Environmental Science and Engineering

Environmental Science

Series Editors

Rod Allan
Ulrich Förstner
Wim Salomons

For further volumes:
http://www.springer.com/series/3234

Lothar Mueller · Abdulla Saparov
Gunnar Lischeid
Editors

Novel Measurement and Assessment Tools for Monitoring and Management of Land and Water Resources in Agricultural Landscapes of Central Asia

Springer

Editors
Lothar Mueller
Gunnar Lischeid
Institute of Landscape Hydrology
Leibniz Centre for Agricultural Landscape Research
Müncheberg
Brandenburg
Germany

Abdulla Saparov
Kazakh Research Institute of Soil Science
Almaty
Kazakhstan

ISSN 1431-6250
ISBN 978-3-319-01016-8 ISBN 978-3-319-01017-5 (eBook)
DOI 10.1007/978-3-319-01017-5
Springer Cham Heidelberg New York Dordrecht London

Library of Congress Control Number: 2013942143

Printed on acid-free paper

Springer is part of Springer Science+Business Media (www.springer.com)

Preface

Origin of the Book

In 2009, during the 18th International Soil Tillage Research Organization (ISTRO) conference in Izmir, Mekhlis Suleimenov from Kazakhstan and Lothar Mueller from Germany came into contact and had their first few talks about soil and water conservation and the situation of agriculture in their countries. They found some common areas of interest, and it seemed to be worthwhile and challenging to explore opportunities for a more intensive exchange of ideas. Abdulla Saparov, Director of the Uspanov Institute for Soil Science in Almaty, and Gunnar Lischeid, Head of the Institute of Landscape Hydrology at the Leibniz Centre for Agricultural Landscape Research in Muencheberg, were very responsive to the suggestion of starting to cooperate and encouraged, inspired and headed this project. A research funding initiative by the German Federal Ministry of Education and Research (BMBF) provided the framework, and our submission was confirmed. In September 2010, the project KAZ 10/001 "Novel Measurement and Assessment Tools for the Monitoring and Management of Water and Soil Resources in Agricultural Landscapes of Central Asia" got started. The International Bureau of the BMBF escorted and administered the project. Over a period of two years, funding has been provided to support short visits by experts, exchanges by young scientists, a workshop in Almaty and the publication of this book. The duration of the project was not enough for extended joint experiments but was sufficient to gain an impression and basic understanding of the status and achievements of research, and of the great potential benefits that longer lasting cooperation would have.

Purpose of the Book

This book is intended to be a source of information for all those dealing with its subject: methods for the characterisation and wise utilisation of water and land resources in Central Asia. There are indications that existing methodologies do not meet international standards and current resource use is not sustainable. The book is to help improve this situation and initiate sustainable developments in Central Asia.

We advocate the role of science and technology in improving our understanding of ecosystem processes and creating monitoring and controlling mechanisms. Reliable data based on advanced, internationally proven and acknowledged methods are required. This implies the exchange of knowledge and the transfer and joint advancement of methods in the scientific community.

The main intended innovation of the book is its focus on methodologies, not on results and facts. Scientific tools will be proposed for measuring, evaluating, modelling and controlling processes in agricultural landscapes. Their application will create a knowledge shift and synergetic effects leading to practical results and conclusions. The book shall act both as a milestone by offering novel tools and ideas, and as a cornerstone by creating lasting research cooperation between scientists and institutes of Eurasia.

Our addressees are people dealing with the development and conservation of land and water in a vast region, where these valuable resources have often been handled wastefully in the recent past. The book mainly addresses scientists, planners, teachers, students and decision makers. It is intended to be a source of information and inspiration for all readers who feel responsible for initiating the sustainable use of resources in Central Asia. This shall help to prepare a secure and better future for the young generation growing up, by preserving the capacities of terrestrial and aquatic ecosystems.

Content and Structure of the Book

The book offers a broad array of methods to measure, assess, forecast, utilise and control land and water resources: laboratory and field measurement methods, methods of resource evaluation, functional mapping and risk assessment and remote sensing methods for monitoring and modelling large areas. It contains methods for data analysis and ecosystem modelling, methods for the bioremediation of soil and water and the field monitoring of soils and methods and technologies for optimising land use systems.

The book has 43 individual chapters in three sections and eight thematic clusters. In order to focus on the scientific value of individual chapters and the expertise of their authors, the editors have decided to keep the structure on a flat level of hierarchy and to allocate the chapters to three parts only. Part I, *Environmental and Societal Framework for the Monitoring and Management of Land and Water Resources*, shall provide an overview of issues related to land and water in Central Asia and prepare the reader for an understanding of the methodological chapters presented in the subsequent two sections. Part I contains 6 chapters analysing the current status and trends. Part II, entitled *Novel Methodologies for the Measurement of Processes and Assessment of Resources* and

Part III, *Applications and Case Studies*, shall provide information about novel methods and give examples of their practical use. Methods that are not yet well known in Central Asia but may have a particular novelty and potential importance are presented in Part II, whilst other new methods and solutions are given in part III. A fourth section, *Executive Summary and Conclusions*, allocates all individual chapters to thematic clusters, reviews them and makes proposals for how they can be applied.

Authors, Readers and their Responsibilities

The authors are inventors and activists behind novel methods, as well as being innovative and experienced scientists. Most of them come from Kazakhstan, Germany, Uzbekistan and Russia, others from different regions of the globe. Not all the authors took part in the project. Many of them were invited to contribute an article afterwards because of the relevance and novelty of their approaches.

Possible divergences between the findings, conclusions and statements of some individual authors are natural. They do not necessarily need to coincide with the particular opinion of the editors. The authors are free to highlight and point out aspects of their study from their typical, individual perspective. The transliteration of local names for rivers, cities or other geographical items or units may also differ from chapter to chapter. All statistical data given in the various chapters of this book may include slight uncertainties, biases and inconsistencies. The editors have made no attempt to harmonize them because this is natural and reflects the different sources and local and temporal scales of the data.

The editors are hopeful that readers will gain sufficient information and inspiration for their own work from this book. However, it is not a cookbook with clear recipes. Readers will become aware of the inconsistencies and deficiencies of some approaches when it comes to measuring and assessing processes in complex ecosystems. They are encouraged to find their individual optimum when drawing conclusions and acting imaginatively.

In some chapters, trade names are used to provide specific information. Mentioning a trade name does not constitute a guarantee of the product by the authors or editors. Neither does it imply an endorsement by the authors or editors of comparable products that are not named.

One brief remark on the book's language. Our aim of providing standard scientific English throughout the book could not be ensured for some chapters. Some constraints restricted this. Despite those deficits we decided to include these chapters because of their relevance and novelty. Though the English is imperfect, in our opinion the content is a valuable contribution to the book. We believe it is preferential and more useful to reach out to some important potential readers in the region by also providing the titles, summaries, figure captions and table headers in

Painting Teris-Asjibulak

Teris-Asjibulak is a small village and correspondent water reservoir (Терс-Ашибулакское водохранилище) about 50 km SW of Taras, Kazakhstan. It was built mainly for irrigation purposes in 1962. The view is from the outlet of the reservoir in south direction towards the Alatau mountain range in Kyrgyzstan where the water comes from. The painting shows a midsummer scene in a typical medium-term dry period. Water shortage occurs since some years. Dry agricultural lands and semi-aquatic vegetation which grew up in the former aquatic area form red–brown belts. Painter: Ute Moritz, 2012. She dedicated the painting to this book edition. Material is oil on canvas. Original size 70*50 cm.

Russian. This information will be available as extra material. Readers feel free to contact the chapter authors or the editors.

Müncheberg and Almaty, May 2013

Lothar Mueller
Abdulla Saparov
Gunnar Lischeid

Acknowledgments

Many people and institutions have supported this publication. We would like to thank the German Federal Ministry of Education and Research for the project funding. The International Bureau of the BMBF, Dr. Kirsten Kienzler (Bonn), managed and controlled the project with expertise and prudence.

Dr. Ralf Dannowski, Dipl. Ing. Ralph Tauschke and Dipl. Ing. (FH) Heike Schäfer (Müncheberg) provided editorial support. Many authors of the book and also Dr. Olga Rukhovich (Moscow) and Prof. Dr. Aleksandr Syso (Novosibirsk) served as reviewers of certain chapters. Mrs. Anne Koth (Dresden), Mrs. Theresa Gehrs (Osnabrück) and Dr. Dmitry Balykin (Barnaul) supported the language correction.

Dipl. Ing. (FH) Ute Moritz dedicated her painting "Teris Asjibulak" to this book edition.

Prof. Dr. Jutta Zeitz (Berlin), Dipl. Ing. Igor Klein (Oberpfaffenhofen), Dr. Rolf Sommer (Nairobi), Dr. Eddy de Pauw (formerly in Aleppo), Dr. Konstantin Pachikin (Almaty), Dr. Azimbay Otarov (Almaty), Prof. Dr. Tobias Meinel (Astana) and Dmitry Chistoprudov (Moscow) provided additional photos and graphics.

The Springer publishing house ensured that the editoral and printing process was smoothly managed and completed.

The editors would like to thank all the contributors for their support and engagement.

Contents

Part II Novel Methodologies for Measurement of Processes and Assessment of Resources

Contributors

Kanat Akshalov Scientific Production Center of Grain Farming named after A.I. Barayev, 021601 Shortandy, Kazakhstan, e-mail: kanatakshalov@mail.ru

Aden Aw-Hassan International Center for Agricultural Research in the Dry Areas, 5466, Aleppo, Syria, e-mail: a.aw-hassan@cgiar.org

Batyrbek Badjanov Kazakh Scientific Research Institute of Water Economy, 12 Koygeldy Street, Taraz, Kazakhstan, e-mail: badg_4@mail.ru

Nurlan Balgabayev Kazakh Scientific Research Institute of Water Economy, 12 Koygeldy Street, Taraz, Kazakhstan, e-mail: kiwr-t@mail.ru

Bruce C. Ball Crop and Soil Systems Research Group, SRUC, West Mains Road, Edinburgh EH9 3JG, UK, e-mail: Bruce.Ball@sruc.ac.uk

Dagmar Balla Leibniz Centre for Agricultural Landscape Research, Eberswalder Strasse 84, 15374 Muencheberg, Germany, e-mail: dballa@zalf.de

Dietmar Barkusky Leibniz Centre for Agricultural Landscape Research, Eberswalder Strasse 84, 15374 Muencheberg, Germany, e-mail: dbarkusky@zalf.de

Axel Behrendt Leibniz Centre for Agricultural Landscape Research (ZALF) e. V., Research Station Paulinenaue, Gutshof 7, 14641 Paulinenaue, Germany, e-mail: abehrendt@zalf.de

Gulzhan B. Beiseyeva Kazakh Research Institute of Soil Science and Agrochemistry named after U. U. Uspanov, Joint Stock company "KazAgroInnovation", Ministry of Agriculture RK, Al Faraby Ave 75b, 050060 Almaty, Kazakhstan, e-mail: beiseeva2009 @mail.ru

Ussen Bekbaev Kazakh Scientific Research Institute of Water Economy, 12 Koygeldy Street, Taraz, Kazakhstan, e-mail: b_ussen@mail.ru

Anne Berger Leibniz Institute for Zoo and Wildlife Research, 10315, Alfred-Kowalke-Straße 17, Berlin, Germany, e-mail: berger@izw-berlin.de

Valentina Chernenok S. Seifullin Kazakh Agro Technical University, Zhenis Prospect 51/1, 010000 Astana, Kazakhstan, e-mail: Chernenok2@mail.ru

Nils Cremer Erftverband, Bergheim, Germany, e-mail: nils.cremer@erftverband.de

Ralf Dannowski Leibniz Centre for Agricultural Landscape Research, Eberswalder Strasse 84, 15374 Muencheberg, Germany, e-mail: danno@zalf.de

Ottfried Dietrich Leibniz Centre for Agricultural Landscape Research, Eberswalder Strasse 84, 15374 Muencheberg, Germany, e-mail: dietrich@zalf.de

Hartwig Drechsler Dr. Drechsler Ingenieurdienst, Hannah-Vogt-Str.1, 37085 Göttingen, Germany, e-mail: post@drechsler-ingenieurdienst.de

Temirbolat D. Dzhalankuzov Kazakh Research Institute of Soil Science and Agrochemistry named after U. U. Uspanov, Al Faraby Ave 75b, 050060 Almaty, Kazakhstan, e-mail: d.temirbolat@mail.ru

Hasi Dzhamantikov Research Institute of Rice Production, Joint Stock company "KazAgroInnovation", Ministry of Agriculture RK, Abai Avenue, 25 b, 120008 Kyzilorda, Kazakhstan, e-mail: pniiaesx@rambler.ru

Nodir Djanibekov Leibniz Institute of Agricultural Development in Central and Eastern Europe, Theodor-Lieser-Straße 2, 06120 Halle (Saale), Germany, e-mail: djanibekov@iamo.de

Wilhelmus H. M. Duijnisveld Federal Institute for Geosciences and Natural Resources, Geozentrum Hannover, Stilleweg 2, 30655 Hannover, Germany, e-mail: Wilhelmus.Duijnisveld@bgr.de

Olga Erokhina Kazakh Research Institute of Soil Science and Agrochemistry named after U. U. Uspanov, Al Faraby Ave 75b, 050060 Almaty, Kazakhstan, e-mail: oerokhina@rambler.ru

Frank Eulenstein Leibniz Centre for Agricultural Landscape Research, Eberswalder Strasse 84, 15374 Muencheberg, Germany, e-mail: feulenstein@zalf.de

Robert G. Evans USDA-ARS-NPA (retired), Northern Plains Agric. Research. Laboratory, 1500 North Central Avenue, Sidney, MT 59270, USA, e-mail: bcevans@pocketinet.com

Andreas Fischer Leibniz Centre for Agricultural Landscape Research, Eberswalder Strasse 84, 15374 Muencheberg, Germany, e-mail: afischer@zalf.de

Shinya Funakawa Graduate School of Agriculture, Kyoto University, Kyoto 606-8502, Japan, e-mail: funakawa@kais.kyoto-u.ac.jp

Roger Funk Leibniz Centre for Agricultural Landscape Research, Eberswalder Strasse 84, 15374 Muencheberg, Germany, e-mail: rfunk@zalf.de

Ursula Gessner German Remote Sensing Data Center, Land Surface Oberpfaffenhofen, German Aerospace Center, 82234 Wessling, Germany, e-mail: ursula.gessner@dlr.de

Levke Godbersen Federal Institute for Geosciences and Natural Resources, Geozentrum Hannover, Stilleweg 2, 30655 Hannover, Germany, e-mail: levke.godbersen@gmail.com

Lars-Christian Grunwald Amazone Company, 49205 Hasbergen, Germany, e-mail: LarsChristian.Grunwald@bbg-leipzig.de

Ahmad Hamidov Faculty of Agriculture and Horticulture, Humboldt University of Berlin, Invalidenstrasse 42, 10117 Berlin, Germany, e-mail: ahmad.hamidov@agrar.hu-berlin.de

Katharina Helming Leibniz Centre for Agricultural Landscape Research, Eberswalder Strasse 84, 15374 Muencheberg, Germany, e-mail: helming@zalf.de

Volker Hennings Federal Institute for Geosciences and Natural Resources, Geozentrum Hannover, Stilleweg 2, 30655 Hannover, Germany, e-mail: volker.hennings@bgr.de

Christian Hertel UMS GmbH, Gmunder Straße 37, 81379 Munich, Germany, e-mail: CH@ums-muc.de

Carsten Hoffmann Leibniz Centre for Agricultural Landscape Research, Eberswalder Strasse 84, 15374 Muencheberg, Germany, e-mail: choffmann@zalf.de

Chunsheng Hu Institute of Genetics and Developmental Biology, The Chinese Academy of Sciences, Center for Agricultural Resources Research, 286 Huaizhong. Rd, Shijiazhuang 050021, Hebei, China, e-mail: cshu@sjziam.ac.cn

Saghit Ibatullin Executive Committee of the International Fund for Saving the Aral Sea, Dostyk Avenue 280, Almaty, Kazakhstan, e-mail: Saghit@inbox.ru

Tursun Ibrayev Kazakh Scientific Research Institute of Water Economy, 12 Koygeldy Street, Taraz, Kazakhstan, e-mail: tursun_ibraev@mail.ru

Thomas Kaiser Leibniz Centre for Agricultural Landscape Research, Eberswalder Strasse 84, 15374 Muencheberg, Germany, e-mail: tkaiser@zalf.de

Erkin K. Kakimjanov Al-Farabi Kazakh National University, Al-Farabi Avenue 71, 050038 Almaty, Kazakhstan

Akmal Karimov IWMI- Central Asia, C/o PFU-CGIAR-ICARDA-CAC, Apt 123, Bldg 6, Osiyo str, Tashkent, Uzbekistan 100000, e-mail: A.Karimov@CGIAR.ORG

Zheksenbai Kaskarbayev Scientific Production Center of Grain Farming named after A.I. Barayev, 021601, Shortandy, Kazakhstan, e-mail: tsenter-zerna@mail.ru

Mukhamadkhan Khamidov Bukhara Branch of Tashkent Institute of Irrigation and Melioration, Gazli street 32, Bukhara, Uzbekistan 200100, e-mail: khamidov_m@mail.ru

Sergey Khudyaev Russian Academy of Sciences, Siberian Branch, Institute of Soil Science and Agrochemistry, Ac. Lavrentieva av., 8, 630090 Novosibirsk, Russian Federation, e-mail: xca_nsk@mail.ru

Igor Klein Institut für Geographie und Geologie, German Aerospace Center, Universität Würzburg, , 82234 Oberpfaffenhofen, Wessling, Germany, e-mail: igor.klein@dlr.de

Polina Koroleva V.V. Dokuchaev Soil Science Institute, Russian Academy of Agricultural Sciences, Pyzhevskii per. 7, Moscow, Russian Federation, e-mail: soilmap@yandex.ru

Farida E. Kozybayeva Kazakh Research Institute of Soil Science and Agrochemistry named after U. U. Uspanov, Joint Stock company "KazAgroInnovation", Ministry of Agriculture RK, Al Faraby Ave 75b, 050060 Almaty, Kazakhstan, e-mail: farida_kozybaeva@mail.ru

Claudia Künzer German Aerospace Center (DLR), German Remote Sensing Data Center, Land Surface Oberpfaffenhofen, 82234 Wessling, Germany, e-mail: claudia.kuenzer@dlr.de

Maira Kussainova Kazakh Research Institute of Soil Science and Agrochemistry named after U. U. Uspanov, Al Faraby Ave 75b, 050060 Almaty, Kazakhstan, e-mail: madgu@inbox.ru

Marcos Lana Leibniz Centre for Agricultural Landscape Research, Eberswalder Strasse 84, 15374 Muencheberg, Germany, e-mail: lana@zalf.de

Marina Li Kazakh Research Institute of Water Economy, 12 Koygeldy Street, Taraz, Kazakhstan, e-mail: limarina76@gmail.com

Gunnar Lischeid Leibniz Centre for Agricultural Landscape Research, Eberswalder Strasse 84, 15374 Muencheberg, Germany, e-mail: lischeid@zalf.de

Sebastian Maassen Leibniz Centre for Agricultural Landscape Research, Eberswalder Strasse 84, 15374 Muencheberg, Germany, e-mail: Sebastian. Maassen@zalf.de

Jessica L. McCarty Michigan Tech Research Institute, Ann Arbor, MI 48105, USA, e-mail: jmccarty@mtu.edu

Tobias Meinel Amazone company, 4740704 Shortandy, Kazakhstan, e-mail: Dr. Tobias.Meinel@amazone.de

Ralph Meissner Department Soil Physics, Helmholtz Centre for Environmental Research-UFZ, Lysimeter Station, Dorfstrasse 55, 39615 Falkenberg, Germany, e-mail: ralph.meissner@ufz.de

Burghard C. Meyer Geographisches Institut, Universität Leipzig, Johannisallee 19a, 04103 Leipzig, Germany, e-mail: burghard.meyer@uni-leipzig.de

Rickmann Michel Ingenieurbüro BODEN und BODENWASSER, Gesundbrunnenstraße 24, 16259 Bad Freienwalde, Germany, e-mail: rjmichel@t-online.de

Alisher Mirzabaev ICARDA, Central Asia and Caucasus Regional Office, Osiyo 6, Tashkent, Uzbekistan, e-mail: almir@uni-bonn.de

Lothar Mueller Leibniz-Centre for Agricultural Landscape Research, Eberswalder St., 84, 15374 Muencheberg, Germany, e-mail: mueller@zalf.de

Khamit Mukhamedjanov Kazakh Scientific Research Institute of Water Economy, 12 Koygeldy Street, Taraz, Kazakhstan, e-mail: mkv_77@mail.ru

Marco Natkhin Johann Heinrich von Thünen-Institut, Institut für Waldökosysteme, 16225 Eberswalde, Germany, e-mail: marco.natkhin@ti.bund.de

Claas Nendel Leibniz Centre for Agricultural Landscape Research, Eberswalder Strasse 84, 15374 Muencheberg, Germany, e-mail: nendel@zalf.de

Andrew D. Noble International Water Management Institute, P.O. Box 2075, Colombo, Sri Lanka, e-mail: A.Noble@cgiar.org

Mohie Omar Nile Research Institute, National Water Research Center, Ministry of Water Resources and Irrigation, El-Quanater El-Khairya, Egypt, e-mail: mohie.omar@hotmail.com

Sylvia Ortmann Leibniz Institute for Zoo and Wildlife Research, Alfred-Kowalke-Straße 17, 10315 Berlin, Germany, e-mail: ortmann@izw-berlin.de

Azimbay Otarov Kazakh Research Institute of Soil Science and Agrochemistry named after U. U. Uspanov, Al Faraby Ave 75b, 050060 Almaty, Kazakhstan, e-mail: azimbay@bk.ru

Konstantin Pachikin Kazakh Research Institute of Soil Science and Agrochemistry named after U. U. Uspanov, Al Faraby Ave 75b, 050060 Almaty, Kazakhstan, e-mail: kpachikin@yahoo.com

Valeria N. Permitina Institute of Botany and Phytointroduction, Ministry of Education and Science Republic Kazakhstan, Timiryzeva Avenue 36d, 050040 Almaty, Kazakhstan, e-mail: v.permitina@mail.ru

Manzoor Qadir Institute for Water, Environment and Health, United Nations University, 175 Longwood Road South, Suite 204, Hamilton Ontario L8P 0A1, CANADA, e-mail: Manzoor.Qadir@unu.edu

Matthias Reiche Leibniz Centre for Agricultural Landscape Research, Eberswalder Strasse 84, 15374 Muencheberg, Germany, e-mail: reiche@zalf.de

Vladimir Romanenkov Department of Geographic Network of Field Experiments, Pryanishnikov All-Russian Institute of Agrochemistry (VNIIA), Russian Academy of Agricultural Sciences, Pryanishnikova St., 31a, 127550 Moscow, Russian Federation, e-mail: geoset@yandex.ru

Dmitry I. Rukhovich V.V. Dokuchaev Soil Science Institute, Pyzhevskii per. 7, Moscow, Russian Federation, e-mail: landmap@yandex.ru

Holger Rupp Department Soil Physics, Helmholtz Centre for Environmental Research-UFZ, Lysimeter Station, Dorfstrasse 55, 39615 Falkenberg, Germany, e-mail: holger.rupp@ufz.de

Abdulla Saparov Kazakh Research Institute of Soil Science and Agrochemistry named after U. U. Uspanov, Al Faraby Ave 75b, 050060 Almaty, Kazakhstan, e-mail: ab.saparov@yahoo.com

Roland Schindler NEW Niederrhein Wasser GmbH, Rektoratstrasse 18, 41747 Viersen, Germany

Uwe Schindler Leibniz Centre for Agricultural Landscape Research, Eberswalder Strasse 84, 15374 Muencheberg, Germany, e-mail: uschindler@zalf.de

Vera Schreiner Institut für Länderkunde, Schongauerstraße 9, 04329 Leipzig, Germany, e-mail: V_Schreiner@ifl-leipzig.de

Manfred Seyfarth UGT-Environmental Measurement Devices Ltd., Eberswalder Str. 58, 15374 Muencheberg, Germany, e-mail: info@ugt-online.de

T. Graham Shepherd BioAgriNomics Ltd., 6 Parata Street, Palmerston North 4410, New Zealand, e-mail: gshepherd@BioAgriNomics.com

Askhad K. Sheudshen Kuban State Agrarian University, 13 Kalinin Str, 350044 Krasnodar, Russian Scientific Research Institute of Rice, Bjelosernoye settlem, 350921 Krasnodar, Russian Federation, e-mail: kubagrohim@mail.ru

Dana K. Shokparova Al-Farabi Kazakh National University, Al-Farabi Avenue 71, 050038 Almaty, Kazakhstan, e-mail: shk.dana@hotmail.com

Boris Smolentsev Siberian Branch, Institute of Soil Science and Agrochemistry (ISSA), Russian Academy of Sciences, Ac.k Lavrentieva av., 8, 630090 Novosibirsk, Russian Federation, e-mail: pedolog@ngs.ru

Elena Smolentseva Institute of Soil Science and Agrochemistry (ISSA), Siberian Branch, Russian Academy of Sciences, Ac. Lavrentieva av., 8, 630090 Novosibirsk, Russian Federation, e-mail: smolentseva@issa.nsc.ru

Rolf Sommer International Center for Tropical Agriculture, c/o ICIPE Complex, off Kasarani Road, Nairobi, Kenya, e-mail: r.sommer@cgiar.org

Jörg Steidl Leibniz Centre for Agricultural Landscape Research, Eberswalder Strasse 84, 15374 Muencheberg, Germany, e-mail: jsteidl@zalf.de

Mekhlis Suleimenov Scientific Production Center of Grain Farming named after A.I. Barayev, 021601 Shortandy, Kazakhstan, e-mail: mekhlis@yahoo.com

Victor G. Sychev Pryanishnikov All-Russian Institute of Agrochemistry (VNIIA), Russian Academy of Agricultural Sciences, Pryanishnikova St. 31a, 127550 Moscow, Russian Federation, e-mail: info@vniia-pr.ru

Marion Tauschke Leibniz Centre for Agricultural Landscape Research, Eberswalder Strasse 84, 15374 Muencheberg, Germany, e-mail: mtauschke@zalf.de

Valeriy A. Tumlert Kazakh Scientific Research Institute of Water Economy, 12 Koygeldy Street, Taraz, Kazakhstan, e-mail: tumlert_e@mail.ru

Georg von Unold UMS GmbH, Gmunder Str. 37, 81379 Munich, Germany, e-mail: gvu@ums-muc.de

Jens Utermann Federal Environment Agency, Wörlitzer Platz 1, 06844 Dessau-Roßlau, Germany, e-mail: Jens.Utermann@uba.de

Frants Vyshpolsky Water Quality Laboratory, Kazakh Scientific Research Institute of Water Economy, 12 Koygeldy Street, Taraz, Kazakhstan

Tulkun Yuldashev ICARDA, Central Asia and Caucasus Regional Office, Osiyo 6, Tashkent, Uzbekistan, e-mail: t.yuldashev@cgiar.org

Nikolai Yushchenko Karaganda Research Institute of Crop Science and Plant Breeding, Karaganda, Kazakhstan, e-mail: 12345680@bk.ru

Part I
Environmental and Societal Framework for Monitoring and Management of Land and Water Resources

Land and Water Resources of Central Asia, Their Utilisation and Ecological Status

Lothar Mueller, Mekhlis Suleimenov, Akmal Karimov, Manzoor Qadir, Abdulla Saparov, Nurlan Balgabayev, Katharina Helming and Gunnar Lischeid

Abstract Central Asia is the global hotspot of a nexus of resources. Land, water and food are key issues in this nexus. We analysed the status of land and water resources and their potential and limitations for agriculture in the five Central Asian Transition States. Agricultural productivity and its impacts on land and water quality were also studied. The ecological status of open waters and soils as dependent on the kind of water and land use was shown. The main sources were information and data from the scientific literature, recent research reports, the statistical databases of the FAO and UNECE, and the results of our own field

L. Mueller (✉) · K. Helming · G. Lischeid
Leibniz-Centre for Agricultural Landscape Research (ZALF) e. V., Eberswalder Str. 84, 15374 Muencheberg, Germany
e-mail: mueller@zalf.de

M. Suleimenov
Scientific Production Center of Grain Farming Named After A.I. Barayev, Shortandy 021601, Kazakhstan
e-mail: mekhlis@yahoo.com

A. Karimov
IWMI- Central Asia, C/o PFU-CGIAR-ICARDA-CAC, Apt 123, Bldg 6, Osiyo Str., Tashkent 100000, Uzbekistan
e-mail: A.Karimov@CGIAR.ORG

M. Qadir
United Nations University - Institute for Water, Environment and Health (UNU-INWEH), 175 Longwood Road South, Suite 204, Hamilton Ontario L8P 0A1, CANADA
e-mail: Manzoor.Qadir@unu.edu

A. Saparov
Kazakh Research Institute of Soil Science and Agrochemistry Named After U. U. Uspanov, Al Faraby Ave 75b 050060 Almaty, Kazakhstan
e-mail: ab.saparov@yahoo.com; ab.saparov@mail.ru

N. Balgabayev
Kazakh Scientific Research Institute of Water Economy, 12 Koygeldy Str., Taraz, Kazakhstan
e-mail: kiwr-t@mail.ruIWRE@bk.ru

L. Mueller et al. (eds.), *Novel Measurement and Assessment Tools for Monitoring and Management of Land and Water Resources in Agricultural Landscapes of Central Asia*, Environmental Science and Engineering, DOI: 10.1007/978-3-319-01017-5_1,

work. Agriculture is crucial for the economy of all Central Asian countries and responsible for about 90 % of their water use. We found that land and water resources may provide their function of food supply, but the agricultural productivity of grassland and cropland is relatively low. Irrigation agriculture is sometimes inefficient and may cause serious detrimental side effects involving soil and water salinisation. Dryland farming, as currently practiced, includes a high risk of wind and water erosion. Water bodies and aquatic, arable and grassland ecosystems are in a critical state with tendencies to accelerated degradation and landscape desertification. Despite all these limitations, agricultural landscapes in Central Asia have great potential for multi-functional use as a source of income for the rural population, tourism and eco-tourism included. The precondition for this is a peaceful environment in which they can be developed. All major rivers and their reservoirs cross borders and involve potential conflict between upstream and downstream riparians. The nexus of resources requires more detailed research, both in the extent of individual elements and processes, and their interactions and cycles. Processes in nature and societies are autocorrelated and intercorrelated, but external disturbances or inputs may also trigger future developments. We emphasise the role of knowledge and technology transfer in recognising and controlling processes. There has been a lot of progress in science and technology over the past ten years, but agri-environmental research and education in Central Asia are still in a crisis. Overcoming this crisis and applying advanced methods in science and technology are key issues for further development. Science and technology may provide an overall knowledge shift when it comes to recognising processes and initiating sustainable development. The following chapters introduce the results of further, more detailed and regional analyses of the status of soil and water. Novel measurement and assessment tools for researching into, monitoring and managing land and water resources will be presented. We will inform future elites, scientists and decision makers on how to deal with them and encourage them to take action.

Keywords Central Asia · Soil · Water · Sustainable development

1 Introduction

The 21st century is characterised by accelerating demands for most natural resource commodities. Natural resource governance faces increasing complexity, especially when the linkages and interdependencies between different resources are considered (Andrews-Speed et al. 2012). This has implications for all global regions including Central Asia. Public documents have stated that there is a "Water and Energy Nexus in Central Asia" (World Bank 2004). Based on the situation that all available water resources in the region are trans-boundary, the key questions in the region are "Who has the right to consume all the water?" and

"Water for food or water for energy?". However, this is only part of the problem; the situation is much more complex. Central Asia is the global hotspot of a struggle for all main resources: food, land, water, energy and minerals. These resource categories are closely interrelated in different spatial and temporal scales and dimensions of cognition. Typical segments and key questions differ.

What is Central Asia? The region of Central Asia can be characterised and defined by terms from different scientific disciplines and perspectives. As a geographic category, it is the centre of the Eurasian continent, consisting of striking landscapes such as the high mountains of Tien Shan and Pamir, deserts such as Kara Kum, Kyzyl Kum, or Taklamakan, the second largest desert of the world, large steppe lowlands (Fig. 1), great internal basins, water bodies such as the Aral Sea, Issyk Kul and Lake Balkhash, the Silk Road, the most famous trade route of the world, and more. As a socio-economic or political category, it includes a number of countries in whole or in part. The term "Central Asia" is frequently used for the territory of five land-locked Transition States of the former USSR: Kazakhstan, Kyrgyzstan, Tajikistan, Turkmenistan and Uzbekistan. This coincides with the FAO's area classification (FAO 2012b).

When dealing with topics related to natural resources such as land and water, their implications for human society, and scientific/technological approaches to resolving problems, a mixed consideration of both natural/geographical and political/territorial aspects can be useful. Here, we refer primarily to the territory of Kazakhstan, Kyrgyzstan, Tajikistan, Turkmenistan and Uzbekistan (calling them Central Asia) with some focus on Kazakhstan. The latter country is much larger than the others, contains very diverse landscapes and, in some geographical and agricultural regions, faces typical problems of land and water monitoring and management. Most of these issues are typical for the other countries, too. They may be also true for a much larger area, based on a broader definition of Central Asia. The regions of Siberia, Mongolia, North-Western China and of other neighbour regions surrounding the territory of the five Post-Soviet Transition

Fig. 1 Grazing sheep on semi-natural pastures in the vicinity of Almaty, Alatau mountains in the background. Photo: Courtesy of Volker Hennings

States face similar problems with natural resources and their management. We believe that Kazakhstan holds as a typical example for most scientific problems related to land and water with implications for agriculture and the environment.

In this book, we focus on novel scientific methodologies for measuring and assessing some crucial properties of land and water resources. This is intended to help monitor and manage processes in agricultural landscapes of Central Asia better. A sustainable use of the resources of land and water must be initiated. First, an analysis is required of the extent and current status of the resources of land and water and of some development trends. This will take place in this and some following chapters.

This chapter will provide an overview of quantities, qualities and productivity of land and water, and agriculture based on them in Central Asia. We have analysed available sources on the status of water and land, trends in their use and developments. The information and data came from the scientific literature, recent research reports, statistical databases and commission reports of United Nations Organizations such as the FAO and UNECE, international research and development projects, and the results of our own field work.

2 Key Elements of a Nexus of Resources: Food Security and Water Consumption

2.1 Food Security

Within a resource nexus, food is a special and fundamental category. Having enough food is a basic human right, and food security is crucial for the stabilization and survival of civilisations. The FAO's definition states, "Food security is a situation that exists when all people, at all times, have physical, social and economic access to sufficient, safe and nutritious food that needs their dietary needs and food preferences for an active and healthy life" (FAO 2002).

Other resources such as land, water, energy and minerals directly or indirectly impact on food security (Fig. 2).

The availability and good condition of land and water are natural preconditions for agriculture, the main basis of food production. The situation of human society and governance regarding the utilisation of all resources determines the framework for agricultural production and food security. Some basic data on the five Central Asian countries are given in Table 1.

Kazakhstan has extensive land and water resources and its national economy is in an acceptable condition, mainly due to oil, gas and mineral resources. The country is sparsely populated. In terms of its gross domestic product (GDP) and the human development index (HDI) it ranks above the world average. Related to the land area, water resources are relatively low and unequally distributed. The dependency ratio of 40.1 means that about forty percent of total available water

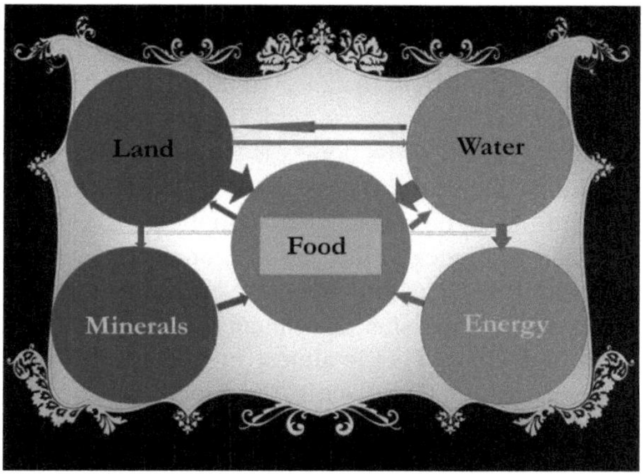

Fig. 2 The resource nexus on the carpet of prosperity and welfare for the countries of Central Asia (simplified, with emphasis on land, water and food)

Table 1 Basic data on land, water and economy in the five Central Asian states

Country	Total land, (M ha)[1]	Water resources (TRWR, km^3/year) and dependency ratio[2]	Population in millions[1]	Economy, GDP $ per capita (rank)[3]	HDI, value (rank)[4]
Kazakhstan	272.490	107.50 (40.1)	16.207	13,000 (95)	0.74 (68)
Kyrgyzstan	19.995	23.62 (1.1)	5.393	2,400 (183)	0.61 (126)
Tajikistan	14.255	21.91 (17.3)	6.977	2,100 (189)	0.60 (127)
Turkmenistan	48.810	24.77 (97.0)	5.105	7,800 (128)	0.67 (102)
Uzbekistan	44.740	48.87 (80.1)	27.760	3,300 (168)	0.62 (115)

[1] Data from FAOSTAT (FAO 2012b), M ha = million hectares, 1ha = 0.01 square kilometres
[2] Data from FAOSTAT (FAO 2012b), *TRWC* Total Renewable Water Resource, dependency ratio = percent of external water resources to TRWC
[3] Data from the CIA World Factbook 2011 (CIA 2011), *GDP* Gross Domestic Product, the rank in parentheses refers to the countries of the world
[4] Data from the Human Development Report (UNDP 2011), *HDI* Human Development Index, mainly based on indexes of income, health and education

resources come from neighbouring countries, in this case mainly from Kyrgyzstan and China.

Kyrgyzstan and Tajikistan are relatively small countries characterised by a poorly developed economy. They have an extremely low GDP and rank well below the world average in terms both of GDP and of HDI. Related to their land area, they have extensive water resources and are largely independent of other countries. Their potential for hydropower is very good.

Turkmenistan and Uzbekistan are extremely dependent on water resources coming from other countries, mainly from Kyrgyzstan and Tajikistan. Though they have some fossil fuel resources, their economy is still weakly developed. Uzbekistan is the most densely populated country in the region. In some oases, the population density is particular high.

The different overall economic situation in these countries is underpinned by their trade balance. In 2011, Kazakhstan, Turkmenistan and Uzbekistan had a positive overall trade balance (merchandize exports higher than imports), whilst the balance of Kyrgyzstan and Tajikistan was negative (WTO 2012). The combination of a low gross domestic product with a negative trade balance is a burden for the development of a prospering economy taking ecology into consideration. These different figures may have implications for the status and utilisation of water and land resources in the various countries.

Overall, the territory of the five Central Asian Transition States Kazakhstan, Kyrgyzstan, Tajikistan, Turkmenistan and Uzbekistan covers about 399 million hectares (M ha). This is almost in the order of the geographic magnitude of the European Union (EU), whilst the population is only about 12 % of that in the EU.

Regional food security is a key issue for human development and social and political stability everywhere. It cannot yet be considered to be ensured in Central Asia. Over the past twenty years, all five countries witnessed an initial decline in agricultural productivity, and currently they have an increase in productivity. The situation is not stable, and there are big differences between countries (Swinnen and Vranken 2010). Food quality is another issue related to food security. Awareness of good food quality has to be based on reliable procedures and data regarding monitoring and control (Tultabaieva 2012).

Sustainable food security in the region does not only depend on food production: there are other, decisive factors influencing the long-term balance between food demand and food availability (Dukhovny and Stulina 2011). Factors controlling food demand include demographic structure, population growth, preferred kinds of food and consumer incomes. Some main factors in food availability are the status of natural resources such as land and water, and policies promoting their utilisation through initiatives, skills and investment.

Currently, there are gaps between the production and consumption of some kinds of basic food products. For grain products, Tajikistan shows a clear gap, whilst Kazakhstan has an over-production. Oils are in deficit in all five countries. In case of vegetables, production and demand coincide. With meat, Tajikistan and Kazakhstan indicate deficits (Fig. 3).

Overall, Tajikistan shows deficits in food security from these data (Dukhovny and Stulina 2011). The factors affecting this are the population's insufficient purchasing power, limited production and very high population growth rates. Food prices increased due to inflation by 10–120 % for many agricultural products in 2010. A law "On Food Security" of 29 Dec. 2010 was one measure used to improve the situation (Bobodzhanova 2012). In Kyrgyzstan, too, the situation has worsened over the past 25 years. In the 1980s the country was self-sufficient and exported food. Now, food security issues are based on the 2008 law "About food safety in the Kyrgyz Republic" (Mansurova 2012). More than half of food commodities need to be imported, amongst them basic products such as bakery products, meat, sugar and oils. In 2009, about one third of the population, the vast majority of them in rural areas, lived below the poverty line (Kulmyrzaev 2012).

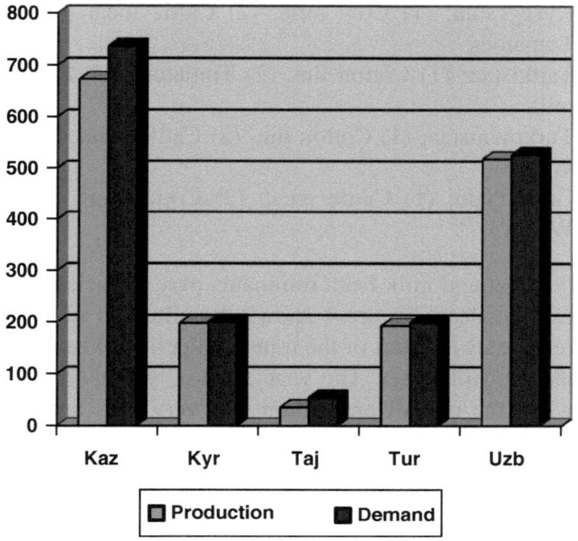

Fig. 3 Production and demand of meat (in millions of tonnes), data from Dukhovny and Stulina 2011

Uzbekistan is self-sufficient but supplying food for the balanced, safe nutrition of a growing population remains a challenge (Payziyeva and Paiziev 2012).

The land-locked situation of Central Asia should be a reason to promote regional markets and to reduce imbalances between production and consumption by prosperous trade between countries, embedded in a tension-free and friendly societal and political environment.

Agriculture is a pillar of the economy of all Central Asian countries. For example, in 1999, agriculture made up 11 % of the gross domestic product (GDP) in Kazakhstan, 19 % in Tajikistan, 27 % in Turkmenistan, 33 % in Uzbekistan, and 38 % in Kyrgyzstan (World Bank 2004). More recent figures from 2008 (Kienzler 2009) are 5, 26, 20, 22 and 19 %, showing a downwards trend due to increases in other economic branches such as the energy and mining industries. In Uzbekistan, agriculture is the largest sector of the economy, providing about 25 % of exports and 31 % of employment (Turayeva 2012). Of the Central Asian countries, Uzbekistan has the highest rural population. However, for the economy and to supply food to the population of all five countries in the region, agriculture remains crucial. Agriculture in a dryland region necessarily means irrigated agriculture. Even in Kazakhstan, where only 5–6 % of cropland is irrigated, about one third of the total food production comes from that land (Ismukhanov and Mukhamedzhanov 2003).

FAO statistics (FAO 2012b) show that in 2010 the top five agricultural products by country were:

Kazakhstan: (1) Cow milk, (2) Cattle meat, (3) Wheat, (4) Sheep meat, (5) Potatoes

Kyrgyzstan: (1) Cow milk, (2) Cattle meat, (3) Potatoes, (4) Sheep meat, (5) Tomatoes
Tajikistan: (1) Cotton lint, (2) Tomatoes, (3) Sheep meat, (4) Potatoes, (5) Cow milk
Turkmenistan: (1) Cotton lint, (2) Cattle meat, (3) Sheep meat, (4) Cow milk, (5) Wheat
Uzbekistan: (1) Cattle meat, (2) Cotton lint, (3) Cow milk, (4) Tomatoes, (5) Wheat

Meat and milk from ruminants play a crucial role as food for the population of Central Asia. Figure 4 shows the situation of milk production. The year 1992 represents the start of the transition period, when old structures in agriculture were largely still intact. The year 2000 is typical for the end of the transition period, when the overall production was very low, and the data for 2010 are the most recent available. It can be seen that all countries were keen to increase the per-capita production of milk. This has been successfully achieved in all countries. The data are probably not very reliable, for Tajikistan in particular. Many poor people breed goats for their subsistence, and those data are obviously not available. However, increasing per-capita production despite a fast-growing population indicates increasing pressure on grasslands. Sustainable grassland management is a precondition for securing these traditional commodities on available lands.

For Kazakhstan, Turkmenistan and Uzbekistan, wheat also belongs to the group of the top five agricultural products. Wheat production in Kazakhstan is shown in Fig. 5. The situation has stabilised in comparison with that in the 1990s but prevailing rainfed cropping wheat production still remains largely weather-dependent.

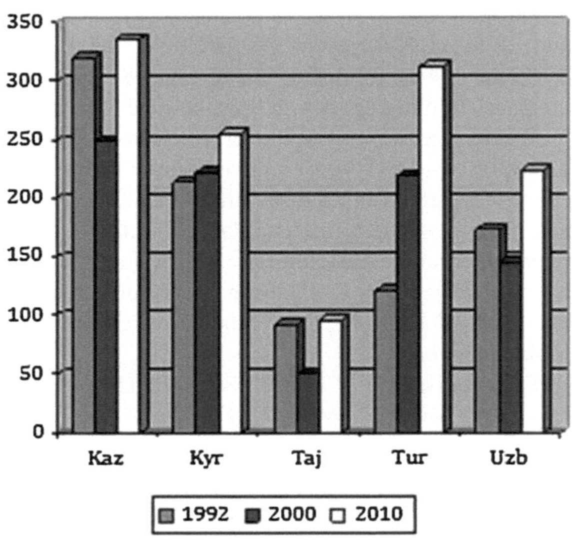

Fig. 4 Production of milk in kg per capita in typical years of three different periods, data from FAOSTAT database (FAO 2012b)

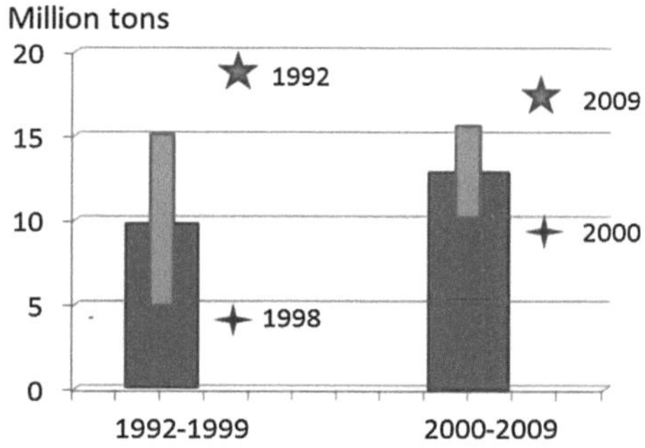

Fig. 5 Annual wheat production of Kazakhstan in two periods: the transition period after independency in the 1990s, and the first 10 years of this century. *Blue bars* indicate the average, small *green bars* the 95 % confidence interval, and *stars* are extremes. The data come from the FAOSTAT database (FAO 2012b)

Table 2 Main imported basic foods and agricultural commodities in 2010 (FAO 2012b)

Country	Top imported basic food commodities	Top exported agricultural commodities
Kazakhstan	Sugar, chicken meat	Wheat, wheat flour, cotton, barley
Kyrgyzstan	Wheat, chicken meat, sugar, sunflower oil	Beans, cotton, cow milk, potatoes
Tajikistan	Wheat flour, wheat, sugar	Cotton, tomatoes
Turkmenistan	Wheat, potatoes, sugar	Cotton
Uzbekistan	Wheat flour, sugar, sunflower oil	Cotton, tomatoes, wheat

From this figure the annual production seems to be remaining relatively stably above 10 million tonnes. However, in 2010 the production was 9.6 million tonnes, in contrast to 17 million tonnes in 2009. The situation remains unstable. Cereal production in other Central Asian countries is mainly irrigated cropping and more stable than rainfed cropping.

Studying imports and exports of agricultural products confirms that wheat and wheat flour are among the main basic food commodities imported into Kyrgyzstan, Tajikistan, Turkmenistan and Uzbekistan (Table 2). All these countries need to import sugar, though sugar beets grow well on salinised soils. On the export side, cotton dominates. Kazakhstan is an important exporter of wheat and barley.

2.2 Water Resources and Water Consumption

2.2.1 Status of Main Rivers and Reservoirs

Water is a dynamic resource. Melting snow and glacier ice in the high mountains of Central Asia, mainly located in Kyrgyzstan and Tajikistan, are the only source of fresh water in the region. The runoffs of the large rivers, such as the Syr Darya, Amu Darya, Ili, Shu, Talas, Zeravshan, Atrek, Karatal, Aksu, Lepsa, etc., originate in the high-altitude mountains. The Syr Darya is the longest river in Central Asia (2,212 km). It originates in the Tien Shan mountains and flows along the borders

Table 3 Main rivers and basins in Central Asia

Basin and sub-basins	Total area (km^2)	Recipient	Riparian countries
Ob	2,972,493 from that KAZ: 734,543	Arctic Ocean (Kara Sea)	RU(MN, CN, KZ via Irtysh)
Irtysh	1,643,000	Ob	RU, KZ, CN, MN
Tobol	426,000	Irtysh	RU, KZ
Ishim	176,000	Irtysh	RU, KZ
Amu Darya	n/a	Aral Sea	AF, KG, TJ, UZ, TM
Surkhan Darya	13,500	Amu Darya	TJ, UZ
Kafirnigan	11,590	Amu Darya	TJ, UZ
Pyanj	113,500	Amu Darya	AF, TJ
Vakhsh	39,100	Amu Darya	KG, TJ
Zeravshan	n/a	Desert sink	TJ, UZ
Syr Darya	n/a	Aral Sea	KZ, KG, TJ, UZ
Naryn	n/a	Syr Darya	KG, UZ
Kara Darya	28,630	Syr Darya	KG, UZ
Chirchik	14,240	Syr Darya	KZ, KG, UZ
Chu	62,500	Desert sink	KZ, KG
Talas	52,700	Desert sink	KZ, KG
Ili	413,000	Lake Balkhash	CN, KZ
Murgab	46,880	Desert sink	AF, TM
Tejen	70,260	Desert sink	AF, IR, TM

Basic data and table from UNECE (2007), modified, n/a- not available, country abbreviations: *AF* Afghanistan, *CN* China, *KZ* Kazakhstan, *KG* Kyrgyzstan, *MN* Mongolia, *RU* Russia, *TM* Turkmenistan, *TJ* Tajikistan, *UZ* Uzbekistan

of or across four states before ending in the North Aral Sea. The Irtysh is the longest river in Kazakhstan, flowing 1,200 km through the country.

Central Asia consists of some large and a number of small river basins (Table 3). The Ob river and its main tributary, the Irtysh, drain the north of Kazakhstan towards the Arctic Ocean. Other rivers remain in Central Asia and feed inner basins.

From the data in the AQUASTAT database (FAO 2012a), the river basins by countries are:

Kazakhstan: the major river basins are the Arctic Ocean via the Ob river (60 % of the outflow); internal basins, Lake Balkhash included (26 %), Caspian Sea (8 %) and Aral Sea (6 %).

Kyrgyzstan: There are six groups of main river basins: the Syr Darya (62 %), the Chu, Talas and Assa (16 %), a group of southeastern small mountainous river basins draining to China, mainly the Aksay, Sary Dzhaz and Kek Suu (14 %), the Amu Darya (4 %), the Issyk–Kul (3 %) and Lake Balkhash (1 %).

Tajikistan: Four major river basins may be distinguished: the Amu Darya (84 %), the Syr Darya (11 %), the Zeravshan (4 %), and a group of northeastern small mountainous river basins draining to China (1 %).

Turkmenistan: Two main groups of river basins are the Amu Darya (68 %), and a group of Murghab, Tedzhen, Atrek and others (32 %). The overall discharge is 1 km^3/year only.

Uzbekistan: Two main river basins forming the Aral Sea basin can be distinguished: the Amu Darya basin in the south (86 %) and the Syr Darya in the north (14 %). The given percentages refer to renewable water resources, e.g. a calculated outflow, not to the area of recharge. The latter would provide a different figure. For example, the territory of Kazakhstan can be divided into 8 main hydro-economic basins: the Aral- Syr Darya basin (345 K km^2), the Balkhash-Alakol basin (353 K km^2 in K), the Ural-Caspian basin (415 K km^2), Shu-Talas (64.3 km^2), the Irtysh basin, the Ishim basin (216 K km^2), the Nura-Sarysu basin, and the Tobol-Turgai basin (214 K km^2), (UNDP 2004). The latter four basins belong to the Ob-Irtysh basin and drain to the Arctic Ocean.

Central Asian countries have a high stock of reservoirs, which were mainly constructed about 40–60 years ago. For example, in the Aral Sea basin there are more than 50 large reservoirs for the storage of water for irrigation and hydropower production (Kamilov 2003). Reservoir siltation is a problem and has to be minimised by reducing erosion in upstream regions. There are 475 reservoirs in Kazakhstan, of which 75 are in the southern regions, with a total capacity of 95.5 km^3 and surface area of over 10,000 km^2. The largest reservoirs in Kazakhstan are the Bukhtarma reservoir on the Irtysh River (49 km^3) and the Kapchagay reservoir on the Ili river (28 km^3). The great majority of large reservoirs are multipurpose, for all sectors of agriculture, including fish production, industry and supplying the population (Ismukhanov and Mukhamedzhanov 2003). The construction of water power station dams has created serious problems for the migration and natural reproduction of native fish.

Though Central Asia is a dryland region, there are many natural lakes of different degrees of salinity. The largest lakes in Kazakhstan are Lake Balkhash (18,000 km^2 area and 112 km^3 volume), Lake Zaisan (5,500 km^2) and Lake Tengiz (1,590 km^2). The overall number of natural lakes in Kazakhstan is more than 17,000, with a total area of about 45,000 km^2 and an estimated total volume of about 190 km^3 of water (FAO 2012a).

All these rivers, reservoirs, lakes and other open waterbodies cannot be considered as a resource to provide, store or drain off water only. They are important ecosystems and must be maintained as habitats for flora, fauna and overall biodiversity. For humans, they provide a number of ecosystem functions and services, including recreation and experiencing natural wildlife. Some regions of Central Asia have good potential for ecotourism based on their unique, diverse and exciting landscape (Fig. 6). Open waters are important elements of this landscape ensemble. The protection of lakes, water cavities, rivers and marsh lands takes top priority for actions in the National Biodiversity Strategies and Action Plan of the Republic of Kasakhstan (National Strategy and Action Plan 1999). Kazakhstan signed the Convention of Biological Diversity (CBD) in 1992 and ratified it in 1994, while the other four countries accessed in the 1990s (CBD Secretariat 2012).

Fig. 6 Impressions of three transboundary rivers. **a** River and meadow of the Talas river near Taraz (Kazakhstan). A few kilometres upstream is the border to Kyrgyzstan. Downstream are irrigated lowlands. The river ends in a desert sink. **b** Ili river near Karakol, at the begin of the delta, about 150 km to the river mouth. The Ili originates in China and feeds extended irrigated lands there before entering Kazakhstan. The Ili is the largest river ending in Lake Balkhash, and is crucial for its hydrological stabilisation. **c** Charyn river, a tributary of the Ili river. The Charyn has formed an impressive canyon landscape (Photo: Courtesy of Azimbay Otarov)

Water resources and water quality issues are also set down on the United Nations' official list of "Millennium Development Goal (MDG) indicators". These are: "Goal 7: Ensure environmental sustainability, Target 7.A: Integrate the principles of sustainable development into country policies and programmes and reverse the loss of environmental resources, indicators 7.4 Proportion of fish stocks within safe biological limits and 7.5 Proportion of total water resources used". Also, "Target 7.B Reduce biodiversity loss..." refers to the functions of land and water to preserve biodiversity (MDG indicators 2012). This may be a further reason to monitor and control water resources carefully. The welfare and future prosperity of the region are linked to the status of aquatic ecosystems (Yessekin et al. 2008).

2.2.2 Water Consumption

Irrigation agriculture is the dominating water user, and evapotranspiration is the largest sink in the landscape water balance. Between 82 % (Kazakhstan) and 98 % (Turkmenistan) of all water withdrawal is needed for agriculture (CIA 2011). This proportion of water withdrawal is clearly above the global average (averaged over countries) for irrigated land, which is 70 % (Siebert et al. 2010).

Drinking water is a most important and valuable food. An OECD study revealed that the quality of drinking water in the region was poor (OECD 2003). More than one third of the population of Central Asia uses drinking water that does not meet quality standards, and this proportion exceeds 50 % in some regions. Microbiological parameters and nitrates are particular problems (OECD 2003). Improvement of the water quality and the supply of clean drinking water to the population take top priority in the national water strategies of all Central Asian countries (FAO 2012a).

2.3 Potential for Water Conflicts

The potential for conflict between riparians may result from water quantity and quality issues. All great rivers and their main tributaries are trans-boundary. Threats to water security exist for the downstream countries of Kazakhstan (Ryabtsev 2011), Uzbekistan (Rakhmatullaev et al. 2009; Turayeva 2012) and Turkmenistan (Allouche 2007).

This includes the potential for water conflicts between countries and kinds of resource users (Giese et al. 2004; Grewlich 2010). The upstream countries (Tajikistan and Kyrgyzstan, also China) where the most water is generated will increasingly claim the water resources for themselves. There are significant "upstream–downstream" issues with hydropower potential upstream and irrigation demands downstream (Granit et al. 2010). Even though there is the potential for conflict, the situation could be turned into a win–win situation for all the riparian states (Giese et al. 2004; Wegerich et al. 2007).

The imbalance between flow generation and water consumption is a main problem of managing water quantity. Control of reservoirs is a permanent issue. The example of the Toktogul reservoir, the largest reservoir in Kyrgyzstan, on the Naryn river—the Syr Darya catchment—may demonstrate the situation. The water regime of the Toktogul reservoir has shifted to a strategy meeting the demands of the hydropower industry, not the demands of agricultural water management of downstream rural areas (Dukhovny and Stulina 2011). On the other hand, the economy of Kyrgyzstan is characterised by an energy crisis. A substantial part of the population has no access to electricity. Kyrgyzstan generates about 90 % of its electricity from hydro-electric power stations (Liu and Pistorius 2012). To produce more electricity in the winter, most water of the reservoir is released in the winter. This causes problems in downstream regions, floods to the Arnasay depression (Uzbekistan) and the region of Kyzyl Orda (Kazakhstan) in the winter and a lack of water in the summer (World Bank 2004).

Central Asia is already prone to natural flood disasters. Man-made floods such as that in the Naryn catchment exacerbate the situation. Floods and earthquakes are the most frequent major natural disasters in the region (Thurman 2011). Reservoirs involve a permanent risk for downstream inhabitants in the case of earthquakes.

Climate change trends in Central Asia will exacerbate problems with water scarcity and water resource management. Central Asia is one of the most vulnerable regions to climate change (Lioubimtseva and Henebry 2009). With increasing climate variability and a warming trend in the region, food and water security issues are becoming even more crucial (Qi et al. 2012). Central Asia is becoming warmer more quickly than the global average (Gupta et al. 2009). Droughts are creating a higher water demand for irrigation. An increase in water consumption of at least 10–15 % due to rising temperatures is expected (Dukhovny and Stulina 2011).

Other important upstream countries claiming water for agriculture in the region are China and Afghanistan. China is going to develop large irrigation projects in the catchments of the Irtysh and the Ili rivers with implications for the downstream riparian Kazakhstan. Climate-induced hazards in mountain areas, such as breakthroughs of glacier lakes, floods, landslides and mudflows, may create problems of freshwater availability. Increasing competition for water will require more effort for economic and political cooperation to avoid conflicts. According to an analysis by Sehring and Giese (2011), hotspots of potential conflict are (a) the Fergana valley, the most densely populated area in Central Asia, belonging partly to Kyrgyzstan, Uzbekistan and Tajikistan, (b) Kyrgyzstan, where a North–South gradient exists, (c) the poor Tajikistan, where the situation is fragile, and (d) the delta of the Amu Darya, where water distribution conflicts between Turkmenistan and Uzbekistan could arise (Sehring and Giese 2011).

3 Natural Conditions for Farming and Rural Development

3.1 Climate and Agro-Climate

Climate is a crucial factor for agriculture and a driver of rural development. Central Asia is a dryland region and has a continental climate with hot summers and cold winters. 90 % of this area has less than 400 mm of precipitation per year, and the average over the total area is 266 mm (de Pauw 2007, in Gupta et al. 2009). The distribution over the region is very different, ranging from extremely high values in the mountains to desert conditions with amounts of less than 100 mm south of the Aral Sea, Lake Balkhash and some smaller regions (Fig. 7). The orographic and climate situation differs between countries. Kyrgyzstan and Tajikistan are typical mountain areas with large climatic differences between the mountains and valleys. Most land area in Turkmenistan is lowland. The Kara Kum Desert covers 80 % of the total area of that country. In Uzbekistan and Kazakhstan, lowlands prevail.

For Kazakhstan, the average annual precipitation is estimated at 344 mm, ranging from less than 100 mm in the Balkhash-Alakol depression in the central-

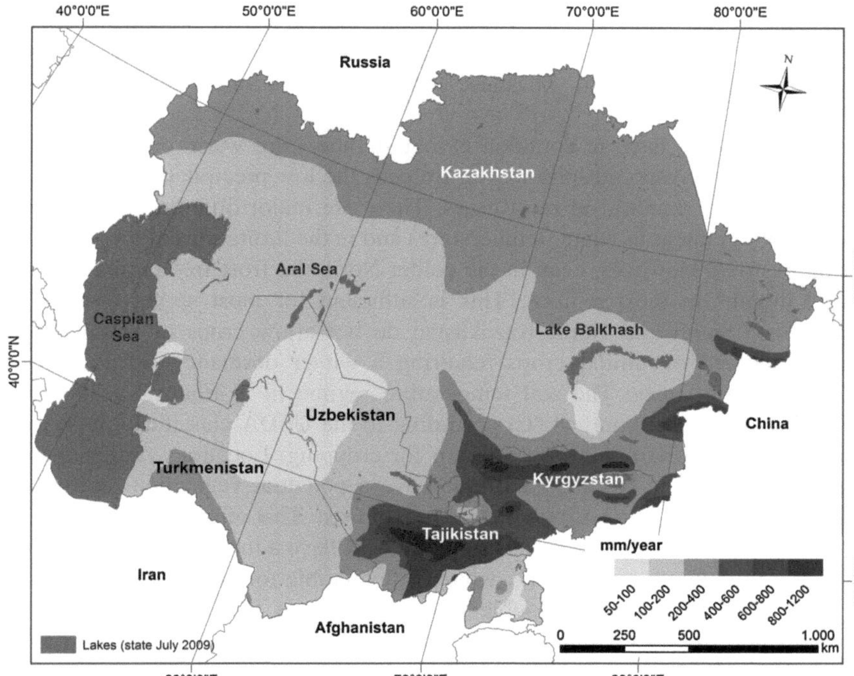

Fig. 7 Distribution of precipitation over Central Asia. Map provided by Igor Klein

eastern part of the country or near the Aral Sea in the south, up to 1,600 mm in the mountain zone in the east and southeast of the country. In Kyrgyzstan, the average annual precipitation is estimated at 533 mm, varying from 150 mm in the Fergana valley to over 1,000 mm in the mountains. Tajikistan also has a continental climate, but in the floodplains of the rivers the climate is characterised by hot, dry summers and mild, warm winters. The average annual precipitation is 691 mm, ranging from less than 100 mm in the southeast up to 2 400 mm on the Fedchenko glacier in the central part of the country. In Turkmenistan the climate is that of a subtropical desert. The average annual precipitation is about 191 mm only, ranging from less than 80 mm in the northeast to 300 mm in the Kopetdag mountain zone in the southwest. The climate of Uzbekistan is arid in over 60 % of the territory. The average annual rainfall is 264 mm, ranging from less than 97 mm in the northwest to 425 mm in the mountainous zone in the middle and southern parts of country (FAO 2012a).

Further climate data from individual stations is brought in here to underpin the agro-climatic situation and big contrasts over the whole region. Table 4 shows climate data from five locations in Kazakhstan: Astana, the capital, located in the centre/north region, Irtyssk, in the Pavlodar region, about 500 km northeast of Astana, Kyzyl Orda on the Syr Darya river, Emba in the western part of the country, and Fort Shevtchenko, also in the west, directly on the Caspian Sea. Further locations are Bishkek, the capital of Kyrgyzstan, Khujand in Northern Tajikistan in the Fergana Valley, Ashgabad, the capital of Turkmenistan, and Tashkent, the capital of Uzbekistan. Urganch, in the Khorezm region, is another example from Uzbekistan. All these locations are at fairly low to moderate altitudes.

One point they have in common is that summers are warm or hot. Also, the annual potential evapotranspiration far exceeds the low precipitation, leading to a classification as semiarid in most cases. There are major differences and a clear north–south gradient in winter temperatures and in the limitations of the vegetation period by frost. However, even in the colder North the frost-free period is longer than 3 months in most regions. This is sufficient for most spring crops. Cold winters in Central and North Kazakhstan do not allow crops of winter wheat, winter rapeseed or similar crops requiring a winter dormancy period, nor the growing of fruit trees. The soil temperature regime in Northern Kazakhstan is "frigid" (soil temperature <8 °C) according to the USDA classification (Keys to Soil Taxonomy 2010) and sub-optimum for cropping. Locations farther than 43 degrees south of the Northern Latitude have a potential vegetation period from April to September which is completely free of frost. This offers great potential for farming and gardening. Most locations in Table 4 have a mesic (>8 °C) or thermic (>12 °C) soil temperature regime, which is favourable for plant root development in soils. The latter enables double cropping systems if water is available. A comparison of the Net Primary Production (NPP) data shows that crop growth is clearly limited by water at all locations.

The high-mountain areas of the Tien-Shan and Pamir are located in the South and South-East of Central Asia. Water from their melting snow and glaciers feeds

Table 4 Climate data for some locations in Central Asia

	Astana, KAZ	Irtyshsk, KAZ	Kyzyl Orda, KAZ	Irgiz, KAZ	Fort Shevtchenko, KAZ
Latitude, Deg. North	51.13	53.35	44.76	48.61	44.55
Longitude, Deg. West	71.36	75.45	65.53	61.26	50.25
Altitude, m	348	94	129	114	−20
Mean annual temp. °C	2.7	1.9	9.8	5.7	11.7
Temp. January, °C	−15.9	−17.4	−8.4	−14.9	−2.3
Temp. July, °C	21.2	21.0	27.5	25.2	25.5
Frost-free months, No.	3	3	5	4	5
Ground frost probability %					
April	40	45	2	22	17
May	5	5	0	0	0
September	6	6	0	1	0
October	45	48	10	34	11
Annual Precipitation, mm	317	290	149	181	181
Potential Evapotranspiration, mm	608	531	1139	737	892
Koeppen climate zone	Dfb	Dfb	BWk	BSk	BWk
NPP_{Temp} g/m^2	808	758	1385	1036	1557
NPP_{Rain} g/m^2	569	526	283	340	339
	Bishkek, KYR	**Khujand, TAJ**	**Ashgabad, TUR**	**Tashkent, UZB**	**Urganch, UZB**
Latitude, Deg. North	42.80	40.21	37.96	41.26	41.56
Longitude, Deg. West	74.50	69.73	58.33	69.26	60.56
Altitude, m	756	414	228	489	99
Mean annual temp. °C	10.6	14.4	16.4	14.2	11.9
Temp. January, °C	−3.6	−0.4	2.2	0.5	−4.9
Temp. July, °C	24.7	28.2	30.8	27.6	27.5
Frost-free months, No.	5	5	7	5	5
Ground frost probability %					
April	0	0	0	0	0
May	0	0	0	0	0
September	0	0	0	0	0
October	14	0	0	3	10
Annual precipitation, mm	441	164	227	426	93
Potential Evapotranspiration, mm	1035	1393	1537	1394	1020
Koeppen climate zone	Dsa	BWk	BWk	BSk	BWk
NPP_{Temp} g/m^2	1463	1794	1958	1775	1573
NPP_{Rain} g/m^2	761	310	420	739	180

Source Database LocClim 1.10 (FAO 2006). Potential evapotranspiration was not measured at some stations; in those cases the data were interpolated from neighbouring stations. *Dfb* humid snow climate with warm summers, *Dsa* snow climate with dry and hot summer, *BSk* arid cold steppe, *BWk* cold desert, NPP_{Temp} and NPP_{Rain} data were also given by LocClim 1.10, calculated by the Miami model developed by Lieth (FAO 2006)

all the major rivers and is a precondition for irrigated agriculture and rural living in the semi-arid and arid lowlands of the region. In the north of Kazakhstan, the climate is semi-arid, and precipitation rates of 200–400 mm per year enable some rainfed cropping on the natural short-grass steppes.

3.2 Land for Cropping and Grazing

Kazakhstan, Kyrgyzstan, Tajikistan, Turkmenistan and Uzbekistan cover a land area of 393 M hectares (Table 5). This is a large resource providing several functions and services in the global ecosystem including the production of food, livestock foraging and fibre for local people and for the market.

Most of the total land in Central Asia is grassland (63 %). Among the top countries by grassland area, Kazakhstan ranks 6th in the world. Due to decades of drought and overgrazing, the productivity is low on pastures, with about 0.1–0.4 t/h of dry matter (Kazakh Ministry 2007). Arable land amounts about 11 % of the agricultural land of Central Asia. Forests are sparse and particularly threatened, covering only about 4 % of the total land area (FAO 2012a) and rapidly declining further on, as has been detected from satellite images (Klein et al. 2012). In terms of croplands, Kazakhstan has more than 20 million hectares of land for rainfed cropping in the northern part of the country. These large areas for spring wheat growing are the result of an immense land conversion initiative in the former Soviet Union in the 1950s. About 42 million hectares of steppe soils were reclaimed for agriculture. More than half of this land (about 25 m ha) was located in Kazakhstan, another 6 M ha in the neighbouring Western Siberia (Meinel 2002). Some land for rainfed cropping is also available in the foothill areas in the south of the country. On rainfed cropping land, wheat is the main crop grown about 80 % in the rotations. Wheat monocultures with included all-year fallow phases prevail, and yields are low and instable.

The type of crops grown on arable land is largely determined by developments in the past. In the Soviet era each republic specialised in producing a specific agricultural commodity according to the prevailing agro-climatic and biophysical resources. Uzbekistan specialised in producing cotton and Kazakhstan in

Table 5 Agricultural land in the 5 transition states of Central Asia in million hectares

Country	Total land	Agricultural land	Grassland	Arable land	Irrigated cropland[1]	Drained land[2]
Kazakhstan	270.0	208.5	185.1	23.4	1.2	0.4 (0.7)
Kyrgyzstan	19.2	10.6	9.3	1.3	1.1	0.1 (0.8)
Tajikistan	14.0	4.7	4.0	0.7	0.7	0.3 (0.6)
Turkmenistan	47.0	32.9	31.0	1.9	2.0	1.0 (1.2)
Uzbekistan	42.5	26.4	22.1	4.3	3.7	2.8 (3.3)
Total	392.7	283.1	251.5	31.6	8.7	4.6 (6.6)

Source AQUASTAT database, FAO 2012a

[1] Total harvested land, full control of irrigation, data for Kazakhstan Tajikistan and Turkmenistan 2006, Uzbekistan 2005 and Kyrgyzstan 1993. More recent data given by Kienzler et al. (2011), based on country statistics, show that the gross irrigated land in 2008 was somewhat higher: 2.08 M ha for Kazakhstan, 1.07 M ha for Kyrgyzstan, 0.72 M ha for Tajikistan, 1.08 M ha for Turkmenistan and 4.21Mha for Uzbekistan

[2] The number in parentheses is land in need of drainage; data from 1993/1994

Fig. 8 Grasslands and croplands. **a** Grasslands dominate Central Asia. Overgrazed and heavily disturbed grasslands in Kyrgyzstan. Photo: Jutta Zeitz. **b** Wheat field in Kazakhstan. Rainfed wheat cropping is often practised with minimum inputs of agrochemicals. This is partly considered environmentally friendly, but nutrient mining may exhaust the soil resources. Competing weeds reduce crop yields

Table 6 Yields of cereals (tonnes per hectare), averaged data from the FAOSTAT database (FAO 2012b)

Country	Average yield in t/ha		Minimum/maximum yield	
	1992–1999	2000–2009	1992–1999	2000–2009
Kazakhstan	0.90	1.01	0.56–1.34	0.94–1.33
Kyrgyzstan	2.27	2.51	1.63–2.77	2.38–3.03
Tajikistan	1.07	1.62	0.88–1.31	1.31–2.78
Turkmenistan	1.91	2.40	0.82–2.61	2.12–3.29
Uzbekistan	1.94	2.94	1.59–2.52	2.44–4.64

producing cereals. After their independence in 1991, all republics had to develop their own independent economies in which agriculture continues to play an important role both for local food needs and for the international market (Gupta et al. 2009) (Fig. 8).

It has been possible to increase cereal yields over recent years (Table 6) but the figures remain relatively low compared with other countries. Kazakhstan produces most cereals under rainfed conditions, whilst in other countries irrigated cereal cropping dominates. The fluctuations of the yield from year to year are also relatively high in some cases even under irrigated conditions. Annual yields lower than 1 t/ha under dryland conditions and lower than 2 t/ha under irrigated conditions should be considered very low and inacceptable. It can be seen that the crop yield level in both Tajikistan and Kazakhstan is low and instable. Uzbekistan has made the most progress in increasing and stabilising the yields of cereals.

4 Performance and Environmental Effects of Irrigated Agriculture

4.1 Cropping Structure and Productivity

Irrigation has been practiced in Central Asia since ancient times. Without irrigation extended oasis regions such as Khorezm and Bukhara would still be desert. Water and irrigation have been important factors for progress, the development of culture and sciences, and the co-operation of people inhabiting Central Asia (Dukhovny and Stulina 2011).

About one third of the arable land in Central Asia is under irrigation, but differences between countries are large. Almost all the arable land in Kyrgyzstan, Tajikistan, Turkmenistan and Uzbekistan is irrigated cropland (Table 5). Large irrigation areas have been installed in the lowlands of all the main rivers in the region (Figs. 9, 10, 11).

Irrigation water is taken from rivers and reservoirs. The current systems have existed since the Soviet era. During that period an extensive irrigation infrastructure was constructed in the form of reservoirs, irrigation canals, pumping stations and field canals (World Bank 2004). The current status of the irrigation systems and structures is a phase of consolidation. The main water courses are being used and maintained, whilst some older and smaller inter-farm transport systems of irrigation water have disintegrated over time (Fig. 12). In some regions irrigation systems have decayed so severely that much of the water never reaches

Fig. 9 Irrigation areas in Central Asia. Fresh water from the main rivers is the basis of extended irrigation. This map was redrawn and modified based on a figure by Bucknall et al. 2003

Fig. 10 Smallholder cotton field near Bukhara, Uzbekistan. Post-harvest situation. The woman picks late-ripened lint. Photo: Courtesy of Dagmar Balla

Fig. 11 Potatoes under sprinkler irrigation

the fields. In Kazakhstan all irrigation and drainage constructions will need rebuilding in due time (Suleimenov et al. 2012). Irrigation technologies are traditional: surface irrigation methods with a high water consumption (Fig. 13). In Kazakhstan, sprinkler irrigation is also common on about 24 % of the irrigated land, mainly in the Northern provinces (FAO 2012a).

The irrigated land of Central Asia, Uzbekistan in particular, is dominated by cotton (*Gossypium hirsutum* L., Payziyeva and Paiziev 2012). Uzbekistan was a main global cotton producer, and cotton remains crucial for the economy of this country. More than 70 % of the irrigated land is used for cotton production. Rice and wheat are also cultivated there, but the production of rice is declining. Because of freshwater shortages, saline water is often used for irrigation. Over the last 15 years the irrigated area has been reduced in Uzbekistan, and it will continue to decline (Turayeva 2012).

Crops grown on irrigated land differ by country. The following FAO data (FAO 2012a) refer to the 1990s, but may characterise the typical status at a later stage. In Kazakhstan, the main irrigated crops were fodder (alfalfa dominating), cereals,

Fig. 12 Technical systems for irrigation. Water transport is provided by hydrotechnical systems constructed 30–40 years ago. **a** Main canal for irrigation water transport. These channels are often not lined, causing water losses, leaching and secondary salinisation of adjacent areas. **b** Central weir for water supply control. **c–d** Inner-farm supply of different maintenance statuses. Some old smaller irrigation systems for cropping land which need some effort for their maintenance, such as artificial ditches made from concrete, are still working (**c**). Many of them have completely disintegrated (**d**)

Fig. 13 Irrigated crops and technology. **a** Experimental rice field in the irrigation system of the Ili river (Almaty region in Kazakhstan). Rice growing is profitable, but has extremely low water use efficiency and causes side-effects of secondary salinisation on adjacent land. **b** Raised-bed cropping of onions and other vegetables and crops in combination with furrow irrigation or low-pressure plastic hoses are emerging technologies. Example of Besagasch research site, near Taraz, Kazakhstan. The research station is operated by the Institute for Water Economy in Taras

cotton, fruits, potatoes and sugar beets. In 1993, irrigated crop yields were 1.8 t/ha for cotton, 1.5 t/ha for wheat, 4.3 t/ha for rice, 3 t/ha for maize, and 2.5 t/ha for grapes. Irrigated farming predominated in the south of Kazakhstan, where cotton is produced on 37 % of the irrigated area, forage crops on 18 % and cereals on 15 %, and vegetables and other cultures on 30 % (Ismukhanov and Mukhamedzhanov 2003). In Kyrgyzstan, the main irrigated crops were fodder and cereals, with wheat dominating. Average wheat yields were 2.2 t/ha. In Tajikistan, the major irrigated crops were cotton, fodder, fruits and grapes, cereals and vegetables. Cotton, fruits and grapes were the most important export crops. In 1994, irrigated crop yields were 1.9 t/ha for cotton, 0.85 t/ha for wheat, 1.7 t/ha for rice and 3 t/ha for grapes. The major irrigated crops in Turkmenistan were cereals (mainly wheat), cotton and fodder. In 1994, irrigated crop yields were 2.3 t/ha for cotton, 1.6 t/ha for wheat, 1.8 t/ha for barley and 2.4 t/ha for rice. In Uzbekistan, cotton dominated, followed by fodder, wheat and fruits. Yields were 2.5 t/ha for cotton, 2.1 t/ha for wheat, 3 t/ha for rice and 4.0 t/ha for grapes (FAO 2012a). Crop yields were low during that period of structural reforms or re-organization in agriculture, a few years after independence. Later, crop yields from irrigated land improved slightly in many cases, but their level remains low or very low in comparison with the leading countries in terms of irrigated agriculture, such as China.

In Uzbekistan, yields of some cash crops such as cotton are sufficient but not stable, while others such as rice remain very low (Abdullaev and Molden 2004). The water productivity of the cotton-growing areas of the Syr Darya basin, one of the largest irrigation systems of Central Asia, is still lower than the world average. The water productivity of rice was 0.21 kg/m^3 water. This is extremely low compared to the world average of 0.7–0.8 kg/m^3 (Abdullaev and Molden 2004). Water productivity is on the increase but is still lower than 1980 (Turayeva 2012). On the other hand, comparing water productivity data with those from other regions can be biased, and results of gross water and solute balancing may be an improper measure of water productivity. The latest developments in lysimetry (Meißner et al. 2010) may lead to better data and a better understanding of the processes.

Over-irrigation may lead to nutrient leaching, water table increases, or runoff and erosion on slightly sloping lands (Reddy et al. 2013). The majority of farms have no crop rotations. The challenge for irrigated agriculture is to increase water productivity, improve and maintain soil fertility and diversify cotton-wheat rotations (Gupta et al. 2009; Turayeva 2012).

4.2 Side-Effects of Irrigated Farming

The long existence of large oases in Central Asia shows that local people managed their irrigation systems under extreme climate conditions in a viable and sustainable way. However, irrigation farming may affect water and land resources adversely in terms of both quality and quantity. During the last 50 years, irrigated areas have been expanded largely without considering resource conservation

(Kienzler et al. 2012). The soils of irrigated land located in former river beds and deserts in the lowlands have a sandy-loam texture and favour losses of nutrients and organic matter (Gupta et al. 2009). Excessive irrigation may lead to the formation of soils with poor physical and chemical properties. Leaching of gypsum, carbonates and organic matter from the soil profile and the accumulation of Na^+ and Mg^{2+} have occurred (Karimov et al. 2009).

Currently, irrigated farming is the main source of water and soil pollution by salinisation. Only a minority of irrigation canals are water proofed. More than 50 % of irrigated soils in Central Asia are salt-affected and/or waterlogged (Qadir et al. 2009). Salinity is a major driver of land desertification (D'Odorico et al. 2012). Waterlogging and salinity problems depend on the position of the area within the river system. They are relatively low in the upper reaches, high in the lower reaches and extremely high in the tail areas.

Salinity is closely related to drainage conditions. About 70–80 % of irrigated areas would need adequate drainage to mitigate problems of waterlogging and salinity, but only about 50 % of irrigated land is drained or equipped for drainage (Table 5). Groundwater tables are too high because of excessive irrigation intensity and insufficient drainage in many cases. For example, in the Khorezm region of Uzbekistan average ground water levels were 1.1–1.4 m at the start of the leaching period and 0.9–1.4 m in July during the growing season. Optimum levels would be 1.4–1.5 m. (Ibrakhimov et al. 2011). Farmers try to adapt to the salinity level by growing salt-tolerant crops. Fodder crops for the winter feeding of livestock are grown on field sites where salinity and poor drainage conditions prevent other crops from being grown (FAO 2012b).

Environmental problems caused by excessive and improperly managed irrigation agriculture are worst in the large downstream countries of Uzbekistan, Turkmenistan and Kazakhstan. A typical example of an inefficient irrigation system is the water loss from the Kara Kum canal, whose banks are unlined. A minimum of 18 % of the total flow is lost through seepage, which has been causing waterlogging and salinisation of the surrounding land. In general, water in the agricultural drainage networks and rivers of Turkmenistan is of poor quality, containing high concentrations of salts and pesticides coming from local and upstream irrigation systems (FAO 2012b). Water is free of charge for agricultural use in Turkmenistan. This promotes wasteful irrigation practices, soil salinity and waterlogging. About half of irrigated land is damaged. Crop yields have declined by 20–30 per cent over the past 10 years (UNECE 2012).

In Uzbekistan, too, land degradation is significant through waterlogging and salinisation of irrigated land, crop diseases and pests due to cotton monoculture (FAO 1997). In rice-growing areas on the tail reaches of the Syr Darya basin the quality of irrigation water is particular low. This can be seen as a barrier to crop diversification, i.e. to shifting from rice to other high-value crops. Improvements in water management in the upper and middle reaches of the basin could lead to improvements in the quality of water delivered to the tail reaches (Abdullaev and Molden 2004).

Meanwhile, groundwater in the tail reaches may be more polluted than surface water (Törnqvist et al. 2011). This means increased health risks for downstream populations when switching to groundwater-based drinking water supplies. Ongoing irrigation expansion has created cumulative health hazards due to high concentrations of copper, arsenic, nitrate and DDT. The Amu Darya delta region is characterised by a very high groundwater salinity in the first shallow, unconfined aquifer. Even in non-irrigated downstream regions of irrigation areas, extremely high salinity levels of 23 g/l were found, higher than in the groundwater of the upstream irrigated region, where the salt content was 3 g/l (Johansson et al. 2009).

4.3 Formation of New Ecosystems: Deserts and Wetlands

Another possible consequence of extended irrigation is the risk of disappearance of the largest freshwater lakes of Central Asia and the formation of new deserts (Figs. 14 and 15). The case of the Aral Sea has reached global public dimensions and awareness. Cai et al. (2003) addressed the Aral Sea disaster as a prime example of unsustainable irrigation development. The sea is located on the territories of Uzbekistan and Kazakhstan. The Aral Sea had an area of about 68,000 km^2 and a volume of 1,061 km^3 in 1960. The water levels were mainly stabilised by annual discharges of 56 km^3 by the Syr Darya and Amu Darya rivers before 1960. As a result of intensive irrigation water consumption until the 1990s, the annual water runoff reaching the Amu Darya and Syr Darya river deltas was reduced to 5 km^3 per year. The Aral Sea lost about two thirds of its volume and surface area and declined 40 m in depth during that time (Kamilov 2003). The sea consists now of some small, separate lakes. Their overall volume is about 90 km^3: the Aral Sea has lost over 90 % of its water, and its salinity has increased about ten to 20-fold (Zavialov 2011).

Meanwhile, the Amu Darya river runs dry before reaching the South Aral Sea. Uzbekistan has started oil exploration in the drying South Aral seabed. Kazakhstan has undertaken some efforts to stabilize the North Aral Sea by building a dam. In 2008, the water level in this lake had risen by 24 m and its salinity had dropped. Some fish stocks have recovered.

Fig. 14 Small Aral Sea, some remaining water, surrounded by a new desert (Photo: Courtesy of Azimbay Otarov)

Fig. 15 Lake Balkhash, currently the largest freshwater lake of Central Asia, is under risk of rapid desiccation and accelerated salinisation (Photos: Courtesy of Azimbay Otarov and Konstantin Pachikin)

The former sea area consists of huge plains covered with salt and toxic chemicals. The consequences include a collapse in the fish industry and sanitation and health problems among the local people (Wikipedia 2012). Technical solutions have been proposed to restore the Aral Sea using water from the rising Caspian Sea (Cathcart and Badescu 2011) but due to the high gradient of about 60 meters (Altitudes: Caspian Sea − 29 m, Aral Sea + 29 m) this would require a great deal of energy: a careful check of the economic and ecological consequences is needed. Taking some flood water from the Ob-Irtysh river system seems to be more realistic (Giese et al. 2004) but would also cause problems due to expected water scarcity in the Irtysh catchment (Hrkal et al. 2006).

The new Aral Kum (Desert) has become a very active dust source over the last three decades (Indoitu et al. 2012).

Groll et al. (2012) measured dust emissions in the desiccated Aral Sea region and found the highest monthly deposition rate in Uzbekistan (up to 56.2 g per m^2). The impact of the Aral Kum as the dominant source of Aeolian dust was significant and limited to a region of approximately 500,000 km^2 surrounding the former Aral Sea. Dust clouds and deposits in Western Uzbekistan come from the drying Aral seabed, and this is often clay mineral agglomerate material and can carry dangerous pollutants (Smalley et al. 2006).

There are many national and international activities for improving or stabilizing the hydrological situation in the Aral Sea region. An International Fund for Saving the Aral Sea (IFAS) has been established. The mission of its Executive Committee (EC IFAS) is "to coordinate cooperation at national and international levels in order to use existing water resources more effectively, and to improve the environmental and socio-economic situation in the Aral Sea Basin" (EC IFAS 2011).

It is important to mitigate the consequences of missing water and to mediate between conflicting interests of stakeholders. However, as long the rainfall does not increase and water evaporates through vegetation and soil on fields instead of flowing to the Aral Sea, the problem will be present. Faster glacier melting could mask the problem for a while.

There are early signs that Lake Balkhash (Fig. 15), currently the largest freshwater lake in the region, could experience a similar destiny. Propastin (2013) calculated the implications of climate change and human activity on the disaster risk for water resources in the Balkhash Lake drainage basin. Water from the Ili River is the most important for the stabilisation of Lake Balkhash. The construction of irrigation systems in the Kazakh part of the drainage basin in the 1970–1990 period led to a significant drop in the water level. As China is planning irrigation systems and intends to reduce the outflow of the Ili river to Kazakhstan, serious consequences for Lake Balkhash are possible. Of three scenarios, two show disaster-like implications for Lake Balkhash (Propastin 2013).

The Aydar-Arnasay lake system, located in Uzbekistan, is an example of a human-made aquatic ecosystem. Its formation is the result of an uncoordinated use of reservoirs. At first it was thought to constitute environmental and economic damage, creating social unrest (Rakhmatullaev et al. 2009). Today, the lake system plays an important role in the regional economy such as fisheries, biodiversity maintenance and conservation, having been designated a RAMSAR site in 2007. Recreation and tourism play also a role (Rodina and Mnatsakanian 2012).

5 Water Quality of Rivers, Lakes and Reservoirs

5.1 Reasons for Diminished Water Quality and Outlook

The quality of all the water bodies in the region is determined both by natural recharge and drainage conditions, and the extent and proportion of sectorial water users, e.g. agriculture, industry or the population, which may also be polluters. Infrastructure, the state of wastewater processing, the strictness with which the authorities control water quality issues and the general environmental awareness of the population are further important factors as regards water quality. Depending on these preconditions, rivers and other water bodies of Central Asia face different risks and perspectives.

The northern and central part of Kazakhstan belong to the catchment of the Ob-Irtysh. This region is characterised by mining and the processing industries. Water and substances in the water drain northwards to Russia and into the Arctic Ocean, which is the final sink. Addressing sources of pollution and taking proper measures may lead to improvements in the river water quality in the long run.

Except the Ob-Irtysh system, all the other main rivers of Central Asia drain into internal basins and are dominated by irrigated agriculture. Improving river water quality is also possible, but substances in the water will accumulate in the internal final sinks: inland lakes, the groundwater of desert sinks and wetlands. Current trends need to be slowed by stopping sectorial users from having an excessive effect on water quality. Monitoring and controlling salinisation is crucial.

Water reservoirs are also important sinks. Soil erosion in upstream areas of Kyrgyzstan and Tajikistan may result in the siltation of reservoirs and water pollution by phosphorus and other soil colloids and solutes. In Tajikistan, the industrial sector, though using much less water than agriculture, is the main source of water pollution with toxic substances (Safarov et al. 2006).

Water pollution by salts and other chemical compounds may have serious implications for aquatic life in waters of all categories. The production of freshwater fish decreases with increasing salt concentration in the water, but further studies are required to discover which water salinity levels are harmful to native fish (Umarov 2003; Djancharov 2003). Serious environmental issues in Central Asia have reduced freshwater supplies to the region and affected the local economy adversely, creating a potential for social unrest. There is a need to develop new technologies to mitigate these problems (Qi and Kulmatov 2008). Water quality standards and norms have been elaborated for all Central Asian countries, as reported by Jumagulov et al. (2009) for Kazakhstan, but need to be monitored and used as a basis for decisions. Examples of the pollution status of some important basins are given below.

5.2 Ob-Irtysh River Basin

The legacy of the past, when environmental standards were low, and recent pollution determine the quality of both main rivers and their tributaries in these watersheds. Industrial activities are the main cause of serious water pollution of the Irtysh river (Hrkal et al. 2006) (Fig. 16). Recharges from contaminated groundwater will be a long-lasting problem. Some locations in the Irtysh river basin, such as the Semipalatinsk nuclear polygon, are among the most ecologically endangered and affected regions on our planet (Hrkal et al. 2006). Extreme water pollution originating from thermal power stations, oil and gas enterprises, from coal mines, and metallurgic enterprises, caused by heavy metals, petroleum

Fig. 16 Black Irtysh in the north of Kazakhstan. It is embedded in an interesting river landscape. The river may carry a high load of toxic substances from industry and mining activities. Stream mud sediments are polluted. Photo: Courtesy of Konstantin Pachikin

products, phenols, nitrates and organic substances, is also reported from the Sarysu River (Karaganda) and the Irtysh-Karaganda canal (Dahl and Kuralbayeva 2001).

Overall, the Irtysh river was one of the most polluted rivers in Kazakhstan at the end of the last century. Excessive water pollution with copper, boron, phenol and cases of extremely high-level pollution with zinc, twice as high as the maximum allowable concentrations, was measured in the Irtysh river or its tributaries (UNECE 2007). The sources of pollution included the metal-processing industry and the discharge of untreated water from mines and other sources. In the Russian Federation, the water quality of the Irtysh falls into the classes "polluted" and "very polluted". Since the start of this century, the water quality has improved (UNECE 2007). The extension of agriculture in these watersheds has implications on possible emissions, which need to be monitored and controlled.

5.3 Aral Sea basin

Crosa et al. (2006) reported on the high salinisation levels of the Amu Darya water, mainly due to sulphates and chlorine. Drainage intensity of agricultural land in the lower catchment and snow and glacier melting in the upper catchment are the main driving forces governing the temporal variation of the salinity of the river water. During low-flow periods salinity is strongly influenced by return water flows from irrigated land (Fig. 17).

Olsson et al. (2013) analysed the Zerafshan River in Uzbekistan at upstream and downstream locations and found an increase in salinity and chemical oxygen demand (COD) concentrations as well as a more sulphate-rich and chloride-rich composition of the downstream waters. A study published by the Scientific Information Center of the Interstate Commission for Water Coordination in Central Asia (SIC ICWC 2011) in the Amu Darya River Basin revealed an increased anthropogenic content over the last 25 years. The development of urban areas, industry, agriculture and insufficient investment in wastewater cleaning

Fig. 17 Ditch near Bukhara, Uzbekistan, highly salinised and wastewater polluted. No fish can live in these waters. Photo: Courtesy of Dagmar Balla

technologies have led to an increased pollution of natural water resources throughout the basin. There was evidence of water pollution by many kinds of industries, including light industry, food, textiles, coal, iron, nonferrous metals, chemicals and more. Wastewater from industrial and municipal sources lead to water quality parameters exceeding the quality targets to a factor of ten, and are finally discharged into surface and ground water bodies (SIC ICWC 2011). When reaching larger rivers, a dilution effect occurs. Thus, high salt contents dominate the monitored water parameters in these rivers.

The salt contents of the Amu Darya range from about 700 mg/l in the upper reaches to 1,200 mg/l in the lower and tail reaches. In the Syr Darya, this range is slightly broader, from 650 to 1,400 mg/l. Sometimes, up to 3 g/l salt were measured as early on as in the Fergana valley. Increased salt contents were associated with significant coliform index increases and higher concentrations of phenols (SIC ICWC 2011).

Contamination by radionuclides has been reported from locations in the Syr Darya river catchment (Skipperud et al. 2012). Polonium-210 was found in the water, and ^{210}Pb and ^{210}Po had accumulated in fish organs from 3 different fish species in the Taboshar Pit Lake, located in the uranium mining area in Tajikistan, and in the Kairakkum Reservoir, Tajikistan. The authors concluded that there was a health risk for the local population through the consumption of fish from the Taboshar Pit Lake.

5.4 Ural River Basin

Dahl and Kuralbayeva (2001) found a significant level of water pollution in the Ural River in Western Kazakhstan. This river drains large regions in Russia. On the other hand, the Ural River has the only remaining spawning habitats in the entire Caspian basin for all sturgeon species. The natural hydrological regime and the ecosystem still seem to be intact (Lagutov 2008). It is argued that legal overfishing is the reason for decreasing fish stocks, not bad water quality.

5.5 Inland Lake Basins

An unknown number of inland lake basins could be partially contaminated by local industries. Two examples are given. Lake Issyk–Kul (Kyrgyzstan), the second largest pristine highland lake of the world, is endangered by radioactive pollution. The source is the abandoned Kadji-Sai field of uranium-bearing brown coal on the southern coast of the lake. Ephemeral streams transport ^{210}Pb, ^{238}U and ^{226}Ra into the lake (Gavshin et al. 2005).

River Nura and its floodplain in Central Kazakhstan face threats by mercury contamination. For several decades, mercury-rich wastewater from an acetaldehyde plant has been discharged largely without treatment. During spring floods highly contaminated silts consisting of fly ash and mercury are transported downstream, leading to a widespread contamination of the river bed and the floodplain (Heaven et al. 2000). River Nura ends in Lake Tengiz, a RAMSAR wetland ecosystem.

6 Land Quality, Land Degradation and Land Use Potential

6.1 Industrial Pollution of Soils

An analysis by Dahl and Kuralbayeva (2001) revealed that the industrial pollution of soil is a serious problem in Kazakhstan. Many areas around major metallurgical, chemical, and energy enterprises have been found to be polluted by toxic substances such as heavy metals, oil and oil products, sulphur oxides, carbon nuclides, and chemical wastes. Almaganbetov and Grigoruk (2008) summarised the situation by stating that the land quality is also adversely affected by soil contamination with oils, particles of heavy metals, radionuclides and other pollutants. Soil contamination with radionuclides is reported from the Semipalatinsk site located in the Irtysh catchment (Hrkal et al. 2006). Areas around uranium mining sites such as the Kurday site in Kazakhstan are also enriched with radionuclides and trace elements (Salbu et al. 2012).

Industrial products for agriculture are also important soil polluters. Nurzhanova et al. (2012) sampled and analysed 80 former pesticide storehouses in the Almaty region of Kazakhstan. They found that soils in and around twenty-four of them were contaminated with organochlorine pesticides residues, showing concentrations higher than the maximum allowable levels.

The pollution of soils by industrial by-products such as heavy metals or persistent chemicals may entail a high risk for food production and biodiversity. On the other hand, that kind of pollution can be frequently traced to its source. Once detected, the area of damage can be localised and measures to eliminate or mitigate the problem can be initiated, based on a risk analysis.

The cases shown and discussed here refer mainly to Kazakhstan, where monitoring systems have been established and research activities in this field are being carried out and published. In Turkmenistan, the status of the soil pollution is unknown, as the country does not provide soil monitoring or soil analyses, or publish state-of-the-environment reports (UNECE 2012).

6.2 Soil Quality for Agriculture

For environmental monitoring, both water and soil quality can be measured using sets of chemical, biological and physical data. In the case of soils, there is a lack of conventions and international standards on the parameters required for measuring. A complete preventive monitoring of all agricultural land would also be an unrealistic goal.

It is more useful and common to measure and evaluate soil quality in terms of its functions for society. For example, the specific role of soil and land in producing plant biomass for humans (productivity function, Mueller et al. 2010) remains crucial. Consequently, higher soil quality means the land has a higher crop yield potential. In this context of agriculture, soil can be considered to be a subcategory of land or even synonymous with it.

Soil quality monitoring by measuring productivity potential may help to answer questions such as: What is the soil quality of a particular land in Central Asia as compared with other regions, and what are main risks or possible reasons for its decline?

Central Asia has many soils which have developed on loess, the source of which is deflated silt or fine sand particles from deserts. This material has favourable physical and chemical properties for agriculture, with a high water and nutrient storage capacity in particular. The best soils formed from loess material and humus content are Chernozems and Kastanozems, located in the northern part of Kazakhstan (brown and red in Fig. 18). Typical Sierozems (Calcisols acc. to WRB 2006, Xerosols in Fig. 18) are also very common in Central Asia, and they

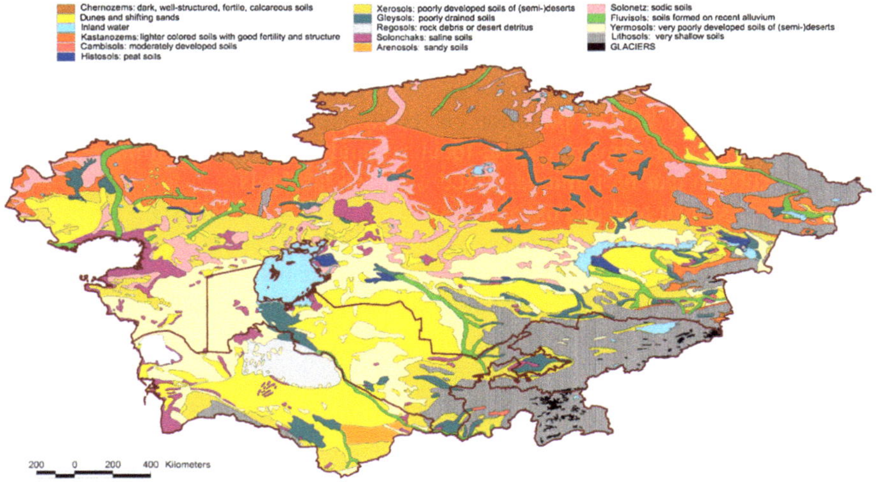

Fig. 18 Map of major soil units of Central Asia. Map was drawn by Rolf Sommer and Eddy de Pauw based on data from (FAO-UNESCO1995). Courtesy of Rolf Sommer and Eddy de Pauw

Fig. 19 Four examples of soil types and soil quality under grassland. **a** Chernozem in the pre-mountain region near Almaty; soil classification acc. to WRB 2006: Haplic Chernozem (Pachic, Siltic). Very good to good soil quality, 81 M-SQR points (Mueller et al. 2010). **b** Alluvial soil in the meadows of the Ili river. Soil classification: Gleyic Fluvisol (Eutric, Siltic). Good soil quality, 62 M-SQR points, **c** Light Kastanozem in the Almaty region. Soil classification: Haplic Calcisol (Eutric, Arenic). Medium soil quality, 43 M-SQR points. Vegetation is degraded by overgrazing, but the vegetation cover may still prevent wind erosion. **d** Brown soil on the high terrace of the Charyn river. Classification: Haplic Calcisol (Eutric, Skeletic). Very low soil quality, 8 M-SQR points. The soil surface is the result of out-blow processes. The stony cover protects the soil surface from further erosion. Extremely dry location

are valuable for rainfed and irrigated agriculture (Turayeva 2012). The majority of these soils have formed on loess or loess-like materials. Loess soils get high scores in basic ratings of agricultural soil quality. Due to climate constraints caused by drought and in some cases by too low temperatures, their overall soil quality and crop yield potential are low to very low on a global scale. Irrigated land has a medium to high overall soil quality (Smolentseva et al. 2011). Figure 19 shows some examples. The overall situation of soil quality in Central Asia and future trends in the context of the Eurasian and global situation does not seem to be clear yet but could be found out by using the Muencheberg Soil Quality Rating (M-SQR, Mueller et al. 2010) to create a strategy of assessing food security for Eurasia or the world.

6.3 Land Degradation by Agriculture

The productivity of available land is often limited by erosion, soil fertility decline, pollution, salinisation and waterlogging (Gupta et al. 2009). Loessy soil substrates have a favourable structure for rooting, but are prone to hydrocollapse and water erosion (Smalley et al. 2006). Land degradation of cropland in rainfed agriculture mostly occurs through soil erosion by both wind and water (Suleimenov et al. 2012). Water erosion induced by furrow irrigation is also a serious problem (Saparov et al. 2013). In the mountainous countries of Tajikistan and Kyrgyzstan, eroded soils dominate agricultural lands, limiting their productivity, whilst

Fig. 20 Two types of salinisation: **a** Primary salinisation due to natural conditions. Charyn river catchment, hummocky landscape, precipitation 180 mm p.a. **b** Secondary salinisation due to leaching from irrigation canals in a rice grown area. Ili river catchment, lowland, precipitation 280 mm p.a. Channel wall in the background

waterlogging, salinisation and sodification are typical problems on irrigated lowlands (Safarov et al. 2006) (Figs. 20, 21 and 22).

Soil degradation through nutrient mining can be a problem too, but is largely reversible with better soil management and fertiliser use (Gupta et al. 2009). Central Asia is a hotspot of wind erosion by nature (Indoitu et al. 2012). Deflation of fine material may lead to skeletic or stony surfaces in deserts and Loess deposits in other regions (Fig. 19d). One positive aspect is that the stony surface protects the brown desert soil underneath and reduces or stops the soil loss. Such mechanisms may be a reason for measured tendencies of decreasing deposition (Indoitu et al. 2012) in Central Asia and offer potential for re-greening the landscape.

Wind erosion may be reach global dimensions in future due to mismanagement of soils by ploughing (BMBF 2011). In this case of permanent rotating by tillage, the soil loses its protective vegetation or its exo-skeletal cover. In the past, the main reasons for wind erosion were overgrazing, logging, the cutting down of shrubs, vehicular traffic, and oil exploration in desert areas. Recent Loess deposition may come from heavily contaminated areas such as the new Aral Kum and may pollute other agricultural sites successively.

Overall, the land area where agriculture is restricted is relatively high in Central Asia (Fig. 23). The total area of land exhibiting significant limitations is 232 million hectares or 60 % of the total land, and 80 % of this problematic land belongs to Kazakhstan (Bot et al. 2000). Sodicity (high pH and Na sorption) and salinity are the dominating processes. They are largely associated with climate conditions, but also caused by land management. A report by Turayeva (2012) revealed that Uzbekistan is characterised by significant land degradation. Three million hectares have been damaged by wind and water erosion. About half of all soils have been damaged by agricultural land use activities: pollution by pesticides, waterlogging and secondary salinisation. Since 1990 the area of salinised soils has increased by one third (Turayeva 2012).

Fig. 21 Slip erosion (landslide erosion) is a common feature on steep deforested land in mountain areas of Central Asia

Fig. 22 Takyr-like soil structures are considered unsuitable for vegetation development. On the other hand, genuine Takyrs are useful landscape elements for rainwater harvesting in deserts

Halting anthropogenically induced land degradation by introducing more sustainable land management is a challenge for all Central Asian countries. This has also implications for the sector of agri-environmental research.

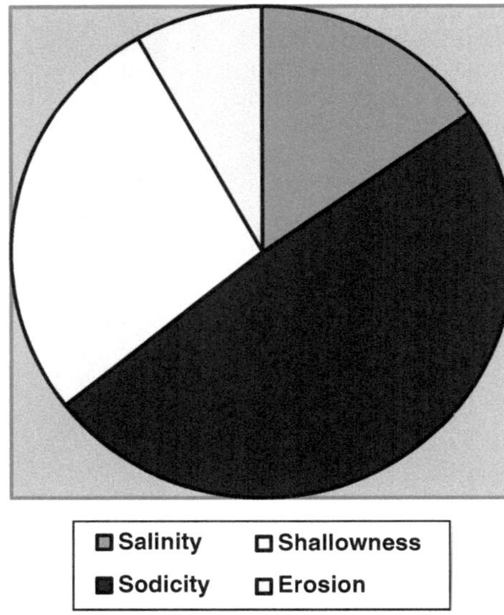

Fig. 23 Types of problematic lands in the five Central Asian Countries. (Data from Bot et al. 2000)

6.4 Ecological Status of Pastures and Rangelands

Grasslands and rangelands are characterised by sparse, varying greening patterns of the vegetation due to weather cycles, and contrasts between locations at different altitudes: valleys and mountain plains. Nomadic culture was a suitable way to utilise the vegetation for livestock grazing. It was part of the lifestyle of the peoples of Central Asia for millennia (Rahimon 2012). Pastoralists followed long-distance migratory routes each season. Seasonal stock movement was essential to avoid degradation (Robinson et al. 2003). The break came in the last century, but in some regions of Kazakhstan some nomadic systems continued with state farm support during the Soviet period (Kerven et al. 2008). The change from nomadic grazing to fixed settlements had some consequences for the vegetation. The current situation of rangelands and pastures is also influenced by the post-Soviet transition phase of restructuring the economy of pastoral systems. Because stock numbers had decreased and based on some field assessments, Robinson et al. (2003) assumed that the rangelands in Kazakhstan were in good condition. Other reports indicate rangeland degradation, but the situation seems to be geographically different.

The degradation of pastures may also be considered a result of poorly managed intensification (Gupta et al. 2009). After the collapse of state farms most pastoralists were constrained to graze their animals in circuits around their homesteads. Severe vegetation and soil degradation has been monitored. There has been some more vegetation degradation around watering points and villages (Rajabov 2009). This can be recognised in spaceborne imagery as brightness belts (Karniely et al. 2008; Bazarbayev and Bayekenova 2008). A number of wells and watering points have disintegrated and reduced the possibilities of grazing. Moreover, in mountain areas, where a lack of water is not as limiting as in the steppe or desert lowlands, grassland degradation has been observed. Based on data collected by the state property registry office in Kyrgyzstan, the country's pasture land was not in best condition in 2005–2006: 2.5 million hectares (27 %) were littered with inedible weeds, 1.7 million hectares (19 %) were eroded, and 3.0 million hectares (33 %) were substantially degraded (USAID 2007). In Uzbekistan, five million hectares of pasture were subject to desertification (Turayeva 2012). Those statistical data need to be considered with care as they may lack background methodology. The occurrence of non-palatable plant species in semi-natural grassland ecosystems is common. It will require more detailed studies to conclude on whether land has been degraded, i.e. on an irreversible loss of ecosystem functions.

There is a lack of fresh, reliable data on the topic of pasture or rangeland degradation. There are also no conventions about methodologies. A loss of plant and wild animal diversity, an increase in unpalatable or toxic plants, a loss of soil fertility and productivity and a decline in livestock production are examples of possible indicators. Scientists of different disciplines or working for different stakeholders have no common basis for measuring and assessing degradation indices (Kerven et al. 2012). Decreased stock capacities, mainly due to the economic recession and adaption phase in the 1990s, provided opportunities for rangeland recovery in some regions (Robinson et al. 2003). Rangeland recovery may comprise palatable biomass, biodiversity and rare species.

Biodiversity is influenced or threatened by several disturbances such as habitat loss, fragmentation of natural communities, over-exploitation such as overgrazing, penetration of non-native species, environmental pollution, climate change, and other elements. On grasslands, overgrazing is a crucial factor. The long-term conservation of vertebrate communities may depend upon the maintenance of ecologically and socially sustainable grazing systems (Sanchez-Zapata et al. 2003). Pictures and trends are unclear, but some trends can be modelled. For example, some endangered bird species have recovered because of abandoned arable fields and ungrazed pristine steppe, whilst others' habitats are associated with the effects of livestock concentration (Kamp et al. 2011). The Saiga Antelope (*Saiga tatarica tatarica*) is a key indicator species of the rangelands in Western Central Asia. Their populations are under threat from habitat loss, poaching, lack of protection and gaps in ecological knowledge (Singh et al. 2010).

Establishing reliable, comparable monitoring systems for Central Asians grasslands is the best way to maintain these valuable ecosystems.

Nevertheless, stocks of grazing animals may give some indication of trends in pasture and rangeland development. Table 7 shows that in Kazakhstan, cattle and sheep stocks collapsed in the 1990s. Sheep stocks also collapsed in Kyrgyzstan, whilst in other countries the decline was more moderate or did not take place at all (Turkmenistan). Stocks of cattle and sheep have been increasing since 2000 in all Central Asian countries. There were still fewer sheep in 2010 than in 1992 in Kazakhstan and Kyrgyzstan, and fewer cattle in Kazakhstan. The steep increase in

Table 7 Stocks of cattle, sheep and goats in million head, rounded data from FAOSTAT (FAO 2012b)

		Kazakhstan	Kyrgyzstan	Tadjikistan	Turkmenistan	Uzbekistan	Overall
Cattle	1992	9.08	1.19	1.39	0.78	5.11	17.55
	2000	4.00	0.95	1.04	1.40	5.27	12.68
	2010	6.10	1.28	1.90	2.20	8.51	19.99
Sheep	1992	33.91	9.22	2.48	5.38	8.27	59.26
	2000	8.72	3.26	1.47	7.50	8.00	28.95
	2010	14.66	3.88	2.62	13.50	12.16	32.16
Goats	1992	0.69	0.30	0.87	0.22	0.92	3.00
	2000	0.93	0.54	0.71	0.50	0.89	3.57
	2010	2.71	0.93	1.58	2.80	2.28	10.30

Fig. 24 Different situations of pastoral grazing in Kazakhstan at the same time in July 2012. **a** Vegetation and soil degradation around watering points in the dry steppe. **b** River lowland with vital vegetation growth, no restrictions on drinking water for the livestock

goat numbers in all countries is remarkable. The goat is the "cow of the poor people". Goats can utilise plants of less nutrition value and some *Artemisia* species. However, they graze relatively aggressively and may eradicate protective bush vegetation. From these facts it is very plausible that land surfaces around villages are now much more degraded than 20 years ago.

A recent study by Vanselow et al. (2012) in the Eastern Pamirs confirmed that pastures close to villages are heavily overgrazed. On summer pastures the grazing pressure is lower, but stock densities are moderate and high, and there is no longer any underuse of grasslands. The grasslands of Turkmenistan suffer from overgrazing and destruction of the vegetation cover due to increased livestock numbers. Rangelands surrounding desert wells have been degraded because of excessive cutting of shrubs for firewood (UNECE 2012). A recent analysis by Suleimenov et al. (2012) confirmed that the most intensive land degradation processes in Kazakhstan have developed on rangelands (Fig. 24).

6.5 Landscape Potential for Recreation of the Urban Population

The global trend of growing population and urbanisation is also valid for Central Asia. More and more people will work and live in cities. They consider and require rural and remote areas as a source of recreation and inspiration (Fig. 25). This is a great opportunity for inhabitants of rural areas and communities to improve their income and to keep pace with the increasing living standard in cities.

Attractive landscapes include lands covered with typical flora and fauna and may also include productive grassland and agricultural ecosystems. Rural landscapes which look like uncontrolled waste disposal sites are not healthy or attractive for visitors. The same holds for rivers, lakes and other open waters. Despite some worsening factors and data regarding the status of land and water, Central Asian landscapes offer great potential for recreation, leisure activities and eco-tourism. This potential needs to be developed and preserved. From their inherent nature and religions, the people of Central Asia have an ecological way of thinking, are inspired by nature and respect its laws. This great human potential is going to be revived.

The protection of the region's great biological diversity including the development of environmental tourism is being addressed and operationalised by National Strategy and Action Plans as a duty in the context of the UNO convention on biodiversity (CBD Secretariat 2012).

The "National Strategy and Action Plans on Conservation and Sustainable Use of Biological diversity in the Republic of Kazakhstan" and the "Long-Term Strategy of the Republic of Kazakhstan up to the Year 2030: Environment and Natural Resources" take the line that "Kazakhstan should become a clean and green country with fresh air and transparent water..." (National Strategy and Action Plan 1999).

All Central Asian countries, and especially the mountainous countries of Kyrgyzstan and Tajikistan, have great potential for environmental tourism. They cover a vast array of different habitats, ranging from polar to temperate and

Fig. 25 Some desert regions such as the Charyn Canyon (Almaty region) have a high potential for tourism (Photos: Maira Kussainova)

Fig. 26 Kirgiz range with *Trollius* in the foreground. Photo: Courtesy of Konstantin Pachikin

Fig. 27 *Tulipa greigii*, a Red Book species, grows in the high herb semi-savanna in West Tien Shan. Photo: Courtesy of Konstantin Pachikin

subtropical ecosystems containing a great richness of species and biological resources (Ministry of Environmental Protection 1998; Safarov et al. 2006) (Figs. 26 and 27). However, in Tajikistan and Kyrgyzstan, the critical state of agriculture, industry and the overall economy is a major constraint for protecting the environment and developing eco-tourism. Thus, the National Environmental

Action Plan of Tajikistan (Safarov et al. 2006) includes planned measures and investment for improving agricultural drainage and irrigation infrastructure, reconstructing a number of drinking water supply and sewerage systems both for rural areas and the industrial sector.

Environmental tourism has many benefits for tourists and for local people. The ecosystems of Central Asia are of great functional, cultural, aesthetic and recreational importance. The agricultural lands and rural population provide a major contribution to maintaining biodiversity and biological resources and must be the main stakeholders in all processes of landscape development. Impact assessment procedures (Helming et al. 2011) are value tools for finding the best solutions for the development of multi-functional landscapes.

7 Research Activities for Initiating Sustainable Resource Management

7.1 The Crisis of Agro-Environmental Research, Education and Monitoring

In the framework of the Central Asian Countries Initiative for Land Management (CACILM), Gupta et al. (2009) revealed a crisis in national agricultural research and education systems in all five countries. It is precisely the branch of the society which needs to be the most innovative and flexible which seems to be suffering the longest from the collapse of the Soviet Union. There are many reasons, such as shortages of research funds, an exodus of staff or an overall lag in technology (Alimgazinova 2009). One key factor should be emphasised: the collapse of the Soviet system took place during a period when the international scientific community switched to a uniform language. English became the only official language at most important international scientific congresses. All leading scientific publications are written in English. This led to a shift in knowledge and technologies worldwide, but to a lesser degree or not at all in Central Asia. The results of the latest research were accessible and beneficial for the English-speaking community, but many good scientists in the former Soviet Union became isolated from the international community and from the benefits of international public goods. Overcoming this crisis in research and education is one of the greatest challenges of the years to come. Scientific cooperation may promote this process (Fig. 28).

A better understanding of interacting processes in agricultural landscapes needs to be gained through research in order to monitor processes, optimise the management of agricultural systems, develop new strategies and find solutions for the sustainable development of rural areas. Knowledge transfer and the site-specific application of available methodological tools can support this work.

There has been a great deal of progress in hydrological and agri-environmental research over recent years that may help to find new and more sustainable

Fig. 28 Education, research and scientific cooperation are key to resolving problems of sustainable resource management. **a** Kyrgyz students practise methods of analysing and assessing soils on a joint cooperation project with Berlin's Humboldt University. Photo: Courtesy of Jutta Zeitz. **b** Researchers from the Uspanov Institute of Soil Science and Agrochemistry sample soils under *Haloxylon* vegetation

solutions for resource management. Most works refer to Uzbekistan and Kazakhstan, and the most progress has been made in these countries (Fig. 29). Some examples will be shown below. The majority of results come from internationally funded projects.

The situation of environmental monitoring, public information and education seems to be critical in Turkmenistan. Water and land monitoring are insufficient. Only two of 16 reservoirs are monitored. Since 1991, no soil analyses have been performed, and no qualitative assessments of land have been carried out since 1998. Analytical equipment is obsolete (UNECE 2012).

Fig. 29 Agri-environmental monitoring station in the meadows of the Ili river. Modern monitoring stations for agrometeorological and environmental data to calculate water, solute and gas exchange of soils and vegetation with the atmosphere are essential for modern research. Kazakh researchers from the Uspanov Institute of Soil Science and Agrochemistry plan to set up the first generation of such stations

7.2 Recent Advances in Research on Food Security and Water Management

7.2.1 Novel Technologies for Measuring, Modelling and Data Processing for Hydrological Systems

In the framework of the Regional Research Network "Water in Central Asia" (CAWa, Echtler et al. 2009), new Remotely Operated Multi-Parameter Stations (ROMPS) for hydrometeorological monitoring in Central Asian headwaters were developed and installed. They not only monitor standard meteorological and hydrological parameters but also deliver GPS data for atmospheric sounding and for tectonic studies (Schöne et al. 2012).

Siegfried et al. (2011) elaborated a coupled climate, land-ice and rainfall runoff model for the Syr Darya catchment to predict the effect of climate changes on hydrological processes. Earlier snow melts affect runoff seasonality. They found a risk for the densely populated, agriculturally productive, and politically unstable Fergana Valley. The seasonal shift in runoff, as projected by their model, is likely to cause serious problems that can only be addressed by constructing new hydrotechnical systems and improving water management.

As open water resources decrease, the utilisation of groundwater becomes more and more important (Turayeva, 2012). Karimov et al. (2010) analysed the storage capacity of aquifers in the Fergana valley and found a limited but not negligible resource that could be taken into account for future projects. Significant water savings were possible using improved *balancing procedures*. The water-saving potential was estimated at about 10 % of the total inflow in the Fergana Valley (Karimov et al. 2012).

Understanding the spatio-temporal patterns of water quality parameters and detecting causal chains in water polluting processes is possible by means of new multi-variate statistical methods of monitoring data. Using those methods, Olsson et al. (2013) quantified hot spots with organic and nutrient pollution due to return flows from industrial effluent and municipal wastewater.

7.2.2 Satellite-Based Monitoring and Modelling of Resources

Areas of glacier retreat, floods and other climate or weather-induced changes in the landscape hydrology can be observed using remote sensing data (Spivak et al. 2004; Bolch et al. 2011). Bai et al. (2012) monitored the surface area changes in inland lakes as an indicator of climate changes and human activities. The area of the inland lakes in Central Asia has shrunk significantly over the past 32 years.

Also, vegetation development processes and human impacts can be studied for complete catchments or regions. Rachkovskaya and Bragina (2012) analysed the patterns of zonal steppe vegetation and constructed a new phyto-geographic map of Kazakhstan which may serve as a basis for the conservation of steppe ecosystems. The Net Primary Production (NPP) for the whole country of Kazakhstan has been calculated and electronically mapped in a recent study by Eisfelder et al. (2012)) in the context of the CAWa project. These data may provide a new quality of land resource monitoring and land use planning. In the same project, Gessner et al. (2012) analysed the relationship between time series of precipitation anomalies and vegetation activity in Central Asia. The results indicate that vegetation is particularly sensitive in areas with 100–400 mm of annual rainfall. Those results could be used to optimise pastoral regimes. Zolotokrylin and Titkova (2011) developed an approach for recognizing the dynamics of desertification centres based on satellite observation data on albedo and surface temperature. They found increasing desertification in the Astrakhan region of Russia and in Western Kazakhstan. Satellite data can be used to model changes in the carbon stocks in the soil and vegetation (Propastin and Kappas 2010).

7.2.3 Modelling Crop and Soil Performance to Optimise Agricultural Systems

Crop models are excellent tools for understanding and quantifying plant growth processes. Sommer et al. (2008) and Kienzler (2009) applied the CropSyst model (Stockle et al. 1994) to optimise nitrogen efficiency in irrigated cotton production in the Khorezm region of Uzbekistan. In combination with field experiments (Ibragimov et al. 2012) optimum N levels of 120–180 kg/ha were quantified for winter wheat depending on available mineralised soil nitrogen. Pereira et al. (2009) developed and tested the irrigation scheduling simulation model ISAREG in the Fergana Valley and found that 20 % of irrigated water percolated out of the root zone. When optimizing irrigation strategies, the contribution of groundwater to the water and solute balance has to be taken into consideration. Sommer et al. (2010) developed the Farm-Level Economic-Ecological Optimization Model (FLEOM), a site-specific integrated model. It optimises land and resource use at farm level by displaying the ecological and economic consequences of different management options.

Carbon stocks of soils and their possible losses causing greenhouse gas emissions are important issues affecting land use strategies with respect to climate change. Soil carbon loss is also associated with a decline in soil fertility and productivity potential when inputs of agrochemicals are limited. Causarano et al. (2010) applied the Environmental Policy Integrated Climate (EPIC) model for estimating soil carbon stocks under different land use and management systems in the semiarid region of central east Kazakhstan. They predicted an increase in soil organic carbon stocks if conservation agriculture were applied, whilst other current systems would lead to a loss of carbon. Modelling approaches such as this are important but need to be validated by measurement data. Ibraeva et al. (2010) measured a significant loss of soil organic carbon under paddy rice cultivation in Southern Kazakhstan. Sommer and de Pauw (2011) used the experimental data given in the literature and parameterised the FAO-UNESCO Soil Map of the

World. Organic carbon stocks in the upper 30 cm of native soils and their declining trends were estimated for the whole of Central Asia.

7.2.4 New Technologies for Land and Water Management at Farm Level

Gupta et al. (2009) list a number of advanced technologies for better agricultural productivity and efficiency at field level. In their studies, laser-assisted land levelling improved water use efficiency in surface-irrigated fields. Irrigation with plastic chutes and re-use of drainage and irrigation water increased crop productivity. Raised-bed technologies for seeding and cropping improved seed germination and provided wheat yields of more than 6 tonnes per hectare. Intercropping of cotton, maize or barley with forage legumes was profitable for farmers. Conservation tillage and planting systems with remaining stubble or mulch on the field increased available soil moisture and reduced erosion. On sloped land, terraces were beneficial (Gupta et al. 2009). Better N fertiliser management cannot close present gaps between the officially recorded yields and those technically achievable but could form one part of optimised irrigated agriculture (Kienzler et al. 2011).

On magnesium-rich soils new technologies have been developed to improve productivity by applying phosphor-gypsum (Vyshpolsky et al. 2008). The accumulation of salts in surface soil layers can be managed using mulch layers, which reduce evaporation from the soil surface. Cotton yield and water productivity under mulching treatments were significantly greater (Bezborodov et al. 2010). New technologies such as drip irrigation led to water savings of 28–42 % in comparison with furrow-irrigated cotton, and the cotton yield increased by 10–19 % (Ibragimov et al. 2007).

All these tools need to be adapted site-specifically based on a thorough diagnosis of crop yield limiting properties and projected economic efficiency and ecological perspectives. Rapid methods such as the application of electromagnetic conductivity meters for a field-scale diagnosis of salinisation (Akramkhanov et al. 2011) can support the process of optimisation. The current project KULUNDA (BMBF 2011) aims to understand, mitigate and prevent processes of land degradation focusing on wind erosion in the Kulunda steppe, a huge region for spring wheat cropping, located in Northern Kazakhstan and Siberia. Within this project, new technological systems are being developed for large farms in this region for conservation tillage in dryland agriculture. Today, no-till and minimum tillage have been adopted as practices on 11.2 million ha of cropland (Suleimenov et al. 2012).

7.2.5 Biotechnologies, Phytoremediation and Biosaline Agriculture

Toderich et al. (2008a) developed agricultural systems based on biosaline crops and livestock to improve people's livelihoods in poor rural areas. This was supported by transferring methodologies developed at the International Centre for Biosaline

Agriculture. The central aspects of these systems are planting valuable halophyte species including bushes and trees to improve the productivity of marginal salt-affected lands, and lowering the groundwater tables by means of biological drainage. They found that some promising species for saline, sandy desert sites were *Haloxylon apphyllum* (saxaul), *Salsola paletzkiana andSalsola richteri* (saltwort). For clay loamy hydromorphic soils, *Atriplex undulata*, *Hippophae rhamnoides* (sea-buckthorn), *Eleagnus angustifolia* (oleaster, Russian olive), *Acacia ampliceps* (salt wattle), *Ulmus pumila* (Siberian elm), *Populus euphratica* (Euphrates poplar) and *Populus nigra* (black poplar) var. *pyramidalis*, *Robinia pseudoacacia* (black locust), *Morus alba* (white mulberry), and *Morus nigra* (black mulberry) were suited. For arable cropping, new varieties of sorghum and pearl millet were tested and introduced. In Turkmenistan, Mamedov et al. (2009) also tested locally adapted plant species for landscape diversification and bioremediation.

Phytoremediation, the use of plants to detoxify pollutants through biological processes, is an effective and ecologically friendly technology to remediate polluted soils (Toderich et al. 2010). For the phytoremediation of sites contaminated with pesticides, Nurzhanova et al. (2012) tested pesticide-tolerant wild plant species. The principle worked with many such plant species, but soil de-contamination is difficult as organochlorine pesticides mainly accumulate in the root system. The bioremediation of marginal or abandoned saline land using halophytes is a promising means of site improvement. It serves multiple purposes: improving livestock feed resources throughout the year, preserving soil and resources, using shrubs as windbreaks to spare the land for other crops and protecting the soil from wind erosion and sand encroachment (Toderich et al. 2008b).

Hbirkou et al. (2011) evaluated the impact of afforestation on soil fertility after 4 years of afforestation. Plantations of *Populus euphratica* and *Ulmus pumila* showed significant levels of reduced soil salinity, increased aggregate stability and improved soil organic carbon stocks.

Fig. 30 River Ili with stands of *Eleagnus* (*left corner*) on the banks. *Eleagnus* (Russian olive) is a preferred plant for afforestation

Fig. 31 Mixed grassland and saksaul (*Haloxylon aphyllum* (Minkwitz)) vegetation in a sandy desert area of the Balkhash river basin. In those regions, saksaul is often the only possible forest and must be protected and re-established. Taproots of *Haloxylon* and some other plants (*Astragalus)* are able to utilise groundwater resources at depths of some meters

Converting degraded cropland to forested areas of *Elaeagnus angustifolia* could be an option if the effects of much lower water consumption and carbon sequestration were taken into account economically (Djanibekov et al. 2012) (Fig. 30). As phosphorus is a growth-limiting element on these soils, P fertilisation improves tree growth significantly (Djumaeva et al. 2012).

In desert regions, saxaul (*Haloxylon aphyllum, H. ammodendron*) is a most important plant (Fig. 31). Restoring and conserving saxaul vegetation is one way to sequester carbon through vegetation in Uzbekistan and Turkmenistan (Thevs et al. 2013). Many other locally adapted plants have potential for technical purposes (Figs. 32 and 33).

Fig. 32 *Ephedra* grows on drylands. It plays a role in traditional medicine. More and more people prefer natural medical products to synthetic ones. In Kazakhstan, ephedrin has resources of 200,000 t of dry raw materials. It would be possible to produce up to 1,000 t annually (National Strategy and Action Plan 1999). Photo: Konstantin Pachikin

Fig. 33 **a** Chu river with extended stands of common reed (*Phragmites australis*) **b** Chee grass (*Achnatherum splendens*). Both species are salt-tolerant and have potential as technical and bioenergy plants

8 Conclusion

Despite much progress in utilising water and land resources in Central Asia more sustainably, many challenges and numerous gaps in knowledge remain. We conclude that there is a need for a shift in scientific methodologies, including measurement methods, evaluation methods, models and technological solutions. Scientists of Central Asia should have access to globally leading technologies for the monitoring and management of resources. They and their work must be embedded in the international scientific community.

The given examples show that scientific cooperation and the inclusion of expertise from foreign partners may help to recognise and resolve problems. A list of further measurement and evaluation methods is available and shall be presented in the next chapters. Some of them will need investment and site-specific studies before they can be applied in Central Asia. Others could be applied immediately and at no cost. This would require awareness and understanding on the part of decision makers and researchers and the introduction of some initiatives.

References

Abdullaev I, Molden D (2004) Spatial and temporal variability of water productivity in the Syr Darya basin, Central Asia. Water Resour Res 2004:1–11

Akramkhanov A, Martius C, Park SJ, Hendrickx JMH (2011) Environmental factors of spatial distribution of soil salinity on flat irrigated terrain. Geoderma 163(2011):55–62

Allouche J (2007) The governance of Central Asian waters: national interests versus regional cooperation. Central Asia at the crossroads. Online source: http://www.unidir.org/pdf/articles/pdf-art2687.pdf. Accessed on 23 Dec 2012

Almaganbetov N, Grigoruk V (2008) Degradation of soil in Kazakhstan: problems and challenges. Soil chemical pollution, risk assessment, remediation and security. NATO Science for Peace and Security Series 2008, pp 309–320

Andrews-Speed P, Bleischwitz R, Boersma T, Johnson C, Kemp G, Van Deveer SD (2012) The global resource nexus. The struggles for land, energy, food, water, and minerals. Transatlantic Academy Washington, DC May 2012, p 100. http://www.gmfus.org/wp-content/blogs.dir/1/files_mf/ta2012report_apr12_web.pdf. Accessed on 6 Dec 2012

Bai J, Chen X, Yang L, Fang H (2012) Monitoring variations of inland lakes in the arid region of Central Asia. Front Earth Sci 6(2):147–156

Bazarbayev AT, Bayekenova MK (2008) The challenge of water provision for pasture lands: a restrictive factor for livestock breeding development in Kazakhstan. Environmental problems of central asia and their economic, social and security impacts. NATO Science for Peace and Security Series C: Environmental Security 2008, pp 201–208

Bezborodov GA, Shadmanov DK, Mirhashimov RT, Yuldashev T, Qureshi AS, Noble AD, Qadir M (2010) Mulching and water quality effects on soil salinity and sodicity dynamics and cotton productivity in Central Asia. Agric Ecosyst Environ 138(2010):95–102

BMBF (2011) KULUNDA—Wie verhindert man die nächste "Global Dust Bowl"?—Ökologische und Ökonomische Strategien zur nachhaltigen Landnutzung in Russischen Steppen: Ein Beitrag zur Anpassung an den Klimawandel. http://www.kooperation-international.de/detail/info/verbundvorhaben-kulunda-wie-verhindert-man-die-naechste-global-dust-bowl-oekologische-und-oekon.html. Accessed on 1 Jan 2013

Bobodzhanova K (2012) Food safety issues in Tajikistan. In: Alpas H, Smith M, Kulmyrzaev A (eds) Strategies for achieving food security in Central Asia. NATO science for Peace and Security Series C: Environmental Security. Springer Science + Business Media B.V. 2012, p 31–42

Bolch T, Peters J, Yegorov A, Pradhan B, Buchroithner M, Blagoveshchensky V (2011) Identification of potentially dangerous glacial lakes in the northern Tien Shan. Nat Hazards 59(3):1691–1714 http://link.springer.com/article/10.1007%2Fs11069-011-9860-2. Accessed 1 Jan 2013

Bot AJ, Nachtergaele FO, Young A (2000) Land resource potential and constraints at regional and country levels. World Soil Res Rep (FAO), No. 90, Rome (Italy) FAO Land and Water Development Division

Bucknall J, Klytchnikova I, Lampietti J, Lundell M, Scatasta M, Thurman M (2003) Irrigation in Central Asia. Social, economic and environmental considerations. The World Bank, February 2003. http://siteresources.worldbank.org/ECAEXT/Resources/publications/Irrigation-in-Central-Asia/Irrigation_in_Central_Asia-Full_Document-English.pdf. Accessed 23 Nov 2012

Cai X, McKinney DC, Rosegrant MW (2003) Sustainability analysis for irrigation water management in the Aral Sea region. Agric Syst 76(2003):1043–1066

Cathcart RB, Badescu V (2011) Aral Sea partial refill with imported Caspian Sea water. In: Badescu V, Cathcart RB (eds) Macro-engineering seawater in unique environments, environmental science and engineering. Springer, Berlin, pp 317–349

Causarano HC, Doraiswamy PC, Muratova N, Pachikin K, McCarty GW, Akhmedov B, Williams JR (2010) Improved modeling of soil organic carbon in a semiarid region of Central East Kazakhstan using EPICAgron Sustain Dev (2011) 31:275–286, doi:10.1051/agro/2010028

CBD Secretariat (2012) Convention of biological diversity. http://www.cbd.int/countries/profile. Accessed 13 Jan 2013

CIA (2011) The world factbook. https://www.cia.gov/library/publications/the-world-factbook/fields/2201.html. Accessed 9 Dec 2012

Crosa G, Froebrich J, Nikolayenko V, Stefani F, Galli P, Calamari D (2006) Spatial and seasonal variations in the water quality of the Amu Darya River (Central Asia). Water Res 40(2006):2237–2245

Dahl C, Kuralbayeva K (2001) Energy and the environment in Kazakhstan. Energy Policy 29(2001):429–440

Djancharov D (2003) The use of irrigation systems for fish production in Kyrgyzstan. In: Petr T (ed) Fisheries in irrigation systems of arid Asia. FAO Fisheries Technical Paper. No 430. Rome, FAO, 2003, 150p, 71–75, ftp://fao.org/docrep/fao/007/y5082e/y5082e00.pdf. Accessed on 23 Nov 2012

Djanibekov U, Khamzina A, Djanibekov N, Lamers JPA (2012) How attractive are short-term CDM forestations in arid regions? The case of irrigated croplands in Uzbekistan. Forest Policy Econ 21(2012):108–117

Djumaeva D, Lamers JPA, Khamzina A, Vlek PLG (2012) The benefits of phosphorus fertilization of trees grown on salinized croplands in the lower reaches of Amu Darya, Uzbekistan. Agroforestry Systems, in print, doi: 10.1007/s10457-012-9576-7

D'Odorico P, Bhattachan A, Davis KF, Ravi S, Runyan CW (2012) Global desertification: drivers and feedbacks. Adv Water Res http://dx.doi.org/10.1016/j.advjwatres.1212.01.013. Accessed on 18 Dec 2012

Dukhovny VA, Stulina G (2011) Water and food security in Central Asia. In: Madramootoo C, Dukhovny V (eds) NATO science for peace and security series C: environmental security, water and food security in Central Asia. 10.1007/978-90-481-9974-7_1, Springer Science + Business Media B.V. 2011

EC IFAS (2011) Online source http://www.ec-ifas.org/about/. Accessed on 18 Dec 2012

Echtler H, Frede H.-G, Küppers AN, Mandychev AN, Merz B, Moldobekov B, Unger-Shayesteh K, Vorogushyn S (2009) Das Forschungsnetzwerk 'Water in Central Asia' (CAWa) – Von Klimawandel bis Wassermanagement.—In: Kramer, M. (eds) Integratives und nachhaltigkeitsorientiertes Wassermanagement : Kooperationspotenziale zwischen Deutschland und Zentralasien, Gabler, 59–68

Eisfelder C, Klein I, Niklaus M, Kuenzer C (2012) Remote sensing based net primary productivity modeling for semi-arid Kazakhstan. In: IGARSS 2012 international geoscience and remote sensing symposium in Munich. Paper. MO3.14.1, Oral presentation on 23 July 2012. http://69.5.195.238/Papers/viewpapers.asp?papernum=2418. Accessed on 18 Dec 18 2012

FAO (2002) The state of food insecurity in the world 2001. FAO Rome http://www.fao.org/docrep/003/y1500e/y1500e00.htm. Accessed on 25 Dec 2012

FAO (2006) New_LocClim 1.10. Local climate estimator. Download–Address:ftp://ext-ftp.fao.org/SD/SDR/Agromet/New_LocClim/ on ext-ftp.fao.org. Accessed on 10 Oct 2010

FAO (2012a) AQUASTAT database, food and agriculture organization of the United Nations (FAO) http://www.fao.org/nr/water/aquastat/main/index.stm. Accessed on 13 Nov 2012

FAO (2012b) FAOSTAT database, http://faostat3.fao.org/home/index.html, Area classification: http://faostat.fao.org/site/371/default.aspx. Accessed on 6 Dec 2012

Gavshin VM, Melgunov MS, Sukhorukov FV, Bobrov VA, Kalugin IA, Klerkx J (2005) Disequilibrium between uranium and its progeny in the Lake Issyk-Kul system (Kyrgyzstan) under a combined effect of natural and manmade processes. J Environ Radioact 83(2005):61–74

Gessner U, Naeimi V, Klein I, Kuenzer C, Klein D, Dech S (2012) The relationship between precipitation anomalies and satellite-derived vegetation activity in Central Asia. Global and Planetary Change, in press, GLOBAL-01897, p 14, doi: http://dx.doi.org/10.1016/j.gloplacha.2012.09.007. Accessed on 18 Dec 2012

Giese E, Sehring J, Trouchine A (2004) Zwischenstaatliche Wassernutzungskonflikte in Zentralasien Nr. 18, Giessen im Mai 2004 http://geb.uni-giessen.de/geb/volltexte/2004/1823/. Accessed on 28 Nov 2012

Granit J, Jägerskog A, Löfgren R, Bullock A, de Gooijer G, Pettigrew S, Lindström A (2010) Regional water intelligence report Central Asia. Stockholm, March 2010 http://www.watergovernance.org/documents/WGF/Reports/Paper-15_RWIR_Aral_Sea.pdf. Accessed on 6 Dec 2012

Grewlich KW (2010) Pipelines, Drogen, Kampf ums Wasser—greift die EU-Zentralasien-Strategie? Neues „Great Game"von Afghanistan bis zum Kaspischen Meer? ZEI discussion paper. Center for European Integration Studies. Rheinische Friedrich-Wilhelms-Universität Bonn 2010. http://www.zei.uni-bonn.de/dateien/discussion-paper/dp_c200_grewlich.pdf. Accessed on 28 Nov 2012

Groll M, Opp C, Aslanov I (2012) Spatial and temporal distribution of the dust deposition in Central Asia—results from a long term monitoring program. Aeolian Research. Available online 29 September 2012, In Press, http://dx.doi.org/10.1016/j.aeolia.2012.08.002. Accessed on 18 Dec 2012

Gupta R, Kienzler K, Martius C, Mirzabaev A, Oweis T, de Pauw E, Qadir M, Shideed K, Sommer R, Thomas R, Sayre K, Carli C, Saparov A, Bekenov M, Sanginov S, Nepesov M, Ikramov R, (2009) Research Prospectus: a vision for sustainable land management research in Central Asia. ICARDA Central Asia and Caucasus Program. Sustainable Agriculture in Central Asia and the Caucasus Series No.1. CGIAR-PFU, Tashkent, Uzbekistan. 84 pp

Heaven S, Ilyushchenko MA, Kamberov IM, Politikov MI, Tanton TW, Ullrich SMU, Yanin EP (2000) Mercury in the River Nura and its floodplain, Central Kazakhstan: II. Floodplain soils and riverbank silt deposits. Sci Total Environ 260(2000):45–55

Hbirkou C, Martius C, Khamzina A, Lamers JPA, Welp G, Amelung W (2011) Reducing topsoil salinity and raising carbon stocks through afforestation in Khorezm, Uzbekistan. J Arid Environ 75(2011):146–155

Helming K, Diehl K, Bach H, Dilly O, König B, Kuhlman T, Perez-Soba M, Sieber S, Tabbush P, Tscherning K, Wascher D, Wiggering H (2011) Ex ante impact assessment of policies affecting land use, Part A: analytical framework. Ecol Soc [Electronic resource].16(1):Art. 27, 1–17, http://www.ecologyandsociety.org/vol16/iss1/art27/. Accessed on 18 Dec 2012

Hrkal Z, Gadalia A, Rigaudiere P (2006) Will the river Irtysh survive the year 2030? Impact of long-term unsuitable land use and water management of the upper stretch of the river catchment (North Kazakhstan). In: Environ Geol 50:717–723, doi: 10.1007/s00254-006-0244-y

Ibraeva MA, Otarov A, Wiłkomirski B, Suska-Malawska M (2010) Humus level in soils of Southern Kazakhstan irrigated massifs and their statistical characteristics. Nat Environ. Monit. Środowiska Przyrodniczego, vol. 11, p. 55–61, Kieleckie Towarzystwo Naukowe, Kielce

Ibragimov N, Evett SR, Esanbekov Y, Kamilov BS, Mirzaev L, Lamers JPA (2007) Water use efficiency of irrigated cotton in Uzbekistan under drip and furrow irrigation. Agric Water Manag 90(1):12–120

Ibragimov N, Djumaniyazova Y, Ruzimov J, Eshchanov R, Scheer C, Kienzler K, Lamers JPA, Bekchanov M (2012) Optimal irrigation and N-fertilizer management for sustainable winter wheat production in Khorezm, Uzbekistan. In: Cotton, water, salts and soums, Springer pp 171–180

Ibrakhimov M, Martius C, Lamers JPA, Tischbein B (2011) The dynamics of groundwater table and salinity over 17 years in Khorezm. Agric Water Manag 101(2011):52–61

Indoitu R, Orlovsky L, Orlovsky N (2012) Dust storms in Central Asia: spatial and temporal variations. J Arid Environ 85:62–70

Ismukhanov K, Mukhamedzhanov V (2003) The use of irrigation systems for sustainable production of agricultural and fish products in the Republic of Kazakhstan. In: Petr T (ed) Fisheries in irrigation systems of arid Asia. FAO Fisheries Technical Paper. No. 430. Rome, FAO, 2003. 150p, 101–114, ftp://fao.org/docrep/fao/007/y5082e/y5082e00.pdf. Accessed on 23 Nov 2012

Johansson O, Aimbetov I, Jarsjö J (2009) Variation of groundwater salinity in the partially irrigated Amudarya River delta, Uzbekistan. J Mar Syst 76(2009):287–295

Jumagulov A, Nikolayenko A, Mirkhashimov I (2009) Water quality standards and norms in the Republic of Kazakhstan. The Regional Environmental Center for Central Asia. Almaty, 2009. 44 p. ISBN 9965-659-89-3

Kamilov B (2003) The use of irrigation systems for sustainable fish production: Uzbekistan. In: Petr T (ed) Fisheries in irrigation systems of arid Asia. FAO Fisheries Technical Paper. No. 430. Rome, FAO, 2003, 150p, 115–126, ftp://fao.org/docrep/fao/007/y5082e/y5082e00.pdf. Accessed on 23 Nov 2012

Kamp J, Urazaliev R, Donald PF, Hölzel N (2011) Post-Soviet agricultural change predicts future declines after recent recovery in Eurasian steppe bird populations. Biol Conserv 144(2011):2607–2614

Karimov A, Qadir M, Noble A, Vyshpolsky F, Anzelm K (2009) Development of magnesium-dominant soils under irrigated agriculture in southern Kazakhstan. Pedosphere 19(3):331–343

Karimov A, Molden D, Khamzina T, Platonov A, Ivanov Y (2012) A water accounting procedure to determine the water savings potential of the Fergana Valley. Agric Water Manag 108(2012):61–72

Karimov A, Smakhtin V, Mavlonov A, Gracheva I (2010) Water 'banking' in Fergana valley aquifers: a solution to water allocation in the Syrdarya river basin? Agric Water Manag 97(2010):1461–1468

Karnieli A, Gilad U, Ponzet M, Svoray T, Mirzadinov R, Fedorina O (2008) Assessing land-cover change and degradation in the Central Asian deserts using satellite image processing and geostatistical methods. J Arid Environ 72(2008):2093–2105

Kazakh Ministry (2007) Kazakh ministry of environmental protection. environmental monitoring and information management system for sustainable land use. Design guidelines for the SKO subsystem, Kazakhstan Ministry of Environmental Protection, Kazakhstan Asian Development Bank: TA-4375; Mott Mc Donald. 11th Nov 2007. http://www2.adb.org/Documents/CACILM/Guidelines-ADB-LandUse.pdf. Accessed on 23 Nov 2012

Kerven C, Shanbaev K, Alimaev I, Smailov A, Smailov K (2008) Livestock mobility and degradation in Kazakhstan's semi-arid rangelands. The Socio-economic causes and consequences of desertification in Central Asia. NATO Sci Peace Secur Ser 2008:113–140

Kerven C, Steimann B, Dear C, Ashley L (2012) Researching the future of pastoralism in Central Asia's mountains: examining development orthodoxies. Mountain Res Develop 32(3):368–377

Keys to Soil Taxonomy (2010) Eleventh edition. ftp://fc.sc.egov.usda.gov/NSSC/Soil_Taxonomy/keys/2010_Keys_to_Soil_Taxonomy.pdf. Accessed on 28 Nov 2012

Kienzler KM (2009) Improving the nitrogen use efficiency and crop quality in the Khorezm region, Uzbekistan. Agrarwissenschaften, Bonn, ZEF/Rheinische Friedrich-Wilhelms-Universität Bonn. p 246. http://hss.ulb.uni-bonn.de/2010/1983/1983.pdf. Accessed on 28 Nov 2012

Kienzler KM, Djanibekov N, Lamers JPA (2011) An agronomic, economic and behavioral analysis of N application to cotton and wheat in post-Soviet Uzbekistan. Agric Syst 104(2011):411–418

Kienzler KM, Lamers JPA, McDonald A, Mirzabaev A, Ibragimov A, Egamberdiev O, Ruzibaev E, Akramkhanov A (2012) Conservation agriculture in Central Asia—What do we know and where do we go from here? Field Crops Res 132(2012):95–105

Klein I, Gessner U, Kuenzer C (2012) Regional land cover mapping and change detection in Central Asia using MODIS time-series. Appl Geogr, 35(1–2):219–234

Kulmyrzaev A (2012) Urgent measures to improve food quality and safety control in the Kyrgyz Republic. In: Alpas H, Smith M, Kulmyrzaev A (eds) Strategies for achieving food security in Central Asia. NATO science for Peace and Security Series C: Environmental Security. Springer Science + Business Media B.V., pp 3–12

Lagutov V (2008) The Ural River sturgeons: population dynamics, catch, reasons for decline and restoration strategies. rescue of sturgeon species in the Ural River Basin. NATO Sci Peace Secur Ser C: Environ Secur 2008:193–276

Lioubimtseva E, Henebry GM (2009) Climate and environmental change in arid Central Asia: Impacts, vulnerability, and adaptations. J Arid Environ 73(2009):963–977

Liu MFM, Pistorius T (2012) Coping with the energy crisis: impact assessment and potentials of non-traditional renewable energy in rural Kyrgyzstan. Energy Policy 44(2012):130–139

Mamedov EY, Esenov PE, Durikov MK, Zverev NE, Tsukanova SK (2009) Experience of halophyte cultivation on saline soils. National Institute of Deserts, Flora and Fauna of the Ministry of Nature Protection of Turkmenistan, Ashgabat, 2009, p 44 http://www.rrcap.ait.asia/library/ca-report/Experience%20of%20Halophyte%20Cultivation%20on%20Saline%20Soils_En.pdf. Accessed on 1 Jan 2013

Mansurova A (2012) Issues of food safety in the Kyrgyz Republic. In: Alpas H, Smith M, Kulmyrzaev A (eds) Strategies for achieving food security in Central Asia. NATO science for Peace and Security Series C: Environmental Security. Springer Science + Business Media B.V. 2012, pp 13–20

MDG indicators (2012) Millenium development goals indicators. The official United Nations site for the MDG indicators. Online source, http://mdgs.un.org/unsd/mdg/host.aspx?Content=indicators/officiallist.htm. Accessed on 23 Dec 2012

Meinel T (2012) Die geoökologischen Folgewirkungen der Steppenumbrüche in den 50-er Jahren in Westsibirien//Diss. Halle 2002. Online: http://sundoc.bibliothek.uni-halle.de/diss-online/02/03H082/t1.pdf. Accessed on 28 Nov 2012

Meißner R, Prasad MNV, Du Laing G, Rinklebe J (2010) Lysimeter application for measuring the water and solute fluxes with high precision. Curr Sci 99(5):601–607

Ministry of Environmental Protection (1998) Kyrgyz republic biodiversity strategy and action plan. MEP, Bishkek. Online source: http://www.cbd.int/doc/world/kg/kg-nbsap-01-en.pdf. Accessed on 14 Jan 2013

Mueller L, Schindler U, Mirschel W, Shepherd TG, Ball BC, Helming K, Rogasik J, Eulenstein F, Wiggering H (2010) Assessing the productivity function of soils. A Rev Agron Sustain Dev 30(3):601–614

National strategy and action plan (1999) National strategy and action plan on conservation and sustainable use of biological diversity in the Republic of Kazakhstan. Ministry of Natural Resources and Protection of Environment of the Republic of Kazakhstan, Kokshetau, 1999, online source: http://www.cbd.int/doc/world/kz/kz-nbsap-01-en.pdf. Accessed on 14 Jan 2013

Nurzhanova A, Kalugin S, Zhambakin K (2012) Obsolete pesticides and application of colonizing plant species for remediation of contaminated soil in Kazakhstan. Environ Sci Pollut Res, in print, doi: 10.1007/s11356-012-1111-x. Springer

OECD (2003) Organisation for economic co-operation and development. Urban Water Reform in Eastern Europe, Caucasus and Central Asia. Pogress since the Almaty Ministerial Conference. Online source: http://www.oecd.org/environment/environmentinemergingandtransitioneconomies/14626557.pdf. Accessed on 16 Dec 2012

Olsson O, Gaßmann M, Manig N, Ikramova M, Wegerich K (2013) Basin efficiency approach and its effect on streamflow quality, Zerafshan River Uzbekistan. J Hydrol 476:128–135

Payziyeva S, Paiziev A (2012) Food security in Uzbekistan. In: Alpas H, Smith M, Kulmyrzaev A (eds) Strategies for achieving food security in Central Asia. NATO science for Peace and Security Series C: environmental Security. Springer Science + Business Media B.V. 2012, pp 43–56

Pauw E de (2007) Principal biomes of Central Asia. pp 3–24 In Climate change and terrestrial carbon sequestration in Central Asia. Lal R, Suleimenov M, Stewart BA, Hansen DO, Doraiswamy P (eds) Taylor and Francis Group, New York, NY, USA

Pereira LS, Paredes P, Eholpankulov ED, Inchenkova OP, Teodoro PR, Horst MG (2009) Irrigation scheduling strategies for cotton to cope with water scarcity in the Fergana Valley Central Asia. Agric Water Manag 96(2009):723–735

Propastin P (2013) Assessment of climate and human induced disaster risk over shared water resources in the Balkhash Lake drainage basin. Climate Change and Disaster Risk Management. Climate Change Manag 2013:41–54

Propastin PA, Kappas M (2010) Modelling carbon Sequestration in drylands of Kazakhstan using remote sensing data and field measurements. Land Degradation and Desertification: Assessment, Mitigation and Remediation, pp 297–306

Qadir M, Noble AD, Qureshi AS, Gupta R, Yuldashev T, Karimov A (2009) Salt-induced land and water degradation in the Aral sea basin: a challenge to sustainable agriculture in Central Asia. Nat Res Forum 33(2):134–149

Qi J, Kulmatov R (2008) An overview of environmental issues in Central Asia. environmental problems of central asia and their economic, social and security impacts. NATO Sci Peace Sec Ser C: Environ Secur 2008:3–14

Qi J, Bobushev TS, Kulmatov R, Groisman P, Gutman G (2012) Addressing global change challenges for Central Asian socio-ecosystems. Front Earth Sci 6(2):115–121

Rachkovskaya EI, Bragina TM (2012) Steppes of Kazakhstan: diversity and present state. Eurasian steppes. Ecological Problems and Livelihoods in a Changing World. Plant Veg 6:103–148

Rahimon RM (2012) Evolution of land use in Pastoral Culture in Central Asia with special reference to Kyrgyzstan and Kazakhstan. In: Victor Squires (Ed) Rangeland Stewardship in Central Asia. balancing improved livelihoods, biodiversity conservation and land protection. ISBN: 978-94-007-5366-2

Rajabov T (2009) Ecological assessment of spatio-temporal changes of vegetation in response to piosphere effects in semi-arid rangelands of Uzbekistan. Online Source: http://www.unulrt.is/static/fellows/document/rajabov-t.pdf. Accessed on 9 March 2013

Rakhmatullaev S, Huneau F, Le Coustumer P, Motelica-Heino M, Bakiev M (2009) Facts and perspectives of water reservoirs in Central Asia: a special focus on Uzbekistan. Author manuscript, published in Water 2, 2 (2010) pp 307–320. DOI: 10.3390/w10x000x, Online: http://hal-insu.archives-ouvertes.fr/docs/00/45/78/24/PDF/Rakhmatullaev-Water-2009.pdf. Accessed on 21 Dec 2012

Reddy JM, Jumaboev K, Matyakubov B, Eshmuratov D (2013) Evaluation of furrow irrigation practices in Fergana valley of Uzbekistan. Agric Water Manag 117(2013):133–144

Robinson S, Milner-Gulland EJ, Alimaev I (2003) Rangeland degradation in Kazakhstan during the Soviet era: re-examining the evidence. J Arid Environ 53:419–439, doi:10.1006/jare.2002.1047

Rodina K, Mnatsakanian R (2012) Spills of the Aral Sea: formation, functions and future development of the Aydar-Arnasay Lakes. Lagutov V (ed) Environmental security in watersheds: the sea of Azov, Chapter 11, 183, NATO Science for Peace and Security Series C: Environmental Security, doi: 10.1007/978-94-007-2460-0_11, Springer Science + Business Media B.V. 2012, pp 183–215

Ryabtsev AD (2011) Threats to water security in the Republic of Kazakhstan: the transboundary context and possible ways to eliminate them. In: Madramootoo C, Dukhovny VA (eds) Water and food security in Central Asia, NATO Science for Peace and Security Series C: Environmental Security, pp 69–75. DOI 10.1007/978-90-481-9974-7_6, Springer Science + Business Media B.V. 2011

Safarov NM, Asrorov IA, Isufov U, Mahmadaliev BU, Novikova TM, Khairulloev R (2006) Republic of Tajikistan. National Environmental Action Plan, Dushanbe 2006, online Source: http://www.unpei.org/PDF/National_Environmental_Action_Plan_eng.pdf. Accessed on 14 Jan 2013

Salbu B, Burkitbaev M, Strømman G, Shishkov I, Kayukov P, Uralbekov B, Rosseland BO (2012) Environmental impact assessment of radionuclides and trace elements at the Kurday U mining site, Kazakhstan. J Environ Radioact, in print, http://www.sciencedirect.com/science/article/pii/S0265931X12001154. Accessed on 1 Jan 2013

Sanchez-Zapata JA, Carrete M, Gravilov A, Sklyarenko S, Ceballos O, Dona′zar JA, Hiraldo F (2003) Land use changes and raptor conservation in steppe habitats of Eastern Kazakhstan. Biol Conserv 111:71–77

Saparov A, Mirzakeev EK, Sharypova TM, Saparov GA, Abuduwaili J (2013) Irrigation erosion of irrigated soils in the foothills of southern Kazakhstan. J Arid Land 5(2):166–171 doi: 10.1007/s40333-013-0157-z

Schöne T, Zech C, Unger-Shayesteh K, RudenkoV, Thoss H, Wetzel H.-U, Zubovich A (2012) A new permanent multi-parameter monitoring network in Central Asian high mountains—from measurements to databases. Geosci Instrum Method Data Syst Discuss. 2:301–336 www.geosci-instrum-method-data-syst-discuss.net/2/301/2012/ doi:10.5194/gid-2-301-2012. Accessed on 5 Dec 2012

Sehring J, Giese E (2011) Global environmental change and conflict potential in Central Asia. In: Brauch HG, Oswald Spring Ú, Mes¬jasz C, Grin J, Kameri-Mbote P, Chourou B, Dunay P, Birkmann J (eds) Coping with global environmental change, disasters and security—threats, challenges, vulnerabilities and risks. Hexagon series on human and environmental security and Peace, vol. 5 Springer, Berlin, pp 525–534

SIC ICWC (2011) Scientific-Information Center, Interstate Commission for Water Coordination in Central Asia (SIC ICWC), United Nations Economic Commission for Europe (UNECE), Central Asian Regional Environmental Center (CAREC), Water quality in the amudarya and syrdarya river basins. Analytical Report. Tashkent 2011. 77 pp. Online source: http://www.cawater-info.net/water_quality_in_ca/files/analytic_report_en.pdf. Accessed on 16 Dec 2012

Siebert S, Burke J, Faures JM, Frenken K, Hoogeveen J, Döll P, Portmann FT (2010) Groundwater use for irrigation—a global inventory. Hydrol Earth Syst Sci 14(2010):1863–1880

Siegfried T, Bernauer T, Guiennet R, Sellars S, Robertson AW, Mankin J, Bauer-Gottwein P, Yakovlev A (2011) Will climate change exacerbate water stress in Central Asia? Climatic change. Springer Science + Business Media B.V. 2011, doi: 10.1007/s10584-011-0253-z, http://www.ib.ethz.ch/docs/Climatic_Change.pdf. Accessed on 1 Jan 2013

Singh NJ, Grachev IA, Bekenov AB, Milner-Gulland EJ (2010) Tracking greenery across a latitudinal gradient in central Asia—the migration of the saiga antelope. Divers Distrib 16:663–675 (Blackwell)

Skipperud L, Jørgensen AG, Heier LS, Salbu B, Rosseland BO (2012) Po-210 and Pb-210 in water and fish from Taboshar uranium mining Pit Lake, Tajikistan. J Environ Radioact, in press, corrected proof http://www.sciencedirect.com/science/article/pii/S0265931X12000793. Accessed on 1 Jan 2013

Smalley IJ, Mavlyanova NG, Rakhmatullaev KL, Shermatov MS, Machalett B, O'Hara Dhand K, Jefferson IF (2006) The formation of loess deposits in the Tashkent region and parts of Central Asia; and problems with irrigation, hydrocollapse and soil erosion. Quatern Int 2006:152–153

Smolentseva EN, Rukhovic O, Lukin S, Suleimenov M, Saparov A, Balgabayev N, Pachikin K, Kussainova MD, Bekbaev U, Schindler U, Mueller L (2011) Ozenka katcestva i potenzial urozhainosti pocv v globalnom masshtabe. Pocvovedenie i Agrochimija, Almaty 4:81–91

Spivak L, Arkhipkin O, Pankratov V, Vitkovskaya I, Sagatdinova G (2004) Space monitoring of floods in Kazakhstan. Math Comput Simul 67:365–370

Sommer R, Kienzler K, Conrad C, Ibragimov N, Lamers JPA, Martius C, Vlek PLG (2008) Evaluation of CropSyst for simulating the potential yield of cotton in Uzbekistan. Agron Sustain Develop 28(2):345–354. doi:10.1051/agro:2008008

Sommer R, Djanibekov N, Salaev O (2010) Optimization of land and resource use at farm-aggregated level in the Aral Sea Basin of Uzbekistan with the integrated model FLEOM—model description and first application, ZEF Discussion Papers On Development Policy No. 139, Center for Development Research, Bonn, July 2010, pp. 102, Online source: http://www.zef.de/fileadmin/webfiles/downloads/zef_dp/ZEF_DP_139.pdf. Accessed on 16 Dec 2012

Sommer R, de Pauw E (2011) Organic carbon in soils of Central Asia—status quo and potentials for sequestration. Plant soil 338:273–288, doi: 10.1007/s11104-010-0479-y

Stockle CO, Martin SA, Campbell GS (1994) CropSyst, a cropping systems simulation model: water/nitrogen budgets and crop yield. Agric Syst 46:335–359

Suleimenov M, Saparov A, Akshalov K, Kaskarbayev Z (2012) Land degradation issues in Kazakhstan and measures to address them: research and adoption. Pedologist 2012:373–381

Swinnen JFM, Vranken L (2010) Reforms and agricultural productivity in Central and Eastern Europe and the former soviet republics: 1989–2005. J Prod Anal 33:241–258. Doi: 10.1007/s11123-009-0162-6

Thevs N, Wucherer W, Buras A (2013) Spatial distribution and carbon stock of the Sexual vegetation of the winter-cold deserts of Middle Asia. J Arid Environ 90:29–35

Thurman M (2011) Natural disaster risks in Central Asia: a synthesis. Online source: http://europeandcis.undp.org/uploads/public1/files/vulnerability/Senior%20Economist%20Web%20site/CA_Disaster_Risks_-_MT_-_D5_11_04.pdf. Accessed on 16 Dec 2012

Toderich K, Tsukatani T, Ismail S, Massino I, Wilhelm M, Yusupov S, Kuliev T, Ruziev S (2008a) Extent of salt affected land in Central Asia: Biosaline agriculture and utilization of the salt-affected resources, KIER Working Papers, 35 pp. Abstract at: http://ideas.repec.org/p/kyo/wpaper/648.html. Accessed on 23 Dec 2012

Toderich KN, Ismail S, Juylova ES, Rabbimov AA, Bekchanov BB, Shyuskaya EV, Gismatullina LG, Osamu K, Radjabov TB (2008b) New approaches for biosaline agriculture development, management and conservation of sandy desert ecosystems. In: Biosaline agriculture and high salinity tolerance. Springer 2008, pp 247–264. http://link.springer.com/chapter/10.1007%2F978-3-7643-8554-5_23?LI=true. Accessed on 23 Dec 2012

Toderich KN., Shuyskaya EV, Khujanazarov TM, Ismail S, Kawabata Y (2010) The structural and functional characteristics of Asiatic desert halophytes for phytostabilization of polluted sites. In: Ashraf M, Ozturk M, Ahmad MSA (eds) Plant adaptation and phytoremediation pp 245–274, doi: 10.1007/978-90-481-9370-7, Springer Science + Business Media B.V. 2010

Törnqvist R, Jarsjö J, Karimov B (2011) Health risks from large-scale water pollution: trends in Central Asia. Environ Int 37(2011):435–442

Tultabaieva T (2012) Raw materials and products in Kazakhstan. In: strategies for achieving food security in Central Asia (2012) NATO science for peace and security series C: environmental security. In: Alpas H, Smith M, Kulmyrzaev A (eds) Strategies for achieving food security in Central Asia. Chapter 7: maintenance of quality of meat and dairy. ISBN: 978-94-007-2501-0

Turayeva S (2012) The problems of management and effective utilization of water and ground resources in Uzbekistan. In: Alpas H, Smith M, Kulmyrzaev A (eds) Strategies for achieving food security in Central Asia. NATO science for Peace and Security Series C: environmental Security. Springer Science + Business Media B.V. 2012. pp 57–66

Umarov P (2003) Uzbekistan irrigation systems and their management potential for fisheries in regional context. In: Petr T (ed) Fisheries in irrigation systems of arid Asia. FAO Fisheries Technical Paper. No. 430. Rome, FAO. 2003. 150p., 125–150, ftp://fao.org/docrep/fao/007/y5082e/y5082e00.pdf Accessed on 23 Nov 2012

UNDP (2004) Water resources of Kazakhstan in the New Millennium Almaty, 2004, A series of UNDP publication in Kazakhstan, # UNDPKAZ 07 http://waterwiki.net/images/a/ad/KazakhstanWater.pdf. Accessed on 12 Jan 2013

UNDP (2011) United Nations Development Program (UNDP). Human Development Report 2011. Country Profiles and International Human Development Indicators. http://hdr.undp.org/en/countries/. Accessed on 16 Dec 2012

UNECE (2007) United Nations, Economic Commission for Europe, convention on the protection and use of transboundary watercourses and international lakes. 1st Assessment of Transboundary rivers, lakes and groundwaters. United Nations, New York and Geneva, 2007, http://www.unece.org/fileadmin/DAM/env/water/blanks/assessment/assessment_full.pdf. Accessed on 10 Dec 2012

UNECE (2012) United Nations Economic Commission For Europe. Environmental performance reviews. Turkmenistan. First review synopsis. UNITED NATIONS. New York and Geneva, 2012. 35 pp. http://www.unece.org/index.php?id=31562. Accessed on 3 Jan 2013

USAID (2007) United States agency international development. Pasture reform. suggestions for improvements to pasture management in the Kyrgyz Republic. Online source. http://pdf.usaid.gov/pdf_docs/PNADN532.pdf. Accessed on 29 Dec 2012

Vanselow KA, Kraudzun T, Samimi C (2012) Grazing practices and pasture tenure in the Eastern Pamirs. The nexus of pasture use, pasture potential, and property rights. Mountain Res Develop 32(3):324–336

Vyshpolsky F, Qadir M, Karimov A, Mukhamedjanov K, Bekbaev U, Paroda R, Aw-Hassan A, Karajeh F (2008) Enhancing the productivity of high-magnesium soil and water resources in Central Asia through the application of phosphosgypsum. Land Degrad Develop 19:45–56

Wegerich K, Olsson O, Froebrich J (2007) Reliving the past in a changed environment: hydropower ambitions, opportunities and constraints in Tajikistan. Energy Policy 35(2007):3815–3825

Wikipedia (2012) Aral Sea. Online source: http://en.wikipedia.org/wiki/Aral_Sea. Accessed on 10 Dec 2012

World Bank (2004) Water energy nexus in Central Asia. Improving regional cooperation in the Syr Darya Basin. The World Bank, Washington DC, January 2004. http://siteresources.worldbank.org/INTUZBEKISTAN/Resources/Water_Energy_Nexus_final.pdf. Accessed 28 Nov 2012

WRB (2006) World reference base for soil resources 2006, a framework for international classification, correlation and communication, FAO Rome, 2006, World Soil Resources Reports 103, p 145
WTO (2012) Trade profiles. Online source: http://stat.wto.org/CountryProfile/WSDBCountryPFHome.aspx?Language=E. Accessed on 13 Jan 2013
Yessekin B, Burlibayev M, Medvedeva N, Stafin S (2008) Water ecosystems of central asia: important factors affecting the environmental-social prosperity of the region. transboundary water resources: a foundation for regional stability in Central Asia. NATO Sci Peace Secur Ser C: Environ Secur 2008:43–64
Zavialov P (2011) Ongoing changes in physical and chemical regimes of the Aral Sea. Quatern Int 279–280:553–554 16 November 2012, XVIII INQUA Congress, 21st–27th July, 2011, Bern, Switzerland
Zolotokrylin N, Titkova TB (2011) A new approach to the monitoring of desertification centers. Arid Ecosyst 1(3):125–130. ISSN 2079_0961 (Pleiades Publishing, Ltd.,)

Soil Resources of the Republic of Kazakhstan: Current Status, Problems and Solutions

Abdulla Saparov

Abstract This chapter includes information on the current status of the soil landscape of Kazakhstan and an analysis of materials based on our own research published during the last two decades, as well as literature on land degradation, salinisation, soil contamination and land resource use. It has been found that a current tendency can be observed towards intensive land degradation and desertification, salinisation (including secondary salinisation and soil contamination by oil, petroleum products, chemicals and radioactive substances), de-humification, decreasing soil fertility and deterioration of the ecological status of soil resources. In Kazakhstan this problem is complex because of the vast territory and variety of both natural conditions and anthropogenic impacts on soil. The sustainable development of agriculture in Kazakhstan and ensuring food and environmental safety are closely related to the rational and efficient use of soil. In this regard, it is necessary to develop appropriate laws and regulations, programmes and related activities and to take measures to prevent rapid land degradation, desertification and deterioration of the environmental situation, and to start to restore soil fertility.

Keywords Kazakhstan · Soil · Degradation · Soil fertility

1 Soil Inventory

The soil surface of Kazakhstan is characterized by a distinct horizontal and vertical zoning which is linked with the significant length of the plains stretching from north to south, high altitudes in the mountains, and changes in the bioclimatic

A. Saparov (✉)
Kazakh Research Institute of Soil Science and Agrochemistry Named After U. U. Uspanov,
Al Faraby Ave 75b, 050060 Almaty, Kazakhstan
e-mail: ab.saparov@yahoo.com; ab.saparov@mail.ru

L. Mueller et al. (eds.), *Novel Measurement and Assessment Tools for Monitoring and Management of Land and Water Resources in Agricultural Landscapes of Central Asia*, Environmental Science and Engineering, DOI: 10.1007/978-3-319-01017-5_2,

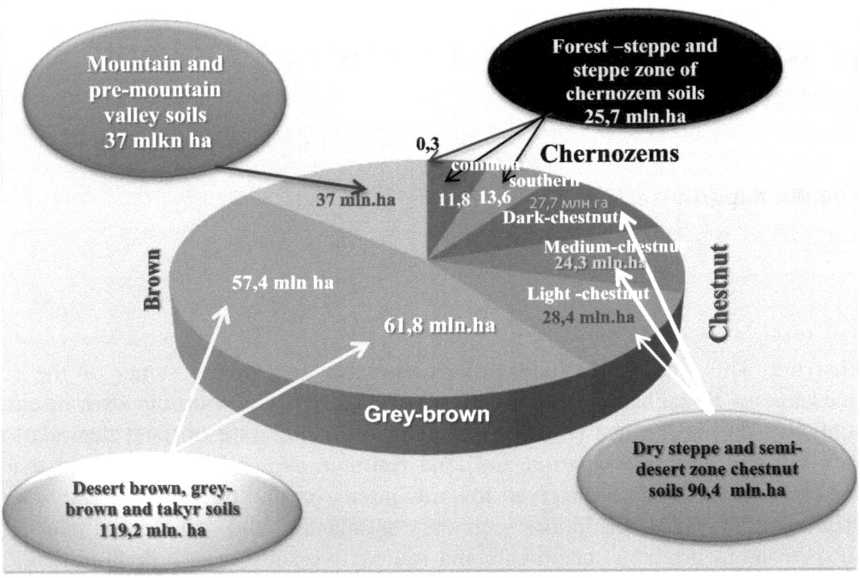

Fig. 1 Soil surface of the Republic of Kazakhstan (Saparov et al. 2006b). Read *comma* as point

conditions of soil formation. The flat area occupies 86 % of the Republic's territory.

In Kazakhstan, there are 25.7 million hectares (M. ha) of black soils (Chernozems), 90.4 M. ha of chestnut soils (Kastanozems), 119.2 M. ha of brown and grey-brown soils (Calcisols), and 37 M. ha of mountain soils (Faizov et al. 2006, Fig. 1).

The flat territory of the Republic is divided into natural zones and subzones:

- Moderately wet forest steppe with grey forest soils.
- Moderately arid and arid steppe zone with black soils.
- Moderately dry, dry and desert-steppe zone with chestnut soils.
- Desert zone with brown and grey-brown soils.

According to bioclimatic indices and basic and auxiliary soil formation processes, the moderately wet forest steppe zone with grey forest soils, and the moderately arid and arid steppe zone with black soils, are divided into three subzones:

- Moderately wet southern forest steppe with grey forest soils, black soils (Chernozems) and meadow-black soils (Meadow Chernozems);
- Moderately arid steppe with common black soils (Ordinary Chernozems);
- Arid steppe with southern black soils (Southern Chernozems).

In the structure of the soil landscape, the black soils and meadow-black soils occupy about 40 %, grey forest soils and salty soils 20, semi-hydromorphic and hydromorphic soils more than 20, salty and saline marsh soils 19 % of the territory. On the background of zonal soils, the common black soil complexes with saline soils comprise about 17 %, semi-hydromorphic and hydromorphic soils 10 %, saline soils 5 %, and saline soils and saline marsh soils 4 % of the area. Zonal southern black soils occupy 65 % of the subzone area, of which 22 % are their complexes with saline marsh soils, 10 % semi-hydromorphic and hydromorphic soils and 3 % saline soils and saline marsh soils.

In the zone of moderately wet forest steppe with grey forest soils and black soils, processes of de-humification and decreases in natural fertility have been observed. In the moderately arid steppe zone with black soils (common and southern) intensive processes of dehumification and erosion are observed, particularly strongly on calcareous soils. The moderately dry, dry and desert-steppe zone with chestnut soils occupies more than 30 % of the territory of Kazakhstan. The soil landscape in this zone is characterized by complex soils, with alkaline, calcareous and saline soils widespread, and more than 40 % is occupied by a variety of complexes of soils with saline soils. The basis of the soil is dark chestnut soils that are spread over the low Kazakh hills, the trans-Ural and pre-Ural plateau and the low mountains of Mugaldzhar and some parts of the Caspian, as well as the Turgay and pre-Ural plateau and the Irtysh plain. This is a zone of rainfed agriculture, i.e. four or five years out of ten are dry, and there is no yield.

The subzone of desert steppe with light chestnut soils occupies the extreme southern level of the chestnut soils zone from the Caspian depression in the west to the foothills of Altay and Tarbagatai in the east. The main background of the soil landscape in the subzone is light-chestnut saline and saline marsh soils. The usual light chestnut, calcareous soils and weakly developed soils are less common. In the structure of the soil landscape, the zonal light chestnut soils occupy 63.3 % of the subzone area, saline soils 20.2 %, semi-hydromorphic soils 5.2 %, hydromorphic soils about 4 %, saline marsh soils 1.6 % and sands 1.2 %.

In the moderately dry, dry and desert-steppe zone of chestnut soils, where the variety of saline soil complexes are widely spread, intense degradation, desertification, salinisation and alkalinisation is observed, and in irrigated areas secondary salinisation is observed. The desert zone of brown and grey-brown soils is the southern level of the latitudinal bioclimatic zones of Kazakhstan, which covers the un-drained areas of the south Caspian lowlands, the plateaus of Mangyshlak, Ustyurt and Betpakdala, the Aral Sea areas, Shu-Moinkum, the Balkhash-Alakol depression, the sloping foothill landscapes of the Tien Shan, Zhongariya, Altay and Saur-Tarbagatai. Their total area makes up 44 % of the country and nearly 15 % of the area of typical desertified land surface (Faizov 1980; Babaev et al. 1986).

Deserts are the most arid regions of Kazakhstan, where soil-forming processes take place in conditions of severe water shortage, and high levels of soil degradation and desertification. The main natural reasons for these processes are a flat terrain, a high degree of arid climate, salinity, carbonate content, a lack of structure and low natural soil fertility (Saparov and Faizov 2006).

The influence of anthropogenic factors is seen almost in all natural landscapes, especially in the Aral Sea region, where degradation and desertification processes are becoming more widespread. In the most fertile delta-alluvial plain of the Syr-Darya river, the area of desertified land is 1.1 M. ha, and in the dried-up bottom of Aral Sea it is 1.5 M. ha, of which saline marsh soils occupy 0.8 M. ha. In contrast to other natural zones, these vast areas are occupied by sands (17.5 M. ha), saline soils (2.6 M. ha) and takyr plains (0.3 M. ha). The total area of saline soils in the desert zone with brown and grey-brown soils exceeds 60 M. ha, and alkaline complexes are present in 22 M. ha.

Mountain soils occupy the mountain ranges of south and south-eastern Kazakhstan, and amount to 13.6 % of the whole territory of Kazakhstan. Moreover, 5 M. ha of mountain soils are located in the low mountain areas of central and western Kazakhstan. The main and most general factor regulating mountain soils is the strong vertical zoning. Mountain soils are formed under different hydrothermal conditions and do not form complexes with saline soils. They are divided into three major regions depending on the combination of soil formation conditions and composition of soil landscape in the vertical zones and belts: Altay, north Tien Shan and west Tien Shan. Each mountain region has its own specific vertical soil zoning structure.

The soil landscape of Kazakhstan is the wealth of the Republic and the basis for ensuring food security. At the same time, some of the country's rich soil resources are not used effectively or efficiently.

2 Problems of Soil Quality

Analysis of the current status of the soil landscape has shown that intensive land degradation and desertification processes can be observed, with deterioration of the soil and environmental conditions of the country taking place. According to the data of the Agency on Land Resources Management, more than 75 % of the territory of Kazakhstan is subjected to degradation and desertification; over 14 % of pastures have reached an extreme degree of degradation or are completely degraded, and it is impossible to use them. Therefore, 15.2 % of lands are classified as "waste" fallow lands. The desertification of large territories is accompanied by soil contamination, waterlogging by surface water and groundwater, and a decrease in general regional biological capacity. According to scientists' preliminary evaluation, the damage caused by the erosion of arable land, secondary soil salinity, land degradation and desertification is estimated as 93 billion tenge, or 6.2 billion USD (RFCA 2010).

On the vast territory of the Republic, there are a number of regions where the combination of various forms of environmentally damaged soils has resulted in a crisis situation. Disastrous environmental conditions can be observed in the Aral

Sea area: zones of intensive soil desertification, salinisation and deflation. The regions of Central and East Kazakhstan, which are the major industrial centres, are the centres of technogenic disturbances and the industrial pollution of soils with toxic chemicals (lead, mercury, chromium, etc.). Every year, about 3–4 M. tonnes of polluting chemicals are emitted to the atmosphere or deposited on the soil surface.

The conditions and use of 59.6 M. ha of land in the Aral Sea area is causing particular concern. As a result of the Aral Sea drying up, major changes have occurred in the current delta of the Syr Darya river and on the dried Aral Sea bed. The negative factor is the process of salt-dust transfer over long distances. At the Kazakhstan part of the Aral Sea, three large sources emitting sand-salt aerosols into the atmosphere have formed and their influence can be observed at a distance of 200–250 km and more. The area of dust distribution and deposition is about 25 M. ha.

Since 1956 researchers at Kazakhstan's U.U. Uspanov Research Institute of Soil Science and Agrochemistry have conducted monitoring studies which have determined that the drying up and desertification of hydromorphic soils at the current delta of the Syr Darya river are accompanied by increased salinisation processes, a sharp decrease in non-saline soils and increases in soil salinity of various degrees, including the formation of saline marsh soils. Saline marsh soils make up more than 50 % of the area.

In the oil regions of Western Kazakhstan and the Torgay plain, on an area of more than 500 thousand ha, there are large sections of soil contaminated with oil and radioactive materials, high levels of salinity with industrial wastewater and technological transformation of the soil landscape, leading to the accumulation of toxic heavy metals (lead, cobalt, nickel, vanadium etc.) and radionuclides (thorium, barium, radium). Lands contaminated with heavy metals and radioactive substances occupy about 21.5 M. ha. These pollutants are spread on 59 % of the area in the Atyrau region, 19 % in Aktobe, 13 % in West Kazakhstan and 9 % in Mangystau.

Scientists from the Institute identified that as a result of oil pollution, deep morphogenetic changes are taking place in the soils, with soils in the zone being transformed and acquiring new properties. In contaminated soils, the most important genetic characteristics are damaged: their natural morphological profile changes along with their chemical and biological properties, and dense bituminous crusts are formed, which are soaked in fuel. Their morphological profile is transformed under the influence of crude oil and their horizon turns a brown-grey and resinous-black colour; it becomes a viscous, sticky and cloddy feature.

After the evaporation of the light fraction of oil, the heavy fractions which remain in the soil, saturated with resins, wax and asphaltene, glue together the granulometric fractions to form a dense mass, and form bituminous crusts ranging from 5–10 to 20–40 cm (Figs. 2, 3): specific technogenic soils with different genetic characteristics than the natural zonal soils are formed. It has been determined that oil-polluted soils are characterized by an increased content of gross and

Fig. 2 Bituminous crusts

mobile forms of lead 1–6 times higher than the maximum allowable concentration (MAC), 7–12 times higher for molybdenum and 2–3 times higher for cobalt.

In some cases, the presence of vanadium and nickel and low concentrations of copper, zinc and cadmium are observed. In the coastal zone, an increased amount of boron was revealed, caused by alkaline and saline sea sediments (Saparov et al. 2006a, b, 2010; Saparov and Mamyshov 2008). The soil is becoming an accumulator and storage medium for toxic chemicals. In addition, a geochemical flow goes into the Caspian Sea, which could cause a global environmental disaster. The industrial zone featuring oilfields is contaminated with hydrogen sulphide, sulphur dioxide and other toxic substances (Figs. 4, 5).

According to the nature of the initial soil salinity, the areas of the Caspian basin are classified as chloride and sulphate-chloride types of salt accumulation, the Aral Sea basin as the chloride-sulphate type and Lake Balkhash as the soda-sulphate type. On some parts of the plateau and layer plains in Mangyshlak, Ustyurt and Betpakdala, there are gypsum deposits. The loose sediments are spread at large distances from the parent rocks and differ in age, genesis and mineral composition.

Everyone knows about the serious consequences of nuclear tests at the Semipalatinsk, "Kapustin Yar" and "Lira" test grounds (among others). At the former Semipalatinsk nuclear test ground, about 2 M. ha of agricultural lands have been subjected to radioactive contamination. In the technogenic area of Shymkent City, soil is contaminated with mobile forms of lead and cadmium, with maximum concentrations of lead from 200 times the maximum allowable concentration to 1,500 MAC. A similar situation is observed for specific elements in the East Kazakhstan, Karaganda and Pavlodar regions.

In Kazakhstan, large areas (more than 30.5 M. ha) are subject to erosion processes. The degree of their occurrence varies and depends on climatic conditions, the physical and physical–mechanical properties of the soil and parent rock, slope,

Fig. 3 Oil pit

Fig. 4 Contamination with sulphur

degree of disturbance of vegetation, and soil cultivation methods. Erosion processes at the vast massifs of Karakum, Kyzylkum, Moinkum, and the Sary-Ishikotraukum sands are very active, as well as in regions where light texture and calcareous soils are common. Areas of land subjected to wind erosion occupy 25.5 M. ha, and those subject to water erosion more than 5 M. ha, of which 1 M. ha are arable land. The largest areas of water erosion can be observed in the South Kazakhstan region, with 958.7 thousand (K) ha, of which eroded arable land makes up 223.6 K. ha. In the Almaty region, there are 801.9 K. ha, in the Mangistau region 802.8 K. ha, and in Akmola 559.4 K. Ha including 286.2 K. ha of eroded arable land. In the Aktobe region, there are 488.3 K. ha, and in East Kazakhstan 419.0 K. ha, of which 134.5 K. ha are eroded arable land.

Fig. 5 Sulphur storage

Erosion processes are also developing intensively in the Akmola, South Kazakhstan and Almaty regions. In 8 districts of the Akmola region, there are slightly eroded soils (where the thickness of the humus horizon has decreased by 30 %), medium-eroded soils (by 50 %), and heavily eroded soils (characterized by the lack of an arable horizon).

In Southern Kazakhstan, in recent years, the processes of erosion on irrigated fields and pastures have developed rapidly: every year 19 million tonnes of soil are washed off with 400 K. tonnes of humus. Every year about 2.5–2.6 million tonnes of manure would be needed to cover these losses.

In the irrigated fields of Mangyshlak consistent fine, strongly spotted and solid secondary soil salinisation can be observed. Progressive salinisation of irrigated land is taking place in the Arys-Turkestan, Tashutkul, Bakanas and other irrigated areas. Secondary salinisation of irrigated lands has increased rapidly.

Intensive use of agricultural lands of the country without consideration of the agro-ecological potential of the territory and scientifically justified land cultivation systems has led to a significant decrease in soil fertility. Monitoring studies carried out by scientists at Kazakhstan's U.U. Uspanov Research Institute of Soil Science and Agrochemistry have determined changes in the content of humus in the major soil types in Kazakhstan and, accordingly, decreases in the potential soil fertility. The loss of humus after the cultivation of virgin and fallow lands was one-third of the original content, or 45–48 % including the most valuable humic acids and hydrolyzed nitrogen, and up to 60 % in irrigated conditions. The annual losses of humus in agriculture in Kazakhstan are 1.2–1.6 t/ha.

These results suggest that the processes of de-humification, classified as stage 1 (low degree) cover 4.5 M. ha of arable land, stage 2 (moderate degree) 5.2 M. ha,

Table 1 Application of mineral fertilizers in the Republic of Kazakhstan

Types of fertilizer	Years						
	1986	1995	2000	2005	2006	2007	2008
Mineral fertilizers (kilotonnes)	1,039	36.2	10.7	37.5	41.4	58.9	30.9
Organic fertilizers	33,196.0	1,141.0	17.5	7.8	13.7	7.8	7.4
Applied mineral fertilizers per 1 ha of arable land, kg NPK	29	1.3	0.7	2.0	2.3	3.1	1.5
Share of arable lands treated with fertilizers, %	47	1.5	0.5	3.5	3.6	4.8	4.9
Total arable area, K. ha	35,500	28,679	15,400	18,400	18,400	18,954	20,119

and stage 3 (strong degree) 1.5 M. ha. On irrigated arable land, 0.7 M. ha are subject to de-humification. Due to these significant losses of humus, the plants are not provided with the proper quantity of nutrients.

The main reasons behind humus loss are the inefficient use of land, failing to observe the laws of interaction between nature and society, and, most importantly, neglecting scientifically justified land cultivation systems.

3 Research Activities on Soil Monitoring and Restoring Soil Functions

One of the major factors in the conservation and improvement of soil fertility is a scientifically justified fertilizer application system. In Kazakhstan, the application of fertilizers has significantly decreased, resulting in a decrease in soil fertility and yields. In 1965, 170.4 K. tonnes of fertilizer were supplied and applied, i.e. 3.6 kg of NPK per 1 ha of arable land. As agriculture was intensively developed in the country, the areas of fields where the fertilizers were applied increased by 47 % of their total area. In 1986, the total volume of deliveries of mineral fertilizers amounted to 1,039 K. tonnes, and the quantity of fertilizers per 1 ha of arable land was 29 kg (Table 1). By 1986, yields had increased by 26 % compared with 1965, spring wheat from 6.1 to 10.1 decitonnes (dt)/ha, rice from 19.1 to 45.1 dt/ha, corn from 20. 8 to 38.8 dt/ha, sugar beets from 235.8 to 288.0 dt/ha, cotton from 17.9 to 25.8 dt/ha, potatoes from 75.0 to 106.2 dt/ha and vegetables from 66.1 to 170.0 dt/ha (Saparov et al. 2011).

Then, in 1987, the production and supply of mineral fertilizers for agriculture sharply decreased, and this situation lasted until 2000. The total amount of fertilizers used in agriculture decreased to 10.7 K. tonnes, and the intensity of their use decreased from 29.0 to 0.7 kg/ha of NPK. This led to a decrease in yields of 8.8–52.1 % depending on the crops. In comparison with the period 1986–1990, the yield capacity of grain crops during 1996–2000 decreased by 14.4 %, and that of wheat by 8.7 %. Particularly high losses were observed in the yields of sunflower, sugar beet, corn, rice, cotton and vegetables.

Since 2001, there has been a steady tendency towards land cultivation growth in Kazakhstan (Table 1). In 2009, 56.4 K. tonnes of mineral and 125.5 K. tonnes of organic fertilizers were applied, i.e. 10 % of the required amount.

Monitoring of the use of fertilizers in Kazakhstan at a long-term station involving experiments with crop rotations has showed that soil fertility can be not only conserved, but also extensively restored using organic and mineral fertilizers. Thus, scientists from the Kazakh Institute of Land Cultivation and Crop Production determined that in an 8-field crop rotation scheme, using 60 t/ha of complete fertilizer and manure once per rotation, an extensive quantity of humus was restored in the soil. Beginning with the third rotation, humus reserves increased both in the upper and lower layers of soil.

By systematically applying complete mineral fertilizer, the concentration of mobile phosphorus in the soil by the end of the first rotation was increased by 13.1 mg/kg of soil and reached optimal values. At the same time, in the fertilized variants in the upper layer an increase in the concentration of mobile phosphorus was observed after the application of not only the optimum level of mineral fertilizers but also high values of combined mineral and organic fertilizers. Fertilizers providing a positive impact on the nutrient regime of irrigated light chestnut soils have enhanced the increase in the productivity of crop rotation with sugar beet. In the long-term cultivation of crops without fertilizers only low yields were harvested, and the quality of crops in a sugar beet crop rotation was low. The systematic application of mineral fertilizers provided an increase in the yield of sugar beets of almost 3 times the level, 1.6 times for wheat grain, 1.3 times for corn, and 1.5 times for alfalfa hay. After the application of mineral and organic fertilizers the yield of sugar beet increased 3 times, that of wheat 1.6 times, that of corn 1.5 times, and that of alfalfa hay 1.6 times. The quality of production improved significantly: the content of crude protein in the wheat grain increased by 0.8–1.2 %, the concentration of sugar in the roots hs increased by 0.9–1.0 %, and the quality of corn grain and alfalfa hay also improved (Saparov et al. 2010).

In a station, experiment conducted by the Kazakh Research Institute of Potato and Vegetable Growing in irrigated dark chestnut soils on the basis of intensive vegetable crop rotation, it was determined that during 18 years (1991–2008) the concentration of humus in the variant without fertilizers decreased by 15.2 %, and in the case of variants with fertilizers (combined use of mineral and organic fertilizers), a tendency was observed for extensive restoration of soil fertility. However, there is an assumption that fertilizers are a potential source of soil contamination with heavy metals.

In view of this, we conducted extended agro-environmental monitoring in the south-east of Kazakhstan from 2001–2005. 11 field experiments applying fertilizers were selected as the object of monitoring, lasting for different durations (15–63 years) and taking place at 6 scientific institutions. Land plots without fertilizer and with full fertilizer application were tested. The soils at these sites were light and common grey soils, meadow-marsh and light chestnut soils, and they differed in the degree of contamination with heavy metals.

Table 2 Inputs of heavy metals (HMs) into the soil through the long-term use of complete mineral fertilizers (1961–2005)

Fertilizer	Application of fertilizers, kg/ha	Input into soil, g/ha						
		Zn	Cu	Fe	Pb	Cd	Cr	Ni
Urea	2,850	37.2	5.0	155	8.0	1.5	–	–
Granulated superphosphate	2,500	253.8	188.0	21,678	559	46	78.9	43
Potassium chloride	3,080	67.4	24.7	2,209	68.5	23	–	–
Input of HM, g/ha of soil	–	358.4	217.7	24,042	635	71	78.9	43
Input of HM, mg/kg of soil	–	0.12	0.07	8.0	0.2	0.02	0.02	0.01

An assessment of the degrees of soil contamination with heavy metals was carried out, identifying the mobile forms in them according to Vazhenin (1987): weak, moderate, medium, increased, high and very high levels of pollution.

To calculate the amount of heavy metals penetrating the soil with the fertilizers, data was used from a long-term station experiment conducted by the Kazakh Institute of Land Cultivation and Crop Production (Table 2).

The research results showed that the highest quantity of heavy metals penetrating the soil, in terms both of the accumulation and the concentration of the impurity, was with phosphorus fertilizers. The figures for potash fertilizers are slightly lower, and penetration of heavy metals in the soil with nitrogen fertilizers is insignificant. The calculation confirms that the application of scientifically justified fertilizer quantities does not lead to the contamination of soil with heavy metals.

There are various ways of solving these problems. A special approach needs to be created for each specific region, taking into account the natural climatic and ecological conditions. To stop and reduce land degradation and desertification, salinisation and pollution it is necessary to implement scientifically justified systems of land cultivation and complex organizational, economic, agrotechnical and hydro-technical activities, forest reclamation schemes etc.:

- Complete inventory of arable lands according to their current state of degradation and desertification, susceptibility to erosion, salinisation, pollution and other negative factors;
- Integrated soil resource management and their protection, introduction of scientifically justified crop rotation schemes, land cultivation systems, industrial and innovative technologies, pasture rotation, surface and radical improvement of soil reclamation and environmental conditions on unproductive degraded agricultural lands, rangelands and their irrigation;
- Rational use of water resources; strengthening efforts to prevent water loss, introducing modern irrigation technologies;
- Establishment of environmental monitoring and improvement of ecological situation by implementing environmental standards and certification in agriculture;
- Development of highly effective technologies for extracting hydrocarbons;
- Conservation of flooded oil wells.

Fig. 6 Forest bands of black saksaul in the Kozjetpes valley

In the Aral Sea area, significant attention is paid to the phyto-reclamation development of the formed land, and in particular soil conservation planting of trees and shrub species (halophytes) which will support the removal of salt, dust and sand, and stop the movement of sand dunes (Figs. 6, 7).

For a rational use of the soil landscape and the improvement of the soil reclamation and environmental status of land resources, one of the major tasks of our time is to develop new, highly relevant methods of soil resource management. In this regard, the relevance and priority of this issue are obvious, and it requires urgent solution and actions.

To solve these problems, it is necessary to develop specific programmes for studying contaminated and disturbed lands, develop laws and make appropriate recommendations for re-cultivation and land restoration. The adopted measures will serve as the basis for sustainable and intensive development of agri-industrial complexes in the Republic and ensuring food security of the country, aimed at implementing a common development policy in the agricultural sector.

Fig. 7 Natural tamarisk plants on sand in the Aral area

References

Babaev AG, Zonn IS, Drozdov NN, Freykip ZG (1986) Deserts (Пустыни) p 318

Faizov KS (1980) The soils of the desert zone of Kazakhstan (Почвы пустынной зоны Казахстана). Nauka, Almaty, p 135

Faizov KS, Kenenbaev SB, Mamutov ZU, Esimbekov MB (2006) Soil geography and ecology of Kazakhstan (География и экология почв Казахстана). Almaty, p 348

RFCA (2010) Analysis of crop production of RK. Analytical service of rating agency RFCA (Анализ растениеводства РК. Аналитическая служба Рейтингового Агентства РФЦА). (http ://www.rfcaratings.kz). Almaty

Saparov AS, Faizov KS (2006) Degradation and soil conservation issues in Kazakhstan (Деградация и проблемы сохранения почв Казахстана. Вестник сельскохозяйственной науки Казахстана). J Agric Sci Kaz 10:22–26, Almaty

Saparov AS, Mamyshov MM (2008) Environmental issues of soil surface of the Republic of Kazakhstan/issues of agricultural chemistry and ecology (Экологические проблемы почвенного покрова Республики Казахстан//Проблемы агрохимии и экологии). Sci-Theor j (Научно-теоретический журнал) 3:23–26

Saparov AS, Faizov KS, Asanbaev IK (2006a) Soil-ecological state of the Caspian oil and gas area and ways to improve them (Почвенно-экологическое состояние Прикаспийского нефтегазового региона и пути их улучшения). Almaty, p 148

Saparov AS, Kozybayeva FE, Saparov GA (2006b) Status of soil ecology in Kazakhstan/current status of potato and vegetable growing and their scientific support (Состояние экологии почв Казахстана/Современное состояние картофелеводства и овощеводства и их научное обеспечение). Almaty, p 694–698

Saparov AS, Tokseitova G, Saparov G (2010) Degradation of soils in Kazakhstan and prevention measures. In: Proceedings of international soil science congress on management of natural resources to sustain soil health quality. http://www.soil2010.omu.edu.tr. p 176

Saparov AS, Eleshev RE, Ramazanova SB, Aitbaev TE, Bazilzhanov EK (2011) Status of agrochemical monitoring of soil fertility and productivity of crops (Состояние агрохимического мониторинга плодородия почв Республики Казахстан и продуктивность сельскохозяйст, венных культур), vol 3. Soil Science and Agricultural Chemistry (Почвоведение и агрохимия), Almaty, pp 45–54

Vazhenin IG (1987) Methodological guidelines for soil surface mapping by levels of pollution by industrial emissions (Методические рекомендации по картографираованию почвенного покрова по уровням загрязненности промышленными выбросами). Moscow, p 18

Long-Term Monitoring and Water Resource Management in the Republic of Kazakhstan

Tursun Ibrayev, Batyrbek Badjanov and Marina Li

Abstract This chapter highlights current issues regarding long-term monitoring and water resource management in the Republic of Kazakhstan. It includes basic information about water resource monitoring systems, features and characteristics of water use in Kazakhstan, and the average costs of water for production. The most urgent water management problems have been outlined, for example that of water distribution from trans-boundary rivers such as the Ertis, Ile, Syrdarya and others, and ensuring the security of water facilities in Kazakhstan. The results of a catastrophic flood in the village of Kyzylagash are shown along with data from operative space imaging of the reservoir. Space-based information is described with a proposal for how it could be applied for the complex assessment of emergency risks in flood-endangered parts of river basins. A technological scheme is developed for Remote Sensing of Earth (RSE) data processing using GIS technology, and an outline is provided of the stages of digital data processing and GIS analysis technology and data interpretation. The proposed methodology is most promising and offers a new approach to evaluating water resources and identifying emergency situations and their consequences.

Keywords Kazakhstan · Water resources · Monitoring · Management

T. Ibrayev (✉) · B. Badjanov · M. Li
Kazakh Scientific Research Institute of Water Economy, 12 Koygeldy Street, Taraz, Kazakhstan
e-mail: tursun_ibraev@mail.ru

B. Badjanov
e-mail: badg_4@mail.ru

M. Li
e-mail: limarina76@gmail.com

L. Mueller et al. (eds.), *Novel Measurement and Assessment Tools for Monitoring and Management of Land and Water Resources in Agricultural Landscapes of Central Asia*, Environmental Science and Engineering, DOI: 10.1007/978-3-319-01017-5_3,

1 Introduction

Global water consumption has increased 12-fold in the 20th century to about 5,000 km^3 per year. This is almost 14 % of the annual flow of all rivers in the world. Rivers are the predominant source of water in the world. About 70 % of global water consumption is in agriculture, 13 % in industry, 10 % for household needs, and 7 % for other needs of the water resource economy (hydropower, transportation, fisheries, etc.) (Shimova 2001). The purpose of the long-term monitoring of water resources is to obtain and evaluate data from repeated observations. This is achieved using up-to-date methodologies for measuring parameters and data collection. It allows scientists to gather information regarding the current status of water resources and assess trends in their changing characteristics, as well as forecasting the limits of possible changes. Water resource monitoring includes the monitoring of available water (surface and underground), water economy systems and facilities, and water use.

2 State Water Monitoring in Kazakhstan

The state monitoring of water is an integral part of the State Monitoring of the Environment and Natural Resources. It involves regular observations of hydrological, hydrogeological, hydrogeochemical, sanitary-chemical, microbiological, parasitological, radiological and toxicological indicators, followed by the collection, processing and transmission of the obtained information. The purpose is to detect negative processes, evaluate and forecast their development, make recommendations on preventing harmful consequences and determine the degree of effectiveness of public water economy activities. This characterization is based on the Water Code of the Republic of Kazakhstan, Article 60.

Water resource monitoring systems provide information on water resource management in the country.

The main characteristics of these monitoring systems are:

- Integrated approach;
- Continuous monitoring in space and time;
- Use of common methodological approaches;
- Organization of GIS-based monitoring system;
- System designed to be open for practical linking with other systems;
- Focus on computer technologies for collection, storage and data processing.

The data sources of the monitoring system are:

- Data on status of water environment (hydrological parameters): "Kazhydromet" State Enterprise;
- Data on water use—Basin inspection by the Committee on Water Resources at the Ministry of Agriculture;

- Data on use of underground water (hydrochemical parameters, salinity): Committee of Geology and Protection of Resources (CPC), Ministry of Energy and Mineral Resources;
- Data on safety of hydraulic structures: Ministry of Emergency Situations;
- Data on water quality in water intakes, sewage and household water: Departments of Water Channels, Sanitation Centres.

The long-term monitoring of water resources in Kazakhstan is characterized by a number of problems: insufficient financing, poor coverage of the territory of the country by a network of observations, data collection that is performed separately and is insufficient, outdated equipment and methods of data collection and analysis, poor equipment at observation units, dissociation of monitoring networks of various agencies, insufficient scientific supply and development of water resource monitoring systems.

To solve the above problems, it is necessary to implement the following activities: (1) identification of the information required on various water users and natural ecosystems; (2) modernization of observation units with modern equipment and technologies for data collecting and analyzing; (3) provision of funding for monitoring of water resources by State and other organizations; (4) creation of a unified system of water resource monitoring based on GIS technology, involving all stakeholders; and (5) improvement of the quality of research work on monitoring systems in the Republic of Kazakhstan.

3 Water Resources in Kazakhstan

The territory of Kazakhstan, most of which is arid and semi-arid, has limited water resources. Therefore, every year the country faces a shortage of fresh and irrigation water. The determining factor for the effective use of limited water resources is the current status of water facilities.

85 % of the water supply to the economic branches is from surface water, the rest partly from ground, sea and sewage water. The water consumption of the Kazakhstan economy is on average 32.5 km^3 per year, of which the largest consumer is agriculture, using 75 %. More than half of this volume (53 %) is used in the Aral-Syr Darya basin—a traditional area of irrigated agriculture.

The largest industrial water users are located in the Ertis basin (38 %), Nura-Sarysu (29 %) and Oral-Caspian (21 %). The household sector consumes about 5 %. Consequently, the major water consumers are located in trans-boundary river basins (Orman 2010).

In terms of the water supply, Kazakhstan occupies one of the last places among the Commonwealth of Independent States (CIS) countries. The specific water supply is 37,000 m^3 per km^2 and 6,000 m^3 per person per year. In comparison, in Kyrgyzstan these data are 245,000 m^3 per km^2 and 11,763 Thous. m^3 per person, respectively.

The average water flow in the republic has decreased by 25.3 km^3 per year over the last 30 years, including a fall of 10.1 km^3 in local flow and 15.2 km^3 in cross-border flow. This phenomenon is due to climate change and anthropogenic impacts on water resources.

Water resources in Kazakhstan are distributed very unevenly. The Eastern region contains 34.5 % of resources, the South-East region 24.1 %, the Southern Region 21.2 %, the Western Region 13.4 %, the Northern Region 4.2 % and the Central Region 2.6 %. The East Kazakhstan Region has the most water resources at 290,000 m^3 per km^2, whilst the Atyrau, Kyzylorda and Mangistau regions face dramatic shortages of water, and there are practically no freshwater sources (Balapanova 2010).

Surface water resources in Kazakhstan amount to 100.5 km^3 in an average year, of which only 56 km^3 have their origin in the territory of the Republic. The remaining 44 km^3 come from neighbouring countries: China (18.9 km^3), Uzbekistan (14.6 km^3), Kyrgyzstan (3.0 km^3) and Russia (7.5 km^3). Seven of the eight major river basins in Kazakhstan are trans-boundary basins. The water economy conditions in the basins of Kazakhstan's trans-boundary rivers are characterized by the following features and tendencies:

- Drop in river flow;
- Threat of floods;
- High level of water pollution;
- Disturbance of the natural balance of ecological systems with negative impacts on the environment;
- Risk for population's health.

4 Water Management Efficiency

The quality of surface water resources is characterized as very unsatisfactory due to the intense economic activity of all Central Asian states. The water from most trans-boundary rivers is not suitable for drinking or fisheries, but it can be used for irrigation if protective measures are taken to prevent the salinisation of irrigated areas. The hydrochemical, hydrobiological, and sanitary regimes of the rivers have substantially changed. The low natural self-purification capacity of the water has resulted in dangerous environmental conditions in the basins of trans-boundary rivers.

In this situation, Kazakhstan, as a downstream water user, is subject to significant damage, economic, social, and environmental, due to the unsatisfactory quality of the incoming river flow. This is the reason for a decrease in yields and the quality of agricultural products, and an increase in water costs by 1.5–2.0 times per production unit due to the need to conduct leaching irrigation regimes to prevent the salinisation of irrigated lands, and other average costs of obtaining additional water volumes (Table 1, Chagay and Seytbembetov 2006).

According to data provided by the experts of the Committee on Water Resources of the Republic of Kazakhstan, 13–17,000 m^3/ha of water is used for irrigation and 26,000 m^3/ha for rice production, while in developed countries, it is 5–7,000 m^3/ha. For one m^3 of water used in the Southern Regions of Kazakhstan there is 0.4–0.8 kg of crop production, while in developed countries it is 2–8 kg. The most rational way to further develop agricultural production is to introduce water-saving irrigation technologies (Orman 2010).

5 Trans-Boundary Water Problems

The issues involved in using water from trans-bordering rivers have remained unresolved between the countries participating in one of the most influential political organizations in Central Asia, the Shanghai Cooperation Organization (SCO). For many years, most of the neighbouring countries failed to use trans-border water resources in compliance with principles and norms of international laws. International laws do not allow a country to make any prohibitive or restrictive provisions on the use of water from the rivers within its own territory. Water relations should be based solely on the basis of mutual agreements.

Currently, Kazakhstan's closest neighbours (China, Uzbekistan and others) have not acceded to two major international agreements, the "Convention on the right of non-navigational use of trans-border water flows" (1997) and the "Convention on conservation and use of trans-border water flows and international lakes" (1992), and they prefer to solve the issue of regulating trans-border waters only through bilateral negotiations which, in practice, provide only a partial solution to this problem. Only 37 countries have signed the UN documents concerning this matter, and only 20 of them have ratified them. China is not among the parties to the agreement, and has never had any agreements with any country on trans-boundary rivers.

Thus, the planned large-scale use of water from the trans-boundary rivers Ertis, Ile, Syrdarya and others will have serious and far-reaching harmful consequences for the national economy of Kazakhstan, and will mostly have negative impacts on

Table 1 Average costs of obtaining 1,000 m^3 of water

No	Methods of obtaining water	Costs, USD
1	Freshening of mineralized waters	1,000–250
2	Rehabilitation of hydro-technical system	800–100
3	Territorial re-distribution	750–200
4	Sewage water purification	120–20
5	Water reservoir regulation	70–20
6	Introduction of water-saving technologies	3–2

the natural ecological balance of the whole Central Asian region. Due to the threatening nature of the problem of cross-border rivers, in recent years, Kazakhstan has undertaken active diplomatic efforts to resolve its water problem. The leadership of Kazakhstan is concerned about the situation and has repeatedly proposed that all interested parties conduct multilateral negotiations on trans-boundary rivers.

Assessments by water economy experts from the Republic of Kazakhstan suggest that in the immediate future the economic potential of the country will grow due to the development of rich resources of minerals, fossil fuels, energy and land. However, there will be a serious problem with the water supply. In this situation, the issue of using water from trans-boundary rivers and sharing it with neighbouring countries according to the principles of international law and mutual cooperation is very important to Kazakhstan.

6 Enabling the Performance and Safety of Water Facilities

Another pressing issue relating to water resources in the Republic of Kazakhstan is to ensure the coordinated performance and safety of water facilities. In Kazakhstan, there are 643 hydro-technical facilities which belong to different departments and owners. This includes 340 hydro-units and hydro-technical facilities operating in the water economy system. At present, 270 reservoirs with a complex of hydro-technical facilities are situated on the territory of Kazakhstan, of which 62 reservoirs are of national importance and 208 of local importance. By the Decree of the President of the Republic of Kazakhstan, 57 reservoirs and 29 water-supporting hydro-technical complexes are included in the list of facilities of special strategic importance. Water reservoirs with long-term flow regulation regimes have a major influence on the redistribution of the annual water flow of the rivers. The largest of them are the Bukhtarma reservoir (on the Ertis river) with a total volume of 49.0 km^3, the Kapshagay reservoir (on the Ile river) with a volume of 28.1 km^3, the Shardara reservoir (on the Syrdariya) with a volume of 5.2 km^3, the Upper Tobol and Karatomarsky reservoirs (on the Tobol) with a capacity of 0.82 and 0.59 km^3, respectively, and the Viacheslavskoye and Sergeevskoye reservoirs (on the Ishim) with a volume of 0.4 and 0.7 km^3 ("Dam Safety in Central Asia" Project 2010).

According to data from the Ministry of Emergency Situations of Kazakhstan (MES RK) 268 hydro-facilities, including 28 large ones, currently need urgent repair. The state owns 24 % of large hydro-technical facilities (61 water reservoirs, 91 hydro-junctions and line channels), and the rest belong to municipal, industrial and agricultural enterprises. One serious problem is that a number of small hydro-technical facilities, some of which are abandoned, have no owners or operational service. Their technical condition is extremely bad. Annual losses caused by their unsatisfactory state of functioning and failure to protect against the harmful effects of floods can be estimated in dozens of millions USD. Besides this, similar damage is caused to water resources. Experience shows that a regular evaluation of the technical conditions of hydro-technical devices and the conducting of repairs would allow damage from the harmful effects of water and possible accidents to be reduced significantly.

It should be noted that the large hydro-technical systems forming the basis of Kazakhstan's water resources, as a rule operated by functioning water management systems, are prepared for floods and emergencies. However, this does not fully apply to the numerous blocking facilities on small rivers, whose owners, even if there is one, are not certified for operation, and whose safety in operating conditions is not always ensured. This can lead to serious flooding accidents (Figs. 1, 2) (MES RK 2010).

7 Remote Sensing for Water Monitoring

These and other water management problems in Kazakhstan can only be addressed through integrated surface-satellite studies. Such studies are a crucial element for gathering primary data to be processed using GIS technologies. The application of space-based remote sensing data has become a major practical and scientific method for investigating processes and phenomena on the Earth's surface and in the atmosphere and hydrosphere. Space-based monitoring is a system of regular observations, data collection, storage and the processing and analysis of

Fig. 1 Flooding disaster in the village of Kyzylagash

Fig. 2 Consequences of flooding disaster in the village of Kyzylagash

information, involving ground control, surveying land and water from space, the provision of targeted information, ground data receipt, initial processing, data storage, final processing and the analysis of information.

The following tasks are directly related to the monitoring of water resources in river basins, and can be solved by means of remote sensing:

- Monitoring the hydrological and hydrochemical characteristics of the river network;
- Monitoring the flood situation of the rivers, flood control, floods of different origin (rain, snow melt, effects of earthquakes, accidents at hydro power plants, etc.);
- Monitoring ice conditions during floods on the rivers;
- Detecting the discharge of pollutants into water reservoirs and seas;
- Monitoring the melting of mountain glaciers;
- Checking sea tidal areas.

The main purpose of monitoring water resources from space is to map water passage lines and flooded areas by means of operational and review monitoring. Operational monitoring reflects the current status and is the result of the operational processing of daily space data. Review monitoring characterizes the development of the water situation during a certain period of time: a decade, month or season. Decade maps of water passage lines and flood zones are based on daily data on water passage lines and flood zones obtained during operative monitoring. They represent total water passage lines during the passage of flood waters and flooded areas during floods over a decade in the study area. Monthly maps of water passage lines and flooded areas are made in a similar way, based on the decadal data, and seasonal maps are produced on their basis. During dam breaks, medium- and high-resolution images are obligatory (Fig. 3 and 4) (NCSRT Report 2011).

River basin areas with signs of flood threats are surveilled using different satellite types and scenes, Table 2 (NCSRT Report 2011). Regular monitoring surveys are conducted twice a year, in spring and fall. The spring monitoring of water bodies is carried out to assess the status of the water collecting area when floods pass through after the melting of snow or spring rains. During spring, monitoring the time and volume of water passage and flooding are recorded and natural climatic conditions before the floods pass through are analyzed, with area indices of the flood territory determined on actual masks of flooded areas during flood passes. Total fall monitoring is conducted to check whether hydro-technical facilities are prepared for winter. By this time, all summer repair work must be completed. Autumn monitoring is carried out using a sample survey of specific hydro-technical facilities on which repair work has been carried out.

8 Processing of Data and GIS Technologies

Currently, Remote Sensing of Earth (RSE) data are the most operative source of geo-information data. Consequently, they are the main source for maintaining relevant GIS data, especially if their factor of relevance is critical (military intelligence, control of natural disasters, environmental monitoring, natural resource exploration, etc.). Therefore, it is expedient to consider this type of technology separately. Because of the importance of geo-information technologies for the processing of RSE data, it should be stressed that there is a trend towards the mutual convergence of GIS technologies and RSE data processing, because in

Fig. 3 Water reservoir near the village of Kyzylagash in 2009 before dam break (Courtesy of the Sultangazin Aerospace Institute, Almaty)

Fig. 4 Satellite image (Landsat) of the water reservoir near the village of Kyzylagash in March 2010 after dam break (Courtesy of the Sultangazin Aerospace Institute, Almaty)

Table 2 Content of satellite information and its use in the complex evaluation of emergency situations

Type of satellite images	Periods	Territory
Terra/Aqua- MODIS (NOAA)	Every day during flood risk season	Rivers: Syrdaria (Koksaray, Shardara), Ertys, Oral, Ile (Balkhash)
Landsat 5 (IRS-P5 and other space data of medium resolution, 15–30 m)	Annual mosaics (especially in flood risk season)	Rivers: Syrdaria (Koksaray, Shardara) (10 scenes), Ertys (11 scenes), Oral (6 scenes), Ile (Balkhash) (12 scenes).
Quick Bird (other space data of super high resolution, 0.5–1 m or aerial photography)	1 scene every year	Rivers: Koksaray, Shardara (dams), Ertys (2 dams), Oral (1 site), Ile (1 site).
RADARSAT-1 (data from other RL satellites)	1 scene every year	Rivers: Syrdaria (Koksaray, Shardara), Ertys, Oral, Ile (Balkhash).

RSE technology, digital image processing techniques play the leading role on the stage of data collection. Figure 5 shows a typical scheme of RSE processing in GIS technologies (NCSRT Report 2011). The stages of processing digital data are specially outlined because at these stages of processing the volume of data for processing is higher than in GIS modelling. In the first stage of processing, data is imported from satellites or scanning. The second stage assumes that this data has been analysed, which is necessary for the subsequent preparation of an image processing plan. In the third stage, the images are rectified. If necessary, they can be transmitted into a given cartographic projection. In the fourth stage, several images can be joined or combined to obtain a complete picture of the object or phenomenon studied. In the next phase, the image synthesized from several different images is processed in order to improve its quality, and different characteristics from the initial images are studied to find their common characteristics and create a single synthesized image. To improve the quality of decoding, and the quality of vectoring of the raster image, image objects are automatically classified, then clustered by property in the sixth stage. This makes the process of organizing attribute data simpler. The next step provides the basis for the application of GIS technologies. During this stage, attribute data structures are formulated in accordance with the requirements of a particular GIS, and the structure of links between positional and attribute data is formed. At the eighth stage, the raster image is vectorised using the classification data and organized links between coordinates and attributes. At this point, the initial data is compressed substantially (by a factor of 2–3) preserving the information ability of the selected objects. At the ninth stage, a digital model is formulated as the basis for data storage and modelling in GIS. At tenth stage the procedures of geo-information modelling are carried out, which can repeatedly include a number of procedures, such as combination of the objects, rectification, classification, etc. The goal of these procedures is a more in-depth study of GIS objects. At the last stage, the results of geoinformation modelling are presented in the form of reports, presentations, references, maps and other documents.

Attention should be drawn to the specifics of the technology considered:

1. In contrast to the classic technologies in which a digital model is formulated on the principle "one map—one digital model", in the technology shown in Fig. 5 other principles of formulation are possible: "several different maps—one model", "several scenes (all scales)—one model", "several maps and several pictures—one digital model".
2. The vectorization process is preceded by a classification process which not only groups the initial objects of vectorization that form the basis for effectively selecting vectorization filters but also establishes the structure of attribute data, transferring the process (as yet only partially) from humans to a program.

This complex technology includes traditional methods of organization and data interpretation:

- Photogrammetric image analysis, cartographic analysis of initial information;
- Automated data processing methods that transfer the load from human to computer or allow factors to be identified that humans cannot identify in a visual analysis;
- Heuristic data analysis methods based on traditional statistical processing and analysis;
- Analysis of temporal data sets (time series), which allow the dynamics of the process or phenomenon to be studied over time.

Remote sensing methods provide a large amount of terrain data. However, the latter still do not represent information on the object and phenomena in a pure form. For the algorithmic processing, the results of RSE, including satellite imagery, should be analyzed, identified and classified. RSE gives a lot of visual information obtained from visual and instrumental surveys, visual surveys, different types of sensing and image shots. A large amount of visual information requires visual analysis of data and their interpretation.

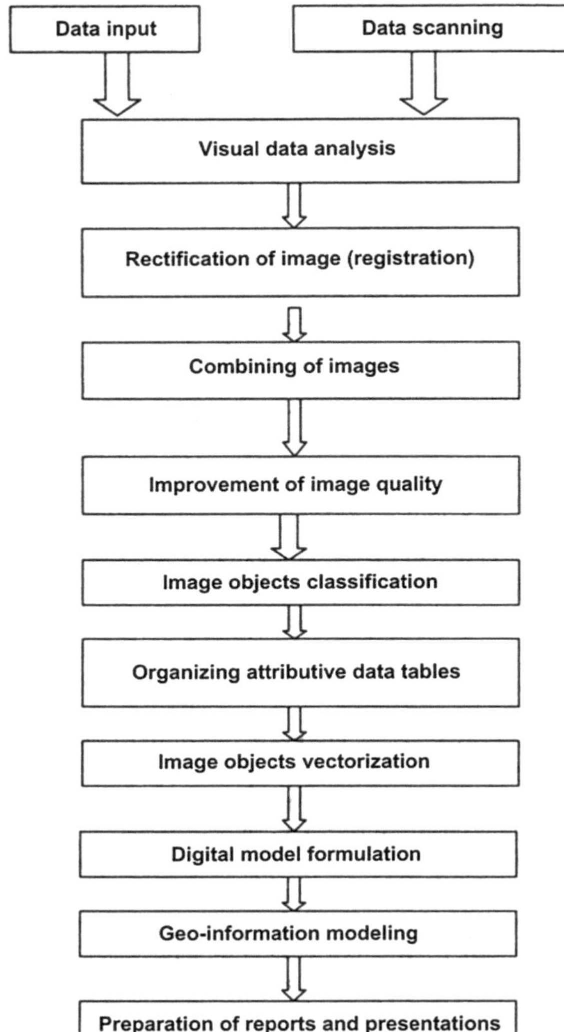

Fig. 5 Technological scheme of RSE processing in GIS technologies

GIS-based analysis is a multidimensional concept. In geoinformatics, analysis is divided into qualitative and quantitative analysis according to the methods and results of processing. Methods of processing can be divided into automated or statistical methods, or the analysis of time series. On the qualitative level, data analysis can be divided into systematic, generalized (sometimes structural), semantic (meaning), and parametric analyses (estimation). The types of analysis considered can form different combinations depending on the impact or value of aspect of the data study. Figure 6 shows a diagram of geo-information analysis technology and data interpretation (NCSRT Report 2011).

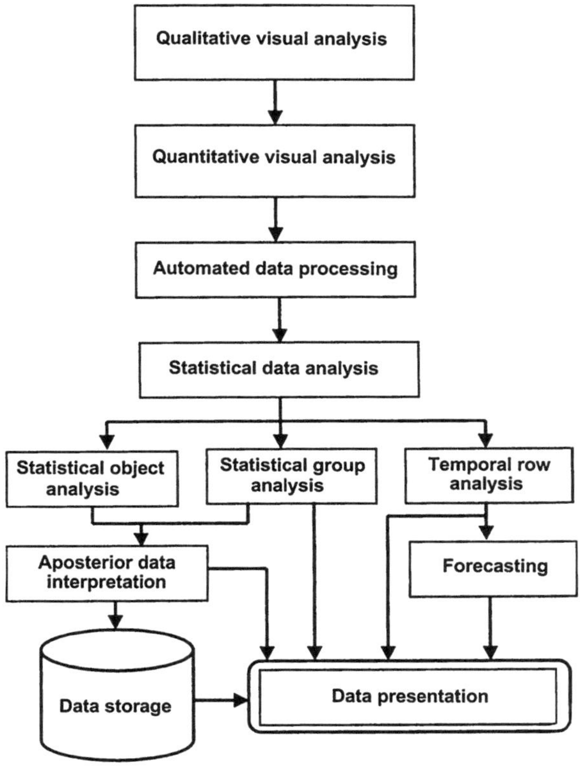

Fig. 6 Scheme of analysis and interpretation of RSE in GIS technologies

One specific feature of this technology is the need for comprehensive processes of analysis, interpretation and data processing. This applies to all GIS data, including RSE. This technology includes a visual qualitative and quantitative data analysis. The next stage includes automated processing, which is essentially a stage of automated analysis. Automated processing provides analysis using well-known algorithmic methods, which is possible and effective in the study and classification of data structures and data classes. Further statistical analyses are conducted, providing an estimate of the groups of study objects or their group characteristics.

The efficiency of the analysis increases substantially when using a database that stores various data obtained at different stages of analysis. The results of visual, automated and statistical analysis allow geospatial a posteriori data interpretation to be carried out. This interpretation serves as a basis for presenting data electronically or on paper. The example below is a scheme of total flood zones and zoning for the fragment of the territory of the Kyzylorda region where flooding is frequent (Figs. 7 and 8, NCSRT Report 2011).

The level of automation of combined GIS technology, including RSE data processing techniques, is higher than in many other GIS technologies. It is most

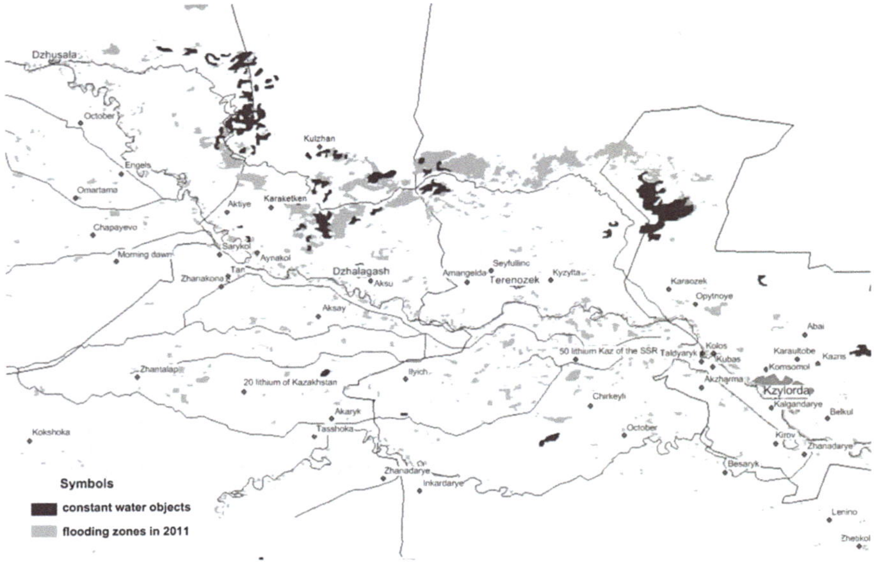

Fig. 7 Total flood zones in the Kyzylorda region (courtesy of the Sultangazin Aerospace Institute, Almaty)

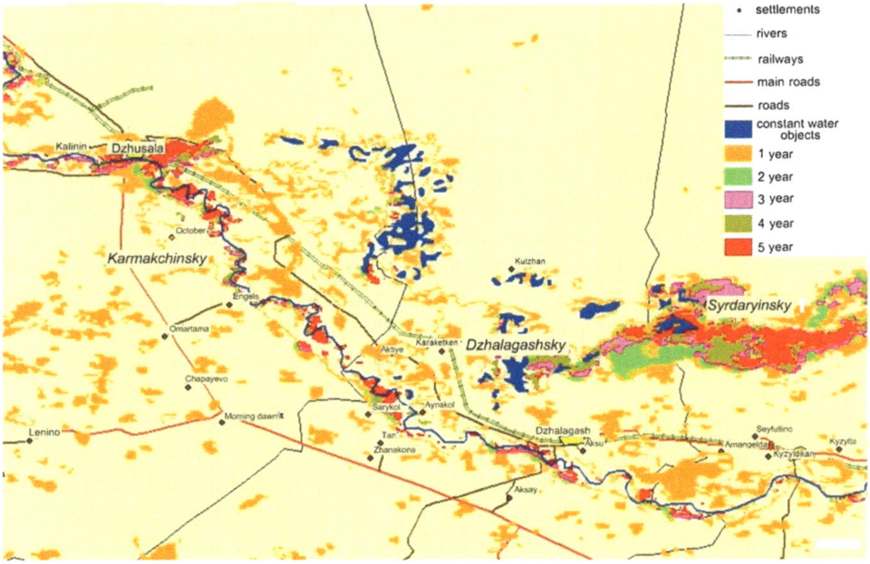

Fig. 8 Zoning of the fragment of the territory of Kyzylorda region where flooding is frequent over 5 years (Courtesy of the Sultangazin Aerospace Institute Almaty)

promising, since it contributes to the capacity building of geo-information systems as processing systems.

Ground-space monitoring of water resources on the basis of remote sensing of Earth from space, together with combined GIS technology, is a contemporary approach facilitating the long-term monitoring of water resources and the identification of emergency situations and phenomena leading to disaster, as well as the evaluation of their impact.

References

Balapanova A (2010) Kazakhstan and the issue of allocation of water resources of transboundary rivers. Bus Kaz 13(210) 09 Apr 2010

Chagay E, Seytbembetov K (2006) The ecological status of water resources and influence of water condition on the environment. UNDP project, a newsletter. Current issues of the Aral-Syr Darya basin. Kyzylorda, pp 56–60

MES RK (2010) Images of rescue services of the ministry for emergency situations of the Republic of Kazakhstan

NCSRT Report (2011) National center for space research and technology. Establishing a complex of space monitoring and risk assessment of emergency situations (floods, dam breaks, glacial lakes, etc.) in the basins of transboundary rivers. Almaty, p 85

Orman A (2010) What determines the well-being of people? Kazakhstanskaya Pravda 231–232:5 13 Sep 2010

Project Dam Safety in Central Asia (2010) Dam safety in central Asia: capacity building and regional cooperation. Second phase. report of the Republic of Kazakhstan. Almaty

Shimova OS (2001) Fundamentals of ecology and environmental economics. BSEU, Minsk, p 367

Trends in the Agriculture of Central Asia and Implications for Rangelands and Croplands

Mekhlis Suleimenov

Abstract During the first years of independence all Central Asian countries focused their policies on a maximum increase of wheat grain production. In Kazakhstan this strengthened farm economies, while in other countries it was a step towards self-sufficiency in wheat grain production. However, this single-crop production strategy had an obvious negative impact on crop diversification, including the production of feed and forage crops, as well as oilseeds, pulses and sugar beets. The governments of most Central Asian countries seemed to be concerned with wheat and cotton production while the livestock industry long remained a neglected sector, left to household plots. In recent year's, the livestock industry has gained more support from the governments, but livestock productivity may only be improved when measures to improve rangeland management and increase forage production are undertaken. In this respect it is the right time to change policies towards supporting rangeland improvement and integrated crop and livestock production. This will also improve long-term soil conservation.

Keywords Central asia · Agriculture · Wheat production · Soil conservation

1 Introduction

Twenty years ago the countries of Central Asia began to develop as independent economies. Before then all their economies, including the agricultural industry, had been developed as part of one large country known as the Soviet Union (SU). Under the conditions of this centrally planned economy, the major parameters of the agricultural industry were determined by a State Planning Committee which in

M. Suleimenov (✉)
Scientific Production Center of Grain Farming Named After A.I. Barayev,
Shortandy 021601, Kazakhstan
e-mail: mekhlis@yahoo.com

L. Mueller et al. (eds.), *Novel Measurement and Assessment Tools for Monitoring and Management of Land and Water Resources in Agricultural Landscapes of Central Asia*, Environmental Science and Engineering, DOI: 10.1007/978-3-319-01017-5_4,

turn developed its strategy based on guidelines approved by the congresses of the Communist Party of the USSR (CPSU). One major feature of those state plans was administering to all Republics those parameters of development which on paper ensured the accomplishment of major global tasks. Using this approach each Republic had its specialization determined by a centrally managed strategy. One of the major mistakes of this strategy for the Republics of Central Asia was that every effort was directed towards achieving goals from approved state plans, neglecting any other more profitable directions of development. The economists in the State Planning Committee believed that if each Republic specialised in producing whatever was planned, this would ensure the better utilization of the natural resources of the whole country.

2 Specialization of Central Asian Countries as Part of Soviet Union Agriculture

Kazakhstan's major target was a grain production of 22 million tonnes per year, including wheat grain production of 14–15 million tonnes. Wheat production was organized mainly in dryland spring wheat-fallow cropping systems in the Northern Plains with an annual precipitation of 250–350 mm. The spring wheat grain yield in the north depends on both winter and summer precipitation. The winter precipitation as snow is a source of moisture accumulation in soil before the planting of spring wheat in May. This is why snow management in the form of snow ridging or leaving tall stubble is an important part of dryland farming. The most important factor for the spring wheat grain yield is rainfall between late June and the beginning of July, which coincides with the period of tillering/jointing of spring wheat. At this time, however, the rainfall is 35–40 mm on average with fluctuation from 5–10 mm up to 80–100 mm. Thus it is impossible to obtain the targeted grain production every year. The planning system did not, however, take into account weather conditions and demanded an annually targeted grain production. This led to an annual increase in the grain production area by ploughing up marginal lands including rangelands of low fertility and solonetzic soils. Finally, by the end of the Soviet Union, the area involved in grain production in Kazakhstan had increased to 35 million ha, of which the arable land area was about 22–23 million ha.

Besides this, the central government set relatively ambitious, often unrealistic tasks for livestock production. For example, in 1974 when sheep population was about 34 million head, the leader of the CPSU, Brezhnev, unexpectedly postulated an absurd target to increase the sheep population up to 50 million head. Now it is well known that nothing actually happened afterwards, and the sheep population in Kazakhstan remained at a level of 34 million up to the dissolution of the SU. The cattle population was maintained at a level of about 10 million head, with very low productivity. It was essentially dairy cattle breeding stock, the culled part of which was used for meat. The average milk productivity of cows in 1992 amounted to only 1,500 L per year. At the same time the milk productivity of cows in Canada amounted to 6,000 L on average. The slaughter weight of cattle in the two countries was 185 and 278 kg, respectively, at that time.

Livestock production and its basis, forage production from grasslands, were disregarded. The grain production of republics, provinces, districts and state farms was the only acknowledged measure of success. Therefore a large number of rangelands were ploughed and used for grain production. Cereals other than wheat were grown where their grain yield was higher than that of wheat. This explains why barley was grown on a huge area of 7–8 million ha. Proso millet and oats also were planted on areas of just under one million hectares each. Kyrgyzstan specialized in sheep production as well as in the production of alfalfa and maize seeds. The country also produced some cotton and sugar beet under irrigation.

Other Central Asian countries, namely Tajikistan, Turkmenistan and Uzbekistan, were administered to produce as much cotton under irrigation and karakul sheep for pelts as possible. In these republics all irrigated lands, including marginal lands, were used for cotton production. This led to the overuse of water resources from rivers flowing to the Aral sea, resulting in the drying off of the sea in the long term. Grain production in these republics was organized primarily under rainfed conditions for local purposes. Livestock production was based on karakul sheep raised on rangelands in desert and semi-desert areas while in mountainous areas fine wool sheep breeds were raised to produce wool. All other agriculture sectors had local importance and were neglected.

3 Development of Crop Production Industry in the Transition Period

The transition period began with great hardship associated with many unfavourable factors. First, all newly emerged states entered an economic crisis which ended in considerable reductions in their production and in the bankruptcy of most industrial enterprises. Secondly, all independent states began to develop a market-oriented economy with no knowledge on how to move from a centrally planned to a market-driven economy. Thirdly, major hardship was created when property ownerships were changed from state to private with no experience of such transformation processes. The first, most difficult step in the establishment of new economies lasted until the end of the last century. Later on, the countries of Central Asia established new economic structures, determined new priorities in agriculture and achieved initial successes, although the productivity of both the crop and livestock industries remained rather low compared to world levels. States changed priorities, especially in the crop production sector (Table 1).

Table 1 Changes in the planting area of various crops in Central Asian countries from 1992–2010 in thousands of ha (FAOSTAT 2010)

Country	1992			2010		
	Wheat	Barley	Oilseeds	Wheat	Barley	Oilseeds
Kazakhstan	13,700	5,600	360	13,100	1,300	1,250
Kyrgyzstan	250	260	2	380	12	15
Tajikistan	183	60	0	343	70	6
Turkmenistan	197	61	0	850	53	0
Uzbekistan	630	304	3	1420	105	16

4 Kazakhstan

In Kazakhstan the cropland area was reduced significantly (by 12 million hectares), the barley area was reduced considerably, and forage production was almost completely stopped. All cropland was used to produce wheat continuously with summer fallow once in 3–4 years. During the first few years of independence, wheat grain promised to be the only product producing profits despite low-input production technologies. This led to grain exports below world price levels, which also influenced international grain prices. In 1999, the wheat-grown area was reduced by 5 million hectares, but later on it began to gradually expand, reaching the level of the first year of independence by 2009–2010. This happened mainly thanks to governmental support for seed production, plant protection, fertilizer application and the purchasing of new equipment. Furthermore, most of the collective enterprises established during the first stage of privatization of Soviet state farms went bankrupt, and companies dealing with grain handling, storing and marketing took management rights on their land. These companies were restructured into integrated grain production, storage and marketing holdings, covering activities on the area of several former state farms. They had the resources to invest in agriculture and availed themselves of some support provided by the government to purchase modern equipment and build new facilities for grain storage. Grain holdings returned to production some land which had previously been abandoned, and at present they control a major amount of grain production and marketing in the country. Beside this, medium-sized farms were founded, which farmed on former state farm areas, as well as smaller, individual farms of various sizes, known as peasant farms.

After a long crisis, a new stage began in the new century and lasted until 2006–2007. This period was characterized by the strengthening of the grain industry. At this stage wheat grain production stabilized at the level of Soviet times. Compared to the Soviet era soils became less fertile because of the increasing time since the grassland had last been developed, and the application of fertilizers fell dramatically. Nevertheless the soil quality did not go down considerably because almost all marginal lands were excluded from grain production. Additionally, grain holdings invested a lot in purchasing modern equipment made by the best international companies, enabling them to introduce modern technologies with minimum and no tillage, using efficient chemicals to control weeds, plant diseases and insect pests. Thereafter wheat grain yield increased by 25 % (Fig. 1).

A certain crop diversification started, primarily by increase of the oilseed crop area. In the Soviet era this was neglected as it produced lower yields compared to wheat and was more productive in other parts of the Soviet Union. Under market economy conditions it became obvious to producers that even with relatively low yields oilseeds can be much more profitable than wheat thanks to the much higher market prices. Oilseed prices fluctuate considerably less than wheat grain prices.

In the Soviet era oilseeds occupied 360,000 ha, primarily of sunflower, in eastern Kazakhstan (Table 1). The sown area went down to 250,000 ha at the first stage, then increased up to 460,000 ha at the second stage and went up to 650,000–750,000 ha in the third stage. This occurred due to a considerable increase in the area planted with sunflower not only in the east but also in the north, where it was not grown before. Rapeseed became the second important oilseed crop, grown until 2004 on a rather small area of about 10,000 ha. By 2010 it already occupied an area of 305,000 ha, mainly in the north. Rapeseed producers obtain yields of 0.6–0.8 t ha^{-1} on average, which is 50–60 % compared to wheat grain yield, and this provides a higher profit margin for the crop producers, especially when wheat grain prices are low. The third oilseed crop became flax, which occupied 58,000 ha in 2009, 225,000 ha in 2010, and 326,000 ha in 2011. The flax seed yield is about 0.8 t/ha on average, which also mainly makes this crop profitable for growing in the north. In spite of rather low yields (0.4–0.5 t ha^{-1}) safflower also is grown on a considerable area, mostly on marginal rain-fed lands in the south. Originally this started in southeastern Kazakhstan but recently it has

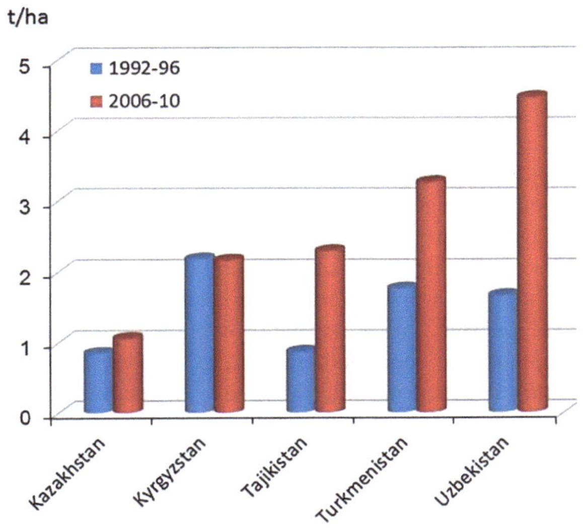

Fig. 1 Changes in wheat grain yield (t ha^{-1}) in Central Asian countries during 1992–2010, (FAO statistics FAOSTAT 2010)

expanded to the southeast and even the north. Its sown area reached 200,000 ha in 2010. Thus, compared to Soviet times the area planted with oilseed has increased fivefold and reached 1,250,000 ha, allowing the country to achieve self-sufficiency in edible plant oil.

During the third period of the transition, grain producers started showing interest in food legumes, starting with dry peas. Chickpeas are also becoming popular, and some farmers are trying growing lentils. These crops are grown on a relatively small area so far. Many producers understand the importance of these crops and have started to replace the summer fallow with dry peas and chickpeas. This trend might continue in future. Food legumes have very good prices on international markets.

5 Uzbekistan

Achieving self-sufficiency in bread wheat grain has been a major feature of agricultural restructuring in this country in recent times. This was supported by government actions. In Soviet times only cotton production was carefully administered, while wheat was grown only under rain-fed conditions on an area of about half a million hectares. The government had two alternatives: to continue the same policy of preferring cotton and buying wheat grain on international markets or to start planting wheat on irrigated land. Wheat production under irrigation instead of cotton seemed to be a less economic strategy. The political decision, however, was in favour of allocating over one million hectares of irrigated lands for wheat production (Table 1). Thus, the area grown with cotton was reduced significantly, and alfalfa and other forage crops were removed from irrigated lands. Cotton-alfalfa rotation was replaced by a cotton-wheat rotation. The soil tillage practice changed thereafter and instead of traditional ploughing after the cotton harvest, wheat was spread into cotton stand and the seeds covered by subsequent shallow harrowing or cultivating. The wheat grain yield was tripled after wheat was grown on irrigated lands (Fig. 1). Achieving grain yields of 4.5–4.7 t ha^{-1}, the total wheat production reached over 6 million tonnes per year. This was enough for self-sufficiency in grain but was not enough to produce the necessary amount of high-quality wheat grain. The point is that wheat grain produced under irrigation very often does not meet the criteria of baking quality for bread wheat. Therefore, there was a need to buy about one million tonnes of wheat flour.

In spite of privatization and though agricultural production was moved to private peasant farms, strong government-administered management methods of cropland use remained. In fact, all producers must use cropland for wheat and cotton production in accordance with established government plans. The government-controlled system provides farmers with subsidies and takes all cotton produced as well as the planned wheat grain harvest. Interestingly, in the case of crop failure the farmer has to buy wheat on the market and deliver it to the governmental buyers at government prices to meet the target plan.

Farmers can grow all other crops on their land according to their own plans, but not much land is usually left for this. As a result, the area under rice, barley and maize has been cut by a factor of three. Potato production has increased considerably. During Soviet time most potatoes were delivered from Belorussia and the Baltic Republics, but were of poor quality. They were very often already rotten when they arrived in the country. The area currently planted with potatoes has increased from 43,000 to 68,000 ha and yields have tripled, resulting in a fivefold increase in production. There is very small area under oilseeds, just 5–10 thousand hectares of sunflower and safflower. The country prefers to buy about 60,000 tonnes of edible sunflower oil a year, paying US$ 1,000 per tonne. This is equal to the sum received for exporting grapes and raisins. As a whole, Uzbekistan has great potential for producing fruits and vegetables for export, but for several reasons it does not do so. Pulses are hardly ever grown, although most families plant a lot of chickpeas on their garden plots. Dry beans are sometimes grown as a double crop after a harvest of wheat, but these areas are probably not included in the statistics. Thus widespread wheat-cotton rotations do not include legumes and grasses, which may result in land degradation.

6 Kyrgyzstan

During Soviet times, Kyrgyzstan specialized in the production of alfalfa seed, potatoes and maize. After independence the country, like others, aimed at self-sufficiency in wheat grain and doubled its wheat-sown area from 250 to 550 thousand hectares by 1997 (Table 1). At the second stage of transition it stabilized at a level of half a million hectares and gradually fell to 360–400 thousand hectares in the current century. The wheat yield did not change notably during the whole period and got left at the level of 2.2 t ha^{-1} (Fig. 1). The wheat yield is not high, as it is winter wheat grown on both rain-fed and irrigated cropland. The so-called irrigated land is not properly managed; the irrigation is limited just to one watering period in the season. In the Soviet era the area sown with barley was the same as that of wheat but under market conditions it was immediately halved.

One of the crops which proved to be profitable under market economy conditions was maize. The sown area under this crop initially fell from 55 to 38 thousand hectares, with the grain yield reduced from 5 to 3 t ha^{-1}. Later on, however, the grain yield went up to 6 t ha^{-1} and the grown area went up to 75,000 ha accordingly. As in other countries of the region the planted area under potatoes tripled.

The most successful story about crop development under market economy conditions is dry beans sown under irrigation. The sown area under this pulse crop increased from 1,000 ha in the Soviet era to 48,000 ha in 2010. The grain yield went up from 0.8 t ha^{-1} to 1.8–1.9 t ha^{-1} in 1999–2005 but at present it has gone down to 1.5–1.7 t ha^{-1}, which might be because of continuous cropping. The area of other pulses such as rain-fed dry peas, is very small.

At the beginning of transition, the safflower sown area increased rapidly from 2 up to 50 thousand hectares but later on it dropped to 15,000 ha. The safflower seed yields are at 0.8–0.9 t ha^{-1}.

7 Tajikistan

Like other Central Asian countries, Tajikistan specialized in cotton production under irrigation in the Soviet era. After independence, the country focused on growing bread wheat under irrigation by replacing some cotton and removing forage crops from irrigated land. From the first years of independence the wheat grown area was doubled and it remains until now at the level of 360 thousand hectares (Table 1). The grain yield level is very low, which can be explained by a shortage of fertilizers, inefficient irrigation methods and a poor seed production system (Fig. 1). It remained below 2 t ha^{-1} and has only increased up to 2.3 t ha^{-1} in the last few years. It is not enough to reach the targeted wheat production for self-sufficiency, so the country imports about 400 hundred tonnes of both wheat grain and flour annually.

All the other crops occupy relatively small areas of land. As in the neighbouring countries, potato production increased considerably during the transition period, tripling the planting area and almost doubling the crop yield from 13 to 24 t ha^{-1}. Still, some 30,000 tonnes of potatoes are imported annually. The area under pulses and under oilseeds is very small.

8 Turkmenistan

The strategy towards increased bread wheat production under irrigation was also shared by Turkmenistan. It produced notable results in wheat grain production, which increased tenfold (Table 1). This result was achieved by increasing the wheat-sown area fourfold. The wheat yield increased significantly (Fig. 1), although it is rather low for irrigated lands. In spite of this considerable increase in wheat production, the country still imports some 300,000 tonnes of wheat grain annually. Barley is placed on 50–60 thousand hectares of rain-fed or irrigated area with limited water resources. During independence the rice-sown area has increased from 28 to 65 thousand hectares, with grain yields remaining at a level of 2.0–2.2 t ha^{-1}. Potato production has increased sevenfold mostly due to increased planted area. Maize production, however, is not a success story. Its sown area has fallen from 43 to 16 thousand hectares because of very low maize grain yields. Pulses and oilseeds are grown on very small areas.

9 Land Degradation Issues

In dryland agriculture, a major feature of the so-called conservation agriculture system widely adopted in Kazakhstan during Soviet times was the replacement of the mouldboard plough by undercutting sweep and blade equipment. This tillage system included tillage operations in fall and in spring resulting in almost no mulch cover after sowing. Moreover, that farming system included summer fallow once in every 4–5 years as a mandatory element. The summer fallow management technology included four to five tillage operations in summer to control weeds. This resulted in wind and water erosion. Presently, conservation agriculture principles include no-till and minimum-tillage methods as well as eliminating summer fallow (Gan et al. 2008). The long-term research data proves that on black soils summer fallow can be replaced by food legumes (Suleimenov et al. 2010). Minimum tillage has become a widely adopted practice in Kazakhstan while no-till systems are used on 10–15 % of cropland. The area under summer fallow was reduced by 1 million hectares in 2008–2011 but it is still quite large, occupying over 3 million hectares every year. Chemical fallow might be a good option to replace the traditional fallow with many tillage operations. There are no adequate statistics but approximately 5 % of summer fallow is chemical, and on 10–15 % of the summer fallow area minimum tillage is practised as a combination of chemical and tillage operations.

In irrigated agriculture, the prevailing technology is the combination of minimum tillage when sowing winter wheat into cotton stubble with traditional ploughing after the harvest of winter wheat (Khusanov 2009). In addition to inefficient cropping systems, excessive irrigation and poorly designed and maintained irrigation and drainage systems have led to severe land and water degradation in Central Asia. Only 50–70 % of the irrigation water used reaches crops, due to water losses in inter-farm and intra-farm irrigation pipes, with 10–25 % of the water lost in the inter-farm irrigation network and 20–30 % lost in intra-farm networks (Bekturova and Romanova 2007).

Due to increased demand for food and forage, many rangeland areas in the region are poorly managed, leading to a lack of feed, land degradation, a loss of plant biodiversity, and expanding desertification. Distant pasturing has been replaced by a daily "home-pasture-home" scheme for most of the herders (Bekturova and Romanova 2007). Open access to communal rangelands near the settlements is leading to land degradation and the loss of pasture species (Iniguez et al. 2004). There is clear evidence of increased rangeland degradation around settlements (Gintzburger 2004; Suleimenov and Thomas 2007).

10 Development of Livestock Industry in the Transition Period

10.1 Livestock Population

Ignoring feed and forage, production in all Central Asian countries can be partly explained by the fact that livestock industry on large-scale farms collapsed during the first few years of transition and was moved to households which had no rangeland and no cropland at all. This resulted in a dramatic fall in animal numbers in countries with more liberal reforms, such as Kazakhstan and Kyrgyzstan (Table 2).

In Kazakhstan, the privatization of land and agriculture happened against the background of a collapse of the whole economy, and government failure to support the agricultural sector. Under the hardships of the transition to a market-oriented economy, the livestock industry could not be profitable because of the low prices at local markets. Furthermore, animals were used by collective enterprises as barter for agricultural input materials and services. The most dramatic drop in the livestock population occurred in small ruminants, whose number fell from 34 to 9 million head by the end of the last century. In the new century it gradually started to recover and reached 17 million head by 2010 (Table 2). Sheep breeds, on the other hand, changed from fine wool to coarse wool breeds, and the share of goats also increased. The cattle population decreased less considerably, because many cows were kept in household farms, and there is hope of a return to 1992 figures thanks to a new government programme supporting the purchase of pedigree breed cattle from overseas. However, the recovery of the cattle industry will not come easy because of rangeland degradation and the collapse of feed and forage production. During the transition period Kazakhstan lost 1.7 million pigs, 26 million chickens, 220 thousand horses and 400 thousand turkeys.

The livestock industry in Uzbekistan was not damaged during the transition period as in its neighbouring country. This can be explained by the fact that there were not many large-scale livestock operation units except karakul sheep herds. Moreover, Uzbekistan in general kept government-controlled collective farms for longer. Presently, cattle are held in numerous small household farms. In this way

Table 2 Changes in livestock population in Central Asian countries in 1992–2010, millions of head (FAOSTAT 2010)

Country	Cattle		Sheep and goats		Pigs		Chicken	
	1992	2010	1992	2010	1992	2010	1992	2010
Kazakhstan	9.1	6.1	34.6	17.4	3.0	1.3	59	33
Kyrgyzstan	1.2	1.3	9.5	4.8	0.4	0.1	12	4
Tajikistan	1.4	1.0	3.4	4.2	0.1	0	7	4
Turkmenistan	0.8	2.2	5.6	16.3	0.2	0	8	15
Uzbekistan	5.1	8.5	9.2	14.5	0.7	0.1	34	35

the population of cattle was maintained at a level of 5 million head and gradually increased up to 8.5 million head (Table 2). The small ruminant population increased up to 14.5 million head with a growing share of coarse wool sheep and goats. Pig numbers fell quickly because of the departure of many Russians and other Slavic people. The number of chickens dropped by two thirds during the first stage of transition but then recovered and returned to a population similar to that in Soviet times. As a whole there is a positive trend in livestock numbers but productivity remains at a low level.

The sheep, pig and poultry production sectors suffered most in the Kyrgyzstan livestock industry (Table 2). The sheep population fell to 3 million at the first stage of transition and later gradually recovered slightly. However, the sheep breeds changed from fine-fleece to rough-wool, fat-tail breeds. As everywhere, goat numbers increased threefold. The dramatic reduction in the pig population can be explained by the departure of the Slavic population. Unlike the situation in Kazakhstan the cattle population was maintained at a certain level. The country exports some milk products to Kazakhstan.

In Tajikistan the livestock population dropped at the first stage but then recovered except that of pigs, which vanished completely (Table 2). The cattle population even increased compared with Soviet times. Fine-fleece sheep breeds were replaced by fat-tail, coarse-wool ones while goat numbers doubled.

The livestock population growth rate has been most rapid in Turkmenistan (Table 2). The most striking feature is a threefold increase in small ruminant numbers. This appears unbelievable against the background of a disastrous drop in the sheep populations of both Kazakhstan and Kyrgyzstan. It is explained by the fact that in Soviet times the country specialized in karakul sheep bred to produce karakul pelts by slaughtering newly born lambs. This market was lost and a decision was made in favour of keeping lambs to produce meat. In this way the sheep population was increased up to 14–15 million head in a very short time. During this time the number of goats increased 13-fold. It could be assumed that the potential of deserts to feed sheep and goats was underestimated in Soviet times, but the sustainability of the current situation needs to be questioned. Unlike that of other countries of the region, the cattle population increased threefold, while the number of chickens doubled.

10.2 Livestock Productivity

While livestock population developments were quite different in various countries, in most cases productivity remained at very low levels or even went down. Before privatization the low animal productivity in the Soviet era was associated with poor livestock management on state-owned farms. At that time livestock productivity in households was notably better. During the privatization of the agricultural sector and the removal of the state planning system all the main branches of animal production vanished very quickly, and they were left only in households. However, the result was the same, sometimes even worse (Table 3).

The only slight increase in cow milk productivity during the transition period was observed in Kazakhstan. In all the other countries it remained at a very low level. Cattle slaughter weights were also at a very low level, in some countries even lower than before. All this can be explained by the fact that only wheat and cotton production have been supported by government policies until now. Moreover, forage crops such as alfalfa have been removed from irrigated lands and replaced by wheat. In all countries feed and forage crops have also been removed from cropland, and livestock only remains in households with no cropland to grow forage. The utilization of rangelands has also become problematic because rangeland improvement is now never attempted. Furthermore most rangelands in Kazakhstan have been privatized, making it difficult for the majority of small holders to use seasonal pastures; they can only use limited, low-quality communal rangelands.

Comparative data on animal productivity shows considerable differences between the countries of Central Asia and Turkey, where livestock is also bred on small farms (Table 3). In 1992 the livestock productivity in Turkey was even lower than in Central Asian countries. During the next 18 years cow milk productivity in Turkey doubled and the cattle slaughter weight increased by half. There is no need to compare cattle productivity with Canada but it is worth noting that in this developed country livestock productivity notably increased during the same time in both the milk and meat sectors.

11 Conclusions and Outlook

(1) During the first few years of independence all Central Asian countries focused their policies on increasing wheat grain production to the maximum. In Kazakhstan this strengthened farm economies, while in other countries it was a step towards self-sufficiency in wheat grain production. However, this single-

Table 3 Changes in cattle productivity in Central Asian countries compared with Canada and Turkey in 1992–2010 (FAOSTAT 2010)

Country	Milk productivity per cow, kg		Slaughter weight, kg	
	1992	2010	1992	2010
Kazakhstan	1,492	2,255	185	157
Kyrgyzstan	1,914	1,983	191	178
Tajikistan	959	633	115	177
Turkmenistan	1,308	1,226	179	185
Uzbekistan	1,736	1,731	202	180
Turkey	1,436	2,847	146	217
Canada	5,952	8,202	278	340

crop production strategy had an obvious negative impact on crop diversification, including the production of feed and forage crops, as well as oilseeds, pulses and sugar beets.

(2) In Kazakhstan in recent years, crop diversification has been successfully started by the wide adoption of oilseed crops. This was influenced by government support as well as by favourable market prices compared to wheat grain prices. Oilseed crops include not only traditional sunflower but also new crops such as rapeseed, flax and safflower. Crop diversification also includes pulses although the sown area under dry peas, chickpeas and lentils is still not large enough to affect soil fertility management notably. Feed and forage production to support government-subsidized cattle stock improvements remains a serious problem. Thus the area under annual and perennial forage crops needs to be increased in both irrigated and rain-fed agriculture. In addition, sugar beet production dropped out of sight due to increasing sugar exports.

(3) In other countries in the region a political decision to achieve self-sufficiency in bread wheat grain by replacing feed and forage crops and some cotton with wheat under irrigation seemed to seriously damage the integrated development of crop and livestock production. The removal of alfalfa from irrigated lands is a major concern in new crop production systems as there is no space for crops other than wheat and cotton. Alfalfa has played a strategic role in crop rotations by maintaining soil fertility as well as producing high-quality forage. The successful introduction of dry bean production into Kyrgyzstan's agriculture demonstrated good opportunities in crop diversification for soil fertility management and strengthening the national economy. In other countries the area under pulses is very low. It could be increased by introducing pulses as a double crop after the harvesting of wheat under irrigation. Cotton seeds seem to be a major source of edible oil but a lot of sunflower oil has been imported by all countries in the region. Therefore safflower could be widely grown on dryland while sowings of sesame could be expanded on irrigated lands as a double crop.

(4) The governments of most Central Asian countries have seemed to care only about wheat and cotton production while livestock industry and forage production have remained a neglected sector. Presently most large agricultural farms are not producing forage crops because there is no livestock on these farms. Livestock is bred on household plots and on communal rangelands which are overstocked and degraded. In recent years the livestock industry has received more support from the governments but livestock productivity might be improved only if measures are undertaken to improve rangeland management and increase forage production. In this respect it is the right time to change policies towards supporting rangeland improvement and integrated crop and livestock production.

(5) Vast rangelands of Central Asia could serve as an important sink for atmospheric CO_2 (Suleimenov and Thomas 2007). Lal (2004) estimated that the potential of soil carbon sequestration in Central Asia is 8 to 24 million tonnes

of C per year (16 ± 8 M t of C per year) for about 50 years. Degraded and desertified rangelands cannot fulfil this function.

(6) In Kazakhstan the area under barley and forage crops has been increasing in recent years in accordance with more government support to livestock industry. In Uzbekistan, one possible option to remedy the fodder scarcity could be planting barley on saline areas abandoned from cotton cultivation in the country, since barley is highly salt-tolerant, more than wheat and cotton.

(7) Land degradation issues in dryland agriculture can be efficiently addressed and prevented by introducing conservation agriculture practices including the reduction of summer fallow area, crop diversification and no-till farming. Many water and land management technologies and practices are available that can substantially improve the efficiency of irrigation water use in Central Asia. Among these are drip irrigation, sprinklers or micro-sprinklers, cutback and alternate furrow irrigation, raised bed planting, field levelling, contour furrows, micro furrows, mulching, plastic chutes for conveying water, the re-use of drainage water for irrigation, and the planting of more water-efficient crops (Akhmadov 2009; Bekenov 2009; Khusanov 2009; Pender et al. 2000).

References

Akhmadov H (2009) Tajikistan report. IFPRI-ICARDA project on economic evaluation of sustainable land management options in central Asia. ICARDA, Tashkent

Bekenov M (2009) Kyrgyzstan report. IFPRI-ICARDA project on economic evaluation of sustainable land management options in Central Asia. ICARDA, Tashkent

Bekturova G, Romanova S (2007) Traditional land management knowledge in Central Asia: resource pack. S-Print, Almaty 86 p

FAOSTAT (2010) Online source: http://faostat.fao.org. Accessed on 5 July 2012

Gan Y, Harker KN, McConkey B, Suleimenov M (2008) Moving towards no-till practices in Northern Eurasia. In: Goddard T, Zoebisch M, Gan Y, Ellis W, Watson A, Sombatpanit S (eds) No-till farming systems. World Association of Soil and Water Conservation Special Publication No. 3. Bangkok, Thailand

Gintzburger G (2004) Agriculture and rangelands in Middle Asian Countries, In: Ryan J, Vlek P, Paroda R (eds), Agriculture in Central Asia: research for development. International Center for Agricultural Research in the Dry Areas (ICARDA), Tashkent, pp 154–175

Iniguez I, Suleimenov M, Yusupov S, Ajibekov A, Kineev M, Kheremov S, Abdusattarov A, Thomas D (2004) Livestock production in Central Asia: constraints and research opportunities. In: Ryan J, Vlek P, Paroda R (eds) Agriculture in Central Asia: research for development. International Center for Agricultural Research in the Dry Areas (ICARDA), Tashkent, pp 278–301

Khusanov R (2009) Uzbekistan report. IFPRI-ICARDA project on economic evaluation of sustainable land management options in Central Asia. ICARDA, Tashkent

Lal R (2004) Carbon sequestration in soils of Central Asia. Land Degrad Dev 15(6):563–572

Pender J, Mirzabaev A, Kato E (2000) Economic analysis of sustainable land management systems in Central Asia. Final Report submitted to Asian Development Bank (ADB)

Suleimenov M, Thomas R (2007) Central Asia: ecosystems and carbon sequestration challenges. In: Lal R, Suleimenov M, Stewart B, Hansen D, Doraiswamy P (eds) Climate Change and Terrestrial Carbon Sequestration in Central Asia. Taylor &Francis Group, London
Suleimenov M, Kiyas A, Kaskarbayev Z (2010) Replacement of summer fallow with oats and food legumes on black soils of northern Kazakhstan. Asian Aust J Plant Sci Biotech 4 (Special issue 1):81–86@2010 Global Science Books

Landscape Hydrology of Rural Areas: Challenges and Tools

Gunnar Lischeid

Abstract Currently the world hosts seven billion people that require food. About one third of the earth's land area is now intensively used by agriculture, and another third extensively. Agriculture inevitably depends on soil quality as well as on water resources. More than 70 % of human water use is due to irrigation. In addition, transpiration from rainfed agriculture comprises a substantial part of the earth's water cycle. Thus, land use both highly depends and affects water availability and is intimately intertwined with water resources management. The term "landscape" is used here to account for a variety of feedback effects between natural resources, human land use, economy and demography. The world population continues to increase. In addition, growth of economic wealth in many countries increased demand for upmarket agricultural products, and there is increasing demand for biofuel production as well which increases pressure on soil and water at a global scale. During the last 50 years cultivated area per capita decreased by half and likely will continue to decrease. Thus there is urgent need for advanced concepts of sustainable use of water and soil resources. Already today groundwater and river water over-exploitation due to increasing irrigation is a matter of concern. In some regions groundwater levels have been decreasing by 1 m per year during the last decades. Even large rivers fell increasingly dry. This has severe implications for water resources further downstream, not to mention biodiversity aspects. For example, the Aral Sea has been shrinking by more than 90 % within a few decades, giving place to a hostile salt desert. Thus, inefficient water management and land degradation are closely connected to each other. Climate change is now considered an increasing threat on water resources. Increasing air temperature is often associated with increasing evapotranspiration and thus increasing utilization of rare water resources. However, this is an over-simplification. On the one hand, warmer air masses can transport more water vapour, and increasing temperature comes along with increasing energy for mass

G. Lischeid (✉)
Leibniz Centre for Agricultural Landscape Research (ZALF) e. V.,
Eberswalder Strasse 84 15374 Muencheberg, Germany
e-mail: lischeid@zalf.de

L. Mueller et al. (eds.), *Novel Measurement and Assessment Tools for Monitoring and Management of Land and Water Resources in Agricultural Landscapes of Central Asia*, Environmental Science and Engineering, DOI: 10.1007/978-3-319-01017-5_5,

transport which could even increase precipitation. In addition, higher CO_2 partial pressure likely will increase water use efficiency of plants and thus reduce water consumption. On the other hand, large scale atmospheric circulation patterns most probably will be affected by climate change and thus change spatial patterns of precipitation which is not trivial to predict. Correspondingly, climate change models are fraught with large uncertainties with respect to precipitation. However, experts agree that in general frequency and intensity of extreme events like floods and drought likely will increase which poses agricultural management and soil resources at increasing risk. Besides, melting of glaciers in mountainous regions currently increases water availability in the lowlands downstream, whereas the opposite is true in the long-term. Thus there is urgent need for an "Integrated Water Resources Management" (IWRM) which has to include soil quality management as well. Various ideas, concepts and methods exist. However, local conditions have to be considered. Thus, experience from different parts of the world needs to be exchanged. This book is intended to contribute to that.

Keywords Landscape hydrology · Water over-exploitation · Integrated water resources management

1 Challenges and Tools

About one third of the earth's land surface is now intensively used for agricultural production (Lambin and Meyfroidt 2011), and another third is extensively used, mainly as pasture. Thus two third of the earth's land surface can be considered rural areas which have more or less intensively be shaped and are used by man (Hurtt et al. 2006; Miegel 2011). Pressure on agricultural land and the need for enhanced agricultural production is even increasing continuously. The world's population recently reached seven billion and continues to rise. Due to growing income the demand for meat and upmarket food is growing as well in many countries of the world. In addition, there is now increasing demand for biomass production on arable land as a renewable source of energy.

The indispensable prerequisites for agricultural production are soils and water. Soils do not only provide a substrate for plant roots, and a stock for nutrients, but also play a major role for water availability. Production of 1 kg rice requires about 3,500 l, and that of 1 kg beef, including water used for fodder production, about 15,000 l (Hoekstra and Chapagain 2008). Thus, about 70 % of the global annual water withdrawal by man is used for agriculture. However, irrigation water comprises only 20 % of the annual total water uptake and transpiration by agricultural plants (Coates et al. 2012). Consequently, agriculture is a major driver of the global water cycle. Agricultural water use competes with discharge to streams and wetlands as hotspots of biodiversity as well as with withdrawal of drinking water and for industrial purposes (Fig. 1).

Fig. 1 Aquatic ecosystems and wetlands are crucial elements of biodiversity. Their monitoring is a mandatory national task in the frame of the UN Biodiversity Convention, Photo: Lothar Mueller

Thus, agricultural management has substantial implications that reach far beyond the local cropland. On the other hand, local agricultural management as well as water availability depend on a multitude of influencing factors comprising natural conditions as well as the socio-economic boundary conditions (Fig. 2). To account for this complex interplay the term "landscape" will used in the following. It is not restricted to a certain spatial scale. However, as regions differ with respect to natural as well as to socio-economic conditions it makes sense to address single regions separately. In the following, I will focus on the hydrological aspects of rural landscapes and agricultural management.

It is assumed that the world population will increase up to 8.3 billion in 2030, and will exceed 9 billion in 2050 (UNDESA 2009). According to Bruinsma (2009), this would imply an increase of food demand by 50 % until 2030, and by 70 % by 2050. Please note that demand for bioenergy likely will add to this. Expansion of agricultural land use might be possible in some regions, although to a

Fig. 2 Creating new solutions for decision support and agricultural water management is a main task of landscape hydrological research, Photo: Lothar Mueller

limited extent. Often water availability is a key constraint. On the other hand, about 0.6 % of farmland is subject to land degradation and drops out from agricultural production every year (Coates et al. 2012). In fact, as a consequence of increasing population as well as of increasing urbanization the area of cultivated land per capita reduced by 50 % from 1961 to 2005 and is now close to 0.2 ha per capita (Coates et al. 2012).

Progress in plant breeding, agricultural management and technologies might provide some of the required increase of agricultural production. On a shorter time scale irrigation is considered the key for increasing crop yield. Irrigated area increased by 80 % from 1970 to 2008 and continues to increase (Coates et al. 2012). Globally, irrigation increases crop yield by 170 % on average (Coates et al. 2012). However, this comes along with over-abstraction of river water and groundwater resources in some regions. Hu et al. (2005) report on groundwater level decrease by 0.7 m per year from 1978 to 2003 at Luancheng in the North China Plain. Satellite data revealed that in the Indian states of Rajasthan, Punjab and Haryana groundwater depletion currently occurs at a rate of 40 mm/year (Rodell et al. 2009). The area of the Aral Sea has been shrinking by more than 90 % from the 1960s, giving place to more than 40,000 km^2 of hostile salt desert, with more than 150,000 t/year of dust and salt blown out in the 1990s (Micklin and Williams 1996).

Thus, over-exploitation of water resources is now considered a global problem. The concept of "virtual water" has gained a lot of interest in the public. It is defined as the ratio of water consumption of the plants over crop yield (Hoekstra 2003). This concept was developed by Allan (1993, 1994) to account for water consumption in extensively irrigated fields in the Middle East. Production and export of, e.g. cotton, thus implies export of virtual water which should be considered in economic assessments of resources allocation in arid regions (Allan 1994; Liu et al. 2009). Applying that concept, however, should differentiate between different sources of water: The term "green water" is used for rainfed agriculture, "blue water" is irrigation water from surface water or groundwater, and "grey water" denotes wastewater irrigation. Problems arise mainly with "blue water", whereas "grey water" use might even be considered a favourable way of water and nutrient recycling as well as of wastewater treatment.

Applying the virtual water concept to "green water", however, is not appropriate. The same holds for the term "water consumption". Plants do not consume water like a car consumes gasoline. Except for a minuscule fraction that becomes part of the plant's tissue, water taken up by plants is transpired, returns to the atmosphere and will rain out later on. In fact up to a fourth of summer precipitation in Europe can be due to internal recycling within the continent (Bisselink and Dolman 2009). Dirmeyer et al. (2009) show that at a global scale within-continents moisture recycling is especially high, e.g., in East Asia. Man does affect moisture recycling by land use management. E.g., Oguntunde et al. (2013) showed that reforestation in West Africa would significantly increase the region's annual precipitation, and agriculture would benefit from that.

At a local scale, replacing existing forests by arable fields would decrease annual evapotranspiration and thus increase groundwater recharge in temperate climates. In contrast, rainfed extensive grasslands in steppe regions transpire water irrespective whether they are grazed or not (Coates et al. 2012).

Climate change is now increasingly considered a major thread for land use in many parts of the world. It is often associated with increasing desertification. Increasing air temperature will increase vapour water deficit which in turn either increases evapotranspiration or water stress. However, this is not necessarily true. In fact, potential evaporation determined by class A pans exhibited significant decreasing trends during the last decades which is ascribed to an increase of cloud cover (Roderick and Farquhar 2002).

On the other hand, it is argued that water use efficiency of the plants will increase due to carbon dioxide enrichment. However, some authors argue that this effect will only partly compensate for increasing evaporative demand (Bunce 2004). Moreover, warmer air masses can transport larger quantities of water vapour and thus could intensify the water cycle, including enhanced precipitation. Climate change likely will affect circulation patterns at larger scales and will change local weather conditions and climate in an unpredictable way. Actual global circulation models have been quite successful with respect to hindcasts of air temperature, but still have considerable problems with respect to precipitation (Lorenz and Kunstmann 2012). In spite of all of these uncertainties, frequency and magnitude of extreme events, that is, storms, heavy rainstorms and droughts are assumed to increase due to higher energy content of the air masses. This will pose increasing threats for agriculture. However, so far land use effects had much larger effects on hydrology and runoff compared to climate change (Piao et al. 2007). But experts agree on the fact that regions that make use of mountain rivers which are fed from glaciers, like in South Kazakhstan, will benefit from increasing glacier melting in the short-term but have to account for decreasing discharge in the long-term.

Beside water quantity soil and water quality effects have to be considered. The nitrate load of major rivers per unit area of basin is highly correlated with population density in the basin (Howarth et al. 1996), which is only partly due to sewage efflux. In addition to nitrogen, intensive agricultural land use usually comes along with substantial phosphorus and pesticide contamination of groundwater, rivers and lakes. About half of the German groundwater resources are considered not to be able to reach a good qualitative state until 2005 (BMU 2008), mainly due to high agricultural contamination. Protecting groundwater resources is closely intertwined with soil protection. Soil degradation by salinization due to non-adequate irrigation, soil compaction, or erosion of the topsoil layer deprive the soil of its potential to protect groundwater resources as well as to warrant high crop yields.

Wise water resources management is an indispensable basis for living and economic wealth (Muller 2012). Managing landscape water resources has to account for a variety of impact factors and interactions. To that end the term "Integrated Water Resources Management" (IWRM) has been introduced. It is defined as “a process which promotes the coordinated development and management of water, land and related resources, in order to maximize the resultant economic and social welfare in an equitable manner without compromising the sustainability of vital ecosystems” (GWP 2012). Integrated Water Resources Management can and should make use of ecosystem services, like the purification effect of intact soils, but without overstressing them. A variety of approaches have been developed and tested in different parts of the world. Thus, many tools are available by now. However, they need to be adapted to local conditions. Wise water resources management can benefit a lot from local, not necessarily formally documented experience. This book is intended to document, communicate and share information and knowledge in order to stimulate and foster our knowledge and ability for wise water and soil resources management to the benefit of our peoples.

References

Allan JA (1993) Fortunately there are substitutes for water otherwise our hydro-political futures would be impossible. In: ODA (ed) Priorities for water resources allocation and management. ODA, London, pp 13–26

Allan JA (1994) Overall perspectives on countries and regions. In: Rogers P, Lydon P (eds) Water in the Arab world: perspectives and prognoses. Harvard University Press, Cambridge, pp 65–100

Bisselink B, Dolman AJ (2009) Recycling of moisture in Europe: contribution of evaporation to variability in very wet and dry years. Hydrol Earth Syst Sci 13:1685–1697

BMU (Bundesministerium für Umwelt, Naturschutz und Reaktorsicherheit; ed.). 2008. Grundwasser in Deutschland. Berlin

Bruinsma J (2009) The resource outlook to 2050: by how much do land, water and crop yields need to increase by 2050? Prepared for the FAO expert meeting on ‘How to Feed the World in 2050′, 24–26 June 2009, Rome

Bunce JA (2004) Carbon dioxide effects on stomatal responses to the environment and water use by crops under field conditions. Oecologia 140:1–10. doi:10.1007/s00442-003-1401-6

Coates D, Connor R, Leclerc L, Rast W, Schumann K, Webber M (2012) Water demand: What drives consumption? In: WWAP (World Water Assessment Programme): The United Nations world water development report 4: managing water under uncertainty and risk. Paris, UNESCO, Chapter 2, pp 44–76

Dirmeyer PA, Brubaker KL, DelSol T (2009) Import and export of atmospheric water vapor between nations. J Hydrol 365:11–22

GWP (Global Water Partnership) (2012) What is IWRM? Downloaded at http://www.gwp.org/The-Challenge/What-is-IWRM/. 21 Jan 2013

Hoekstra AY (2003) Virtual water: an introduction. In: Hoekstra AY (ed) Virtual water trade. Proceedings of the international expert meeting on virtual water trade. Value of water research report series No. 12. UNESCO-IHE, Delft, pp 13–23

Hoekstra AY, Chapagain AK (2008) Globalization of water: Sharing the planet’s freshwater resources. Blackwell Publishing, Oxford

Howarth RW, Billen G, Swaney D, Townsend A, Jaworski N, Lajtha K, Downing JA, Elmgren R, Caraco N, Jordan T, Berendse F, Freney J, Kudeyarov V, Murdoch P, Zhao-Liang Z (1996) Regional nitrogen budgets and riverine N and P fluxes for the drainages to the North Atlantic Ocean: natural and human influences. Biogeochemistry 35:75–139

Hu C, Delgado JA, Zhang X, Ma L (2005) Assessment of groundwater use by wheat (*Triticum aestivum* L.) in the Luancheng Xian region and potential implications for water conservation in the northwestern North China Plain. J Soil Water Conserv 60:80–88

Hurtt GC, Frolking S, Fearon MG, Moore B, Shevliakovas E, Malyshev S, Pacala SW, Houghton RA (2006) The underpinnings of land-use history: three centuries of global gridded land-use transitions, wood-harvest activity, and resulting secondary lands. Glob Change Biol 12:1208–1229

Lambin EF, Meyfroidt P (2011) Global land use change, economic globalization, and the looming land scarcity. PNAS 108(9):3465–3472, doi/10.1073/pnas.1100480108

Liu J, Zehnder A, Yang H (2009) Global consumptive water use for crop production: the importance of green water and virtual water. Water Resour Res 45:W05428. doi:10.1029/2007WR006051

Lorenz C, Kunstmann H (2012). The hydrological cycle in three state-of-the-art reanalyses: intercomparison and performance analysis. J Hydrometeorol, 120402130714000 doi:10.1175/JHM-D-11-088.1

Micklin P, Williams WD (eds) (1996) The aral sea basin. NATO ASI Series 2: Environment, vol. 12. Berlin, Springer, pp 5–6

Miegel M (2011) Exit. Wohlstand ohne Wachstum. List Taschenbuch, Ullstein Buchverlage GmbH, Berlin, ISBN 978-3-548-61031-3

Muller M (2012) From raw data to informed decisions. In: WWAP (World Water Assessment Programme): The United Nations world water development report 4: managing water under uncertainty and risk. Paris, UNESCO, Chapter 6, p 157–173

Oguntunde PG, Abiodun BJ, Lischeid G (2013) Modeling the impacts of reforestation on the projected hydroclimatology of Niger River Basin, West Africa. Ecohydrology (in press), doi:10.1002/eco.1343

Piao S, Friedlingstein P, Ciais P, Noblet-Ducoudré N, Labat D, Zaehle S (2007) Changes in climate and land use have a larger direct impact than rising CO_2 on global river runoff trends. Proc Natl Acad Sci 104:15242–15247

Rodell M, Velicogna I, Famiglietti JS (2009) Satellite-based estimates of groundwater depletion in India. Nature 460:999–1003. doi:10.1038/nature08238

Roderick ML, Farquhar GD (2002) The cause of decreased pan evaporation over the past 50 years. Science 298(5597):1410–1411

UNDESA (United Nations Department of Economic and Social Affairs, Population Division) (2009) World population prospects: the 2008 revision, highlights, working paper no. ESA/P/WP.210. New York, UN

Productivity Potentials of the Global Land Resource for Cropping and Grazing

Lothar Mueller, Uwe Schindler, Bruce C. Ball, Elena Smolentseva, Victor G. Sychev, T. Graham Shepherd, Manzoor Qadir, Katharina Helming, Axel Behrendt and Frank Eulenstein

Abstract The chapter gives an overview of global land potentials, crop yields and their limiting factors, and of methods to evaluate the productivity potential of land. Maintaining the capacity of the global land resource to produce plant biomass which can be used for humans is one of the most challenging issues of the 21st

L. Mueller (✉) · U. Schindler · K. Helming · F. Eulenstein
Leibniz-Centre for Agricultural Landscape Research (ZALF) e. V., Eberswalder Str. 84
15374 Muencheberg, Germany
e-mail: mueller@zalf.de

B. C. Ball
Crop and Soil Systems Research Group, SRUC, West Mains Road, Edinburgh EH9 3JG, UK
e-mail: Bruce.Ball@sruc.ac.uk

E. Smolentseva
Russian Academy of Sciences, Siberian Branch, Institute of Soil Science and Agrochemistry (ISSA), Ac. Lavrentieva av., 8, 630090 Novosibirsk, Russian Federation
e-mail: smolentseva@issa.nsc.ru

V. G. Sychev
Russian Academy of Agricultural Sciences, Pryanishnikov All-Russian Institute of Agrochemistry (VNIIA), Pryanishnikova St., 31a, 127550 Moscow, Russian Federation
e-mail: info@vniia-pr.ru

T. G. Shepherd
BioAgriNomics Ltd, 6 Parata Street, Palmerston North 4410, New Zealand
e-mail: gshepherd@BioAgriNomics.com

M. Qadir
Institute for Water, Environment and Health (UNU-INWEH), United Nations University, 175 Longwood Road South, Suite 204, Hamilton, ON L8P 0A1, Canada
e-mail: Manzoor.Qadir@unu.edu

A. Behrendt
Leibniz Centre for Agricultural Landscape Research (ZALF) e. V, Research Station Paulinenaue, Gutshof 7 14641 Paulinenaue, Germany
e-mail: abehrendt@zalf.de

L. Mueller et al. (eds.), *Novel Measurement and Assessment Tools for Monitoring and Management of Land and Water Resources in Agricultural Landscapes of Central Asia*, Environmental Science and Engineering, DOI: 10.1007/978-3-319-01017-5_6,

century. We need methodologies to observe and control the status of the potential productivity of agricultural and other lands. Methods of overall soil quality assessment which include the most significant factors and indicators relevant to soil productivity potentials can be useful tools for monitoring and managing the global soil resource sustainably. The aim was to find a common basis for soil productivity evaluation, as required by a global community of land users to allow achievement of high productivity in the context of a sustainable multifunctional use of landscapes. Results showed that soil types or reference groups in most existing soil classifications are largely defined on pedogenetic criteria and provide insufficient information to assess soil functionality. Traditional specific soil and land evaluation schemes already exist at national levels. They are based on different concepts of soil fertility or quality, local soil properties and the types of land use and management that prevail in the region or country. Their soil data inputs differ, ratings are not transferable and not applicable in transnational studies. At a transnational level, methods like agro-ecological zoning or ecosystem and crop models provide reliable assessments of land productivity potentials. Such methods are not intended for a field scale application to detect main soil constraints or to derive soil management recommendations in situ. A comparative analysis of several soil and land evaluation methods revealed the usefulness of indicator-based approaches applicable reliably, simply and consistently over different scales, from field level to large regions (aided by soil maps). Basic soil survey methods, including visual tactile soil structure assessment, are useful diagnostic tools for the recognition of productivity limiting soil attributes and estimation of indicator values. We advocate a straightforward indicator-based soil functional assessment system supplementing the current WRB (2006) classification or the coming Universal Soil Classification. It operates as a useful tool for monitoring, planning and management decisions based on soil quality (SQ) by detecting properties and limitation of soils for cropping and grazing and by providing estimates of attainable crop yields over different scales. The Muencheberg Soil Quality Rating (M-SQR), described in a chapter of Part II, has the potential to serve as a global reference assessment method of soil productivity potentials consistently over different scales. It combines visual methods of soil assessment (methods of soil survey, visual assessment of soil structure) with climate data in expert-based evaluation, classification and ranking schemes. M-SQR has been successfully tested in most agricultural regions worldwide. It provides concrete results about soil quality but also a frame for further research towards sustainable agricultural practices.

Keywords Soil functions · Soil quality · Crop yield · Sustainable agriculture · Land rating

1 Introduction: The need for Information on the Functional Status of Land Resources

The landmass of our planet is a proper habitat for human life. Land includes soils, the fragile skin of the earth. Soils provide plant growth and are the basis of our highly-evolved civilisation. Their performance to produce food, fibre, clear water, energy and further essential goods for humans will be decisive for our survival. This source is largely non-renewable and must be handled with care. Increasing world population, limited fossil fuel resources, changing climate and other global issues raise pressure on existing or potential agricultural land.

Soon, feeding about 10 Billion people will be one of the greatest challenges of our century. Dynamic agricultural development as demanded by Borlaug (2007) has constraints given by rules of nature and humanity. FAO prospects from, for example, Nachtergaele et al. (2011) may imply an optimistic view on the trends of global food security. However, expected gains are mainly due to the expansion of the areas of agricultural land and of irrigation agriculture. This poses a risk to other scarce and fragile ecosystems and resources.

Land use has impacts on global carbon, water and nutrient cycles and on biodiversity. Whilst natural ecosystems are converted into agricultural land, urban sprawl seals over the best soils worldwide at a high daily rate. Management of soils by societies must be sustainable to maintain the function of all global ecosystems. Land use strategies including development of agriculture have to be imbedded in balanced strategies to develop multi-functional landscapes (Helming et al. 2008). Crucial ecosystem services must be maintained by linking agricultural intensification with biodiversity conservation (Tscharntke et al. 2012).

Soils have to provide several ecological and social functions. Based on a definition of Blum (1993) a key soil functions is "Food and other biomass production". Mueller et al. (2010) called this the "Productivity function" of soil. This productivity function is related to the most common definition of soil quality as "the capacity of a specific kind of soil to function, within natural or managed ecosystem boundaries, to sustain plant and animal productivity, maintain or enhance water and air quality, and support human health and habitation" (Karlen et al. 1997).

Good land is rare and very unequally distributed over the globe. However, what is good land and what is bad land? This simple question is scale-dependent and depends on agrarian policies. In some regions of Central Europe, solar plants have been installed recently on relatively "bad land" which otherwise has a potential for producing 4–8 t/ha of wheat under rainfed conditions. In other regions like in Central Asia, some "good land" provides a potential for 1–2 t/ha wheat at a high risk of soil degradation.

Maintaining and improving agricultural soil quality and halting soil degradation are essential. Currently, the preconditions for achieving this goal are not available. These are standards and frameworks for the taxonomic and functional classification and allocation of soils. At the 19th World Congress of Soil Science in

Brisbane the creation of a Universal Soil Classification System was declared (Golden et al. 2010). This is a first step towards a common language of soil scientists, but this goal is focusing on taxonomy, e.g. soils will be re-named following standardised rules. This will not provide progress towards an international functional soil classification. Our chapter deals with the requirements of such a functional soil evaluation system and a comparison of existing approaches to meet those requirements.

2 The Global Land Resource for Cropping and Grazing and its Current Productivity

According to statistical data of the FAO, the global land resource amounts about 130 Million square kilometres. Most of this land is covered by ice and snow, deserts or rocks and is not suitable for agriculture. Agricultural land is about 49 Million square kilometres or 38 % of global land. Most agricultural land of the world is sparsely used grassland. Arable land (cropland) amounts to 15 Million square kilometres (31 % of agricultural land) and land grown with cereals is less than the half of this, about 7 Million square kilometres (Fig. 1). Cereals are staple food and play a most important role for global food security. Predominant cereals are wheat, maize and rice. Most important areas for wheat production are Northern China, the EU states and the North American Great Plains.

The productivity in terms of grain yields of wheat and maize in different countries is shown in Fig. 2. Yields of wheat are very high in the EU and in Northern China. They are relatively low in other big wheat producing countries like Russia, Kazakhstan and Australia. Grain yields of maize are extremely high in the USA due to irrigated cropping systems. Much maize is grown worldwide for livestock feeding with silage and for biogas plants.

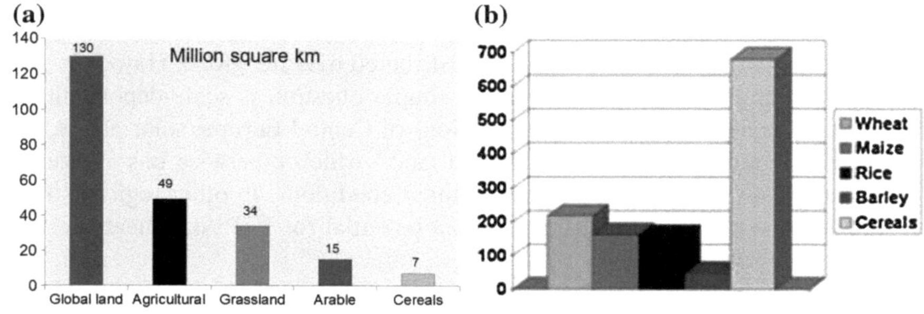

Fig. 1 **a** Global land resource for grazing and cropping. Million square kilometers (Mkm^2), based on AQUASTAT database (FAO 2012). 1 km^2 = 100 ha, **b** area under different cereal types in the Million hectares (data from Töpfer International 2012)

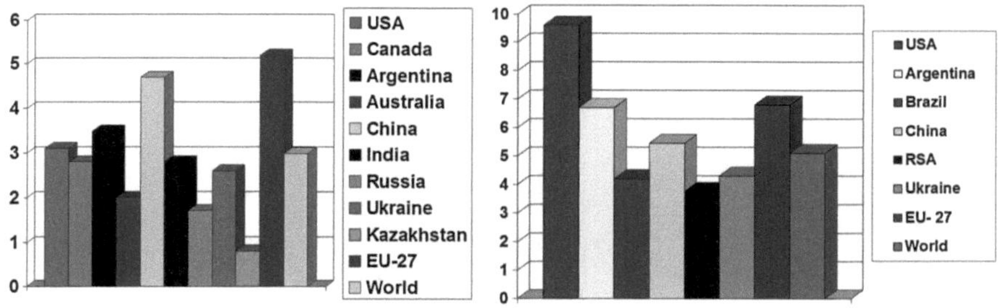

Fig. 2 Wheat yields, left hand, and Maize yields, right hand (Tons/ha) in 2011 by country (data from Töpfer International 2012)

Unfavourable natural conditions of climate and soil clearly have an influence on yields of cereals and all crops. Some crops are limited to specific regions. However, exact quantification scales are not available. This underpins the usefulness of a uniform evaluation frame of land quality.

Meat and milk of ruminants is a traditional basic food of many humans living in steppe and desert regions. Grasslands provide forage for lifestock but have much more functions in the global ecosystem like carbon storage and biodiversity preservation (White et al. 2000). Grasslands are very complex ecosystems of a high biodiversity dependent on their extent of managemente. In a wider sense, e.g. including rangelands and partly shrubs, they are among the largest ecosystems in the world having an estimated area of 52.5 Million square kilometers (Suttie et al. 2005).

Data about grassland productivity and reasons for local and temporal variability in terms of palatable biomass are relatively rare. Whilst some grass species in temperate grasslands may provide yields of more than 10 t/ha dry matter (Mueller et al. 2005), it is about 1 t/ha in the cool steppe of Mongolia (TERC 2005). The productivity in semi-arid and arid regions is lower than 1 t/ha. On pastures in Kazakhstan it was between 0.1 and 0.4 t/h (Kazakh Ministry of Environmental Protection 2007). Artemisia-ephemeral rangelands have low productivity and vegetation biomass with annual production varying from 0.15 to 0.36 t dry matter per ha (Morozova 1959, cited by Rajabov 2009).

Stocking rates and the resulting degradation status of the vegetation have a crucial influence on the productivity and quality of forage from grasslands (Fig. 3). On rangelands in Balochistan, Pakistan, above ground dry biomass production varies from 40 to 200 kg/ha in open areas as compared to 200 to 865 kg/ha in protected (non-grazed) areas (Ahmad et al. 2012).

Table 1 shows the influence on grazing intensity on the change of the composition of species in Western Kazakhstan. The palatable biomass decreases with the status of grazing intensity and may reach values of less than 100 kg/ha for cattle and sheep in overgrazed arid regions. The total biomass is often higher due to deep rooting secondary vegetation like dwarf bushes or unpalatable aliens. On the other hand, the encroachment of pastures with unpalatable bush species has benefits for protecting the soil against erosion and for carbon sequestration (Gupta

Fig. 3 Examples of overgrazed grasslands in cold steppes. Left hand: Xilingere river catchment, Inner Mongolia, China. Precipitation 420 mm/year. Serious overgrazing around farms and livestock stations results in *Stipa* became dominant and replacing *Leymus* communities. Right hand: Ili river catchment, Kazakhstan. Precipitation 260 mm/year. Long-term extreme overgrazing results in grasses being almost completely replaced by *Artemisia*, *Salsola* and other largely unpalatable species

et al. 2009). Grasslands are under stress from overgrazing, and many areas may change to deserts in dryland regions in the near future.

3 Constraints to Soil Productivity

"It is possible to modify almost any site to produce a particular crop or to build a structure. However, in practice, the extent of any modification depends on the land and climate characteristics, the cost of modifying them, the benefit gained, the availability of capital, and considerations of the long-term risk of failure" (Shepherd and Jessen 2002).

Plants need light, CO_2, adequate temperature, water and nutrients for photosynthesis. They need a proper substratum for germination, emergence, rooting and growth. They do not need soil for their metabolism. However, soils are a naturally effective reactor, transformer, catalyst and buffer for plant growth and other functions. They facilitate plant growth and biomass formation if further conditions

Table 1 The effect of grazing intensity on grassland changes on dark chestnut soils in Western Kazakhstan (Boonman and Mikhalev 2005; based on Larin 1956)

Grazing intensity	Vegetation species
Absent	*Stipa* spp., *Festuca sulcata; herbs; Agropyron* spp. *and Bromus* spp.frequent
Weak	*Stipa lessingiana, Festuca sulcata, Stipa capillata, Artemisia austriaca*
Moderate	*Festuca sulcata, Koeleria gracilis, Stipa capillata, Artemisia austriaca*
Intensive	*Artemisia austriaca, Poa bulbosa, Euphorbia virgata*
Excessive	*Polygonum aviculare, Agropyron* spp., *Ceratocarpus arenarius, Bassia* spp.

Fig. 4 Examples of site-specific limitations to cropping at different scales. Left hand: Oder river catchment, Germany where extreme contrasts of sunflower growth are due to the heterogeneity of the soil substrate with poor stands on medium sand of low water storage capacity, Gleyic Fluvisol (Eutric, Arenic, Drainic). Good stands are on loamy clay, Gleyic Fluvisol (Eutric, Clayic, Drainic). Right hand: Ili river catchment, Kazakhstan, where uniform sparse stands of spring wheat occur on dryland due to restricted water availability, soil management and inputs, Haplic Calcisols (Siltic)

are suitable, mainly an appropriate climate, providing radiation, temperature and precipitation. Nutrients are present in the soil and can be given by fertilisation. Plant production on agricultural land is based on human activity which may provide high yields for selected species dependent on inputs.

Natural constraints to soil productivity for agriculture exist at different scales (Fig. 4). They can be classified into three major groups. The first group includes the thermal and moisture regimes of soils. Plants require appropriate temperatures and moisture of the soil for their growth. For most soils, thermal and moisture regimes directly depend on climatic conditions. They define the boundaries of limitations like drought, wetness or a too short vegetation period. The American Soil Taxonomy (Keys to Soil Taxonomy 2010) has included climatic conditions and topography in the definition of soil types. Constraints to soil productivity may be characterised in terms of soil properties. This simple concept is useful for assessing productivity potentials. Soil moisture is the main limiting factor in most agricultural systems worldwide. Drylands cover more than 50 % of the global land surface (Asner and Heidebrecht 2005).

The availability of a deep soil substratum having physical, chemical and biological properties suitable for rooting, water uptake and nutrition of plants is a second precondition for soil productivity. Shallow soils, rock outcrops, stoniness or adverse chemistry like salinity, sodicity, acidity or contamination may cause most severe restrictions to plant growth or the capability to manage the land.

The third group includes suitable terrain, sometimes considered as an external soil property, which helps to prevent soil erosion and to provide accessibility for management.

Human activity in terms of management inputs also has a clear effect on plant growth and crop yields. It can range widely from unmanaged largely natural grasslands to high–input cropping systems, and from native, site-adapted plants or rotations, to monocultures. As a consequence, a straightforward quantification of soil productivity potentials (soil quality scores or classes in terms of poor, moderate or good), requires definition of input- and management levels on the one hand, and of plants or rotations on the other hand. The scale of soil quality assessment should be valid for the area in question.

4 Defining an Optimum of Soil Quality: Loess Soils and Their Suitability for Cereal Cropping

About 10 % of the land surface of our planet is covered by Loess. Loess is an aeolian sediment of silty texture. Different types of soils have developed on loess sediments (Fig. 5). Loess soils are very fertile and play a crucial role for agriculture and human development worldwide. They have been a basis for the emergence of ancient and current highly-evolved civilisations like those in China, Europe and the United States. Suitable texture and mineral composition of loess ensure a good supply of plant-available water and nutrients, potassium in particular, good soil aeration, extensive penetration by plant roots and easy cultivation and seedbed production (Catt 2001). Loess occurs mainly in the mid latitudes. According to climate conditions, some typical soil types or reference groups such as Calcisols, Kastanozems, Chernozems, Phaeozems, Cambisols or Luvisols have been developed on Loess substrates. Chernozems, soils of the tall grass steppe, are the best known group and are characterised by a deep mollic horizon. Mollic horizons readily develop on Loess. Optimum crumb structure of those horizons can be observed on Chernozems and Phaeozems developed under Loess and under grassland vegetation.

Very high grain yields >10 t/ha in single cropping systems and more than 15 t/ha in maize-wheat double cropping systems can be achieved in the Northern China Plain on Cambisols developed on loess. The humus content of these soils is low (1 %). High rates of irrigation and mineral N- fertiliser are a precondition for high yields. Luvisols on loess in a humid climate of Western Europe provide wheat grain yields of 10 t/ha in relatively dry years at very high mineral fertilisation (240 kg N/ha).

A not too harsh climate providing cropping of winter wheat and favourable soil properties are preconditions for high crop yields. However, on Loess soils, favourable soil properties and high yields are not for free and not for ever. Inadequate soil management has been leading to degradation by accelerated wind and water erosion, humus depletion and other human impacts. Extended irrigation and high inputs of agrochemicals have exhausted and degraded aquifers under Loess in China and elsewhere.

Fig. 5 Two examples of soils from Loess. Left hand: Chernozem in the Ob- catchment, Western Siberia, Haplic Chernozem (Molliglossic, Siltic), cold temperature regime, precipitation 450 mm. The soil structure is close to optimum. Grain yields of spring wheat are 2–4 t/ha. Right hand: Stagnosol in the Rhine catchment, Germany, Luvic Stagnosol (Eutric, Siltic), mesic temperature regime, precipitation 830 mm. The soil structure is of medium quality and the soil is prone to wetness. The grain yield of winter wheat or winter barley is typically 6–8 t/ha due to sufficient rain in the vegetation period, temperate climate and high inputs of farming

5 Processes that Degrade Soils and Reduce Their Productivity Potential

Soils are threatened by degradation and desertification that limit their functions in the global ecosystems and their productivity in particular. About one-third of the global land surface is subject to desertification (Eswaran et al. 2001). Semiarid and arid regions are threatened by global warming induced problems of soil degradation and desertification risk (Lobell et al. 2008).

Climate change is expected to limit the agronomic productivity and sustainability in most currently dense populated regions (European Commission 2009) and may also aggravate problems of soil degradation in these areas. A basic reason for land degradation is over-exploitation of soils (Lal 2009). At EU and global levels, some processes have been identified as main threats to soil functions including productivity (Louwagie et al. 2009). Erosion, compaction and organic matter decline are all examples of human-induced negative side-effects of many farming systems. Erosion (Figs. 6, 7 and 8) affects soil productivity in several

Fig. 6 Different forms of water erosion: Gully erosion (**a**, **b**) and slip erosion (**c**). In cold steppes, huge gullies may form at springtime snowmelt. Examples from the steppe of Kazakhstan (**a**) and a canola field in Germany (**b**). In humid landscapes, gully erosion occurs on ploughland and may pollute adjacent water bodies. Slip erosion is typical on steep land in humid landscapes. The example is from Manawatu, New Zealand. The natural vegetation of forest would normally prevent slip erosion. Tillage erosion is also typical for arable farming on sloped and steep lands. (**d**) The example on the right is from the Palouse region, Washington, USA

Fig. 7 Typical steppe landscape under cropping and prone to wind erosion. This example is from Oregon, USA, where there is no protecting vegetation cover or shelter belts. Part of the dust is being mobilised by tillage operations. Storm events may create dust bowls

Fig. 8 Example of high wind erosion intensity on a fine sandy soil at a cropping site at Gu Yuan, Hebei, at the border to Inner Mongolia, Northern China. The left side shows a field, currently grown with linseed (*Linum usitatissimum*) and intensively ploughed and cropped for 40 years. The apparent wall at the right side of the photo is a windbreak. The depth of the deposition accumulated over 40 years is 1.2 m, an average deposition rate of 30 mm/year

ways. Topsoil removal diminishes plant emergence, initial development and rooting depth, mainly because of reduced water and nutrient storage capacities. In humid regions of Europe, productivity of land is not likely to be heavily reduced by soil erosion, (Bakker et al. 2007), but the offsite- damage of these processes is most significant. By contrast at limited inputs, the threat of declines in productivity due to erosion is much greater. Crop yield reductions of more than 60 % or even almost total failures have been reported from African studies (Oyedele and Aina 2006; Salako et al. 2007; Mairura et al. 2008).

As a global process, erosion damage should be evaluated more specifically. In some regions it has probably positive effects for carbon sequestration due to the building up a new A-horizon at eroded parts on the one hand and conservation of buried organic matter in the colluvium on the other hand.

Slip erosion is typical on steep land in humid landscapes. The example is from Manawatu, New Zealand (Fig. 6c). The natural vegetation of forest would normally prevent slip erosion. Tillage erosion is also typical for arable farming on sloped and steep lands.

Accelerated compaction of soils is a consequence of agricultural mechanisation and may lead to complex limitations of soil ecological functions (Lipiec et al. 2003). Compaction of soils alters their structure and affects ecological processes adversely. These include reduction in rooting depth and nutrient uptake from the soils, biochemical changes due to denitrification and a reduction in nitrification. Stress to crops limits their productivity.

Part of this damage can be compensated by increased inputs of fertilisers or water, but those may not be possible in some farming systems or are ineffective. In organic farming, a non-compacted soil having favourable structure properties is essential for high productivity (Munkholm et al. 2003) because mineral nitrogen fertilisers cannot be applied. Compaction lowers the efficiency of water and radiation use by plants (Sadras et al. 2004). Soils in many countries of the humid temperate zone are severely compacted (Jones et al. 2003). Subsoil compaction is very persistent and must be accepted in most cases. Technical solutions to diminish or alleviate compaction have been available over recent decades (Spoor 2006), but there are no clear, thorough strategies to stabilise and maintain recovered soil structure. A main reason for targeting adaption strategies for compaction instead of prevention and recovery strategies is the difficulty of quantifying the ecological effects of compaction (Lipiec et al. 2003). As the compaction status of fields has increased over the past years so too has the crop yield level so that the negative impact of compaction on crop yields often remains hidden. As long as the weight of agricultural machinery increases, the damaging processes of compaction will continue (Mueller et al. 2011a).

Decline in soil organic matter (SOM) mainly occurs in response to soil management like tillage, rotation and fertilisation. The conversion from natural vegetation of forest or grassland to cropland has already led to a marked SOM decline. Crop residue removal has negative consequences for the soil carbon balance and productivity (Blanco-Canqui and Lal 2009). The process of SOM decline can be aggravated by climate change. Lal (2009) stated that, regardless of the region and farming system, maintenance or increase of SOM levels is not only essential for the carbon balance and GHG related issues but also for sustaining productivity. Under favourable conditions of climate and soils, for example on Loess soils in China, high crop yields can be achieved also with very low SOM contents, but in soils of the humid tropics yields decline dramatically with SOM losses (Lal 2006).

Like other soil degradation processes, salinisation and sodification or acidification are natural processes which may be made worse by land management. Monitoring and control of these processes requires international guidelines and evaluation scales. Table 2 shows an example of the assessment of sodification at EU level by measuring Exchangeable Sodium Percentage (ESP). Another indicator of this process at field level could be pH measurement. pH-values greater than 8.2 may indicate the threat of sodification. For salinisation, which is often combined with sodification in many arid areas, the electrical conductivity may serve as a good indicator (Varallyay, in Huber et al. 2008). However, plant genera and single crop types have very specific salt tolerance (Fig. 9).

Table 2 Thresholds of relative fertility of soil due to sodium hazard (Varallyay, in Huber et al. 2008)

Relative fertility of soil	ESP* (%)	Remark
100	<5	Non-sodic soil
60–75	10–15	Moderately sodic soil
20–30	25–30	Sodic soil
0	>50	Strongly sodic soil

*ESP = Exchangeable sodium percentage

Fig. 9 Examples of different salt tolerance of plants near Nanpi, Hebei province, China. The sugar beets (*left photo*) and cotton (*right photo, background*) are relatively tolerant to salinization whereas sunflowers (right foreground) show clear growth depressions

Long-term use of soils as cropland does not nessessarily lead to degradation of soil quality and productivity potentials. Sustainable management and direct soil improvements e.g. by land drainage or amelioration of the soil substrate, may also lead to an aggradation of soils in the long run (Fig. 10).

Finally, the complete loss of soil, mainly by sealing in urban areas, is the most severe process of land degradation worldwide. Measures of landscaping and urban agriculture may maintain productivity potentials of those lands to a certain degree, but those ecosystems are artificial and vulnerable. Producing food on urban land requires high inputs and faces an increasing risk to human health but is likely to be very important for future food security.

6 Do Soil Taxonomic Classification Systems Contain Information on Crop Yield Potentials?

Soil classification systems consider a combination of different principles, mainly of recognition and performance. Attributes used for classification may reflect both. Pedogenetic criteria dominate at higher levels of soil classification systems but functional criteria are also part of most soil classifications. In some cases pedogenetic and functional criteria are combined, and genetic soil types provide information about soil productivity potentials.

For example, it is common knowledge that the soil type Chernozem, which has developed mainly from loess material and contains a mollic epipedon, rich in humus, has a high crop yield potential, whilst the soil type Leptosol is a shallow soil of low productivity. Podzols are leached sandy soils lacking nutrients and water storage capacity. These examples show the soil type is associated with typical substrate and climate conditions, from which one may make conclusions on functional properties.

Fig. 10 Examples of soil degradation and aggradation. Human impacts on soils may lead to degradation or aggradation of productivity potentials. On the left is a Chernozem degraded by wind erosion in the Kulunda Steppe, Russia. Outblow of fine material and organic matter has led to an irreversible alteration of soil texture and structure with negative impacts on the soil moisture regime and emergence of crops. On the right is a former Podsol aggraded by adding organic material (*grass sods*) in the past, location in Northern Germany. This specific type of Anthrosol (*Plaggenesch*) has improved its water and nutrient status and provides high crop yields

Apart from those extremes, functional information from higher level soil classifications is relatively minor (Table 3). Some soil types or reference groups of WRB 2006, (examples: Cambisols, Fluvisols, Regosols) may be have developed from different soil substrates in different climate environments. In those cases, certain information about soil productivity at a local or regional scale will be provided if the classification includes further soil attributes like texture, organic matter, eutrification status and pH.

Overall, the information value on soil functionality and productivity of most soil classifications including the world reference base (WRB 2006) is relatively low. Higher coefficients of determination in Table 3 show that information on the thermal and moisture regime of soils, which is a component of the US taxonomy but not of the WRB, may improve the information content on soil productivity potentials considerably. However, only if a genuine soil or land rating procedure like the Muencheberg Soil Quality Rating (Mueller et al. 2007a) is included, can a satisfactory level of information about soil productive potentials (B = 0.8) be provided. Depending on land use intensity and the quality of indicator

Table 3 Coefficient of determination (B) of multiple regressions between grain yield of cereals and soil parameters (Mueller et al. 2009b)

Reference soil groups (RSG) and land use	Variant of classification	B
All RSG, high-input farming (n = 352)	(1) Soil attributes only (Qualifiers of WRB 2006)	0,20**
	(2) Soil attributes and thermal and moisture regime	0,61***
	(3) M-SQR-score (Muencheberg soil quality rating)	0,78***
Phaeozems, Chernozems und Kastanozems, high-input farming (n = 54)	(1) Soil attributes only (Qualifiers of WRB 2006)	0,00
	(2) Soil attributes and thermal and moisture regime	0,65***
	(3) M-SQR-score (Muencheberg soil quality rating)	0,74***
All RSG, organic and low-input farming (n = 43)	(1) Soil attributes only (Qualifiers of WRB 2006)	0,36**
	(2) Soil attributes and thermal and moisture regime	0,63***
	(3) M-SQR-score (Muencheberg soil quality rating)	0,87***

parameterisation, final M-SQR scores may explain between 50 and 80 % of the crop yield variability.

7 Soil Structure as a Criterion of Management Induced Agricultural Soil Quality

Soil structure is a key to soil biological, chemical and physical processes, and visual soil structure assessment is a useful diagnostic tool for their recognition. Methods of visual soil structure examination enable semi- quantitative information of soil quality for use in extension and monitoring. Effects of compaction and differences in soil management can be recognised by visual structure criteria (Batey and McKenzie 2006; Mueller et al. 2009a).

Several methods have been developed over the past five decades. One of the oldest but most accepted methods is that of Peerlkamp (1967). A quantitative comparison of some easily-applied methods and their correlations with measured physical parameters of soil structure revealed that many methods provided similar results after standardizing data (Mueller et al. 2009a). Types and sizes of aggregates (Fig. 11) and abundance of biological macropores were the most reliable criteria as related to measurement data and crop yields. Differences in soil management could be recognised by visual structure criteria.

Methods based on or supplemented by illustrations have clear advantages for a reliable allocation of visual diagnostic criteria to rating numbers. The latest update

Fig. 11 Different features of soil aggregates in New Zealand (Structure indicators of the VSA drop-shatter procedure of Shepherd 2000). On the left is Manawatu silt loam Haplic Fluvisol (Eutric, Siltic, Drainic) with moderately good structure, rating of 1.5 (0 = worst, 2 = best) and on the right is Kairanga clay Fluvic Gleysol (Eutric, Clayic, Drainic) with poor structure, rating of 0

of the Peerlkamp method provided by Ball et al. (2007) and the method of Brunotte et al. (2011) are well illustrated. The New Zealand Visual Soil Assessment (VSA, Shepherd 2000, 2009) as an illustrated multi-criteria method, enabling reliable assessments of the soil structure status. It is currently recommended by the FAO for controlling the soil structure in African farming systems (FAO 2008). These methods are feasible tools for soil structure monitoring and management recommendations at field or local scales.

8 Comparison of Available Methods to Assessing Soil Productivity Potentials

Over the past 100 years, specific soil and land evaluation systems targeting soil quality for crop yields have been developed in almost all countries. They differ in purpose and performance and their results are rarely comparable or transferable into other regions.

A global assessment framework of the overall soil quality and productivity potential is useful and needs to meet the following requirements (Mueller et al. 2010):

- Monitoring tool of the functional status of the soil
- Precise in operation, based on indicators and thresholds of the most functionally relevant attributes

- Consistently applicable over different scales
- Potential for suitability and capability classifications
- Straightforward for use in extension
- Potential as a crop yield estimator and thus acceptable to farmers.

A comparison of available methods regarding these criteria (scale of validity, field method capability, reliability, relation of soil and climate data, plant suitability and others) is shown in Fig. 12. This analysis by a multi-dimensional scaling procedure visualises the similarity of methods. The plot is a computed distance map based on a performance matrix of some well known methods. Wide separations indicate dissimilarities of methods. The closer the distance to the center, the better is the suitability for a global assessment methodology.

This procedure shows clear separations between traditional soil ratings and visual assessments of soil structure quality (VSA, Shepherd 2000). Traditional soil ratings include the Storie Index Rating (Storie 1978) and the German and Austrian Soil Rating (Rothkegel 1950; Bodenaufnahmesysteme 2001). Also the soil quality rating system of the former Soviet Union (Gavrilyuk 1974; Tjumenzev 1975) would be located at this corner.

Soil texture is correlated with other important functional attributes like water and nutrient storage capacity and thus has become a dominant criterion of all existing functional classification systems. However, at the supra-regional or global scale, qualifiers of soil internal attributes like texture may only explain less than 50 % of the crop yield variability (Table 3). Variations in the soil's moisture and

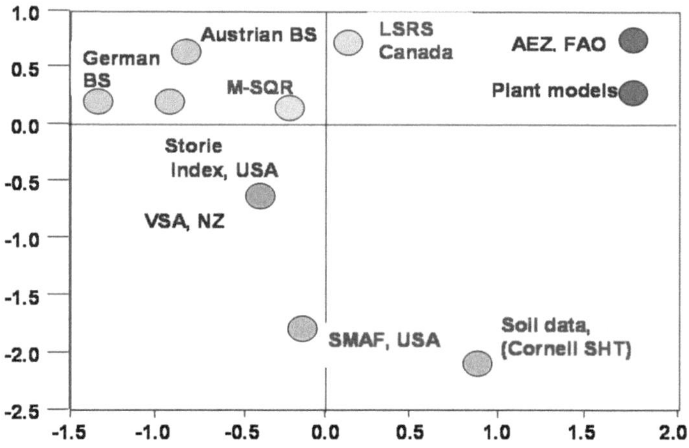

Fig. 12 Similarity plot (Euclidean distance model) of some evaluation schemes relevant to soil productivity (Mueller et al. 2010). Key: German BS (German soil rating, Rothkegel 1950) Austrian BS (Austrian soil rating, Bodenaufnahmesysteme in Österreich 2001) Muencheberg soil quality rating (M-SQR, Mueller et al. 2007b), visual soil assessment (VSA, Shepherd 2009), LSRS Canada (Land suitability rating system, Agronomic Interpretations Working Group 1995), Agro-ecological zoning (AEZ, Fischer et al. 2002), soil management assessment framework (SMAF, Andrews et al. 2004), Cornell soil health test (Cornell SHT, Schindelbeck et al. 2008)

thermal regimes are most crop-yield relevant at those scales and have to be taken into consideration.

Crop models and the AEZ methodology (Fischer et al. 2002) may reflect these climate factors well. They are similar both in purpose and in results. They are located far from the centre because these procedures are not field methods of soil assessment and are mainly based on climate information. Soil data sets and the Soil Management Assessment Framework (Andrews et al. 2004) also occupy isolated positions as, although they contain detailed soil information, they do not include climate information and are based partly on laboratory analyses. More recent procedures like the Canadian Land Suitability Rating System (LSRS), and the Muencheberg Soil Quality Rating (M-SQR) which include more crop yield relevant parameters (Climate, soil structure) are intermediate and closer to the center. The rating procedures differ but inputs are similar. LSRS has been validated with EPIC simulations in Canada (Bremer and Ellert 2004) and M-SQR indicator ratings are expert-based and validated with crop yield data from Germany, Russia, China and the USA. Also, the sophisticated VSA method of the revised 2009 edition (Shepherd 2009) is an excellent tool for the field and regional scale. The Muencheberg Soil Quality Rating meets most criteria of a global framework as specified above as it contains information on climate and soil properties relevant to crop yield, and soil structure in particular. It has potential for consistent ratings of the soil productivity function on a global scale (Smolentseva et al. 2011) and could serve as a functional supplement of the WRB 2006 classification or the upcoming Universal Soil Classification.

9 Overall Agricultural Soil Quality, Crop Yields and Water use Efficiency

The Muencheberg Soil Quality Rating has been tested on many agricultural sites over the world and this has confirmed its usefulness for the characterisation of overall soil quality, yield limiting soil properties and crop yield potentials. Research fields and lysimeters have been the preferred subjects of testing as they provide reliable data of crop yields and water balance. Figure 13 shows an example for the grassland productivity of peatlands in Germany. The management strategy of peatlands is a big issue because of conflicting interests of stakeholders in utilising soil functions. In areas subject to re-wetting, trade-offs between gains (e.g. climate effects) and losses (e.g. yield) are preconditions for decisions. Soil quality on peatlands is controlled by soil substrate conditions and water table management. M-SQR is able to reflect the impacts of these factors on forage yields of grasslands in a simple way. The rating scheme has revealed that the overall soil quality and the input of nitrogen fertiliser nitrogen have crucial influence on the yield of reseeded and well managed peat grassland. If both managed and more natural or re-wetted plots are considered in the same pool of data, a significant

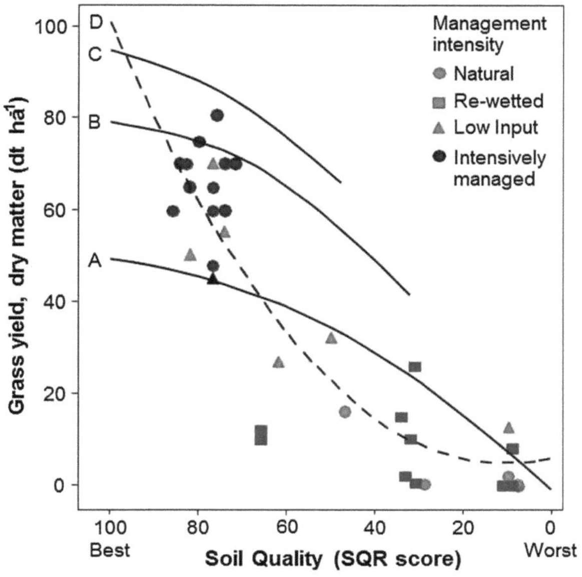

Fig. 13 Overall quality and productivity of peatland soils under grass in Germany, (Mueller et al. 2008, modified). *Lines A*, *B* and *C* refer to different intensities of nitrogen fertilisation based on trials. *Line D* is a regression line based on plotted data of experimental fields. Grass yield refers to the palatable biomass. Single measurement points are not shown in this graph

correlation is found. The influence of management intensity and fertilisation leads to a positive increase of palatable biomass at higher soil quality (M-SQR scores greater than 60).

Regressions for seeded grassland were:

A ► 0 kg N ha^{-1}, $y = 49.15 - 0.109\,(100 - x) - 0.004\,(100 - x)^2$, $n = 83$, $r^2 = 0.82$ ***

B ► 100 kg N ha^{-1}, $y = 78.09 - 0.018\,(100 - x) - 0.007\,(100 - x)^2$, $n = 83$, $r^2 = 0.82$ ***

C ► 200 kg N ha^{-1}, $y = 93.99 - 0.259\,(100 - x) - 0.005\,(100 - x)^2$, $n = 83$, $r^2 = 0.79$***

The regression for semi-natural and reseeded grassland, for current effective field grass yield, palatable plants only, was:

D ► $y = 100.75 - 2.144\,(100 - x) + 0.012\,(100 - x)^2$, $n = 45$, $r^2 = 0.85$***

The efficiency of water use by crops, rotations, and plants grown in different irrigation or tillage systems is a decisive criterion of productivity in agriculture worldwide. In most cases, different crops or management systems have been compared at the same sites. On a groundwater influenced sandy soil rich in humus (Lysimeter station Paulinenaue, Germany) the correlation of above-ground biomass with annual evapotranspiration was linear, but plant-specific. Water use efficiency of the C4-plant maize (*Zea maize*) was significantly higher than that of the mesophile C3-perennial ryegrass (*Lolium perenne*). The hydrophile C3-reed canary grass (*Phalaris arundinacea*) had the lowest water use efficiency, WUE, (Mueller et al. 2005).

Rose et al. (2012) found that management influenced grassland biomass and water balance primarily through fertilisation. A grassland experiment in a temperate climate revealed fertilisation with 180 kg N/ha increased the WUE by 20–30 %.

WUE may also vary from location to location and is clearly associated with soil quality (Fig. 14) and management inputs. Soil quality scores of M-SQR include effects of soil properties and climate. On better soils, plants produce more vegetation cover and protect the soil surface from evaporation. Figure 14 is based on data from 24 lysimeter stations across the world. In the group of very good soils (M-SQR scores >80) about 1.5 kg/m^2 biomass or 0.8 kg/m^2 grains were produced by 535 mm (kg/m^2) water. On soils of poor quality, mainly due to frequent drought stress, 400 mm water were necessary to produce 0.5 kg/m^2 dry aboveground biomass or about 0.2 kg/m^2 of grain. Postel (2000) reported that for the production of 1 unit of grain, a thousandfold mass of water will be required. Our data in Fig. 14 confirm this in general. However, WUE depends largely on the quality of land cereals are grown on. The efficiency can be much lower on poor quality soils (factor up to 2000) and much higher on good soils (factor down 670).

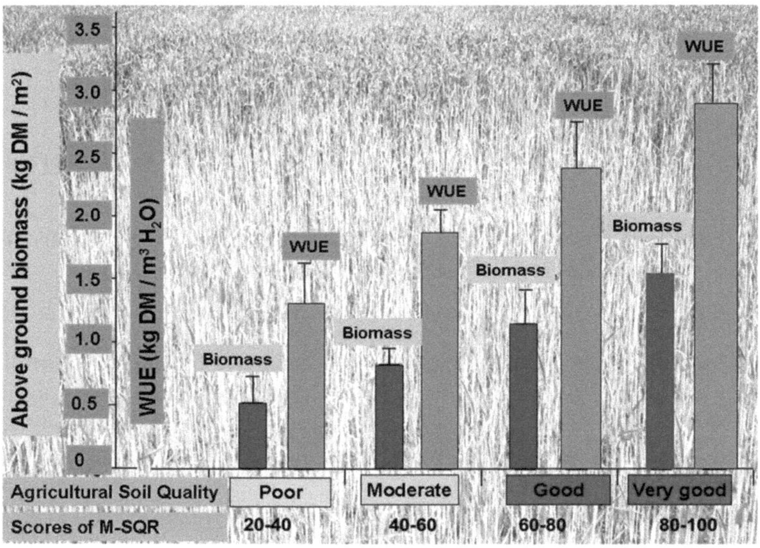

Fig. 14 SQR score and water use efficiency(WUE) of small grain cereals (Mueller et al. 2011b, modified) WUE refers to the whole year and to the total annual above-ground biomass in single cropping systems

10 Soils and Their Functions in the Context of Real Landscapes

Landscapes and their soils as structural and functional components may provide a number of functions like maintaining biodiversity, the water and carbon cycle, source of minerals, ground for built structures, archives of natural and cultural history and others.

Highly evolved civilisations and big cities have historically developed in regions of fertile soils (Fig. 15). The impacts pressure on soils are particularly high in urban regions. But also in more traditional agricultural landscapes, trade-offs between soil functions may arise. For example, agricultural crop production and groundwater recharge may compete in sub-humid landscapes (Wessolek and Asseng 2006). Traditional commodities of agriculture like grain could be replaced in future by commodities like water or carbon. On the other hand, some apparent conflicts between agricultural productivity and other soil functions could disappear in the light of new results of research. Due to limited water supplies, increases in crop production must come from rainfed cropping systems which account for more than 80 % of the cropland worldwide (Lal 2008). Increasing crop yields can also be associated with soil aggradation and benefits for the carbon cycle. More root and stubble residues need to remain on the soil in order to enhance its carbon pool and productivity (Lal 2008). These aspects need to be tested by more measurements and experiments.

Better measurement and assessment tools are required for understanding processes and their cycles and monitoring changes in cropland and grassland ecosystems. M-SQR is a tool providing concrete results about soil quality but may also serve also a frame for further research towards the optimisation of agricultural systems at an optimum level of ecological intensity. Holistic experiments under optimum to extreme environmental conditions (Fig. 16) could help to get

Fig. 15 Competition for land at the local scale in China, northern part of the Loess Plateau. The winner is visible in the background. These soils are among the best soils worldwide (M-SQR score >80, grain yields 10–15 t/ha wheat plus maize). Urban sprawl is a threat for fertile soils worldwide. Soils are sealed and lose many functions irreversibly. Also soils in the neighbourhood of big cities face risks of pollution

Fig. 16 At Dryland research station in Lind, Washington, USA, experiments with cropping under extreme dry conditions (mean precipitation 266 mm/year) have been successfully performed for many years (Schillinger 2011). The *right photo* shows spring wheat stubbles serving as a trap for snow to enrich the soil with moisture for the next and overnext season. Soils are formed on Loess substrate and have a good Basic rating of M-SQR and a high potential for water storage. Due to extreme drought, the overall rating reveals an allocation into the groop of poor soils (M-SQR score 20). Crop yield is 2–2.5 t/ha spring wheat every second year

information about the efficiency of agricultural practices. This is the basis for sustainable land management by new technologies.

11 The Perspective of Eurasia for Global Food Production

Eurasia has potentials for a sufficient food production based on high technological and environmental standards. This is mainly referring to al kinds of rain-fed agriculture, cereal cropping in particular. Those prospections are supported by climate change scenarios and economic analyses. Globally, due to climate change, rainfed rice and wheat yield will increase in developed countries, whilst yield in irrigated cropping and developing countries decreases (Jaggard et al. 2010). Grain production in Russia, Ukraine and Kazakhstan is likely to increase due to a combination of winter temperature increase, extension of the growing season, and natural fertilisation by increased carbon dioxide concentrations (Lioubimtseva 2010).

Agriculture in this region is an important target of investments by international agroholdings. Apparently "virgin land" in Siberia and the northern part of Kazakhstan plays an important role in processes of land versatility (Deininger et al. 2011; Petrick et al. 2012). Superfarms are key driver of increases in grain production (Deininger et al. 2011) and rural development (Petrick et al. 2012). Amongst countries of Eurasia, Russia, Kazakhstan and the Ukraine could become major players in global grain markets besides the EU and China. They have most current and potential land for wheat production (Fig. 17). Russia alone possesses more potential cropland than the sum of other leading wheat producing countries

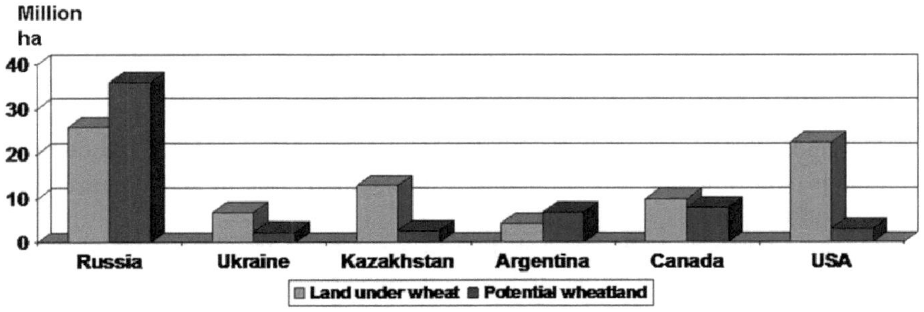

Fig. 17 Current and potential land for wheat production (Million hectares) in some countries of Eurasia and America. Figure bases on data of Petrick et al. (2012)

shown in this graph. Most of this potential land is located in the steppe zone and forest steppe zone.

Whilst suitable land seems to be available, the yield gap and the risk of resource degradation are high. This is clearly demonstrated in other chapters of this book. Ploughing steppe land cannot be considered as a sustainable way of rainfed cereal cropping. No- tillage as a practicable option is presented in chapters of Suleimenov and Meinel et al. Broad adoption of no-tillage on more than 105 Million hectares worldwide shows the great adaptability of the system to all kinds of climates, soils and cropping conditions (Derpsch et al. 2010).

Prospections for rainfall are generally uncertain for Eurasia, whilst the increase of temperatures is signigicant. Aridity is expected to increase in most regions of Central Asia, that are now already too dry for rainfed cropping. These are the western parts of Turkmenistan, Uzbekistan, and Kazakhstan (Lioubimtseva and Henebry 2009). Current and future performance of grasslands and rangelands of Central Asia is uncertain.

EU countries are going to develop criteria and frameworks of soil resource evaluation (Huber et al. 2008). In Central Asia, exact inventories and analyses of yield potentials of crop lands and grasslands are missing due to missing methodical frames and guidelines. Thorough status analyses of agricultural land, assessing its productivity potentials and risks of resource degradation, based on internationally acknowledged methods, are required but missing. Further chapters of this book present tools and solutions for this.

12 Conclusions

1. The function of soil to sustainably produce plant biomass is one of the main global challenges of the 21st century and embraces food security, environmental quality and climate change. Global land and soil resources are limited, largely non-renewable and prone to degradation.

2. There is a lack of a methodical, universal platform for assessing soil productivity potentials for the growing global community of stakeholders to achieve sustainable use of the soil resource. Existing soil and land evaluation and classification systems operate on a regional or national basis. A recommended or standardised internationally applicable method providing soil productivity ratings over different scales, ranging from the field scale to global assessments is required.
3. The Muencheberg SQR has potential to serve as a straightforward indicator-based soil functional rating and classification system. It is suitable for sustainable land use and soil conservation and rehabilitation assessments at an internationally comparable scale. It could also serve as a universal soil productivity estimator providing fast appraisals of attainable crop yields over different scales.
4. Based on M-SQR and other transferable tools presented in this book, the performance of the land resource of Eurasia for cropping and grazing can be measured and evaluated.
5. Those monitoring data based on international conventions and standards may provide the development of sustainable management strategies of the global land resource.

References

Agronomic Interpretations Working Group (1995) Land suitability rating system for agricultural crops: 1. Spring-seeded small grains. In: Pettapiece WW (ed.) Agriculture and agri-food Canada, Centre for Land and Biological Resources Research, Tech. Bull. 1995-6E, Ottawa, p 90

Ahmad S, Islam M, Mirza SN (2012) Rangeland degradation and management approaches in Balochistan, Pakistan. Pak J Bot 44:127–136 (Special Issue)

Andrews SS, Karlen DL, Cambardella CA (2004) The soil management assessment framework: a quantitative soil quality evaluation method. Soil Sci Soc Am J 68:1945–1962

Asner GP, Heidebrecht KB (2005) Desertification alters regional ecosystem- climate Interactions. Glob Change Biol 11:182–194

Bakker MM, Govers G, Jones RA, Rounsevell MDA (2007) The effect of soil erosion on Europe‘s crop yields. Ecosystems 10:1209–1219

Ball BC, Batey T, Munkholm LJ (2007) Field assessment of soil structural quality—a development of the Peerlkamp test. Soil Use Manag 23:329–337

Batey T, McKenzie DC (2006) Soil compaction: identification directly in the field. Soil Use Manag 22:123–131

Blanco-Canqui H, Lal R (2009) Crop residue removal impacts on soil productivity and environmental quality. Crit Rev Plant Sci 28:139–163

Blum WEH (1993) Soil protection concept of the council of Europe and integrated soil research. In: Eijsackers HJP, Hamers T (eds) Integrated soil and sediment research: a basis for proper protection. Soil and environment, vol 1. Kluwer Academic Publishers, Dordrecht, pp 37–47

Bodenaufnahmesysteme in Österreich (2001) Bodeninformationen für Land-, Forst-, Wasser- und Abfallwirtschaft, Naturschutz-, Landschafts-, Landes- und Raumplanung, Agrarstrukurelle Planung, Bodensanierung und -regeneration sowie Universitäten, Schulen und Bürger.

Mitteilungen der Österreichischen Bodenkundlichen Gesellschaft Heft 62, zugleich eine Publikation des Umweltbundesamtes Wien, 2001, p 221. Online http://www.umweltbundesamt.at/fileadmin/site/umweltthemen/boden/9_-_Schlussartikel.pdf. Accessed on 12 Feb 2010

Boonman JG, Mikhalev SS (2005) Chapter 10, the Russian steppe, In: Suttie JM, Reynolds SG, Batello C (eds) Grasslands of the world. Plant production and protection series No. 34, Food And Agriculture Organization of the United Nations, Rome 2005. http://www.fao.org/docrep/008/y8344e/y8344e00.htm. Accessed on 28 Nov 2012

Borlaug N (2007) Feeding a hungry world. Science 318:359

Bremer E, Ellert K (2004) Soil quality indicators: a review with implications for agricultural ecosystems in Alberta. Report prepared for: Alberta environmentally sustainable agriculture soil quality program, Alberta agriculture, food and rural development 2004. Online http://www1.agric.gov.ab.ca/$department/deptdocs.nsf/all/aesa8681. Accessed on 12 Feb 2010

Brunotte J, Senger M, Haaren V, Heyn M, Brandhuber J, Voßhenrich R, Epperlein H, Vorderbrügge J, Ortmeier T, Lorenz B, Kyas M (2011) Einfache Feldgefügeansprache für den Praktiker., Johann Heinrich von Thünen Institut (vTI) für Agrartechnologie Braunschweig, and Gesellschaft für Konservierende Bodenbearbeitung e. V. (GKB) Neuenhagen, 2. Auflage, p 2

Catt JA (2001) The agricultural importance of loess. Earth Sci Rev 54:213–229 (Hindawi Publishing Corporation)

Deininger K, Byerlee D, Lindsay J, Norton A, Selod H, Stickler M (2011) Rising global interest in farmland: Can it yield sustainable and equitable benefits? The World Bank, Washington, DC, doi: 10.1596/978-0-8213-8591-3, p 214. http://siteresources.worldbank.org/INTARD/Resources/ESW_Sept7_final_final.pdf. Accessed on 12 Feb 2013

Derpsch R, Friedrich T, Kassam A, Li H (2010) Current status of adoption of no-till farming in the world and some of its main benefits. Int J Agric Biol Eng 3(1):1–25. Open Access at http://www.ijabe.org, http://www.fao.org/ag/ca/CA-publications/China_IJABE.pdf. Accessed on 12 Jan 2013

Eswaran H, Lal R, Reich PF (2001) Land degradation: an overview. In: Bridges EM, Hannam ID, Oldeman LR, Pening FW, de Vries T, Scherr SJ, Sompatpanit S (eds) Responses to land degradation. Proceedings of 2nd international conference on land degradation and desertification, Khon Kaen, Thailand, Oxford Press, New Delhi, India. Online http://soils.usda.gov/use/worldsoils/papers/land-degradation-overview.html. Accessed on 12 Feb 2010

European Commission (2009) New challenges for agricultural research: climate change, food security, rural development, agricultural knowledge systems, 2nd scar foresight exercise. Directorate-General for Research, Food, Agriculture and Fisheries, and Biotechology, p 104. http://ec.europa.eu/research/agriculture/scar/pdf/scar_2nd_foresight_exercise_en.pdf. Accessed on 12 Feb 2010

FAO (2008) Visual soil assessment (VSA). Field guides. Food and Agriculture Organization of the United Nations (FAO), Rome. http://www.fao.org/docrep/010/i0007e/i0007e00.htm. Accessed on 28 Nov 2012

FAO (2012) AQUASTAT database. Food and Agriculture Organization of the United Nations (FAO). Website accessed on 13 Nov 2012 9:33

Fischer G, van Velthuizen H, Shah M, Nachtergaele F (2002) Global agro-ecological assessment for agriculture in the 21st century: methodology and results. International Institute for Applied Systems Analysis, Laxenburg, Austria, p 154

Gavrilyuk FY (1974) Bonitirovka pochv. Moskva.: Vysshaya shkola, p 270

Golden M, Micheli E, Ditzler C, Eswaran H, Owens P, Zhang G, McBratney A, Hempel J, Montanarella L, Schad P (2010) Time for a universal soil classification system. In: Gilkes RJ, Prakongkep N (eds) Proceedings of 19th world congress of soil science. Soil solutions for a changing world. Published on DVD, pp 48–51

Gupta R, Kienzler K, Martius C, Mirzabaev A, Oweis T, de Pauw E, Qadir M, Shideed K, Sommer R, Thomas R, Sayre K, Carli C, Saparov A, Bekenov M, Sanginov S, Nepesov M, Ikramov R (2009) Research prospectus: a vision for sustainable land management research in

Central Asia. ICARDA Central Asia and Caucasus program. Sustainable agriculture in Central Asia and the Caucasus series No.1. CGIAR-PFU, Tashkent, Uzbekistan, p 84
Helming K, Tscherning K, König B, Sieber S, Wiggering H, Kuhlman T, Wascher D, Perez-Soba M, Smeets P, Tabbush P, Dilly O, Hüttl RF, Bach H (2008) Ex ante impact assessment of land use change in European regions : the SENSOR approach. In: Helming K, Pérez-Soba M, Tabbush P (eds) Sustainability impact assessment of land use changes. Springer, Berlin, pp 77–105
Huber S, Prokop G, Arrouays D, Banko G, Bispo A, Jones RJA, Kibblewhite MG, Lexer W, Möller A, Rickson RJ, Shishkov T, Stephens M, Toth G, Van den Akker JJH, Varallyay G, Verheijen FGA, Jones AR (eds) (2008) Environmental assessment of soil for monitoring: volume I indicators and criteria. EUR 23490 EN/1. Office for the Official Publications of the European Communities, Luxembourg, p 339
Jaggard KW, Qi A, Ober ES (2010) Possible changes to arable crop yields by 2050. Phil Trans R Soc B 365:2835–2851. doi:10.1098/rstb.2010.0153 (Review paper)
Jones RJA, Spoor G, Thomasson AJ (2003) Vulnerability of subsoils in Europe to compaction: a preliminary analysis. Soil Tillage Res 73:131–143
Karlen DL, Mausbach MJ, Doran JW, Cline RG, Harris RF, Schuman GE (1997) Soil quality: a concept, definition and framework for evaluation. Soil Sci Soc Am J 61(1):4–10
Kazakh Ministry of Environmental Protection (2007) Environmental monitoring and information management system for sustainable land use design guidelines for the SKO subsystem. KazakhstanMinistry of Environmental Protection, Kazakhstan Asian Development Bank: TA-4375; Mott Mc Donald 11th Nov 2007. http://www2.adb.org/Documents/CACILM/Guidelines-ADB-LandUse.pdf. Accessed on 23 Nov 2012
Keys to Soil Taxonomy (2010) Eleventh edition. ftp-fc.sc.egov.usda.gov/NSSC/Soil_Taxonomy/keys/2010_Keys_to_Soil_Taxonomy.pdf. Accessed on 28 Nov 2012
Lal R (2006) Enhancing crop yield in the developing countries through restoration of soil organic carbon pool in agricultural lands. Land Degrad Dev 17:187–209
Lal R (2008) Soils and sustainable agriculture. A Rev Agron Sustain Dev 28(2008):57–64
Lal R (2009) Soils and food sufficiency. A Rev Agron Sustain Dev 29(2009):113–133
Larin IV (1956) Lugovodstvo i pastbishchnoe khozyaistvo [Grassland husbandry]. Selkhozgiz, Leningrad and Moscow, Russia
Lioubimtseva E (2010) Global food security and grain production trends in Central Eurasia: Do models predict a new window of opportunity? Online source http://www.energy-101.org/wp/wp-content/uploads/2012/01/NSSA-Paper-2010.pdf. Accessed on 12 Feb 2013
Lioubimtseva E, Henebry GM (2009) Climate and environmental change in arid Central Asia: impacts, vulnerability, and adaptations. J Arid Environ 73(2009):963–977
Lipiec J, Arvidsson J, Murer E (2003) Review of modelling crop growth, movement of water and chemicals in relation to topsoil and subsoil compaction. Soil Tillage Res 73:15–29
Lobell DB, Burke MB, Tebaldi C, Mastrandea MD, Falcon WP, Naylor RL (2008) Prioritizing climate change adaptation needs for food security in 2030. Science 319:607–610
Louwagie G, Gay SH, Burrell E (eds) (2009) Addressing soil degradation in EU agriculture: relevant processes, practices and policies report on the project 'sustainable agriculture and soil conservation (SoCo)' JRC scientific and technical reports. Online http://eusoils.jrc.ec.europa.eu/ESDB_Archive/eusoils_docs/other/EUR23767.pdf. Accessed on 12 Feb 2010
Mairura FS, Mugendi DN, Mwanje JI, Ramisch JJ, Mbugua PK, Chianu JN (2008) Scientific evaluation of smallholder land use knowledge in Central Kenya. Land Degrad Develop 19:77–90
Mueller L, Behrendt A, Schalitz G, Schindler U (2005) Above ground biomass and water use efficiency of crops at shallow water tables in a temperate climate. Agric Water Manag 75(2):117–136
Mueller L, Schindler U, Behrendt A, Eulenstein F, Dannowski R (2007a) Das Muencheberger soil quality rating (SQR): ein einfaches Verfahren zur Bewertung der Eignung von Boeden als Farmland. Mitteilungen der Deutschen Bodenkundlichen Gesellschaft 110(2):515–516

Mueller L, Schindler U, Behrendt A, Eulenstein F, Dannowski R, (2007b) The Muencheberg soil quality rating (SQR). Field manual for detecting and assessing properties and limitations of soils for cropping and grazing. Online Manuscript http://www.zalf.de/de/forschung/institute/lwh/mitarbeiter/lmueller/Documents/field_mueller.pdf. Accessed on 28 Nov 2012

Mueller L, Drösler M, Schindler U, Behrendt A, Höper H, Eulenstein F, Kantelhardt J, Sommer M (2008) Implications of soil properties, vegetation and management intensity for peatland quality. In: After wise use—the future of peatlands : Proceedings of the 13th international peat congress; Tullamore, Ireland 8–13 June 2008: vol 1. International Peat Society, Jyväskylä, pp 616–619

Mueller L, Kay BD, Hu C, Li Y, Schindler U, Behrendt A, Shepherd TG, Ball BC (2009a) Visual assessment of soil structure: evaluation of methodologies on sites in Canada, China and Germany: part I: comparing visual methods and linking them with soil physical data and grain yield of cereals. Soil Tillage Res 103(1):178–187

Mueller L, Schindler U, Behrendt A, Smolentseva E, Hennings V, Schad P, Hu C, Ball BC, Schlindwein S, Eulenstein F, Helming K, Schlindwein S (2009b) Bodenklassifikation und mittlere Ernteerträge in einigen Agrarlandschaften Eurasiens. Jahrestagung der DBG: Böden—eine endliche Ressource; Kommission VIII, September 2009, Bonn: 1–4; Deutsche Bodenkundliche Gesellschaft, Oldenburg. http://eprints.dbges.de/270/. Accessed on 28 Nov 2012

Mueller L, Schindler U, Mirschel W, Shepherd TG, Ball BC, Helming K, Rogasik J, Eulenstein F, Wiggering H (2010) Assessing the productivity function of soils. A Rev Agron Sustain Dev 30(3):601–614

Mueller L, Lipiec J, Kornecki TS, Gebhardt S (2011a) Trafficability and workability of soils. In: Gliński J, Horabik J, Lipiec J (eds) Encyclopedia of agrophysics. Springer Science+Business Media B.V, Dordrecht, pp 912–924

Mueller L, Schindler U, Behrendt A, Fank J (2011b) Konzept zur Bewertung von Ertragspotentialen für Lysimeterstandorte. In: Bericht/14. Gumpensteiner Lysimetertagung : am 3. und 4. Mai 2011 am LFZ Raumberg-Gumpenstein. Lehr- und Forschungszentrum für Landwirtschaft Raumberg-Gumpenstein, Irdning, pp 137–142

Mueller L, Schindler U, Shepherd TG, Ball BC, Smolentseva E, Hu C, Hennings V, Schad P, Rogasik J, Zeitz J, Schlindwein SL, Behrendt A, Helming K, Eulenstein F (2012) A framework for assessing agricultural soil quality on a global scale. Arch Agron Soil Sci 58(Supplement 1):S76–S82

Munkholm LJ, Schjonning P, Rasmussen KJ, Tanderup K (2003) Spatial and temporal effects of direct drilling on soil structure in the seedling environment. Soil Tillage Res 71:163–173

Nachtergaele F, Bruinsma J, Valbo-Jorgensen J, Bartley D (2011) The state of land and water resources for food and agriculture anticipated trends in the use of global land and water resources SOLAW background thematic report—TR01. http://typo3.fao.org/fileadmin/templates/solaw/files/thematic_reports/TR_01_web.pdf. Accessed on 24 Aug 2012

Oyedele DJ, Aina PO (2006) Response of soil properties and maize yield to simulated erosion by artificial topsoil removal. Soil Plant 284:365–373

Peerlkamp PK (1967) Visual estimation of soil structure. In: de Boodt M, de Leenherr DE, Frese H, Low AJ, Peerlkamp PK (eds) West European methods for soil structure determination, vol 2(11). State Faculty Agricultural Science, Ghent, pp 216–223

Petrick M, Wandel J, Karsten K (2012) Rediscovering the Virgin Lands: agricultural investment and rural livelihoods in a Eurasian frontier area, World Development (in press). http://dx.doi.org/10.1016/j.worlddev.2012.09.015. Accessed on 12 Jan 2013

Postel S (2000) Redesigning irrigated agriculture, in state of the world 2000, vol 272, 1st edn. The Worldwatch Institute, New york, pp 39–58

Rajabov T (2009) Ecological assessment of spatio-temporal changes of vegetation in response to piosphere effects in semi-arid rangelands of Uzbekistan. Online Source http://www.unulrt.is/static/fellows/document/rajabov-t.pdf. Accessed on 9 Mar 2013

Rose L, Coners H, Leuschner C (2012) Effects of fertilization and cutting frequency on the water balance of a temperate grassland. Ecohydrology 5(1):64–75

Rothkegel W (1950) Geschichtliche Entwicklung der Bodenbonitierungen und Wesen und Bedeutung der deutschen Bodenschätzung. Stuttgart, Ulmer, p 147
Sadras VO, O'Leary GJ, Roget DK (2004) Crop responses to compacted soil: capture and efficiency in the use of water and radiation. Field Crops Res 91(2–3):131–148
Salako FK, Dada PO, Adejuyigbe CO, Adedire MO, Martins O, Akwuebu CA, Williams OE (2007) Soil strength and maize yield after topsoil removal and application of nutrient amendments on a gravelly Alfisol toposequence. Soil Tillage Res 94:21–35
Schillinger WF (2011) Practical lessons for successful long-term cropping systems experiments. Renewable Agric Food Syst 26:1–3
Schindelbeck RS, van Es HM, Abawi GS, Wolfe DW,Whitlow TL, Gugino BK, Idowu OJ, Moebius-Clune BN (2008) Comprehensive assessment of soil quality for landscape and urban management. Landscape Urban Plan 88 (2–4):73–80. doi:10.1016/j.landurbplan.2008.08.006. Accessed on 12 Feb 2010
Shepherd TG (2000) Visual soil assessment, volume 1. Field guide for cropping and pastoral grazing on flat to rolling country. Horizons.mw/Landcare Research, Palmerston North, p 84
Shepherd TG (2009) Visual soil assessment, volume 1. Field guide for pastoral grazing and cropping on flat to rolling country, 2nd edn. Horizons Regional Council, Palmerston North, New Zealand, p 118
Shepherd TG, Jessen M (2002) Land versatility for food and fibre production and urban use in the information area of the western Bay of Plenty subregion, for the SmartGrowth consortium. In: Graham Shepherd T, Jessen MR (eds) Landcare research contract report: LC0102/114
Smolentseva EN, Rukhovic O, Lukin S, Suleimenov M, Saparov A, Balgabayev N, Pachikin K, Kussainova MD, Bekbaev U, Schindler U, Mueller L (2011) Ozenka katcestva i potenzial urozhainosti pocv v globalnom masshtabe. Pocvovedenie i Agrochimija, Almaty. 4:81–91
Spoor G (2006) Alleviation of soil compaction: requirements, equipment and techniques. Soil Use Manag 22:113–122
Storie RE (1978) Storie index soil rating (Revised). Special publication 3203. Division of agricultural science. University of California, Berkeley, CA. http://anrcatalog.ucdavis.edu/pdf/3203.pdf. Accessed on 12 Feb 2010
Suttie JM, Reynolds SG, Batello C (eds) (2005) Grasslands of the world. Plant production and protection series No. 34, Food and Agriculture Organization of the United Nations, Rome. http://www.fao.org/docrep/. Accessed on 28 Nov 2012
TERC (2005) Productivity of grassland in central Mongolia. Terrestrial Environment Research Center Tsukuba University. In: Proceedings of the 3rd international workshop on terrestrial change in Mongolia—joint workshop of AMPEX, IORGC and RAISE projects. http://www.suiri.tsukuba.ac.jp/publication/tercbull/tercbull05s/. Accessed on 2 Dec 2012
Tjumenzev NF (1975) Susvtnost bonitirovki pocv na genetiko-proisvodstvennoij osnove. Isdatelstvo Nauka. Sibirskoje otdelenije, Novosibirsk, p 286
Töpfer International (2012) http://www.acti.de/en/. Accessed on 24 Aug 2012
Tscharntke T, Clough Y, Wanger TC, Jackson L, Motzke I, Perfecto I, Vandermeer J, Whitbread A (2012) Global food security, biodiversity conservation and the future of agricultural intensification. Biol Conserv 151(2012):53–59
Wessolek G, Asseng S (2006) Trade-off between wheat yield and drainage under current and climate change conditions in Northeast Germany. Europ J Agron 24(2006):333–342
White R, Murray S, Rohweder M (2000) Pilot analysis of global ecosystems (PAGE): grassland ecosystems. World Resources Institute, Washington, DC, Nov 2000/paperback/ISBN 1-56973-461-5/. http://sustentabilidad.uai.edu.ar/pdf/info/page_grasslands.pdf. Accessed on 28 Nov 2012
WRB (2006) World reference base for soil resources 2006, a framework for international classification, correlation and communication, FAO Rome, World soil resources reports 103, p 145

Part II
Novel Methodologies for Measurement of Processes and Assessment of Resources

A Novel Method for Quantifying Soil Hydraulic Properties

Uwe Schindler

Abstract A knowledge of soil hydraulic properties—the water retention curve and unsaturated hydraulic conductivity—is required for soil water modelling and for various studies of soil hydrology. Taking measurements using traditional techniques is time consuming, the equipment is costly and the results can be uncertain. The evaporation method is frequently used for the simultaneous determination of hydraulic functions of unsaturated soil samples, i.e. the water retention curve and hydraulic conductivity function. Due to the limited range of common tensiometers, all the methodological variations of the evaporation method suffer from the limitation that the hydraulic functions can only be determined to a maximum of 70 kPa. The extended evaporation method (EEM) overcomes this restriction. Using new cavitation tensiometers and setting the air-entry pressure of the tensiometer's porous ceramic cup as a final tension value allow both hydraulic functions to be quantified close to the wilting point. Additionally, soil shrinkage dynamics as well as soil water hysteresis can be quantified. Here, the HYPROP system was selected, a commercial device with vertically aligned tensiometers optimised to perform evaporation measurements. The HYPROP software was developed for recording, calculating, evaluating, fitting and exporting hydrological data. A good match between the results of soil hydraulic functions was shown when those obtained from traditional methods and the extended evaporation method were compared. Systematic deviations were not found.

Keywords Soil · Hydraulic properties · Measurement · Extended evaporation method (EEM)

U. Schindler (✉)
Leibniz Centre for Agricultural Landscape Research (ZALF) e. V., Eberswalder Strasse 84
15374 Muencheberg, Germany
e-mail: uschindler@zalf.de

L. Mueller et al. (eds.), *Novel Measurement and Assessment Tools for Monitoring and Management of Land and Water Resources in Agricultural Landscapes of Central Asia*, Environmental Science and Engineering, DOI: 10.1007/978-3-319-01017-5_7,

1 Introduction

Classical determination of soil hydraulic properties—the water retention curve and unsaturated hydraulic conductivity characteristics—has been carried out using various methods and procedures. Depending on the desired measuring range, different methods and devices are available to determine the soil water retention curve. In the low tension range, between 0 and 10 kPa, the sand box (Cresswell et al. 2008) is the common method for quantifying the water retention data points. The sand/kaolin box is mainly used in the tension range between 10 and 50 kPa. For higher tensions (100–1,500 kPa), the pressure plate extractor is applied (Dane and Hopmans 2002).

In addition, various methods are available to estimate the unsaturated soil hydraulic conductivity function of soil samples. The steady state pressure membrane procedure (Henseler and Renger 1969; Boels et al. 1978; Schindler et al. 2010a) and the tension disc infiltrometer method (Reynolds and Elrick 1991) allow the measurement of hydraulic conductivity values only in the low tension range. Hot-air methods (Arya 2002; Tyner et al. 2006) and centrifugation techniques (Nimmo et al. 2002) enable soil water diffusivity and unsaturated hydraulic conductivity to be measured rapidly. The measurement conditions, however, differ markedly from the natural conditions for all these methods. The one-step (Kool et al. 1985) and especially the multistep outflow method (Hopmans et al. 2002; Fujimaki and Mitsuhiro 2003; Durner and Iden 2011) produce reliable hydraulic conductivity data and are widely in use.

Evaluation of traditional methods:

- Expensive;
- Time consuming (several months to quantify both functions);
- High level of uncertainty;
- Artificial process;
- Inflexible—transporting the equipment requires a lorry and instrumentation takes several weeks or months.

The evaporation method (Wind 1966) allows the simultaneous determination of both the water retention curve and the hydraulic conductivity function. Some modifications have been developed as described by Becher (1970), Schindler (1980), Klute and Dirksen (1986), Plagge (1991), Halbertsma (1996), Wendroth et al. (1993) Bertuzzi et al. (1999), Šimunek et al. (1999)and Schindler and Müller (2006). Measurement time and cost of the equipment are much less than when using the traditional methods. However, all variations of the evaporation method suffer from one limitation, namely the measurement limit of about 70 kPa of the tensiometers. Unfortunately, most hydrological studies require exact soil hydraulic properties at higher tensions.

The extended evaporation method (EEM) described by Schindler et al. (2010a, b) overcomes this limitation. Using new cavitation tensiometers and setting the air entry value of the tensiometer's ceramic cup allows the range to be extended to close to wilting point.

Following the EEM method, the measurement device HYPROP (**HY**draulic **PROP**erty Analyser) and procedures are described, and measurements are presented. The results of a comparison between traditional and evaporation measurements are also shown.

2 Evaporation Method

2.1 Description, Based on Versions of Schindler (1980) and Schindler et al. (2010a, b)

Two tensiometers are installed at depths of 1.25 and 3.75 cm in a soil sample (250 cm^3, height 5 cm) (Fig. 1). The sample is saturated with water from the bottom, sealed at the bottom and placed on a balance. Its surface remains open to free evaporation. Tensions (Ψ) and sample mass (m) are recorded at consecutive times. Single points of the water retention curve are calculated on the basis of the water loss per volume of the sample at time t and the geometric mean tension of the sample at that time. The hydraulic conductivity (K) is calculated according to modified Darcy-Buckingham's Law (Eq. 1) where the evaporated water volume per time interval relates to half the sample height versus hydraulic gradient as determined by the tensiometers (Schindler 1980). The flux (v) is derived from the soil water volume difference ΔV (1 cm^3 of water = 1 g) per surface area (A) and time unit (Δt). The mean hydraulic gradient (i_m) is calculated on the basis of the mean tensions in time intervals

$$K(\overline{\Psi}) = \frac{\Delta V}{2A\Delta t i_m} \tag{1}$$

where $\overline{\Psi}$ is the mean tension geometric averaged over the upper and the lower tensiometer and the time interval, ΔV is the evaporated soil water volume [= mass difference (Δm)] during the interval, A is the cross-sectional area of the sample, Δt is time interval and i_m is the mean hydraulic gradient in the interval.

At the end of the measurement, the residual amount of storage water is derived from water loss through drying in the oven (105 °C). The initial water content is determined by total water loss (evaporation part plus residual amount) related to core volume. Dry bulk density is derived from dry soil mass divided by core volume. For this reason, the volume of the tensiometer holes (1 cm^3) is subtracted from the core.

Assumptions for the validity of Eq. 1 are: (1) that water flow out of the core can be treated as a "succession of steady states" where the flux and hydraulic gradient are effectively constant within each time interval and (2) linear decreasing water content difference across the sample height in the measuring interval. Accordingly, the flux through the measuring layer is half of the total flux and can be calculated from the total evaporative soil water volume (mass) difference in the time interval.

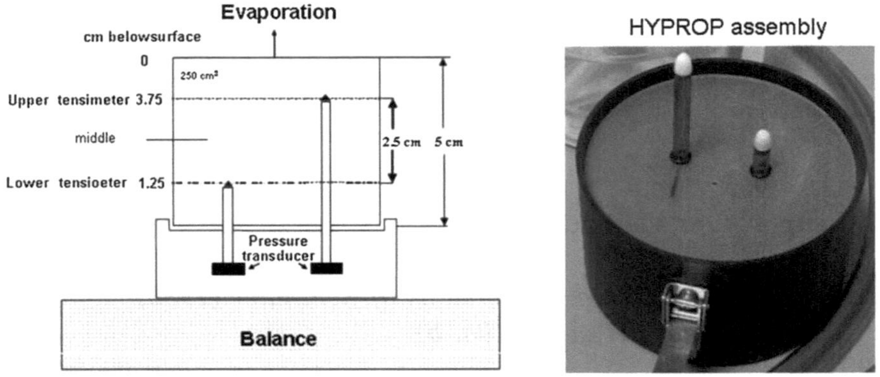

Fig. 1 Schematic illustration of the evaporation experiment and photo of the HYPROP device

The assumptions were found to be valid for sand, silt, loam and peat soils in the original measurement range between 0 and 60 kPa (Schindler and Müller 2006). Schindler et al. (2010b) reported that linearisation in space led to only minor errors, even in the late stage of evaporation where strongly nonlinear tension profiles emerge.

2.2 Procedure

Intact soil cores are placed in stainless steel cylinders (8 cm diameter, 5 cm height) with a sharpened leading edge to minimise soil disturbance during insertion. Cores are slowly saturated in the laboratory by placing them in a pan of water (tensiometer holes at the top) (Fig. 2). To minimise air entrapment, the water table should be increased slowly and in steps from an initial height of about 0.5 cm above the bottom of the sample to a final height of about 0.2 mm below the sample surface. Holes are vertically excavated for the tensiometers at the bottom of the core using a special auger and a template.

After saturation, tensiometers are prepared, and inserted in the core. The core is still in the pan at that time. The configuration is connected to the core (Fig. 3), it is removed from the pan (Fig. 4), inverted (Fig. 5) and the core ring is clamped to the tensiometer. This procedure prevents water drainage and evaporation out of the base of the core. All excess water is removed from the device, and the sample surface is sealed with a lid. The equipment is connected to the computer, and tensions are read. Hydraulic equilibrium is assumed when the tension difference between the upper and the lower tensiometer is 0.25 kPa. The whole device is placed on a balance, connected to the computer, the soil surface is exposed for evaporation and the measurement starts (Fig. 6).

Fig. 2 Saturation of samples and tensiometer holes

Fig. 3 Connection of the HYPROP device to the core

Tension (Ψ) and sample mass (m) are recorded consecutively using tensioVIEW software (http://www.ums-muc.de 2012). The HYPROP-Fit software was developed for calculating, evaluating, fitting and exporting the hydraulic functions (http://www.ums-muc.de 2012). In the remainder of this paper, the difference between atmospheric pressure and water pressure inside a tensiometer is referred to as tension (Ψ) and is expressed as a positive quantity in hPa or kPa.

Fig. 4 Removing the device and the core from the pan

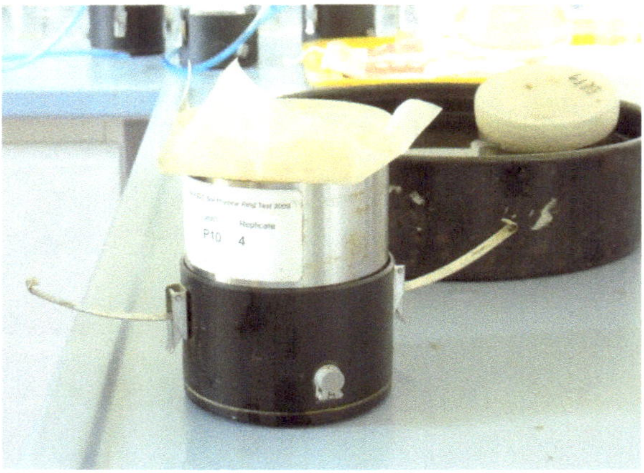

Fig. 5 Inverted device

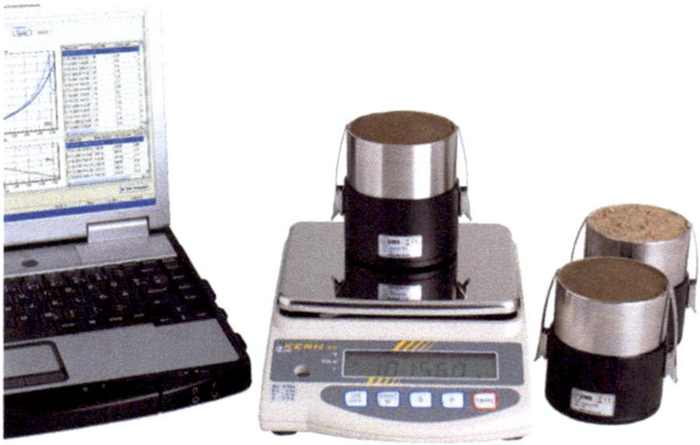

Fig. 6 Sample is clamped, placed on a balance and connected to a computer (photo courtesy of UMS GmbH Munich)

3 Extending the Range of Measurement

3.1 Tensiometric Measurements in a Drying Soil

A top-of-the-range tensiometer with highly reproducible measuring characteristics is a prerequisite for extending the range of measurement (Schindler et al. 2010a). These tensiometers consist of three basic interconnected elements: (1) a semi-permeable porous cup (2) a water reservoir, and (3) a measurement gauge or pressure transducer. Pressure equilibrium between the water in the tensiometer and the surrounding soil is achieved through water movement across the porous tensiometer cup. If the tension of the soil water exceeds the air-entry pressure, the cup drains and becomes air-permeable. Air enters the tensiometer and its internal tension recedes. The tensiometer's ceramic cup is therefore configured to ensure that its air-entry pressure is greater than the highest soil water tension (>100 kPa) that has to be measured.

The dynamics of a tensiometric measurement in a drying soil can be divided into three distinct stages (Fig. 7).

In the first stage, the measured tension reflects the matric potential of the surrounding soil. For most tensiometers, the upper limit of the tensiometer method is approximately 80 kPa (Young and Sisson 2002). For optimal performance, the water inside the tensiometers is free of dissolved gas. If dissolved gas is present, a small gas bubble will form that expands continually during the drying process and yields a slightly corrupted tensiometric measurement (Durner and Or 2005). This can be avoided by checking that the tensiometer is functioning properly before installation, as described in Schindler et al. (2010a).

The second stage is the vapour pressure stage. If the absolute soil water pressure is decreased to below the liquid's vapour pressure, the water inside the tensiometer will start to boil. The pressure inside the tensiometer will equilibrate to the vapour pressure, which is close to vacuum. Water in contact with the porous cup will flow through it into the surrounding soil, whilst the vapour bubble inside the cup will expand continually. As a consequence, the soil in the immediate vicinity of the cup

Fig. 7 Tension stages

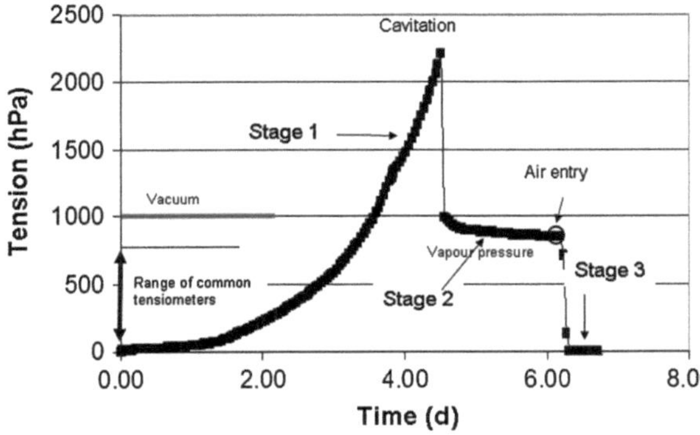

will be less dry (i.e. have a lower tension) than it would be without the presence of the tensiometer. The tensiometer readings in this stage are no longer representative of the soil water's matric potential. The initiation of stage 2, however, can be delayed if boiling retardation occurs. With a suitable tensiometer design, reliable tension values >400 kPa can be measured before cavitation occurs and the pressure inside the tensiometer collapses to the liquid's vapour pressure (Schindler et al. 2010a). The third and final stage can be called the "air-entry stage". It occurs when the tension in the surrounding soil exceeds the air-entry pressure of the ceramic material. The largest continuous pore of the ceramic cup drains and air enters from the soil into the tensiometer. At this moment, the measured tension collapses towards zero, which is easily visible in the tensiometer reading.

3.2 Principle of the Extension of the Measurements

The basic concept for extending the measurement range is to use the ceramic cup's air-entry pressure at the exact moment of the collapse in tension, i.e. at the initiation of stage three, as an additional measurement of the soil's matric potential. If this assumption is valid (which will be discussed later), an interpolation of the tension from the last reliable values of stage 1 to the initiation point of stage 3 can be performed. Figure 8 shows the basic values and in Fig. 9 the interpolation is demonstrated. Any smooth function with higher-order continuity, such as polynomial functions or Hermitian spline interpolation, can be used for interpolation with relatively little uncertainty. Applying this procedure to both tensiometers extends the data evaluation into the dry range (Schindler et al. 2010b).

As well as the general pre-condition that the matric potential of the tensiometer cup is in equilibrium with the soil with which it is in contact, the validity of the proposed method depends on the following points: (1) the air-entry pressure is well defined and reproducible, and (2) the water loss from the tensiometer to the surrounding soil during stage 2 does not affect the soil's tension at the beginning of stage 3. It is helpful, however, but not absolutely necessary, to achieve the

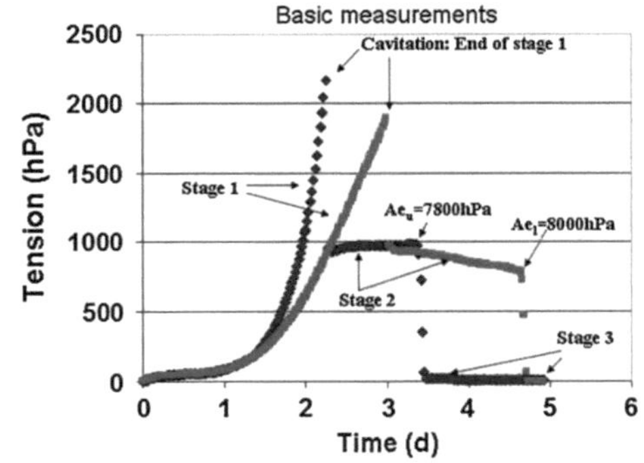

Fig. 8 Principle of tension interpolation—basic measurements

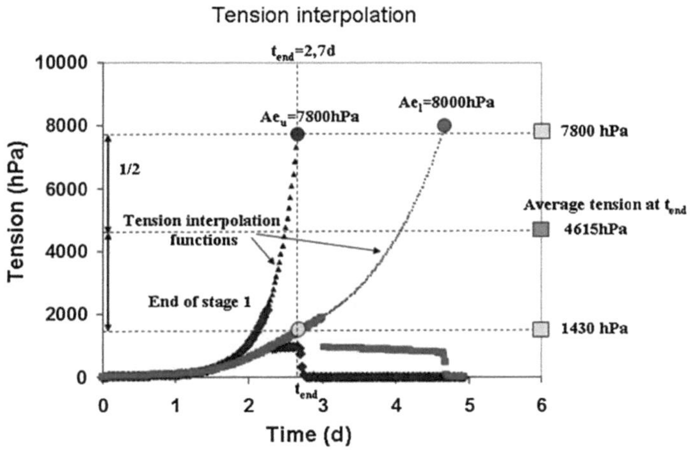

Fig. 9 Principle of tension interpolation

cavitation range for extending the range using the tensiometer's air entry value. The first assumption can be tested empirically by repeatedly determining the air-entry pressure of the tensiometer cup material, as demonstrated in Schindler et al. (2010a). The second assumption depends on a variety of factors. Most important amongst them are (1) the speed of drying the soil, (2) the unsaturated hydraulic conductivity of the surrounding soil material, (3) the size of the contact area between tensiometer cup and the soil and, most importantly, (4) the amount of water loss from inside the tensiometer into the surrounding soil. The latter is directly related to the void space inside the tensiometer, but also to the alignment of the instrument. To investigate the bias in the tension measurements due to water loss from the tensiometer, HYDRUS-2D software was used to numerically simulate the drying process of the soil with an embedded tensiometer (Šimůnek et al. 1999). The HYPROP® (UMS Munich) system was used in the experiment. This is a commercial piece of apparatus with vertical aligned tensiometers that is optimised to perform evaporation measurements. Additionally, the effect of horizontally embedded tensiometers was simulated as this is typical in traditional laboratory design. It was found that the wetting effect is of negligible importance for the accuracy of hydraulic functions when the tensiometers used are designed and vertically embedded as in the HYPROP equipment (Schindler et al. 2010b).

3.3 Determining the Air-Entry Value of the Tensiometer's Ceramic Material

To determine the air-entry pressure of the tensiometer cups, a procedure was developed (Fig. 10) that produces reliable and reproducible results (Schindler et al.

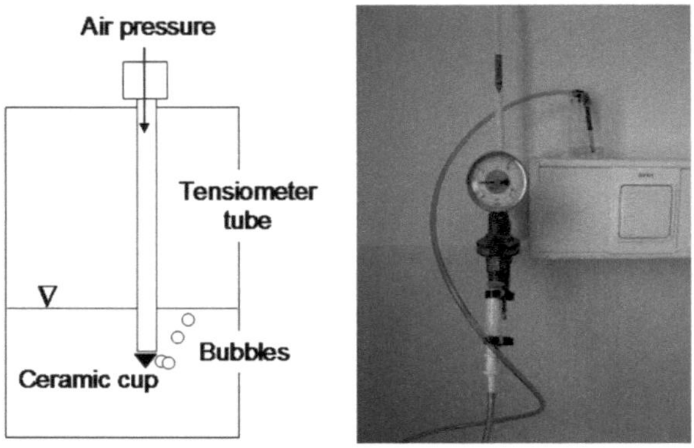

Fig. 10 Determination of the tensiometer's air-entry value

2010b). It is recommended that measurements are taken after those for evaporation. The protocol is as follows:

(1) The procedure starts with the saturation of the ceramics, which follows the standard tensiometer preparation protocol before use. Saturate the ceramic cup carefully with de-ionised water under vacuum. This step takes approximately 1 h.
(2) Empty the tensiometer and place it in a glass filled with de-ionised water.
(3) Apply a positive pressure within the tensiometer tube by connecting it to a compressor or compressed air bottle. Increase the pressure up to the expected air-entry pressure p_1 (material characteristic—provided by the ceramic's manufacturer) and wait approximately 30 min (base check for tightness). If no visible bubbles form at the ceramic cup, increase the pressure slowly in 20 kPa steps and pause at each step for about 2 min. It is advisable to use a loupe for the observation.
(4) Stop the experiment when bubbles leave the ceramic cup and take the pressure reading (p_2). To verify this air-entry pressure, repeat the procedure as described in the next step.
(5) Saturate the ceramic cup again and start the experiment with pressure $p_3 = (p_1 + p_2)/2$. Wait approximately 30 min.
(6) If no bubbles leave the ceramic cup, increase the pressure from p_3 in 10 kPa steps and wait at each step for approximately 30 min. The experiment is complete (specified air-entry pressure) when bubbles leave the ceramic cup.
(7) If bubbles form already in the repetition cycle at p_1, start the procedure at pressure p_1 minus 100 kPa, and repeat the procedure as described.

3.4 Extended Hydraulic Functions

Figures 11 and 12 show exemplary results for extended hydraulic functions for a sandy soil sample. The interpolation between representative measurements (stage

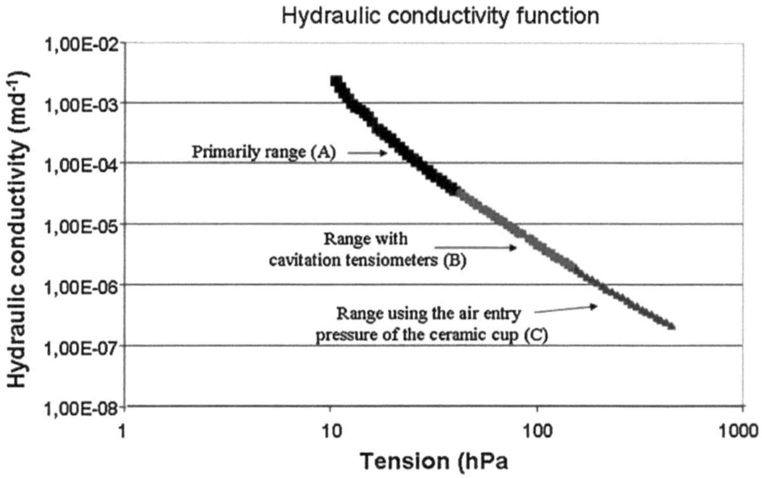

Fig. 11 Extended hydraulic conductivity function measured using the HYPROP system, sand, Al horizon, Muencheberg site

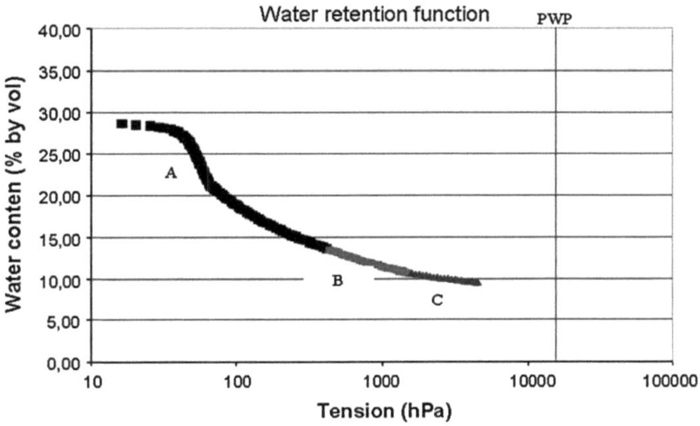

Fig. 12 Extended water retention function measured using the HYPROP system, sand, Al horizon, Muencheberg site

1) and air-entry pressure extends the range of the hydraulic functions almost by one order of magnitude, close to the permanent wilting point.

The actual range for the extension depends on (1) the air-entry pressure of the ceramic material and (ii) the tension difference between the tensiometers at the end of measurement. The HYPROP tensiometer's ceramic material is very uniform. Generally, air enters the tensiometer interior between 8 and 9 bar. The procedure described above (Schindler et al. 2010b) enables this to be determined accurately. The tension difference at the end of measurements depends on the soil and the evaporation rate. Generally, sand and clay soils have high tension differences, whereas the differences between peat and silt soil are rather small. Thus the extension ranges between 4.5 and 7.5 bar.

The measurement time depends on the soil and the evaporation rate. Under laboratory conditions with evaporation rates between 2 and 5 mmd^{-1} generally the total measurement varied between 3 and 10 days.

4 Conclusions

1. The evaporation method allows the water retention curve and the hydraulic conductivity function to be measured simultaneously.
2. Measurement time ranges between 2 and 10 days. The measurement of multiple samples can be achieved at the same time.
3. Applying evaporation functions reduces costs whilst retaining the accuracy of the measurement (Schindler et al. 2006).
4. Using specially designed tensiometers and setting the air entry pressure of the ceramic cup allows the extension of the measurement range to close to wilting point.
5. The HYPROP software tensioVIEW was developed for data recording, and HYPROP-Fit for calculating, evaluating, fitting and exporting hydraulic functions (UMS GmbH Munich (2012), http://www.ums-muc.de 2012).
6. Water retention measurements using traditional methods (CLM) corroborated EEM results. Systematic deviations between the methods were not found.

Acknowledgments Author thanks the HYPROP developer group of the UMS GmbH Munich (Dipl. Ing. Georg von Unold, Dipl. Ing. Thomas Pertasek, Dipl. Ing. Andreas Steins), Prof. Dr. Wolfgang Durner (TU Braunschweig), and Dr. Andre Peters (TU Berlin) for their cooperation.

References

Arya LM (2002) Wind and hot air method. In: Dane JH, Topp EC (eds) Methods of soil analysis part 4: physical methods, SSSA Book Ser. 5, SSSA, Madison, WI, pp 916–926

Becher HH (1970) Ein Verfahren zur Messung der ungesättigten Wasserleitfähigkeit. Z. Pflanzenern. Bodenkd. 128(1):1–12

Bertuzzi P, Mohrath D, Bruckler L, Gaudu JC, Bourlet M (1999). Wind's evaporation method: experimental equipment and error analysis. In: van Genuchten MTh, Leij FJ, Wu L (eds) Proceedings of international workshop on characterization and measurement of the hydraulic properties of unsaturated porous media, 22–24 Oct 1997, University of California, Riverside, CA, pp 323–328

Boels D, van Gils JBHM, Veerman GJ, Wit KE (1978) Theory and system of automatic determination of soil moisture characteristics and unsaturated hydraulic conductivities. Soil Sci 126:191–199

Cresswell HP, Green TW, McKenzie NJ (2008) The adequacy of pressure plate apparatus for determining soil water retention. Soil Sci Soc Am J 55(72):41–49

Dane JH, Hopmans JW (2002) Pressure plate extractor. In: Dane JH, Topp EC (eds) Methods of soil analysis part 4: physical methods, SSSA Book Ser. 5, SSSA, Madison, WI, pp 688–690

Durner W, Iden S (2011) Extended multistep outflow method for the accurate determination of soil hydraulic properties near water saturation. Water Resour Res 47:W08526. doi:10.1029/2011WR010632

Durner W, Or D (2005) Chapter 73: soil water potential measurement. In: Anderson MG, McDonnell JJ (eds) Encyclopedia of hydrological sciences, Chapter 73, Wiley, London, pp 1089–1102

Fujimaki H, Mitsuhiro I (2003) Reevaluation of the multistep outflow method for determining unsaturated hydraulic conductivity. Vadose Zone J 2:409–415

Halbertsma J (1996) Wind's evaporation method, determination of the water retention characteristics and unsaturated hydraulic conductivity of soil samples Possibilities, advantages and disadvantages. In: Durner W, Halbertsma J, Cislerova M (eds) European workshop on advanced methods to determine hydraulic properties of soils, Thurnau, Germany, 10–12 June 1996, Department of Hydrology, University of Bayreuth, pp 55–58

Henseler KL, Renger M (1969) Die Bestimmung der Wasserdurchlässigkeit im wasserungesättigten Boden mit der Doppelmembran-Druckapparatur. Z. Pflanzenernähr. Bodenkd. 122:220–228

Hopmans JW, Simunek J, Romano N, Durner W (2002) Inverse methods. In: Dane JH, Topp EC (eds) Methods of soil analysis part 4: physical methods, SSSA Book Ser. 5, SSSA, Madison, WI, pp 763–1008

HYPROP-Fit software (2012) (http://www.ums-muc.de)

Klute A, Dirksen C (1986) Hydraulic conductivity and diffusivity: laboratory methods, 687–734. In: Klute A (ed) Methods of soil analysis. Part 1, 2nd edn, Agron. Monogr. 9. ASA and SSSA, Madison, WI

Kool B, Parker JC, van Genuchten MTh (1985) Determining soil hydraulic properties from one-step outflow experiments by parameter estimation: I. Theory and numerical studies. Soil Sci Soc Am J 49:1348–1354

UMS GmbH Munich (2012) HYPROP©—Laboratory evaporation method for the determination of pF-curves and unsaturated conductivity, Online: http://www.ums-muc.de/en/products/soil_laboratory.html

Nimmo JR, Perkins KS, Lewis AM (2002) Steady-state centrifuge, 903–916. In: Dane JH, Topp GC (ed) Methods of soil analysis. Part 4. Physical methods. SSSA Book Ser. 5. SSSA, Madison, WI

Plagge R (1991) Bestimmung der ungesättigten hydraulischen Leitfähigkeit im Boden PhD Thesis. Technical University Berlin, Institute of Ecology, Department of Soil Science, p 152

Reynolds WD, Elrick DE (1991) Determination of hydraulic conductivity using a tension infiltrometer. Soil Sci Soc Am J 55:633–639

Schindler U (1980) Ein Schnellverfahren zur Messung der Wasserleitfähigkeit im teilgesättigten Boden an Stechzylinderproben Arch. Acker- u. Pflanzenbau u. Bodenkd, Berlin 24, 1, 1–7

Schindler U, Mueller L (2006) Simplifying the evaporation method for quantifying soil hydraulic properties. J Plant Nutr Soil Sci 169(5):169623–169629

Schindler U, von Durner W, Unold G, Mueller L (2010a) Evaporation method for measuring unsaturated hydraulic properties of soils: extending the range. Soil Sci Soc Am J 74:1071–1083

Schindler U, Durner W, von Unold G, Mueller L, Wieland R (2010b) The evaporation method—extending the measurement range of soil hydraulic properties using the air-entry pressure of the ceramic cup. J Plant Nutr Soil Sci 173:563–572

Šimůnek J, Šejna M, van Genuchten MTh (1999) The HYDRUS-2D software package for simulating the two-dimensional movement of water, heat, and multiple solutes in variably-saturated media, Version 2.0. U.S. Salinity Laboratory, Riverside, California

TensioVIEW Software (2012) http://www.ums-muc.de

Tyner JS, Arya LM, Wright WC (2006) The dual gravimetric hot-air method for measuring soil water diffusivity. Vadose Zone J 5:1281–1286

Wendroth O, Ehlers W, Hopmans JW, Klage H, Halbertsma J, Woesten JHM (1993) Reevaluation of the evaporation method for determining hydraulic functions in unsaturated soils. Soil Sci Soc Am J 57:1436–1443

Wind GP (1966) Capillary conductivity data estimated by a simple method. In: Proceedings of UNESCO/IASH Symp. Water in the unsaturated zone Wageningen, the Netherlands, pp 181–191

Young MH, Sisson JB (2002) Tensiometry. In: Dane JH, Topp EC (eds) Methods of soil analysis part 4: physical methods. SSSA Book Ser. 5, SSSA, Madison, WI, pp 575–608

Advanced Technologies in Lysimetry

Ralph Meissner, Holger Rupp and Manfred Seyfarth

Abstract The ability to quantify soil water flow is a prerequisite for the accurate prediction of solute transfer within the unsaturated zone. Monitoring these fluxes is a challenge because the results are required for answering not only scientific but also practical questions regarding the protection of groundwater, the sustainable management of agricultural, forestry, mining or set aside industrial areas, the reduction of leachate loss from landfills, and for explaining the fate of environmentally harmful substances. Both direct and indirect methods exist for estimating water flux rates; these have been applied with varying success. In Europe, the use of direct lysimetry methods for measuring water and solute fluxes in soils has increased significantly in recent years. Although this technique generates reliable drainage data, it involves relatively high investment and maintenance costs. New lysimeter techniques have been developed to tackle this problem. It is now possible to collect large monolithic soil columns and to measure the soil water balance of these monoliths (surface area 0.03–2 m^2 and depth to 3 m) with a high degree of precision (±20 g). Furthermore, progress in lysimetry enables us to ascertain experimentally the mass input of dew and to calculate actual evapotranspiration, precipitation and seepage rates. Weighable groundwater lysimeters have been developed in addition to gravitation lysimeters. Different lysimeter types and their usage will be presented and explained using practical examples.

R. Meissner (✉) · H. Rupp
Department Soil Physics, Helmholtz Centre for Environmental Research—UFZ, Lysimeter Station, Dorfstrasse 55 39615 Falkenberg, Germany
e-mail: ralph.meissner@ufz.de

H. Rupp
e-mail: holger.rupp@ufz.de

M. Seyfarth
UGT-Environmental Measurement Devices Ltd, Eberswalder Str. 58 15374 Muencheberg, Germany
e-mail: info@ugt-online.de
URL: www.ugt-online.de

L. Mueller et al. (eds.), *Novel Measurement and Assessment Tools for Monitoring and Management of Land and Water Resources in Agricultural Landscapes of Central Asia*, Environmental Science and Engineering, DOI: 10.1007/978-3-319-01017-5_8,

Keywords Lysimeter · Percolation · Solute flux · Soil hydrology · Soil monolith extraction · Weighing precision

1 Objectives

Exact information about the soil water balance is required in order to quantify solute transfer within the unsaturated zone. A number of methods for measuring water and solute flux in and below the root zone have been developed and critically reviewed (Gee and Hillel 1988; Führ et al. 1998; Meissner et al. 2010). However, there is no standard method available for measuring soil water flux (Gee et al. 2009). A state-of-the-art overview and evaluation of insitu soil water extraction methods is given by Weihermueller et al. (2007), Weihermueller (2012) and Schlotter et al. (2012). In addition to discussing the advantages and disadvantages of the respective methods, they explain that the decisions taken in favour of the appropriate system depend on the scientific objective and the potential limitations in terms of installation effort and accuracy, maintenance time and financial budget. Gee et al. (2002) suggest the need to differentiate between indirect methods, which involve measuring specific soil characteristics and calculating the water flux rate using data from tensiometers, time domain reflectometry (TDR), frequency domain reflectometry (FDR) or heat pulse probes, and direct methods, which involve measuring drainage water using a buried device such as drainage lysimeters or wick-based water samplers.

In the international literature, the term "lysimeter" is used for different measuring devices (Weihermueller et al. 2007; Schlotter et al. 2012). As we understand it, the use of lysimeters is one of the methods for directly measuring water and solute fluxes in soil. In Europe, the use of direct drainage lysimetry methods for measuring water and solute fluxes in soils has increased in recent years (Lanthaler and Fank 2005). The German Industrial Standard DIN 4049-3 (1994) defines a lysimeter as a device for collecting drainage water for mass and solute balances in relation to soil, parent rock, vegetation, local climate and other site conditions. In general, it consists of a square or round vessel filled with soil and a mechanism to collect and quantify the amount of water discharged at the bottom. Only lysimeters enable the quantity of water that percolates through a soil profile and the type and quantity of solutes contained in the water to be determined directly. Hence, lysimeters are much more reliable at calculating solute loads carried towards groundwater than any other method (Allen et al. 1991; Klocke et al. 1993; Gee et al. 2009). If the lysimeter is weighable, actual evapotranspiration can be calculated from its weight (mass) change. Due to these characteristics, lysimeters are an excellent tool for deriving or calibrating water and solute transport models (Wriedt 2004).

A wide range of lysimeters have been developed and used in the past, ranging from small, free-draining pan lysimeters or tension-controlled lysimeters that often capture only a small portion of the drainage water, to large drainage lysimeters that limit divergence, capturing most or all of the drainage water within a specific area (Aboukhaled et al. 1982; Meissner et al. 2000, 2008). The main difference between the lysimeter types used are:

- soil-filling procedure (disturbed–undisturbed)
- weighability (weighable or non-weighable)
- lysimeter size (depends on the scientific question and scale of observation)
- lower boundary conditions (free drainage or suction-controlled drainage).

The design of a lysimeter (surface and length required) depends mainly on the scientific question, the way in which the vessel is filled (disturbed or undisturbed), the lower boundary, and the location of the installation. Small-scale heterogeneity of a site is averaged using a larger lysimeter base area. Furthermore, lysimeters with vegetation should represent natural crop inventory, and the maximum root penetration depth should be taken into account. With the exception of generating well-defined recurrences of the same soil conditions, it is recommended to fill the lysimeter vessel monolithically. According to our knowledge, a large weighable lysimeter is the best method for obtaining reliable data about the quantity and quality of seepage water. However, it is expensive to construct and maintain large drainage lysimeters (especially the weighing type) (Meissner et al. 2008). To solve these problems, new lysimeter techniques have been developed and applied in various countries (Weihermueller et al. 2007; Meissner et al. 2008; Weihermueller 2012).

The objectives of this paper are to inform readers about the advances made in lysimeter techniques and technology, in particular (i) the methods for obtaining undisturbed soil monoliths, (ii) a technique for storing lysimeter vessels in a containerised station, (iii) newly developed types of lysimeter and (iv) examples of their practical application.

2 Materials and Methods

2.1 Soil Monolith Extraction

In order to establish flow and transport conditions that are comparable to natural field conditions, it is critically important that the soil monolith undergoes minimal disturbance during the extraction and subsequent filling of the lysimeter vessel. In the past, several methods were used to extract and fill lysimeter vessels vertically—including hand digging, employing sets of trihedral scaffolds with lifting blocks and ballast, or using heavy-duty excavators that could shear and cut large blocks of soil. More recently, technologies have been developed to extract cylindrical soil monoliths by using ramming equipment or screw presses. One of the great disadvantages of the aforementioned methods is the compaction or settling of soil that occurs during the "hammering" or "pushing".

For this reason, a new technology was developed that cuts the outline of the soil monolith using a rotary cutting system (Meissner et al. 2007). The principal scheme of this technology is shown in Fig. 1. The newly developed cutting tool enables soil monoliths to be cut out with a high degree of precision. The soil monolith remains intact during the cutting process, affecting the extraction site only minimally. A tripod frame is used to bring the lysimeter vessel into a vertical position and to keep it vertical during the cutting process. The vessel is made of stainless steel, and can be coated on the inside with an inert protective surface. At the top of the frame is a hydraulic cylinder which, in conjunction with guard and adjustable slip rails, guides the lysimeter vessel during the cutting process. The bottom of the vessel features a rotary cutting tool, driven by a small hydraulic motor. This motor, also located at the bottom of the vessel, uses a chain and sprocket arrangement. The cutting tool can be fitted with various types of chisels, enabling it to be adjusted to different soil and site conditions.

When it rotates, the cutting tool carves out the soil some 4 cm wider than the diameter of the lysimeter vessel, i.e., it leaves an excess of 2 cm of soil all around the rim of the vessel. With its own mass as the driving force, the vessel concurrently penetrates into the carved soil and shears off the aforementioned excess in the process. If necessary, additional force can be applied by the hydraulic cylinder on top of the frame. Since the vessel slides over a soil core that is slightly larger than itself, there is a tight fit between the soil and the vessel. This precludes gaps, which may act as preferential flow paths.

A pit is dug on one side of the vessel. This pit, 20 cm wider, 40 cm longer and some 10 cm deeper than the diameter of the lysimeter vessel, eventually penetrates the soil. This is necessary because the metal plate and accompanying hydraulic

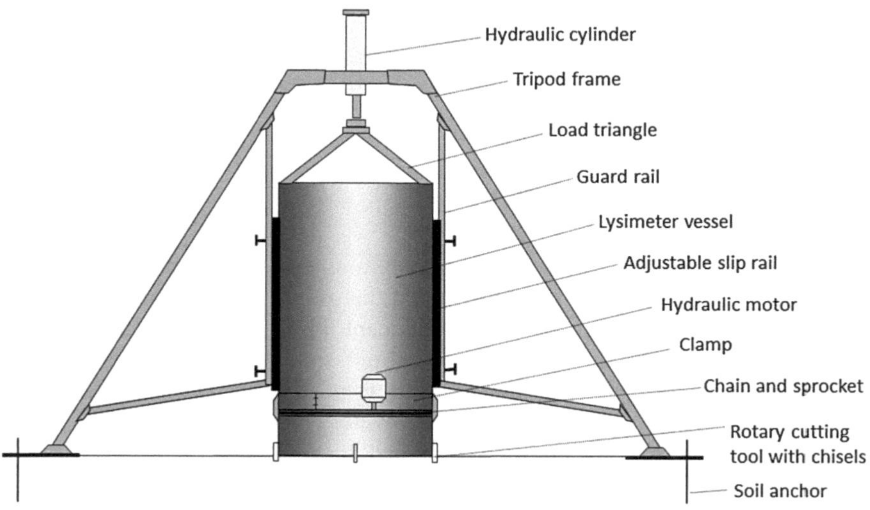

Fig. 1 Schematic of apparatus for the vertical collection of a lysimeter vessel with an undisturbed soil monolith

pushing device for cutting the base of the monolith need to be accommodated. Furthermore, peelings from the cutting process are discarded into the pit, although a much smaller pit would suffice for this purpose. Once the desired depth has been reached, the cutting tool stops rotating and the chisels are detached. Next, the monolith is severed at the bottom and the cutting plate is left attached to the bottom of the vessel. A crane is then employed to lift the whole assembly out of the pit. Smaller monoliths (e.g., surface area <0.5 m^2 and depth <1.0 m) can be lifted by the aforementioned hydraulic cylinder, without requiring any additional lifting device. Once they have reached the soil surface, the rotary cutting tool and the hydraulic motor that drives it are removed. A lid is then fixed to the top of the vessel, which is turned upside down. Now the lower 15 cm layer of soil in the monolith is removed and replaced with a graded filter layer of quartz sand and gravel (0.1–0.5, 0.71–1.25 and 3.15–5.6 mm in diameter). The time required to collect a soil monolith depends on the soil, site conditions and size of the soil column. It usually takes one day for the whole procedure of obtaining a large undisturbed soil monolith. This technology has been used successfully for different soil types (from gravel to sand to clay) and different lysimeter sizes (surface area 0.03–2 m^2 and depth to 3 m). Different types of cutting tools are available for cutting out the soil monoliths; Fig. 2 shows the most important tools. Over 400 monoliths (from sand to gravel to clay) have already been extracted using this method. No preferential flow occurred in any of these cases.

2.2 Containerised Lysimeter Station

Lysimeters are usually located in a special lysimeter station offering access for functional inspection and for the accommodation of measurement, control and weighing devices. In most cases, such stations feature an expensive steel or concrete cellar. To reduce costs and to secure mobility, a containerised polyethylene (PE-HD) lysimeter station was developed. The principle scheme of the PE-HD lysimeter station, where four lysimeter vessels are located in a clover-type arrangement around a central access hatch, is shown in Fig. 3 (Meissner et al. 2008). Variations in the quantity and arrangement of the lysimeter vessels are possible, as demonstrated in Fig. 4. Containerisation enables lysimeters to be established virtually anywhere, e.g., at the extraction site.

Fig. 2 Types of cutting tool available for obtaining undisturbed soil monoliths of different sizes and from various sites (starting from *top left*—surface area of: 2, 1, 0.5, 0.07, 0.03 and 0.03 m^2 for peat soil)

3 Types of Lysimeter

3.1 Weighable Gravitation Lysimeter

Figure 5 shows a schematic of a weighable gravitation lysimeter equipped to measure water and solute flux and to calculate actual evapotranspiration. This type of lysimeter can be produced with surface areas from 0.5 to 2.0 m^2 and total depths of 1.0–3.0 m; in Germany, the lysimeter type with 1.0 m^2 surface area and a total depth of 2.0 m is often used. The lysimeter vessel was extracted from the investigation site as an undisturbed soil monolith using the aforementioned collection technique. As described above, a 15 cm thick filter layer (sand over coarse sand over gravel) was placed at the bottom of the lysimeter to minimise natural flux disturbances.

Instead of using a mechanical weighing system, which is currently the most widely used technology, our lysimeters are equipped with three shear stress cells,

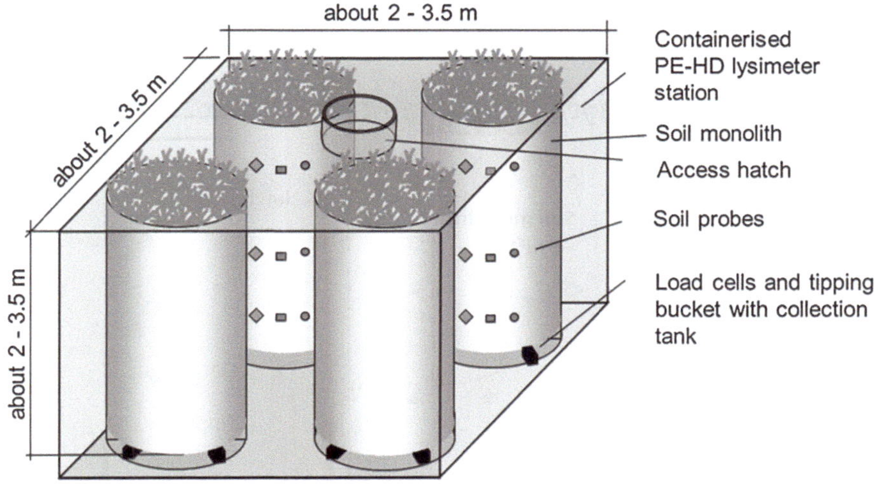

Fig. 3 Schematic of a containerised polyethylene (PE-HD) lysimeter station with *four lysimeters* in a clover arrangement and an access hatch at the centre

Fig. 4 Alternative designs of containerised PE-HD lysimeter stations (from *left to right*—4-fold, 2-fold and 1-fold lysimeter station during installation)

placed on top of aluminium pedestals. Even with a total lysimeter mass of 4,000–4,500 kg, this weighing system can register mass changes of ±20 g. Tensiometers, time domain reflectrometry (TDR) probes, thermometers and suction cups are installed at depths of 0.30, 0.90 and 1.50 m. Measured values are consolidated and stored in a data logger at recording intervals selected by the user. It permits a very high temporal resolution (<1 min). The quantity of seepage water is measured using a tipping bucket (values are stored by a data logger) and collected in a storage container from which water samples can be taken for chemical analysis. The tube leading from the bottom of the lysimeter to the tipping bucket has a large diameter and is open to the atmosphere to allow free drainage out of the lysimeter. There is no hanging water column. Computer software is

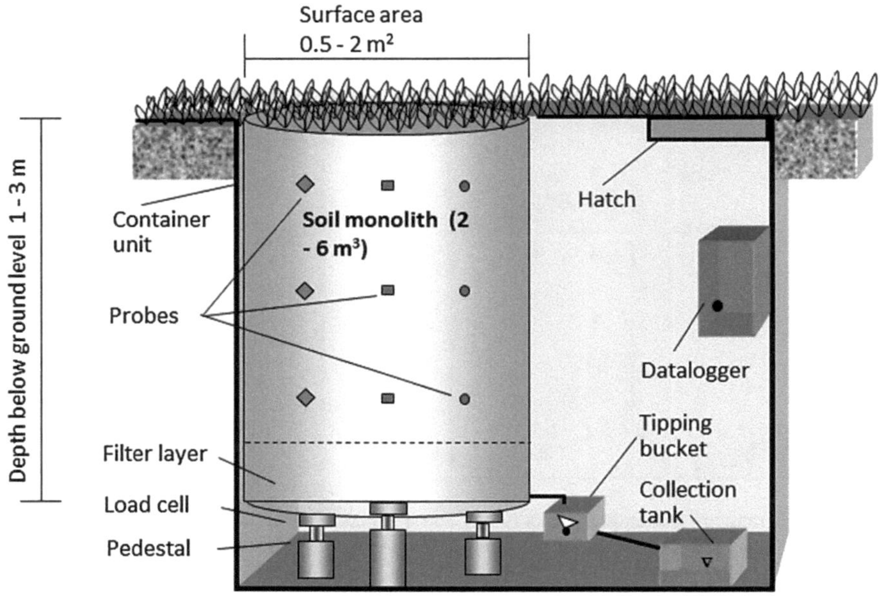

Fig. 5 Schematic of a weighable gravitation lysimeter

installed, enabling all of the measured parameters to be presented in various ways (e.g., average, minimum, maximum or all values).

Based on the measured parameters, actual evapotranspiration (ETa in mm) can be derived using the following equation:

$$\mathrm{ETa} = \mathrm{P} - \mathrm{S} \pm \Delta\mathrm{W} \tag{1}$$

where

P precipitation (mm),
S quantity of seepage water (mm) and
ΔW change in the quantity of water stored (mm), as determined from the mass change of the lysimeter over time (1 kg $\approx$ 1 l m^{-2} = 1 mm)

If the water balance is calculated correctly, the solute load (L in mg m^{-2}) can be determined very accurately from the following relationship:

$$\mathrm{L} = \mathrm{C_s} * \mathrm{S} \tag{2}$$

where

C_s solute concentration in the seepage water (mg l^{-1}) and
S quantity of seepage water (l m^{-2} = mm)

Vegetation in the lysimeter and how it is managed can be matched to the specifics at the extraction site. However, it is also possible to choose any other vegetation type or management practice.

3.2 Weighable Groundwater Lysimeter

The dynamics of soil water have a substantial impact on groundwater recharge as well as on the concurrent process of solute movement. Knowledge of these processes is essential for long-term planning with regard to floodplains that serve as a source for obtaining drinking water from bank filtrate. In the event of flooding, parameters that influence groundwater dynamics need to be quantified as extensively as possible to enable the water absorption capacity of adjacent floodplains to be predicted more accurately. Since it is very difficult to measure in detail the pedohydrological parameters in the actual floodplain, a weighable groundwater lysimeter was developed (Bethge-Steffens et al. 2004). This lysimeter type is the basis for recording water balance quantities such as precipitation, evapotranspiration, groundwater recharge, capillary rise, and interaction with the water course.

The water balance equation can be specially adapted for plane floodplain sites. Surface runoff can be neglected. Outflow and inflow rates can be recorded in the saturated zone. During a flood, floodwater is an additional input parameter. The modified water balance equation for groundwater-influenced and temporarily water-logged sites reads:

$$\mathrm{P} + \mathrm{Pond} = \mathrm{ETa} + (\mathrm{R}_{\mathrm{out}} - \mathrm{R}_{\mathrm{in}}) \pm \Delta\mathrm{S} \quad (3)$$

P	precipitation
Pond	surface floodwater
ETa	actual evapotranspiration
R_{out}	groundwater outflow
R_{in}	groundwater inflow
ΔS	change of storage capacity.

The principle scheme of a groundwater lysimeter is shown in Fig. 6. It usually has a surface area of 1 m^2 and is 2 m deep. This lysimeter type was also filled monolithically using the excavation technique described above and located in a PE-HD container lysimeter station. As above, for the groundwater lysimeter a 15 cm thick filter layer (sand over coarse sand over gravel) was also placed at the bottom of the lysimeter to minimise natural flux disturbances.

The groundwater lysimeter is weighable and equipped with the aforementioned load cells. Weight is recorded every minute. In order to balance groundwater inflow and outflow, mass changes need to be recorded with as little delay as possible. Measurements are condensed to 15-min averages to reduce any effects,

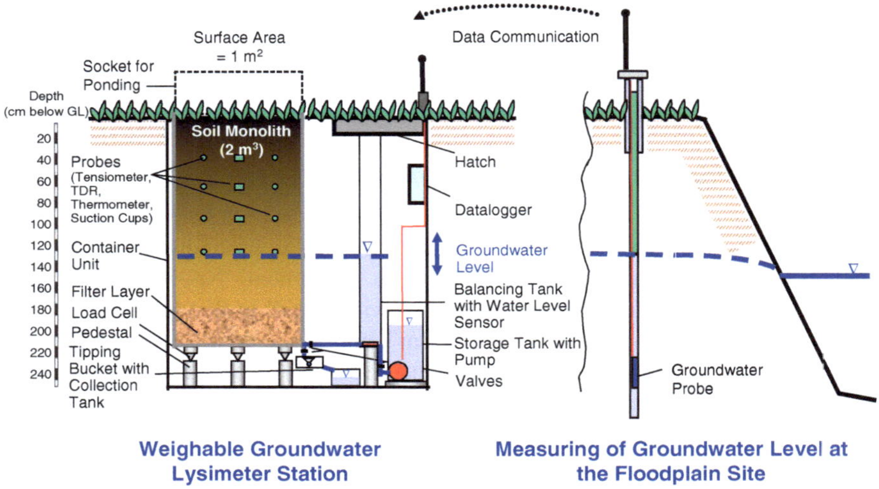

Fig. 6 Schematic of a weighable groundwater lysimeter with a groundwater control system and radio data transmission

for example, caused by wind (mass oscillation) or brief mass increases (due to passing animals etc.).

TDR probes, tensiometers and thermometers that record hourly values are installed at depths of 0.30, 0.60, 0.90 and 1.20 m beneath the surface. Furthermore, suction cups are installed at the same depths to enable soil water samples to be taken. A tipping bucket with a storage tank is installed for measuring the quality and quantity of seepage water at times when groundwater is below the bottom of the lysimeter. This enables the groundwater lysimeter to be used additionally as a gravitation lysimeter. Vegetation height is matched to the rest of the vegetation in the floodplain by mowing, since it has a decisive influence on evapotranspiration (especially transpiration and interception).

The groundwater lysimeter is connected to an immediately adjacent balancing tank using the principle of communicating pipes. The balancing tank is fitted with a water level sensor. If the water level in the balancing tank deviates from the target water level by more than 1 cm, the valve to the lysimeter closes and the water level is raised or lowered by a pump. A storage tank serves to provide the water required or to absorb surplus water. The valve between the balancing tank and the lysimeter then opens, and the water level is adjusted. This process is repeated until the target water level has been reached in both the lysimeter and the balancing tank.

An automatic groundwater control system was developed that allows groundwater levels measured in the floodplain to be recreated as quickly as possible in the lysimeter. In this way, the natural course of capillary rise and groundwater recharge are gauged as realistically as possible in the lysimeter experiments. Groundwater level data from the floodplain are transmitted once a day by modem to the lysimeter site and regarded as the target water level.

The differences in water level required to achieve the target water level in the balancing tank are added in order to determine the quantity of water that flowed into or out of the lysimeter during the period concerned. The balance is calculated using the following equation:

$$V_{efft=m} = A_{AG} * [(B_{t=m} - B_{t=0}) - (W_{t=m} - W_{t=0})] \quad (4)$$

where

$V_{efft=m}$ volume of water that flowed into/out of the lysimeter at $t = m$
A_{AG} base area of the balancing tank
$B_{t=0}$ value of balancing at the start of the adjustment process
$B_{t=m}$ value of balancing at the end of the adjustment process, comprising $B_{t=0}$ plus/minus the changes to the water level in the balancing tank during the period concerned
$W_{t=0}$ water level in the balancing tank at the start of the adjustment process
$W_{t=m}$ water level in the balancing tank at the end of the adjustment process.

4 Results and Discussion

4.1 Weighing Precision of a Gravitation Lysimeter

In water balance studies, lysimeters are typically used to quantify rainfall, actual evapotranspiration and drainage. However, if weighing precision is sufficiently high, as in the case of the lysimeters presented here, precipitation in the form of dew, fog and frost can be measured accurately (Meissner et al. 2007; Xiao et al. 2009). These new types of lysimeter are ideal for developing and testing models for soil hydrologic processes, because they enable detailed measurements to be taken with very high temporal resolution.

As an example of the high precision of the new weighing technique, Fig. 7 shows a chart of the lysimeter mass (the mass change allows us to calculate the change in the amount of water stored in the soil column) recorded in northern Germany over a five-day period in April 2010. There was no rainfall from 16 April to the evening of 18 April, meaning that the lysimeter mass decreased due to evapotranspiration. In the early morning of 17 April, dew formation was visible because the lysimeter mass increased slightly. The rising sun's radiation leads to increasing evapotranspiration with a typical day-night rhythm. In the late evening of 18 April, a rain event occurred, which led to an increased mass change in the lysimeter. Nine further rain events with different quantities of precipitation were registered until the afternoon of 19 April. A total of 5.5 mm precipitation was measured, leading to an increased mass of 5.5 kg.

This example gives an overview of the various ways in which this lysimeter type can be used. It enables measurements to be taken with very high temporal

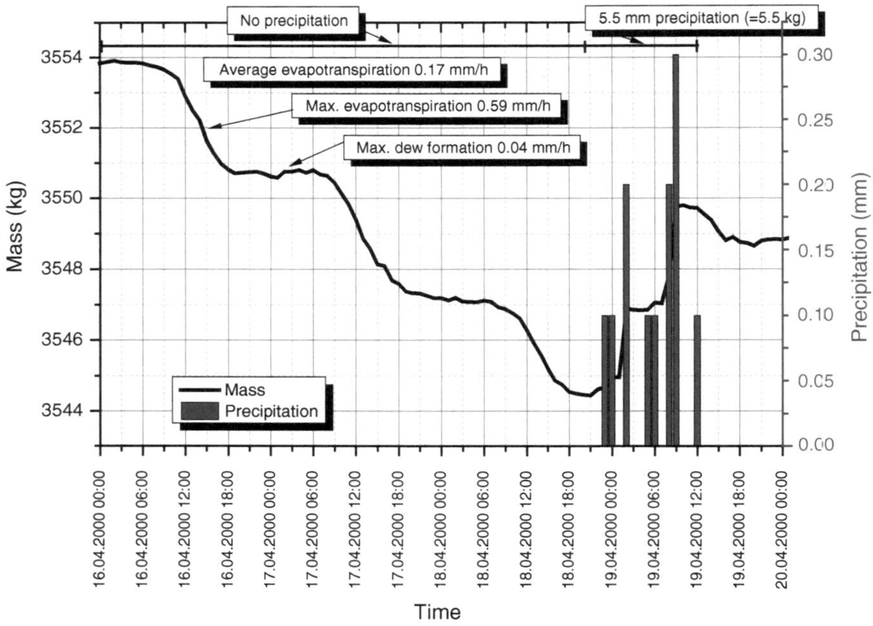

Fig. 7 Example of the diurnal mass change of a weighable gravitation lysimeter planted with grass

resolution; it gives an "inside view" of hydrological processes; it is an essential tool for developing and testing hydrological models. In Europe, this newly developed lysimeter type is currently being used for investigating and monitoring water and solute balances in agricultural, post-mining and forested areas. It is also used in scientific studies in which the movement of water and chemicals (nutrients, pesticides, important contaminants) through the soil profile is measured.

4.2 Water Balance of a Groundwater Lysimeter

Initial estimates of individual quantities of the soil water balance at floodplain sites were made using the soil hydrological parameters measured in a groundwater lysimeter in February 2010 (Fig. 8). In unadjusted periods (time intervals where the groundwater level is unaffected by groundwater control, i.e., by pumping processes), precipitation and current evapotranspiration can be directly quantified on the basis of recorded mass changes in the lysimeter monolith. These data are supplemented by groundwater control values.

Using the modified water balance Eq. (3), the following balance can be defined for this period:

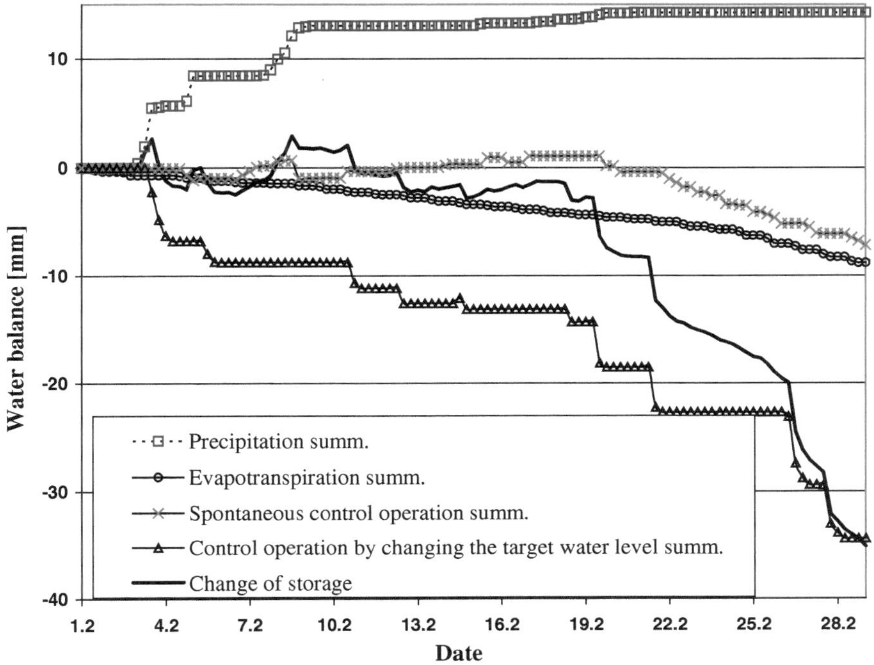

Fig. 8 Water balance parameters (cumulative) in February 2010 based on measurements using the weighable groundwater lysimeter

$$\begin{aligned} &P + \text{Pond} = \text{ETa} + (R_{\text{out}} - R_{\text{in}}) \pm \Delta S \\ &15\ \text{mm} + 0\ \text{mm} = 9\ \text{mm} + (7\ \text{mm} + 34\ \text{mm} - 0\ \text{mm}) - 35\ \text{mm} \end{aligned} \tag{5}$$

The quantity of precipitation during the study period was 15 mm, which occurred as snow or sleet mainly at the beginning of the month. It led to a short-term and slight replenishment of soil water storage.

Depending on the course of the weather, daily values of evapotranspiration varied between 0.1 and 0.7 mm per day. In February, the total actual evapotranspiration of 9 mm had only a small effect on the reduction in soil water storage.

The water level in the balancing tank is constantly compared with the target water level. If the water level in the lysimeter deviates from the specified target water level by more than ±1 cm, an adjustment is necessary. This type of adjustment process is called **spontaneous control operation**, classified as groundwater recharge or capillary rise. In order to maintain target water levels, limited adjustments were necessary until 20 February 2010. As of 20 February 2010, increased seepage was observed in the lower soil layers. The resulting parameter “groundwater recharge” generated a total of 7 mm by the end of the month.

The other types of adjustment processes are external changes in the target water level for groundwater control in the floodplain and unchanged water levels in the corresponding lysimeter. It can be assumed that this change in groundwater level in the floodplain was caused by interaction with the water course. This kind of adjustment is called **control operation by changing the target water level**, classified as infiltration or exfiltration.

During the investigation period, however, "control operations by changing the target water level" constituted the largest proportion of groundwater outflow. The considerable decline in the groundwater level caused a groundwater outflow of 34 mm. This quantity is classified as "exfiltration into the water course". Hence, the findings of the lysimeter experiments can be used to determine the scale at which groundwater interacts with river water in the study area.

The quantities of evapotranspiration, groundwater recharge and exfiltration into the water course resulted in a 35 mm reduction in soil water storage that could not be compensated by precipitation.

In addition to the described use of this lysimeter type, new weighable groundwater lysimeters with a surface area of 1 m^2 and a depth of 2 to 3 m have been constructed to investigate hydrological processes in post-mining overburden heaps and in a swamp area with a fluctuating groundwater table.

5 Conclusions

In Europe, the use of direct drainage lysimetry methods for measuring water and solute fluxes in soils has increased in recent years. Although this technique generates reliable drainage data, it involves relatively high investment and maintenance costs. There is a tendency to develop new lysimeter techniques with the aim of improving measuring accuracy and reducing costs. Progress is visible in a new technology to obtain large undisturbed soil monoliths and in the development of mobile PE-HD lysimeter stations. A high-resolution weighing technology and an automatic groundwater control system enable detailed investigations to be carried out into the water balance, forming the basis for very accurate solute balance calculations and for the modelling of hydrological processes.

In the future, lysimeter studies will also be an essential tool for scaling up results achieved in small-scale experiments to larger geographical units. Combining lysimeter studies with in situ measurements in the field or catchments allows a direct comparison to be made of relevant soil hydrological parameters. Furthermore, the newly developed experimental setup allows a scenario simulation of typical climatic and hydrologic questions, e.g., global warming and the impact it has on the water and solute balance, the influence of dew and fog on the establishment of a vegetation cover in arid areas, and the transport of contaminants during heavy rainfall following a severe drying up of the soil profile.

References

Aboukhaled A, Alfaro A, Smith M (1982) Lysimeters, FAO Irrigation and Drainage, Paper 39, Food and Agriculture Organization of the United Nations, Rome Italy, p 68

Allen RG, Howell TA, Pruitt WO, Walter IA, Jensen ME (1991) Lysimeters for evapotranspiration and environmental measurements. In: Proceedings of the International Symposium on Lysimetry, American Society of Civil Engineers, New York, NY, pp 79–84

Bethge-Steffens D, Meissner R, Rupp H (2004) Development and practical test of a weighable groundwater lysimeter for floodplain sites. J Plant Nutr Soil Sci 167(4):516–524

DIN 4049-3 (1994) Hydrologie—Teil 3: Begriffe zur quantitativen Hydrologie. DIN Deutsches Institut für Normung e. V. (Herausgeber), Oktober 1994. Beuth Verlag GmbH, Berlin, p 80

Führ F, Hance RJ, Plimmer JR, Nelson JO (1998) Comprehensive tracer studies on the environmental behavior of pesticides: the lysimeter concept. In: Führ F, Hance RJ, Plimmer JR, Nelson JO (eds.) American Chemical Society, Washington, DC

Gee GW, Hillel D (1988) Groundwater recharge in arid regions: review and critique of estimation methods. Hydrol Process 2(3):255–266

Gee GW, Ward AL, Caldwell TG, Ritter JC (2002) A vadose zone water fluxmeter with divergence control. Water Resour Res 38(8):16-1–16-7

Gee GW, Newman BD, Green SR, Meissner R, Rupp H, Zhang ZF, Keller JM, Waugh WJ, van der Velde M, Salazar J (2009) Passive wick fluxmeters: design considerations and field applications. Water Resour Res 45:1–18

Klocke NL, Todd RW, Hergert GW, Watts DG, Parkhurst AM (1993) Design, installation, and performance of percolation lysimeters for water-quality sampling. Transact ASAE 6:429–435

Lanthaler C, Fank J (2005) Lysimeter stations and soil hydrology measuring sites in Europe: results of a 2004 survey. In: Proceedings of the 11th Lysimeter conference, Raumberg-Gumpenstein, Austria, 19–24

Meissner R, Rupp H, Schubert M (2000) Novel lysimeter techniques: a basis for the improved investigation of water, gas, and solute transport in soils. J Plant Nutr Soil Sci 163(6):603–608

Meissner R, Seeger J, Rupp H, Seyfarth M, Borg H (2007) Measurement of dew, fog, and rime with a high-precision gravitation lysimeter. J Plant Nutr Soil Sci 170(3):335–344

Meissner R, Rupp H, Seyfarth M (2008) Advances in out door lysimeter techniques. Water Air Soil Pollut Focus 8:217–225

Meissner R, Rupp H, Seeger J, Ollesch G, Gee GW (2010) A comparison of water flux measurements: passive wick-samplers versus drainage lysimeters. Eur J Soil Sci 61:609–621

Schlotter D, Schack-Kirchner H, Hildebrand EE, von Wilpert K (2012) Equivalence or complementarity of soil-solution extraction methods. J Plant Nutr Soil Sci 175(2):236–244

Weihermueller L (2012) Saugkerze. In: DWA Merkblatt 905. Gewinnung von Bodenlösung. Beprobungssysteme und Einflussgrößen. Hennef, Deutsche Vereinigung für Wasserwirtschaft, Abwasser und Abfall e.V. -DWA-, p 36

Weihermueller L, Siemens J, Deurer M, Knoblauch S, Rupp H, Goettlein A, Puetz T (2007) In situ soil water extraction: a review. J Environ Qual 36:1735–1748

Wriedt G (2004) Modelling of nitrogen transport and turnover during soil and groundwater passage in a small lowland catchment of Northern Germany. Doctoral Thesis, Potsdam University, Germany

Xiao H, Meissner R, Seeger J, Rupp H, Borg H (2009) Testing the precision of a weighable gravitation lysimeter. J Plant Nutr Soil Sci 172:194–200

Third-Generation Lysimeters: Scientific Engineered Monitoring Systems

Christian Hertel and Georg von Unold

Abstract Third-generation lysimeters meet the needs of 21st century environmental research and monitoring. High-resolution sensors, field-replicating hydraulic and thermal conditions and an excavation method showing the soil inside the lysimeter cylinder are important features in their quality process chain. Older lysimeter systems are unable to provide such detailed information about the soil water budget and all linked processes. Due to an increasing awareness of climate change, water management, agronomy and soil science issues, it was essential to upgrade lysimeter systems in order to gather more detailed information about processes and fluxes. Different lysimeter station layouts were developed for specific requirements and to increase their fields of application for particular tasks such as fertilisation treatments or irrigation and reproducing identical climatic conditions. Additionally, highly engineered third-generation bespoke lysimeter types are available to support particular projects. As an example, the meteorological lysimeter precisely measures precipitation, evapotranspiration and leachate. For this, the lysimeter weighs to the nearest gramme range a surface of 1 m^2 and supplies results to an accuracy of 0.01 mm for water input such as rain, dew, frost or snow and water output by evapotranspiration and leachate as well as reporting the change of soil water content. Combined with additionally measured meteorological data, this enables water balance models to be developed and potential evapotranspiration can be determined.

Keywords Lysimeter · Types · Evapotranspiration · Filtration · Accuracy

C. Hertel (✉) · G. von Unold
UMS GmbH, Gmunder Str. 37 81379 Münich, Germany
e-mail: CH@ums-muc.de

G. von Unold
e-mail: gvu@ums-muc.de

L. Mueller et al. (eds.), *Novel Measurement and Assessment Tools for Monitoring and Management of Land and Water Resources in Agricultural Landscapes of Central Asia*, Environmental Science and Engineering, DOI: 10.1007/978-3-319-01017-5_9,

1 Introduction

Formerly, lysimeters were only used for measuring leachate. Over the past 20 years, innovative technology has developed lysimeters as high-precision tools for environmental monitoring research projects such as climate research, water management, agronomy, soil science and site remediation. Today, lysimeters are more than a standard tool for evapotranspiration, and will be described in the following paper. The word *lysimeter* originates from the Greek—"lysis" means dissolution or movement and "metron" is to measure (Aboukhaled et al. 1982). It is clear that the word "lysimeter" means the measurement of the percolation of water in soils (Howell et al. 1991). The first lysimeter investigations were performed by Philippe de La Hire in 1688 and focused on determining the origins of springs. Following various studies, de La Hire found out that lysimeters in grass evaporate more water than those in fallow ground (Kohnke et al. 1940). Historically, the design of lysimeters was generally a simple container to define the water movement through a soil boundary (Howell et al. 1991). According to Kohnke et al. (1940), lysimeters can be classified into two groups: those systems that can weigh and those that cannot. In the early beginnings of lysimetry, weighing was an often stated problem. Lysimeters with weighing systems are able to determine evapotranspiration directly through the mass loss of the water within the lysimeter. In contrast, non-weighing systems determine evapotranspiration indirectly through volume balance (Howell et al. 1991). By predicting runoff and lateral flux, lysimeters can provide data for water fluxes and because of the substance concentration in relation to the water balance, the mass balances too, as shown in Fig. 1.

Modern-day lysimeters invented by UMS Muenchen (http://www.ums-muc.de) monitor water fluxes according to the hydraulic and thermal conditions of the

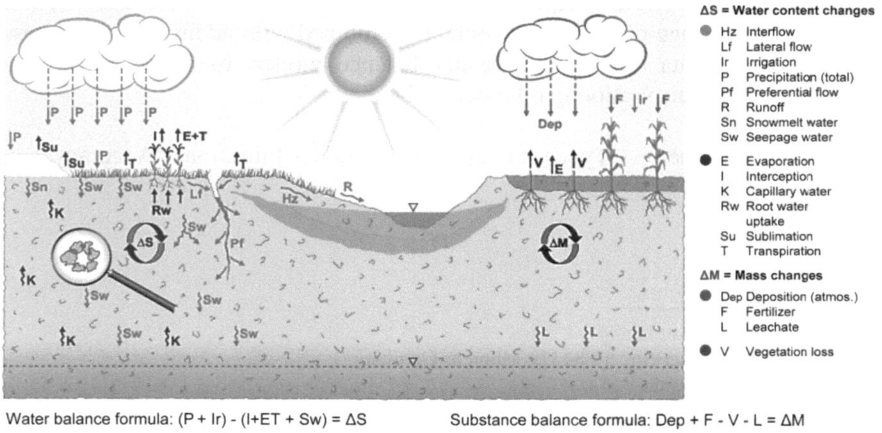

Fig. 1 Precise water and substance balances. Soil water as well as input and output parameters of water and substance balance

surrounding field. Their sophisticated measuring, sampling, controlling and regulating instrumentation monitor the real field situation. Soil water is the decisive parameter for mass balance and any flux study. For studies on groundwater recharge or mass transport or for metabolism research, it is crucial to have a field-replicating water regime inside the lysimeter. Lysimeters offer thermodynamic comparability between field and lysimeter specially designed for climate research or for the analysis of microbial activities. Both features are important for many tasks, as well as aspects of research into conventional cultivation, snow coverage during winter operations and high substance conversion after snow melt.

Scientific engineering lysimeters use a precision weighing system that measures water fluxes in soils by weighing and calculating water input and output over time and can weigh lysimeter columns with a mass of 50 up to 6,000 kg. They can measure soil columns to a high level of precision and resolution. This absolute precision depends on the total weight of the lysimeter. The precise resolution can be even 10 g or 0.01 mm precipitation. Thus, lysimeters are exceptionally suitable for the measurement of all types of precipitation including rain, dew, frost and snow. And above all, they can measure precipitation as it occurs in the surrounding field—right on the soil surface. High-resolution data loggers store short-interval readings over extended periods of time to demonstrate ongoing processes, mass conversion or solute movements. Loggers store readings internally or on USB sticks. As an option, the logger can transmit the data to an Ethernet or Web server network via GSM or GPRS. Thus, data can be combined from numerous lysimeter sites, and this is then available as a common and substantial database for research networks. When investigating climate change, the mesocosmos containing soil, vegetation and micro-organisms is significant as it is exposed to a natural yet different climate. Thus, it reveals symptoms of climate change on soil, plants, water balance, CO_2 and O_2 dynamics, long before they may become apparent elsewhere. The following chapter describes various systems and their respective applications in this area of research.

2 Materials and Methods

Stations with several lysimeters increase the measuring capacity or compare the effects of various treatments of fertilisation or irrigation under identical climatic conditions. In order to determine various treatments at one site, two or more lysimeters are connected to one logger and one service well (Fig. 2a). They can compare diverse cultivation methods and crop rotations, as well as conventional and organic farming. For comparison studies on the same soil but with different treatments (fertilisation, irrigation or CO_2 treatment), up to six lysimeters can be connected to one logger and service well. The advantages of a tetragonal layout are an installed service well in the corner of four fields and the flexible length of the connection pipes, so the lysimeters can be placed far inside the field to minimise vegetative island effects (Fig. 2b). Another possibility is a linear layout to compare various crop rotations or variations from Fig. 2c A to F. A linear alignment and parcelling around the lysimeter is recommended as this is less disruptive to cultivation. Furthermore, lysimeters in a linear arrangement are preferable for autonomous robots or automatic systems for irrigation and tracer application or when gas treatment hoods are used. Another option is the hexagonal layout. For studies into a comparison of soils under changed climatic conditions, lysimeters can be arranged as a ring. Then, the soils can be brought in from different locations and be exposed and surveyed under changed climatic conditions. Due to the geometric similarity of the hexagonal layout, the lysimeters are exposed to equal conditions, which is especially advantageous for comparative studies. Then, lysimeters of Fig. 2d A to F are filled with a different mesocosmos from various locations, or represent the variability of one ecosystem.

Innovative solutions for delivering the best quality assurance in cutting and conserving lifting and rotation methods are important for lysimeter systems operations. As each soil is unique, it is essential to have experience in the

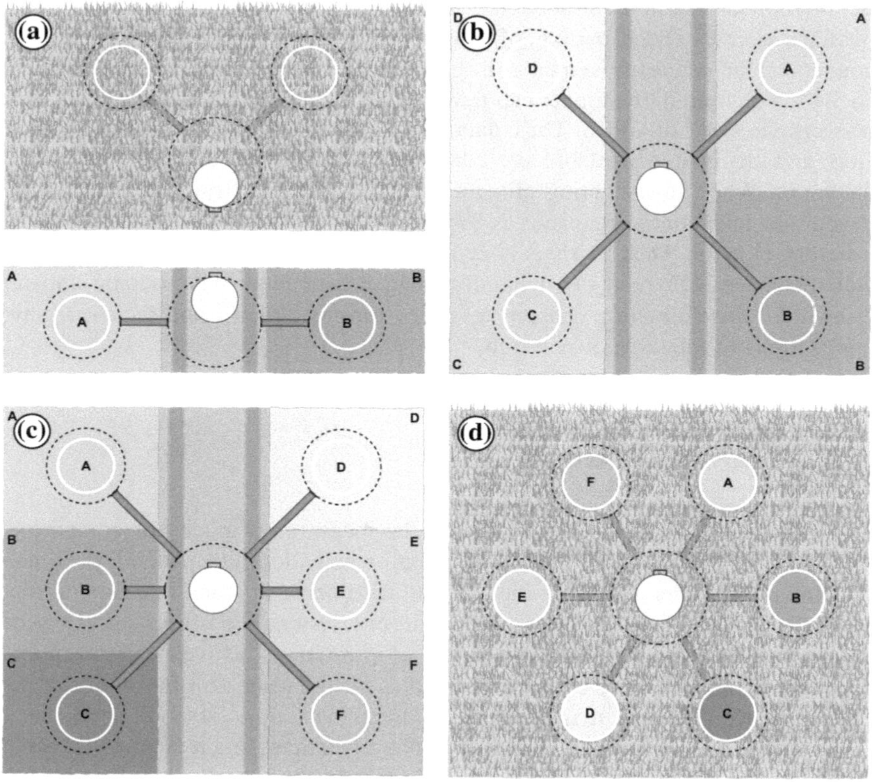

Fig. 2 Various third-generation lysimeter layouts. **a** 90° layout and duplex field layout. **b** Tetragonal layout. **c** Hexagonal linear layout. **d** Hexagonal circular layout

excavation of lysimeter soil columns. Special cutting methods were developed to ensure quality. This method prevents soil compaction and gaps and during the cutting process shows exactly the soil that will be later inside the lysimeter. The first step is inspection borings to determine if rocks, roots, cavities or other disturbances might require preventive measures. Ideally, the site will have trenches dug with an excavator at least one metre deeper than the last lysimeter to show the exact origin of the soil in its texture, structure and any peculiarities. Furthermore, continuous observation of the cutting process by experts is crucial. A specially developed tool for cutting precise perpendicular under protection by the utility model is being used for third-generation lysimeters. Its cutting edges are made from special steel which minimises friction and prevents soil compaction inside the lysimeter column. Finally, stones or roots need to be removed by hand beneath the cutting edge so they do not create cavities or grind grooves onto the soil column. As soon as the soil is inside the lysimeter cylinder, it becomes a black box, so care needs to be taken during the cutting process to ensure its quality. The last step in cutting a lysimeter soil column is a precise shear-off procedure. A polished cutting plate with specially shaped cutting edges is driven by hydraulics to assure a careful and accurate shearing. After the cutting process, the soil column is lifted and turned upside down. The lifting force is applied evenly and close to the balance point to prevent column deformation. The soil body is kept safe and free from deformation. Third-generation lysimeter design assures that the cylinder is not deformed when lifted, as the induced load torque is reduced to a minimum. The load is carefully distributed away from the short bolts over the large welded-on base plate. The filled lysimeters are transported on special anti-shock trucks, if required. The special air suspension of the vehicles ensures that the soil column remains undisturbed. The lysimeter's innovative lifting and rotating mechanism prevents a deformation of the soil body and eliminates the development of gaps inside the cylinder—an essential quality criterion for obtaining substantial data. This prevents the need for heavy weight clamps around the lysimeter and invisible preferential gaps due to handling. After cutting and lifting, the next step in the lysimeter process is the installation and the adjustment of the field-identical water regime.

The basic design of a third-generation lysimeter has a cylindrical stainless steel surface of 1 m^2 and a monolith depth of 1 m. A range of different sensors are available and can be selected according to the lysimeter's task of providing a view into the black box of the soil monolith. Thanks to highly resolved weighing systems, measurements of water fractions with a resolution of 10 g can be achieved (for lysimeters with a surface of 1 m^2 and a length of 1.5 m). Drainage water measurements by a second cell enable high resolution (von Unold and Fank 2008). Tipping buckets for drainage measurements are applicable too, but lead to losses in resolution, depending on the tip volume. The bottom of the monolithic pillar is formed as a matric potential area to achieve field comparable fluxes. The difference between the water regime in the field and inside the lysimeter has always been criticised as a weakness in design. In conventional gravity lysimeters, the leachate simply drains out of the bottom of the lysimeter. In reality, variable matric potentials, which are the main determinants for water flow, occur in the field. In rainy seasons or wet periods, the field's soil water is pulled towards the lower groundwater. This does not happen inside the lysimeter, as the potential on the very bottom is zero. As a result, the lysimeter has more moisture than the field. In dry periods, depending on the soil type, capillary suction causes the water in the field to rise. As this is not the case in the conventional lysimeter, the soil in the lysimeter remains drier than the soil in the field. The described third-generation lysimeter has solved this problem by measuring and comparing the matric potential in both the bottom of the lysimeter and in the field at the same depth. If the lysimeter has more moisture, then water is sucked out through the rake of the suction cups. If the lysimeter is drier, then water is injected. In this way, the old problem with lysimeter bottom plates was overcome, enabling the monitoring of field water fluxes, which was exactly what was wanted.

3 Lysimeter Designs

Third-generation lysimeters depend on different application requirements. Special lysimeter designs are described in detail below in terms of their specific field applications.

3.1 Hydro-Lysimeter

The Hydro-Lysimeter in its basic design (Fig. 3) has been devised to solve the water balance equation (von Unold and Fank 2008). It measures the weight of the lysimeter monolith as well as drainage. The standard lysimeter height is 1.5 m with a surface of 1 m^2. Precipitation and evapotranspiration are determined from the change in the lysimeter's weight over time. The precipitation evaluation and drainage measurement enable the determination of ground water recharge capacities (von Unold and Fank 2008). With the Hydro-Lysimeter, it is possible to obtain soil water parameter measurements and their interfaces to atmosphere and aquifer. The controlling input parameters for the lysimeter are the measured water reaching the surface and the water equivalent of snow. To prevent problems in snow measurements, the snow bridge crossing the gap between the lysimeter and its surrounding area is cut to maintain correct weight measurements. On other sites, the output parameters are real evapotranspiration and drainage rates. In addition, an identical temperature regime is available within the lysimeter field.

The design includes a lysimeter cylinder with silicon carbide porous cups' rake and a precision weighing system. Moreover, the field tensiometer measures the field situation and the lysimeter tensiometer steers the controller to keep the lysimeter at field conditions.

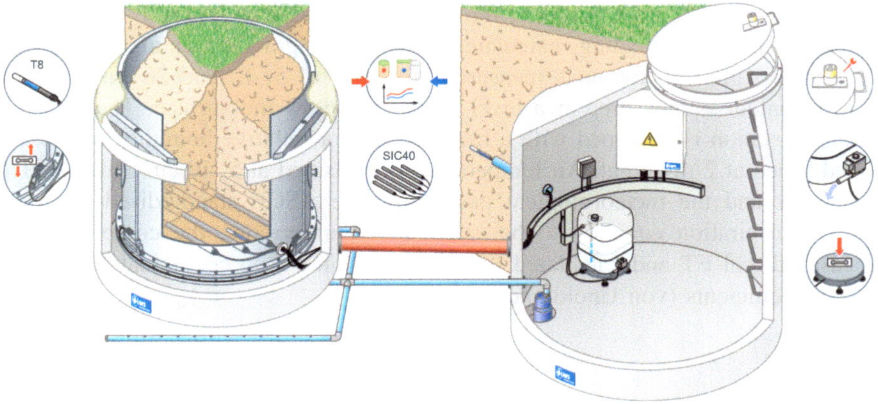

Fig. 3 The Hydro-Lysimeter with service well

3.2 Meteo-Lysimeter

The meteorological lysimeter, as shown in Fig. 4, is specially designed for determining evapotranspiration (ET) and grass reference evapotranspiration over short time intervals (ET_0); drainage and precipitation from lysimeter weighing data and measured peripheral meteorological data.

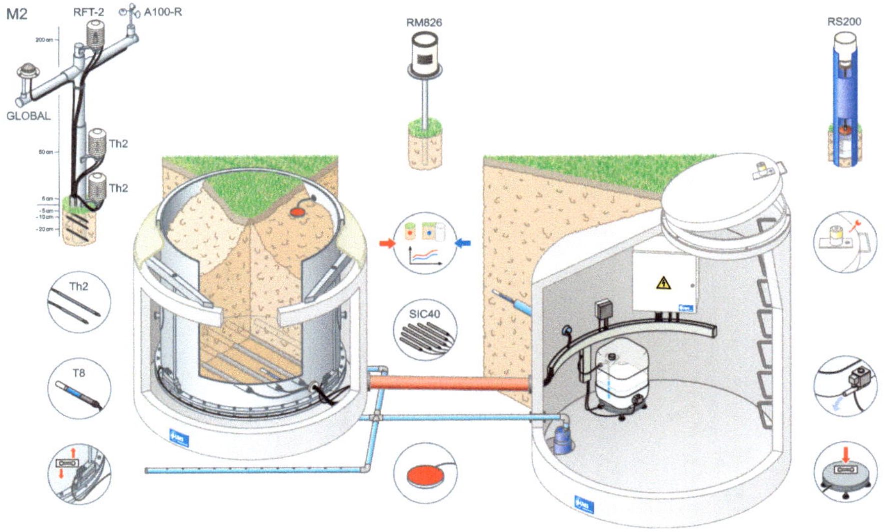

Fig. 4 The Meteo-Lysimeter with weather station and service well

As with the Hydro-Lysimeter, the Meteo-Lysimeter measures ET_0 as a continuously recorded reading, calculated by a 12 cm grass cover algorithm. Furthermore, soil temperature is measured at depths of 5, 10 and 20 cm. The soil water content is measured at a level of 10 cm below the surface. The included weather station is equipped with a radiation sensor, relative humidity sensor and wind speed at 2 m height. Air temperature is measured at 5, 50 cm and 2 m. The lysimeter and the meteorological data allow the ASCE standardised reference evapotranspiration equation to be calculated. Furthermore, the combination of measured real ET and calculated ET_0 can determine crop coefficients and water stress coefficients (von Unold and Fank 2008).

3.3 *Scientific Field-Lysimeter*

This lysimeter, designed especially for scientific applications, has a standard depth of 2 m (Fig. 5). It has been designed as a configurable system for investigating specific properties of soil, soil utility and conservation. It combines the advantages of laboratory and field investigations, as it provides laboratory precision even under rough field conditions. Water samples and readings inside the monolith are taken from 10 cm down to 180 cm. Through its ability to supply the additional measurements of volumetric water content, matric potential and soil temperature, a detailed description of soil water fluxes within the monolith can be generated. For quality assurance, sensors are also installed in the undisturbed field to compare the significance of the lysimeter investigations with natural field conditions (von Unold and Fank 2008). For more detailed research tasks, a combination of suction cups and a tension controlled vacuum for pore water extraction can be installed,

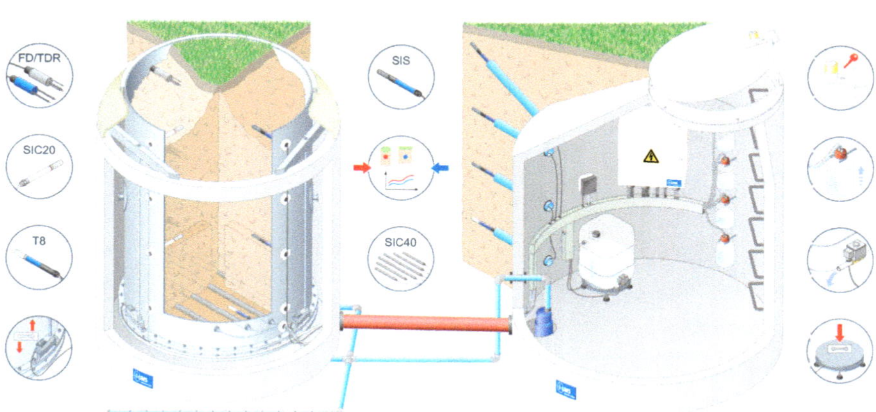

Fig. 5 The science field-lysimeter with various types of instrumentation and reference sensors

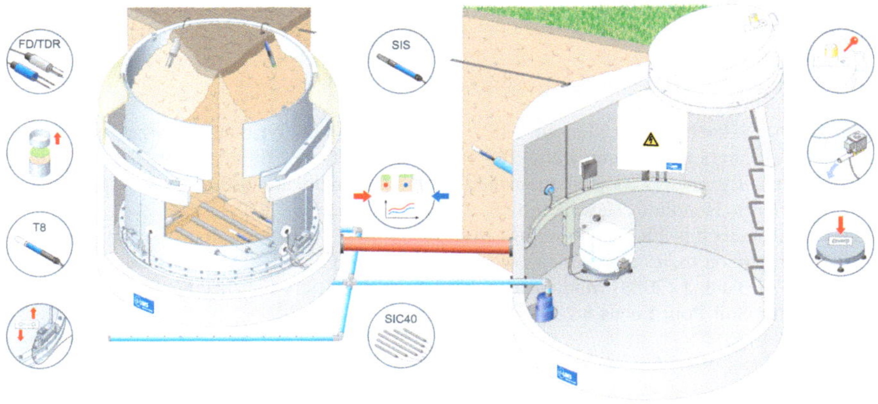

Fig. 6 The agro-lysimeter with various types of instrumentation and reference sensors

giving detailed information about metabolisms or contamination inside the monolith.

3.4 Agro-Lysimeter

The Agro-Lysimeter (Fig. 6) is adjusted to measure variables such as root water stress, drainage water and solute fluxes in cultivated field investigations. The specifications of the lysimeter are 1 m^2 with a depth of 1 m. The special construction of the removable upper ring means that the field can be cultivated and it can be replaced after tillage. The field matric potential is transferred into the lysimeter by the suction cup rake. The Agro-Lysimeter is not weighable and additional sensors should be adjusted in the upper soil horizon. Furthermore, seepage water measurements are detected by using a 0.1 mm capsulated tipping bucket (von Unold and Fank 2008). Its ideal application area is the recurrent monitoring of ground water recharge and solute leaching from arable land.

4 Conclusion

Lysimeters have been used for centuries, but only the instrumentation and innovation of the past 20 years has reduced the large water potential errors. To avoid problems in lysimetry, a detailed review is crucial to find out which configuration and layout are best suited to resolving scientific questions. The chapter above has described general operations and technical concepts to show the flexibility of lysimeters for different tasks and applications.

References

Aboukhaled A, Alfaro JF, Smith M (1982) Lysimeters. FAQ Irrigation and Drainage Paper, no 39. Rome

Howell TA, Schneider AD, Jensen ME (1991) History of lysimeter design and use for evapotranspiration measurements. In: Proceedings of the International Symposium on Lysimetry: Lysimeters for evapotranspiration and environmental measurements, pp 1–9, July 23–25, 1991, Honolulu, Hawaii. ASCE, 345 E. 47th St., NY, 1017–2398

Kohnke H, Dreibelbis FR, Davidson JM (1940) A survey and discussion of lysimeters and a bibliography on their construction and performance, vol 372. USDA, Washington, DC, p 68

von Unold G, Fank J (2008) Modular design of field lysimeters for specific application needs. Water Air Soil Poll: Focus 8:233–242

A Field Method for Quantifying Deep Seepage and Solute Leaching

Uwe Schindler

Abstract Quantifying groundwater recharge to aquifers is necessary for agricultural and environmental research. Lysimeters are the main devices for monitoring and quantifying soil water and solute balances and transport processes. However, certain characteristics of lysimeters, high costs and limited flexibility act as restricting arguments. There is a lack of effective and reliable methods for quantifying deep seepage under undisturbed soil and for managing conditions in the field. Soil hydrology measurements provide an alternative way of analysing the soil water and matter status in situ. The method presented here aims to estimate deep seepage and solute leaching in the field based on soil hydrological measurements below the zero flux plane and a calibrated hydraulic conductivity function. This method offers simple handling, flexibility, and costs less than lysimeters. The required soil hydraulic properties are only derived from tension and water content field recordings within the measurement depth. After calibration, no further information about soil properties, weather, management and land use data is required, nor is any other data. Since 1994, the method has been successfully applied at many sites in northeast Germany. A comparison between lysimeter discharge measurements and discharge calculations has confirmed the validity of this method.

Keywords Soil · Deep seepage · Solute leaching · Measurement · Field method

---

U. Schindler (✉)
Leibniz Centre for Agricultural Landscape Research (ZALF) e. V., Eberswalder Strasse 84
15374 Muencheberg, Germany
e-mail: uschindler@zalf.de

L. Mueller et al. (eds.), *Novel Measurement and Assessment Tools for Monitoring and Management of Land and Water Resources in Agricultural Landscapes of Central Asia*, Environmental Science and Engineering, DOI: 10.1007/978-3-319-01017-5_10,

1 Introduction

Quantifying groundwater recharge to aquifers is necessary for tackling numerous economic and environmental problems such as providing non-polluted water for different uses, protecting drinking water and determining safe yields from aquifers. Soil type, land use and land management practices and their relationship with climatic conditions are decisive factors in seepage flow and solute leaching (Benson et al. 2006; Köhler et al. 2006). Efficient water use and intelligent water management are essential for environmental protection, especially in dry regions with an annual water balance deficit. Climate change scenarios predict a shortfall in the water budget in many parts of the world. As a consequence of the significant rise in temperatures, potentially, evapotranspiration will increase and rainfall will decrease during the vegetation period. Throughout the regions likely to be affected, there is a demand for measures to support groundwater recharge as well as wetland conservation. For the time being, this will result in maximum water retention in catchment areas which will involve reducing any non-productive discharge (surface runoff, artificial drainage flow) and assuring that groundwater resources are replenished by deep seepage. To meet these requirements, there is a need for research into suitable land use and land management systems that provide sustained seepage flow.

For centuries, lysimeters have been the main devices for monitoring and quantifying soil water and solute balances and transport processes. Goss and Ehlers (2009), present an overview of the development of lysimeter techniques. The first quantitative experiments were performed in the 17th century by Johan Baptista von Helmont. John Law constructed the first lysimeters with natural soil structure and soil stratification in Rothamsted, England in the 1880s. Monitoring hydrological processes with weighable lysimeters was first carried out by the German, von Seelhorst, in 1902. Since that time, many lysimeters have been constructed and used for agricultural and environmental research (Edwards 1986; Schindler et al. 2001; Goss and Ehlers 2009; Meissner et al. 2010; Haferkorn et al. 2011; Prasuhn et al. 2011; Puetz et al. 2011). Lysimeters differ in their construction and management. They can be: (1) weighable or not weighable); (2) in terms of size, between a few cm^2 to several m^2; (3) their depth can range from 50 to 500 cm, (4) structure of the soil material can be monolithic or refilled soil; (5) the lysimeter material can be either steel or plastic; (6) the boundary conditions can be either free draining or suction at the base; (7) the location can be either in or outside of the field, and so on. Lysimeters are useful for quantifying soil hydrological processes. Depending on the instrument's construction and management, results are influenced by special lysimeter effects (Schindler et al. 2008). Besides this weakness, lysimeters are expensive and inflexible.

Soil water content and tension are important hydrological variables that reflect the effects of land surface processes. In spite of its importance in land use planning, agriculture and drought monitoring, information about soil moisture is not widely available. There is a lack of effective and reliable methods for quantifying deep seepage under undisturbed soil and managing conditions in the field. There are also few models for predicting the effect of land use change on the soil water and solute status. Evaluating the impact of land use and management practices on soil water dynamics and solute leaching turns out to be difficult because only a few techniques are capable of monitoring the quantity and quality of soil water flow below the root zone in the field without disturbing the soil profile and affecting natural flow processes. Generally, underground lysimeter techniques are applied for measuring the seepage flow through the soil matrix by intercepting water and then either manually or automatically sampling the accumulated volume of water (Brye et al. 2001; Albright et al. 2004; Masarik et al. 2004; Gee et al. 2009).

Soil hydrology measurements provide an alternative way of analysing the soil water and matter status in situ under undisturbed soil conditions. Compared with lysimeters, these methods are less expensive and more flexible. However, taking measurements at various sites and with many sensors across the whole profile is also expensive and is not a feasible solution. The development of a simpler method requires, in the first instance, investigations into providing knowledge about the processes dynamics of the soil water and solute status under different conditions of soil, land use, climate and agricultural management. Utilising this knowledge for transmitting quantitative results requires, as the next step, the development and application of an effective field method for quantifying deep seepage and solute leaching. Soil hydrology measurements may provide the qualified basis for validating models for transmitting quantitative results within a particular area (Mölders et al. 2003; Miao et al. 2003; Bryant et al. 2006; Köhler et al. 2006; Perkins et al. 2011).

2 Method

2.1 Process Analysis

Tensions and water content time series at a selected profile at the Muencheberg site (Schindler et al. 2010) were used to show the typical behaviour of variables in the field. Muencheberg is located in a Pleistocene region. Sandy and week loamy soils dominate. The Muencheberg site's soil type is classified as Haplic Albeluvisol (Drainic, Arenic) (WRB 2006). A sandy Ap (0–35 cm) and Al Horizon (35–60 cm) is underlain by a Bt Horizon at a depth between 60 and 80 cm, below which is a sandy C Horizon. Figures 1 and 2 demonstrate the tension dynamics at different depths between 60 and 300 cm over a four year period from 2008 to 2012. Tensions show the same behaviour at all depths. However, when depth increased, redistribution processes occurred, tension signals were damped and the hydraulic conditions became more uniform. The variability of tensions decreased markedly and its temporal fluctuations occurred slowly and continually.

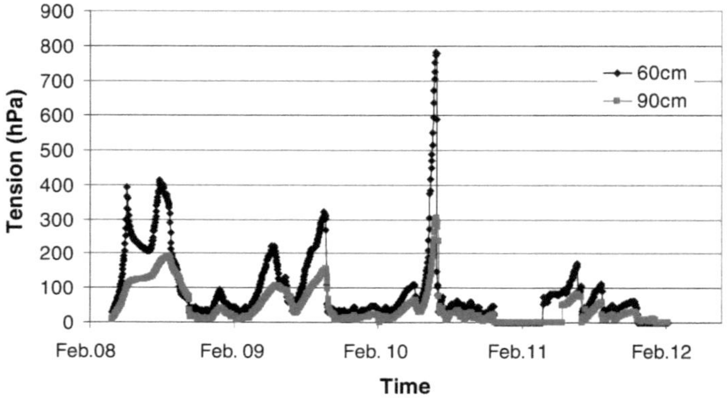

Fig. 1 Tension time series at 60 and 90 cm depths, Haplic Albeluvisol, Muencheberg site

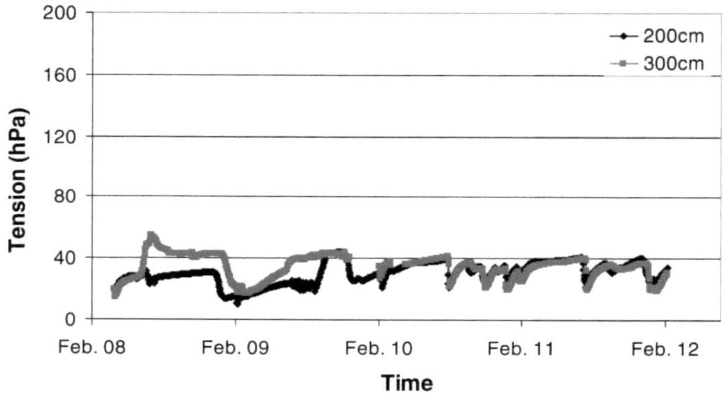

Fig. 2 Tension time series at 200 and 300 cm depth, Haplic Albeluvisol, Muencheberg site

High tensions and positive as well as negative hydraulic gradients show alternating flow conditions at depths down to 90 cm. Soil water conditions at these depths were influenced on the one hand by plant water uptake and capillary rise (positive hydraulic gradients) and on the other by negative hydraulic gradients demonstrating downward water movement. The hydrological conditions at greater depths were completely different. Between 200 and 300 cm negative hydraulic gradients were dominant. The hydraulic gradient varied by around −1 (unit gradient). Water only moved downwards when driven by gravity (Fig. 3).

Time series of water content values confirmed the findings of damping behaviour and lessening variability as depth increased (Figs. 4 and 5). From 120 cm depths, daily water content changes become small. As $d\Theta/dt \cong 0$ was true during the recording intervals of 1d, "steady-state conditions" could be assumed for quantifying soil water fluxes (Schindler and Müller 1998).

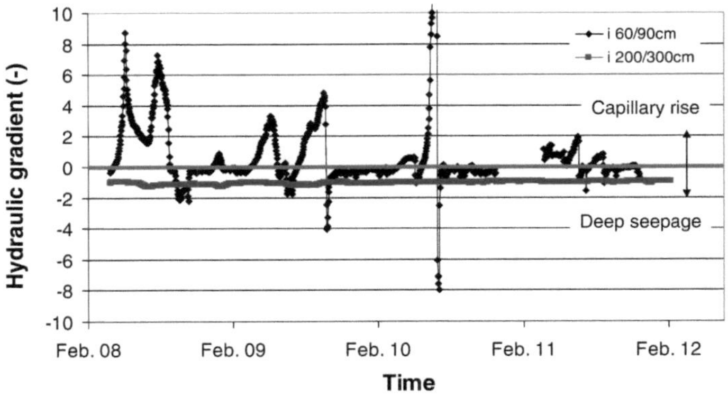

Fig. 3 Hydraulic gradient (1) between 60 and 90 cm and 200 and 300 cm depth at the Muencheberg site

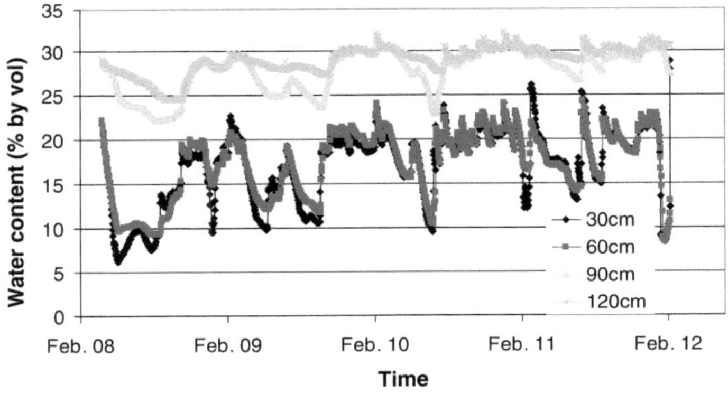

Fig. 4 Water content time series between 30 and 120 cm depths at the Muencheberg site

2.2 Method for Quantifying Deep Seepage from Soil Hydrology Field Measurements (Schindler and Müller 1998)

2.2.1 Concept

Below the zero flux plane, water flow is driven by gravity only. Over time, the soil in that zone can be considered as being a pipe with various fillings. As an indicator of filling levels, the soil water content is taken. All changes of water content or tension are induced by changing seepage conditions. Transformation of the filling level (soil water content) into flux (seepage flow rate v) can be undertaken using the Darcy-Buckingham equation, containing a non-linear scaling factor, where the

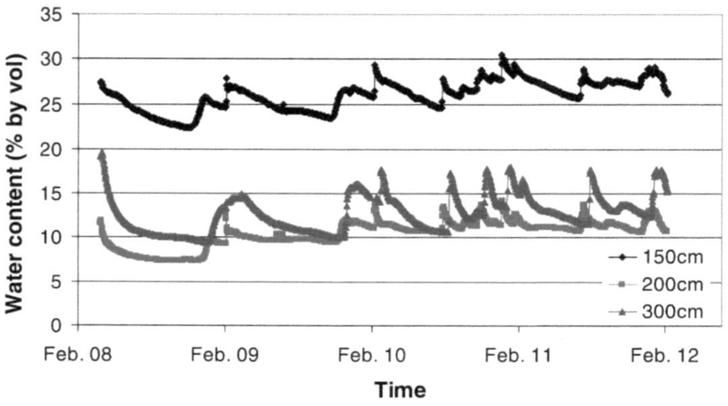

Fig. 5 Water content time series between 150 and 300 cm depths at the Muencheberg site

hydraulic conductivity function (K(Θ)) is dependent on water content (Θ), and the hydraulic gradient (1) is the driving force. "Steady-state" conditions are assumed at daily intervals and the unit hydraulic gradient is considered to be valid (Schindler and Müller 1998). This method is feasible on soils without ground water influence. Daily deep seepage rates (v) are calculated based on water content measurements below the zero flux plane, and an unsaturated hydraulic conductivity function K(Θ) calibrated to the water balance. Further information about soil properties, land management, weather data and other data is not necessary. Prerequisites are high precision and temporally stable tensiometers and TDR probes with no drift or temperature influence and temporally constant flow pathways (Fig. 6).

Fig. 6 Methodological concept

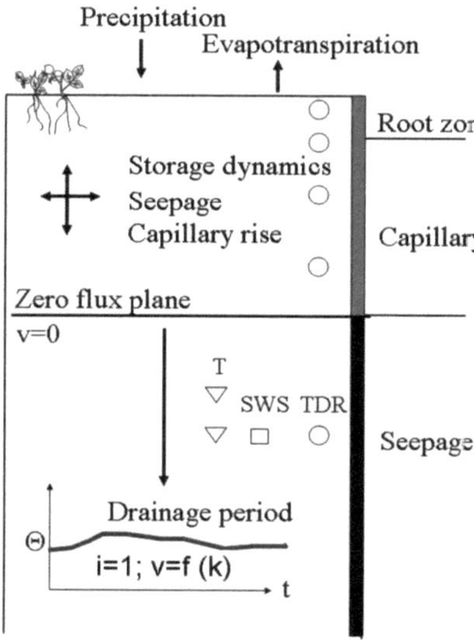

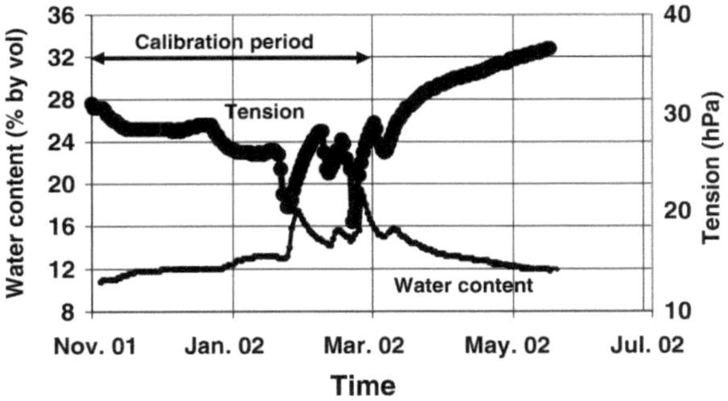

Fig. 7 Recording of tensions and water content values below the zero flux plane

$$v = K(\Theta) * i \tag{1}$$

2.2.2 Procedure

The method consists of two main parts:

(1) The measurement of soil hydrological variables (tension and water content) below the zero flux plane, and the calculation of fictitious fluxes based on water content time series and a relative hydraulic conductivity function.
(2) To provide reliable fluxes, the relative hydraulic conductivity function has to be calibrated to the water balance over a calibration period and then transformed to a reliable level.

Procedure:

Water content and tensions are recorded with a high temporal resolution below the zero flux plane (Fig. 7). The recommended measurement depth at arable sites in a humid climate is 3 m, and at forest sites, 5 m.

In the second step, the van Genuchten model Eq. 2 (van Genuchtens 1980) is fitted to the relationship between tension and water content, providing a field retention function (Fig. 8),

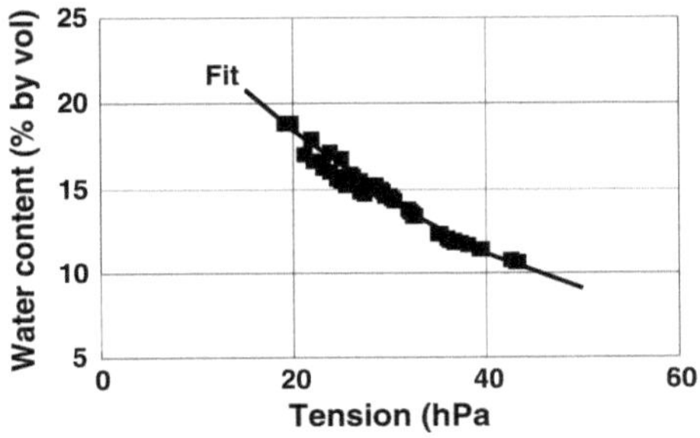

Fig. 8 Fitted water retention function

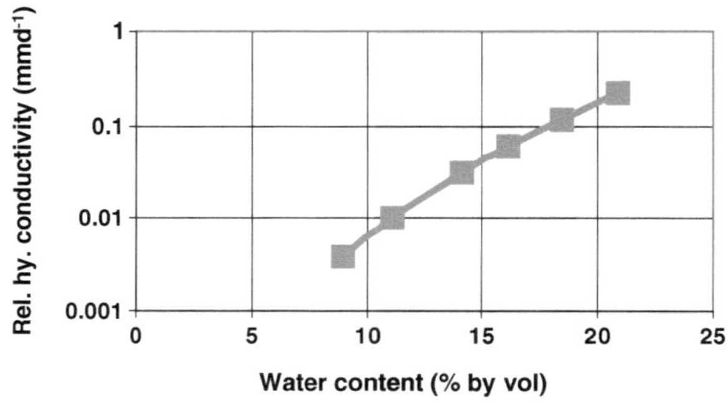

Fig. 9 Relative hydraulic conductivity function K(Θ)

$$\Theta(\Psi) = \Theta_r + \frac{\Theta_s - \Theta_r}{[1 + (\alpha\Psi)^n]^m} \tag{2}$$

with

Θ_s saturated water content

Θ_r residual water content

α, n, m (m = 1-1/n) parameters

and the relative hydraulic conductivity function $K_r(\Psi)$ is predicted Eq. (3) (Mualem 1976).

$$\frac{K(\Psi)}{K_s} = K_r(\Psi) = \frac{\left\{1 - (\alpha\Psi)^{n-1}[1 + (\alpha\Psi)^n]^{-m}\right\}^2}{[1 + (\alpha\Psi)^n]^{\frac{m}{2}}} \tag{3}$$

with

K(Ψ) hydraulic conductivity dependent on tension

K_s saturated hydraulic conductivity

$K_r(\Psi)$ relative hydraulic conductivity

Substituting Ψ in Eq. 3 by Eq. 4 gives the hydraulic conductivity as a function of water content Θ (Fig. 9). The use of K(Θ) minimises hysteresis effects. Any extrapolation of K(Θ) beyond field recordings is not permitted.

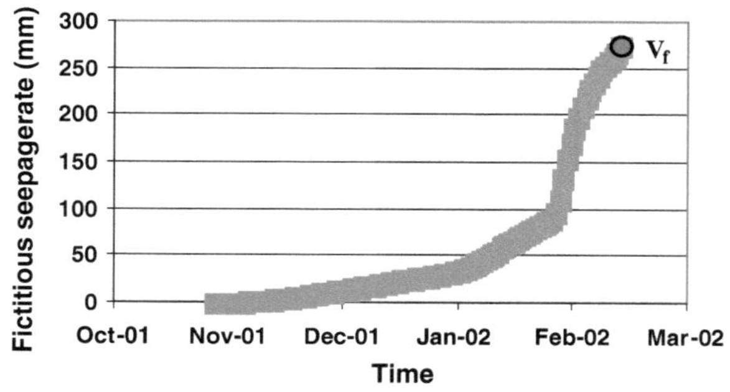

Fig. 10 Fictitious seepage rate

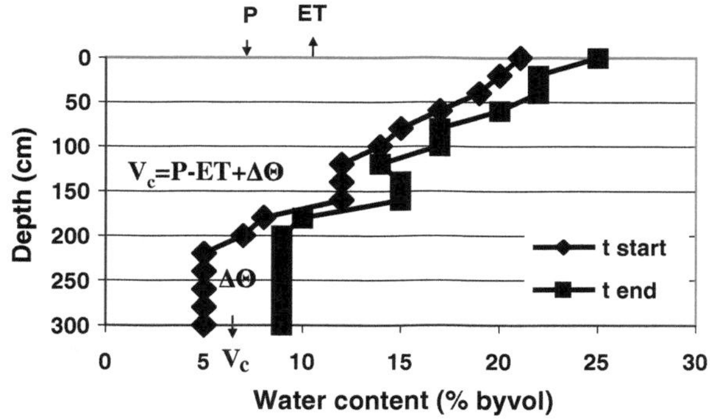

Fig. 11 Determining water balance

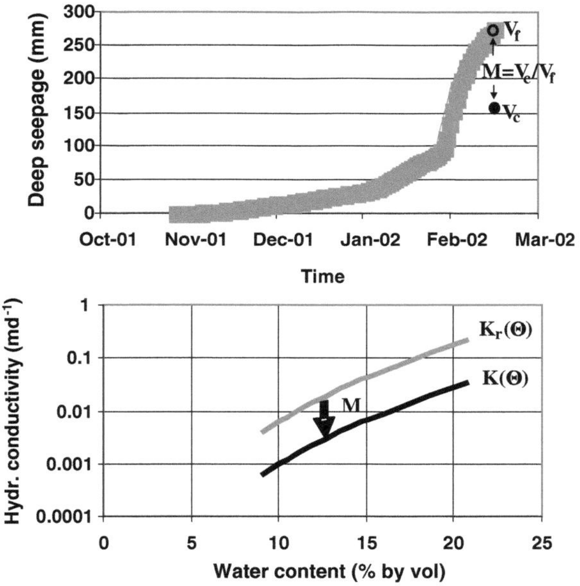

Fig. 12 Matching the K function

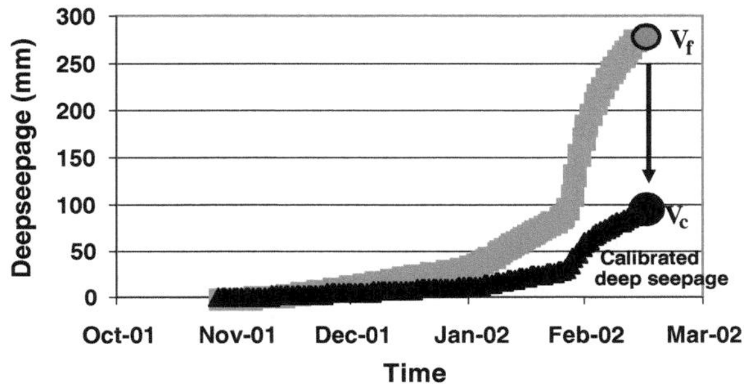

Fig. 13 Transforming fictitious seepage rates to a reliable level

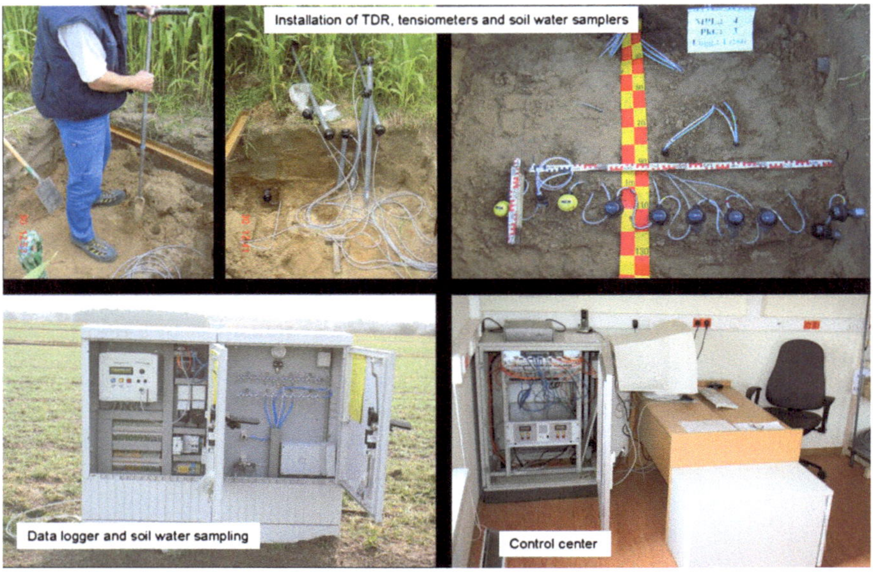

Fig. 14 Installation of TDR probes, tensiometers and soil water samplers below the plough horizon at the Muencheberg experimental site

$$h = f(\Theta) = \left\{ \left[\left(\frac{\Theta_s - \Theta_r}{\Theta - \Theta_r} \right)^{\frac{1}{m}} - 1 \right]^{\frac{1}{n}} \right\}^{-\alpha} \tag{4}$$

In the next step, fictitious seepage rates (v_r) are calculated (Eq. 1) based on the recorded water content values (Θ) and the relative hydraulic conductivity function $K_r(\Theta)$ (Fig. 10).

A unit gradient $i = 1$ is assumed. This assumption has been validated (Schindler and Müller 1998), and prompts only small uncertainties. Inaccuracies in tension measurements, however, can produce larger errors. However, when using the unit gradient tension, uncertainties can be negated for calculating fluxes.

To provide reliable seepage rates (V_c), the relative hydraulic conductivity function has to be converted to a reliable level. First, the water balance (Eq. 5) is determined for a frost-free autumn/winter period (Fig. 11). An autumn/winter period should be chosen because it minimises uncertainties for calculating evapotranspiration. Precipitation (P) should be measured at the site of the experiment. The soil water storage difference ($\Delta\Theta$) is derived from water content measurements in the profile from the beginning and the end of the calibration period. The ratio V_c/V_f reveals M as the matching factor (Eq. 6) for transforming the $Kr(\Theta)$ function to a reliable level ($K(\Theta)$) (Fig. 12). Finally, Eq. 5 is used for calculating reliable daily seepage rates (Fig. 13). The term $f(\Theta)$ in Eq. 7 substitutes Ψ according to Eq. 4.

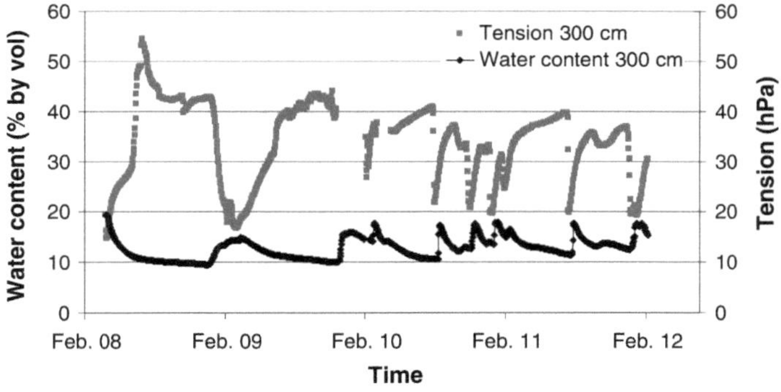

Fig. 15 Time series of water content and tension at a depth of 3 m at the Muencheberg site

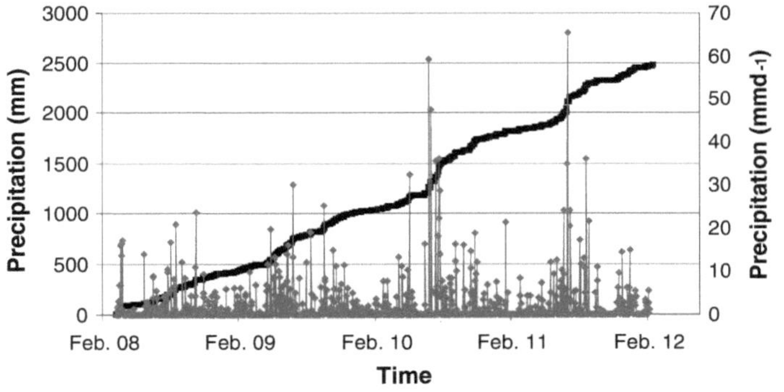

Fig. 16 Precipitation at the Muencheberg site

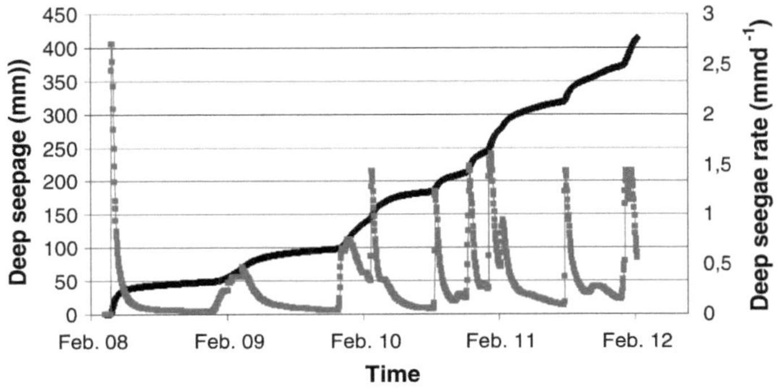

Fig. 17 Deep seepage at the Muencheberg site

$$V_k = P - ET + \Delta\Theta \tag{5}$$

$$M = \frac{V_c}{V_f} \tag{6}$$

$$v = M \cdot K_r(\Theta) = \frac{V_c}{V_f} \frac{\left\{1 - (\alpha \cdot f(\Theta))^{n-1}[1 + (\alpha \cdot f(\Theta))^n]^{-m}\right\}^2}{[1 + (\alpha \cdot f(\Theta))^n]^{\frac{m}{2}}} \tag{7}$$

2.2.3 Instruments and installation

Figure 14 provides an exemplary overview of the installation practice and the instruments used. All devices (tensiometers, TDR probes, soil water samplers) are installed below the plough horizon. Tractors and agricultural machinery are able to move over the plots as normal. The aim of this activity, to collect data, is not corrupted by artificial management effects. All data are recorded by data loggers (4 plots to one data logger) and forwarded to a central computer.

3 Results

The results from the Muencheberg site are shown in the diagram below. Figure 15 illustrates the time series of tension and water content at a depth of 3 m. Both variables show comparable dynamics. Where water content increased, the tension decreased, and the reverse applied. These dynamics correspond with those of precipitation as shown in Fig. 16 for the observation period 2008 to 2012.

A total of about 2,500 mm precipitation with maximum daily events of 65 mm was observed. 2010 (747 mm) and 2011 (626 mm) were wet in both summer and in winter. This observation corresponds with the deep seepage findings. Overall, 414 mm of deep seepage was calculated, with annual rates of: 2008: 49 mm; 2009: 56 mm; 2010: 134 mm; 2011: 131 mm; 2012: 44 mm (Fig. 17).

4 Conclusion

Since 1994, the soil hydrology method for quantifying deep seepage and solute leaching has been successfully applied for various land use and management systems in northeast Germany (Schindler et al. 2010).

Advantages:

- The method is simple to handle, is flexible and is less expensive than lysimeters.
- Measurements of soil hydraulic properties in the laboratory or in the field are not necessary. Soil hydraulic properties are derived from field recordings of tension and water content.

- After calibration, no further information about soil properties, weather, management, land use data or anything else is required.
- If information about precipitation and soil water storage changes, it is possible to estimate evapotranspiration. This procedure is suitable only for longer periods (>1 month).
- The quality of calibration is important for the accuracy of absolute flow rates. The temporal changes and relationships between fluxes, however, are not affected by the calibration and its accuracy.

Limitations and difficulties:

- This method is feasible only at sites with no groundwater influence. Preferential flow cannot be considered. However, at depths of 3 m and deeper, in general, preferential flow is of low importance.
- The demands placed on water content accuracy are high. However, recently manufactured TDR and FDR devices are able to measure water content values to that standard.
- The method is sensitive to water content errors and especially to measurement drift.
- Whenever water content sensors are changed, a re-calibration is required.

The method was found to be valid in comparison with lysimeter discharge measurements (Schindler and Mueller 2004; Schindler et al. 2009).

References

Albright WH, Benson CH, Gee GW, Roesler AC, Abichou T, Apiwantragoon P, Lyles BF, Rock SA (2004) Field water balance of landfill final covers. J Environ Qual 33:2317–2332

Benson VS, van Leeuwen JA, Sanchez J, Dohoo IR, Somers GH (2006) Spatial Analysis of land use impact on ground water nitrate concentrations. J Environ Qual 35:421–432

Bryant RB, Gburek WJ, Veith TL, Hively WD (2006) Perspectives on the potential for hydropedology to improve watershed modeling of phosphorus loss. Geoderma, 131(3–4):299–307

Brye KR, Norman JM, Bundy LG, Gower ST (2001) Nitrogen and carbon leaching in agroecosystems and their role in denitrification potential. J Environ Qual 30:58–70

Edwards WRN (1986) Precision weighing lysimetry for trees, using a simplified tared-balance design. Tree Physiol I:127–144

Gee GW, Newman BD, Green SR, Meissner R, Rupp H, Zhang ZF, Keller JM, Waugh WJ, van der Velde M, Salazar J (2009) Passive wick fluxmeters: design considerations and field applications. Water Resour Res 45:W04420

Goss MJ, Ehlers W (2009) The role of lysimeters in the development of our understanding for soil water and nutrient dynamics in ecosystems. Soil Use Manag 25:213–223

Haferkorn U, Winkler A, Bernsdorf S (2011) Lysimetermessungen und pflanzenbauliches Monitoring zum Nachweis der bodenverbessernden Wirkung von Klärschlammkompost am Beispiel der Braunkohlekippe Witznitz II.: 14. Gumpensteiner Lysimetertagung. 103–110

Köhler K, Duijnisveld WHM, Böttcher J (2006) Nitrogen fertilization and nitrate leaching into groundwater on arable sandy soils. J Plant Nutr Soil Sci 169:185–195

Masarik KC, Norman JM, Brye KR, Baker JM (2004) Improvements to measuring water flux in the vadose zone. J Environ Qual 33:1152–1158
Meissner R, Rupp H, Seeger J, Ollesch G, Gee GW (2010) A comparison of water flux measurements: passive wick-samplers versus drainage lysimeters. European J Soil Sci, 61(4):609–621
Miao Z, Cheplick MJ, Williams MW, Trevisan M, Padovani L, Gennari M, Ferrero A, Vidotto F, Capri E (2003) Simulating pesticide leaching and runoff in rice paddies with the RICEWQ–VADOFT Model. J Environ Qual 32:2189–2199
Mölders N, Haferkorn U, Döring J, Kramm G (2003) Long-term investigations on the water budget quantities predicted by the hydro-thermodynamic soil vegetation scheme (HTSVS)—Part I: Description of the model and impact of long-wave radiation, roots, snow, and soil frost. Meteorol Atmos Phys 84(1/2):115–135
Mualem Y (1976) A new model for predicting the hydraulic conductivity of unsaturated porous media. Water Resour Res 12:513–522
Perkins KS, Nimmo JR, Rose CE, Coupe RH (2011) Field tracer investigation of unsaturated zone flow paths and mechanisms in agricultural soils of northwestern Mississippi, USA. J Hydrol 396(1–2):1–11
Prasuhn V, Spiess E, Humphrys C (2011) Sickerwasser und Nitratauswaschung—erste Ergebnisse der neuen Lysimeteranlage Zürich-Reckenholz. Proceedings 14. Gumpensteiner Lysimetertagung 2011, 207–210
Puetz Th, Kiese R, Zacharias ST, Bogena H, Priesack U, Wollschlaeger U, Schwank M, Papen H, von Unold G, Vereecken H (2011) TERENO-SOILCan—Ein Lysimeter Netzwerk in Deutschland. In: Proceedings: 14. Gumpensteiner Lysimetertagung, 5–10
Schindler U, Mueller L (2004) Comparison of deep seepage estimations of a virtual with a real lysimeter by means of TDR-measurements. Int Agrophysics Lublin 19(1):69–73
Schindler U, Müller L (1998) Calculating deep seepage from water content and tension measurements in the vadose zone at sandy and loamy soils in north-east Germany. Arch Agron Soil Sci 43:233–243
Schindler U, Wolff M, Kuehn G (2001) Lysimeterstudie zum Einfluss von Düngung und Bewirtschaftung auf die Ertragsbildung, den Wasserhaushalt und die Nährstoffauswaschung im Trockengebiet der Uckermark. J Plant Nutrition Soil Sci 169:697–703
Schindler U, Verch G, Wolff M, Müller L (2008) Bedeutung der unteren Randbedingung in Lysimetern für die Sickerwassermessung. Proceedings: Seminar "Wasserkreislaufparameter" Schloss Seggau. http://www.lysimeter.com/seminar2008/Seminarband_Seggau2008_web.pdf
Schindler U, Fank J, Müller L (2009) Alternativlösung zur Quantifizierung der Tiefensickerung in situ.—In: Bericht/13. Gumpensteiner Lysimetertagung : am 21. und 22. April 2009 am LFZ Raumberg-Gumpenstein: 93–98; Irdning (Lehr- und Forschungszentrum für Landwirtschaft Raumberg-Gumpenstein)
Schindler U, Müller L, Dannowski R, Barkusky D, Francis G (2010) Long-term measurements to quantify the impact of arable management practices on deep seepage and nitrate leaching. In: Long-term ecological research : between theory and application: 243–252; Dordrecht (Springer Science+Business Media B.V)
van Genuchten MTh (1980) A closed-form equation for predicting the hydraulic conductivity of unsaturated soils. Soil Sci Soc Am J 44:892–898
WRB World Reference Base for Soil Resources (2006) A Framework for International Classification, Correlation and Communication, FAO Rome, 2006, World Soil Resources Reports 103, 145 p

Simple Field Methods for Measurement and Evaluation of Grassland Quality

Lothar Mueller, Axel Behrendt, T. Graham Shepherd, Uwe Schindler, Bruce C. Ball, Sergey Khudyaev, Thomas Kaiser, Ralf Dannowski and Frank Eulenstein

Abstract Grasslands and rangelands are very important ecosystems influencing natural cycles and human existence and well-being. Their functional status can be greatly affected by soil and water management. Grasslands are prone to degradation, but comprehensive frameworks and objective criteria for their monitoring are largely absent. Simple field methods of measurement and visual rating may help to detect properties and processes limiting the function of grasslands, and the results used to develop criteria and thresholds of soil and vegetation quality. Methods characterising aspects of the physical, chemical and biological status of grasslands in conjunction with soil survey data are presented here. Soil strength measured with penetrometers, sink cones and shear testers may characterise spatio-temporal alterations of soil resistance conditions best. Important attributes of the soil water status can be measured by TDR probes, field tensiometers and simplified infiltration equipment. Experience and care are necessary when interpreting field

L. Mueller (✉) · U. Schindler · T. Kaiser · R. Dannowski · F. Eulenstein
Leibniz-Centre for Agricultural Landscape Research (ZALF) e. V., Eberswalder Str. 84
15374 Muencheberg, Germany
e-mail: mueller@zalf.de

A. Behrendt
Leibniz Centre for Agricultural Landscape Research (ZALF) e. V., Research Station Paulinenaue, Gutshof 7 14641 Paulinenaue, Germany
e-mail: abehrendt@zalf.de

T. G. Shepherd
BioAgriNomics Ltd, 6 Parata Street, Palmerston North 4410, New Zealand
e-mail: gshepherd@BioAgriNomics.com

B. C. Ball
Crop and Soil Systems Research Group, SRUC, West Mains Road, Edinburgh EH9 3JG, UK
e-mail: Bruce.Ball@sruc.ac.uk

S. Khudyaev
Russian Academy of Sciences, Siberian Branch, Institute of Soil Science and Agrochemistry (ISSA), Ac. Lavrentieva av., 8, Novosibirsk 630090, Russian Federation
e-mail: xca_nsk@mail.ru

L. Mueller et al. (eds.), *Novel Measurement and Assessment Tools for Monitoring and Management of Land and Water Resources in Agricultural Landscapes of Central Asia*, Environmental Science and Engineering, DOI: 10.1007/978-3-319-01017-5_11,

measurement data. A number of expert-based approaches for estimating site properties by visual-tactile methods and by bio-indicative vegetation analyses are available and should be utilized more. Some of these methods can be applied solely to study particular aspects of grassland quality like trampling effects under different animals, machinery and grazing systems or trafficability of meadows and measure the compaction status of soil, the quality of soil structure and pasture quality. In many cases a set of methods in combination with an overall assessment of soil quality (Visual Soil Assessment, VSA, and Muencheberg Soil Quality Rating, M-SQR) will provide a reliable assessment of the status of grasslands and rangelands. Methods presented here should be considered and proposed to be used as possible standard components of frameworks for assessing the functional status of grasslands by uniform methodologies over Eurasia.

Keywords Grassland ecosystems · Soil structure · Soil quality · Vegetation · Indicators · Muencheberg Soil Quality Rating

1 Introduction

Grasslands produce forage for livestock. Additionally, grasslands and rangelands provide numerous ecosystem services beneficial for humans. They play a major role for the maintenance of global biodiversity, carbon sequestration, water cycling and other functions. Improperly managed, they can be a source of significant greenhouse gas emissions (Ball et al. 2012). Pasture farming is efficient at capturing solar energy and is likely to become more important in future as fertiliser becomes more expensive and scarce e.g. Salatin (2010). Central Asia is a grassland region that differs from grasslands in temperate zones by climate, soils, vegetation and management. They are more prone to degradation and currently are already largely degraded (Saparov 2013). Overgrazing and intensive trampling are common reasons for grassland damage by erosion (Zhou et al. 2010) or initiation of degradation. Recovery of soil structure and site-specific vegetation are then difficult or impossible. Grassland monitoring is an urgent need.

Soils under grassland can be greatly affected by soil and water management. Whilst the conditions of grasslands differ between Europe and Asia, measurement and assessment methods could be based on similar, uniform frameworks and principles of measurement and assessment. Simple field methods may help to detect properties and processes limiting grassland use and functions. Assessment of overall soil quality for grassland can be done by common soil survey methods, visual methods of soil evaluation and information about climate (Shepherd 2009; Mueller et al. 2013). The framework of the Muencheberg Soil Quality Rating (M-SQR) offers some options to include more detailed information about physical, chemical and biological soil properties into rating tables and procedures. Simple but reliable field methods may provide this. This chapter gives information about simple field methods that have been successfully developed and/or applied in studies on grasslands of the temperate zone.

We have developed and tested methods characterising aspects of the physical, chemical and biological status of grasslands in conjunction with soil survey data. Some of these methods can be applied solely to study particular aspects of grassland quality like trampling effects under different animals and grazing systems or trafficability of meadows, thereby providing information on the compaction status of soil, the quality of soil structure and vegetation. In many cases a set of methods in combination with an overall assessment of soil quality will be useful for a reliable assessment of the status of soil in spatio-temporal studies.

Our hypothesis is that the majority of our methods has potential for applications in studies on grasslands in Central Asia and worldwide. Those studies could be useful for a better understanding and monitoring of degradation processes and managing soils under grasslands including rangelands more sustainably.

2 Measurement Methods Characterising the Soil Physical Status

2.1 Soil Strength

Any use and management of soil requires a minimum strength of ground and sward. Soil strength is a property of a soil to withstand external forces. It can be easily measured in the field by penetrometers, sink cones, shear testers or other devices. Figure 1 shows examples of devices used on grasslands of North–East Germany.

Fig. 1 Devices for measuring soil strength. **a** Penetrometer. **b** Sink cone. **c** Vane shear tester

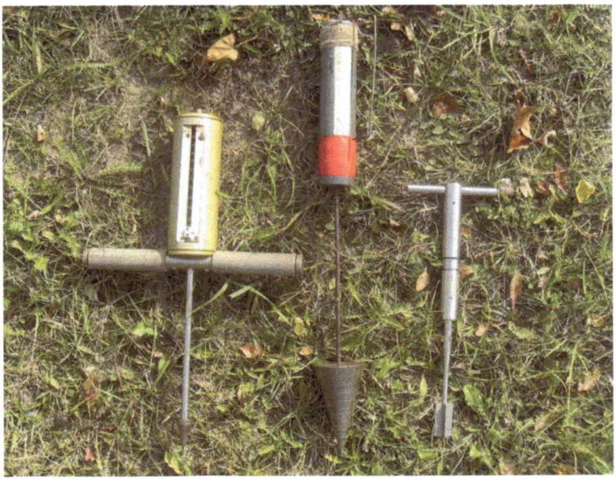

Their measurement values provide fast and reliable information about relative soil strengths across and down soil profiles and soil surfaces. Typical reasons for measuring soil strength are to assess soil compaction by machinery, livestock or natural processes. Theory and principles of soil strength measurements including the application of penetrometers and shear testers are contained in textbooks in soil mechanics like (Schultze and Muhs 1967; Kezdi 1969; Terzaghi et al. 1996). Penetrometers are very common for measuring soil strength properties (Hartge and Bachmann 2004) at depths of about 0.1–1 m. It needs to be borne in mind that the size and shape of the penetrometer cone or cone cylinder affects absolute measured values inconstantly over a range of soils (Mittelstedt and Mueller 1989). Thus, information about cone properties has to be given for all measurements, and standardised devices should be applied.

In our studies on relatively wet soils we measured penetration resistance by hand held penetrometers having a 12–16 mm diameter tip. On drier soils a 10 mm tip or smaller is better. An important factor in penetrometer use is to avoid friction at the penetrometer shaft by employing a tip diameter 3-8 mm greater than the shaft diameter. Vertically operating penetrometers are commonly used (Fig. 2).

The application of penetrometers for characterising soil surface strength is sub-optimal. Measuring the sinking depth of a cone of defined weight and dimensions and calculating the cone resistance is more relevant to soil surface processes like stability against hooves of animals, sinking of tires or stability of crusts. We successfully used a large 30° sink cone combined with a penetrometer for characterising aspects of soil trafficability (Mueller et al. 1990, 2004). On grasslands, this is a good indicator of soil stability against livestock trampling. The combination of the cone with a penetrometer-like device is useful for the application of a defined load to the soil.

Vane shear tests are another option to characterise soil strength. Kraschinski et al. (2001) used them for testing the stability of very soft grassland soils. They detected the significant effects of plant root systems for increasing the resistance of

Fig. 2 Vertical penetrometer for testing soil strength on a red-deer pasture in the research station Paulinenaue, Germany

Table 1 Field measurement methods of soil strength

Method	Principle	Preferred unit	Reference*
Penetrometer resistance	Pushing a relatively small cone or cone cylinder through the soil	MPa	Mittelstedt and Mueller (1989)
Cone resistance (sink or drop cone)	Pressing a large cone on the soil surface, measuring the sinking depth	MPa	Mueller et al. (1990)
Vane shear test	Shearing a defined soil volume by turning a shear vane, reading maximum value of torque at soil disruption	MPa	Mueller et al. (2007c)

* Reference refers to tests by authors, not to the first description in literature

peatlands to agricultural traffic. We applied the Eijkelkamp light vane shear tester (Fig. 1) of vane size 16 mm diameter × 42 mm length on cropland and grassland. Vane shear testers are preferably applied on soils of low strength like in peatlands, on wet soils or on tilled soils. Under drier or stony conditions problems with damage to the shear vane may occur.

Soil strength is closely correlated with both density and moisture status of soil. Neglecting the latter aspect may lead to wrong conclusions on limiting factors. On sandy soils at field capacity, soil strength data provides good information about aspects of soil structure. On soils rich in clay, penetrometer resistance and other soil strength parameters do not provide information about soil structure (Mueller et al. 2013). The grassland sward consists of plants and soils. Their strength is largely influenced by the mechanical stability of living plants. This is a reason for the application of methods of Table 1, cone resistance and vane shear tests in particular. Ball and O'Sullivan (1982) compared cone resistance and vane shear strength with bulk density and produced limiting values for restricting plant emergence.

Though all measurements of Table 1 result in metric values of a pressure unit like MPa, the values may be very different due to the differing specific procedures. Data values obtained by all three methods are normally distributed over a number of measurements, and the number of replications at a single point can thus be relatively low. About 4–8 replications may provide reliable averages if classical statistical methods are being used.

2.2 Soil Moisture and Density Status

Water content or moisture content is the quantity of water in a soil, expressed as a ratio or percentage relative to dry soil. It can be given on a mass (gravimetric) or volumetric basis. For many practical purposes in soil hydrology, e.g. water flow through soils, the volumetric basis is of interest. For other purposes, for example, engineering behaviour of soil, the mass basis is more common.

Dry bulk density is another important soil physical parameter, related to the density or compaction status of soil. The simple formula

$$w_{vol} = w_{grav} * DBD, \tag{1}$$

combines all three parameters. Where w_{vol} is the volumetric water content in m^3/100 m^3, DBD is the dry bulk density in t/m^3 and w_{grav} is the gravimetric water content in t/100 t.

Knowing the water retention curve of a soil, measured water contents can be related to the energy status of soil (e.g. the soil water content at given pressure heads or suctions), to the pore size distribution and plant ecological states of soil like water excess by lack of air (anaerobism) or drought stress by lack of soil water (Kutilek and Nielsen 1994; Schindler et al. 2010).

Volumetric water content can be measured simply by methods of time-domain reflectrometry (TDR). Hand-held field probes developed by the Institute of Agrophysics in Lublin, Poland, are examples of reliable instruments Easy Test (2012, Fig. 3).

A complication in spatio-temporal field studies is the non-availability of fast field methods for measuring DBD and gravimetric water content. This would require oven-drying of samples in a laboratory. To overcome this problem, a hand-held TDR-probe and measuring the wet bulk density (WBD) by a standard cylinder of calibrated volume and known mass, and an electronic balance can be applied. WBD is the net mass of a wet volumetric sample (wf) in gram divided by the soil volume (V) of this cylinder in cm^3.

$$WBD = wf/V \tag{2}$$

The calculation of the dry bulk density can be done by the formula

$$DBD = WBD - w_{vol}/100 \tag{3}$$

Fig. 3 Field tensiometer (Tensio 100 of UGT Muencheberg) and field TDR probe (Easy—Test, IPAN Lublin) may provide information rapidly on the soil water status

DBD is the dry bulk density in g/cm^3 (equals t/m^3), WBD is the wet bulk density (g/cm^3) and w_{vol} is the volumetric water content (g/100 g) measured by the TDR probe.

The gravimetric water content can then be calculated by using (1). This method has some potential for errors, thus well- calibrated TDR-probes and sampling cylinders no smaller than 250 cm^3 at four replications are required. An advantage of this procedure is the estimation of DBD and gravimetric moisture content in the field without laboratory equipment.

Hand-held field tensiometers may provide measurements of the energy status of the soil water, at least in a relatively moist range of about −0.01—−700 hPa suction. Measured values of suction may be related to field capacity of soils. Field capacity of soils ranges between about −60 hPa in sandy soils to −300 hPa in clayey soils. Measured values greater than −60 hPa or even positive values indicate wetness, and values outside the measurement range (less than −700 hPa) indicate drought tendencies. However, exact thresholds depend on the water retention curve and the soil depth under consideration. Modern tensiometers may provide a broader range of measurements of suction towards drier conditions (Schindler et al. 2010), but currently these high-tech devices are not yet applicable as portable field devices. Application of tensiometers in groundwater influenced soils provides information about hydrological states and processes (Fig. 4). It is particular important to know that such sites have very limited gravitational drainage. At a water table of 60 cm as shown in Fig. 4, there is no air in the subsoil of clay soils, and hydraulic gradients for seepage and salt leaching are also too low.

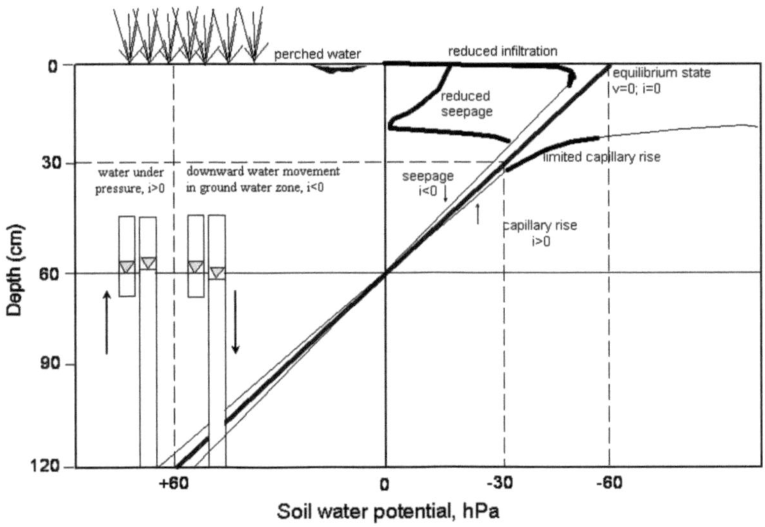

Fig. 4 Soil water potentials as indicators of hydrological processes in soils with a shallow water table (Schindler et al. 2003)

2.3 Water Infiltration

Water infiltration is an extensively studied property of soil. There is much literature about steady-state infiltration capacity of soils e.g. Jarvis et al. 2008; Moret-Fernández and González-Cebollada 2009. Sophisticated devices like the Guelph infiltrometer or disc infiltrometers provide site-specific values for soil characterisation and modelling of hydrological processes.

For more practical purposes of grassland and cropland evaluation, it is important to know the infiltration properties at different locations both between and within fields at approximately the same time, for example, after a heavy rain or snowmelt. For those spatio-temporal studies, the application of steady-state infiltration methods or the determination of a final, constant infiltration rate are not possible and do not adequately reflect natural processes. The use of rainfall simulators (Dimanche and Hoogmoed 2002; Kato et al. 2009) comes closest to real processes of infiltration during a rain but is relatively expensive in use.

We applied two simpler and faster methods for getting information on the initial infiltration rate by ponding infiltration at different field points on the same day. The initial infiltration rate was measured with simple infiltrometers. Standard thin core cylinders (common DBD-cylinders of 250 or 100 cm^3) were pushed 30–45 mm vertically into the soil (Fig. 5). The rings were then filled with 30 mm water. Differences in water levels and elapsed time were measured with a ruler and a stopwatch. As infiltration measurement values are not normally-distributed, more than six replications are necessary when comparing means with statistical methods.

Another simple ponded infiltration method can be applied to evaluate the infiltration of water beneath the soil surface (Fig. 6). This is useful in combination with an assessment of vertical biogenic pores. No infiltration rings are required. The procedure is applicable in cohesive soils. Lower parts of the sidewalls of the soil pit are sealed by soil of soft consistency to prevent lateral water flow off. The size of these pits should be about 0.1–0.2 m^2. We found clear relationships between infiltration rates, water table depth and area of biogenic macropores.

Fig. 5 Water infiltration on cattle trampling pathways as compared with adjacent parts of the paddock

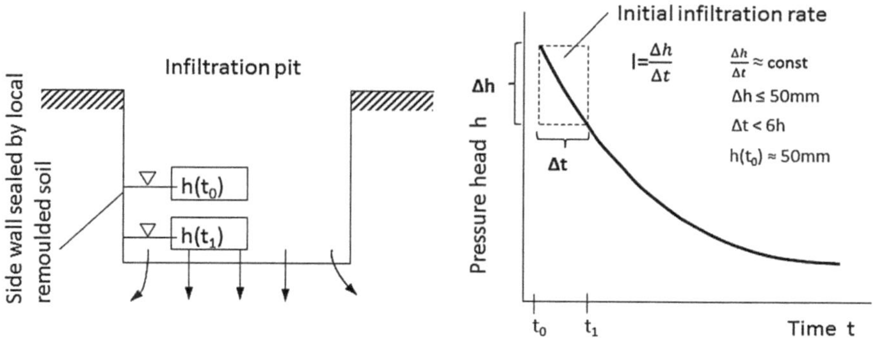

Fig. 6 Soil pit-infiltration for measuring the subsoil infiltration capacity of cohesive soils (Mueller 1988)

Table 2 Field measurement of water infiltration

Method	Principle	Unit	Reference
Initial infiltration rate at soil surface	Ponded infiltration of about 30 mm through the soil surface, sidewalls single rings (common DBD cylinders)	m/d	Mueller et al. (2009)
Infiltration rate of subsoil ("Soil pit infiltration")	Ponded infiltration of about 30-50 mm water through a soil layer beneath the surface, sidewall of sealed soil	m/d	Mueller (1988)

2.4 Detecting Soil Layering and Sampling of Soils

Root limiting layers within the upper 1.5 m of croplands and 0.8 m of grasslands may affect soil functions and crop yield potentials adversely. Those layers may be detected and evaluated by digging, augering or penetration resistance of thin steel probes. Layers within the topsoil (0–30 cm) can be detected using the Visual Evaluation of Soil Structure (Ball et al. 2007). All these methods are useful for ground-truthing of non-destructive indirect measurements for exploring the structure of soil landscapes. Rill probes of the Pürckhauer type are very common. Amongst field augers, the Edelman auger 7 cm (Fig. 7) (product of Eijkelkamp company) may be used as a standard device for detecting soil layering or a shallow water table and for any kind of soil survey. A shallow water table is a very common barrier to rooting due to anoxic conditions for valuable grassland plants. We also used the Edelman auger for soil sampling and analysing of nitrogen in soil profiles down to 5 m depth (Eulenstein et al. 2003). The third method for detecting inhomogeneities in soils, steel probe penetration, is very effective in some lowland grasslands. This method can be applied in spatio-temporal studies of grasslands on

Fig. 7 The Edelman field auger is best suited for detecting root limiting layers and for most kinds of soil survey and sampling on all soils. It can also be suitable for the installation of temporary groundwater installation tubes in lowland soils

peat soils and river lowland soils underlain by sand or gravel (Mueller et al. 2007b). Using well-constructed, thin steel probes, a physically fit person is able to detect the minèral base of peatlands down to 16 m and the thickness of lowland Holocene clay layers down to 7 m. In combination with a leveling device, those measurements permit rapid reconstruction of pre-holocene landscape structures in the field scale.

For sampling peat soils and sediments in semi-aquatic areas, the Dutch sampler, the Wardenaar sampler (Fig. 8), the Vrijwit auger and others are practicable, but every different devices suit different specific conditions. They allow taking of semi-disturbed samples. Cylinder core samples for DBD can be taken from big soil samples of the Wardenaar sampler.

Fig. 8 Dutch sampler and Wardenaar sampler in use on peatlands and under semi-aquatic conditions. Both samplers are commercial products of the Eijkelkamp company, Giesbeek, The Netherlands

3 Visual Soil Structure Assessment Methods

Visual soil structure assessment methods may deliver much information on soil properties in the field relevant to plant growth with a minimum of equipment. Procedures provide information on the feature and function of soils from evaluation of macro-morphological characteristics of the soil structure. The procedure includes generally (1) primary recognition and description of soil structure features (2) classification, evaluation and parameterisation of visual soil structure, and sometimes (3) conclusions on the functional status of soil (Mueller 2011). Type and size of aggregates and number and size of biogenic pores are reliable criteria of assessment. Soil structural features meet the farmer's perception on soil quality (Shepherd 2000) and are correlated with measured data of physical soil quality and crop yield (Mueller et al. 2009). Over the past decades, several methods have been evolved. Most of them differ in several important ways including depth of the soil under consideration, handling the soil prior to assessment, emphasis placed on particular features of soil structure, and application of size, increments and direction of scoring scales. One of the most accepted methods is that of Peerlkamp, cited in Ball et al. (2007). It has a conjoint scale referring to type and size of aggregates and pores. The main advantages of this method are speed and minor soil disturbance, providing comparative statistical analyses both in large fields and also in small plots of long-term trials. However, the scoring scheme has potential for subjective errors. Illustrated methods like the updated Peerlkamp method, called VESS (Ball et al. 2007, Table 3, Fig. 9) and the Visual Soil Assessment (Shepherd 2009, Figs. 10 and 11) leading to ordinally scaled scores are particularly well and reliable in handling. Unfavourable visual structure scores were associated with increased dry bulk density, higher soil strength and lower infiltration rate but correlations were site-specific (Mueller et al. 2009). Visual soil structure assessment is a feasible tool for structure monitoring and management recommendations. Overall soil quality rating schemes like the Muencheberg Soil Quality Rating (Mueller et al. 2007d) include visual soil structure indicators. Techniques such as VESS also allows for the identification and assessment of limiting layers in the topsoil and may guide depth of sampling for core measurements of soil physical properties (Ball et al. 2007).

Table 3 Practicable methods of visual soil structure assessment

Method	Principle	Reference
Visual Soil Assessment (VSA)	Digging a small soil pit, taking a cube of soil and dropping it and assessing aggregates, pores, colour and smell of soil, worms and other parameters by using a manual, calculation, classification and evaluation of a soil score and a plant score	Shepherd (2000), (2009)
Visual examination of soil structure (VESS)	Digging a small pit, taking a spadeful soil, assessing shape of aggregates, pores and rooting by an illustrated conjoint scale	Ball et al. (2007)

Structure quality	Ease of break up (moist soil)	Size and appearance of aggregates	Visible porosity	Roots	Appearance after break-up: various soils	Appearance after break-up: same soil different tillage	Distinguishing feature
Sq1 Friable (tends to fall off the spade)	Aggregates readily crumble with fingers	Mostly < 6 mm after crumbling	Highly porous	Roots throughout the soil			Fine aggregates
Sq2 Intact (most is retained on the spade)	Aggregates easy to break with one hand	A mixture of porous, rounded aggregates from 2mm - 7 cm. No clods present	Most aggregates are porous	Roots throughout the soil			High aggregate porosity
Sq3 Firm	Most aggregates break with one hand	A mixture of porous aggregates from 2mm -10 cm; less than 30% are <1 cm. Some angular, non-porous aggregates (clods) may be present	Macropores and cracks present. Some porosity within aggregates shown as pores or roots.	Most roots are around aggregates			Low aggregate porosity
Sq4 Compact	Requires considerable effort to break aggregates with one hand	Mostly large > 10 cm and sub-angular non-porous; horizontal/platy also possible; less than 30% are <7 cm	Few macropores and cracks	All roots are clustered in macropores and around aggregates			Distinct macropores
Sq5 Very compact	Difficult	Mostly large > 10 cm, very few < 7 cm, angular and non-porous	Very low; macropores may be present; may contain anaerobic zones	Few, if any, restricted to cracks			Grey-blue colour

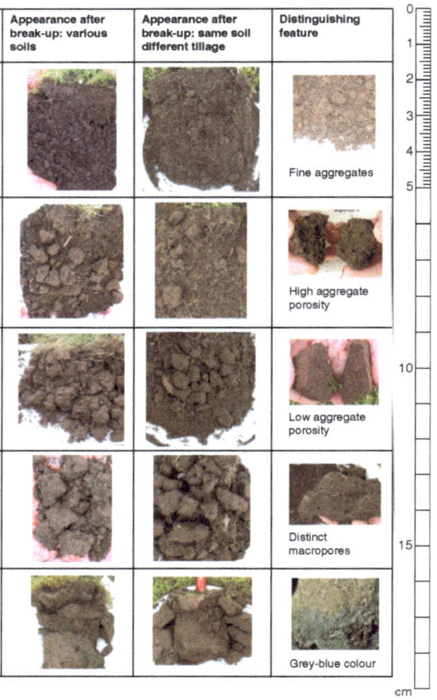

Fig. 9 Assessment scale of visual soil structure by the VESS procedure (Ball et al. 2007). Photo: Bruce C. Ball

Fig. 10 Visual scoring of soil porosity under pasture (Shepherd 2009, p. 18)

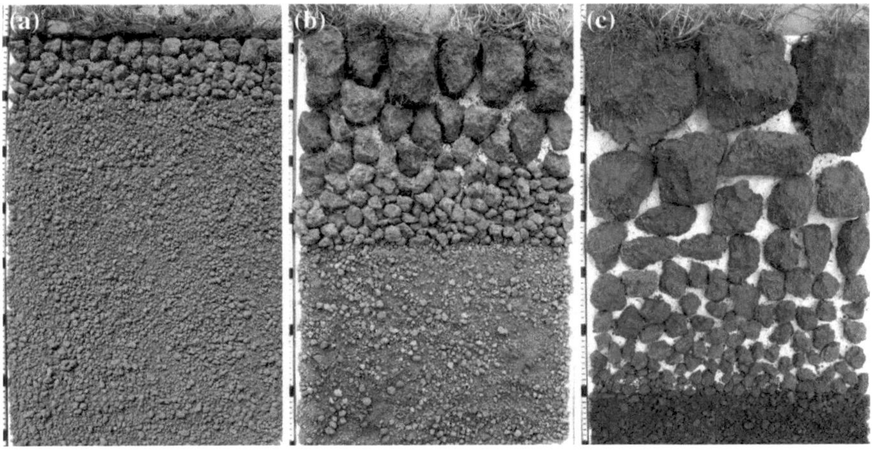

Fig. 11 Soil structure scores of VSA (Shepherd 2009). Aggregates were sorted after a drop-shatter test. **a** Good conditions, score = 2. **b** Moderate conditions, score = 1. **c** Poor conditions, score = 0. Photos: T. Graham Shepherd 2009

4 Methods Characterising Chemical Properties of Soil and Water

Lime content of soil, soil reaction (pH) and electrical conductivity are crucial properties of agricultural sites. Their estimation is part of routine soil survey and is essential for any taxonomic and functional classification of soils. Simple field test methods exist and should be part of monitoring programs to assess the status of grassland soils. Lime content can be estimated by dropping 0.1 n hydrochloric acid on a sample. The degree of effervescence is a measure of lime content and can be assessed by a scale (Boden 2005).

Electrical conductivity and pH should be measured both in soils and adjoining ground and surface waters. This is relevant to many grasslands because they are often adjacent to ponds, rivers or other water bodies or are completely located in lowlands and wetlands. Grassland is the most common land use of wetlands in Eurasia. Their possible use and land quality depends largely on their inundation or groundwater regime. Water table and salinity monitoring are important preconditions for land use optimisation and for soil and water management (Fig. 12).

For measurements of shallow water table height and water quality, a borehole is drilled down to the water table. Conductivity and pH are important parameters of salinity and sodicity classifications of soil and water (Withers et al. 1978; Mueller et al. 2007d).

Besides the ground water level, crucial properties of the soil and ground-water like pH and electric conductivity can be measured within auger holes at each sampling point by simple test kits and instruments. Those test kits are currently

Fig. 12 Soil structure deterioration and salinisation due to stock trampling. Small ponds or wetlands are prone to salinisation even in sub-humid climates. High stocking densities and livestock trampling contribute to structure deterioration and salinisation. Puddling effects decrease soil structure. Urine of cattle contributes to increasing salt concentrations. Both effects lead to adverse living conditions for earthworms and deep rooting plants, which restrict the creation of biological topsoil draining biogene macropores

accepted as reliable and effective screening tools for point-scale assessment of soil quality by providing accurate and precise data over a range of soil conditions (Liebig et al. 1996; USDA/NRCS 2001).

We used a combined pocket meter MultiLine P3 pH/LF-SET in most of our spatio-temporal grassland studies, though separate instruments for pH and conductivity can also be used (Fig. 13). pH can be measured directly by dunking the tip of the measurement electrode into the water. If the soil is too dry, a saturation extract can be created by putting the soil into a small vessel, adding de-ionised water and mixing a saturation extract. For getting reliable readings, it is important to calibrate the probes before daily use. In general, even in the driest region of Germany in the vicinity of Berlin, electric conductivity of soil and water was lower than 1 ms/cm. In some peatlands or organic clays, the common threshold of salinity (2 ms/cm) is already being exceeded locally (Mueller et al. 2007a). Measured values of topsoil electrical conductivity and pH should be assessed by evaluation of scales of salinity, sodicity or acidification. Values given in Tables 4 and 5 are crop-yield relevant and indicate reduced overall grassland soil quality which can be confirmed by performing the Muencheberg Soil Quality Rating.

The procedure for performing of the Muencheberg Soil Quality Rating will be explained in another chapter of this book.

Fig. 13 Measuring water salinity and pH by pocket devices

Table 4 Classification of measured electrical conductivity of topsoil and consequences for the Muencheberg Soil Quality Rating (Mueller et al. 2007d, mod.)

Electrical conductivity EC (mS/cm)[a]	Degree of salinisation	Score	Multiplier grassland[b]
<2	None	2	3
2–4	Low	1.5	2.6–s3
4–8	Moderate	1	2–2.6
8–16	Strong	0.5	1.5–2
>16	Very strong	0	<1.5

[a] Saturation extract of topsoil (mS/cm = mmho/cm = dS/m). If EC is measured in 1:5 solution, conversion is necessary according to Guidelines 2006
[b] Multiplier for the basic score. The final SQR score (0–100) = Basic score* active multiplier. The Basic score ranks about between 10 and 34 (Mueller et al. 2007d)

Table 5 Classification of measured pH of topsoil and consequences for the Muencheberg Soil Quality Rating (Mueller et al. 2007d, mod.)

pH	Degree of acidification or sodification	Score	Multiplier grassland
<3.3	Extreme acidification	**0**	<2.5
3.3–4	High to very high acidification	**0.5**	2.5–2.7
4–4.5	Moderate to high acidification	**1**	2.7–3
4.5–5.2	Low to moderate acidification	**1.5**	3
5.2–8.2	No acidification, no sodification	**2**	3
8.2–8.4	Low to moderate sodification	**1.5**	2.7–2.9
8.4–8.6	Moderate to high sodification	**1**	2.5–2.7
8.6–8.8	High to very high sodification	**0.5**	2–2.5
>8.8	Extreme sodification	**0**	<2

5 Assessing the Ecological Status of Sites by Vegetation

Ellenberg et al. (1991) developed a helpful system of grassland site bio-indication by vegetation. First, the plant species composition is recorded on sample plots, for example by application of a common scale (Kaiser et al. 2005). Then, ecological ranking numbers for moisture, soil reaction, nitrogen content, temperature, continentality and salt content are allocated and means can be computed for mapping units or plots. Ellenberg's ecological rank numbers are currently available for most vascular plant species of Central Europe and have already been tested in drier climates (Böhling 2004). Sometimes a local modification is necessary, but in general, the usefulness and accuracy of this system has been proven and confirmed by many authors, e.g. Schaffers and Sýkora (2000) and Pykälä (2005). Numbers are well correlated with soil properties, for example, the moisture number with watertable depth or the nitrogen number with soil nitrogen contents. Indicator values can be handled as quasi-metric data in many cases. Moisture numbers (Table 6) are particularly well studied and are reliable bio-indicators that can be also transformed to other scales of soil moisture conditions (Kaiser and Käding 2005). We used the Ellenberg system of site bio-indication in grassland studies in river lowlands, on peatlands and on sandy grasslands in Germany (Mueller et al. 2003; Kaiser et al. 2005). This system has started to be extended to other regions of Eurasia and worldwide. It should be borne in mind that all airborne methods of exploring grassland quality are very sensitive to vegetation pattern. Also, results of terrestrial soil survey and soil physical, chemical and biological measurements are closely related to vegetation. Thus, vegetation composition and its ecological value should be always assessed in all studies dealing with grassland quality. Examples are given in Figs. 14 and 15. It would be useful to test the applicability of Ellenberg's bio-indicator system over Eurasia and to consider typical species of other climate zones.

Sites of Fig. 14 are semi-natural grasslands in Northeast Germany in a sub-humid climate. Prime land use functions are nature protection/biodiversity (a,b,d) and flood protection (c). Agriculture is necessary to maintain ecosystems, but is a secondary land use function and underlies restrictions. Management intensity is low. Figure 14(a) shows steppe vegetation with *Adonis vernalis, Potentilla argentea* and *Stipa capillata* on sandy soils and southern exposure. It has a medium Ellenberg moisture number (mF) of 2.5 (dry). Biomass is 1–2 t DM/ha, and effective grassland yield without unpalatable species (EGY) is 0.5–1 t/ha. The land is used as zero-input sheep pasture for landscape maintenance. Figure 14(b) shows meadow steppe vegetation of high floral diversity with *Dactylis* spec., *Campanula* spec., *Hieracium* spec., *Plantago media, Onobrychis* spec. The Ellenberg moisture number mF is 4 (dry to moist), the biomass is 2–3 t DM/ha, and EGY is 1.8–2.5 t/ha. Land use is also zero-input sheep pasture for landscape maintenance. Figure 14(c) shows a river lowland with *Alopecurus pratensis*, *Phalaris arundenacea, Calamagrostis epigejos*

Table 6 Ellenberg's scale of moisture numbers and some examples of grassland species (Ellenberg et al. 1991, mod.)

F number	Description	Examples of species[a]
1	"Starktrockniszeiger". Indicator of extreme dryness, restricted to soils which often dry out for a certain time	
2	Intermediate between 1 and 3	*Sedum acre* *Artemisia campestris*
3	"Trockniszeiger". Dry-site indicator, more often found on dry ground than on moist places, never on damp soil	*Agropyron intermedium* *Chondrilla juncea* *Centaurea scabiosa*
4	Intermediate between 3 and 5	*Convolvulus arvensis* *Hypericum perforatum* *Medicago sativa*
5	"Frischezeiger". Moist-site indicator, mainly on fresh soils of average dampness, absent from both wet and dry ground	*Dactylis glomerata* *Plantago major* *Taraxacum officinale*
6	Intermediate between 5 and 7	*Alopecurus pratensis* *Artemisia vulgaris* *Holcus lanatus*
7	"Feuchtezeiger". Dampness indicator, mainly on constantly moist or damp, but not on wet soils	*Calamagrostis epigejos* *Cirsium oleraceum*
8	Intermediate between 7 and 9	*Polygonum lapathifolium*
9	"Nässezeiger" Wet-site indicator, often on water-saturated, badly aerated soils	*Caltha palustris* *Cicuta virosa*
10	"Wechselwasserzeiger". Indicator of sites occasionally flooded, but free from flooding for long periods	*Carex elata* *Phragmites australis* *Typha latifolia*
11	"Wasserpflanze". Plants rooting under water, but at least for a time exposed above, or plants floating on the surface	*Polygonum amphibium* *Schoenoplectus lacustris*
12	"Unterwasserpflanze". Submerged plant, permanently or almost constantly under water	

Additional symbols are:
~ "Zeiger für starken Wechsel". Indicator of a very fluctuating water regime
= "Überschwemmungszeiger". Indicator of flooding and inundation
[a] Species are part of the grassland flora of Northeast Germany and Central Asia

and invasive *Bromus hordeaceus*. Ellenberg's moisture number mF is 6.5 (damp). Biomass is 2–5 t DM/ha and EGY 1.8–4 t/ha. Haymaking and cattle grazing are common. Figure 14(d) is a peat lowland with *Carex* spec., *Polygonum bistorta*, *Caltha palustris* and *Dactylorhiza maculata*. The mF is 8.5 (wet), biomass is 6.5 t DM/ha, and EGY 1.5 t/ha. It is non grazed land used for occasional haymaking.

Fig. 14 Examples of grasslands, subject to functional assessment including bio-indication

Fig. 15 Examples of potential bio-indicator plants. **a** Rushes (*Juncus spec.*) are indicators of a poor aerated soil due to compaction and/or wetness. **b** *Achnatherum splendens* is typical for some saline soils in lowlands and depressions of Asia

6 Assessing the Quantity and Quality of Vegetation

Above-ground biomass of pastures is a very good indicator of soil conditions. It can be estimated by several methods. Test cuts of random 1 m^2 samples by throwing a stable quadratic meter frame on the surface are traditional and reliable. Also, non-destructive methods ranging from visual estimation classes to complex electronic instruments are available. Examples are pasture rulers, plate discs, devices based on optical principles, and electronic capacitance meters (Lopez Diaz and Gonzalez-Rodriguez 2003). As the architecture of swards depends on the botanical composition, all these non-destructive methods require good calibration according to local conditions and do not work well on transfer to other environments. Meanwhile, a great number of remote sensing methods have been developed for estimating grass biomass consistently over large regions (Schellberg et al. 2008). These methods need accurate ground-truthing, which should be preferably done by harvesting test plots. We preferred traditional test harvests in all our grassland studies.

Sometimes, for example in wetlands, the biomass can be very high, but animals do not graze it because the plant species are of very low forage quality or even poisonous. The quality of vegetation can be assessed by separating out plant species of cut samples and estimating their proportion or weighing them separately.

Palatability values can be allocated to species. The scale of palatability values created by Klapp et al. (1953) is ten-stage: -1 = poisonous, 0 = no palatability value, 1 = very low palatability value, 2 = low palatability value, 4 = medium palatability value, 6 = high palatability value, 8 = very high palatability value. Briemle converted this scale into a common 9-stage scale, ranging from 1 = poisonous to 9 = very high (Briemle et al. 2003).

Eliminating the proportions of non-palatable species (weeds of values <2) results in an Effective Grass Yield (EGY, Mueller et al. 2008). EGY is better correlated with soil quality scores of VSA or M-SQR than the total biomass.

There is much knowledge available about ecological behaviour and palatability values of Central Asian plants, for example Gintzburger et al. (2003); Inam and Maselli (2012). A comparison of Klapp's palatability values (10-stage scale) with rank numbers of palatability given in the Herders manual edited by Inam and Maselli (2012), which has a 5-stage scale, indicate a correlation at *species* level (Table 7). Also, at *genus* level, correlations exist. However, normalised palatability rank numbers in the Herder's manual are higher indicating a definition and scaling problem of those empirical scales. A possible reason is that the available biomass per animal is higher on pastures in Europe. In well- managed grasslands, the percentage of species have very high palatability is also higher. Cattle can eat more selectively, disregarding species of medium value.

Besides climate and soil properties, the degree of grassland management is very important for determining the botanical composition and yield potentials on grasslands. In some regions of Eurasia, estimating plants and their potential biomass is difficult or practically impossible by ad-hoc methods because of permanent, excessive grazing intensity (Fig. 16). Installation of fenced test plots as practiced by researchers in Inner Mongolia of China, Mongolia (Wesche et al. 2010; Sasaki et al. 2013) and Iran (Mofidi et al. 2013) is an useful method to test the local gene pool and vegetation recovery in overgrazed areas. However, this is expensive and cannot be provided for common grassland monitoring and inventory. Knowledge of overall soil quality (M-SQR, Mueller et al. 2007d) or VSA ratings (Shepherd 2009) in combination with defined degrees and classes of

Table 7 Palatability values of some grassland species in two regions of Eurasia

Species	Palatability values	
	Northeast Germany, Klapp scale, (−1–9)	**Western Pamir**, Scale of Herders' manual, (1–5)
Alopecurus pratensis	7	5
Artemisia vulgaris	1	2
Calamagrostis epigejos	2	3
Convolvulus arvensis	3	3
Dactylis glomerata	7	5
Festuca rubra	5	5
Hypericum perforatum	1	2
Medicago sativa	7	5
Plantago lanceolata	6	4
Plantago major	2	1
Polygonum amphibium	1	4
Polygonum aviculare	1	3
Setaria viridis	3	3
Taraxacum officinale	5	4
Trifolium pratense	7	5

Fig. 16 Overgrazed semi-arid landscape in Asia. In some regions of Eurasia, permanent overgrazing is a common practise. This has altered the vegetation composition and grassland productivity. A complication is that under these conditions a status analysis is difficult. Vegetation analyses cannot be reliably performed nor grassland productivity be measured. In this case, test plots have to be fenced before

management could provide estimates of grassland yields. However, worldwide acknowledged assessment scales of grassland management intensity do not exist. In subhumid to humid climates in Europe, where overgrazing is less common, the degree of management can be classified according to the criteria of Bockholt et al. (1996). Classes range from 1 = no use or extremely low intensity of use to 5 = high intensity. In our studies, it was both necessary and sufficient to introduce an additional class for damaged sward under stock and wheel tracks (Mueller et al. 2007c). Developing a common 9-stage scale of management classes for grasslands of Eurasia would by useful.

7 Conclusions

1. We offer a variety of simple field methods for measuring some crucial grassland properties and evaluating the capacity and performance of grasslands for biomass production.
2. Methods presented here have been proven in grassland studies in temperate zones but could be considered for applications to studies of Central Asian grasslands as well.
3. Some focus should be on expert-based visual or bio-indicator methods, which are feasible and reliable.
4. Procedures of Visual Soil Assessment and Muencheberg Soil Quality Rating in conjunction with physical, chemical and pasture quality field methods provide frameworks and parameters for grassland inventories over Eurasia by using uniform methodologies.
5. As there is currently no internationally acknowledged methodology, we suggest that our procedures may be adopted for the assessment of grassland and rangeland functions and the status of soil quality and degradation.

References

Boden AG (2005) Bodenkundliche Kartieranleitung (KA5), 5th edn. Hannover, Germany, p 432

Ball BC, O'Sullivan MF (1982) Soil strength and crop emergence in direct drilled and ploughed cereal seedbeds in seven field experiments. J Soil Sci 33:609–622

Ball BC, Batey T, Munkholm LJ (2007) Field assessment of soil structural quality: a development of the Peerlkamp test. Soil Use Manag 23:329–337

Ball BC, Cameron KC, Di HJ, Moore S (2012) Effects of trampling of a wet dairy pasture on soil porosity, and mitigation of nitrous oxide emissions by a nitrification inhibitor, dicyandiamide. Soil Use Manag 28(2):194–201

Bockholt R, Fuhrmann U, Briemle G (1996) Anleitung zur korrekten Einschaetzung von Intensitaetsstufen der Gruenlandnutzung. In: Natur und Landschaft 71(Nr. 6):249–251

Böhling N (2004) Southern aegean indicator values: derivation, application and perspectives. In: Rhodes, Greece, Arianoutsou and Papanastasis (eds) Proceedings 10th MEDECOS conference, April 25–May 1, ISBN 90 5966 016 1

Briemle G, Nitsche S, Nitsche L (2003) Nutzungswertzahlen für Gefäßpflanzen des Grünlandes. Schriftenreihe für Vegetationskunde 38/03, Bonn-Bad Godesberg, pp 203–225

Dimanche P-H, Hoogmoed WB (2002) Soil tillage and water infiltration in semi-arid Morocco: the role of surface and sub-surface soil conditions. Soil Tillage Res 66(1):13–21

Easy Test (2012) http://Easytest.Ipan.Lublin.Pl/Company_Profile/Index.Html. Accessed 18 Oct 2012

Ellenberg H, Weber HE, Düll R, Wirth V, Werner W, Paulißen D (1991) Zeigerwerte von Pflanzen in Mitteleuropa. Scripta Geobotanica 18

Eulenstein F, Schindler U, Mueller L (2003) Feldskalige Variabilität von Frühjahrsfeuchte und mineralischem Stickstoff in der ungesättigten Zone sandiger Standorte Nordostdeutschlands. Arch Agron Soil Sci 49(2):187–201

Gintzburger G, Toderich KN, Mardanov BK, Mahmudov MM (2003) Rangelands of the arid and semi-arid zones in Uzbekistan, 1st edn. CIRAD, ICARDA 2003, p 426
Guidelines (2006) Guidelines for soil description, 4th edn. FAO, Rome, p 95
Hartge KH, Bachmann J (2004) Ermittlung des Spannungszustandes von Böden aus Werten des Eindringwiderstandes von Sonden. J Plant Nutr Soil Sci 167(3):303–308
Inam R, Maselli D (2012) Herders' manual for Western. In: Pamir B, Inam R, Maselli D (eds) University of Central Asia.—B. 2012, p 124. http://www.ucentralasia.org/downloads/herders_manual_english.pdf. Accessed 9 Feb 2013
Jarvis N, Etana A, Stagnitti F (2008) Water repellency, near-saturated infiltration and preferential solute transport in a macroporous clay soil. Geoderma 143(3–4):223–230
Kaiser T, Käding H (2005) Proposal for a transformation scale between bioindicatively determined water supply levels of grassland sites and mean moisture indicator values according to Ellenberg. Arch Agron Soil Sci 51(3):241–246
Kaiser T, Käding H, Kiesel J, Mueller L, Hierold W, Behrendt A (2005) The derivation of grassland vegetation types on the basis of site and land use characteristics. Arch Agron Soil Sci 51(4):405–416
Kato H, Onda Y, Tanaka Y, Asano M (2009) Field measurement of infiltration rate using an oscillating nozzle rainfall simulator in the cold, semiarid grassland of Mongolia. Catena 76(3):173–181
Kezdi A (1969) Handbuch der Bodenmechanik. Band 1, Bodenphysik, pp 259, Band 3, Bodenmechanisches Versuchswesen, 1th ed. VEB Verlag für Bauwesen Berlin/Kiado Budapest, pp 274
Klapp E, Boeker P, Koenig F, Staehlin A (1953) Wertzahlen der Gruenlandpflanzen. Das Gruenland 2:38–40
Kraschinski S, Prochnow A, Tölle R, Zeitz J (2001) Zur Befahrbarkeit von Niedermoorgrünland. Zeitschrift für Kulturtechnik und Landentwicklung 42(6):277–283
Kutilek M, Nielsen DR (1994) Soil hydrology. GeoEcology textbook, Catena Verlag Cremlingen: Destedt, p 370
Liebig MA, Doran JW, Gardner JC (1996) Evaluation of a field test kit for measuring selected soil quality indicators. Agron J 88:683–686
Lopez Diaz JE, Gonzalez-Rodriguez A (2003) Measuring grass yield by non-destructive methods. In: Kirilov A, Todorov N, Katerov I (eds) Optimal forage systems for animal production and the environment. Proceedings of the 12th symposium of the European grassland federation, May 2003, vol 8. Pleven, Bulgaria, Grassland Science in Europe, pp 569–572
Mittelstedt R, Mueller L (1989) Kräfteanalyse an Penetrometerspitzen in alluvialen Tonböden als Grundlage für die Gestaltung und den Einsatz von Meliorationswerkzeugen. Arch. Acker-Pflanzenbau Bodenkd. Berlin 33(2): 111–120
Mofidi M, Jafari M, Tavili A, Rashtbari M, Alijanpour A (2013) Grazing exclusion effect on soil and vegetation properties in Imam Kandi Rangelands Iran. Arid Land Res Manag 27(1):32–40. doi:10.1080/15324982.2012.719575
Moret-Fernández D, González-Cebollada C (2009) New method for monitoring soil water infiltration rates applied to a disc infiltrometer. J Hydrol 379(3–4):315–322
Mueller L (1988) Efficiency of subsoiling and subsurface drainage in heavy alluvial soils of the G.D.R. Soil Tillage Res 12:121–134 Amsterdam
Mueller L (2011) Soil structure, visual assessment. In: Encyclopedia of Agrophysics. Dordrecht (Springer Science+Business Media B.V), pp 783–786
Mueller L, Tille P, Kretschmer H (1990) Trafficability and workability of alluvial clay soils in response to drainage status. Soil Tillage Res 16:273–287 Amsterdam
Mueller L, Schindler U, Kaiser T, Behrendt A, Frielinghaus Mo (2003) Analyse von Standortbedingungen in der Flußaue der Oder als Grundlage für einen schonenden Umgang mit Boden und Landschaft. Mitteilungen der Deutschen Bodenkundlichen Gesellschaft (102.2): 787–788

Mueller L, Behrendt A, Schindler U (2004) Strukturaspekte der Bodendecke und Bodeneigenschaften zweier Niederungsstandorte in Nordostdeutschland. Arch Agron Soil Sci 50(3):289–307

Mueller L, Wirth S, Schulz E, Behrendt A, Höhn A, Schindler U (2007a) Implications of soil substrate and land use for properties of fen soils in North-East Germany. Part I: basic soil conditions, chemical and biological properties of topsoils. Arch Agron Soil Sci 53(2):113–126

Mueller L, Schindler U, Behrendt A, Shepherd TG, Eulenstein F (2007b) Implications of soil substrate and land use for properties of fen soils in North-East Germany. Part II: aspects of structure in the peat soil landscape. Arch Agron Soil Sci 53(2):127–136

Mueller L, Behrendt A, Shepherd TG, Schindler U, Kaiser T (2007c) Implications of soil substrate and land use for properties of fen soils in North-East Germany. Part III: soil quality for grassland use. Arch Agron Soil Sci 53(2):137–146

Mueller L, Schindler U, Behrendt A, Eulenstein F, Dannowski R, Schlindwein SL, Shepherd TG, Smolentseva E, Rogasik J (2007d) The Muencheberg Soil Quality Rating (SQR). Field manual for detecting and assessing properties and limitations of soils for cropping and grazing, draft, Nov. 2007, p 102. Online: http://www.zalf.de/home_zalf/institute/lwh/lwh_e/mitarbeiter/mueller_l/pdf/field_mueller.pdf

Mueller L, Drösler M, Schindler U, Behrendt A, Höper H, Eulenstein F, Kantelhardt J, Sommer M (2008) Implications of soil properties, vegetation and management intensity for peatland quality. In: After wise use: the future of peatlands. Proceedings of the 13th international peat congress vol 1. Tullamore, Ireland 8–13 June 2008, pp 616–619

Mueller L, Kay BD, Hu C, Li Y, Schindler U, Behrendt A, Shepherd TG, Ball BC (2009) Visual assessment of soil structure: Evaluation of methodologies on sites in Canada, China and Germany: part I: comparing visual methods and linking them with soil physical data and grain yield of cereals. Soil Tillage Res 103(1):178–187

Mueller L, Shepherd G, Schindler U, Ball BC, Munkholm LJ, Hennings V, Smolentseva E, Rukhovic O, Lukin S, Hu C (2013) Evaluation of soil structure in the framework of an overall soil quality rating. Soil Tillage Res 127:74–84

Pykälä J (2005) Plant species responses to cattle grazing in mesic semi-natural grassland. Agric Ecosyst Environ 108:109–117

Salatin J (2010) The sheer ecstasy of being a lunatic farmer. Polyface, Virginia, p 300

Saparov AS (2013) Soil resources of the Republic of Kazakhstan: current status, problems and solutions, chapter of this book

Sasaki T, Ohkuro T, Kakinuma K, Okayasu T, Jamsran U, Takeuchi K (2013) Vegetation in a post-ecological threshold state may not recover after short-term livestock exclusion in Mongolian rangelands. Arid Land Res Manag 27(1):101–110. doi:10.1080/15324982.2012.719568

Schaffers AP, Sýkora KV (2000) Reliability of Ellenberg indicator values for moisture, nitrogen and soil reaction: a comparison with field measurements. J Veg Sci 11(2):225–244

Schellberg J, Hill MJ, Gerhards R, Rothmund M, Braun M (2008) Precision agriculture on grassland: applications, perspectives and constraints. Eur J Agron 29:59–71

Schindler U, Behrendt A, Müller L (2003) Change of soil hydrological properties of fens as a result of soil development. J Plant Nutr Soil Sci 166(3):357–363

Schindler U, Durner W, von Unold G, Müller L (2010) Evaporation method for measuring unsaturated hydraulic properties of soils: extending the measurement range. Soil Sci Soc Am J 74(4):1071–1083

Schultze E, Muhs H (1967) Bodenuntersuchungen für Ingenieurbauten, 2nd edn. Springer, Heidelberg, p 722

Shepherd TG (2000) Visual soil assessment, vol 1. Field guide for cropping and pastoral grazing on flat to rolling country, Horizons.mw/Landcare Research. Palmerston North, p 84

Shepherd TG (2009) Visual soil assessment, vol 1. Field guide for pastoral grazing and cropping on flat to rolling country, 2nd ed. Horizons regional council. Palmerston North, New Zealand, p 118

Terzaghi K, Peck RB, Mesri G (1996) Soil mechanics in engineering practice, 3rd edn. Wiley, New York
USDA/NRCS (2001) Soil quality test kit guide. United States department of agriculture. Natural resource conservation service. Soil quality test kit. Online source: http://soils.usda.gov/sqi/assessment/test_kit.html. Accessed 14 Jan 2013
Wesche K, Ronnenberg K, Retzer V, Miehe G (2010) Effects of large herbivore exclusion on southern Mongolian desert steppes. Acta Oecologica 36:234–241
Withers B, Vipond S, Lecher K (1978) Bewaesserung, 1st edn. Verlag Paul Parey, Berlin und Hamburg, p 241
Zhou ZC, Gan ZT, Shangguan ZP, Dong ZB (2010) Effects of grazing on soil physical properties and soil erodibility in semiarid grassland of the Northern Loess Plateau (China). CATENA 82(2):87–91

Impact Assessment for Multifunctional Land Use

Katharina Helming

Abstract Land is a scarce resource that has to fulfill many functions in addition to the production of food and fiber. Functions include water regulation, carbon sequestration, biodiversity pool, area for settlement and infrastructure, human health and recreation, employment and space for economic activities. In many regions of the world these functions have to be fulfilled simultaneously in space and time. This paper outlines the role of multifunctional land use for sustainable development and how ex-ante impact assessment can support decision making and management on land use towards a balanced performance of the land use functions portfolio. Latest scientific concepts are presented and a case examples given.

Keywords Multifunctional land use · Impact assessment · Land use functions

1 Introduction

World-wide, land use is characterized by the ascending demand of the growing world population for biomass based food, energy and fiber products. By 2050, food production has to satisfy the needs of 9 billion people, which is estimated to amount to about 170 % of the current production figures (FAO 2011). However, the need for an increasing food production is challenged by resources scarcities, demographic changes and the global financial crisis. Foresight studies such as the FAO study 2011, the Agrimonde study of France (Paillard et al. 2012), the UK study of the Government office for Science London (2011) and the 3rd foresight report of the European Commission Standing Committee for Agricultural Research (Freibaur et al. 2011) all point in the direction of the need for drastically

K. Helming (✉)
Leibniz Centre for Agricultural Landscape Research (ZALF) e. V., Eberswalder Strasse 84
15374 Muencheberg, Germany
e-mail: helming@zalf.de

L. Mueller et al. (eds.), *Novel Measurement and Assessment Tools for Monitoring and Management of Land and Water Resources in Agricultural Landscapes of Central Asia*, Environmental Science and Engineering, DOI: 10.1007/978-3-319-01017-5_12,

increasing agricultural production to feed the growing population that is anticipated to increase its livestock based diet, and to provide bio-energy and other non-food products for chemical engineering and replacement of non-renewable sources. The trend for reinforced agricultural production can already be observed and is further triggered by constantly increasing factor prices and land prices worldwide.

In addition to the anthropogenic dynamics, a number of natural factors challenge agricultural production. The first factor is climatic change, which is anticipated to lead to a reduction in water availability for plant production in many world regions accompanied by an increased frequency of severe weather events (droughts, storms). The latter may increase volatility of harvests at long term also in those regions that otherwise may benefit from climate changes (Iglesias et al. 2009). A second factor lies in land degradation processes such as soil erosion, salinization, soil organic matter decline and other processes, which may dramatically reduce the land productivity and which may partly be irreversible. The third factor lies in the diminishing availability of natural resources required for agricultural production, including water, energy, phosphorous. The fourth factor lies in a decreasing area share of land available for agricultural production resulting from a growing demand of other land uses including urbanization, infrastructure, mining, flood control and water retention, nature conservation, and tourism development (PBL 2011). Consequently, an increasing demand for agricultural commodities has to be satisfied on a decreasing land area, the natural productivity of which is declining in many parts of the world.

To cope with these challenges, forward looking decision making is required at the level of the agricultural production entity (farming system) and at the levels of regional, national and international governance and policy making. Decision making has to steer the intensifying agricultural production on a restricted land area and with increasing volatility of production factors and yield uncertainties. Thereby, short term and long term environmental constraints such as efficient use of natural resources (water, fertilisers, energy), soil and biodiversity conservation have to be considered. Long term fertility and productivity of land resources also needs to be guaranteed. Further, social issues (employment, income, human health) and socio-cultural aspects (cultural heritage, rural development) have to be integrated with economic targets of agricultural production and land use. All these aspects call for an integrated, forward looking approach to decision making that considers the multiple functions of land use and that steers into the direction of sustainable development. Research has to stand up to support decision makers through the provision of operational, science based evidence.

This paper outlines how research can support decision making on multifunctional land use by utilizing the approach of ex-ante impact assessment for sustainable development.

2 Multifunctional Land Use

Land is the main asset of rural areas on which competitive development can be grounded. Land use includes agriculture, forestry, settlement, transport- and energy infrastructure, nature conservation and tourism. The diversity of functions and services of land use are partly bound to the economic activities, partly cross-cutting. This is because they include commodities (food, fibre, energy, timber, housing space, infrastructure) contributing to economic performance as well as non-commodities (habitats, biodiversity, cleaning of water and air, greenhouse gas mitigation, buffering of weather extremes, cooling, flood control, cultural assets, recreational and human health assets) contributing to social and environmental performance. The exact regional portfolio of functions and services depends on the relative shares of land uses, its patterns, and its intensities. To characterize this interaction, the term multifunctional land use was coined (Helming et al. 2008). The underlying rationale for multifunctional land use is to consider social, economic, and environmental effects of any land use action simultaneously and interactively, to include those of commodity production and those of negative and/or positive externalities. These effects are linked to spatially explicit geophysical and socio-cultural conditions of landscapes to provide "functions" or "services" in the landscape context (De Groot et al. 2002). They include the provision of abiotic and biotic resources (water, soil air, biotic integrity), the production of food, fibre and other biomass related products, the regulation, transformation, buffering and storage of energy and matter fluxes, the support of health, education and spiritual values including cultural heritage and recreation, and last but not least the basis for economic growth and social welfare. The multifunctionality of any land use action then lies in the degree to which land use affects the ability of the landscape to perform these various functions interactively and simultaneously in space and time (Helming et al. 2008).

To make the concept of multifunctional land use operational for impact assessment and decision making, the Land Use Function (LUF) framework was developed (Perez-Soba et al. 2008). It allows for the assessment and governance of land use towards sustainable development. Developed for scientific purposes (Helming et al. 2011a) the LUF approach was also implemented in a European typology for planning and decision making purposes (ESPON 2012). The concept goes beyond the agricultural perspective in that it integrates other land use sectors and links them with socio-economic and geophysical properties of landscapes that are affected by land use. It emphasises the spatially explicit interaction between land use on the one side, and the condition, structures and processes of landscape on the other side. Consequently, the LUF approach considers all rural land use sectors namely agriculture, forestry, tourism, energy, nature conservation and transport.

Building upon the concepts of ecosystem services and landscape functions on the one hand, and multifunctionality on the other hand, Land Use Functions were defined as "the private and public goods and services provided by the different land uses that summarise the most relevant economic, environmental and societal aspects of a region" (Perez-Soba et al. 2008). Nine land use functions were identified, considering direct and indirect land use impacts on environment, society and economy (Table 1).

Table 1 Land use functions (taken from Helming et al. 2011a)

Mainly Societal Land Use Functions
1. Provision of work: employment provision for all activities based on natural resources, quality of jobs, job security and location of jobs (constraints, e.g. daily commuting)
2. Human health and recreation (spiritual and physical): access to health and recreational services, factors that influence service quality
3. Cultural: landscape aesthetics and quality, and values associated with local culture
Mainly Economic Land Use Functions
4. Residential and land independent production: provision of space where residential, social and productive human activity takes place in a concentrated mode. The utilisation of the space is largely irreversible due to the nature of the activities
5. Land-based production: provision of land for production activities that do not result in irreversible change, e.g. agriculture, forestry, renewable energy and land-based industries such as mining
6. Transport infrastructure: provision of space used for roads, railways and public transport services, involving development that is largely irreversible
Mainly Environmental Land Use Functions
7. Provision of abiotic resources: the role of land in regulating the supply and quality of air, water and minerals
8. Provision of habitat: factors affecting the capacity of the land to support biodiversity, in the form of the genetic diversity of organisms and the diversity of habitats
9. Maintenance of ecosystem processes: the role of land in ecological supporting functions such as soil formation and energy buffering

Corresponding to Land Use Functions, the concept of soil functions have been developed for assessing the value of soil properties and processes for society. Bouma (2010) distinguished seven soil functions:

1. Production of food and other biomass
2. Storing, filtering, and transformation of compounds
3. Providing habitat and gene pool
4. Providing the physical and cultural environment for humankind
5. Source of raw materials
6. Acting as a carbon pool
7. Archive of geological and archeological heritage.

Compared to the Land Use Functions the soil functions are characterized in more fundamental terms and less directly relatable to economic and or societal services for humankind. However, the similarity lies in the potential of both functional concepts to express the value of land use and soil for society, and to specify what land use and soil use means in the context of sustainable development, long term resilience and intergenerational equity.

3 Sustainable Development

Already two decades ago, since the UN Conference on Environment and Development in 1992, sustainable development has been raised to a comprehensive, pioneering programme for policy making to cope with the common future of humankind. This also implies relevancy to the future steering and management of land use and agricultural production systems. The significance of the sustainability concept in international debates can be attributed to its use in the Brundtland Commission's report, Our Common Future (WCED 1987). This report emphasised the economic aspects of sustainability by defining sustainable development as "economic development that meets the needs of the present generation without compromising the ability of future generations to meet their own needs." For the case of agriculture, the term was further defined in the mission statement of the Consultative Group on International Agricultural Research (CGIAR) as "successful management of resources for agriculture to satisfy changing human needs while maintaining or enhancing the quality of the environment and conserving natural resources" (CGIAR 1995). The terminology implies a strong normative component, which makes it attractive for policy makers, but at the same time challenges scientific analysis. While in scientific analyses, indicators and parameters determining environmental, social and economic impacts of any decision are in the focus, normative perceptions on interregional and intergenerational equity have to be considered for substantiating and implementing sustainable development. In this regard, sustainable development is interpreted as a procedural concept, in which societal debates on sustainable development targets are substantial features. This is also manifested in the European Commissions Sustainable Development Strategy (Council of the European Union 2006).

With regards to land use, sustainable development implies a balanced consideration of the multifunctional portfolio of social, economic, and environmental services provided by the land uses in a certain landscape (Wiggering et al. 2006). It also implies a careful consideration of long term attributes of resilience and robustness that are to maintain underlying ecosystem processes. The characteristic features of sustainable land use might considerably vary from region to region as do their natural, political and social characteristics. Also, the question of whether or not certain land use options are sustainable or not, depends not only on the specific characteristics of the respective region, but also on land use options in other regions. If for example, many regions took the same measure of sustainable development, on a larger scale, some key elements of sustainability might vanish due to synergisms and trade-offs between the system components. Therefore, sustainable land development requires a comprehensive landscape approach, but needs to be embedded in a wider spatial and geographical system (Fig. 1).

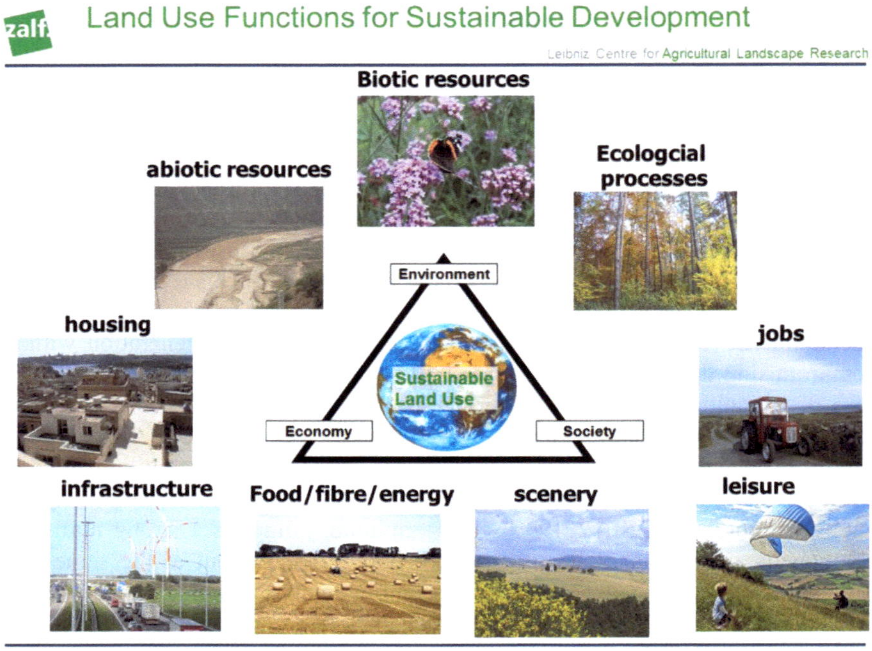

Fig. 1 Land Use Functions for sustainable development [taken from Helming et al. (2008)]

4 Impact Assessment and Decision Support

To provide scientific evidence for land use decision making in the light of sustainable development ex-ante impact assessment is a powerful concept. For researchers, impact assessment is a means to structure the analysis of human-environment interactions. For policy makers, impact assessment is a means to better target policy decisions towards sustainable development. The integration of both provides scientific evidence to decision making, but requires a mutual understanding of the respective objectives and operational restrictions within the scientific and decision making domains (Helming et al. 2011a).

Ex-ante impact assessment for policy decision making includes a series of steps (Table 2) that were formalised e.g. in the Impact Assessment Guidelines of the European Commission (CEC 2009). There, impact assessment is a mandatory process in policy decision making. After having identified the policy problem, the objectives are defined and the main policy options are developed. For every option, the intended and unintended impacts on social, economic and environmental system variables are analysed and compared. Each policy decision making process starts with a thorough identification and analysis of the problem (step 1). This is the basis for a decision to be taken and the justification of the policy intervention. The procedure continues with the definition of the policy objectives (step 2). Step

Table 2 The six steps of impact assessment (adapted from impact assessment guidelines of the European Commission, CEC 2009)

1. Identifying the policy problem
2. Defining the objectives
3. Developing the main policy options
4. Analyzing their impacts
5. Comparing the options
6. Outlining policy monitoring and evaluation

3 covers the development of a choice of alternative policy options for reaching these objectives (step 3). This involves a baseline scenario describing the development without policy intervention (inaction). The alternative policy options may include different levels of intervention (in budgetary terms), different levels of regulation (mandatory, incentive-based, awareness increasing), or different jurisdictional levels of implementation (European, national, regional). In step 4, the impacts of the policy options are to be analysed with respect to the three pillars of sustainable development. This includes intended and unintended impacts, short term and long term impacts, direct and indirect impacts. For the case of land use, production figures have to be analyzed as well as environmental effects on water quality and quantity, biodiversity and soil degradation, and social effects such as employment, rural development and human health. Based on the analyzed indicators a comparison of the policy options is condcted against the baseline scenario (step 5). The last step involves the outline of a set of recommendations for indicators, monitoring procedures and the evaluation of policy implementation (step 6). The latter finally provides for the transition from ex-ante to the ex-post assessment of the policy.

An integrated assessment of land use policies implies simultaneous consideration of all spatially relevant aspects of economic sectors and human activities that are linked to land (Helming et al. 2008). These include agriculture and forestry as the main economic sectors, nature conservation and rural tourism as land conserving activities, and settlement, transport and energy infrastructure as urbanised land uses. All of these sectors and activities compete for land resources, so any policy change affecting one land use has the potential to induce changes in the others.

Particularly at the steps four and five of the impact assessment procedure, scientific evidence can be used in the form of data, models and tools. Indeed, impact assessment has been understood as a process that can best utilise research based knowledge for informed decision making (De Smedt 2010). However, the development of scientific tools and methods for policy and management decision making is a research field that is only recently emerging and that requires robust frameworks and procedural concepts in order to be operational.

5 Scientific Support to Impact Assessment of Multifunctional Land Use

Several interdisciplinary, ex-ante assessment studies have emerged over the last decade in the field of land use. Comprehensive reviews of different types of scenario tools are provided by Bezlepkina et al. (2011), Helming et al. (2012). Generally, the conceptual framework of impact assessment of land use changes can be explained with the DPSIR framework (Gabrielsen and Bosch 2003). Driving forces (D) are exogenous trends that describe social, demographic, technological, policy-making and economic developments leading to land use changes. Via forward looking scenarios, driving forces are translated into land use changes (Pressures) by using land use models or knowledge rules. The spatial changes in land use induce alterations of the bio-geophysical and socio-economic settings of the system (States) that are usually determined by indicators. They can include indicators for soil properties, water status, biodiversity, farmers income, economic figures to name just a few. These changes in the state conditions may affect the area's quality of life and sustainable development (Impacts), as determined by aggregation methods such as multi-criteria analysis and by monetary or non-monetary valuation. Finally, anticipation of these changing conditions may lead to policy and/or management reactions (responses). These responses may again affect any part of the DPSIR chain so that the cycle is closed. At the science-policy interface, this approach is used to help to structure the analysis of policy effects on human activities and the environment and to jointly conceptualise research problems and integrate disciplinary viewpoints (Tscherning et al. 2012).

Figure 2 illustrates an adaption of the DPSIR framework to the case of land use. Driving forces are exogenous forces that include development of land prices, factor prices, technological advancement such as new breeds, innovative irrigation system, or trade and marketing concepts. Via scenario storylines, those driving forces are translated into land use changes (pressures), such as an increase in the share of arable land on the costs of grassland, or the expansion of irrigation agriculture. Those land use and management changes induce alterations of the regional water regime, soil properties, employment rate, marginal farm revenues and other geo-physical and socio-economic parameters. These state changes may alter the regional socio-economic situation and the conditions for human well-being. Finally, management and/or policy reactions can be induced (responses).

The conceptual framework described above may be applied to a variety of issues related to land use and agricultural management. It needs to be specified for each particular case and decision option, and it needs to be specified for the specific geographical region as well as the spatial and temporal scale of analysis. An example for a European policy case is given below.

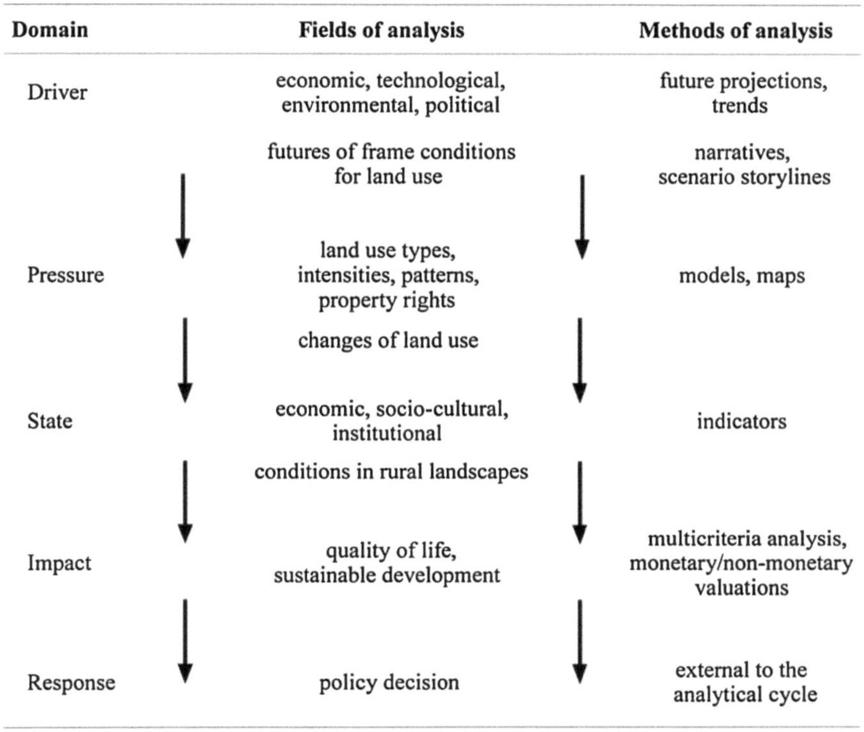

Fig. 2 DPSIR research framework (Gabrielsen and Bosch 2003) adapted to the case of ex-ante land use change assessment [taken from Helming et al. (2012)]

6 Case Examples for Sustainability Impact Assessment of Multifunctional Land Use

SENSOR (Sustainability Impact Assessment Tools for environmental, social and economic effects of multifunctional land use in European Regions) was a large integrated research project funded by the European Commissions 6th framework programme for research. Ex-ante sustainability impact assessment tools were developed to support European policy making in the area of land use (Helming et al. 2008). The approach integrated the relevant land use sectors agriculture, forestry, transport and energy infrastructure, nature conservation and rural tourism.

By using a trend extrapolation of external driving forces (oil price, commodity demand, technological change), a baseline scenario was constructed for the year 2025. It served as counterfactual for a number of policy cases related to the European Common Agricultural Policy (CAP), bio-energy and environmental policies. With a modelling cascade of macro-econometric, sectoral and land use models, scenarios were translated into land use change projections for the EU27 area at regional scale. The next step was to analyse impact indicators with regards to changes in social, economic and environmental conditions. The concept of Land Use Functions was employed to valuate the changes within the greater context of sustainable development.

The DPSIR framework was applied to integrate the research parts of the projects and to weave the logical thread through future oriented scenarios, anticipated land use changes and its expected environmental, social and economic impacts. Two different scenarios regarding reform of the Common Agricultural Policy (CAP) were analysed for the European Union member states: complete discontinuation of both, agricultural market support and direct farm income support, as well as a reduction of CAP support were compared to a baseline scenario (counterfactual: no change in CAP). Each scenario led to different shares of land use in Europe. Based on the DPSIR framework, the land use change was linked to indicators (social, economic and environmental) to obtain the impacts. The resulting scenarios and their impacts were presented, discussed and evaluated by experts and decision-makers at a sub-national scale. The Land Use Function concept was used for aggregating the information and making different scenarios comparable. Through the integration of knowledge from experts at the sub-national level, region-specific impacts and outcomes of political decisions could be presented. The results showed that at aggregated level, land use in Europe is less sensitive to changes in the Common Agricultural Policy than initially expected by experts and policy makers. However, in the context of regions there can be significant impacts on the functions of land use. This finding also confirmed the need for a multi-scale approach: while policy makers need information on aggregated, national levels for decision making, relevant impacts can sometimes only be detected at disaggregated spatial levels. The framework also offered the potential to present decision options to policymakers that were still comparable and based on similar indicators. This additionally augmented the acceptance of the project outcomes at a sub-national level (Helming 2011b).

7 Conclusion

Although challenged by the global financial crisis, sustainable development is still accepted as general standard for the integration of environmental, social and economic concerns as well as interregional and intergenerational equity into forward looking decision making. Land use decision making is particularly devoted to sustainable development because land use is at the heart of human environment interactions. The implementation can be achieved through the concept of multifunctional land use in the sense that all land use functions are considered simultaneously in space and time. This allows for multi-criteria analysis and trade-off identification. Forward looking decision making is facilitated by ex-ante impact assessment that integrates across disciplines and sectors by employing the multifunctionality concept. The approach is complex and requires an efficient interface between research and decision making.

References

Bezlepkina I, Reidsma P, Sieber S, Helming K (2011) Integrated assessment of sustainability of agricultural systems and land use: methods, tools and applications. Special issue. Agric Syst 104

Bouma J (2010) Implications of the knowledge paradox for soil science. In: Donald LS (ed) Advances in agronomy, vol 106. Academic Press, Burlington, pp 143–171

CEC (Commission of the European Communities) (2009) Impact assessment guidelines. European Commission, Brussels, Belgium. http://ec.europa.eu/governance/impact/commission_guidelines/commission_guidelines_en.htm

Council of the European Union (2006) Review of the EU sustainable development strategy—renewed strategy. 10917/06. Brussels, Belgium

Consultative Group on International Agricultural Research, CGIAR (1995) Report of the Task Force on Sustainable Agriculture. Document No. MTMi95110. Nairobi, Kenya. Washington, D.C

De Groot R, Wilson MA, Boumans RMJ (2002) A typology for the classification, description and valuation of ecosystem functions, goods and services. Ecol Econ 41:393–408

De Smedt P (2010) The use of impact assessment tools to support sustainable policy objectives in Europe. Ecol Soc 15(4):30

ESPON (2012) EU-LUPA European Land Use Pattern. Applied Research 2013/1/8. Draft Final Report (Part B) I Version 06/June/2012, p 57

FAO (2011) How to feed the world in 2050

Foresight (2011) The future of food and farming. Executive summary. The Government Office for Science, London

Freibaur A, Mathijs E, Brunoir G, Damianova Z, Faroult E, Girona i Gomis J, O'Brian L, Treyer S (2011) Sustainable food production and consumption in a resource-constrained world. European Commission—Standing Committee on Agricultural Research (SCAR). The 3rd Foresight Exercise

Gabrielson P, Bosch P (2003) Environmental indicators: typology and use in reporting. Internal working paper, European Environmental Agency, Copenhagen, Denmark

Helming K, Perez-Soba M, Tabbush P (Eds). (2008) Sustainability impact assessment of land use changes, Springer, Berlin, p 507

Helming K, Diehl K, Bach H, Dilly O, König B, Kuhlman T, Pérez-Soba M, Sieber S, Tabbush P, Tscherning K, Wascher D, Wiggering H (2011a) Ex Ante impact assessment of policies affecting land use, Part A: analytical framework. Ecol Soc 16(1):27

Helming K, Diehl K, Kuhlman T, Jansson T, Verburg PH, Bakker M, Perez-Soba M, Jones L, Verkerk PJ, Tabbush P, Breton Morris J, Drillet Z, Farrington J, LeMouël P, Zagame P, Stuczynski T, Siebielec G, Sieber S, Wiggering H (2011b) Ex Ante impact assessment of policies affecting land use, Part B: application of the analytical framework. Ecol Soc 16(1):29

Helming K, De la Flor I, Diehl K (2012) Integrated Approaches for ex-ante impact assessment tools—the example of land use. In: Raggamby A. von and F. Rubik (eds.): Sustainable development, evaluation and policy-making. Edward Elgar, Cheltenham, 91–110

Iglesias A, Garrote L, Quiroga S, Moneo M (2009) Impacts of climate change in agriculture in Europe. PESETA-Agriculture study, JRC 55386

Paillard S, Treyer S, Dorin B (2012) Agrimonde Scenarios and challenges for feeding the world in 2050. INRA, Paris

PBL (2011) Getting into the right lande for 2050. Netherlands Environmental Assessment Agency (PBL) and Stockholm Resilience Centre

Perez-Soba M, Petit S, Jones L, Bertrand N, Briquel V, Omodei-Zorini L, Contini C, Helming K, Farrington JH, Tinacci Mossello M, Wascher D, Kienast F, De Groot R (2008) Land use functions : a multifunctionality approach to assess the impact of land use changes on land use sustainability. In: Helming K, Pérez-Soba M, Tabbush P (eds) Sustainability impact assessment of land use changes, Springer, Berlin, pp 375–404

Tscherning K, Helming K, Krippner B, Sieber S, Gomez y Paloma S (2012) Does research applying the DPSIR framework support decision making? Land Use Policy 29:102–110

Wiggering H, Dalchow C, Glemnitz M, Helming K, Müller K, Schulz A, Stachow U, Zander P (2006) Indicators for multifunctional land use: Linking socio-economic requirements with landscape potentials. Ecol Ind 6:238–249

World Commission on Environment and Development WCED (1987) Our common future. Oxford University Press, New York

The Muencheberg Soil Quality Rating for Assessing the Quality of Global Farmland

Lothar Mueller, Uwe Schindler, T. Graham Shepherd, Bruce C. Ball, Elena Smolentseva, Konstantin Pachikin, Chunsheng Hu, Volker Hennings, Askhad K. Sheudshen, Axel Behrendt, Frank Eulenstein and Ralf Dannowski

L. Mueller (✉) · U. Schindler · A. Behrendt · F. Eulenstein · R. Dannowski
Leibniz-Centre for Agricultural Landscape Research (ZALF) e. V., Eberswalder Str. 84
15374 Muencheberg, Germany
e-mail: mueller@zalf.de

T. G. Shepherd
BioAgriNomics Ltd, 6 Parata Street, Palmerston North 4410, New Zealand
e-mail: gshepherd@BioAgriNomics.com

B. C. Ball
Crop and Soil Systems Research Group, SRUC, West Mains Road, Edinburgh EH9 3JG, UK
e-mail: Bruce.Ball@sruc.ac.uk

E. Smolentseva
Russian Academy of Sciences, Siberian Branch, Institute of Soil Science and Agrochemistry (ISSA), Ac. Lavrentieva avenue, 8, 630090 Novosibirsk, Russian Federation
e-mail: smolentseva@issa.nsc.ru

K. Pachikin
Kazakh Research Institute of Soil Science and Agrochemistry named after U. U. Uspanov, Al Faraby Avenue 75b 050060 Almaty, Kazakhstan
e-mail: kpachikin@yahoo.com

C. Hu
Center for Agricultural Resources Research, Institute of Genetics and Developmental Biology, The Chinese Academy of Sciences, 286 Huaizhong Road, Shijiazhuang 050021 Hebei, China
e-mail: cshu@sjziam.ac.cn

V. Hennings
Federal Institute for Geosciences and Natural Resources, Geozentrum Hannover, Stilleweg 2 30655 Hannover, Germany
e-mail: volker.hennings@bgr.de

A. K. Sheudshen
Kuban State Agrarian University, 13 Kalinin Str., 350044 Krasnodar, Russian Federation
e-mail: kubagrohim@mail.ru

L. Mueller et al. (eds.), *Novel Measurement and Assessment Tools for Monitoring and Management of Land and Water Resources in Agricultural Landscapes of Central Asia*, Environmental Science and Engineering, DOI: 10.1007/978-3-319-01017-5_13,

Abstract Sustainable use of soils is a vital issue in the 21st century to meet global challenges of food security, demands for energy and water, climate change and biodiversity. Eurasia has reasons to tackle and solve these problems soon. It covers the largest landmass and has the highest population density of the earth. Tools for reliable, simple and consistent evaluation of the status of the soil over a wide range of scales can help to assess suitability for crop growth and yield potentials. We explain the Muencheberg Soil Quality Rating (M-SQR) for analysing soil properties that limit crop yields and crop productivity potentials consistently over large regions. The approach is based on 8 Basic Indicators and at least 12 Hazard indicators. Ratings of soil quality are made during normal soil survey mainly by applying visual methods of soil evaluation. A field manual provides rating tables based on response curves and thresholds for different hazard indicators (such as risk of drought). Finally, overall soil quality rating scores ranging from 0 (worst) to 100 (best) characterise crop yield potentials. The current approach is valid for grassland and cropland. Field tests in Eurasia confirmed the practicability and reliability of this approach. We conclude that the Muencheberg SQR has the potential to serve as a global functional reference framework for agricultural soil quality of cropland and grassland. We anticipate the creation of comparable soil functional maps of the whole of Eurasia by the use of this method.

Keywords Soil quality · Indicators · Crop yield · Muencheberg soil quality rating

1 Introduction

The capacity of soil to produce plant biomass is closely related to key global issues of the 21st century like food security (Borlaug 2007), climate change and environmental quality (Lal 2008). This productivity potential of land is dependent on maintaining good soil quality (Karlen et al. 2001). A standardised methodology and framework for assessing agricultural soil quality is likely to be demanded by a growing international community of land users and stakeholders for achieving sustainable high soil productivity. This framework has to meet the following criteria: precise in operation, based on indicators and thresholds of soils, consistently applicable over different scales, potential for use in suitability and land capability classifications, adequately relevant to crop yield potential and capable of being integrated into new land evaluation frameworks of the 21st century (Mueller et al. 2010). In particular, it needs to be relevant to the proposed European Union soil strategy. We start from the hypothesis that the status of the Eurasian soil resource for cropping and grazing could be monitored and controlled by a uniform rating method. This paper aims to explain the multi-indicator-based Muencheberg Soil Quality Rating (M-SQR) and how this was tested in different regions of Eurasia. The magnitude and variability of SQ-Indicators and their underlying data are quantified at a number of agricultural sites across the globe, and correlations of ratings with crop yields are computed and shown.

2 The Muencheberg Soil Quality Rating

The Muencheberg Soil Quality Rating (M-SQR, Mueller et al. 2007a) is based on productivity-relevant indicator scoring which provides a functional rating of soils. The underlying concept is that most terrestrial crops require appropriate seedbed conditions and optimum soil quality for a deep and well-established rooting zone. The M-SQR approach includes both indicators of inherent (soil substrate) and dynamic (soil structure) agricultural soil quality, of topography in terms of slope and of climate in terms of the soil thermal and moisture regimes (Fig. 1).

Two types of indicator have been identified and defined by scoring tables. The first are basic indicators and relate mainly to soil textural and structural properties

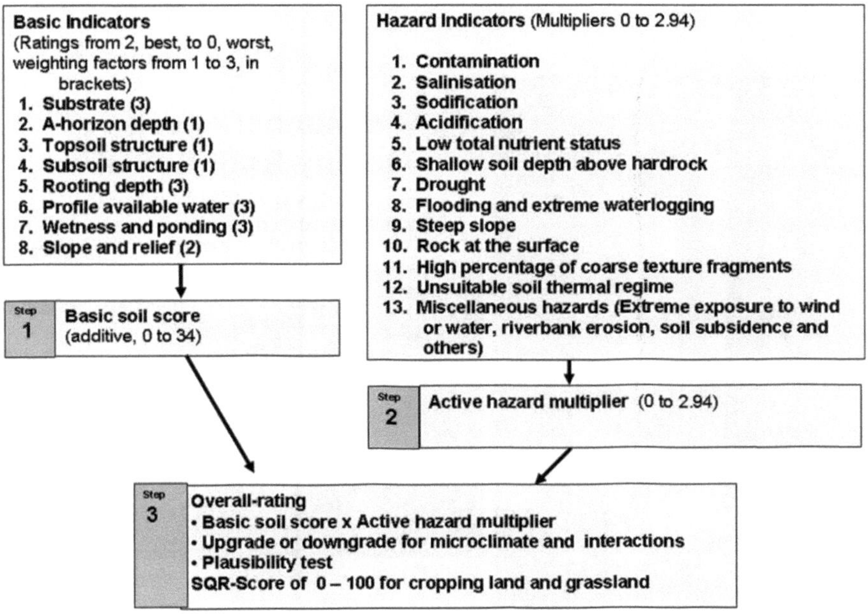

Fig. 1 Rating scheme of M-SQR (Mueller et al. 2007a) First, each of the 8 Basic Indicators is rated on a scale from 2 (best) to 0 (worst), multiplied by a weighting factor from 1 to 3 and then summed. Then the occurrence of Hazard factors is checked and summed as necessary to give a similar rating. The most crop yield limiting Hazard Indicator will be used to estimate a multiplier which may range from 0 to 2.94. The Basic score times the active multiplier will yield an overall M-SQR rating between 0 and 100. Over 100 agricultural research sites worldwide have been rated and classified

relevant to plant growth. These are soil substrate, depth and characteristics of the A horizon, size and shape of topsoil aggregates, features of subsoil structure and compaction, depth of rooting, water supply, wetness and ponding, land slope and relief. These are weighted with extra weight given to rooting and water factors and summed. The second type of indicator is hazard, relating to the most severe restrictions of soil function identified at the site. These are weighted by multipliers according to their relevance. All indicators are rated on a 5-point scale before weighting. The product weighted basic indicator ratings and the most severe (active) hazard indicator provides an overall soil quality rating index (M-SQR score). This M-SQR score may range from 0 (useless for agriculture) to 100 (best soil on a global scale). Indicator ratings are allocated according to a field manual (Mueller et al. 2007b) and utilise soil survey classifications (AG Boden 2005; FAO 2006a), soil structure diagnosis tools and local or regional climate data.

The field procedure for M-SQR consists of digging a small pit to 1 m and augering a hole down to 1.5 m to detect any layering or a shallow watertable. Then the soil profile is scanned to assess the set of indicators shown in Fig. 1 using visual tactile examination, expert-based knowledge and minimum equipment (Fig. 2). A spade, knife and rule are essential for the evaluation of soil textural and

Fig. 2 Essentials for the fieldwork of soil rating. The spade is used for digging the pit and for the spade test of soil structure analysis. The knife is used for checking soil strength and coherence of aggregates. The auger is important for identifying soil texture and other properties like a shallow watertable from 1 to 1.5 m depth. The field manual contains all rating tables. In order to assess the drought risk, monthly data of precipitation and potential evapotranspiration are required. Data can be taken from the local climate estimator New_LocClim 1.10 (FAO 2006b) before starting fieldwork

structural properties. Other instruments like a GPS for marking the exact location and common soil surveying equipment (0.1 n HCL, pH test strips, Munsell colour charts, FAO and/or national guidelines for soil description and others) are useful if the soil has never been surveyed. A basic rating score ranging from 2 (best) to 0 (poorest) will be given for every indicator by the aid of scoring tables related to soil attributes.

The philosophy of the rating procedure is to provide a result based on a minimum of data, but to utilise more detailed information if available. Data need to be allocated to scoring tables, suggested values and sample photographs of the field manual (Mueller et al. 2007b). If for example, analyses of soil density or plant-available water are available and plausible, they should be used instead of the suggested values given in the manual. In the majority of cases, the exact location of the pit for field rating will be clear before field work starts. Where available, use soil and topographic maps, analytical data and other existing information about occurrence of wetness or drought to enhance the reliability of ratings. Monthly climatic data of precipitation and potential evapotranspiration in the main vegetation period are important for assessing the expected drought risk of a site. Some potential evapotranspiration data are inconsistent and unreliable, depending on the method of calculation. Thus, the FAO-Penman–Monteith reference evapotranspiration (Allen et al. 1998) should be used. Climate data of the local climate estimator New Loc_Clim 1.10 (FAO 2006b) are reliable in flat to undulating areas.

The time required for the field rating procedure depends largely on the experience and skills of the expert, but also on the study site and availability of support data. It may range from about 40 min to several minutes. If information about the type of soil is not yet available, it is recommended first to classify the soil by the World Reference Base of Soil Resources (WRB 2006).

3 Important Soil Attributes of the Basic Rating Procedure

The field manual (Mueller et al. 2007b) contains an explanation of indicators, all rating tables, thresholds and orientation guides in detail. Important soil attributes relevant to scoring of Basic indicators are given in Table 1.

On many agricultural sites in temperate climate, this set of attributes and indicators, when summed, is enough to provide productivity-relevant ratings of soil quality at a field scale. The Basic score of M-SQR includes weighting factors which have been developed based on correlations with the crop yield of small grain cereals. The sums of the Basic ratings may range from 0 to 34. They can be converted into a 100-point scale (Upgraded Basic score, UBS) by a multiplier of 2.94. The Upgraded Basic score is often correlated with scores of traditional rating systems, which do not consider drought risk or other hazards.

Table 1 Main soil attributes used for the basic rating

Indicator	Main attributes of scoring	Additional attributes for modifying the score	Relevant depth cm
B1. Soil substrate (WF[a]3)	Soil texture class, parent material	Strong gradients of texture (layering), content of coarse material, low SOM, proportion of artefacts	0–80 (crop land), 0–50 (grassland)
B2. Depth of A- horizon and depth of humic soil (WF1)	Depth of A horizon	Abrupt boundary between topsoil and subsoil, SOM content <4 % (grassland)	0–25
B3. Topsoil structure (WF1)	Type and size of aggregates and pores	Redoximorphic feature	0–25
B4. Subsoil structure (WF1)	Type and size of aggregates and pores, increased soil strength or density	Redoximorphic feature	25–50
B5. Rooting depth and depth of biological activity (WF3)	Occurrence of roots, effective rooting depth	Barriers to rooting and their intensity	150 (cropland), 80 (grassland)
B6. Profile available water (WF3)	Field capacity minus wilting point, rooting depth, capillary supply	Soil texture, stoniness	Rooting zone (<=150)
B7. Wetness and ponding (WF3)	Ponding, depth of ground or perched water table, redoximorphic features, vegetation	Soil position in a depression, wetness due to a perched water table	
B8. Slope and relief (WF2)	Slope at the pedon position	Microrelief and slope aspect at the profile position	

[a] WF = weighting factor of indicator, relevant to crop yield of small grain cereals

4 Important Soil Attributes of the Hazard Rating Procedure

Less fertile regions, widespread worldwide, are characterised by severe restrictions to soil potentials for cropping and grazing. Those hazard factors have the potential to reveal that a soil which looks good with high Basic scores as useless for agriculture. If for example, a fertile Loess soil undergoes severe drought over decades or permanently, no more agricultural cropland use will be possible. There are a number of different potential Hazard factors, but most of them are specific for certain regions. They are combined together into a single index. The reliability of SQ ratings over large regions will be depending on estimating the most critical Hazard factor at the site and its rating. One of the important crop yield limiting factors worldwide is the common site-specific drought risk in the main vegetation period (H7). It may be relevant to almost all regions of the world including humid areas.

The following Table 2 may give an orientation about the relevance of Hazard factors and their recognition.

Hazard indicator ratings are very important for the overall rating result of M-SQR. Having identified the most serious hazard indicator, a multiplier will be derived, which may range from 0 to 2.94. Table 3 gives orientation values of multipliers for drought, which is the factor most limiting soil productivity potentials worldwide. Recommendation values of multipliers consider the number of Hazard indicators with sub-optimum ratings. The number of other hazard factors may also influence the final rating considerably. Additional sub-optimum Hazard indicators like too cool climate (indicator 12) or sodification (indicator 3) are relevant and degrade overall soil quality additionally. If for example, drought is the dominating Hazard indicator at a typical location and has been rated by 1.25 on the basis of the handbook, and additionally Hazard indicators 3 and 12 are also less than 2, the multiplier has to be downgraded to 1.7 using Table 3. To achieve homogenous rating results over larger regions, it is important to estimate the multipliers of drought on the basis of reliable climate data.

The current final rating procedure proposes to check the plausibility of the results and to upgrade or downgrade the result by about 3–15 points, but within the limit of 100 points. Reasons for up- or downgrades are interactions between Hazard indicators, meso- and microclimate and the temporal uniformity of the soil moisture regime within the upper 10 cm.

5 Examples of Field Sites

An example of ratings of four locations in different regions of Eurasia is given in Table 4. Location Muencheberg has sandy soils and the other locations have more favourable Loess soils. Muencheberg and Luancheng are located in moderate climate zones, whilst Ust Kamenka is located in a harsh climate. M-SQR ratings

Table 2 Checklist of Hazard Indicators and criteria for identification (from M-SQR field guide, Mueller et al. 2007b, slightly modified)

Indicator[a]	Thresholds for orientation		
	Direct soil parameters	Indirect parameters of vegetation, climate or others	Reference soil groups (RSG) or qualifiers of WRB (2006)
1. Contamination	Specific for each pollutant according to international thresholds	High risk areas: cities, waste affected soils, vicinity of industrial plants, floodplains	Toxic, (Garbic, Spolic)
2. Salinisation	EC > 4 mS/cm in topsoil	White crusts on soil aggregates or surface, occurrence of halophytes, S-number acc. to Ellenberg >3	Salic, Hypersalic, Puffic, Chloridic
3. Sodification	ESP > 15 % (SAR > 13), pH > 8.2 in topsoil	High pH indicating plants, R-number acc. to Ellenberg of 9	Sodic, Alcalic, Natric
4. Acidification	pH < 5.2 (cropping) or <4.5 (grassland) in topsoil	Low pH indicating plants, R-number acc. to Ellenberg of 3 or lower	Hyperdystric, Hyperthionic
5. Low total nutrient status	Clear deficit of nutrients, cannot be compensated by fertilisation within one year		Hypergypsic, Hypercalcic
6. Soil depth above hard rock	Hardrock or permafrost <120 cm (arable land) or <70 cm (grassland)		Leptic, Lithic, Petric
7. Drought	Water budget in the main vegetation period of 4 month <500 mm, ustic, xeric or aridic soil water regime, total soil water balance in the main vegetation period <50 mm	Climatic water balance in the main vegetation period of 4 months <−100 mm, probability of the occurrence of a dry month >10 %, aridity index acc. to De Martonne <30, benefit of irrigation for cereals	Aridic
8. Flooding and extreme waterlogging	Flooding probability >5 %, peraquic soil water regime	Delay of beginning of farming on cropping land >20d, Grassland mF (Ellenberg) >8, clear benefit of land drainage	Floatic, Gelistagnic, Subaquatic, Tidalic

(continued)

Table 2 (continued)

Indicator[a]	Thresholds for orientation		
	Direct soil parameters	Indirect parameters of vegetation, climate or others	Reference soil groups (RSG) or qualifiers of WRB (2006)
9. Steep slope	Arable land gradient >12 %, grassland gradient >30 %		
10. Rock at the surface	Rock outcrop on arable land >0.01 %, on grassland >0.05 %		Leptosols; Ekranic, Hyperskeletic
11. High percentage of coarse soil texture fragments	Coarse fragments (>2 mm) on arable land >15 % by mass of in topsoil, grassland >30 %		Leptosols; Hyperskeletic, Skeletic
12. Unsuitable soil thermal regime	Cryic or pergelic soil thermal regime, Frigid regime with mean annual temperatures <5 °C	Tundra regions Duration of the frost free period <140 d	Cryosols; Cryic, Glacic

[a] An important characteristic of all indicators is they have rising response curves with crop yields (viz. Higher rating is better). We avoided indicators where response curves have an inner optimum or minimum. For example, if Hazard Indicators 3 (Sodification) and 4 (Acidification) would be combined into an indicator "Soil reaction, pH", the overall rating procedure would work as good as before. However, the approach would lose its potential for the definition of capability classes (for example: Acid soils of moderate productivity potential) as both soils of low and high pH would get low ratings. The functional coding would be not clear without ambiguity

Table 3 Orientation values for ratings and multipliers of hazard indicator 7 « drought»

Rating of drought risk	Orientation value of multiplier for number of H factors, viz. Hazard indicators with values <2[a]			
	0	1	2	3
2 (None)	2.94			
1.75 (None to low)		2.8	2.4	2.1
1.5 (Low)		2.6	2.3	2.0
1.25 (Low to medium)		2.1	1.9	1.7
1 (Medium)		1.8	1.6	1.5
0.75 (Medium to high)		1.5	1.3	1.1
0.5 (High)		1	0.8	0.6

[a] Number of Hazard indicators having ratings <2

consider both aspects. Two rating scores are particularly important. These are the Upgraded Basic score and the overall M-SQR-score. The Upgraded Basic score characterises mainly the physical status of soil for cropping at a certain location. For the locations in Table 4 it reflects the more suitable conditions of Loess soils for plant growth. The overall SQR score considers climate and other factors as specified by Hazard Indicators and that can degrade the soil productivity function seriously. At location Ust Kamenka it is largely determined by unfavourable soil moisture and thermal regimes (Hazard indicators 7 and 12 in Fig. 1). Drought was the dominating

Table 4 Rating examples of four different locations (Smolentseva et al. 2011, modified)

	Muencheberg.D, rainfed	Ust-Kamenka. RU, rainfed	Luancheng.CN, irrigated	Besagasch.KZ, irrigated
Geo-Position[a]	52.7/14.1/37	55.0/83.8/265	37.9/114.7/53	42.8/71.4/620
Climate[b]	540/8.5	514/0.1	537/12.2	330/8,9
Soils[c]	Albeluvisols	Phaeozems	Cambisols	Calcisols
Parent material	Glacial till	Loess	Loess	Loess, re-deposited
Dominant texture class	Sand	Silt loam	Silt loam	Silt loam
Most variable Basic indicators[d]	1,2,5	2, 3	None	5
Upgraded basic score[e]	65 (8)	88 (7)	94 (0)	74
Most limiting H indicator	Drought	Too		
cold+Drought	None	Drought		
Overall M-SQR score[f]	48 (13)	34 (12)	94 (0)	57

[a] Latitude North/Longitude East/Altitude masl
[b] Precipitation in mm/Temperature in °C
[c] Main Reference Soil Group (RSG) of WRB 2006
[d] Basic Indicators 1 = Substrate, 2 = A horizon depth, 3 = Topsoil structure, 4 = Subsoil structure, 5 = Rooting depth, 6 = Profile available water, 7 = Wetness and ponding, 8 = Slope and relief
[e] Basic score in a 100-point scale (100 = best). Mean, (Standard deviation)
[f] 100-point scale: 100 = best for small grain cereals, classes are <20 very poor, 20–40 poor, 40–60 medium, 60–80 good, >80 very good

Table 5 Classes of M-SQR scores and examples of some typical soils

Class of overall soil quality	M-SQR score	Typical examples of soil locations in Eurasia
Very poor	0–20	Soils affected by one or more hazard factors: Solonchaks, Extremely shallow soils, soils at steep slopes, non-irrigated soils in semi-deserts and deserts
Poor	20–40	Loess soils in harsh, too cold and dry climates (cryic or frigid soil temperature regime), soils in Northern Kazakhstan
Moderate	40–60	Sandy and dense loamy soils in temperate subhumid climates of Europe (slight drought risk), loess-like soils in the European part of Russia (frigid temperature regime), some irrigated soils in Central Asia
Good	60–80	Sandy, loamy and clayey soils in temperate humid climates in Central Europe, silty soils in Eastern Europe and Northern China, irrigated soils in Eurasia
Very good	80–100	Loess soils in temperate humid and subhumid climates, in China, Central and Western Europe, irrigated soils in temperate climates

limiting factor of overall soil quality on most sites rated so far. At most locations, the climatic water balance deficit during the main vegetation period could not be compensated by the profile available water. Finally, Luancheng site has a very good overall SQR and crop yield potential, whilst the other three sites have medium or poor conditions for cropping. At local scales, for monitoring and controlling SQR indicators of basic rating like the Upgraded Basic score, the VSA method (Shepherd 2009) or other straightforward evaluation methods of soil structure like VESS (Ball et al. 2007) can be used (Mueller et al. 2012a).

Table 5 gives examples of the classification of soil productivity potentials based on M-SQR scores. It reveals the dominance of the soil thermal and moisture regimes which are largely climate-controlled.

Richter et al. (2009) have shown that these five soil quality classes provide a mapping of agricultural soil quality over larger regions based on available soil and climate data.

6 Overall Rating Scores and Crop Yields

The overall M-SQR score is well correlated with crop yield at a global scale. Figure 3 shows the relationship between M-SQR scores and current yields of small grain cereals at two input levels. The lower linear curve shows crop yields at low and moderate inputs (<100 kg N/ha) and in more ecological farming practices. The regression line is $y = 0.7x$ with a highly significant degree of determination of 0.79. For reasons of readability of this graph, trend lines only are shown. The upper curve is based on high management intensity and high inputs of agro-chemicals (100–450 kg N ha^{-1}). At soils of high and very high quality (M-SQR

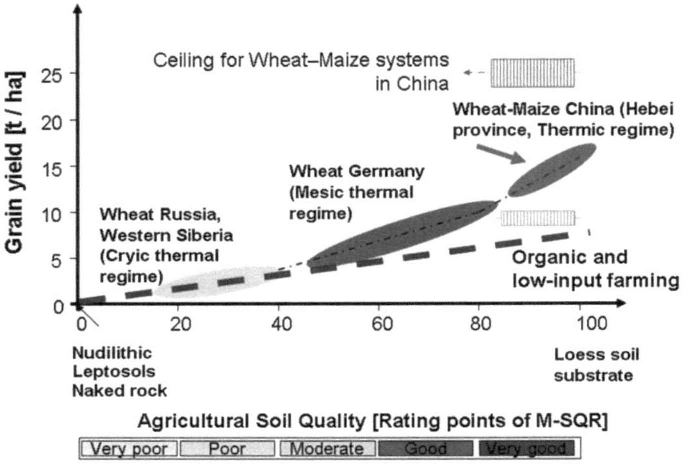

Fig. 3 Agricultural soil quality versus soil productivity (Mueller et al. 2012b, slightly modified). The graph is based on 540 rated soils from different regions of Eurasia. Single points are omitted in this graph. The lower quasi-linear curve is a baseline of M-SQR-Ratings ($y = 0.7x$, $n = 180$, $B = 0.79^{***}$). The upper non-linear curve is also significantly related to yield but reflects high inputs

scores >60) the difference between the curves becomes more marked, showing a high possible response to intensification. Crop yields are higher, the risk of failure is lower and more profit is likely from a given level of inputs.

Crop yields are a measure of productivity and include not only the effects of different soil quality, but also the impact of human activity, skills and management input. At a high level of inputs and management intensity, high crop yields may be achieved even at limited agricultural soil quality, but the risk of resource degradation (mainly water, air and excess fossil energy input) becomes very high. Thus, frameworks for assessing agricultural soil quality are better correlated with crop yields at a low or moderate level of farming intensity.

7 Outlook: Extension of Rating Scales

Currently available evaluation scales of cropland focus on rotations dominated by small grain cereals, reflecting the situation of cropland quality in Central Europe well. However, in some regions other specific crops with different requirements to soil and climate conditions are grown. To cover the majority of cropping and land conditions worldwide, the development of rating scales for other cropping systems is useful and is in progress (Table 6).

Maize and other thermophile grasses like Sorghum are more and more grown in short rotations of Eurasia. Their C-4 metabolism provides high biomass creation in

Table 6 Plant- and management-specific scales of M-SQR

Scale name	Type of dominant land use or crop	Focus of scale	Remark
Grassland scale	Grassland	Grazing for cattle or sheep, pasture, hayland, rangeland	Available
Cropland scale	Small grain cereals dominated rotations	Small grain cereals	Available
Maize scale	Thermophile cereals	Maize or other thermophile cereals, bioenergy crops on cropland	Test phase
Low-input scale	Subsistence cropland agriculture	Inherent soil fertility, organic carbon, zero to low inputs of agrochemicals	In preparation
Multi-cropping scale	Cropland rotations	Limitations of multi-cropping by climate	In preparation

a short growing period at significantly higher water use efficiency as compared to mesophile grasses (Mueller et al. 2005). Increasing temperatures due to climate change promote their extension. A Maize scale is easily created by modification of the thresholds of the Hazard indicator 12 (Soil Thermal Regime).

The usefulness of a Low-input scale is based on the insight that cropping systems worldwide do not follow common rules and conventions about sustainability. Mineral fertilization cannot be provided in many cases of subsistence farming. Under those conditions, the mineral composition of soil and the organic matter content define productivity potential. Creation of a Low-input scale will be also possible in the current framework of M-SQR, but requires some new indicator thresholds and their calibration with crop yields of zero-plots of fertilisation experiments.

The current cropland scale is based on a single cropping approach, which is typical for most areas of cereal-cropping. This could mean a potential overestimation of crop yield potentials on sites having a mesic soil thermal regime as compared to sites having a thermic regime or a mesic regime with mild, frost-free winters. A Multi-cropping scale could be also useful to consider the dominating influence of climate conditions on crop yield potentials.

8 Conclusions

The Muencheberg Soil Quality Rating has potential for assessing the quality of farmland and crop yield potentials on a global scale. More guidance on usage of M-SQR at different scales is needed and can be provided.

It is a realistic goal to create consistent soil functional maps of whole Eurasia by the use of this methodology. These will prove invaluable to land use planners and governments in their strive to improve food security. The Rating can also be used at a smaller scale to help producers get the best from their land.

References

Allen RG, Pereira LS, Raes D, Smith M (1998) Crop evapotranspiration—guidelines for computing crop water requirements. FAO irrigation and drainage paper 56, Food and Agriculture Organization of the United Nations, Rome, Italy, p 300

Ball BC, Batey T, Munkholm LJ (2007) Field assessment of soil structural quality: a development of the Peerlkamp test. Soil Use Manag 23:329–337

AG Boden (2005) Bodenkundliche Kartieranleitung (KA5), 5th edition, Hannover, p 432

Borlaug N (2007) Feeding a hungry world. Science 318:359

FAO (2006). Guidelines for soil description, Rome, 4th edition, p 95

FAO (2006) New_LocClim 1.10. Local climate estimator. Download–Address. ftp://ext-ftp.fao.org/SD/SDR/Agromet/New_LocClim/onext-ftp.fao.org. Accessed on 10 Oct 2010

Karlen DL, Andrews SS, Doran JW (2001) Soil quality: current concepts and applications. Adv Agron 74:1–40

Lal R (2008) Soils and sustainable agriculture. Rev Agron Sustain Dev 28(2008):57–64

Mueller L, Behrendt A, Schalitz G, Schindler U (2005) Above ground biomass and water use efficiency of crops at shallow water tables in a temperate climate. Agric Water Manag 75(2):117–136

Mueller L, Schindler U, Behrendt A, Eulenstein F, Dannowski R (2007) Das Muencheberger Soil Quality Rating (SQR): Ein einfaches Verfahren zur Bewertung der Eignung von Boeden als Farmland. Mitteilungen der Deutschen Bodenkundlichen Gesellschaft 110(2): 515–516

Mueller L, Schindler U, Behrendt A, Eulenstein F, Dannowski R, Schlindwein SL, Shepherd TG, Smolentseva E, Rogasik J (2007) The Muencheberg Soil Quality Rating (SQR). Field Manual for Detecting and Assessing Properties and Limitations of Soils for Cropping and Grazing, Draft, Nov 2007, p 10. http://www.zalf.de/home_zalf/institute/lwh/lwh_e/mitarbeiter/mueller_l/pdf/field_mueller.pdf. Accessed on 6 Aug 2011

Mueller L, Schindler U, Mirschel W, Shepherd TG, Ball BC, Helming K, Rogasik J, Eulenstein F, Wiggering H (2010) Assessing the productivity function of soils. A Review. Agron Sustain Dev 30(3): 601–614

Mueller L, Shepherd G, Schindler U, Ball BC, Munkholm LJ, Hennings V, Smolentseva E, Rukhovic O, Lukin S, Hu C (2012) Evaluation of soil structure in the framework of an overall soil quality rating. Soil Tillage Res. doi:10.1016/j.still.2012.03.002

Mueller L, Schindler U, Shepherd TG, Ball BC, Smolentseva E, Hu C, Hennings V, Schad P, Rogasik J, Zeitz J, Schlindwein SL, Behrendt A, Helming K, Eulenstein F (2012b) A framework for assessing agricultural soil quality on a global scale. Arch Agron Soil Sci 58(Supplement 1):S76–S82

Richter A, Hennings V, Müller L (2009) Anwendung des Müncheberger Soil Quality Ratings (SQR) auf bodenkundliche Grundlagenkarten. In: Jahrestagung der DBG : Böden—eine endliche Ressource, Kommission V, September 2009, Bonn: 1-4; Oldenburg (Deutsche Bodenkundliche Gesellschaft). http://eprints.dbges.de/243/. Visited 6, Aug 2011

Shepherd TG (2009) Visual soil assessment. Volume 1. Field guide for pastoral grazing and cropping on flat to rolling country, 2nd ed., Horizons Regional Council, Palmerston North, New Zealand, p 118

Smolentseva E, Sulejmenov MK, Saparov AS, Pachikin KM, Balgabaev NN, Kusainova MD, Bekbaev UK, Rukhovich OV, Lukin SM, Schindler U, Mueller L (2011) Ocenka kachestva i potencial'noj urozhajnosti pochv v global'nom masshtabe. Pochvovedenie i Agrokhimija (4): 81–91

WRB (2006) World reference base for soil resources. A framework for international classification, correlation and communication, World Soil Resources Reports, FAO Rome, pp 103, 145

Use of Pedotransfer Functions for Land Evaluation: Mapping Groundwater Recharge Rates Under Semi-Arid Conditions

Volker Hennings

Abstract The methodology developed for the new Hydrological Atlas of Germany was applied to derive (hydro-) pedotransfer functions to estimate annual percolation rates from available information on climate, soil characteristics and land use in the Arab region. For this purpose the FAO56 concept was applied and the CLIMWAT database and the CROPWAT model were used, based on the single crop coefficient approach. The first step was to carry out simulations for eight countries, three kinds of land use and varying soil hydrological properties. The second step was to use meteorological data from Syria to carry out simulations for six land use scenarios and varying soil hydrological properties. The resulting country-specific regression equations and nomograms are presented as well as the general magnitude of groundwater recharge under typical crops of the Eastern Mediterranean environment. Prediction results are compared with simulation results for the dual crop coefficient approach and practical information is provided about the accuracy of these types of estimation method.

Keywords Land evaluation · Groundwater recharge · Mapping · Pedotransfer function

1 Introduction

For quantitative water resource management in an arid environment, knowledge of the groundwater recharge rate is essential. From the soil scientist's perspective, percolation beyond the lower boundary of the root zone equals groundwater recharge. Under field conditions deep percolation can be measured directly using

V. Hennings (✉)
Federal Institute for Geosciences and Natural Resources, Geozentrum Hannover,
Stilleweg 2 30655 Hannover, Germany
e-mail: volker.hennings@bgr.de

L. Mueller et al. (eds.), *Novel Measurement and Assessment Tools for Monitoring and Management of Land and Water Resources in Agricultural Landscapes of Central Asia*, Environmental Science and Engineering, DOI: 10.1007/978-3-319-01017-5_14,

lysimeters or water fluxes can be calculated from neutron and TDR probe results. An extensive overview of all available methods for determining groundwater recharge is available from Arbeitskreis Grundwasserneubildung der Fachsektion Hydrogeologie der Deutschen Geologischen Gesellschaft (1977) or particularly for semi-arid climates from Kinzelbach et al. (2002). Information on deep percolation can also be obtained by applying simulation models of the soil water balance. A classification scheme of adequate models is given by Wagenet et al. (1991). Models are generally categorized by the manner in which they represent basic processes of water flow. At one extreme are mechanistic models, which have been developed based on the recognition that potential energy gradients are the driving force for water flow. This kind of model is termed mechanistic, as it is presumed to be representative of the best possible understanding of the basic mechanisms of water flow (Wagenet et al. 1991). At the opposite extreme are functional models, which greatly simplify physical processes in order to reduce data demands and computational time (Wagenet et al. 1991). Both are different in terms of input data requirements and representation of soil physical processes. On a regional scale, to avoid expensive field measurements and limitations in the availability of model input parameters, methods are needed which even more simplifying and robust. This type of model, which includes empirical equations and nomograms, involves (hydro-) pedotransfer functions (HPTFs). They are based on input variables that can be determined easily or are available from existing databases.

2 Previous Work Within the Framework of the Hydrological Atlas of Germany

One example from Germany is the development by Wessolek et al. (2008) of HPTFs to predict the long-term percolation rate from soil. The authors used a simulation model of the soil water balance to calculate actual evapotranspiration and percolation for different climatic regions, soils and land use classes. The spectrum of site variations included four soils with different water storage capacities, six groundwater levels, sixteen climate stations whose climate parameter values can be viewed as representative of the climate regions of Germany, and four land use types [cropland (with the typical succession of grain and root crops), grassland, coniferous forest, and deciduous forest]. These conditions resulted in simulation runs for 57,600 years on a daily basis. The results of all the scenarios were analyzed by multiple regression statistics and equations were derived which could be used to make reliable estimates of the target variable. These HPTFs were used to compile a nationwide map of the annual percolation rate from the soil within the framework of the new Hydrological Atlas of Germany (HAD) (BMU 1998, 2001, 2003). To verify the applied method, the mean rate of percolation plus surface runoff in 106 catchment areas of different sizes, land cover, soil properties, and geomorphological and climatic conditions was compared with the runoff measured by gauging stations. Regression analysis shows a good correlation between the measured and calculated values, with the calculated values averaging about 8 % higher than the measured values. A detailed comparison of the results is given by Wessolek et al. (2008).

2.1 Applied Methodology: The FAO56 Approach

The modelling concept applied to simulate the soil water balance is based on some simple fundamentals:

- actual evapotranspiration is calculated by employing empirical crop coefficients,
- all algorithms originate in FAO Irrigation and Drainage Paper No. 56 (Allen et al. 1998),
- the soil is regarded as a one-dimensional storage pool,
- deep percolation is calculated as the remaining term of the soil water budget after evaporation and transpiration demands have been satisfied.

The FAO56 approach uses crop coefficients to modify the reference evapotranspiration to suit crop-specific conditions. The reference surface is a hypothetical grass crop with specific characteristics. The concept of reference evapotranspiration was introduced to study the evaporative demand of the atmosphere independently of crop type, crop development and management practices. Soil factors do not affect ETo. The most simplified version represents the single crop coefficient concept, where differences in the crop canopy and aerodynamic resistance relative to the reference crop of the FAO Penman–Monteith method are accounted for within the crop coefficient, Kc. The Kc coefficient serves as a lumped parameter for the physical and physiological differences between crops. This approach is realized in the CROPWAT model (Clarke et al. 1998).

In a second, more sophisticated approach Kc is split into two factors that separately describe the evaporation (Ke) and transpiration (Kcb) components. The soil evaporation coefficient Ke is maximal when the topsoil is wet, following rain or irrigation. The largest difference between Kc and Kcb is found in the initial growth stage, where evapotranspiration is predominantly in the form of soil evaporation and crop transpiration is still low. Because crop canopies are near or at full ground cover during the mid-season stage, soil evaporation beneath the canopy has less effect on crop evapotranspiration and the value for Kcb in the mid-season stage will be nearly the same as Kc. Depending on the frequency with which the crop is irrigated during the late season stage, Kcb will be similar to (if infrequently irrigated) or less than the Kc value. Figure 1 presents typical shapes for the Kcb, Ke and single Kc curves. The Kcb curve in the figure represents the minimum Kc for conditions of adequate soil water and dry soil surface. All relevant algorithms were published as part of the FAO's Irrigation and Drainage Paper No. 56, "Crop Evapotranspiration" (Allen et al. 1998).

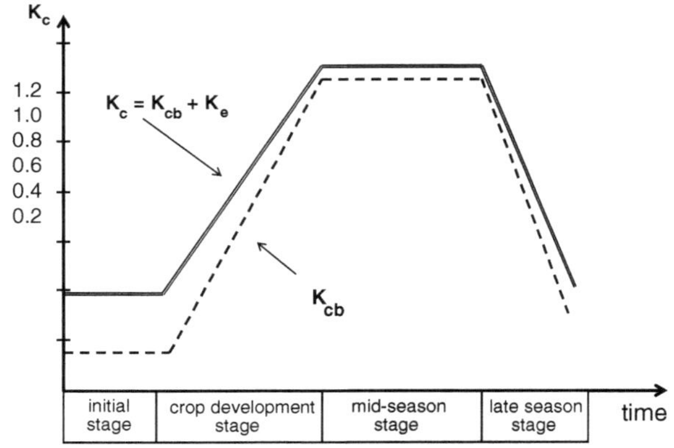

Fig. 1 Crop coefficient curves showing the basal Kcb (*thick line*), soil evaporation Ke (*thin line*) and the corresponding Kc curve (*dashed line*) (based on Allen et al. 1998)

The FAO56 dual crop coefficient approach has been incorporated into a program which has been available since 2011 as part of the WEAP system. WEAP ("Water Evaluation and Planning") is a decision support system for quantitative water resource management (Stockholm Environment Institute 2005). Since 2012, the FAO56 dual crop coefficient approach has also been available as a software tool called SIMDualKc, developed by the Institute of Agronomy from the Technical University of Lisbon (Rosa et al. 2012a).

2.2 *Applied Methodology: CROPWAT Model and CLIMWAT Database*

The methodology applied for the HAD project was used to develop similar nomograms or empirical equations for the countries of the Arab region. For this purpose, the most simplifying functional model (CROPWAT; Clarke et al. 1998) and the most easily accessible database of climatic information (CLIMWAT) were used. Both the CROPWAT software and the CLIMWAT database can be downloaded from the FAO website.

CROPWAT is a decision support system developed by the Land and Water Development Division of the FAO for the planning and management of irrigation. Procedures for calculating the crop water requirements and irrigation requirements are based on the single crop coefficient concept. All calculations are based on the daily soil water balance using various options for water supply and irrigation management conditions.

CLIMWAT 2.0 provides observed agroclimatic data from over 5,000 stations distributed all over the world. CLIMWAT provides long-term monthly mean values for seven climatic parameters required for CROPWAT applications. For this investigation, data were used from 188 meteorological stations from eight Arabic countries (Morocco, Algeria, Tunisia, Libya, Egypt, Jordan, Lebanon,

Syria). Only data from countries with a Mediterranean type of climate were incorporated, while Arabic countries where the main precipitation occurs in summer (such as Yemen) were omitted. Monthly sums of precipitation had to be disaggregated to obtain daily values; in our study an internal algorithm was chosen that distributes rainfall evenly over several days per month, with rain occurring every 5 days. Outside the period of growth, evaporation was calculated using empirical crop coefficients for fallow. Evaporation from bare soil was allowed to take place before and after the period of growth and within its initial phase. Typical kinds of land use in the Mediterranean environment were taken into consideration. Generally, irrigation was simulated up to maximum demand without yield reduction. Five soils were compared, varying in the total available water capacity of the uppermost metre (40, 80, 120, 160, 200 mm). The available water capacity is the pF interval from 2.5 to 4.2. For all simulation runs a fixed amount of surface runoff is assumed to occur, i.e. precipitation in Figs. 2, 3, 4, 5, 6 means effective precipitation or net input.

First, simulation runs were carried out and pedotransfer functions were derived for the entire Arab region; next, country-specific HPTFs were derived for Syria. Six land use scenarios were taken into account: three crops under rainfed agriculture (wheat, barley, olives) were compared with three crops under irrigation (wheat, citrus trees, peas as an example for small vegetables). Therefore, CROPWAT-internal parameter settings were replaced with region-specific plant coefficients, and expert knowledge was incorporated about region-specific agricultural practices (sowing dates, irrigation practices, etc.). The general conditions of the Syrian case study are presented in Table 1.

Table 1 General conditions and summarized results of the country-specific case study

Land use type	Length of growing season	Sowing/ planting date	Period of potential evaporation	Statistically significant input variables of HPTFs for estimating GWR
Winter wheat - no irrigation -	240 days	1Nov	155 days	Prec, ETpot, awc
Barley - no irrigation -	160 days	1Dec	225 days	Prec, ETpot, awc
Olives - no irrigation -	270 days	1March	125 days	Prec, awc
Winter wheat under irrigation	240 days	1Nov	160 days	Prec, ETpot, Prec+Irrig
Citrus trees under irrigation	365 days	(whole year)	60 days	Prec, ETpot, Irrig, awc
Small vegetables under irrigation	100 days	1April	285 days	Prec, ETpot, Prec+Irrig, awc

(Prec = mean annual precipitation; ETpot = mean annual potential evapotranspiration; Irrig = amount of irrigation; GWR = groundwater recharge; awc = soil available water capacity)

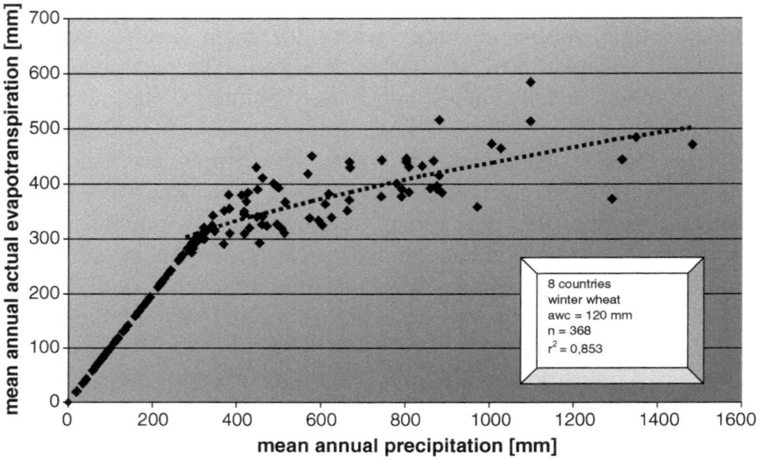

Fig. 2 Annual rate of actual evapotranspiration as a function of annual precipitation for "winter wheat" land use type and derived regression function

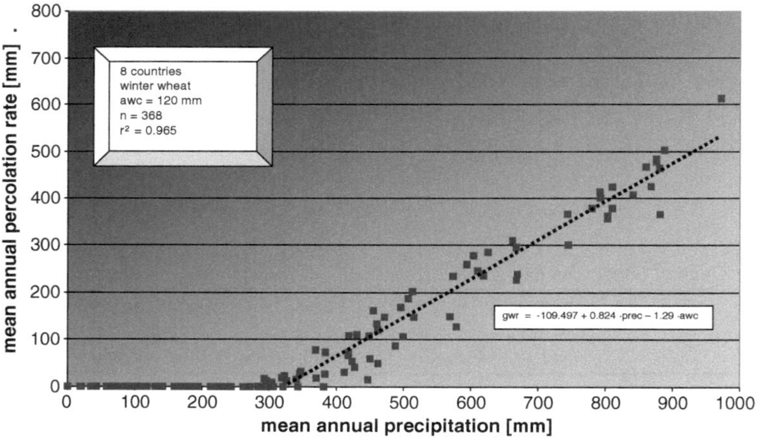

Fig. 3 Annual percolation rate as a function of annual precipitation for "winter wheat" land use type and derived regression function

3 Results and Discussion

3.1 Pedotransfer Functions for Arabic Countries with a Mediterranean Type of Climate

The main characteristics of the soil water balance in a Mediterranean environment, e.g. the Zabadani Basin in the Antilebanon mountains in Syria, are known from field observations and model simulations within the framework of a technical cooperation project between the Arab Center for Studies in Arid Zones and Dry Lands (ACSAD, Damascus, Syria) and the Federal Institute for Geosciences and Natural Resources (BGR, Hannover, Germany) (Droubi et al. 2008). In typical years soil water is stored during the rainy season in winter until the soil water contents are close to field capacity. This status is usually reached in mid-January. As a consequence, groundwater recharge does not start earlier than this. At the beginning of the vegetation period soil water storage completely supplies the needs of plants. During the subsequent weeks the soil water deficit increases; under natural conditions actual evapotranspiration reaches its maximum around 10 May. At this time irrigation usually starts. Under typical conditions (locations close to Barada Spring meteorological station, deeply developed clay loam soils, fruit trees, approx. 650 mm gross precipitation, approx. 12 % surface runoff) without irrigation actual evapotranspiration would not exceed approximately 275 mm and groundwater recharge amounts to approximately 300 mm. In the case of "optimal" irrigation, i.e. where the irrigation amount is always equal to the soil moisture deficit and no yield reduction occurs, actual evapotranspiration increases to approx. 675 mm. Local soils are not completely run dry at the end of the vegetation period; groundwater recharge already starts in November and increases by approx. 95 mm compared to the non-irrigated case (almost 400 mm in total). Under natural conditions recharge is restricted to the winter months of January and February, a period when transpiration is minimal and all incoming water is available for percolation.

In order to develop HPTFs for semi-arid conditions and various land use types, model runs were carried out for 5,640 scenarios from eight Arabic countries. Some examples of results for winter wheat without irrigation and a specific available water capacity are presented in Figs. 2 and 3. ETact increases linearly until a certain amount of annual precipitation is reached, in this case approximately 300 mm (Fig. 2). This amount of rainfall marks the threshold for the occurrence of groundwater recharge. For that reason the fitting procedure of the mathematical model can be restricted to a specific range of precipitation when HPTFs for predicting annual evapotranspiration rates are required (Fig. 2). Figure 3 illustrates the opposite case for percolation rates. Again, the performance of the HPTF can be improved if locations without groundwater recharge are excluded from the regression statistics. When HPTFs for predicting annual percolation rates are required, simple models based on precipitation and available water capacities suffice and provide estimates of high accuracy (Fig. 3). In contrast to the conditions of Central Europe, where rainfall occurs throughout the year, under the climatic conditions of the Middle East the amount of groundwater recharge can be predicted using HPTFs based on simple linear models. In the case shown, groundwater recharge from percolating water does not take place if effective precipitation is below a threshold of $\approx$ 300 mm.

3.2 Country-Specific Pedotransfer Functions for Syria

The results of 1,320 simulation runs were used to derive country-specific HPTFs for Syria, which is represented by 44 meteorological stations in the CLIMWAT database. Regression equations for all land use scenarios are listed in Table 2. Nomograms were developed for specific locations and varying kinds of land use, for specific crops and varying soil properties, and for specific crops and varying irrigation practices. Examples are shown in Figs. 4, 5 and 6.

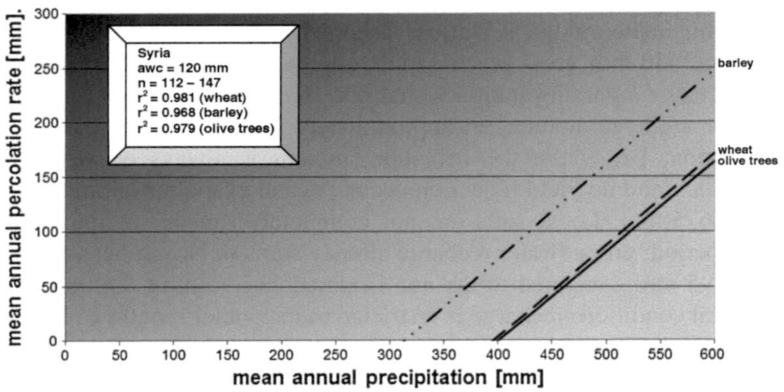

Fig. 4 Nomogram for estimating groundwater recharge in the case of different crops

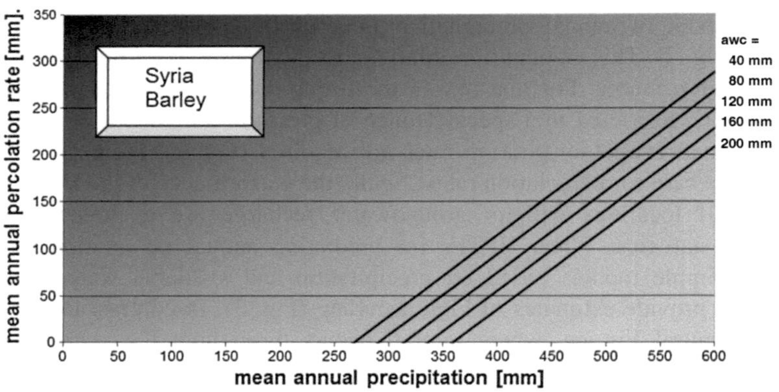

Fig. 5 Nomogram for estimating groundwater recharge in the case of different soil properties

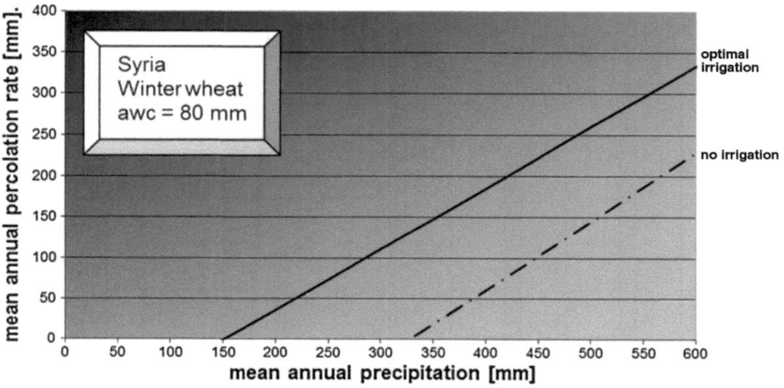

Fig. 6 Nomogram for estimating groundwater recharge in the case of varying irrigation practices

Based on the correlation coefficients in Table 2, the accuracy of these HPTFs is generally high. Accuracy in this context refers to correspondence between the simulation model and pedotransfer function. Simple prediction models as presented here are particularly advantageous where there is a lack of detailed meteorological data of high temporal resolution. (Hydro-) pedotransfer functions that require only easily available input data enable users to avoid time-consuming simulations and can serve as a useful tool to provide reliable estimates of the target variable.

Figures 4, 5, 6 can be used to derive some general rules: annual groundwater recharge rates are higher under crops with shallow rooting depths and short vegetation periods such as small vegetables or barley than under perennial crops with deep root zones such as olives (Fig. 4), rates are higher under shallow, stony or sandy soils with a limited available water capacity than under deeply developed, loamy soils (Fig. 5), and rates are higher in the case of rainfed agriculture than

Table 2 (Hydro-) pedotransfer functions to predict annual groundwater recharge in Syria

Land use type	Equation	Correlation coefficient
Wheat	GWR = −19.39 + 0.807 • Prec – 0.0821 • ETpot – 1.424 • awc	$r^2 = 0.98$
Barley	GWR = −115.8 + 0.845 • Prec – 0.056 • ETpot – 0.495 • awc	$r^2 = 0.97$
Olive trees	GWR = −139.9 + 0.823 • Prec – 1.583 • awc	$r^2 = 0.98$
Irrigated wheat	GWR = −90.8–0.0258 • Prec – 0.332 • ETpot + 0.949 • (Prec+Irrig)	$r^2 = 0.99$
Irrigated vegetables	GWR = −61 + 0.245 • Prec – 0.255 • ETpot – 0.296 • awc + 0.684 • (Prec+Irrig)	$r^2 = 0.99$
Irrigated citrus trees	GWR = −26.275 + 1.02 • Prec – 0.909 • ETpot + 0.026 • awc + 1.093 • Irrig	$r^2 = 0.99$

(Prec = mean annual precipitation; ETpot = mean annual potential evapotranspiration; Irrig = amount of irrigation; GWR = groundwater recharge; awc = soil available water capacity)

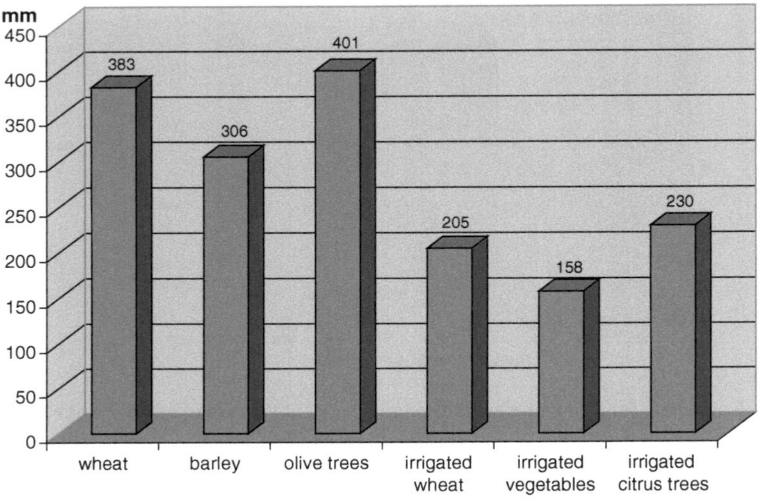

Fig. 7 Minimum annual precipitation necessary for groundwater recharge for six agricultural crops in the case of moderate soil available water capacity

Fig. 8 Predicted groundwater recharge rate [mm] in 2009/10 in the Aleppo Sub-basin when arable land is used for the cultivation of non-irrigated winter wheat

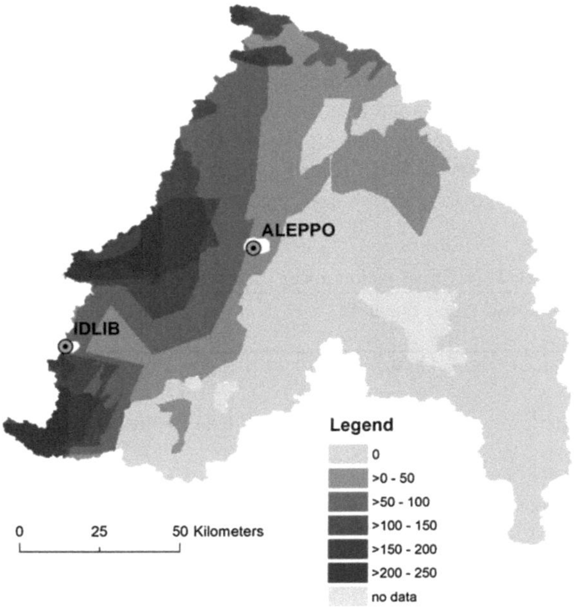

under conditions of irrigation agriculture. The last conclusion is illustrated by Fig. 6: thresholds for the beginning of groundwater recharge in terms of annual precipitation are generally higher for non-irrigated crops than for crops that depend on irrigation in summer. Against this background the following order of mean annual groundwater recharge under standardized conditions along land use types can be derived from model simulations: olive trees < wheat < barley < irrigated citrus trees < irrigated wheat < irrigated vegetables (Fig. 7). When isoline maps of annual precipitation and potential evapotranspiration are available for a specific region as well as soil maps, the prediction equations from Table 2 can be used for land evaluation and thematic maps of land qualities can be compiled. For Fig. 8 information on precipitation in 2009/10 was derived from data originating from NASA's "Tropical Rainfall Measuring Mission" (TRMM) (Schlote 2011), and information on local soil properties was taken from a 1:1,000,000 soil map, published by ACSAD (1985).

4 Practical Information: Accuracy of Estimation Results

The validity of the FAO56 single and dual crop coefficient approach, respectively has been tested and evaluated by several authors (Bodner et al. 2007; López-Urrea et al. 2009; Liu and Luo 2010; Rosa et al. 2012b). Most of the examples illustrate the robustness of the approach, describe the broadness of application options and classify it as a reliable modelling tool to provide accurate results. Liu and Luo (2010) evaluate the approach against lysimeter data from the North China plain and report that the method effectively simulated the quantity of seasonal evapotranspiration for winter wheat but calculated the peak values inaccurately. Rosa et al. (Rosa et al. 2012b), who use rainfed wheat from Northern Syria as one regional example, observe good agreement between the available soil water content observed in the field and that predicted by the model. They furthermore report that the calibrated model does not tend to over- or underestimate available soil water in the course of a season, and that the model also performed relatively well prior to calibration and using standard values for many parameters.

A real validation of estimation results obtained by (hydro-) pedotransfer functions requires measurement data from field experiments in the area of interest. Those data are available from several studies, e.g. a case study on wheat at ICARDA's research farm at Tel Hadya, 30 km south of Aleppo (Rosa et al. 2012b). It is located in the area shown in Fig. 8. Cited results refer to the time interval of 1992/93, which was a comparatively dry year with an annual precipitation of approximately 280 mm. All incoming water was used for evaporation or water consumption by crops and no deep percolation occurred. This result was also predicted by the first equation from Table 2. As a consequence, the dry conditions of 1992/93 do not allow any real validation. Another chance to evaluate estimation results using HPTFs is provided by discharge data from the Quaik River, registered within the framework of the Syrian—German Technical Cooperation Project "Advisory Service for the Ministry of Irrigation in the Geo-Environmental Sector" (Schlote 2011). In developing its capacities for applying a standardized methodology to calculate annual water balances for Syrian water basins, the project is supported by the

Table 3 Comparison of predicted percolation rates under barley at Tel Hadya, 2003/04, for different modelling approaches, varying in details of the FAO56 concept and the temporal resolution of rainfall records

	Conditions of modelling		
Approach	Single crop coefficient	Dual crop coefficient	Dual crop coefficient
Climatic variables	Daily values disaggregated from monthly data	Daily values disaggregated from monthly data	Original daily records
Evaporation		202 mm	173 mm
Transpiration	329 mm	197 mm	189 mm
Percolation	71 mm	1 mm	38 mm

Ministry of Irrigation (Schlote 2011). The discharge data refer to the western part of the Aleppo Sub-basin as shown in Fig. 8, and the time of gauging also corresponds to the period of time used for Fig. 8. The general magnitude of estimated groundwater recharge is confirmed by local discharge data. The discharge of the Quaik River, however, aggregates recharge rates from different soil and land use classes and can only be compared with aggregates from other pedotransfer functions. Against this background the validation of a single HPTF, designed for a specific crop, is not possible on the basis of locally available data.

The accuracy of prediction models such as those shown in Table 2 also depends on the dataset of meteorological variables used for the development of pedotransfer functions, and particularly on their degree of temporal resolution. The methodology applied to develop empirical equations for Syria is based on the single crop coefficient approach. It uses the monthly means of all required climatic parameters, because they are easily accessible via the CLIMWAT database. The limited degree of temporal resolution requires the application of disaggregation techniques, i.e. monthly sums of precipitation have to be distributed over several days. The results of functional models based on the FAO56 concept differ considerably when (1) the dual crop coefficient approach is applied instead of the single coefficient, and when (2) real data of daily temporal resolution are used for modelling. The impact on the modelling results is illustrated on climatic data recorded at ICARDA's research farm at Tel Hadya and kindly made available for this study. For a comparison of three approaches (Table 3) the hydrological year 2003/04 was selected, which was comparatively wet (annual precipitation of approximately 400 mm). The conditions used were for non-irrigated winter wheat and barley as documented in Table 1 and a moderate available water capacity of 120 mm.

The results for deep percolation under wheat are not presented in Table 3 because an annual precipitation of 400 mm is close to the minimum recharge requirements (see Figs. 4 and 7) and therefore the estimated recharge rates differ only very slightly. In the case of barley, differences in the general conditions lead to more varying results. When the single crop coefficient approach is applied and monthly precipitation data are used (left column), 71 mm of recharge are calculated, and this result corresponds to predictions using HPTFs. The application of the dual crop coefficient approach and use of monthly data reduces percolation considerably. In this case precipitation is distributed evenly, rainfall occurs every 5 days, the Ke coefficient from Fig. 1 reaches maximum values of $\approx$ 1.2 every five days, evaporation from bare soil occurs every day, and total evaporation is overestimated. If original daily records are used (right column) evaporation from bare soil is again reduced, and the magnitude of evaporation probably comes closer to reality. If the general conditions of the right column represent the most realistic approach, it has to be assumed that the application of empirical equations from Table 2 tends to overestimate percolation rates from the soil.

In general (hydro-) pedotransfer functions that require only easily available soil, crop and climate information provide reliable estimates of the groundwater recharge rate for most typical land use types in Syria. As a consequence of empirical knowledge about the FAO56 approach and the results from Table 3, the accuracy of HPTFs as listed in Table 2 can be roughly assessed thus: annual groundwater recharge rates in the Arab region or Syria respectively will be overestimated by 30–40 mm. A more detailed validation of prediction results can only be carried out on the basis of available measurements from test sites and field experiments.

References

Arab Center for the Studies of Arid Zones and Dry Lands (ACSAD) (1985) Soil Map of Arab Countries, Soil Map of Syria & Lebanon 1:1.000.000. Prepared by Ilaiwi M. Damaskus. http://eusoils.jrc.ec.europa.eu/esdb_archive/EuDASM/asia/lists/csy.htm

Arbeitskreis Grundwasserneubildung der Fachsektion Hydrogeologie der Deutschen Geologischen Gesellschaft (1977) Methoden zur Bestimmung der Grundwasserneubildungsrate. Geol Jb C 19: 3–98, Hannover

Allen RG, Pereira, LS, Raes D, Smith M (1998) Crop evapotranspiration: Guidelines for computing crop water requirements. FAO Irrigation and Drainage Paper, 56, Rome

Bodner G, Loiskandl W, Kaul H-P (2007) Cover crop evapotranspiration under semi-arid conditions using FAO dual crop coefficient method with water stress compensation. Agric Water Manage 93:85–98

Bundesministerium für Umwelt, Naturschutz und Reaktorsicherheit (BMU) (ed) (1998, 2001, 2003) Hydrologischer Atlas von Deutschland (Hydrological Atlas of Germany). 1st Delivery 1998, 2nd Delivery 2001, 3rd Delivery 2003, Bonn

Clarke D, Smith M, El-Askari KH (1998) CropWat for Windows: user Guide. Land and Water Development Division of FAO, Institute of Irrigation and Development Studies (IIDS) at Southampton University & National Water Research Center (NWRC) of Egypt. http://www.fao.org/waicent/faoinfo/agricult/agl/aglw/cropwat.htm

Droubi AM, Al-Sibai M, Abdallah A, Zahra S, Obeissi M, Wolfer J, Huber M, Hennings V, Schelkes K (2008). A decision support system (DSS) for water resources management: design and results from a Pilot Study in Syria. In: Zereini F, Hötzl H (eds) Climatic changes and water resources in the middle east and north africa. Environmental Science and Engineering, vol 20. Springer-Verlag, Berlin, pp 199–225

Kinzelbach W et al (eds) (2002). A survey of methods for groundwater recharge in arid and semi-arid regions. Early Warning and Assessment Report Series, UNEP/DEWA/RS.02-2. United Nations Environment Programme, Nairobi, Kenya

Liu YJ, Luo Y (2010) A consolidated evaluation of the FAO-56 dual crop coefficient approach using the lysimeter data in the North China Plain. Agric Water Manage 97:31–40

López-Urrea R, Martin de Santa-Olalla F, Montoro A, López-Fuster P (2009) Single and dual crop coefficients and water requirements for onion (Allium cepa L.) under semiarid conditions. Agric Water Manage 96:1031–1036

Rosa RD, Paredes P, Rodrigues GC, Alves I, Fernando RM, Pereira LS, Allen RG (2012a) Implementing the dual crop coefficient approach in interactive software. 1. Background and computational strategy. Agric Water Manage 103:8–24

Rosa RD, Paredes P, Rodrigues GC, Alves I, Fernando RM, Pereira LS, Allen RG (2012b) Implementing the dual crop coefficient approach in interactive software. 2. Model testing. Agric Water Manage 103:62–77

Schlote A (2011) WEAP-Component for Aleppo basin water balance, Hydrological Year 2009/10. Project-internal report of the BMZ-funded project "Advisory Service to the [Syrian] Ministry of Irrigation in the Geo-Environmental Sector". Federal Institute for Geosciences and Natural Resources, Hannover

Stockholm Environment Institute (2005) WEAP: water evaluation and planning system. User Guide for WEAP21, Stockholm

Wagenet RJ, Bouma J, Grossman RB (1991) Minimum data sets for use of soil survey information in soil interpretive models. In: Mausbach MJ, Wilding LP (eds). Spatial variability of soils and landforms. SSSA Special Publication No., vol 28, pp 161–182, Madison, Wisconsin

Wessolek G, Duijnisveld WHM, Trinks S (2008) Hydro-pedotransfer functions (HPTFs) for predicting annual percolation rate on a regional scale. J Hydrol 356:7–27

Nutrient Balances in Agriculture: A Basis for the Efficiency Survey of Agricultural Groundwater Conservation Measures

Frank Eulenstein, Marion Tauschke, Marcos Lana, Askhad K. Sheudshen, Ralf Dannowski, Roland Schindler and Hartwig Drechsler

Abstract In nutrient balances, additions and removals of nutrients are assessed to identify the remaining concentrations of nutrients in soil. Balances can be performed using operational records of nutrients applications and other agronomic information (crops, yields, weather, etc.) at farm or even field level. The aim of performing nutrient balances is to obtain an overview of nutrient levels, in particular to prevent surpluses that could lead to environmental problems such as groundwater contamination, water eutrophication, air pollution and an increase in greenhouse gas emissions. This chapter will provide an overview of methods used to assess soil nutrient levels at farm and field level. The methods described here can be used by farmers, landscape planners, environmentalists, politicians and other stakeholders as a basis for taking agricultural groundwater conservation measures. The procedures and recommendations specified in this chapter are in accordance with the guidelines of the DWA—German association for water, wastewater and waste: "Efficiency of measures to control land use for groundwater conservation—the example of nitrogen" (DWA-M 911 (2013): Effizienzkontrolle von Maßnahmen zur grundwasserschonenden Bodennutzung am Beispiel des Stickstoffs).

F. Eulenstein (✉) · M. Tauschke · M. Lana · R. Dannowski
Leibniz Centre for Agricultural Landscape Research (ZALF) e. V., Eberswalder Strasse 84
15374 Muencheberg, Germany
e-mail: feulenstein@zalf.de

R. Schindler
NEW Niederrhein Wasser GmbH, Rektoratstrasse 18 41747 Viersen, Germany

H. Drechsler
Dr. Drechsler Ingenieurdienst, Hannah-Vogt-Str.1 37085 Göttingen, Germany

A. K. Sheudshen
Kuban State Agrarian University, 13 Kalinin Str., 350044 Krasnodar, Russian Federation
e-mail: kubagrohim@mail.ru

L. Mueller et al. (eds.), *Novel Measurement and Assessment Tools for Monitoring and Management of Land and Water Resources in Agricultural Landscapes of Central Asia*, Environmental Science and Engineering, DOI: 10.1007/978-3-319-01017-5_15,

Keywords Agriculture · Groundwater conservation · Nutrient balance

1 Nutrients in Agro-Ecosystems

Agriculture is a multifunctional activity that goes way beyond the mere production of food and fibre. It acts as a source of fuels and raw materials, giving it a cultural value. In addition, agriculture increasingly plays an important role in agro-ecosystem conservation and recuperation. One of the tools that are relevant for the above-mentioned aspects are nutrient balances (Sutton et al. 2013).

Nutrient balances constitute an instrument to basically evaluate the nutrient-related situation of agro-ecosystems. They are also capable of evaluating the environmental safety of different cropping systems. This evaluation is crucial for providing an overview of nutrient surpluses in intensive agricultural regions. Trends in nutrient consumption, particularly nitrogen and phosphate, are relatively stable in developed countries, despite being used in elevated quantities. The major increase in nutrient consumption, therefore, will occur in emergent economies, where improvements in living conditions, together with the growing population, triggers a growing use of nutrients. This potential increase in nutrient use is particularly high in regions of Africa, Latin America and parts of Asia (especially Central Asia), where there are still large regions in which few nutrients are used (FAO 2008; Sutton et al. 2013). The growth in fertiliser use—especially highly soluble ones—goes hand in hand with an increase in the risk of contaminating soil, water bodies and groundwater.

When nutrients are introduced into the environment in quantities exceeding the agro-ecosystems' extraction or absorption capacity, they become polluting agents. As an example, nitrogen and phosphate, which are very important agricultural nutrients that are also commonly associated with pollution, can contribute to a number of environmental concerns, including water eutrophication, air pollution, groundwater contamination and greenhouse gas emissions.

Regulations governing nutrient levels in soils or water are increasingly being used by government regulatory agencies as a parameter for compliance with environmental regulations. The requirements for documentation of agricultural production processes are currently very high: this applies, for example, to the documentation terms within fertiliser regulations, the EU's Water Framework Directive, or in connection with local environment acts governing drinking water protection areas or nature protection areas. At the same time, the demands placed on traders are rising: in addition to the traded goods, negotiators demand documentation on the methods used for plant production.

When the agronomic management of plant production (which is associated with a flow of nutrients) is properly recorded and documented, it is easy to perform nutrient balances.

In nutrient balances, nutrient sources and sinks are compared in a system. This can be carried out in form of complete operational balance sheets at farm or field level, depending on how detailed the available information is.

In order to prepare nutrient balance sheets, at least three years of farm/field monitoring or assessment are required. This is a mandatory requirement for properly identifying the contribution made by nutrients and agricultural management, enabling the impact of weather on nutrient dynamics and balance to be identified. After 3 years of monitoring, moving averages are used as an indicator of the desired parameters.

2 Control of Nutrient Cycles to Protect Water Resources

Nutrient balances are acknowledged tools for estimating nutrient loadings of nitrogen and phosphate to agricultural soils. European Community fertilisation law, for example, requires the preparation of an operational nutrient comparison for nitrogen and phosphate for the previous agricultural year. This comparison can alternatively be performed by completing an "area balance sheet" or an aggregated field balance, which is then summarised in a long-term nutrient comparison. For this purpose, according to §6 of the German Directive of Fertilisation (*Düngemittelverordnung*, DüV 2007), the maximum nutrient surplus of phosphate is 20 kg/ha per year). Since 2011, a maximum surplus level of 60 kg/ha per year has been permitted for nitrogen. These surplus levels are extremely high, exceeding the overall consumption levels of most agricultural regions worldwide. In other regions of the world, however, permissible nutrient surplus levels may differ to those in the EU and its Member States, or are non-existent. Whilst thresholds differ, the same methods can be applied to detect and control them.

Some direct and indirect methods of measuring, balancing and modelling nutrient cycles and recognising harmful quantities of nutrients for water bodies are acknowledged (Eulenstein et al. 2006, 2008; DWA M-911 2013; Fig. 1). This chapter focuses on nutrient balances.

The nutrient balance, structured like an area balance sheet, is performed by calculating all of the nutrient inputs and those that have been removed from the land. However, the accurate estimation of nutrient balances aiming at efficiency control and refined operational nitrogen management can be very complex, particularly in agro-ecosystems involving livestock production.

Two main approaches regarding nutrient balances can be used: the farm balance and the field balance, both of which feature advantages and disadvantages, which will be discussed below. Farm balances are usually easier to conduct since it is easy to gather the information required for the nutrient balance. Field balances, on the other hand, require greater precision in order to identify the quantity of inputs applied to the field, and hence outputs.

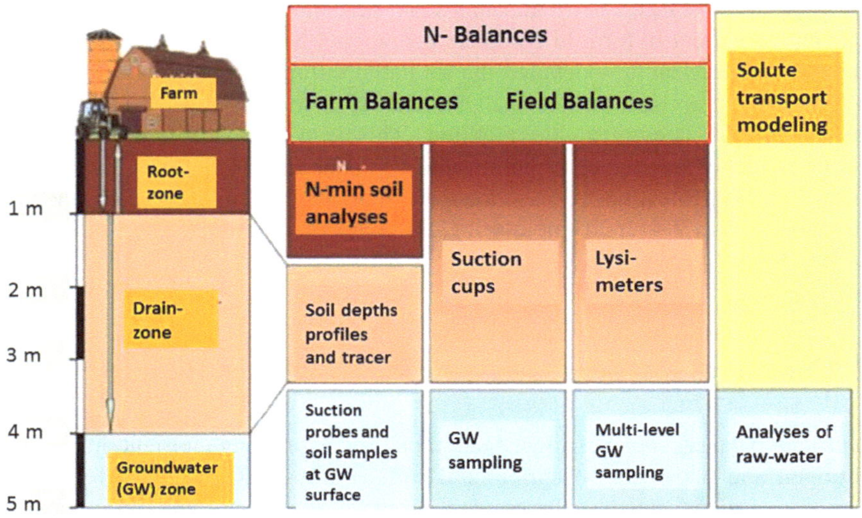

Fig. 1 Acknowledged methods for estimating agricultural nutrient surpluses for the purpose of groundwater protection (example of nitrogen) (DWA-M 911 2013)

3 Farm Balances

With farm balances, the input of nutrients for the whole establishment is compared with the amount of nutrients that leave the establishment from (animal and vegetable) production or residues (of fertilisers, fermenting remains, and so on). Gaseous nitrogen losses from livestock production or fertiliser storage (especially ammoniac) can be described and considered in the nitrogen balance as having been removed from the farm. When adequately described, individual balance sheets are a good representation of the nutrient balance, providing insight into the nitrogen-disperse potential over the soil, water and air of the entire business. Table 1 shows typical examples of inputs and outputs of nutrients; Eq. 1 is used to calculate the average annual nutrient surplus in kg/ha for a farm.

Table 1 Calculation of the nutrient balance at farm level

Nutrient imports due to:	Nutrient exports due to:
Mineral fertiliser	Plant products
Animal manure	Animal products
Digestate, clearing sludge and compost	Animal manure, crop residues
N-fixation by legumes	Losses (gas)
Animals	
Feedstock	

$$\text{Nutrient surplus} = \left(\sum_{i=1}^{n} \left(imports_{kg.year^{-1}} \right) - \sum_{i=1}^{n} \left(exports_{kg.year^{-1}} \right) \right) \Big/ Farmarea(ha) \tag{1}$$

Using supporting reference information related to the location of the farm, the type of soil and other agronomic characteristics, the contamination potential can be evaluated and reported to decision-makers, who can take action to avoid contamination, if required.

3.1 Advantages of Farm Balances

The advantage of farm balances is that calculations of input and output factors are simple to perform. The results are also very accurate at farm level, providing an assessment of nutrient surpluses on a farm or field. Comparing different years, management strategies or agro-ecosystem settings enables potential improvements for nutrient use to be highlighted.

In addition, farm balances enable farms' inputs and outputs and their dynamic results to be compared. This feature also leads to the identification of the most successful practices, which can possibly be adopted by other establishments.

3.2 Disadvantages of Farm Balances

Nitrogen-surplus results cannot easily be upscaled and attributed to similar regions.

The nitrogen surplus does not always provide absolute values, due to the complex temporal and spatial interactions of the nutrients and environmental factors. The nitrogen surplus cannot be used to determine the application of animal manure for crops, especially because it does not provide any information at the level of accuracy required for crop fertilisation, nor discern within-field spatial variability. For the above-mentioned reasons, nutrient balances at farm level cannot provide the adequate information required for water preservation purposes. In cases where governmental regulations determine average levels per farm, the potentiality of the method is also reduced due a relatively low degree of accuracy. Another shortcoming of nutrient balances is the data quality of inputs and outputs: nutrient concentrations of the manure used for fertilisation, yield and residues are usually estimated, whereas data concerning livestock production are defined arbitrarily. Both field and laboratory records are necessary in order to obtain a well-controlled system, which can increase monitoring costs significantly.

3.3 Suitability of Farm Balances for the Efficiency Control of Groundwater Protection

This method is especially suitable for developing a first-level risk assessment of the whole farm, providing information to analyse the nitrogen-surplus risk, enabling the prediction of potential action to be taken in order to mitigate the risk. In this case, the farm balance is to be considered as a component of the farm environment assessment. It is, at first instance, merely a method for internal farming process control, not intended as a substitute for more detailed inspections usually conducted by agents from regulatory agencies. The results of nutrient balances can be validated by using field samples to compare expected and observed nutrient levels. The same procedure can also help to calibrate the nutrient balance by reporting the initial soil conditions. In combination with the standard sampling procedure, nutrient balances assist the long-term evaluation of farms or fields. The information yielded supports farm self-management, providing information about nutrient inputs and outputs.

4 Field Balances of Nitrogen

Nitrogen balances provide information about the management of individual fields. In this case, the amount of nitrogen fertilisers applied, and the use of leguminous crops are accounted for as nitrogen input. Additional inputs are also considered, such as organic and inorganic sources of nitrogen. In the case of organic fertilisation, no distinction is made between the various N fractions, i.e. 100 % of the nitrogen content of the source accounts for the balance. Gaseous losses are not considered in this case. For output calculations, all harvested yield or other removed materials must be taken into account. The nitrogen content of each fraction must be known in order to quantify the amount of nitrogen exported. To avoid errors, strict control of the location and the size of the area is mandatory in this approach.

4.1 Advantages of Field Balances

Nitrogen balances at field level enable the eutrophication risk considering different soils, crops and agronomic management practices to be appraised. The information generated can be used in agronomic decision-making, leading to the correction of management errors that could otherwise cause inefficient nutrient use or contamination.

4.2 Disadvantages of Field Balances

As with farm balances, the temporal connection between nitrogen application and nitrogen demand is inadequately considered in field balances. As an example, nitrogen sources such as manure, biogas sludge, cover crops or other residues can release their nutrients after the cropping season. In this case, off-season nitrogen mineralisation can occur, even during the winter months, which could particularly affect groundwater. Again, the input data quality plays a crucial role in the nutrient balance: nitrogen levels of different nutrient sources—such as animal manure, digestate, clearing sludge and compost—are often assumed based on averages. In the same way, the concentration of nitrogen in harvested products is estimated, leading to probable inaccuracies in the balancing of accounts. In addition, the spatial balance of nitrogen requires records for each field, including inputs, agronomic operations and yield, as well as the production and removal of residues. This implies high efforts for data collection and recording, which may call for human and financial resources.

4.3 Suitability of Farm Balances for the Efficiency Control of Groundwater Protection

For the small-scale assessment and consideration of soil and climatic conditions, this balance can provide information to enable models to be run that calculate the nitrate-exportation potential and the potential nitrate concentration in groundwater. In this case, depending on the quality and scale of application of the models, results can be very accurate, provided that the input data is sufficiently accurate. The validity and accuracy of field balances increases with the length of the balancing period. In order to carry out comparable analyses, it is recommendable to execute field balances over a period of at least one crop rotation (average of 3 years) in combination with additional farm balances for the same period. Figure 2 shows the schematic presentation of nutrient dynamics in an agro-ecosystem, also describing nutrient imports and exports. In order to appraise the nitrogen balance of a field, the parameters presented in Table 2 can be used, applying the same equation as that for farm level assessment.

On the whole, balances can be used as a guide for agricultural farm management. The informative value of these balances basically depends on the accuracy of the input data reflecting the quality of farm management. Thus, farms that are critical from a water management perspective can often only be appraised inadequately by field balances alone. Field balances are not recommended for large-scale analyses (e.g. entire water catchments) due to the large amount of work required for their parameterisation and the potential lack of accuracy involved.

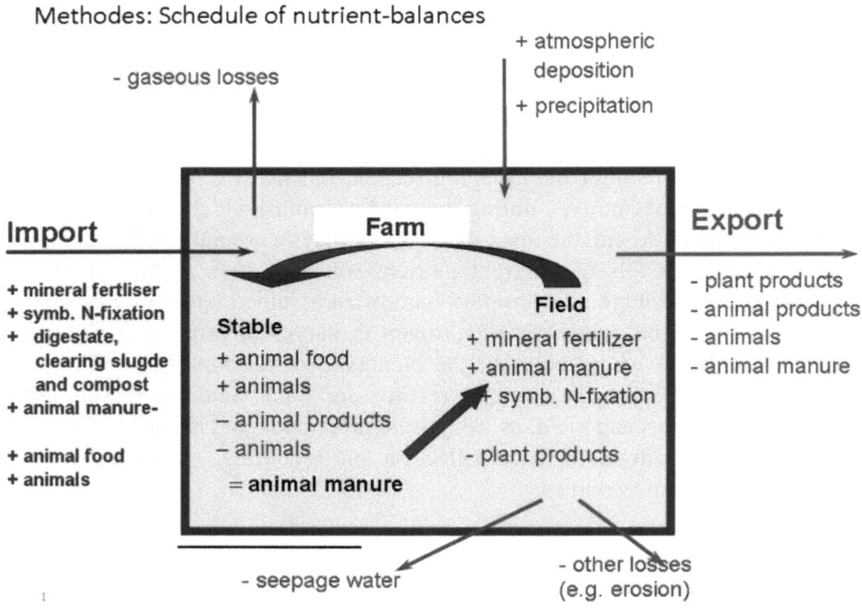

Fig. 2 Elements of nutrient balances

Table 2 Examples of sources and extractors of nutrients for nitrogen balances at field level

Nutrient imports due to:	Nutrient exports due to:
Mineral fertilizer	Plant products
Animal manure	
Digestate, clearing sludge and compost	
N-fixation by legumes	
Quantity of imports	Quantity of exports
Nutrient surplus from field balance = quantity of imports – quantity of exports (kg/yr)	

5 Phosphate Field Balances as an Indicator of Organic Nitrogen Supply

The application of organic fertilisers, such as animal manure, digestate, clearing sludge and compost, harbours the risk of over-fertilising the soil over time. Limiting the quantity of nutrients must not focus on nitrogen alone, but must also involve phosphate. This problem is particularly important for farms where livestock production is performed alongside agriculture. For this reason, in addition to nitrogen, the adequate supply of soils with phosphate, in particular, has to be controlled according to the extraction capacity of this nutrient. When soil levels of phosphate are already sufficient or high, fertilisation should be calculated in order to maintain or reduce these levels. For this reason, soil samples must be taken and analysed prior to applying more fertilisers or other nutrient sources.

Pig manure, pig liquid manure and chicken dung contain high concentrations of phosphate, limiting the amount of these materials that can be applied to fields. For this reason, fertilisation based on one or a few organic sources is usually not balanced in terms of phosphate and nitrogen: the level of the first nutrient is usually achieved due to a lack of the second. As a general recommendation, phosphate (in the form of P_2O_5) could be applied with a surplus of 20 kg/ha over a six-year period. For this reason, calculations to balance nitrogen and phosphate are more complex, and will usually include other nutrient sources in order to reach the desired nutrient levels for the field. Phosphate balances can also be performed using the same procedure as for nitrogen balances.

5.1 Advantages of Phosphate Field Balances

Phosphate field balances are easy to establish, provided that the nutrient concentrations in the source materials are determined before. For calculations, it can be assumed that 100 % of the phosphate in mineral and organic fertilisers is available. Losses by leaching or gaseous losses are usually not considered due to the relatively slow mobility of this nutrient (compared to nitrogen). Phosphate reference levels in the different exports (yield, residues) are relatively stable.

5.2 Disadvantages of Phosphate Field Balances

As with nitrogen balances, phosphate balances also rely on the accuracy of input information to generate assessments with a higher degree of accuracy. The phosphate levels of organic fertilisers must also be known prior to application to the field. Unfortunately, the phosphate levels of inputs are usually estimated, causing inaccuracy.

5.3 Suitability of Phosphate Field Balances for the Efficiency Control of Groundwater Protection

On the basis of phosphate balance sheets, supplemented with data from soil analyses, farm managers can easily evaluate whether the field has been overcharged with phosphate in previous years or cropping seasons. Exaggeratedly high levels of phosphate in the soil usually indicate that organic fertilisation was used, albeit in amounts that surpassed the adequate levels, at least for phosphate. This suggests that the next applications of fertiliser must be calculated in order to meet the demand for nutrients that are below the optimal level—nitrogen, for example—and avoid over-applying nutrients that already are in excess. This procedure can help to avoid the pollution and contamination of the soil and groundwater, preventing potential phytotoxicity problems. Another advantage of phosphate field balances is that fertilisation costs can be rationalised.

6 Discussion

Major elements of the nitrogen balance, such as leaching, were studied in detail in parts of Europe several years ago (Eulenstein and Drechsler 1992; Mueller et al. 2001, 2005; Eulenstein et al. 2003; Schindler et al. 2008). Thus, water flow pathways and data, and orientation values of nutrients can be used for balancing procedures in these regions.

Nutrient balancing methods can be transferred to other regions such as Central Asia, although more research is required to quantify balance input parameters there. Different values were found in sodic and saline soils under irrigation conditions (Devkota et al. 2013). In semi-arid climates, very small nutrient surpluses become harmful agents for water bodies. Thus, carefully established balances in combination with advanced methods of field monitoring, such as lysimeters or soil hydrological set-ups, will provide improved information for ascertaining and controlling nutrient cycles in agro-ecosystems of Central Asia.

References

Devkota KP, Manschadi A, Lamers JPA, Devkota M, Vlek PLG (2013) Mineral nitrogen dynamics in irrigated rice–wheat system under different irrigation and establishment methods and residue levels in arid drylands of Central Asia. Eur J Agron 47:65–76

DWA-M 911 (2013) "Möglichkeiten der Effizienzkontrolle von Maßnahmen zur grundwasserschonenden Bodennutzung am Beispiel des Stickstoffs". Deutsche Vereinigung für Wasserwirtschaft, Abwasser und Abfall e. V., Hennef

DüV (2007) Düngeverordnung Verordnung über die Anwendung von Düngemitteln, Bodenhilfsstoffen, Kultursubstraten und Pflanzenhilfsmitteln nach den Grundsätzen der guten fachlichen Praxis beim Düngen in der Fassung der Bekanntmachung vom 27. Februar 2007, BGBl. I S. 221. Stand: zuletzt geändert durch Artikel 18 des Gesetzes vom 31. Juli 2009, BGBl. I S. 2585

Eulenstein F, Drechsler H (1992) Ursachen, Differenzierung und Steuerung der Nitratkonzentration im Grundwasser überwiegend agrarisch genutzter Wassereinzugsgebiete. Institut für Bodenwissenschaft der Georg-August-Universität Göttingen. Eigenverlag, Göttingen, p 269

Eulenstein F, Schindler U, Müller L (2003) Feldskalige Variabilität von Frühjahrsfeuchte und mineralischem Stickstoff in der ungesättigten Zone sandiger Standorte Nordostdeutschlands. Arch Agron Soil Sci 49(2):187–201

Eulenstein F, Olejnik J, Willms M, Kersebaum K-C (2006) Bestimmung tolerierbarer Stickstoff-Überschüsse in Agrarlandschaften mit Hilfe von Simulationsmodellen. In: Land- und

Ernährungswirtschaft im Wandel—Aufgaben und Herausforderungen für die Agrar- und Umweltinformatik. Referate der 26. GIL Jahrestagung 2006, 57-60. Gesellschaft für Informatik, Bonn

Eulenstein F, Werner A, Willms M, Juszczak R, Schlindwein SL, Chojnicki BH, Olejnik J (2008) Model based scenario studies to optimize the regional nitrogen balance and reduce leaching of nitrate and sulfate of an agriculturally used water catchment. Nutr Cycl Agroecosyst 82(1):33–49

FAO (2008) Current world fertilizer trends and outlook to 2011/12. FOOD AND AGRICULTURE ORGANIZATION OF THE UNITED NATIONS, Rome, 2008, p 57. ftp.fao.org/agl/agll/docs/cwfto11.pdf, accessed on 01.04.2013

Mueller L, Augustin J, Eulenstein F, Seeger J, Meissner R, Behrendt A, Schindler U (2001) Einfluss unterschiedlicher Böden auf die Ausnutzung von Wasser und Stickstoff durch Weidelgras, Sommergerste und Mais auf einigen Grundwasserstandorten. Archiv für Acker- und Pflanzenbau und Bodenkunde 47:277–292

Mueller L, Wirth S, Augustin J, Tauschke M, Schindler U, Behrendt A, Eulenstein F, Meißner R (2005) Verlagerung von Agrochemikalien und Enzymen durch Dränsysteme (Lysimeterexperimente). Mitteilungen der Deutschen Bodenkundlichen Gesellschaft (106):45–46

Schindler U, Müller L, Eulenstein F, Dannowski R (2008) A long-term hydrological soil study on the effects of soil and land use on deep seepage dynamics in Northeast Germany. Arch Agron Soil Sci 54(5):451–463

Sutton MA, Bleeker A, Howard CM, Bekunda M, Grizzetti B, de Vries W, van Grinsven HJM, Abrol YP, Adhya TK, Billen G, Davidson EA, Datta A, Diaz R, Erisman JW, Liu XJ, Oenema O, Palm C, Raghuram N, Reis S, Scholz RW, Sims T, Westhoek H, Zhang FS, with contributions from Ayyappan S, Bouwman AF, Bustamante M, Fowler D, Galloway JN, Gavito ME, Garnier J, Greenwood S, Hellums DT, Holland M, Hoysall C, Jaramillo VJ, Klimont Z, Ometto JP, Pathak H, Plocq Fichelet V, Powlson D, Ramakrishna K, Roy A, Sanders K, Sharma C, Singh B, Singh U, Yan XY, Zhang Y (2013) Our nutrient world: the challenge to produce more food and energy with less pollution. Global Overview of Nutrient Management. Centre for Ecology and Hydrology, Edinburgh on behalf of the Global Partnership on Nutrient Management and the International Nitrogen Initiative. http://www.gpa.unep.org/downloads/doc_download/255-our-nutrient-world.html (accessed on 03.04.2013)

Methods of In Situ Groundwater Quality Monitoring: Basis for the Efficiency Survey of Agricultural Groundwater Conservation Measures

Ralf Dannowski, Roland Schindler, Nils Cremer and Frank Eulenstein

Abstract Against the background of the paramount importance of fresh water resources for human well-being in Europe's industrialised societies, the protection of groundwater bodies is one of the top themes of the environmental ambitions of the European Union. This holds true especially for areas under predominantly agricultural use against nitrates from plant production. A catalogue of agricultural measures of groundwater conservation is available in general. It is to be supplemented by a methodology for in situ monitoring of the groundwater quality as the basis for surveying the efficiency of those measures. General characteristics, the benefits, and the disadvantages of recent monitoring methods are presented, summarised and rated under the term "appropriateness for efficiency survey".

Keywords Groundwater · Quality · Monitoring

This chapter bases largely on the guidelines from the DWA—German association for water, wastewater and waste: "Efficiency of measures to control land use for groundwater conservation—the example of nitrogen" (DWA-M 911 (2013): Effizienzkontrolle von Maßnahmen zur grundwasserschonenden Bodennutzung am Beispiel des Stickstoffs). Roland Schindler was the responsible author of the topic of groundwater quality monitoring.

R. Dannowski (✉) · F. Eulenstein
Leibniz Centre for Agricultural Landscape Research (ZALF) e. V.,
Eberswalder Strasse 84 15374 Muencheberg, Germany
e-mail: danno@zalf.de

F. Eulenstein
e-mail: feulenstein@zalf.de

R. Schindler
NEW Niederrhein Wasser GmbH, Rektoratstrasse 18 41747 Viersen, Germany

N. Cremer
Erftverband, Bergheim, Germany
e-mail: nils.cremer@erftverband.de

L. Mueller et al. (eds.), *Novel Measurement and Assessment Tools for Monitoring and Management of Land and Water Resources in Agricultural Landscapes of Central Asia*, Environmental Science and Engineering, DOI: 10.1007/978-3-319-01017-5_16,

1 Introduction

Stewardship of the fresh water resources is of paramount importance throughout Europe and for all EU environmental policies now and in the future. This finds prominent expression in the European Water Framework Directive (EU WFD) that entered into force in December, 2000 (EU WFD 2000), flanked by the Groundwater Directive (EU GWD 2006) that has been subsequently developed in response to the requirements of Article 17 of the EU WFD and formulating the EU policy on groundwater protection. Since then, the quantitative and qualitative conditions of surface and groundwater bodies have been of exceptional relevance for most of the water-related concepts and measures throughout the European countries. By 2015, to achieve a good quantitative and qualitative state of all European surface water and groundwater bodies is stipulated by the EU WFD.

Abundant fresh, non-salty groundwater resources in particular are among the most valuable environmental goods ensuring life, ecosystems health and prosperity. Therefore, groundwater conservation and protection is a distinct environmental concern. In the case of countries with higher industrial level, the most prevalent impact on groundwater quantity and quality is exerted by agricultural plant production. Hence, conversely, agriculture is the best-suitable field of action for implementing area-wide, large-scale measures of groundwater conservation.

The aim of all the agricultural groundwater-protecting or -conserving measures is essentially—besides the quantitative aspects of preventing overuse—to reduce the amount of harmful substances introduced from agricultural land use into the groundwater, such as nutrients, pesticides, solutes, and pollutants in general. In doing so, systematically to analyse the groundwater quality is an indispensable means and precondition for the observation, inspection, and evaluation—in other words, the efficiency survey—of these agricultural measures. Especially in the shallow, near-surface groundwater, the concentration of dissolved substances rather instantly mirrors the effect of agriculture on the aquatic environment. It can be considered a kind of an early-warning system before impairing the surface or drinking waters quality. This survey is preferably focused on nitrate, as an important substance for plant nutrition and known to behave as a conservative tracer below the root zone carried with the subsurface water (Merz et al. 2009; Behrendt and Dannowski 2005).

Efficiency survey should be organised following the most recent rules and methodology at hand. These are outlined in the present contribution based on the German Technical Guideline DWA-M 911 (2013) entitled “Efficiency of measures to control land use for groundwater conservation—the example of nitrogen”. Though tailored here on groundwater-protecting and -conserving measures, the methodology is suitable also for the more general task of the standard groundwater quality monitoring. To benefit from the following information, the reader should have some basic hydrogeological knowledge at his/her disposal.

2 Methods of In Situ Groundwater Quality Monitoring

2.1 Monitoring the Groundwater Quality of an Aquifer in the Pre-Development Phase

In principle, any extraction well constructed and operated for drinking water supply or crop irrigation equally could be used for groundwater quality monitoring. However, drawing groundwater from those production wells will provide mixed water of averaged quality for the related groundwater catchment, without the possibility of further temporal or spatial differentiation. Therefore, regular groundwater sampling at various places within a subsurface catchment will constitute more deepened information. In the unconsolidated rock region, which the following is referred to, in principle this should be practicable at reasonable expenses.

Prior to planning a monitoring program, the spatial (flow paths) and temporal (residence times) interrelationships—the hydrodynamic situation—within the groundwater catchment ought to be unveiled. Thus, in preliminary investigations, the aquifer structure (stratigraphy, thicknesses, permeabilities), as well as the hydrogeological characteristics (flow directions, flow velocities, depths to the groundwater, spatial extents of the aquifers and where appropriate the aquitards, properties of confining layers, rate of groundwater recharge, seepage dynamics, leakage conditions) are to be determined. Only then a relationship between the implemented groundwater conserving measures and their effect on the groundwater body can be established.

In the case of an allocated water protection area, as a rule, the necessary data will exist, since the delineation of the protection zones is based on this information. Also for groundwater development projects whose water use permits were granted recently, this data should have been assembled in the course of the planning procedure.

2.2 Monitoring Beneath the Groundwater Table

2.2.1 Groundwater Sampling by Means of a Suction Lance

In regions with depths to the groundwater of less than 6 m, one is enabled to obtain relatively short-dated information about the impact of agricultural measures on groundwater quality by taking samples from the near-surface groundwater (Fig. 1a).

By means of a drill hammer with push rod, the seepage zone is penetrated, and a lance completed by a suction cup is inserted down to max. 20 cm below the water table (which can be recognised from being saturated with water). This lance is joined to a suction unit via a hose. Application of low pressure causes extraction of

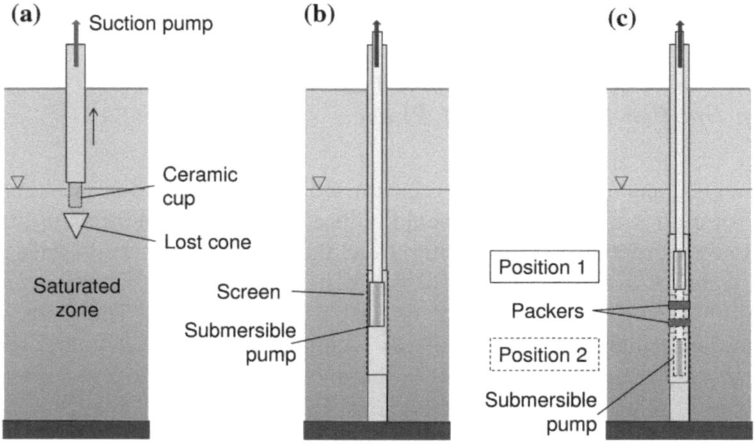

Fig. 1 Schematic of selected methods of groundwater sampling. **a** Suction lance. **b** Monitoring well. **c** Multi-level sampling

water from the groundwater surface. This procedure is repeated until a filter cake has built up around the suction cup to keep suspended materials out. Now a sample of relatively clear water can be abstracted to be analysed in a water laboratory in terms of its main ingredients and some physical–chemical parameters. In the case of peaty soils and sites with higher content of fine silt, no effectual filter structure may develop outside the cup. Those samples are to be pressure-filtered prior to laboratory analysis.

In the best case, with regard to the probable temporal variability of solute concentrations over the seepage period and the effect of the lateral groundwater movement, sampling beneath the water table should be repeated several times during the period of seepage flow. If this might not be feasible, water abstraction at least should be executed once a year by the end of the seepage period. Results obtained at that time at selected sampling points are best comparable with current solute inputs from the land surface because of the short transit time. At areas characterised by limited recharge, in the case of water sampling after prolonged periods of drought, and at higher lateral flow velocities in groundwater, the local displacement of solutes by groundwater flow must be taken into account while interpreting the results of the water analyses. Therefore, it is also suggested to choose groundwater sampling points in a certain distance downstream of the particular area of survey.

Benefits:

This relatively simple procedure does not necessitate any higher material and personnel expenses. Depending on the depth to the groundwater and the texture of the substrate, up to six samples can be taken within an 8-h day. Analysing the obtained water samples, compared with the analysis of soil samples (Sect. 2.2.2), is much simpler and less time- and cost-consuming. The impact of land use and agricultural protective measures on groundwater quality can be assessed at a relatively early stage.

Disadvantages:

In the case of depths to the groundwater of more than 4 m, the procedure takes more and more time, at depths to the groundwater of more than 6 m the technical low pressure will not suffice to abstract water enough to be analysed. Occasionally, water must be filtered prior to analysis, which can lead to errors in determining particular substances.

Appropriateness for efficiency survey:

By taking water samples from beneath the water table, the quality of the recently constituted groundwater is examined. Thus, the results of water analyses can be related to matter inputs in the vicinity of the sampling point. Provided a reasonably fine sampling grid is established, hot spots of solute inputs can be identified within a groundwater catchment. The load balance of such matter inputs can be calculated from concentration, as far as the rate of groundwater recharge is known. Regular area-wide sampling repeated at an interval of one or two years can be used for the efficiency assessment of agricultural measures in groundwater conservation. Compared with supervising based on the groundwater quality from deeper monitoring wells, where the interpretation is complicated by processes of solute displacement, mixing, and transformation in the related groundwater catchment, the described procedure is appropriate in providing results that are allowed to be definitely attributed to a particular cause-effect relationship within an overseeable sub-catchment.

2.2.2 Soil Sampling Beneath the Groundwater Table

Also this method based on taking soil samples from the near-saturation zone, as an alternative to the procedure described in Sect. 2.2.1, enables one to draw conclusions about the effect of agricultural measures to groundwater conservation.

Using a drill hammer with push rod, the unsaturated zone is penetrated down to the water table. Substrate material is sampled in the notch of a trenched rod. Knocking watchful against the rod makes the saturated zone visible (free water will appear), and the uppermost section of about 15 cm is taken from the saturated substrate. This way, per inspected unit (e. g., a field block), four or five soundings are executed and the substrate material is collected in one homogeneously mixed sample. The latter is analysed for nitrate and chloride in the laboratory after elution considering the water content (in Germany, ATV-DVWK 2004).

Benefits:

Relatively simple soil sampling supports area-wide detection of nitrogen inputs at the groundwater table of a whole domain. By means of assembling mixed samples per inspected unit, a field block, the expenditures for analysis are minimised.

Disadvantages:

At the present state, application of the procedure is restricted to unconsolidated sediments and fine-grained substrates (silt, sand); coarse gravel material is out of scope. At depths to the groundwater of more than 6 m, taking soil samples proves to become more and more time- and cost-consuming. In the case of a longer seepage period combined with higher recharge rates, the mean solute concentration of the recently constituted groundwater will tend to be under-estimated as a consequence of dilution.

Appropriateness for efficiency survey:

Because of the fact that water eluted from the soil samples originates from a certain range of depth (ca. 15 cm), results from chemical analyses represent the nitrate input into the groundwater averaged over a period of up to several months. Compared with N_{min} soil analyses from beneath the root zone, they are less subject to short-term fluctuations caused by specific agricultural measures or weather events. Collecting mixed samples offers the measurement-based detection of area-related nitrogen inputs into the groundwater. Just as taking water samples from the near-surface groundwater (Sect. 2.2.1), usage of the procedure indicates effects of agricultural protective measures on the groundwater quality at an early stage.

2.3 Aquifer Monitoring

2.3.1 Groundwater Sampling by the Direct-Push Method

At areas sparsely equipped with monitoring wells, in recent years, groundwater soundings are progressively executed by means of the direct-push method (EPA 2005; Hannappel and Braun 2010). This procedure offers one the direct in-field and depth-specific gathering of groundwater samples. Likewise, an area-specific monitoring of the groundwater quality is feasible, also in the case of a deeper water table.

Direct-push groundwater sampling is performed by hammer-driven ramming a probe head connected to a liner rod down to the groundwater zone. In complex terrain, a small-sized drilling unit is adopted mounted on an off-road vehicle. After the intentional depth is reached, the liner rod is lifted a few decimetres, pulling a filter cone free contained in the liner. This screen, as a rule, is 10–100 cm long and may consist of various materials (e. g., high grade steel, PVC/HDPE, ceramics). Alternatively, also an open filtering system can be inserted. Afterwards, the height of the water table can be measured inside the liner by means of a light plummet. Finally, the groundwater sample is withdrawn via the liner rod by a pump system adequately to the depth.

In the case of depth-specific groundwater sampling, the whole system is subsequently raised to the next position, and the sampling operation is executed as depicted before.

Benefits:

As compared with conventional procedures, direct-push groundwater sampling facilitates the faster and spatially more flexible survey of the geologic and hydrochemical underground conditions combined with diversified local logging. Usage of lightweight automotive equipment provides the accessibility of sampling points also in heavy terrain and supports fast relocation. Because of no permanent installation of groundwater observation points is needed, no additional construction costs will be incurred.

Disadvantages:

Because of the fact that no regular groundwater observation wells are installed, the direct-push method generates survey points for only one-time usage. Repeated groundwater monitoring by use of the same installation is not possible.

Hammer-driven ramming for direct-push groundwater sampling is not everywhere feasible. Difficulties will occur depending on the underground conditions, especially for larger sampling depths and very deep water tables (>20 m), as well as in the case of stony or strongly cohesive sediments. To estimate the depths approachable for sounding and sampling, previous knowledge from well-informed and skilled staff is needed being familiar with the hydrogeological situation. Under certain circumstances, preliminary investigations may be pre-requisite. In the case of no filter cake is built up around the filter cone, filtration-in-place of groundwater samples may be required.

Appropriateness for efficiency survey:

Since the direct-push sampling method offers considerable mobility, it is appropriate especially for the application at hydrogeologically well-perceived areas with nonetheless inadequate density of observation wells, for which extra area- or usage-specific investigations into the groundwater quality are required. This is of concern, e. g., for groundwater bodies under scope of the EU WFD, in the status of ‘at risk’, i. e. where reaching the objectives of water quality is actually unclear or unlikely (EU WFD 2000). The procedure enables the investigator to specify the groundwater quality on a plot-by-plot basis. Knowledge obtained this way may provide important completion of an existing groundwater monitoring program. In particular, potential locations of new points of inspection can be derived and measures to improve groundwater quality can be planned more selectively. By means of the depth-specific sampling provided by the direct-push method, the source and fate of pollutants can be traced back in a more reasonable way. Further on, processes of matter (nitrate) transformation in the groundwater zone can be identified.

2.3.2 Groundwater Sampling from Observation Wells

In the catchments of groundwater abstraction wells for the public water supply, as a rule, observation wells are placed especially in the vicinity of the production wells (Fig. 1b). Based on those water level observations, or on pumping tests in the course of the pre-development phase of the well group, the shared groundwater catchment had been delineated, the hydrogeological parameters of the aquifer had been determined, and supervision of the hydrograph is further ensured on a permanent basis. Such observation wells also should have been occasionally installed for the large-scale assignment of the subterranean catchment of a river or a canal. In addition, suppliers of potable water or water management authorities erect special observation networks to monitor the groundwater quality development, e. g., upstream of a well field or downstream of any polluters (Fig. 2).

These categories of observation wells, though installed for different purposes, all are suitable for investigations into the efficiency of agricultural groundwater-protecting measures provided they meet some requirements with regard to the site conditions and the level of their performance. In particular, this is important for assessing nitrogen inputs.

These requirements are actually:

- Filter lining (screen) be placed in the upper groundwater zone
- Short filter lines
- Unique attribution of the related catchment area be given

For interpreting the measurement results, requirements on the construction and design of observation wells and monitoring networks must be taken into consideration following the technical rules (in Germany, DVGW W 121 (2003) and DVGW W 108 (2003)).

By systematic inspection of the results from observation wells, the effects of groundwater-protecting measures can be detected on the medium to long term. For

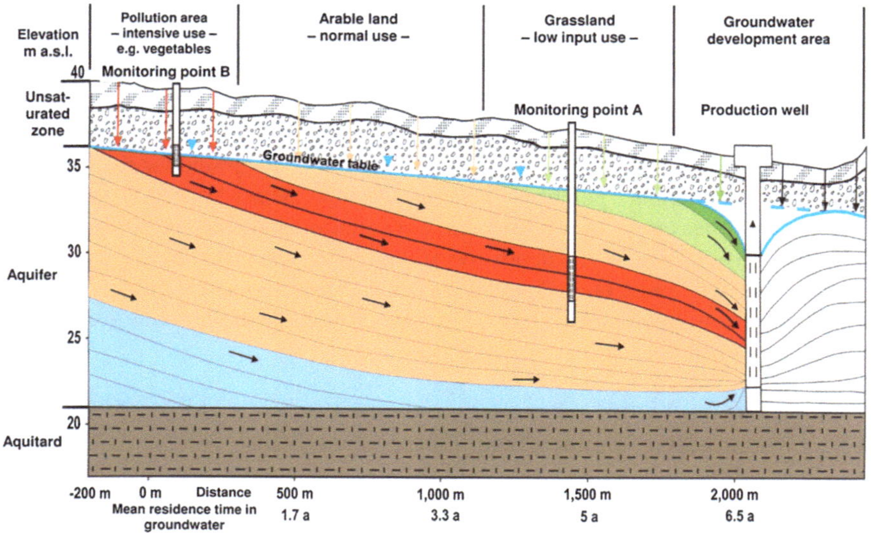

Fig. 2 Groundwater quality monitoring by means of observation wells in the upstream zone of a drinking water production well

this, the contributing subsurface catchment is to be attributed to the observation well according to the depth of the filter screen. This may be complicated in special cases, and sometimes the catchment boundary may vary with time. To increase the trustworthiness of the efficiency assessment, if possible, several observation points should be used (Fig. 2).

Fingerprint parameters of the groundwater age can make sense for the interpretation of the analyses. In many cases, the identification of so-called environmental tracers (^{2}H, ^{3}H, ^{3}He, ^{18}O, ^{85}Kr, CFC, SF_6) has proven to be useful for groundwater age determination. To examine the origin of nitrates and to explore the quantitative schemes of their decay, also radioisotope-hydrologic measurements (^{15}N-NO_3, ^{18}O-NO_3, ^{34}S-SO_4, ^{18}O-SO_4) are appropriate (Schulenberg et al. 1990; Stadtwerke Viersen GmbH 1999). Recently, these methods are increasingly applied in examining the efficiency of agricultural measures to mitigate groundwater pollution (Osenbrück et al. 2000; Schöpel 2000).

Groundwater quality should be monitored at selected observation wells over a longer period at adequate, regular intervals paying attention to the groundwater flow velocity or residence time. Groundwater samples should be taken with moderate pumping power to minimise any water table drawdown and, thus, mixing with water from the adjacent (higher and lower) aquifer zones, which is indispensable in case of the regularly screened observation wells (in Germany, following DWA-A 909 (2011)). Additionally, in interpreting the measuring results to narrow down the activity period of a source, the transit time across the seepage and groundwater zones upstream of an observation well is to be taken into consideration (Fig. 2).

A particular implementation of an observation well is the so-called multi-level well; its function and purpose are explained in Sect. 2.4.

Benefits:

At waterworks catchment areas, as a rule, a groundwater observation network is found for which long-term records of groundwater quality already exist. Optimising such a network for the purposes of efficiency survey permits to capture the quality status of the upper groundwater layer at an acceptable expense.

Disadvantages:

Groundwater quality in the range of a well screen is dependent on a multitude of interacting factors, such like the spatial and temporal patterns of groundwater recharge, the flow velocity and solute transport across the aquifer including dispersion, the matter decay and transformation. Thus, the spatial and temporal attribution of groundwater quality data to the mostly non-point, distributed ('diffuse') sources will render rather erroneous and complicated.

Appropriateness for efficiency survey:

Groundwater observation wells are principally suitable for the efficiency and impact evaluation of agricultural groundwater-protective measures. To be used for relatively short-term information on groundwater quality, however, they should be compatible for certain conditions. Their filter construction (screen) should not be placed too deep below the mean water level, and they must not be biased by larger-scale effects from the catchment. For the assessment of the long-term groundwater quality development it is favourable to include regional main (background) parameters into the monitoring concept. Additionally, the residence time of the sampled groundwater should be estimated. As a consequence from the uncertainties described above in the spatial and temporal allocation of matter inputs, if possible, the results should be re-evaluated and supplemented by independent detection methods.

Observation wells screened in the deeper range of the upper aquifer or in deeper groundwater layers do not provide trustworthy information about the impact of agricultural measures and are not suitable for the purpose of efficiency survey.

2.4 Groundwater Sampling from Multi-Level Observation Wells

Multi-level observation wells (Fig. 1c) are especially appropriate for the hydrochemical survey of thick aquifers or a system of aquifers hydraulically separated from each other by non- or semi-permeable confining layers. In general, they allow for the vertical differentiation of the hydrochemical conditions within one or between several groundwater-bearing units. In the context of the efficiency survey they are suitable for analysing the dynamics of the vertical matter exchange also in the near-surface groundwater zone. In recent times, multi-level observation wells have been often supplemented by probes located in the unsaturated zone at various depths.

In the case of the multi-level observation well "Ruhr University System", dependent on the permeability and the thickness of the aquifer and the specific intention of the groundwater hydrologist, several minifilters are fixed at a guide tube consisting of plastics (50 mm in diameter) by plastics clamps in arbitrary intervals. Each of the minifilters has its own riser tube (14 mm in diameter) leading through to the measuring equipment placed at the soil surface or in a well house. The single minifilters are simultaneously pumped off by means of a rotary pump. This minimises vertical mixing between different aquifer zones and guarantees the intended depth-specific water sampling, provided the permeabilities of the single layers induce an approximately horizontal flow towards the minifilters. For water sampling from probes >7 m below the surface not accessible by suction pumps, so-called injectors can be mounted next to the minifilters. For water level measuring in a selected aquifer zone, a 1 m long filter tube is placed in the required depth within the 50 mm plastics guide tube.

While taking groundwater samples, the pumped discharge has to be adjusted to the water yield of the investigated aquifer zone. To avoid the hydraulic equilibrium over the whole depth of the aquifer be disturbed too much during pumping, a certain water volume (ruled in DWA-A 909 (2011)) drawn up before the actual sampling should not be exceeded.

As stated above with direct push, especially in the case of multi-level observation wells, it is indispensable to recognise the hydrogeological background within the groundwater catchment to be analysed previous to drilling. An important task is to determine the groundwater flow direction. Therefore, besides the extensive and costly multi-level equipment, also a sufficient number of regular observation wells must be installed within the catchment area.

The results of measuring the groundwater quality at single dates are assembled in a concentration–depth profile. Provided the hydrogeological conditions within a given groundwater catchment are reasonably homogeneous, the depth-specific groundwater samples obtained from the multi-level wells can be allocated to their respective areas of origin, under consideration of the rate of groundwater recharge, the effective flow velocity, and the specific yield (actual porosity). As well, deeper aquifer zones can be evaluated with respect to their denitrification potential, taking into account the obligatory regional background parameters such as NO_3^-, Cl^-, SO_4^{2-}, HCO_3^-, O_2. Additionally, isotope-hydrologic methods can be used.

Benefits:

The operation of multi-level observation wells provides vertical higher-resolution information about the hydrochemical conditions of a pore aquifer (or even a system of aquifers and aquitards). Combined with the collection and analysis of additional hydrologic, hydrogeological, hydrochemical, and land use data, also various impacts on groundwater quality, as well as processes of matter transformation within the groundwater zone can be detected.

Disadvantages:

Both, the construction and operation of multi-level observation wells are very extensive and cost-consuming in comparison with regular groundwater observation wells. To avoid imperfections in groundwater sampling (evoking hydraulic short-circuits), only skilled personnel should be commissioned.

Appropriateness for efficiency survey:

Multi-level observation wells provide an important tool for the efficiency evaluation of agricultural groundwater-protective measures, as well as an instrument for the long-term prognosis of the hydrochemical conditions in an existing drinking water protection zone. They are suitable for the comprehensive analysis and monitoring of the hydrochemical processes characterising the pore aquifer. Combined with compatible observation equipment installed in the overlying unsaturated zone, the effects of agricultural groundwater-conserving measures can be studied in their temporal and spatial development. This is of exceptional importance for groundwater catchments marked by approaching or exceeding the limit concentration of nitrate in the upper groundwater layer or where concerns have been raised that the denitrification potential could decline and be exhausted within a relatively short period of time. Prior to establishing a multi-level observation well, however, the hydrogeological conditions of the groundwater catchment must be adequately perceived to be able to localise the appropriate place.

2.5 Quality Monitoring of Raw Water from Production Wells

Effects of agricultural measures on diminishing nitrate loads often are quantified from time series of the untreated groundwater quality at drinking water works, especially as this method offers a specified set of target variables easily to be determined. Interpreting the time series of raw water quality data, the following conditions ought to be reflected (Fig. 2):

- Water abstraction exhibits a markedly accelerating intervention into the natural groundwater dynamics of a catchment.
- In most of the water works, water abstraction takes place via several production wells. Their water quality will differ in general, based on the related sub-catchments characterised by differing groundwater quality parameters.
- Often the production wells have filter constructions located in different depths of an aquifer or even in different groundwater units.
- In larger-sized water works, the untreated groundwater from different well groups is mixed prior to treatment.

Therefore, in the case of conclusions to be drawn about the effectiveness of agricultural measures for groundwater conservation from the evolution of single species concentrations in untreated groundwater, it is mandatory that the water be originated at least from one particular groundwater catchment with stable ratios of water from different aquifers/depths throughout the monitoring period.

As a rule, however, it is preferable to monitor the raw water quality of selected production wells with definitely separated groundwater catchments. But again it must kept in mind that at deeper wells water is mixed from different recharge periods with different, non-natural residence times. Thus, the short-term development (less than 5 years) of matter concentrations in raw water is hardly to be spatially or temporally apportioned to specific agricultural measures. This is true in particular for spacious groundwater catchments with high aquifer thicknesses.

Benefits:

Long-term records of the untreated groundwater quality, in any case, exist in responsibility of the suppliers of potable water as regulated by law. These time series of critical substances, such as nitrates, pesticides, sulphates are suitable for the effect of agricultural groundwater-protective measures being inexpensively assessed.

Disadvantages:

Analysing untreated groundwater samples from production wells does not offer the intended hydraulically undisturbed, highly-differentiated spatial or temporal attribution of the observed water quality to agricultural measures of groundwater conservation.

Appropriateness for efficiency survey:

The evolution of substance concentrations in raw water from selected production wells can be used to evaluate the efficiency under certain circumstances provided the substances are distinctly attributable to a particular groundwater catchment, and agricultural water protection measures have been practiced over a longer period of time. The duration of the monitoring period should last for about one third of the averaged time of groundwater renewal (the quotient of the extractable groundwater volume and the annual recharge rate) in the developed aquifer, plus the mean transit time between the soil surface and the groundwater table. This method, however, will provide only rough information about larger-scale measures that cannot be locally and temporally differentiated.

Drinking water quality monitoring alone, as a rule, is not appropriate for efficiency assessment. Even long-term records are hardly to be interpreted, as they are usually superimposed by the operational impacts of water extraction and groundwater treatment.

3 Conclusions

There are several methods of in situ groundwater quality monitoring suitable for the efficiency survey of agricultural groundwater-protecting measures. Besides their general characteristics, the benefits and disadvantages of the single methods were discussed and rated here under the summarised term "appropriateness for efficiency survey". The basics are outlined in the recently released German Technical Guideline DWA-M 911 (2013), "Efficiency of measures to control land use for groundwater conservation—the example of nitrogen" which was adopted for this contribution.

It can be stated that not one single method exists that would manage all the practical problems at low costs at the same place and time. Instead, a thorough selection or even better a combination of the various methods seems to be required and worthwhile depending on, e. g., the agricultural, pedological, geological, climatic and hydrologic site conditions. Some of the methods provide relatively simple short-term indication for the local efficiency of specific agricultural groundwater-protective measures. Others are more suitable for revealing and understanding the system behaviour of an entire complex of groundwater units in spacious catchments under anthropogenic impact.

Efficiency survey should be related as close as possible to the actual agricultural area under reference to preclude incorrect or vague deductions. This also will raise the farmers' acceptance for conservation measures, as they are immediately informed and can understand the effect of their efforts on the groundwater resources under concern.

References

ATV-DVWK (2004) Weiterentwicklung einer Methode zur Messung des Stickstoffaustrages landwirtschaftlich genutzter Flächen—Teile 1 und 2. ATV-DVWK-Themen. ATV-DVWK-Arbeitsgruppe GB-6.2 "Effizienzkontrolle". Deutsche Vereinigung für Wasserwirtschaft, Abwasser und Abfall e. V. (ATV-DVWK), Hennef

Behrendt H, Dannowski R (eds) (2005) Nutrients and heavy metals in the Odra river system: emissions from point and diffuse sources, their loads, and scenario calculations on possible changes. Weißensee Verlag, Berlin, p 353. (ISBN 3-89998-046-8)

DVGW W 108 (2003) Messnetze zur Überwachung der Grundwasserbeschaffenheit in Wassergewinnungsgebieten. Arbeitsblatt

DVGW W 121 (2003) Bau und Ausbau von Grundwassermessstellen. Arbeitsblatt

DWA-A 909 (2011) Grundsätze der Grundwasserprobenahme aus Grundwassermessstellen. Arbeitsblatt

DWA-M 911 (2013) Möglichkeiten der Effizienzkontrolle von Maßnahmen zur Grundwasser schonenden Bodennutzung am Beispiel des Stickstoffs, Deutsche Vereinigung für Wasserwirtschaft, Abwasser und Abfall e. V., Hennef

EPA (2005) Groundwater Sampling and Monitoring with Direct Push Technologies. U.S. Environmental Protection Agency, Office of Solid Waste and Emergency Response, Washington, DC, OSWER No. 9200.1-51, EPA 540/R-04/005, Aug 2005, p 78. http://www.clu-in.org/download/char/540r04005.pdf

EU WFD (2000) Directive 2000/60/EC of the European Parliament and of the Council establishing a framework for the Community action in the field of water policy, OJ L 327, 22.12.2000, 1–73; http://eur-lex.europa.eu/LexUriServ/LexUriServ.do?uri=CONSLEG:2000L0060:20090625:EN:PDF

EU GWD (2006) Directive 2006/118/EC of the European Parliament and of the Council of 12 December 2006 on protection of groundwater against pollution and deterioration, OJ L 372, 27.12.2006, 19–31; http://eur-lex.europa.eu/LexUriServ/LexUriServ.do?uri=OJ:L:2006:372:0019:0031:EN:PDF

Hannappel S, Braun P (2010) Planung und Durchführung eines Grundwasser-Monitorings mit Direct-Push in Nordbrandenburg. In: Grundwasser Bd. 15, H. 3, 191–201

Merz C, Steidl J, Dannowski R (2009) Parameterization and regionalization of redox based denitrification for GIS-embedded nitrate transport modeling in Pleistocene aquifer systems. Environ Geol 58(7):1587–1599

Osenbrück K, Heinrichs G, Göppel M, Eichinger L (2000) Nitratsanierung von Grundwassererschließungen: Konzept und Zielprognose durch isotopenhydrologische Untersuchungen. In: bbr 3/2000, 20–25

Schöpel M (2000) Ergebnisse des Kooperativen Gewässerschutzes am Beispiel der Kooperation Holstershausen/Üfter Mark. In: Ber. IWW, Bd. 29, Mühlheim a. d. Ruhr

Schulenberg S, Schindler R, Rauert W, Wolf M, Weise S, Voerkelius S (1990) Hydrogeologische, hydrochemische und isotopenhydrologische Untersuchungen in den Grundwassereinzugsgebieten der Stadtwerke Viersen GmbH. Stadtwerke Viersen, p 45

Stadtwerke Viersen [Hrsg.] (1999) Hydrochemische und isotopenhydrologische Untersuchungen in den Grundwassereinzugsgebieten der Stadtwerke Viersen GmbH. Abschlussbericht

Methods in the Exploratory Risk Assessment of Trace Elements in the Soil-Groundwater Pathway

Levke Godbersen, Jens Utermann and Wilhelmus H. M. Duijnisveld

Abstract Groundwater protection starts with soil protection. When soils are contaminated there is always a chance that the groundwater is or will be affected by this contamination as well. Analytical risk assessment is often very time-consuming and expensive. The goal of exploratory investigations is to investigate the risk of groundwater contamination with the minimum possible effort. Exploratory investigations concerning trace elements can be carried out by analyzing soil material, percolation water or groundwater (LABO 2003). While percolation water and groundwater analysis can only confirm a contamination, information gathered by soil analysis permits the current and future contamination risk for the groundwater to be estimated and additionally allows precautionary measures to be planned in order to protect the groundwater. The risk of contamination can only be judged when standards are available for a good quality of soil or percolation water. Local or regional background values of trace elements in the soil or percolation water can serve as adequate standards. This chapter will present and compare the analytical methods applied in exploratory investigations for assessing risks to the soil-groundwater pathway. We start with a little material science regarding trace element analysis in soil science and continue by listing some common methods used to analyze the total content of trace elements in the soil, presenting the conversion functions which allow results from different digestion agents to be translated. Subsequently an aqueous batch extraction method is presented which is routinely used to estimate the soluble fraction of

L. Godbersen (✉) · W. H. M. Duijnisveld
Federal Institute for Geosciences and Natural Resources, Geozentrum Hannover, Stilleweg 2
30655 Hannover, Germany
e-mail: levke.godbersen@gmail.com

W. H. M. Duijnisveld
e-mail: wilhelmus.duijnisveld@bgr.de

J. Utermann
Umweltbundesamt, Federal Environment Agency, Wörlitzer Platz 1 06844 Dessau-Roßlau, Germany
e-mail: jens.utermann@uba.de

L. Mueller et al. (eds.), *Novel Measurement and Assessment Tools for Monitoring and Management of Land and Water Resources in Agricultural Landscapes of Central Asia*, Environmental Science and Engineering, DOI: 10.1007/978-3-319-01017-5_17,

trace elements in the soil for exploratory purposes, and its results are compared with results gained from direct percolation water sampling and a method which adjusts one parameter to in situ conditions.

Keywords Soil · Groundwater · Trace elements · Risk assessment

1 Introduction

Groundwater protection starts with soil protection. When soils are contaminated there is always a chance that the groundwater is or will be affected by this contamination as well. Analytical risk assessment is often very time-consuming and expensive. The goal of exploratory investigations is to investigate the risk of groundwater contamination with the minimum possible effort. Exploratory investigations concerning trace elements can be carried out by analyzing soil material, percolation water or groundwater (LABO 2003). While percolation water and groundwater analysis can only confirm a contamination, information gathered by soil analysis permits the current and future contamination risk for the groundwater to be estimated and additionally allows precautionary measures to be planned in order to protect the groundwater. The risk of contamination can only be judged when standards are available for a good quality of soil or percolation water. Local or regional background values[1] of trace elements in the soil or percolation water can serve as adequate standards.

Figure 1 shows steps which can be taken for an exploratory investigation of suspected sites concerning the soil-groundwater pathway. The first step will be an inventory of the total trace element content in the soil to assess the quantity and quality of the contamination. Comparing the total contents measured in the soil with local background values helps to decide whether the site is contaminated at all. If the total content of trace elements is not significantly higher than the local background content, the investigation stops, because the site under investigation is obviously not a source of contamination.

If, however, the trace element content proves to be significantly elevated compared to the background content, the investigation process continues: the mobility of the trace elements are assessed. The mobility of the contaminant needs

[1] The Guidance on the Determination of Background Values (ISO 19258) defines background values as the "statistical characteristic [...] of the background content". The background content is the "content of a substance in a soil resulting from both natural geological and pedological processes and including diffuse source inputs." Because we usually identify the amount of substances such as trace elements in relation to a defined volume of solid material (e.g. mg/kg) or liquid (e.g. mg/l, μg/l) deviating from the ISO definition, we identify the background value as the statistical characteristics of background concentrations in a soil or percolation water resulting from both natural geological and pedological processes and including diffuse source inputs.

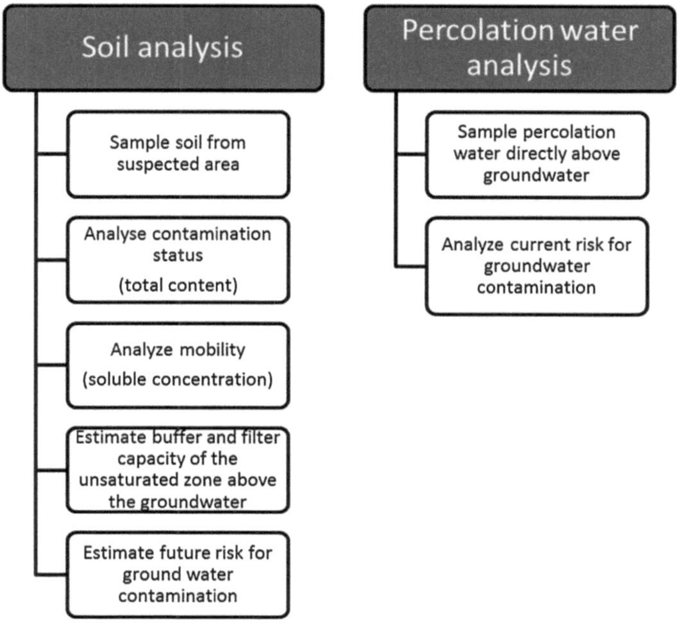

Fig. 1 Workflow in exploratory investigations of suspected contamination with trace elements on the site itself as suggested by the German Federal and State Working Group for soil protection (based on LABO 2003, modified)

to be assessed to judge the risk of groundwater contamination emerging from the suspected site. In Germany, legislation provides threshold values regarding the soluble fraction of trace elements to assess their potential risk regarding groundwater contamination. Local background values of soluble trace elements can fulfil the same function. If the mobility of the trace element gives rise to the suspicion that the groundwater is endangered, the protective effectiveness of the remaining pathway between the contamination and the groundwater needs to be assessed. The judgment is based on parameters such as percolation water rate, thickness of the percolation zone, sorption capacity and sorption isotherms, porosity and permeability. When measured on sites with loamy soils, for example, the same soluble concentration of trace elements in the upper part of the soil profile is much less threatening for the groundwater than when measured in sandy soils, because of the much higher sorption capacity of loamy soils.

If the assessment at that point is inconclusive or if the immediate risk of groundwater pollution is the focus of attention, percolation water can be sampled directly at the transition from the unsaturated to the saturated zone. If the exploratory investigation leads to the conclusion that the groundwater may be threatened by substances emerging from the suspected site, according to German legislation a detailed investigation must be ordered.

This chapter will present and compare the analytical methods applied in exploratory investigations assessing risks to the soil-groundwater pathway. We start with a little material science regarding trace element analysis in soil science and continue by listing some common methods used to analyze the total content of trace elements in the soil, presenting conversion functions allowing results from different digestion agents to be translated. Subsequently, an aqueous batch extraction method is presented which is routinely used to estimate the soluble fraction of trace elements in the soil for exploratory purposes, and its results are compared with results gained from direct percolation water sampling and a method which adjusts one parameter to in situ conditions.

Background values are an important instrument in risk assessment. Knowledge about local background values enables decision-makers to appropriately classify the risk to the environment. For naturally occurring substances such as trace elements, especially, environmental tolerance can change based on the naturally occurring concentrations. Trace elements are relatively immobile and only a very small proportion of the total amount of trace elements is dissolved in the percolation water. Gooddy et al. (1995) report field-based partition coefficients of 2.4–3 log Kd dm^3/kg for trace elements such as Cd or Pb. Consequently, trace elements can only be found in the percolation water in very low background concentrations. Sampling and analyzing percolation water for background values is therefore prone to error due to contamination. In the course of this chapter we also present an improved sampling method for percolation water for the purpose of measuring and/or monitoring background values.

2 Materials for Trace Element Analysis in Soil Science

(Cross-) contamination is one of the main sources of error in inorganic trace element analysis. During sampling the use of metal instruments should be avoided whenever possible. Drilling, however, is nearly impossible without metal probes. Samples from percussion coring tubes should be taken very carefully from the centre of the core. As an alternative, a thick-walled Plexiglas tube can be used instead of a metal probe to sample unconsolidated soil material. Soil and water samples should always be stored, transported and handled with non-metal materials. However, not all plastic materials are free of metals. Some plastic materials have a higher sorption capacity for trace elements, which makes them equally unsuitable for trace element analysis. According to the literature and our own research, fluoropolymers (e.g. polytetrafluoroethylene [PTFE], polychlorotrifluoroethylene [PCTFE], fluorinated ethylene-propylene copolymers [FEPs] or perfluoroalkoxy copolymer [PFA]) are the most suitable materials in trace element analysis (Grossmann 1988; Koch 1993; Morrison 1982). The less expensive alternative mentioned in Reimann et al. (1999), high-density polyethylene (HD PE), is produced under the influence of metal compounds. It can thus not be fully excluded that this material might become the source of contamination during the

Table 1 Typical blank values of trace element analysis

	As	Ba	Cd	Co	Cr	Cu	Mo	Ni	Pb	Sb	V	Zn
(μg/l)	0.5	0.04	0.03	0.04	0.5	0.5	0.05	0.07	0.3	0.01	0.04	3.5

analytical process. PTFE and FEB showed the best properties in terms of mechanical and thermal stress capacity, chemical resistance and gas-tightness (Hellerich et al. 2001; Oberbach 2001). For filtration, cellulose-acetate or nylon are both suitable, where nylon is the better choice when organic carbon is also being analyzed. In order to avoid contamination, disposable gloves need to be worn at all times. Powder-free vinyl-polymer causes the least contamination. All other equipment which is in contact with the sample needs to be cleaned carefully. A standard procedure of rinsing for at least 2 h with 2 % HNO_3 solution and subsequently rinsing twice for at least 2 h each time with ultrapure water produced good results (Table 1).

3 Total Content

An inventory of the contamination is necessary regardless which measure (soil analysis or percolation water analysis) is taken afterwards. Aqua regia digestion (DIN ISO 11466) or pressure-assisted HF-HNO_3 digestion are well-established methods for extracting the total content of trace elements (e.g. Alloway 1995; Mester and Sturgeon 2003; LABO 2003; Arruda 2007; Smith and Cresser 2003). Because HF is very toxic and explicitly dangerous to handle, aqua regia is usually the more popular extracting agent to assess the total content of trace elements in soil materials. Strictly speaking aqua regia does not extract the total content of trace elements and results are thus often referred to as "so-called" totals as opposed to the real totals extracted using HF-HNO_3 digestion (Lewandowski et al. 1997). Ad-hoc AG (2005) offer transfer functions to translate so-called totals of Cd, Cu, Ni, Pb, Zn, Fe, Cr, Co, Mg, Mn, Hg and V into real totals and vice versa (Table 2).

When real total or aqua-regia-extractable concentrations are not significantly higher than local background or baseline concentrations the soil is evidently not contaminated. Thus no risk for the groundwater emanates from the suspected site. Consequently, exploratory investigation can stop here. If, however, the total or aqua-regia-extractable concentrations are significantly higher than the local background a risk of groundwater contamination cannot be excluded and further analysis is necessary.

4 Soluble Fraction

The mobile or soluble fraction of trace elements in the soil can be assessed either by aqueous batch or column extraction from soil material or by analyzing percolation water. Percolation water can be sampled ex situ or in situ. Some examples

Table 2 Coefficients of transfer functions to estimate total trace element concentrations (HF) from aqua regia (AR) extractable trace element concentrations (Ad-hoc AG Boden 2005)

Me(HF) = a + b * Me(AR)				Range of validity (mg/kg in aqua regia extract)
Element	**a**	**b**	R^2	
Cd	0.03487	1.17568	0.923	0.01–1.0
Cu	1.25438	1.00499	0.985	0.05–60
Ni	0.81593	1.09784	0.994	0.3–300
Pb	4.54623	0.9736	0.979	1.2–120
Zn	3.21993	1.10426	0.96	3.8–169
Fe	0.098	0.957	1	0.8–87
Cr	0.205	1.032	0.9	2.3–89
Co	0.077	0.986	0.98	0.3–33
Mg	0.095	0.981	0.98	0.16–55
Mn	0.204	0.937	0.99	7.2–1,900
Hg	0.036	0.992	0.95	0.004–1.7
V	0.208	1.017	0.91	1.9–230
Me(AR) = c + **d** * *Me(HF)*				Range of validity (mg/kg in HF extract)
Element	**c**	**d**	R^2	
Cd	−0.00976	0.74827	0.914	0.01–1.2
Cu	−0.89096	0.95858	0.984	0.8–60
Ni	−0.53833	0.90095	0.994	0.4–320
Pb	−3.95028	0.99022	0.979	2.4–130
Zn	−0.29025	0.8586	0.959	4.4–180
Fe	−0.097	1.039	1	1.2–87
Cr	−0.046	0.87	0.9	2.9–170
Co	−0.064	0.993	0.98	0.46–35
Mg	−0.092	1.002	0.98	0.19–57
Mn	−0.197	1.059	0.99	8.1–1,900
Hg	−0.112	0.956	0.95	0.003–1.8
V	−0.067	0.896	0.91	2.8–220

of ex-situ methods are centrifugation, extraction with positive or negative gauge pressure, the basal cup method or centrifugation with inert solutions. In-situ sampling methods include negative pressure lysimeters, such as suction cups or plates; gravitation lysimeters, and dialysis and diffusion techniques. Among the lysimeters, suction cups are the easiest to install and the least invasive. Although often more expensive and protracted, in situ sampling has serious advantages over the ex-situ sampling of percolation water or the extraction of soil material. The advantages arise mostly because fewer preparation steps are necessary, which reduces the chances of contaminating the sample and of affecting the solubility of the trace elements by accidently changing the physicochemical milieu. In-situ methods yield percolation water samples which can be analyzed almost directly.

Soil sampling is usually cheaper and quicker than sampling in situ percolation water, thus the former is often preferred to the latter. Column extraction methods usually produce results closer to reality, i.e. concentrations more similar to trace element concentrations in percolation water, and yield more liquid sample material than batch extracts (Grathwohl and Susset 2009). However, for exploratory purposes column extraction is too laborious. In Germany the most popular method until recently was the so-called S4-extract, which is an aqueous batch extract with a liquid/solid ratio of 10 l/kg (DIN 38414-4). Batch extraction methods in general are simple to conduct and one advantage of the S4 is the large amounts of extract gained by this method. The relatively wide liquid/solid ratio, however, causes a dilution effect, resulting in an underestimation of soluble trace element concentrations (Schroers 2002; HLUG 2011). Because a wider liquid/solid ratio also affects the equilibrium (Grathwohl and Susset 2009), concentrations measured in extracts from different methods cannot be transferred mathematically. Recently, an aqueous extraction method with a liquid/solid ratio of 2 l/kg has become increasingly popular and has thus been standardized in Germany (DIN 19529 2009). We will present this extraction method in more detail and compare it with a similar batch extraction method developed to estimate the dissolved trace element concentrations in experiments to derive sorption isotherms (Heidkamp 2005; Utermann et al. 2005). The latter method extracts the water-soluble fraction of trace elements in the soil at a liquid/solid ratio of 5 l/kg while the ionic strength is adjusted to in situ conditions.

The instrumentation shown in the following is an example. The volumes of the samples and extracting agents can be increased or decreased as long as the liquid/solid ratio stays the same. There are, however, limitations concerning the amount of filtrate per filter. Reproducible results from batch extraction can only be achieved by ensuring that the ratio of filter size to fluid volume is approx. 1.3, otherwise the filter cake influences the extraction results for trace elements significantly.

5 Batch Extraction of 2 l/kg

5.1 Instrumentation, Equipment and Materials

- Overhead shaker
- Centrifuge
- Disposable gloves
- 50 ml centrifuge tubes (PE)
- Syringes
- Syringe tip filters (nylon or cellulose acetate membrane, pore size <0.45 μm, diameter: 50 mm)
- ultrapure water (EC < 0.1 μS/cm)
- 2 % HNO_3(ultrapure)
- 65 % HNO_3(ultrapure)

5.2 Batch Extraction

15 g of dry soil material are mixed with 30 ml of ultrapure water in a 50 ml polypropylene centrifuge tube. After slowly (<4 rpm) shaking for at least 24 h leave the mixture to settle for 10 min. 30 min of centrifugation at 18 °C with 3,000 g separates the liquid from the solid phase. In order to omit colloidal-bound trace elements, the decanted liquid sample needs to be filtered <0.45 μm. Syringe filters with a nylon membrane yielded the lowest blank values of trace elements and organic substances. For exclusive trace element analysis the cheaper syringe filters with a cellulose acetate membrane produce equally good results. The extract is stabilized until measurement by adding 0.1 ml of ultrapure 65 % HNO to 10 ml of the filtered extract.

6 Batch Extraction of 5 l/kg with Adjusted Ionic Strength

6.1 Instrumentation, Equipment and Materials

- Overhead shaker
- Centrifuge
- Disposable gloves
- 50 ml centrifuge tubes (PE)
- Syringes
- Syringe tip filters (nylon or cellulose acetate membrane, pore size <0.45 μm, diameter: 50 mm)
- ultrapure water (EC < 0.1 μS/cm)
- $Ca(NO_3)_2$ (ultrapure)
- 2 % HNO_3(ultrapure)
- 65 % HNO_3(ultrapure)

6.2 Adjusting Ionic Strength to in Situ Conditions

To estimate the ionic strength in situ we measured the electrical conductivity in suspensions made with ultrapure water of three different liquid/solution ratios (LS) and used the results to extrapolate the electrical conductivity in situ (Fig. 3). The LS in situ can be estimated based on the particle size distribution and the bulk density of the soil, assuming water saturation. Conversion tables to estimate pore volume based on soil texture classes and bulk density classes are available e.g. in the German soil mapping guide (Ad hoc AG 2005, Table 70, p. 344). The liquid/ solid ratio in situ can then be calculated using the equation

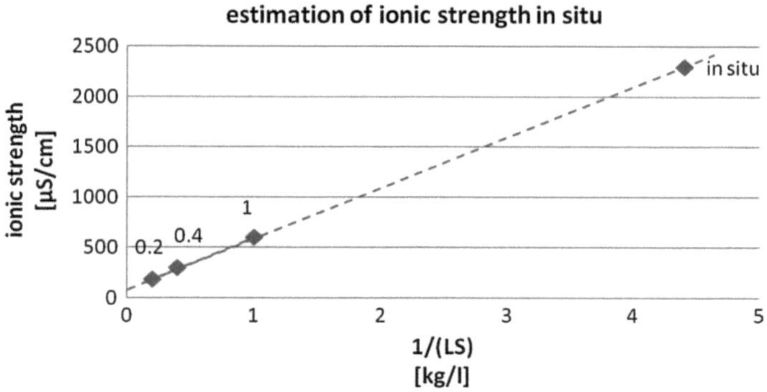

Fig. 2 Ionic strength of one soil for different liquid/solid ratios

$$LS_{insitu} = \frac{AC + eFC}{(100 - AC - eFC) * K}$$

LS_{insitu}	liquid/solid ratio in situ (l/kg)
AC	air capacity % (pores > 50 µm, pF < 1.8)
eFC	effective field capacity % (pores 0.2–50 µm, pF4.2–pF1.8)
K	specific density of the soil material (kg/dm^3)

The necessary amount of $Ca(NO_3)_2$ to adjust the ionic strength of the extracting agent to in situ conditions can be estimated using the following equation following Utermann et al. (2005): (Fig. 2)

$$c(Ca(NO_3)_2) = 0.\frac{013}{3}\left[\frac{\text{mmol}}{1} * \frac{\text{cm}}{\mu\text{s}}\right] * (EC_{insitu})$$

$c(Ca(NO_3)_2)$	concentration of $Ca(NO_3)_2$ needed to adjust ionic strength to in situ conditions
EC_{insitu}	electrical conductivity in situ (µS/cm)

This equation is based on the fact that 1 mmol/l of $Ca(NO_3)_2$ combines 1 mmol/l Ca^{2+} and 2 mmol/l NO_3^- to an ionic strength of 3 mmol/l. $Ca(NO_3)_2$ was chosen as the background electrolyte because Ca^{2+} is abundant in percolation water and in direct competition with the mostly divalent trace elements. NO_3^- builds only weak complexes with trace elements and is thus less likely to affect equilibrium than for example Cl^-.

6.3 Batch Extraction

We mixed 7 g of dry soil material with 35 ml of extracting agent with adjusted ionic strength in a 50 ml polypropylene centrifuge tube. For the comparison of two methods presented later on in this chapter we strictly followed the analytical protocol described by Heidkamp (2005) and applied a rotation speed of 20 rpm. Compared to the 2 l/kg batch extraction this is relatively fast. Faster rotation can mechanically affect the soil material and creates new surfaces, so can increase the amount of soluble trace elements. To avoid increased mobilization due to mechanical stress on the soil particles we suggest rotation speeds of <4 rpm. After shaking the suspension we separated the phases by means of centrifugation (3,000 g, 30 min, and 18 °C). Subsequently the decanted solute was filtered 0.45 μm and stabilized with HNO_3 (0.1 ml of 65 % HNO3 per 10 ml of filtrate).

7 In-Situ Percolation Water Sampling

The ecological risk arising from the soluble amount of trace elements in soils can only be adequately judged if either the threshold values or background concentrations are known. Since background concentrations are usually very low, samples are prone to error due to incorrect handling or contamination. Thus the sampling method for the purpose of deriving background values needs to be exact and liquid samples need to be available for analysis with as few working steps in between sampling and measurement as possible. Duijnisveld et al. (2008) developed a refined suction cup lysimeter system that can be installed by one person within a few hours. Depending on the hydraulic conductivity the system yields enough sample material within 6–36 h for a full physicochemical characterization of the percolation water. Since the filtration step is automatically included in the sampling process the sample can be directly analyzed. The sample is stored within the sampling device, allowing sampling at all depths. Percolation water can be sampled with this suction cup probe in the saturated and in the unsaturated zone. Achieving hydraulic connectivity is sometimes difficult in the unsaturated zone. The system is limited to unconsolidated materials because this is the only case ion which hydraulic connectivity can be achieved. The suction cup, the filter membrane and the sample storage section are made entirely of low sorption materials such as PTFE, nylon and PE. The sampling device is a combination of two main components: the suction cup with probe shaft and stopper and the sample storage section with its recovery device (Fig. 3). The sampling system can be elongated using extension shafts. A construction manual for the sampling device can be found in the Appendix to this chapter.

7.1 Installing the Probe and Sampling Percolation Water

To sample percolation water a hole is drilled down to approx. 10–20 cm above the designated sampling depth. The remaining 10–20 cm of soil material is removed by pressing a Plexiglas tube (80 mm outer diameter, 10 mm wall thickness) down into the soil material. The tube with the soil is lifted up, taking care that the soil material does not fall back into the borehole (e.g. by using a packer technique). Next the borehole is cleaned and the suction cup is installed, assuring hydraulic connectivity. Subsequently the silicone tube coming from the probe shaft is connected with the vacuum line to apply the required vacuum to sample the percolation water (Fig. 3).

After the sampling the internal sample collection device can be withdrawn with the recovery device as described below. The probe shaft with the suction cup can either be recovered as well or be left installed for future sampling events, e.g. for monitoring purposes.

Unsteady physicochemical parameters such as the redox potential, temperature or pH should be analyzed from an aliquot immediately after withdrawing the collection vessel in a field laboratory. It is also best to prepare other aliquots, e.g. for trace element or DOC analysis, as soon as possible after the sample is taken and store them at <4 °C until analyzed.

Before the sampling device is used to sample percolation water on another site, it should be carefully cleaned and prepared (see appendix).

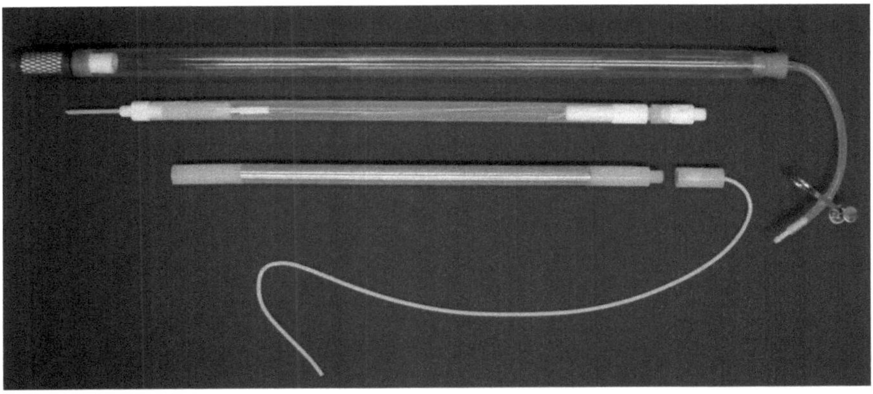

Fig. 3 Suction cup with probe shaft, sample collection vessel and recovery device (from *top to bottom*)

8 Comparison of Batch Extraction with in Situ Percolation Water Samples

All the methods described above aim for the same soluble fraction of trace elements in the soil. Since the methods differ to varying extents in various aspects, the results naturally differ to some extent as well. In the following we will use examples to describe where the results are typically at variance and what strengths and weaknesses each of methods show. We thereby assume that the "real" properties of the percolation water are to be found in the percolation water sampled in situ.

8.1 Physicochemical Parameters Affecting Extractability

The electronic conductivity (EC) of the 2 l/kg extract differs considerably from what can be found in situ. Due to the adjustment, the ECs of the 5 l/kg extracts correspond well with those in situ (Fig. 4). Amongst other things the ionic strength influences the pH value, which is one of the predominant factors governing the solubility of trace elements. Solutions with lower ionic strength showed lower pH values, because the deionised water used in the 2 l/kg extraction added a negative charge to the system, resulting in lower H^+ activity (Appelo and Postma 1993). As a consequence, pH-sensitive anionic elements such as As, Mo or V are extracted to a larger extent by the 2 l/kg extract than in situ or by the 5 l/kg extract (Godbersen et al. 2011).

The dilution effect resulting from the wider LS ratio often surpassed the pH effect for cationic elements in the 5 l/kg extract, especially at low concentration levels. Thus, instead of showing higher concentrations due to the lower pH value,

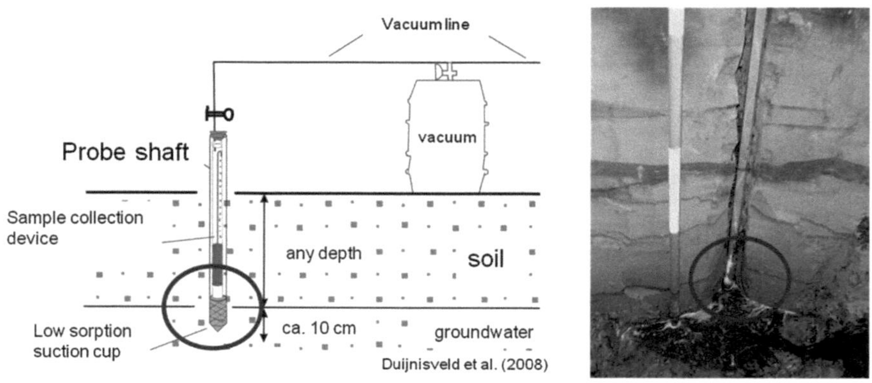

Fig. 4 Diagram and photo of an installed sampling device

the dilution effect caused the concentration of cationic trace elements such as Cd or Ni in the 5 l/kg extract to be lower than in situ but at the same concentration level as in the 2 l/kg extract (Godbersen et al. 2011).

The dilution effect additionally increases the solubility of organic carbon and of elements which tend to build metal–organic complexes such as Cr, Cu or Pb (Godbersen et al. 2011). Temminghoff (1998) for example found that Ca^{2+} acts as a bridge between organic molecules, leading to larger coagulates. Since the 2 l/kg extract shows a weaker ionic strength than in situ, fewer cationic ions such as Ca^{2+} are available, leading to smaller aggregates and thus higher solubility of organic carbon-associated elements. This effect is predominantly observable for soils with higher contents of organic carbon and can affect the solubility of most trace elements (Utermann 2011 in Bachmann et al. ed. 2011).[2] Using batch extraction methods it is very difficult to estimate trace elements whose speciation and solubility is affected by multiple parameters, such as Cr. Directly comparing Cr solubility in aqueous batch extracts and in situ yielded coefficients of determinationof $r^2 < 0.2$ (Godbersen et al. 2011).

8.2 Rotation Speed

The rotation speed influences the solubility of trace elements in batch extracts. A higher rotation speed increases the solubility of trace elements due to higher mechanical stress on the soil aggregates. Table 3 shows the results of a Wilcoxon test comparing rank differences of paired samples, subtracting concentrations measured in samples made at a higher rotation speed from those made at a lower rotation speed (Me_{Slow}–Me_{Fast}). Z is the standardized test statistic. The *p* value is a measure for the statistical significance of the result. Negative Z values show that in most cases the concentration in the extract produced at a higher rotation speed is higher compared to those extracts produced at a lower rotation speed. The effect of the rotation speed is slightly more pronounced for samples with higher DOC concentrations, leading to greater differences in the concentrations of trace elements which tend to build metal–organic complexes (Table 2). Batch extraction with adjusted ionic strength is less susceptible to the rotation speed than batch extracts made solely with ultrapure water, because the Ca^{2+} in the background electrolyte stabilizes the aggregates of organic carbon. As a result the solubility of trace elements is less affected by the rotation speed in 5 l/kg $Ca(NO_3)_2$ extracts extracting samples with high organic content.

[2] German legislation plans to take this observation into account by allowing higher trigger values for 2 l/kg extracts from soil materials with >1% of organic carbon (BMU 2011).

Table 3 Wilcoxon test comparing rank differences of paired samples, subtracting concentrations measured in samples made at a higher rotation speed from those made at a lower rotation speed

Batch extract with **ultrapure water**										
Samples with **higher DOC** concentration (>10 mg/l) in the corresponding percolation water (N = 14)										
	$Cd_{Slow}-Cd_{Fast}$	$Co_{Slow}-Co_{Fast}$	$Mo_{Slow}-Mo_{Fast}$	$Ni_{Slow}-Ni_{Fast}$	$Sb_{Slow}-Sb_{Fast}$	$Zn_{Slow}-Zn_{Fast}$	$Cr_{Slow}-Cr_{Fast}$	$Cu_{Slow}-Cu_{Fast}$	$Pb_{Slow}-Pb_{Fast}$	$V_{Slow}-V_{Fast}$
Z	−2.3	−3.0	−0.85	−3.0	−2.5	−2.8	−2.7	−2.5	−2.6	−3.3
p-value	0.022	0.003	0.397	0.003	0.013	0.005	0.008	0.013	0.009	0.001
Samples with **lower DOC** concentration (>10 mg/l) in the corresponding percolation water (N = 22)										
	$Cd_{Slow}-Cd_{Fast}$	$Co_{Slow}-Co_{Fast}$	$Mo_{Slow}-Mo_{Fast}$	$Ni_{Slow}-Ni_{Fast}$	$Sb_{Slow}-Sb_{Fast}$	$Zn_{Slow}-Zn_{Fast}$	$Cr_{Slow}-Cr_{Fast}$	$Cu_{Slow}-Cu_{Fast}$	$Pb_{Slow}-Pb_{Fast}$	$V_{Slow}-V_{Fast}$
Z	−2.4	−2.8	−4.1	−3.7	−3.4	−1.2	−2.6	−1.9	−2.0	−3.8
p-value	0.016	0.005	0.000	0.000	0.001	0.223	0.008	0.062	0.058	0.000
Batch extract with **adjusted ionic strength**										
Samples with **higher DOC** concentration (>10 mg/l) in the corresponding percolation water (N = 14)										
	$Cd_{Slow}-Cd_{Fast}$	$Co_{Slow}-Co_{Fast}$	$Mo_{Slow}-Mo_{Fast}$	$Ni_{Slow}-Ni_{Fast}$	$Sb_{Slow}-Sb_{Fast}$	$Zn_{Slow}-Zn_{Fast}$	$Cr_{Slow}-Cr_{Fast}$	$Cu_{Slow}-Cu_{Fast}$	$Pb_{Slow}-Pb_{Fast}$	$V_{Slow}-V_{Fast}$
Z	−.031	−0.91	−1.5	−0.91	−0.97	−0.16	−1.3	−0.47	−1.5	−1.3
p-value	0.975	0.363	0.124	0.363	0.331	0.875	0.198	0.638	0.140	0.198
Samples with **lower DOC** concentration (>10 mg/l) in the corresponding percolation water (N = 22)										
	$Cd_{Slow}-Cd_{Fast}$	$Co_{Slow}-Co_{Fast}$	$Mo_{Slow}-Mo_{Fast}$	$Ni_{Slow}-Ni_{Fast}$	$Sb_{Slow}-Sb_{Fast}$	$Zn_{Slow}-Zn_{Fast}$	$Cr_{Slow}-Cr_{Fast}$	$Cu_{Slow}-Cu_{Fast}$	$Pb_{Slow}-Pb_{Fast}$	$V_{Slow}-V_{Fast}$
Z	−1.4	−2.4	−3.7	−3.9	−4.1	−1.3	−3.1	−.49	0.000	−4.0
p-value	0.162	0.015	0.000	0.000	0.000	0.201	0.002	0.627	1.000	0.000

8.3 Variability and Uncertainty

Heterogeneity is an undeniable aspect of nature. Thus all estimations are accompanied by some uncertainty. Uncertainty can, for example, be attributed to spatiotemporal variability in the field, to variability of the sample itself or to differences in analytical protocol. The variability of trace element concentrations in the field can be as high as two orders of magnitude within 90 m (Godbersen et al. 2012). The variability of trace element concentrations in batch extracts made with the same method from aliquots of the same soil sample can range from a COV (coefficient of variation) of 6 % for Cd or Sb to a COV of 28 % for Cr. The deviance between trace element concentrations in situ and in aqueous batch extracts is often greater than 0.5 orders of magnitude, even if soil and percolation water samples originate in exactly the same space (Godbersen et al. 2011).

Appendix

1.1 Construction Manual for Low Sorption Suction Cup Lysimeter with Internal Sample Collection Vessel

The suction cup probe is built for trace element analysis. Thus, carefully deburring and cleaning each item after mechanical manufacturing and before assembly is of vital importance for good results.

The numbers after each item refer to Figs. 5, 6, 7, 8, 9 and Tables 4 and 5 in the appendix.

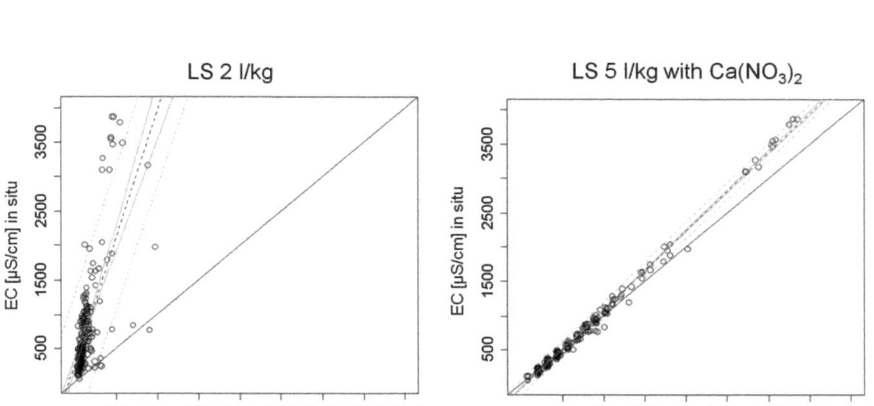

Fig. 5 Electronic conductivity (EC) in extracts compared to EC in situ

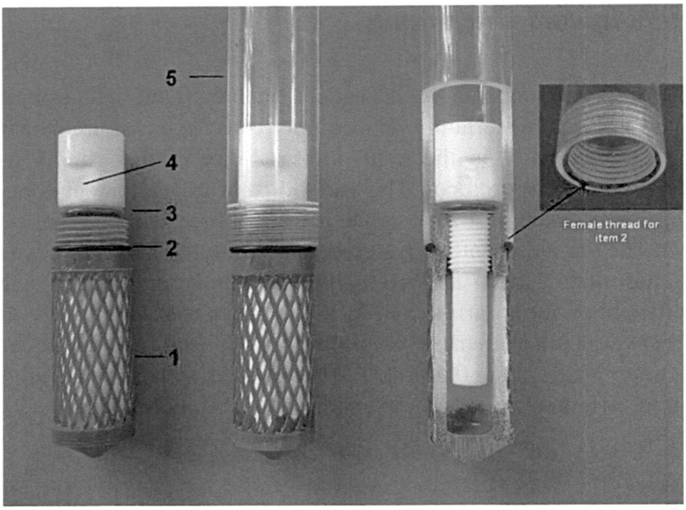

Fig. 6 Suction cup and probe shaft

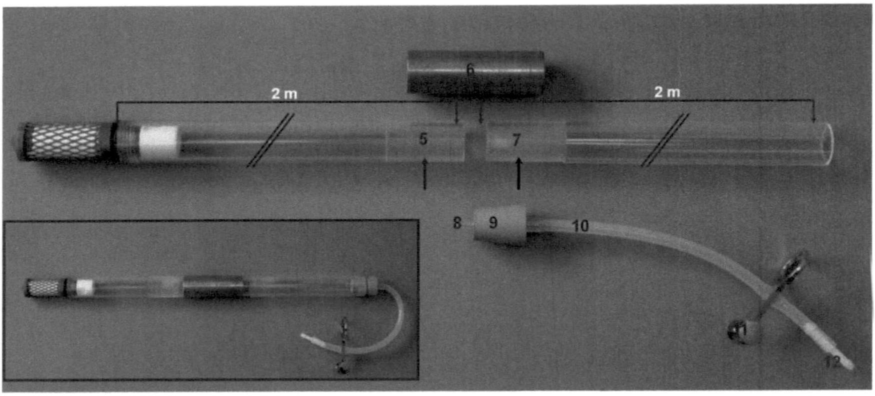

Fig. 7 Suction cup probe and extension shaft

1.2 Suction Cup and Probe Shaft

First a female thread is cut into the suction cup (1) to accommodate the suction cup filling (4). The sealing ring (2) is then placed around the male thread of the suction cup (1). Next the male thread is cut into the filling (4) and a silicone sealing ring (3) is placed over the male thread of the filling (4). Afterwards the filling (3–4) and cup (1–2) are bolted together. Then a female thread is cut into the lower end of the Plexiglas tube for the probe shaft (5) to accommodate the male thread of the suction cup (1). Subsequently the probe shaft (5) is joined to the suction cup (1–4).

Fig. 8 Separator

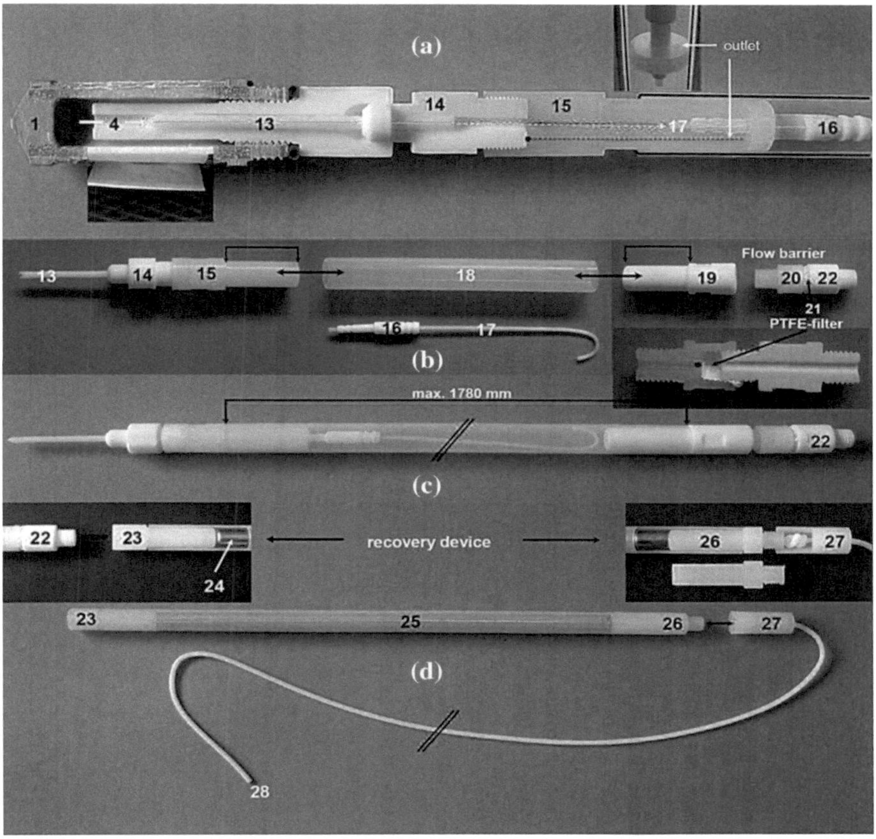

Fig. 9 Components of sample collection vessel (a–c) and of recovery device (d)

Table 4 Material list for low sorption suction cup lysimeter with internal sample collection vessel (items no. 1–12)

Item no.	Materials	Amount	Notes	Source
1	**Low sorption suction cup with nylon membrane on porous PE-supporting body**	1 pc	Delivered without female thread	Umwelt-Geräte-Technik GmbH Müncheberg
2	**Sealing ring (O-ring)** 18 × 1.5 mm	1 pc	Wear part replacement required at greater intervals.	Technical supply
3	**Silicone O-ring, covered with seamless FEP** (FEP = Tetrafluorethylene-perfluoropropylene) 12 × 3.0 mm (Inner diameter x thickness)	1 pc	/	Beichler and Grünenwald GmbH Löchgau
4	**Suction cup filling** connecting component between the suction cup and sample collection vessel PTFE rod, diameter 17 mm available in lengths of 1–2 metres	**See note**	**Custom-made by own precision mechanics workshop** Wear part replacement required at greater intervals	Plastic supplier
5 + 7	**Basic shaft and extension shaft** Plexiglas XT tubes outer diameter: 25 mm/Inner diameter: 19 mm available in standard length: 2,000 mm	1 pc each **see notes**	**Custom-made by own precision mechanics workshop** **Basic tube:** - Attaching female thread for the suction cup –Trimming of approach surface for item 6 **Extension tube:** Trimming of approach surface for item at both ends Additionally: Reworking the thread at basic tube chamfering of basic and extension tube	Plastic supplier

(continued)

Table 4 (continued)

Item no.	Materials	Amount	Notes	Source
6	**Connection sleeve** Brass tube, drawn (CuZn37/Ms63) Outer diameter 26 mm/inner diameter 24 mm All lengths available	1 × 80 mm		Metal supplier
8	**Plexiglas capillary** Outer diameter 5 mm/inner diameter 3 mm Supplied in length: 2,000 mm	1 pc	**Cut to length of 50 mm, chamfering on both ends**	Plastic supplier
9	**Rubber plug** Type "D&N", para grey, onehole, dimensions:17 × 22 × 25 mm	1 pc		Laboratory supplier
10	**Silicone tube** Outer diameter: 7 mm/inner diameter: 4 mm	Approx. 200 mm		Technical suppliers
11	**Pinchcock** By Mohr, brass, nickel plated, elastic	1 pc	**Minimum leg length: 30 mm**	Laboratory supplier
12	**Hose barb** "NORMA" hose connector, GS 4, straight	1 pc		Laboratory or technical supplier
	Liquid nitrogen	Approx 250 ml	**Observe safety rules for handling and transportation!**	Linde AG, Hamburg

Table 5 Material list for low sorption suction cup lysimeter with internal sample collection vessel (items no. 13–28)

Item no.	Materials	Amount	Notes	Source
13	**Trunk**	1 pc	**Identification by manufacturer:**	Beichler and Grünenwald GmbH Löchgau
	FEP capillary		FEP-"Schlauchabschnitt, gerade"	
	Outer diameter 3.5 mm/inner diameter 1.5 mm, length: 80 mm			
14	**Valve**	1 pc each	**Custom-made by own precision mechanics workshop**	Fluoropolymers supplier
19	**Finishing plug**		Refining: deburring, cleaning	
22	**Connecting tube** (flow barrier) made from PTFE rod, diameter 17 mm			Laboratory or technical supplier
15	**Base plug** (sample collection vessel)	1 pc each	**Custom-made by own precision mechanics workshop**	Beichler and Grünenwald GmbH Löchgau
20	**Fixation screw** (flow barrier)		Refining: deburring, cleaning	
23	**Bottom stopper** (recovery device Part 1)			
26	**Upper stopper** (recovery device Part 1)			
27	**Union nut** (recovery device part 2) made from FEP rod (FEP "Natur"), diameter 17 + 0.1 mm, length: 400 mm			
16	Reducing nozzle	1 pc		Laboratory or technical supplier
	PTFE tubing connection, length 45 mm, reduction from 6.8 to 4.5 mm, inner diameter 2 mm			
17	**Uptake tube**	See notes	**Made by thermal deformation**	Novodirect GmbH, Kehl am Rhein
	PTFE micro tube		Length according to following rule of thumb: total length of item 18 – 30 mm	
	Nominal dimensions: outer diameter 1.93 mm, inner diameter, 1.32 mm			

(continued)

Table 5 (continued)

Item no.	Materials	Amount	Notes	Source
18	**Sample collection tube** (for sample collection vessel)	1 pc	**Identification by manufacturer:**	Beichler and Grünenwald GmbH, Löchgau
25	**Tube** (recovery device Part 1)	See notes	FEP-"Schlauchabschnitt, gerade" cutting length for standard probe of 2 m is 1780 mm	
	FEP tube, outer diameter 17 mm (±0.2 mm), inner diameter 13 mm (±0.2 mm), length: 1,800 mm		Cut to desired length	
21	**PTFE filter as a flow barrier**	See note	**Exchange possibly required at fairly long time intervals**	Laboratory supplier
	Diameter 90 mm			
	Pore size 0.2 μm			
24	**Drop weight**	See notes	**To increase weight of recovery device** (part 1 = items 23–26)	Steel supplier
	V2A-rod		Inserted into item 25 (FEP tube) and remains there	
	Material: 1.4301			
	Diameter: 12 mm			
	Length: 300 mm			
28	**Towing Rope**	See note	**Towing rope for part 2 of the recovery device** (items 27 and 28)	Manufacturers of monofilament lines and ropes
	Monofilament nylon rope:		Is required together with part 1 at depths greater than 2 meters	
	Diameter: 2.5 mm		Reworking:	
	Not coloured		Cut to required length	
	UV stabilized		Fixation in item 27	
	PTFE tape	See note	**Exclusively use tapes made of pure PTFE without additives or colour**	Chesterton GmbH Ismaning
	Thickness: 0.1 mm		Optimum thickness: 0.1 mm	
	Max 0.25 mm			
	Width: 12 mm			

It is very important that the sealing is perfect. Sharp edges on the components might cut into the sealing rings and cause leaks. To avoid leakages the edges need to be smoothed. A leaching test is mandatory in every case.

1.3 Probe Extension

The probe shaft can be extended with additional Plexiglas extension tubes (7). The connection is made with a brass sleeve (6). In order to connect the probe and extension tubes the diameters of the tubes are reduced by dipping the end of the probe shaft (5) in liquid nitrogen for approximately 30 s. The tube is then inserted halfway into the brass sleeve (6). Subsequently the diameter of the extension tube (7) is reduced in the same manner and then the tube is inserted into the other side of the brass sleeve. This joint does not need any further sealing. To add further extensions the above procedure is repeated.

To separate the extension tubes an empty, bottomless bottle whose neck fits snugly onto the tube ("separator", Fig. 8) is placed around the brass sleeve and filled with liquid nitrogen. The nitrogen reduces the diameter of the extension tube and the extension tube can be pulled out of the brass sleeve. Next the brass sleeve can be removed from the lower (extension) tube.

1.4 Finishing Plug

To assemble the finishing plug (8–12) a Plexiglas capillary (8) is inserted through a hole into the plug (9). As shown in Fig. 7, 10 mm of the capillary (8) extend beyond each end of the plug (9). Then the silicone hose (10) is connected with the outer end of the Plexiglas capillary (8). Next the hose barb (12) is inserted into the end of the silicone hose. The pinchcock (11) is then adjusted to the hose. Finally the assembled finishing plug is plugged into the far end of the (extended) probe. To test for leakages 100 mbar of negative pressure are applied to the system. A change in pressure, e.g. measured with a high-sensitivity manometer, is a sign of a leakage.

1.5 Sample Collection Vessel

First, the following parts need to be manufactured by drilling bores and cutting threads into them so that they resemble the items in Fig. 9: valve (14), base plug (15), finishing plug (19), fixation screw and connecting tube (22). All the manufactured components are then carefully deburred and cleaned. Next the uptake tube (17) is prepared by cutting the material into appropriate lengths. The tubes are then straightened by heating the material to release stress from the material. To do so, some pieces of the uptake tube are threaded into a glass tube with a suitable inner diameter (approximately 2.2 mm) and heated for 2 h at 180 °C in an oven. Heating and cooling the material causes the molecules in the material to rearrange. Stress is thereby released and the material is thus stabilized. To prepare the vessel (18) the FEP tubes are cut to appropriate lengths. All the manufactured items are cleaned so they are ready to use for trace element analysis.

Assembly starts by inserting the trunk (13) approx. 20 mm deep into the valve (14) as shown in Fig. 9a. Then the reducing nozzle (16) is applied at full length onto the uptake tube (17). Subsequently the nozzle (16) is inserted into the bore of the base plug (15). Next the far end of the uptake tube (17) is heated to approx. 180 °C to make it flexible and then bent with the smallest possible inner radius. The thin wall capillaries are easily buckled, so it should be bent very carefully. Next the FEP tube (18) is applied by inserting the uptake tube (17) into the FEP tube (18) and then pressing the FEP tube (18) over the assembled basis stopper (13–17). After that, the finishing plug (19) is inserted into the opposite end of the sample collection FEP tube (18). The diameters of the components are calculated and manufactured with as little play as possible to ensure that the sample collection vessel is mechanically resilient and sealed tight.

1.6 Flow Barrier

The bolt of the male thread of the connecting tube (22) is made a few mm longer than the female thread of the fixation screw (20). First the PTFE filter membrane (21) is gently pulled taut over the bolt of the connecting tube (22). Next the fixation screw (20) is softly screwed onto the connecting tube and then immediately released. Two or three layers of PTFE tape are wrapped around it to fix the membrane (21) onto the male thread of the connecting tube (22). Finally, the connection tube and the fixation screw are screwed back together. The PTFE membrane is permeable to air, but impermeable to water. This way the flow barrier prevents the sample collection vessel from overflowing.

To mount the flow barrier (20–22) with the sample collection vessel the male thread of the fixation screw (20) is screwed into the female thread in the finishing plug. Finally, the valve (14) is screwed into the base plug (15) to finish assembling the sample collection device. If new components are used, the connection between the valve and base plug is leakproof. Frequently opening and closing the valve, however, can widen the play of the thread, possibly creating a leak. A few layers of PTFE tape wrapped around the male thread usually stop the leak.

1.7 Recovery Device

To assemble the ballast of the recovery device a steel rod (24) is inserted into a 40 cm long piece of FEP tube (25) and lower (23) and upper (26) stoppers are plugged into each end. The steel rod should be locked tight by the stoppers. A female thread links the lower stopper (23) to the connection tube (22) of the flow barrier. The additional ballast of the steel rod is needed to pull the nylon rope (which is otherwise too inflexible) straight when the recovery device is lowered through the extension shaft. The nylon rope (28) is connected with the ballast by a union nut. To fasten the rope the nylon rope is inserted through the bore into the union nut (27) and subsequently knotted and pulled tight so the knot is hidden in the union nut (Fig. 9d). To prevent the knot from bursting when the device is used to recover a full sample collection vessel, the knot is heated with a heat gun. Finally, the union nut (27) is tightly screwed onto the upper stopper (26) of the ballast.

To recover a sample collection vessel with the recovery device, the device is slowly lowered through the probe shaft until the lower stopper (23) hits the connection tube (22). The recovery device is the screwed onto the sampling device and pulled up to collect the sample collection vessel.

To install the sampling collection vessel in the suction cup probe, the sampling device is linked with the recovery device and lowered through the probe shaft. The vessel is dropped very lightly for the last few centimetres. If all components are manufactured precisely enough, the trunk (13) should find the bore of the suction cup filling (4) and the valve (14) makes a tight connection with the filling (Fig. 9a). Check whether the sample collection vessel has properly connected with the suction cup by softly pulling on the rope. If it has properly connected some resistance can be felt.

1.8 Preparing a Newly Built Device for Sampling

Before using the sampling device in the field, carefully clean it. A blank check is conducted for every device with ultrapure water. Before using the device all parts that may have contact with the percolation water sample are conditioned with 2 % HNO_3 for at least 2 h and subsequently rinsed twice with ultrapure water. These parts are then stored in ultrapure water until use. The suction cup and the sample collection vessel are protected during transportation against contamination with a plastic cap (e.g. PE bottle with 100 ml wide neck) or PE plastic foil. The protective caps are discarded after each sampling event.

1.9 Cleaning After Sampling

After sampling the device is disassembled into the suction cup (1 and 2), cup filling (3 and 4), valve (13 and 14), finishing plug (19), basic shaft (5) and sample collection vessel (15–19). All components are rinsed with distilled water until no particles are left. The bore of the valve is cleaned mechanically e.g. with a pipe cleaner (non-metallic!). Then each part is thoroughly rinsed with ultrapure water. Next the suction cup, the cup filling, the valve and the finishing plug are stored in containers with 2 % HNO_3. The sample collection vessel is rinsed twice with ultrapure water and subsequently filled completely with 2 % HNO_3. The basic shaft does not need to be stored in nitric acid since it does not come into contact with the sample. If spare parts need to be applied they are cleaned and prepared in the same manner before installing. A blank check is conducted for the entire system before every use by sampling ultrapure water with the assembled device and storing it for at least 12 h in the device before analyzing it.

References

Ad-hoc-AG B (2005) Leitfaden zum Umrechnungsprogramm "Methodenvergleich Gesamtgehalte Haupt- und Spurenelemente" (http://www.bgr.bund.de/DE/Themen/Boden/Netzwerke/Adhocag/Downloads/methodenvergleich_ad_hoc_ag.pdf?__blob=publicationFile&v=2)

Alloway BJ (1995) Heavy metals in soils. Blackie Academic and Professional

Appelo CAJ, Postma D (1993) Geochemistry, groundwater and pollution. Balkema, Rotterdam, p 536

Arruda M (2007) Trends in sample preparation. Nova Science Publishers, New York

BMU (2011) The federal environment ministry. Arbeitsentwurf der Mantelverordnung Grundwasser/Ersatzbaustoffe/Bodenschutz. Arbeitsentwurf vom, 06 Jan 2011, p 116. http://www.bmu.de/files/pdfs/allgemein/application/pdf/entw_mantelverordnung.pdf

DIN 19529 (2009) Elution von Feststoffen - Schüttelverfahren zur Untersuchung des Elutionsverfahrens von anorganischen Schadstoffen mit einem Wasser/Feststoffverhältnis von 2 l/kg. Beuth Verlag

Duijnisveld WM, Godbersen L, Dilling J, Gäbler H-E, Utermann J, Klump G, Scheeder G (2008) Ermittlung flächenrepräsentativer Hintergrundkonzentrationen prioritärer Schadstoffe im Bodensickerwasser. Endbericht UBA-Forschungsvorhaben 204 72 264, Bundesanstalt für Geowissenschaften und Rohstoffe. http://www.bgr.bund.de/DE/Themen/Boden/Produkte/Schriften/Downloads/2008_Hintergrundkonz_Schadstoffe_in _SIWA.pdf?_blob = publicationFile&v = 3

Godbersen L, Utermann J, Duijnisveld WHM, Altfelder S, Kuhnt G, Böttcher J (2011) Comparing concurrent in situ and batch-extracted trace element concentrations. J Plant Nutr Soil Sci 174:587–595. doi:10.1002/jpln.201000274

Godbersen L, Duijnisveld WHM, Utermann J, Gäbler H-E, Kuhnt G, Böttcher J (2012) Application of groundwater thresholds for trace elements on percolation water—a case study on percolation water from Northern German Lowlands. J Environ Qual (First Look). https://www.crops.org/publications/jeq/view/first-look/q11-0218.pdf

Gooddy D, Shand P, Kinniburgh D, van Riemsdijk W (1995) Field-based partition coefficients for trace elements in soil solutions. Eur J Soil Sci 46:265–285. doi:10.1111/j.1365-2389.1995.tb01835.x

Grathwohl P, Susset B (2009) Comparison if percolation to batch and sequentional leaching tests: theory and data. Waste Manage (Oxford) 29:2681–2688

Grossmann J (1988) Physikalische und chemische Prozesse bei der Probenahme von Sickerwasser mittels Saugsonden. Dissertation, Technische Universität München, Fakultät für Chemie, Biologie und Geowissenschaften

Heidkamp A (2005) Pedotransfer-Funktionen zur Sorption von Schwermetallen in Waldböden. Der andere Verlag

Hellerich W, Harsch G, Haenle S (2001) Werkstoff-Führer Kunststoffe, Eigenschaften-Prüfungen-Kennwerte. Carl Hanser Verlag

Hessisches Landesamt für Umwelt und Geologie (HLUG) (2011) Überarbeitung des Handbuchs Altlasten Band 3 Teil 3, "Untersuchung und Beurteilung des Wirkungspfades Boden-Grundwasser (Sickerwasserprognose)". http://www.hlug.de/fileadmin/ dokumente/altlasten/ Elutionsverfahren-Hinweise_HLUG_SiWaPro_neu_4-2011_.pdf

Koch D (1993) Erfassung und Bewertung der Schwermetallmobilität über Sickerwässer aus Böden hoher geogener Anreicherung und zusätzlicher Belastung. Dissertation, Göttingen, p 149

LABO (2003) Bund/Länder Arbeitsgemeinschaft Bodenschutz: Arbeitshilfe Sickerwasserprognose bei Orientierenden Untersuchungen. http://www.labo-deutschland.de/ documents/SiWaPrognose-120903_91 f.pdf

Lewandowski J, Leitschuh S, Koß V (1997) Schadstoffe im Boden: Eine Einführung in Analytik und Bewertung. Springer, Berlin, Heidelberg, New York

Mester Z, Sturgeon R (2003) Sample preparation for trace element analysis. Elsevier, Amsterdam

Morrison RD (1982) A modified vacuum-pressure lysimeter for soil water sampling. Soil Sci 134(3):206–210

Oberbach K (2001) Saechtling, Kunststoff Taschenbuch. 28. Ausgabe, 1020 S., Carl Hanser Verlag

Reimann C, Siewerd U, Skarphagen H, Banks D (1999) Does bottle type and acid-washing influence trace element analyses by ICP-MS on water samples? Sci Total Environ 239:111–130

Schroers S (2002) Methoden zur Elution von organischen und anorganischen Stoffen aus Böden. "Sickerwasserprognose in Theorie und Praxis" am 16./17.01.2002 im BEW Duisburg

Smith K, Cresser M (2003) Soil and environmental analysis. Marcel Dekker, New York and Basel

Temminghoff E, van der Zee SEATM, de Haan FAM (1998) Effects of dissolved organic matter on the mobility of copper in a contaminated sandy soil. Wageningen University and Research Center Publications. http://library.wur.nl/oai

Utermann J (2011) In: Rosenkranz, Bachmann, König, Einsele: Bodenschutz, Ergänzbares Handbuch (Loseblattsammlung) 1540, V/11, Erich Schmidt Verlag

Utermann J, Meyenburg G, Altfelder S, Gäbler H-E, Duijnisveld WHM (2005) Entwicklung eines Verfahrens zur Quantifizierung von Stoffkonzentrationen im Sickerwasser auf der Grundlage chemischer und physikalischer Pedotransferfunktionen. Bundesanstalt für Geowissenschaften und Rohstoffe; Bundesministerium für Bildung und Forschung

Methods for Quantifying Wind Erosion in Steppe Regions

Roger Funk, Carsten Hoffmann and Matthias Reiche

Abstract Wind erosion has become an important soil degradation process in the steppe regions around the world caused predominantly by overgrazing and by transforming steppe into arable land. Soils, formed by aeolian processes over centuries, are now at risk of being destroyed by the same processes within a short space of time. The main problem with wind erosion is how it is perceived. Although heavy sand and dust storms occasionally attract attention, erosive processes usually go unnoticed. Annual average soil losses up to 40 t ha^{-1} are possible without any visible sign of erosion or deposition. The following chapter introduces common methods for assessing wind erosion and for quantifying the soil losses involved. Consideration of the special characteristics of steppe regions is discussed to enable methods to be applied successfully. Quantification of wind erosion is based on measurements of horizontal fluxes, which can additionally be used to derive soil losses/dust emissions or the deposition of transported particles. The thickness and extent of depositions can be measured to calculate the relocated volume or mass. Losses of fine particles and organic matter can be derived by comparing the grain size distribution and organic matter content of the original soil and depositions. The fallout radionuclide method (FRN, especially ^{137}Cs) is suitable for identifying wind erosion and dust deposition patterns at larger spatial and temporal scales. Remote sensing and GIS procedures are finally used to present wind erosion and dust deposition areas for large landscape units.

Keywords Steppe regions · Wind erosion · Quantification

R. Funk (✉) · C. Hoffmann · M. Reiche
Leibniz Centre for Agricultural Landscape Research (ZALF) e. V.,
Eberswalder Strasse 84, 15374 Muencheberg, Germany
e-mail: rfunk@zalf.de

L. Mueller et al. (eds.), *Novel Measurement and Assessment Tools for Monitoring and Management of Land and Water Resources in Agricultural Landscapes of Central Asia*, Environmental Science and Engineering, DOI: 10.1007/978-3-319-01017-5_18,

1 Introduction

Wind erosion occurs in many arid, semiarid and agriculturally used areas around the world, and is influenced by geological and climatic factors, as well as human activities. Wind erosion relocates huge amounts of soil constituents, especially the finest and most valuable soil particles, leaving behind the less fertile sand fraction (Chepil 1957a, b). It has been estimated that approximately 2,000 Mt of particulates are emitted to the atmosphere each year, predominantly consisting of silt and clay fractions. Of this amount, 1,500 Mt are deposited as dust on land surfaces and 500 Mt in oceans (Shao et al. 2011). Even more soil is redistributed close to the ground by saltation, which mainly affects particles in the sand fraction. Wind erosion can therefore also be regarded as an important process of land evolution and land degradation (Funk and Reuter 2006).

The main problem with wind erosion is how it is perceived. Although heavy sand and dust storms occasionally attract attention, erosive processes usually go unnoticed. Years ago, Chepil (1957b) pointed out that annual average soil losses up to 40 tha^{-1} are possible without any visible sign of erosion. Both erosion and deposition processes take place across large areas, making them difficult to identify. In contrast to water erosion, where the eroded material follows determined paths, wind-eroded material is widely dispersed over the landscape. Furthermore, the direction of transport is subject to change, sometimes in the completely opposite direction, as are the areas of erosion and deposition (Funk and Reuter 2006).

Deserts are a major global source of dust. However, the surrounding semi-arid steppe ecosystems have also changed from former sinks into substantial source regions in recent decades. The causes are always inappropriate land use as the vegetation is destroyed by overgrazing, or the conversion of grazing land to cropland. Since steppe soils represent one of the largest carbon stocks, their degradation has a significant influence on the global carbon balance. To understand the importance of land use change-induced wind erosion and soil degradation to the global carbon balance, it is necessary to identify and quantify the corresponding processes in steppe ecosystems. Especially in grasslands, a clear differentiation between good conditions of the vegetation cover and a critical state for wind erosion is complicated by the continuous transition between all states (Hoffmann et al. 2008a). The processes of wind erosion are determined by the coverage of the soil by vegetation, which is highly variable in space and time. For this reason, methods are required to objectively determine the surface features of grassland and how it influences the processes of wind erosion and dust deposition.

2 Methods for Quantifying Wind Erosion

2.1 Measurements of Horizontal Fluxes

Generally, wind erosion initiates three transport modes of particle motion: creeping and saltation, which mainly affects sand particles, and suspension of silt and clay particles. Since each mode is related to specific grain sizes, wind erosion is also a very effective sorting process for particles that differ in size and density. Generally, two quantities must be measured within an erosion event: the horizontal saltation flux and the vertical dust flux. Reliable measurements of both fluxes are the most problematic procedures in aeolian research (Goossens et al. 2000). Usually, 99 % of the total amount is transported at a height of less than 1 metre by creep and saltation (Zobeck et al. 2003). Especially in semi-arid regions, however, dust storms carry enormous amounts of particles in the suspension mode, resulting in completely different vertical transport profiles (Hoffmann et al. 2008b) (see Fig. 1).

Most of the measurements focus on quantifying wind erosion, which is dominated by the horizontal saltation flux. Although methods for measuring soil losses by wind erosion in steppe regions do not differ from those for agricultural land, some specifics have to be considered. In general, the affected areas are much larger, the time scales of erosion and/or deposition are longer, and the heterogeneity of vegetation (grasses, bushes) have to be considered. In this connection, the transport forms of wind erosion, creep, saltation and suspension occur with some modifications. In landscapes with homogeneous vegetation cover, topography effects are also more relevant, as reflected by the typical deposition patterns for

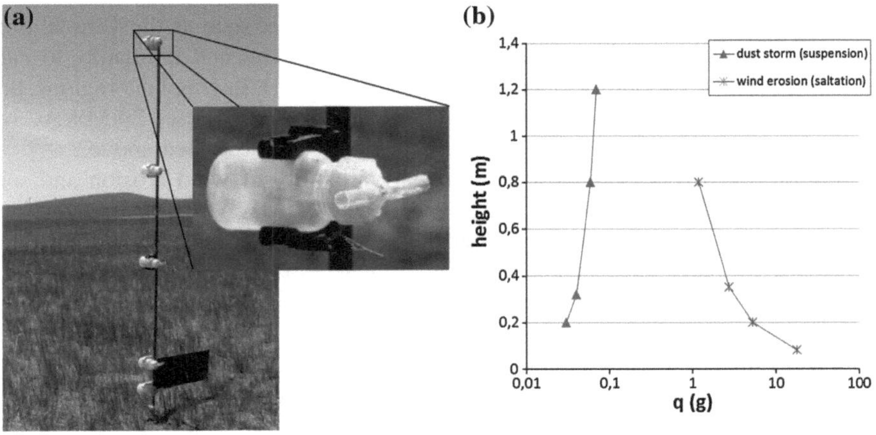

Fig. 1 **a** Modified Wilson and Cooke sediment trap (MWAC, *left*) as a set of five traps on a pole with a wind vane that always aligns the traps to the wind direction. **b** Measured transport profiles of a dust storm on grassland (*blue triangles*) and of a wind erosion event on arable land (*red crosses*)

dust on summits from windward or leeward oriented slopes. As steppe soils are often formed from aeolian deposits, particles smaller than 200 μm predominate to a disproportionately large extent (Hoffmann et al. 2008a).

Two widely-used types of passive samplers are the Big Spring Number Eight (BSNE, Fryrear 1986) and the Modified Wilson and Cooke (MWAC) catcher (Kuntze et al. 1990). The latter is also very efficient for trapping dust particles (Goossens and Offer 2000). One of the most common methods for measuring wind erosion in field studies are sets of BSNE or MWAC to measure the horizontal soil transport at different heights. Generally, more than three traps are arranged vertically at a pole between heights of about 0.05–2.0 m. The trapped amounts of transported soil particles can be used to calculate the entire transport profile and to integrate the sediment over a certain range of height. MWAC have good trapping efficiency for both sand and dust particles, regardless of the wind speed. They have been used in many projects, featuring a well-described trapping efficiency for a wide range of soil textures (Goossens et al. 2000; Goossens and Offer 2000; Funk 1995; Sterk and Stein 1997; Mendez et al. 2011). In addition, the trap is inexpensive, easy to construct, it can be reconstructed using locally available material, and takes measures reliably—all of which are reasons for its preferred use in remote areas and under harsh conditions.

Measurements of wind erosion or dust deposition across an area or along a transect require the installation of several, at least two, traps—one at the incoming and one at the outgoing side of the area under investigation. A certain number of sediment traps are required to balance the spatial distribution of soil losses or gains on a measuring field. Sterk and Stein (1997) used 21 traps on a field spanning 0.24 ha; Funk et al. (2004) set 15 traps on 2.25 ha; Visser et al. (2004) placed 17 traps on a 1.6 ha field in regular and irregular grids. This high trap density is necessary on arable land to derive important transport parameters, their spatial distribution and dependency on the transported distance, such as the vertical flux density, particle composition and organic matter content, and ultimately to calculate the spatial variability of soil loss (Funk et al. 2004). In grassland regions, the distance between traps may be wider. Hoffmann et al. (2008a) used 20 MWAC on a 65 ha grazing site to calculate the same parameter as described above.

Figure 2 gives an example of the spatial variability of wind erosion and soil losses on a sandy soil based on measurements with 15 MWAC traps arranged in a regular grid on an agriculturally used field. The left-hand side shows the total transported eroded soil calculated and interpolated spatially from measured vertical profiles. The characteristics of wind erosion on a sandy soil are the rapid increase of the transported amount, and saturation after a relatively short distance. The wind already reaches the maximum transport capacity at 50 m, which remains constant over the rest of the field. The consequences for the spatial variability of soil loss are: if a distinct soil loss only appears in the windward half of the measuring field, then erosion and deposition are in equilibrium or even deposition prevails locally. Wind erosion therefore not causes only soil losses, it also increases the soil heterogeneity of a field by temporarily depositing coarser fractions on the field and blowing out the finer ones.

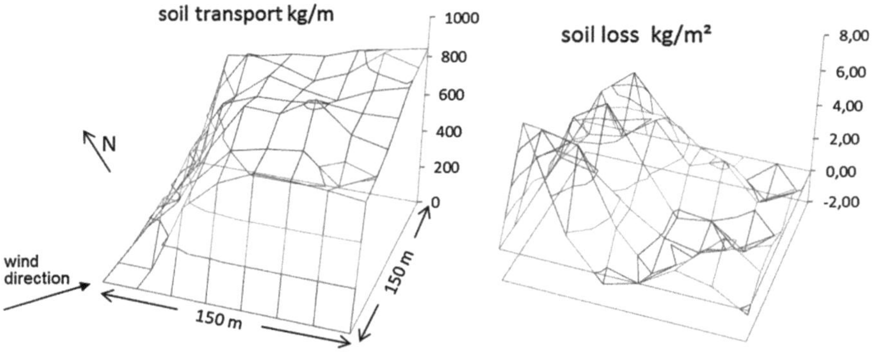

Fig. 2 Spatial distribution of wind erosion on a 150 × 150 m measuring field; *left* total transported soil per m width; *right* soil loss per m^2 (negative values = deposition)

Further techniques to measure horizontal fluxes are electronic impact sensors, such as Saltiphon or Sensit (Spaan and van den Abeele 1991; SENSIT 2012) or recording traps such as SUSTRA (Janssen and Tetzlaff 1991). Together with a meteorological station, they all enable an estimate to be made of the threshold wind or friction velocity, the correlation between wind speed and transport intensity, and information to be obtained about the duration and temporal variability of an event.

2.2 Measurements of Vertical Fluxes

Vertical fluxes are generally measured as dry deposition flux. In this case, dust traps are arranged separately or vertically in a cluster up to heights of several metres (Goossens and Rajot 2008). Traps can be simple boxes or glass jars, funnels or more aerodynamically shaped bowls (VDI 1972; Hall et al. 1994; Goossens 2005). Some have a lattice or a layer of glass marbles on top to prevent the remobilisation of the trapped material (Shao 2008; Groll et al. 2013). The trapping efficiency of these samplers is 50 % less than from an open water surface (Goossens 2005).

Another way to calculate vertical dust flux is correlate vertical to horizontal fluxes (Gillette 1977). Vertical dust fluxes can be derived based on measurements of horizontal fluxes, as shown in Fig. 1b. Using dust concentrations from at least two different heights and the logarithmic wind profile, vertical flux F (upward or downward) can be calculated using Eq. 1.

$$F = \frac{-ku_*(C_2 - C_1)}{\text{In}\left(^{Z_2}/_{Z_1}\right)} \tag{1}$$

With

F	vertical dust flux, $gm^{-2}s^{-1}$
k	von Karman constant 0.4, dimensionless
u_*	friction velocity, ms^{-1}
$C_{1,2}$	concentrations at height z_1 and z_2, gm^{-3}

The direction of vertical fluxes is generally from the higher to the lower concentration. In the example of Fig. 1b, both cases are shown—deposition (vertical downward flux) from a dust storm crossing grassland, and dust emission (vertical upward flux) caused by local wind erosion on cropland.

The determining factor for deposition or emission at an area is aerodynamic roughness. In grassland, for example, it is determined by the height and density of the vegetation, which is influenced by grazing intensity (Fig. 3). Vegetation parameters can be used to express the land use intensity in physically measurable units, which are related to the calculated fluxes. In addition to the dimensions of the vegetation (height, soil cover, leaf area), the effect on surface wind is a necessary parameter, particularly if the vegetation is sparse. This aspect, which can be calculated from wind profile measurements close to the ground, is expressed by aerodynamic roughness length z_0, where $u_{z0} = 0$, calculated from the logarithmic wind profile [Eq. 2].

$$u_2 - u_1 = \frac{u_*}{\kappa} \mathrm{In}\left(\frac{Z_2}{Z_1}\right) \tag{2}$$

The different grazing intensities correspond to the following aerodynamic roughness lengths:

–	heavily grazed (HG)	→	$z_0 < 0.5$ cm
–	moderately grazed (MG)	→	z_0 0.21–1.2 cm
–	lightly grazed (LG)	→	z_0 0.4–1.5 cm
–	ungrazed (UG)	→	$z_0 > 1.9$ cm

In steppe regions, the processes of long-range suspension transport and local wind erosion often occur simultaneously, and are difficult to distinguish from one another. For sustainable grazing management, however, a parameter is required to differentiate between sink and source areas for dust. A correlation between vegetation parameters and vertical fluxes can be drawn using MWAC traps on a wide range of grazing intensities (Fig. 4). In addition, it was possible to derive a transition point, or critical vegetation height in terms of net loss of dust particles (as shown in this example). Thus, all areas where the vegetation height is below 9 cm are endangered by wind erosion, and grazing management should be geared

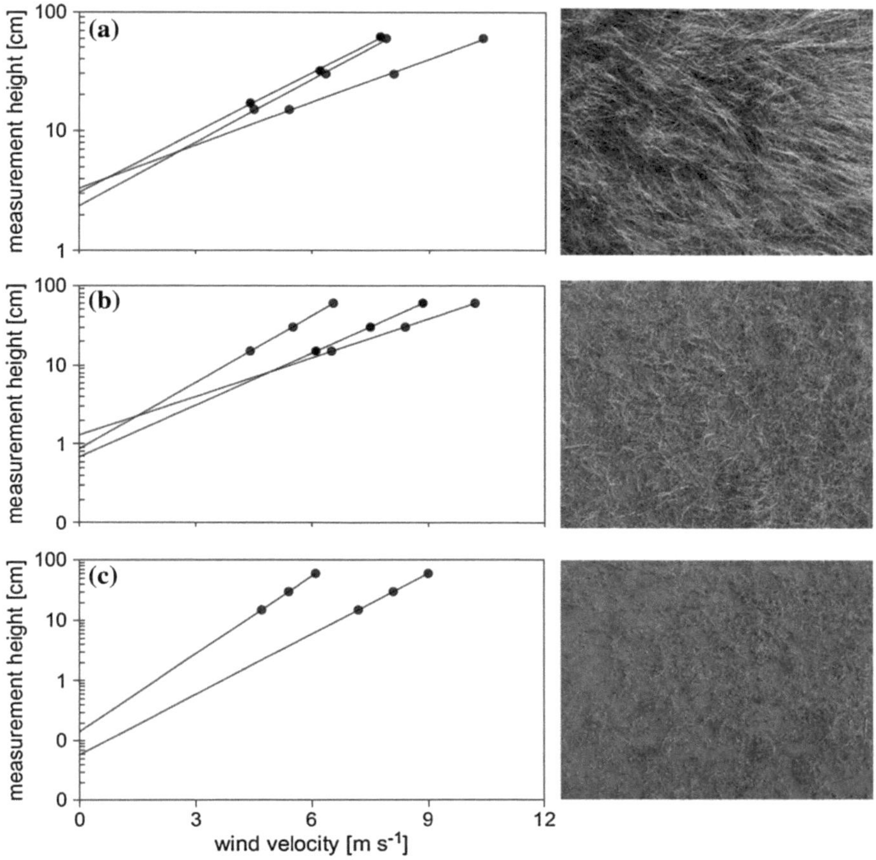

Fig. 3 Wind profiles and top view photos of vegetation on ungrazed (**a**), moderately (**b**) and overgrazed (**c**) plots in a grass steppe (Hoffmann et al. 2008b)

towards grass is at least this height. Even moderate grazing intensity can lead to wind erosion and creeping soil degradation in such grazing areas.

2.3 *Mass Balances*

Another method to quantify wind erosion, and in particular severe erosion events, is the quantification of eroded and deposited material by mass balances. The basic principle is to measure height differences of a surface before and after erosion. The tools required are simple, and soil relocation can be measured using erosion sticks (erosion and deposition) or a ruler (deposition only). The sorting process of material transported by wind makes identification of depositions quite easy. These

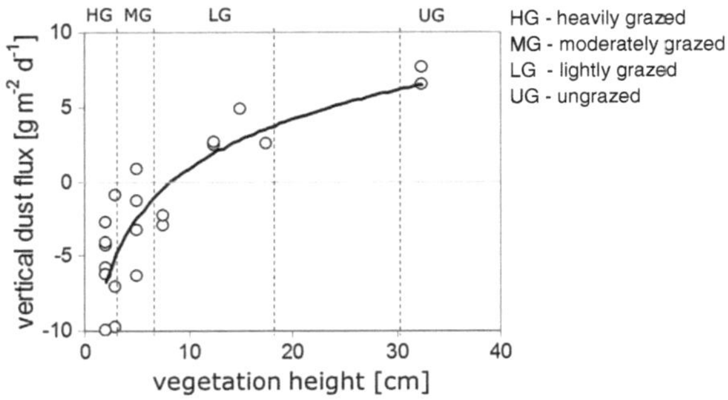

Fig. 4 Vertical dust fluxes on grassland with different grazing intensities (negative values—erosion and dust emission, positive values—deposition)

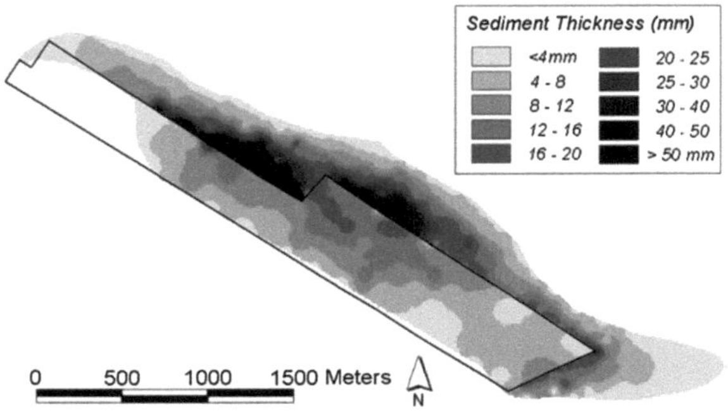

Fig. 5 Thickness of sedimentation, from arable land to adjacent grassland, caused by wind from west corresponding to a soil relocation of 45,900 m^3

are generally composed of a well-sorted, lighter layer above the original soil. The volume and mass of the eroded soil can be calculated from the bulk density of the soil, and the thickness and dimensions of the deposition. Comparing grain size distributions and soil organic carbon contents of the soil and depositions also allows a qualitative analysis to be made of dust emissions and carbon losses (Hoffmann et al. 2011). This method, however, is limited to wind erosion with a clearly defined source area and a manageable area of depositions (Fig. 5).

3 Identification of Long-Term and Large-Scale Erosion or Deposition Patterns

3.1 Long-Term Estimation of Wind Erosion by Radionuclides (^{137}Cs)

Generally, if not ploughed, steppe regions were subject to a gradual change, for better (soil formation—centuries) or for worse (soil degradation—decades). Hence longer periods must be considered and assessed using appropriate methodologies. Fallout environmental radionuclides, in particular ^{137}Cs, are useful tracers and chronometers in soil erosion and sedimentation, as shown in many studies in different regions of the world (Zapata et al. 2002; Ritchie and Ritchie 2007). Compared to conventional monitoring studies, the ^{137}Cs technique enables scientists to estimate soil redistribution processes with relatively little effort (Zapata 2003; Mabit et al. 2008). ^{137}Cs originates from nuclear weapon tests from the 1950s to the early 1970s. It is distributed across the land surface in uniform patterns, depending on latitude and regional rainfall (Davis 1963). It is strongly adsorbed by fine soil particles and nearly non-exchangeable (Tamura 1964). Therefore, any later movement of ^{137}Cs across the landscape can be related to the movement of these fine soil particles. The ^{137}Cs technique can be used to document erosion rates over the last 50 years, beginning with 1950s nuclear bomb testing in the atmosphere. This date corresponds with the start of land use change and management intensification in most Asian grasslands, which was intensified at the same time period. The comparison of the measured ^{137}Cs inventory of a sampling point with a reference inventory indicates either erosion, if it is smaller, or deposition, if it is higher than the reference value. However, reference sites in steppe regions should be selected very carefully because dust generally contains a higher concentration of ^{137}Cs than soil, and is deposited practically anywhere. For this reason, Funk et al. (2012) suggested the following requirements for a reference site in grasslands based on fluid dynamical principles:

1. Summit positions with low dust deposition because of the local increase in wind speed.
2. A dense vegetation or litter cover to prevent soil losses by wind erosion.
3. A downwind hill influenced by a dust-shadow effect by an upwind hill.
4. Plane or only slightly sloped to exclude additional influences of water erosion.

Providing that steppe soils are undisturbed (not ploughed), the "Profile-Distribution" model of Walling et al. (2006) can be used to calculate the annual average soil loss or gain, assuming that the depth distribution of ^{137}Cs follows an exponential function. The estimated net dust emission/deposition of a site can then be related to land use or vegetation parameter.

The ^{137}Cs technique enables annual averages of soil redistribution rates to be estimated, as well as, sometimes, the intensity of the controlling factors that have changed over time. In the case of wind erosion/dust deposition studies, a correlation to dust storm frequency may be useful. Figure 6 shows the relative dust deposition rates in a steppe region of China between 1962 and 2001. These rates correspond very well with the measured depth distribution of ^{137}Cs and the currently measured deposition rates of 110–200 gm^{-2} (Li et al. 2004).

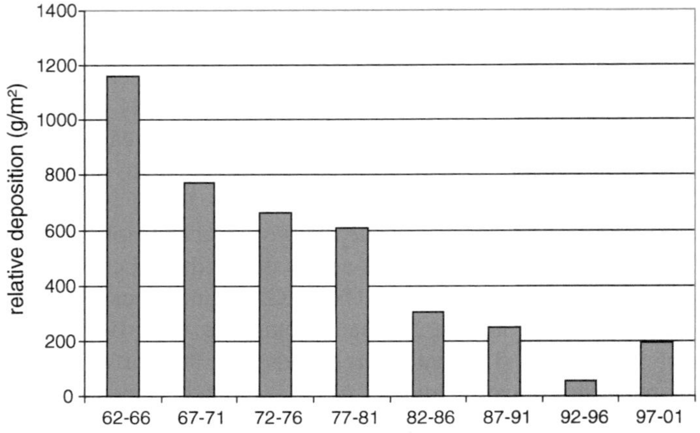

Fig. 6 Relative deposition rates in a steppe region of China from 1962 to 2001

3.2 Large-Scale Estimation of Wind Erosion Using GIS

The large-scale effects of seasonal changing and stable parameters concerning wind erosion need to be considered (Reiche et al. 2012). The former are usually vegetation conditions influenced by seasonal growth and land use intensity. Stable parameters of dust emission/deposition patterns are caused by topography and a dominating wind direction. Satellite images (ASTER, RapidEye) can be used to classify vegetation/land use at regional scales. In steppe regions, where grass residues are often poor or dead, a soil-adjusted vegetation index (SAVI) has a number of advantages over the generally used Normalized Difference Vegetation Index (NDVI) (Huete 1988). If simultaneously measured ground truth data are available, the setup of training areas for supervised classification reduces uncertainties. The correlation of classified vegetation and land use parameters with vertical dust fluxes also enables areas susceptible to wind erosion or dust deposition areas to be designated (Fig. 7).

As already shown, topography is an important factor influencing the spatial pattern of wind erosion and dust deposition (Goossens 1988; Goossens and Offer 1997; Goossens 2006; Hoffmann et al. 2008b). A global data set of a digital elevation model (DEM) is freely available from the Shuttle Radar Topography Mission (SRTM) in $90 \times 90\ m^2$ resolution, which is sufficiently accurate for large-scale analysis. The DEM can be used to identify leeward sheltered areas for a specific wind direction by setting a virtual shadow with the length of 40 times the height of an elevation (Reiche et al. 2012). In cases with different wind directions that are relevant for wind erosion or dust transport, this calculation can be performed for each direction.

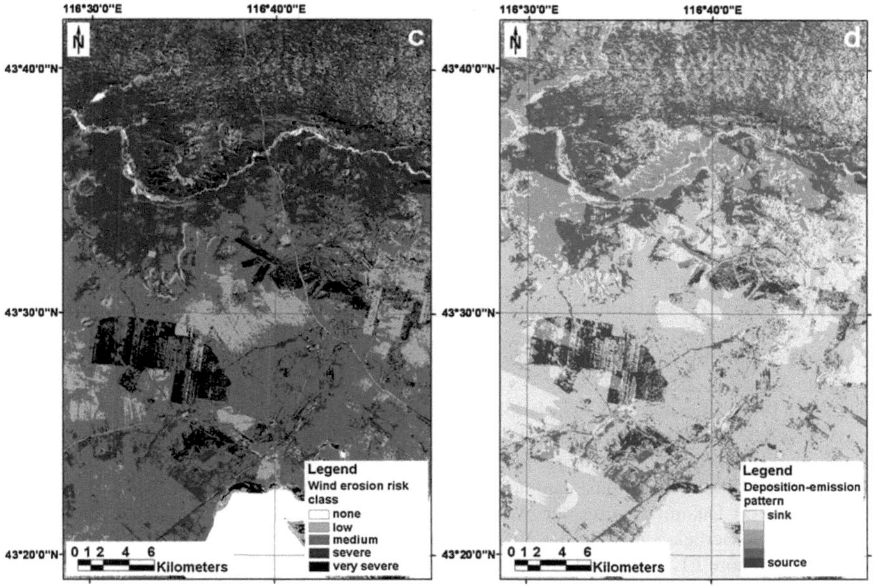

Fig. 7 Maps of wind erosion risk (*left*) and dust emission or dust deposition areas in a steppe region (Reiche et al. 2012)

4 Conclusions

There are various methods available to estimate wind erosion and dust depositions in steppe regions, from single point measurements to the large-scale mapping of processes. Measuring horizontal fluxes at different heights is the most recommended method because it is easy to implement and enables further important parameters to be derived. Most of the samplers used have been tested and calibrated in numerous studies under various conditions. In addition to the quantification of wind erosion, the reliability and comparability of measurements are also important criteria. A comprehensive summary of the basic requirements for wind erosion measurements can be found in Zobeck et al. (2003).

The authors listed important details that need to be considered in field studies. These include:

1. field data: location and dimensions, direction of tillage, occurrence and description of any upwind obstructions;
2. sediment sampler data: number, type, placement, efficiency, sampling frequency, fetch distance, time of sampling;
3. soil surface data: soil type and classification, texture, soil moisture, organic matter and calcium carbonate content, random and oriented roughness, dry aggregate size distribution, presence of crust and estimate of stability and loose erodible material, soil cover type and amount;

4. meteorological data: wind speed and direction with averaging times during storms, aerodynamic roughness, friction velocity, duration of storms, antecedent rainfall, relative humidity, solar radiation, air temperature, and wind direction variability during storm sampling.

References

Chepil WS (1957a) Sedimentary characteristics of dust storms. 1. Sorting of wind-eroded soil material. Am J Sci 255:12–22

Chepil WS (1957b) Sedimentary characteristics of dust storms. 3. Composition of suspended dust. Am J Sci 255:206–213

Davis JJ (1963) Caesium and its relationship to potassium in ecology. In: Schultz V, Klement AW (eds) Radioecology: proceedings of the 1st National symposium on radioecology. Reinhold Pub. Corp., New York, pp 539–556

Fryrear DW (1986) A field dust sampler. J Soil Water Conserv 41(2):117–120

Funk R (1995) Quantifizierung der Winderosion auf einem Sandstandort Brandenburgs unter besonderer Berücksichtigung der Vegetationswirkung. ZALF-Bericht Nr. 16, Müncheberg

Funk R, Skidmore EL, Hagen LJ (2004) Comparison of wind erosion measurements in Germany with simulated soil losses by WEPS. Environ Model Software 19:177–183

Funk R, Reuter HI (2006) Wind erosion in Europe. In: Poesen J, Boardman J (eds) Soil erosion in Europe. Wiley, London, pp 563–582

Funk R, Li Y, Hoffmann C, Reiche M, Zhang Z, Li J, Sommer M (2012) Using ^{137}Cs to estimate wind erosion and dust deposition on grassland in Inner Mongolia—selection of a reference site and description of the temporal variability. Plant Soil 351:293–307

Gillette DA (1977) Fine particulate-emissions due to wind erosion. Trans ASAE 20:890–897

Goossens D (1988) The effect of surface curvature on the deposition of loess—a physical model. Catena 15:179–194

Goossens D, Offer ZY (1997) Aeolian dust erosion on different types of hills in a rocky desert: wind tunnel simulations and field measurements. J Arid Environ 37:209–229

Goossens D, Offer ZY (2000) Wind tunnel and field calibration of six aeolian dust samplers. Atmos Environ 34:1043–1057

Goossens D (2005) Quantification of the dry aeolian deposition of dust on horizontal surfaces: an experimental comparison of theory and measurements. Sedimentology 52:859–873

Goossens D (2006) Aeolian deposition of dust over hills: the effect of dust grain size on the deposition pattern. Earth Surf Proc Land 31:762–776

Goossens D, Offer ZY, London G (2000) Wind tunnel and field calibration of five aeolian sand traps. Geomorphology 35:233–252

Goossens D, Rajot JL (2008) Techniques to measure the dry aeolian deposition of dust in arid and semi-arid landscapes: a comparative study in West Niger. Earth Surf Process Landforms 33:178–195

Groll M, Opp C, Aslanow I (2013) Spatial and temporal distribution of the dust deposition in Central Asia—results from a long term monitoring program. Aeolian Research 19:49–62. doi.org/10.1016/j.aeolia.2012.08.002

Hall DJ, Upton SL, Marsland GW (1994) Designs for a deposition gauge and a flux gauge for monitoring ambient dust. Atmos Environ 28:2963–2979

Hoffmann C, Funk R, Wieland R, Li Y, Sommer M (2008a) Effect of grazing on wind driven carbon and nitrogen ratios in the grasslands of Inner Mongolia. Catena 75:182–190

Hoffmann C, Funk R, Wieland R, Li Y, Sommer M (2008b) Effects of grazing and topography on dust flux and deposition in the Xilingele grassland, Inner Mongolia. J Arid Environ 72:792–807

Hoffmann C, Funk R, Reiche M, Li Y (2011) Assessment of extreme wind erosion in Inner Mongolia, China. Aeolian Res 3:343–351

Huete AR (1988) A soil adjusted vegetation index (SAVI). Remote Sens Environ 25:295–309

Janssen W, Tetzlaff G (1991) Entwicklung und Eichung einer registrierenden Suspensionsfalle. Z. Kulturtech. Landesentwickl. 32:167–180

Kuntze H, Beinhauer R, Tetzlaff G (1990) Quantification of soil erosion by wind, I. Final Report of the BMFT project. Project No. 0339058 A, B, C. Institute of Meteorology and Climatology, University of Hannover, Germany

Li FR, Zhao LY, Zhang H, Zhang TH, Shirato Y (2004) Wind erosion and air-borne dust deposition in farmland during spring in the Horqin sandy land of eastern Inner Mongolia, China. Soil Till Res 75:121–130

Mabit L, Benmansour M, Walling DE (2008) Comparative advantages and limitations of the fallout radionuclides ^{137}Cs, ^{210}Pbex and ^{7}Be for assessing soil erosion and sedimentation. J Env Rad 99:1799–1807

Mendez M, Funk R, Buschiazzo DE (2011) Field wind erosion measurements with big spring number eight (BSNE) and modified Wilson and Cook (MWAC) samplers. Geomorphology 129:43–48

Reiche M, Funk R, Zhang Z, Hoffmann C, Reiche J, Wehrhan M, Li Y, Sommer M (2012) Application of satellite remote sensing for mapping wind erosion risk and dust emission-deposition in Inner Mongolia grassland, China. Grassland Sci 58:8–19

Ritchie JC, Ritchie CA (2007) Bibliography of publications of ^{137}Cs studies related to erosion and sediment deposition. http://www.ars.usda.gov/Main/docs.htm?docid=15237

Sensit (2012) http://www.sensit.com. Accessed on 12 Jan 2013

Shao Y (2008) Physics and modelling of wind erosion, 2nd edn. Kluwer Academic Press, Dordrecht

Shao Y, Wyrwoll KH, Chappell A, Huang J, Lin Z, McTainsh GH, Mikami M, Tanaka TY, Wang X, Yoon S (2011) Dust cycle: an emerging core in Earth system science. Aeolian Res 2:181–204

Spaan WP, van den Abeele GD (1991) Wind borne particle measurements with acoustic sensors. Soil Technol 4:51–63

Sterk G, Stein A (1997) Mapping wind-blown mass transport by modelling variability in space and time. Soil Sci Soc Am J 61:232–239

Tamura T (1964) Consequences of activity release: selective sorption reactions of caesium with soil minerals. Nucl Safety 5:262–268

VDI 2119. VDI-Richtlinie Blatt 2 (1972) Messung partikelförmiger Niederschläge; Bestimmung des partikelförmigen Niederschlages mit dem Bergerhoff-Gerät (Standardverfahren)

Visser SM, Sterk G, Snepvangers J (2004) Spatial variation in wind-blown sediment transport in geomorphic units in northern Burkina Faso using geostatistical mapping. Geoderma 120:95–107

Walling DE, Zhang Y, He Q (2006) Models for converting radionuclide inventories (^{137}Cs, Excess ^{210}Pb, and ^{7}Be) to estimates of soil erosion and deposition rates (Including Software for Model Implementation). Department of Geography, University of Exeter, Exeter, U.K. report

Zapata F (2003) The use of environmental radionuclides as tracers in soil erosion and sedimentation investigations: recent advances and future developments. Soil Till Res 69:3–13

Zapata F, Garcia-Agundo E, Ritchie JC, Appleby PG (2002) Introduction. In: Zapata F (ed) Handbook for the assessment of soil erosion and sedimentation using environmental radionuclides. IAEA, Austria, pp 1–13

Zobeck TM, Sterk G, Funk R, Rajot JL, Stout JE, van Pelt RS (2003) Measurement and data analysis methods for field-scale wind erosion studies and model validation. Earth Surf Process Landforms 28:1163–1188

Generation of Up to Date Land Cover Maps for Central Asia

Igor Klein, Ursula Gessner and Claudia Künzer

Abstract Human activity and climate variability has always changed the Earth's surface and both will mainly contribute to future alteration in land cover and land use changes. In this chapter we demonstrate a land cover and land use classification approach for Central Asia addressing regional characteristics of the study area. With the aim of regional classification map for Central Asia a specific classification scheme based on the Land Cover Classification System (LCCS) of the Food and Agriculture Organisation of the United Nations Environment Programme (FAO-UNEP) was developed. The classification was performed by using a supervised classification method applied on metrics, which were derived from Moderate Resolution Imaging Spectroradiometer (MODIS) data with 250 m spatial resolution. The metrics were derived from annual time-series of red and near-infrared reflectance as well as from Normalized Difference Vegetation Index (NDVI) and thus reflect the temporal behavior of different land cover types. Reference data required for a supervised classification approach were collected from several high resolution satellite imagery distributed all over the study area. The overall accuracy results for performed classification of the year 2001 and 2009 are 91.2 and 91.3 %. The comparison of both classification maps shows significant alterations for different classes. Water bodies such as Shardara Water Reservoir and Aral Sea have changed in their extent. Whereby, the size of the Shardara Water Reservoir is very dynamic from year to year due to water management and the eastern lobe of southern Aral Sea has decreased because of the lack of inflow

I. Klein (✉)
Universität Würzburg, Institut für Geographie und Geologie, German Aerospace Center (DLR), Oberpfaffenhofen 82234 Wessling, Germany
e-mail: igor.klein@dlr.de

U. Gessner · C. Künzer
German Aerospace Center (DLR), German Remote Sensing Data Center, Land Surface Oberpfaffenhofen 82234 Wessling, Germany
e-mail: ursula.gessner@dlr.de

C. Künzer
e-mail: claudia.kuenzer@dlr.de

L. Mueller et al. (eds.), *Novel Measurement and Assessment Tools for Monitoring and Management of Land and Water Resources in Agricultural Landscapes of Central Asia*, Environmental Science and Engineering, DOI: 10.1007/978-3-319-01017-5_19,

from Amu Darja. Furthermore, some large scale changes were detected in sparsely vegetated areas in Turkmenistan, where spring precipitation mainly affects the vegetation density. In the north of Kazakhstan significant forest losses caused by forest fires and logging were detected. The presented classification approach is a suitable tool for monitoring land cover and land use in Central Asia. Such independent information is important for accurate assessment of water and land recourses.

Keywords Central Asia · Land cover maps · MODIS data

1 Introduction

Central Asia has been facing serious land cover and land use changes in the past few decades. The transformation of the natural steppe to cropland during the former Soviet Union (Lioubimtseva and Henebry 2009) and inefficient use of natural water recourses are the most famous examples. Since the political change in this region some land was abandoned by farmers (Propastin et al. 2008a). The change in livestock and livestock practice leads to degradation of land due to overgrazing on the one hand and rehabilitation of drylands due to undergrazing on the other (Lioubimtseva et al. 2005). Furthermore, in context of climate change Central Asia undergoes multiple changes such as glacier loss and salinization of soil. Therefore, a permanent land cover and land use (LCLU) monitoring is required to enable an accurate assessment of the actual processes at this wide area. Information on LCLU and connected alterations are important to evaluate the impact of man and climate change on environment (Meyer and Turner 1992). For regional modeling applications for example in climatology, hydrology and carbon storage, information on LCLU are significant as well as for various regional management questions (Giri et al. 2005).

In this chapter we describe a regional classification approach based on the published study of Klein et al. (2012) with more specific focus on the methodology for the performed classification which is described in detail in section two. Within the study a classification scheme for Central Asia based on Land Cover Classification System (LCCS) was derived and an appropriate reference data was collected. Pixels enclosed by the training polygons were used to derive specific temporal and spectral signatures for the respective LCLU classes (Abd El Kawy et al. 2011), whereby the training polygons were specifically collected for the purpose of 250 m resolution classification map in Central Asia considering regional characteristics. Annual time series of the Moderate Resolution Imaging Spectroradiometer (MODIS) data with 250 m spatial resolution were used for the classification approach. The classification was performed with decision tree classifier based on C5.0 algorithm (Quinlan 1993). In section three, we present the main results including accuracy assessment as well as some exemplary changes between 2001 and 2009 to demonstrate the short-term variability of LCLU in Central Asia on regional scale and to emphasize the need of regularly regional land cover monitoring.

2 Methodology and Data

2.1 Regional Classification Scheme

A regionally adapted, hierarchical classification legend was developed considering both the requirements of land and water resource management and the information content of MODIS data. The legend was designed based on the principles of the Land Cover Classification System (LCCS) of the Food and Agriculture Organisation of the United Nations Environment Programme (FAO-UNEP). This standardized a priori classification system has a range of advantages as it is independent from data, region or thematic (Jansen and Di Gregorio 2002; Di Gregorio 2005). Most important is the comparability and conformability of the classification scheme as all classes within LCCS are based on standardized thresholds (e.g. of life form and canopy cover). Figure 1 shows the developed classification legend for Central Asia. At the first level (L1), vegetated and non-vegetated areas are distinguished. The second level (L2) separates terrestrial areas from aquatic or regularly flooded areas. At the third level (L3), classes are subdivided according to management aspects and artificiality of the land cover. At the fourth level (L4), 13 land cover classes are distinguished based on a combination of several predefined pure land cover classifiers and attributes. Detailed information on the class definition is given in Table 1.

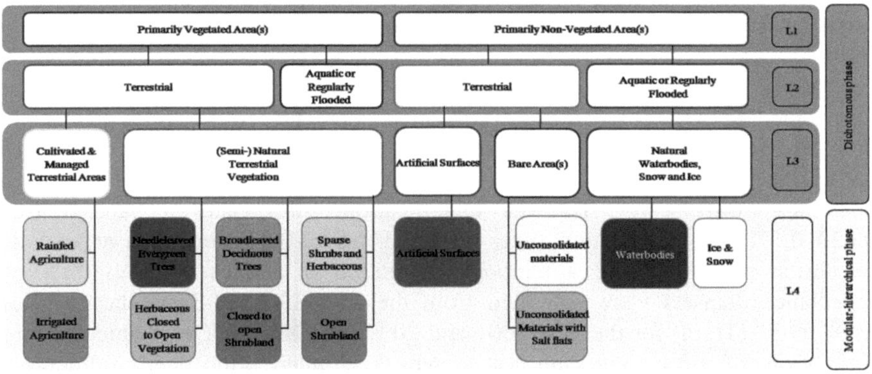

Fig. 1 Classification scheme for Central Asia developed according to LCCS standards. Modified after Klein et al. (2012)

Table 1 Description of the land cover/land use classes of the regionally adapted legend

Class name	Definition after LCCS
Natrural and semi-natural terrestrial vegetation	
Needleleaved trees	Needleleaved evergreen trees, main layer: trees >65 %
Broadleaved trees	Broadleaved deciduous trees, main layer: trees >65 %
Sparse vegetation	Sparse shrubs (5–15 % 30 cm to 3 m) and sparse herbaceous: 5–15 % (30 cm to 3 m)
Grassland	Herbaceous closed to open vegetation: main layer: herbaceous: 15–100 % (3 cm to 3 m)
Closed shrubland	Closed medium high shrubland, main layer: shrubs: >65 % (50 cm to 3 m)
Open shrubland	Open medium high shrubland: main layer: shrubs: 15–65 % (50 cm to 3 m)
Cultivated and managed terrestrial areas	
Rain-fed agriculture	Rain-fed agriculture
Irrigated agriculture	Water supply mainly by irrigation
Bare areas	Unconsolidated material(s), less than 4 % vegetative cover
Bare areas with salt flats	Unconsolidated material(s) with salt flats, less than 4 % vegetative cover
Water bodies and snow	
Water	Artificial and natural
Ice and snow	Artificial and natural
Artificial surfaces and associated areas	
Artificial	Built up and sealed areas

2.2 Data

Main data source of the land cover classification are annual time series of optical remote sensing data acquired by MODIS. This sensor is operating on board of the satellites Terra (in orbit since 1999) and Aqua (in orbit since 2002). The entire Earth surface is covered every two days in 36 spectral bands and three spatial resolutions 250, 500 and 1,000 m. The U.S. Geological Survey's Land Processes Distributed Active Archive Center (USGS LP DAAC) offers several products with daily, 8-day, 16-day, monthly, quarterly and yearly temporal resolutions (USGS LP DAAC 2012). The datasets are distributed in a sinusoidal grid tilling system and tiles of 10 × 10°. For our classification, reflectance of band 1 (red, 0,620–0,670 μm) and band 2 (near infra red, 0.841–0.876 μm) data at a spatial resolution of 250 m and a temporal resolution of 8 days were used. These reflectance datasets were extracted from the MODIS level-3 standard product MOD09Q1 (Terra) for the year 2001 and 2009. The MOD09Q1 product contains estimations of surface reflectance at ground level where atmospheric influence is eliminated. For each pixel, the datasets contain the best (optimal) observation acquired during an 8-day period. The best observation is selected considering the aspects such as low view angle, absence of clouds and cloud shadow (USGS LP DAAC 2012). Furthermore, the MODIS Normalized Difference Vegetation Index (NDVI) product MOD13Q1 with 250 m spatial resolution and 16-day temporal resolution was involved in the classification process.

For an entire coverage of the Central Asian land surface, nine MODIS tiles need to be processed. Figure 2 shows the coverage of these tiles in red.

Besides the medium resolution MODIS data, Landsat data was used for training data selection in the classification process. With a spatial resolution of 30 m and the six spectral bands, data of the Landsat TM and ETM + sensors (GLCF 2011) qualifies for sample selection. In this comparatively high resolution images, it is easier to distinguish between the classes of the regionally adapted legend (Sect. 2.1). For the region of Central Asia, 20 Landsat images of May or June 2001 and 2009 (covering app. 8 % of the study region) were selected for sample selection (Fig. 2). In May/June the vegetation is supposed to be fully developed and therefore those images are adequate for class assignment as they represent the maximum state of vegetation of vegetation development as requested by LCCS. The selected images are distributed equally over the study area and cover all land cover types allowing for a representative sampling.

In addition to optical remote sensing data, the digital elevation model of the Shuttle Radar Topography Mission (SRTM) with a spatial resolution of 90 m

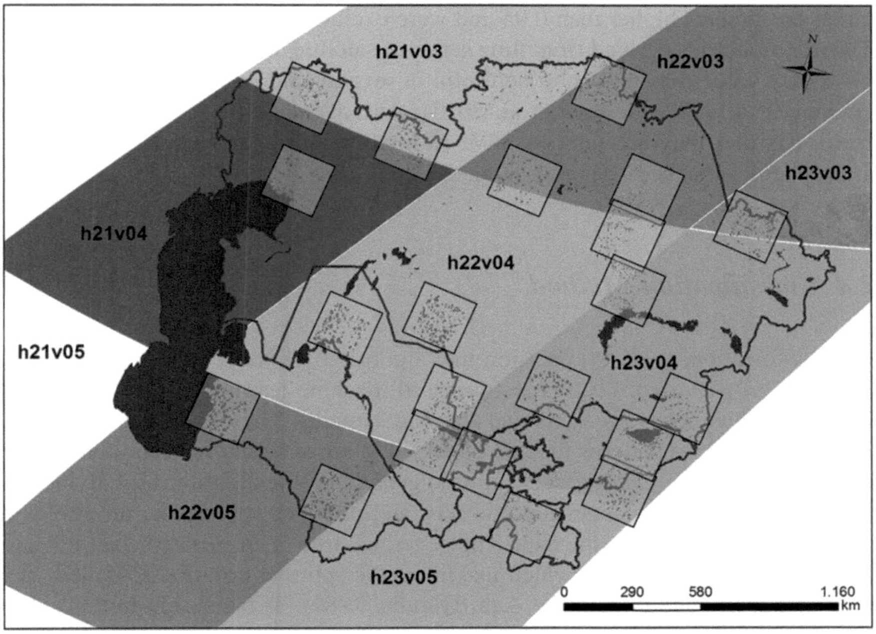

Fig. 2 The coverage of MODIS tiles (*coloured tiles with ID*) of Central Asia (*purple line*) and the distribution of used Landsat images (*grey transparent*)

(Rabus et al. 2003) was used in the post-classification process. Furthermore, re-analysed precipitation data of the Global Precipitation Climatology Centre (GPCC) with a spatial resolution of 0.5° lat/lon (Rudolf et al. 2010) was involved. The datasets were rescaled to the MODIS pixel grid of 250 m.

2.3 Pre-processing of MODIS Data

The MODIS products MOD09Q1 and MOD13Q1 provide a number of 46 and 23 datasets per year, respectively. From these annual time series, descriptive statistics were calculated for defined temporal sections (periods). Besides the full annual cycle from January to December we defined further three periods according to the climatic characteristics of the study region: he winter/spring period as the time period before growing season (Jan–Mar), the summer section as the growing season (Apr–Sep) and the autumn section as the post growing period (Oct–Dec). For these temporal sections we calculated median, maximum, minimum and amplitude of the red and near-infrared reflectance and NDVI. The resulting 48 metrics were cross correlated to identify possible redundancy in information content. Six metrics (full year maximum of NDVI, winter/spring minimum of NDVI, autumn/winter minimum of NDVI, summer minimum of red band, winter/spring maximum of red band, autumn/winter maximum of red band) had correlation coefficients higher than 0.95 and were excluded from classification process. The use of metrics derived from time series of satellite data versus using weekly or bi-weekly values has proven to be useful in several studies (e.g. Reed et al. 1994; Hansen et al. 2002; Gessner et al. 2013). An advantage of such multi-temporal metrics is that they are less sensitive to inter-annual phenological variations or atmospheric influences (Hansen et al. 2002).

2.4 Classification Method

For classification, the C5.0 decision tree algorithm introduced by Quinlan (1987, 1993, 1996) was used. C5.0 is a supervised machine learning system (Kotsiantis 2007) and thus needs pre-classified training samples. These samples are used to extract informative patterns that are collected in a tree structured classifier, the so-called decision tree. During decision tree generation the algorithm subdivides the training pixels into homogeneous subsets, and provides appropriate thresholds of all attributes for splitting the data. The generated tree is univariate because it uses only one attribute (the one, which maximize the splitting criterion so-called 'gain ratio') at each node to split the data (Quinlan 1987). The decision tree is subsequently transformed into a set of concise if–then rules. This transformation improves the prediction of unseen cases because dispensable brunches and splits are removed (Quinlan 1987). The rule-set classifier is finally used for classification of the entire region. The advantage of decision tree and delineated rule-set classifiers is that they are able to reveal nonlinear and hierarchical relationships between the training cases (Hansen et al. 1996). C5.0 is fast in computing speed and rather insensitive to noise in the training data (Quinlan 1996) and is well qualified for classification of wide areas such as Central Asia (Klein et al. 2012).

2.5 Collection of Training Samples

As decision tree algorithms are empirical learning systems they need pre-classified cases or so-called training samples. There are many different approaches to train empirical learning systems for example based on areas which belong to the same class in different harmonized global maps (Herold et al. 2008). Such approach is based on many generalized global maps and those inherent misclassifications to the training processes. To enable a more accurate and less generalized regional classification, we decided to train the C5.0 algorithm based on comprehensive and high quality reference data derived from higher resolution Landsat imagery. The selection of reference data is the only manual process in presented classification approach. This process is done visually by an expert with knowledge of the study area and image interpretation. To support an appropriate derivation of training polygons from Landsat images we first applied a multi-resolution segmentation to all selected Landsat scenes. Multi-resolution segmentation is a clustering of neighboring and spectral similar pixels. It extracts image objects based on spectral and textural characteristics. Size and shape of these objects are based on spectral heterogeneity (sum of standard deviation of spectral values of all assigned pixels in this segment) and shape heterogeneity of the object (defined by smoothness and compactness) (Baatz and Schaepe 2000; Schiewe 2002). A pre-set threshold defines when the clustering of spatially homogeneous pixel stops (Aguirre-Gutierrez et al. 2012).

Training polygons were selected from the image objects that resulted from segmentation. A training sample had to fulfill certain criteria. It needs to show a minimum area of nine MODIS pixels. Furthermore, the polygon must be spatially homogeneous and cover only one land cover type (based on the legend described in Sect. 2.1). Furthermore, the polygon needs to delineate an area where land cover did not change between 2001 and 2009. For an appropriate classification a minimum number of training samples per class are required. We decided to pre-classify at least 120 polygons for each class. These manually assigned polygons meet all requirements defined in former three steps. The assignment of samples was done by visual interpretation using expert knowledge and a priori information, high resolution information from GoogleEarth (Kurtz et al. 2010). Finally, the polygons were morphologically eroded in order to remove potential mixed pixel at the border of the polygons and thus to improve the homogeneity of the samples. By adjusting the severity of morphological erosion for each class individually, the area of training data was adjusted for each class to the approximate proportion of the class in the study area (Fig. 3).

The final reference dataset for the classification includes more than 2,000 polygons comprising about 220,000 MODIS pixel. To put this number in a context, we mention the global 500 m resolution MODIS Collection 5 Global Land Cover product, which were based on 704 sites and 14,136 pixel for whole Asia (Friedl et al. 2010). The temporal consistency of our samples between 2001 and 2009 allows using the reference dataset for classification of both years. The described sampling approach exempts our classification from using samples based on global products and thus from generalized coarse resolution maps.

Two thirds of the reference dataset were used for training the C5.0 algorithm. Whereby, the set of training samples was selected randomly. For each year the C5.0 rule-sets were created independently from the MODIS metrics of the respective year resulting in an individual decision trees for each year. Figure 3 shows the overall workflow for building a comprehensive reference dataset.

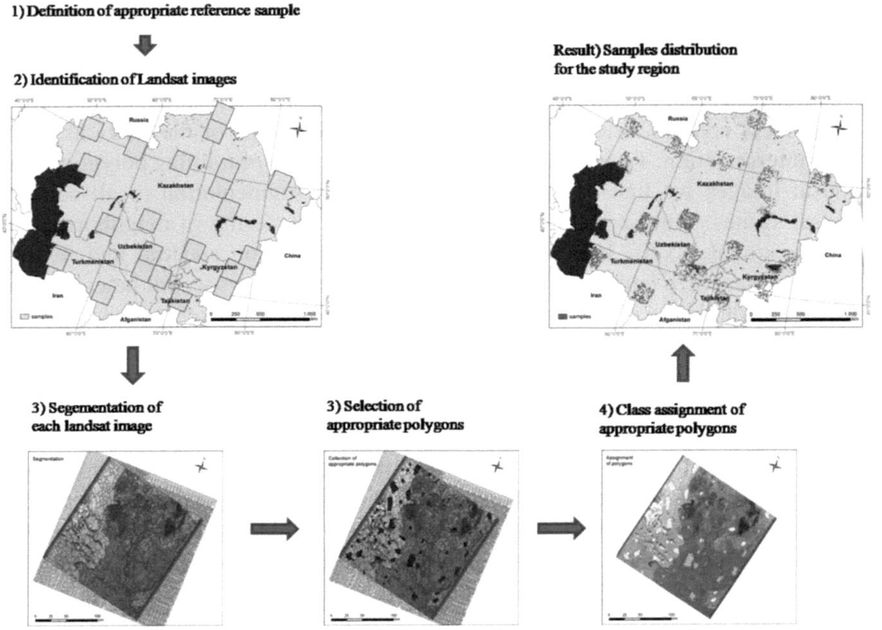

Fig. 3 Workflow for building a comprehensive reference dataset

2.6 Post-classification

Like all supervised classifiers, the C5.0 algorithm is based on two main assumptions. It supposes that the training data is representative for all pixels and that the used spectral-temporal information is able to differentiate between the target classes (Friedl et al. 2010). In this context, the spectral-temporal information content of MODIS and biases which are inherent to tree-based models could limit the quality of the results. In some cases, the classification results need to be adjusted due to errors and biases from uneven representation of classes and inadequate class separation in remote sensing data (Sulla-Menashe et al. 2011; Herold et al. 2008). Therefore, we allied some post-classification steps to improve the results especially of spectral ambiguous classes. The post-classification approach is based on basic, logical rules considering topographic land features required for the accumulation of water and on climatological conditions required for rain-fed agriculture.

Misclassifications between 'water bodies' and shadowed slopes in mountainous regions were removed by applying a threshold value of >0 % of inclination as derived from the SRTM DGM. Furthermore, spectral ambiguous classes such as 'ice and snow' and 'bare areas with salt flats' are corrected by taking terrain elevation into account. All pixels below 2,000 m above mean sea level (AMSL)

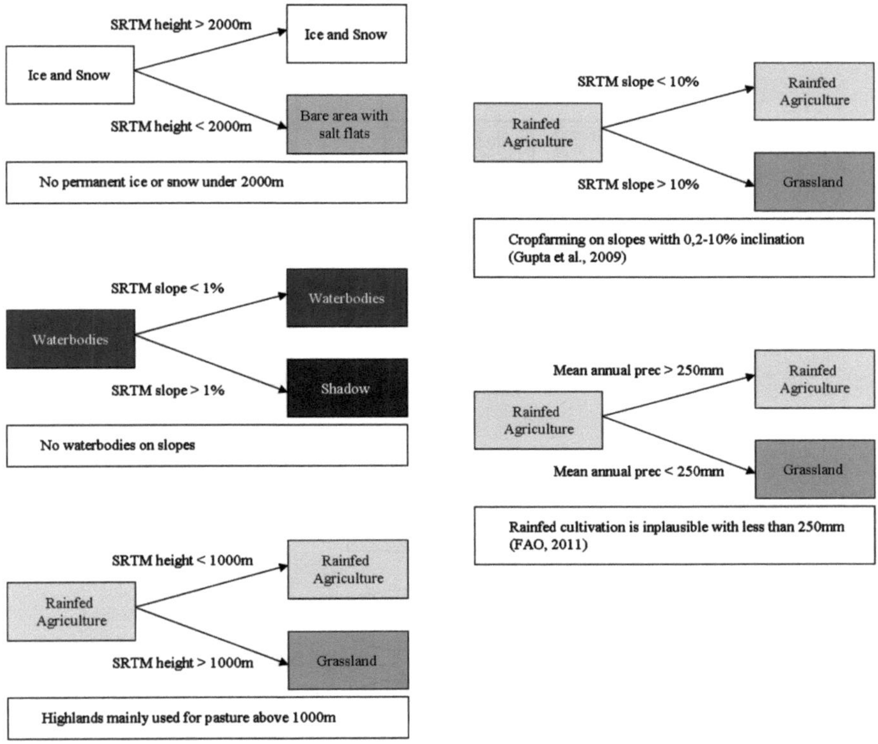

Fig. 4 Post-classification steps applied to improve the classification results. Modified after Klein et al. (2012)

that had been classified as 'ice and snow' were reassigned to 'bare areas with salt flats' (Klein et al. 2012). Furthermore, the confusion between classes 'rain-fed agriculture' and 'grassland' can be removed by taking into account the long term annual precipitation. As rain-fed agriculture in Central Asia is practised in regions with annual precipitation between 250 and 400 mm (Gupta et al. 2009), all areas classified as 'rain-fed agriculture' but averaging less than 250 mm of annual rainfall were reclassified to 'grassland'. Furthermore, Gupta et al. (2009) pick out that land above 1000 m AMSL is mainly used as pasture and crop farming is practised on slopes with inclination between 0.2 and 10 %. Therefore, we reclassified areas classified as 'rain-fed agriculture' above 1,000 m and areas classified as 'rain-fed agriculture' on slopes stepper than 10 % to 'grassland'. Such mountainous areas mainly in Tajikistan and Kyrgyzstan are mainly used for pasture (Propastin et al. 2008a). All post-classification rules applied for the LCLU mapping in this study are summarized in Fig. 4.

Figure 5 illustrates all steps within the methodological process described in this section from data acquisition till resulted maps.

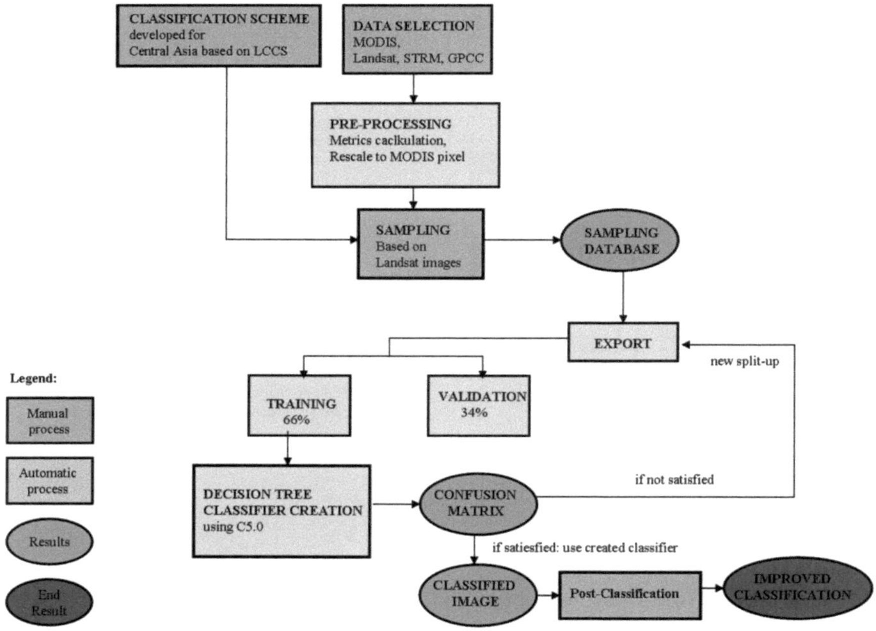

Fig. 5 Classification chain applied for mapping of LCLU in Central Asia. Modified after Klein et al. (2012)

3 Results

3.1 Land Cover Map

The generated classification for the year 2009 is depicted in Fig. 6. In the northern parts of Kazakhstan, wide areas of 'rain-fed agriculture' are present. The rain-fed areas transmit into Kazakhstan's steppes classified as 'grassland' further south. Classes 'bare areas', 'bare areas with salt flats' and 'sparse vegetation' represent Central Asia's deserts and semi-deserts in the south of the study region. Also the wide areas of irrigated agriculture of the Fergana Valley, Khorezm, Kyzyl Orda and Mary province in Turkmenistan are well depicted. The mountainous regions of Tien Shan, Alai and Pamir are mainly represented by the classes 'ice and snow',

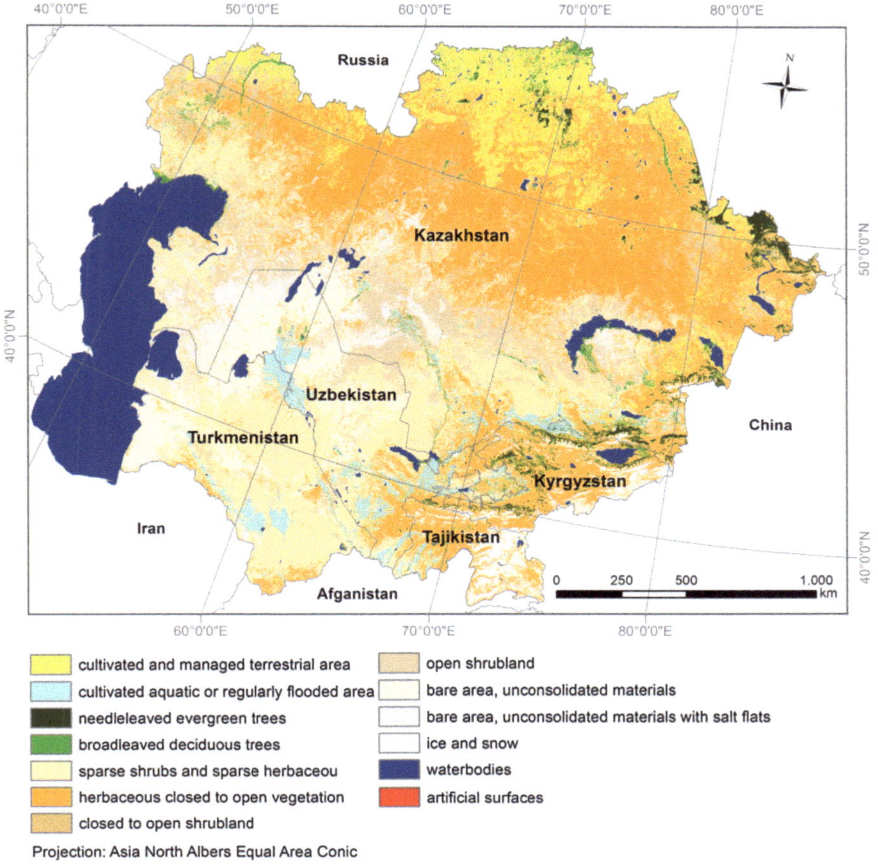

Fig. 6 LCLU Classification map of the year 2009 for Central Asia

'needleleaved trees' and 'grassland'. The mixed vegetation class 'open shrubland' is located in between the steppes and deserts and semi-deserts forming a gradual transition zone. The classes 'broadleaved deciduous trees' and 'closed shrubland' are found adjacent to rivers and water bodies and cover only small areas within Central Asia. Furthermore, the patches of 'needleleaved trees' in the North-East of Kazakhstan are part of for the Taiga.

For information about the quality of maps derived from remote sensing the so called error matrix has been widely used (Foody 2002). An error matrix provides accuracy measures such as overall accuracy (sum of all correctly classified pixel divided by total number of pixel), producer's accuracy (total number of correct classified pixel of class A divided by total number of class A pixel identified in reference data) and user's accuracy (correct classified pixel of class A divided by total number of pixel classified as class A) (Congalton and Green 2009; Jung et al. 2006). The overall accuracy of the classification 2009 was 91,26 % and for year 2001 91,22 %. However, the user and producer accuracy differs between single classes. For example the classes 'ice and snow', 'water' 'bare area', 'bare areas with salt flats' and 'sparse vegetation' hold high user and producer accuracies above 90 %. Classes such as 'closed shrubland', 'open shrubland' and 'broadleave trees' show lower accuracies between 70 and 80 % with lowest user accuracy for the class 'closed shrubland' with 41.07 %. The reason for such quality difference in the accuracy assessment is due to limited number of samples because those classes cover very small areas of the study region. Another reason is the over and under prediction of supervised classification methods which causes further misclassification (Friedl et al. 2010). Furthermore, such classes are mistaken due to their spectral-temporal ambiguities (Table 2).

Table 2 Accuracies for the 2001 and 2009 LCLU maps

Class names	User accuracy 2001	Producer accuracy 2001	User accuracy 2009	Producer accuracy 2009
Rainfed	84.78	88.41	83.49	87.50
Irrigated	86.45	88.83	82.23	87.91
Needleleaved	89.38	94.22	93.43	92.04
Broadleaved	73.45	83.33	76.47	73.27
Grassland	86.74	87.24	86.53	85.79
Closed shrubs	39.06	62.50	41.07	60.53
Open shrubs	67.83	79.76	71.49	71.20
Sparse Veget	93.34	90.88	93.43	90.23
Bare areas	95.68	93.47	95.10	95.79
Bare areas with salt flats	92.68	92.83	96.14	95.79
Ice and snow	96.30	100	91.67	100
Waterbodies	99.69	99.84	92.44	95.78
Overall accuracy		*91.22*		*91.26*

3.2 Detected Exemplary Changes

In the following section some of the detected land cover changes are exemplary discussed. Figure 7 illustrates the relative change from 2001 to 2009 of each class, where the situation of 2001 is considered as the basis status.

The class ‘bare areas’ has decreased dramatically and changed to ‘sparse vegetation’. This change can be explained by the high interannual variability of precipitation as the annual vegetation activity and cover of both classes are very sensitive to water availability in arid to semi-arid regions (Gessner et al. 2012; Yang et al. 1998; Richard and Poccard 1998). The inter-annual response of vegetation to precipitation was discussed by Propastin et al. 2008b; Wang et al. 2003; Nicholson and Farrar 1994; Nezlin et al. 2005. The seasonal precipitation differences between 2001 and 2009 are depicted in Fig. 8. Particularly strong changes

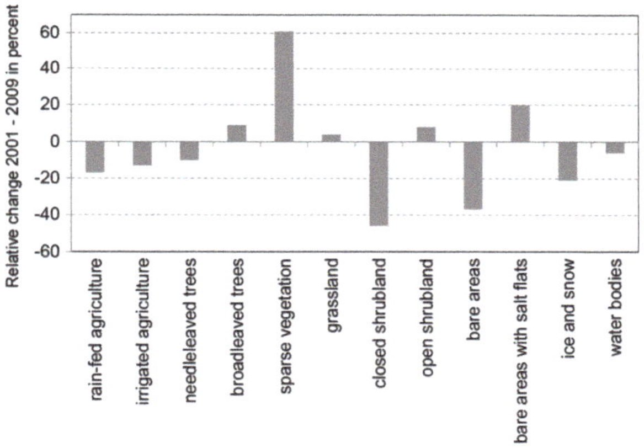

Fig. 7 Relative classified land cover change of 2009 compared to classification of 2001 (0 % line)

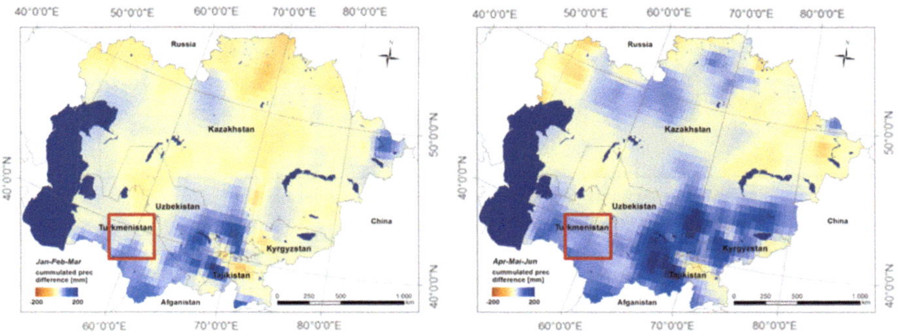

Fig. 8 Difference in precipitation for three months intervals (Jan to Mar and Apr to Jun) between 2001 and 2009. Positive values indicate higher precipitation in year 2009. Precipitation data were derived from GPCC, 2011. Water bodies characterize the situation of July 2009

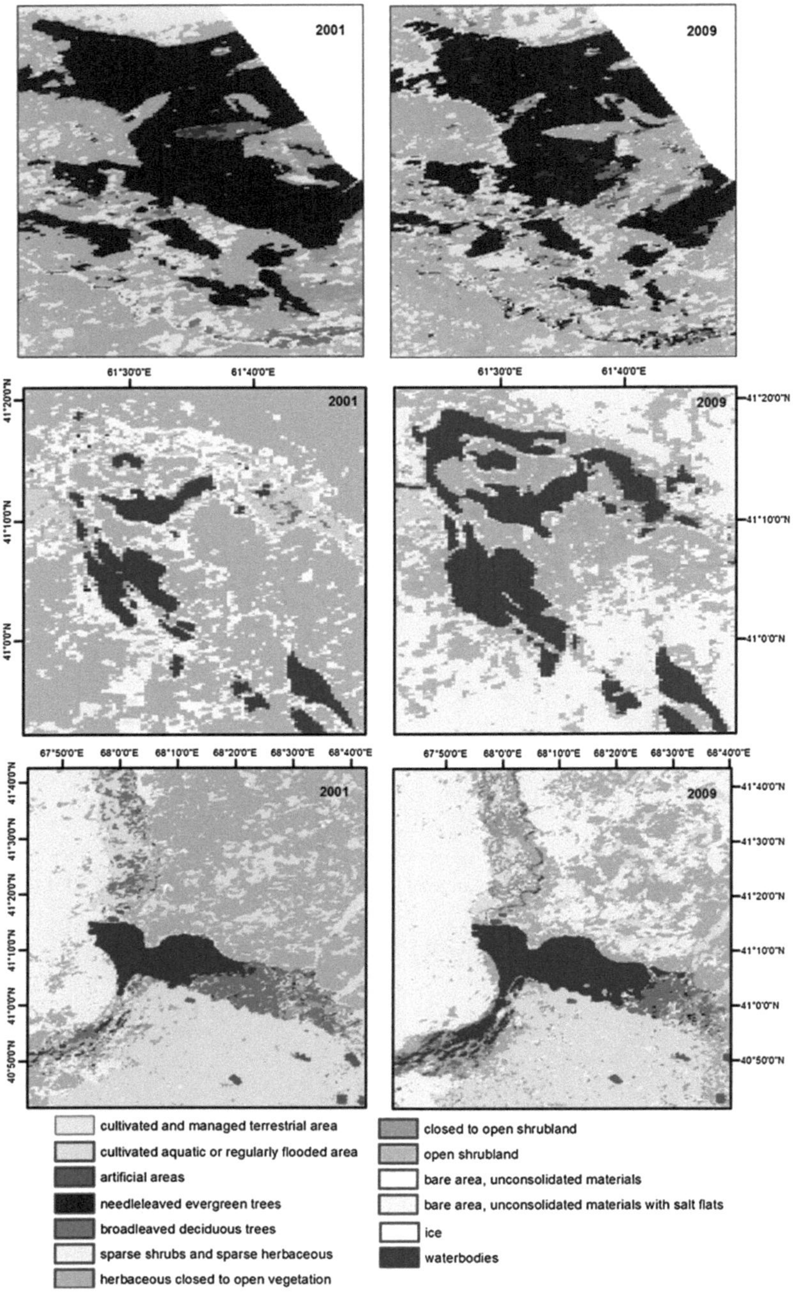

Fig. 9 Comparison of exemplary changes between 2001 and 2009 classification. Projection lat/lon; WGS84

from 'bare areas' to 'sparse vegetation' occurred in central Turkmenistan (red framed area in Fig. 8). These areas show moister conditions in spring 2009 than in 2001. Positive anomalies of precipitation lead to higher foliage amount and vegetation coverage in desert to semi-desert regions. In our classification results this is reflected in a transition to a class with denser vegetation cover. This is however not a permanent land cover change but a natural fluctuation due to meteorological conditions. Another reason for the decrease of 'bare areas' can be found in some regional studies discussing the rehabilitation of drylands due to decreased livestock (Karnieli et al. 2008; Lioubimtseva et al. 2005; De Beurs and Henebry 2004).

Figure 9 illustrates further exemplary changes between 2001 and 2009. The upper example shows the forest lost in northern Kazakhstan, which occurred due to forest fires and logging. This area is very vulnerable to forest fires as the danger of fires is classified as very high (Arkhipov et al. 2000). The second example demonstrates the difference in flooded and irrigated areas which apparently varies from year to year depending on water availability. The third examples shows one of the biggest water reservoirs in Central Asia, the Shardara Reservoir in south Kazakhstan close to Uzbek and Kyrgyz borders. The water body in July 2009 is significantly larger compared to July 2001. The amount of water in this reservoir is mainly controlled by human regulation and water management (Shemratov 2004). However, in some cases there was uncontrolled growing of the water body due to increase water discharge into the reservoir caused by snow melt. In such cases serious winter and spring floods occurred downstream of the reservoir, especially as the water discharge to Arnasai depression onto Uzbeks territory was reduced by construction of dams due to hydro power interests (Ryabtsev 2008).

4 Conclusion

Accurate LCLU maps are very important for a range of applications and as information data base as there is still a lack of accessible data. In this report we described our classification approach for Central Asia. The approach is based on a comprehensive reference data which were collected specifically for Central Asia. The reference data base was collected with the aim of a multi year classification based on temporal metrics derived from MODIS data with 250 m resolution. For this study we classified the years 2001 and 2009. The results confirm with zonal distribution of different land cover types known for Central Asia. Considering land cover changes between both years, there were some significant alteration detected such as forest loss in the north of Kazakhstan due to forest fires, changes in aerial extent of water bodies due to water management decisions and regulation, changes between natural classes from bare to sparse vegetated areas due to precipitation variation. The comparison of both maps indicates high variability of natural and semi-natural land cover classes due to water availability. The presented spatially higher resolution maps might be more suited for regional hydrologic, climatologic and carbon storage modeling studies than global products and used as improved data source for regional applications. Furthermore, the presented approach is repeatable, temporally transferable and therefore enables a multi-temporal classification to capture and understand the dynamics and changes of different LCLU classes. The approach enables an annual independent LCLU monitoring for Central Asia which can be transferred to other regions in Asia by adding further suitable classes and reference data.

Acknowledgments This study was performed within the Regional Research Network CAWa (Water in Central Asia). We thank the German Federal Foreign Office (AA) for funding CAWa and this research. We appreciate that MODIS data, Landsat images and SRTM data were provided free of charge by the NOAA/USGS and GPCC Full Data Reanalysis Precipitation Data by the Global Precipitation Climatology Centre hosted at Deutscher Wetterdienst (DWD). Furthermore, we would like to thank CAIAG for providing us with vector datasets of the study region. We grateful thank the editors and ZALF to give us the opportunity to publish our results in this book.

References

Abd El-Kawy OR, Rod JK, Ismail HA, Suliman AS (2011) Land use and land cover change detection in the western Nile delta of Egypt using remote sensing data. Appl Geogr 31:483–494

Aguirre-Gutierrez J, Seijmonsbergen AC, Duivenvoorden JF (2012) Opimizing land cover classification accuracy for change detection, a combined pixel-based and object-based approach in a mountainous area in Mexico. Appl Geogr 34:29–37

Arkhipov V, Moukanov BM, Khaidarov K, Goldammer JG (2000) Overview on forest fires in Kazakhstan. Int Forest Fires News IFFN 22:40–48

Baatz M, Schaepe A (2000) Multiresolution segmentation—an optimization approach for high quality multi-scale image segmentation. J Photogrammetry Remote Sens 58(3–4):12–23

Congalton RG, Green K (2009) Assessing the accuracy of remotely sensed data: principles and practices. Lewis Publishers, Boca Raton

De Beurs K, Henebry GM (2004) Land surface phenology, climatic variation, and institutional change: analyzing agricultural land cover change in Kazakhstan. Remote Sens Environ 89(4):497–509

Di Gregorio A (2005) Land cover classification system (LCCS), classification concepts and user manual, Software version 2. Food and Agriculture Organization (FAO) of the United Nations, Rome

FAO (2011) Countries. http://www.fao.org/countries/55528/en/kgz/ (last accessed: 17/10/2011)

Foody GM (2002) Status of land cover classification accuracy assessment. Remote Sens Environ 80:185–201

Friedl MA, Sulla-Menashe D, Tan B, Schneider A, Ramankutty N, Sibley A, Huang X (2010) MODIS collection 5 global land cover: algorithm refinements and characterization of new datsets. Remote Sens Environ 114:168–182

Gessner U, Machwitz M, Conrad C, Dech S (2013) Estimating the fractional cover of growth forms and bare surface in savannas. A multi-resolution approach based on regression tree ensembles. Remote Sens Environ 129:90–102

Gessner U, Naeimi V, Klein I, Kuenzer C, Klein D, Dech S (2012) The relationship between precipitation anomalies and satellite-derived vegetation activity in Central Asia. Global and Planetary Change, http://dx.doi.org/10.1016/j.gloplacha.2012.09.007

Giri C, Zhu G, Reed B (2005) A comparative analysis of the Global Land Cover 2000 and MODIS land cover data set. Remote Sens Environ 94:123–132

GLCF (2011) Landsat imagery. Available at: http://glcf.umiacs.umd.edu/data/landsat/ (last accessed: 10/11/2011)

Gupta R, Kienzler K, Martius C, Mirzabaev A, Oweis T, de Pauw E, Qadir M, Shideed K, Sommer R, Thomas R, Sayre K, Carli C, Saparov A, Bekenov M, Sanginov S, Nepesov M, Ikramov R (2009) Research prospectus: a vision for sustainable land management research in Central Asia. In: ICARDA Central Asia and Caucasus Program. Sustainable agriculture in Central Asia and the Caucasus Series 1. CGIAR-PFU, Tashkent, Uzbekistan, p 84

GPCC (2011) The global precipitation climatology centre hosted at Deutscher Wetterdienst (DWD). http://gpcc.dwd.de (last accessed: 10/11/2011)

Hansen MC, Dubayah R, DeFries RS (1996) Classification trees: an alternative to traditional land cover classifiers. Int J Remote Sens 17:1075–1081

Hansen MC, DeFries RS, Townshend JRG, Sohlberg R, Dimiceli C, Carroll M (2002) Towards an operational MODIS continuous field of percent tree cover algorithm: examples using AVHRR and MODIS data. Remote Sens Environ 83:303–319

Herold D, Mayaux P, Woodcock CE, Baccini A, Schmullius C (2008) Some challenges in global land cover mapping: an assessment of agreement and accuracy in existing 1 km datasets. Remote Sens Environ 112:2538–2556

Jansen LJM, Di Gregorio A (2002) Parametric land cover and land-use classifications as tools for environmental change detection. Agric Ecosyst Environ 91:89–100

Jung M, Henkel K, Herold M, Churkina G (2006) Exploiting synergies of global land cover products for carbon cycle modeling. Remote Sens Environ 101:534–553

Karnieli A, Gilad U, Ponzet M, Svoray T, Mirzadinov R, Fedorina O (2008) Assessing land-cover change and degradation in the Central Asian deserts using satellite image processing and geostatistical methods. J Arid Environ 72:2093–2105

Klein I, Gessner U, Kuenzer C (2012) Regional land cover mapping and change detection in Central Asia using MODIS time series. Appl Geogr 35:219

Kotsiantis SB (2007) Supervised machine learning: a review of classification techniques. Informatica 31:249–268

Kurtz DB, Schellberg J, Braun M (2010) Ground and satellite based assessment of rangeland management in sub-tropical Argentina. Appl Geogr 30:210–220

Lioubimtseva E, Cole R, Adams JM, Kapustin G (2005) Impacts of climate and land-cover change in arid lands of Central Asia. J Arid Environ 62:285–308

Lioubimtseva E Henebry GM (2009) Climate and environmental change in arid Central Asia: impacts, vulnerability, and adaptations. J Arid Environ 73:963–977

Meyer WB Turner II BL (1992) Human Population Growth and Global Land Use/Land Cover Change. Ann Rev Ecold Syst 23:39–61

Nezlin NP, Kostianoy AG, Li B-L (2005) Inter-annual variability and interaction of remote-sensed vegetation index and atmospheric precipitation in the Aral Sea region. J Arid Environ 62:677–700

Nicholson SE, Farrar TJ (1994) The influence of soil type on the relationships between NDVI, rainfall and soil moisture in Semiarid Botswana. I. NDVI response to rainfall. Remote Sens Environ 50:107–120

Propastin PA, Kappas M, Muratova NR (2008a) A remote sensing based monitoring system for discrimination between climate and human-induced vegetation change in Central Asia. Manage Environ Qual Int J 19(5):579–596

Propastin PA, Kappas M, Muratova NR (2008b) Inter-annual changes in vegetation activities and their relationship to temperture and precipitation in Central Asia from 1982 to 2003. J Environ Inf 12(2):75–87

Rabus B, Eineder M, Roth A, Bamler R (2003) The shuttle radar topography mission—a new class of digital elevation models acquired by spaceborne radar. Photogrammetry Remote Sens 57:241–262

Reed BC, Brown JF, VanderZee D, Loveland TR, Merchant JW, Ohlen DO (1994) Measuring phonological variability from satellite imagery. J Veg Sci 5:703–714
Richard Y, Poccard I (1998) A statistical study of NDVI sensitivity to seasonal and inter-annual rainfall variations in southern Africa. Int J Remote Sens 19:2907–2920
Ryabtsev AD (2008) Threats to water security in the Republics of Kazakhstan: the transboundary context and possible ways to eleminate them. In: Madramootoo CA, Dukhovny VA (eds) Water and food security in Central Asia. Springer, Netherland, p 69
Rudolf B, Becker A, Schneider U, Meyer-Christoffer A, Ziese M (2010) GPCC full data reanalysis Version 5. GPCC Status Report
Schiewe J (2002) Segmentation of high-resolution remotely sensed data—concepts, application and problems. In: Symposium on geospatial theory, Processing and applications, Ottawa
Shemratov D (2004) Will Koksarai save the Shardara reservoir? (21/4/2004). In: Gazeta KZ (ed) http://engarticles.gazeta.kz/art.asp?aid=43843 (last accessed 11/12/2011)
Sulla-Menashe D, Friedl MA, Krankina O, Baccini A, Woodcock CE, Sibley A, Sun G, Kharuk V, Elsakov V (2011) Hierarchical mapping of Northern Eurosian land cover using MODIS data. Remote Sens Environ 115:392–403
U.S. Geological Survey—Land Processes Distributed Active Archive Center (USGS, LP DAAC) (2012) https://lpdaac.usgs.gov/products/modis_overview (last accessed: 29/5/2012)
Wang J, Rich PM, Price KP (2003) Temporal response of NDVI to precipitation and temperature in the central Great Plains, USA. Int J Remote Sens 24:2345–2364
Quinlan JR (1996) Improved use of continuous attributes in C4.5. J Artif Intell Res 4:77–90
Quinlan JR (1993) C4.5 programs for machine learning. Morgan Kaufmann Publishers Inc., San Mateo
Quinlan JR (1987) Generating production rules from decision trees. In: Proceedings of the 10th international joint conference on artificial intelligence, pp 304–307
Yang L, Wylie B, Tieszen LL, Reed BC (1998) An analysis of relationship among climate forcing and time-integrated NDVI of grassland over the U.S. Northern and Central Great Plains. Remote Sens Environ 56:25–37

Estimating Black Carbon Emissions from Agricultural Burning

Vladimir Romanenkov, Dmitry Rukhovich, Polina Koroleva and Jessica L. McCarty

Abstract High sensitivity of the Arctic region to short-lived climate forcers, including black carbon (BC), makes crop residue burning an important source of emissions. A high to moderate uncertainty in cropland burning emission estimates from remote sensing-based analyses currently exists and is problematic for establishing baseline estimates of black carbon emissions from global remote sensing products. Straw burning and possible BC emissions were estimated at the oblast level for Russia for years 2003 through 2010. A study was based on 1 km Moderate Resolution Spectroradiometer (MODIS) Active Fire Product, oblast level agricultural statistics, 1:25,000–1:50,000 scale GIS vector field maps and developing algorithms for calculating the size and intensity of fires as well as testing the accuracy of the predictions in areas with contrast land use. Both Active Fire Product and statistics methods demonstrated consistent results, including increasing fire activity in the years with additional straw surplus and the highest absolute values for vast territories with quite intensive grain production, mainly in European Russia. Straw burning can be a source of at least 1/3 total BC emissions from agriculture and grassland fires and does not appear to be the main source of total BC emissions for the Russian Federation. For regions with small number of cropland fires, the accuracy of existing remote sensing-based land cover products

V. Romanenkov (✉)
Department of Geographic Network of Field Experiments, Pryanishnikov All-Russian Institute of Agrochemistry (VNIIA), Pryanishnikova St., 31a 127550 Moscow, Russian Federation
e-mail: geoset@yandex.ru

D. Rukhovich · P. Koroleva
V.V. Dokuchaev Soil Science Institute, Pyzhevskii per. 7, Moscow, Russian Federation
e-mail: landmap@yandex.ru

P. Koroleva
e-mail: soilmap@yandex.ru

J. L. McCarty
Michigan Tech Research Institute, Ann Arbor, MI 48105, USA
e-mail: jmccarty@mtu.edu

L. Mueller et al. (eds.), *Novel Measurement and Assessment Tools for Monitoring and Management of Land and Water Resources in Agricultural Landscapes of Central Asia*, Environmental Science and Engineering, DOI: 10.1007/978-3-319-01017-5_20,

is insufficient for reliable classification of agricultural fires from satellite products. Incorrect classification of agricultural fires may exceed 25 %, increasing for the northern part of the country where forests are the predominant land cover. An improved method would be to calculate BC emissions from burned area using high resolution field masks and ground validation of fire sources in cropland areas.

Keywords Cropland burning · Remote sensing · GIS · Black carbon · Straw burning · Russia

1 Introduction

The burning of agricultural areas within Russia is a well-documented phenomenon (Soja et al. 2004), which began to receive attention within Russia as early as the 1980s (Derevyagin 1987). Local oblast-level regulations and acts which banned the field burning of crop residues were introduced in the 2000s.

Cropland burning not only results in serious environmental issues and wastes biological resources which can be used as organic fertilizers but is also an important source of emissions critical for the Arctic region with its high sensitivity to short-lived climate forcers such as black carbon (BC), which is a product of incomplete biomass combustion during crop residue burning. However, high uncertainty in emission estimates from cropland burning comparison of different remote sensing products means an analysis based on global remote sensing land cover and fire products is problematic.

For example, a recent study using different statistical and remote sensing analyses found the range of average annual BC emissions from cropland burning in the Russian Federation was 2.49–22.2 Gg for the time period of 2003–2009 (McCarty et al. 2012). This figure seems a more reliable estimation in comparison with the 2009 special report by the Clean Air Task Force (CATF), where average spring emissions assumed as high as 38.9 Gg in 2004–2007 (CATF 2009). If all straw surpluses in Russia as estimated from official agricultural reports were assumed to burn, only 24 Gg BC would be released for the above mentioned period. In reality annual burning of unused straw can produce about 6–9 Gg BC, less than 30 % of potential total BC emissions. In comparison, the total open burning consumption of straw in China is estimated to produce 10.3–10.8 Gg BC annually or about 23 % of the total output crop straw (Cao et al. 2008).

High uncertainty in these estimates is connected with the fact that few ground monitoring data are available for confirmation. Land use in Russia was subjected the largest change of the 20th century in the northern Hemisphere (Lyuri et al. 2010) when after 1990 approximately 43 million ha of agricultural lands (including 30.2 million ha of arable land) were abandoned according to official agricultural statistics. Absence of biomass removal and cultivation allows for the classification of fires on these abandoned lands as open burning but not cropland burning, and hence, uncontrolled burning. The mass scale of agricultural land abandonment and necessity of land cover product validation and update is one of many important methodological challenges to investigate crop burning in Russia. Non-uniformity of crop yields, agricultural practices, and economic incentives within all agricultural areas of Russia as well as absence of official statistical data on the total amount of crop straw burned do not allow for a management and/or policy explanation of the seasonal and interannual variations in agricultural fire activity.

Crop straw can be used as fodder and in animal husbandry, for compost production, and for household and industrial uses. At present, the proportion of the straw of grain and grain-leguminous crops used for household and fodder has substantially decreased and currently represents a minor use of crop straw. The analysis of incoming and expenditure items of straw balance shows that about 50 % of gross harvest of straw is a surplus (Rusakova 2006). Straw surplus increased from 51.4 % in 1990–1994 to 60.0 % in 2000–2004 and 63.8 % in 2005–2009, demonstrating only in 2010 a tendency to decrease in a bad harvest year resulting from the long drought in grain-production regions.

As a result, burning in the fields is a common agricultural practice used during the harvesting, postharvesting, or preplanting periods (Fig. 1).

Seasonal BC emissions from cropland burning on the territory of Russia, the Baltic countries, Belarus, Ukraine and Kazakhstan were generally highest in spring and fall, which connected with field preparations triggered by snowmelting (with a peak of more than 800 detections per day) and after harvesting before fall tillage (Fig. 2; Korontzi et al. 2006; Stohl et al. 2007).

But is straw burning the main source of agricultural fires in Russia? How can high resolution field masks be used to decrease the uncertainty of remote sensing data? How can variation in agricultural practices and economic incentives for farmers in different regions be accounted for in calculations? The present study was undertaken to answer these questions.

Fig. 1 Partial stubble burning in Krasnodarskij Kraj (**a**) and burning straw in windrows in Belgorodskaya Oblast (**b**) after harvesting. Figure **a** courtesy of Lothar Mueller and **b** courtesy of Dmitry Chistoprudov

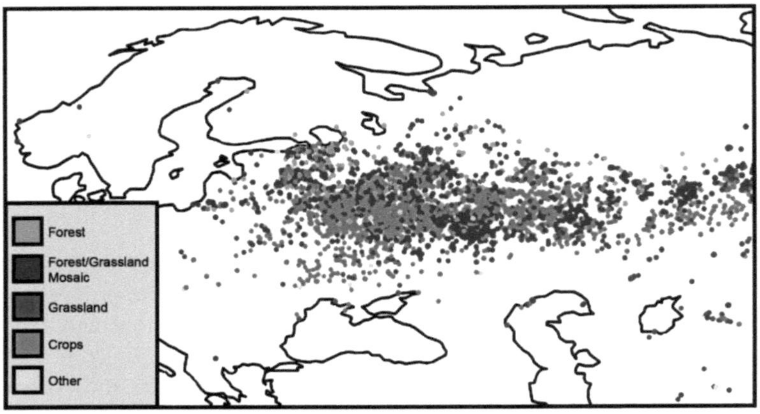

Fig. 2 MODIS active fire detections during spring 2006 classified by general land cover type across Eurasia. Late spring snow melt triggered intense field burning in April in Russia, the Baltic countries, Belarus, Ukraine and Kazakhstan

2 Data and Methods

2.1 Fire Data

The MODIS sensors, onboard the sun-synchronous polar-orbiting satellites Terra and Aqua, acquire four near-global observations daily at 1030 and 2230 (Terra) and 0130 and 1330 (Aqua), equatorial local time since 1999. The MODIS Level 2 fire product, commonly referred to as the MODIS Active Fire Data, is collected daily at 1 km resolution and includes the latitude, longitude, fire power radiation, and confidence of the fire detection. For this research, each discrete active fire point was considered to represent a burned area of 1 km^2, a common assumption that burned area is proportional to simple counts of fire pixels (Giglio et al. 2006).

2.2 Land Cover Data

Global land cover products were also considered, including the 1 km (MODIS 1 km Land Cover dataset) and 300 m (MERIS 300 m GlobCover v2.2). Cropland classifications from the MERIS GlobCover product were compared with the high resolution field boundaries for Rostovskaya, Moskovskaya, and Kostromskaya oblasts. Coordinates of active fires detected by MODIS on agricultural lands in 2003–2010 have been compared all three of these cropland land cover products (see McCarty et al. 2012).

Like the International Geosphere-Biosphere Programme (IGBP) classification schema used by the 1 km MODIS land cover dataset, the Land Cover

Classification System (LCCS) schema of the 300 m MERIS GlobCover product permits mosaic classes. LCSS land cover represents a more detailed classification compared with the IGBP as it permits 20–50, 50–70 % and >70 % of mosaic cropland within a class.

2.3 The Pilot Study Based on High Resolution GIS Field Maps

Refinements were achieved with the manual identification of croplands based on 1:25,000–1:50,000 scale field maps. GIS vector field maps have the advantage of allowing for different land cover classes in one map pixel. As a result, more precise dataset of cropland burning can be derived from the satellite fire data.

The choice of these pilot areas is connected with the fact that they are representative territories of different farming and/or cropping intensity in the absence of field masks evenly distributed for the whole agricultural area within Russia. Rostovskaya Oblast specialization is grain production based on intensive agriculture, Moskovskaya Oblast has different crop specialization with intensive technologies, and Kostromskaya Oblast is mainly a forested area and agriculture is represented by traditional crops and technologies. Rostovskaya oblast can be considered as a typical territory of the south of European Russia with 80 % of cropland from the total area and a typical field size more than 1 km^2. Moskovskaya oblast represents a region in the Central European Russia with about 50 % of arable land and a typical field size less than 1 km^2. Kostromskaya oblast is situated in the north of the European Russia with a mosaic of small fields (less than 0.5 km^2) and cropland area is about 20 % for the entire oblast.

The burned area assumption of 1 km^2 of MODIS active fire counts (Sect. 2.1) was assessed using a contemporary GIS cropland field mask. If an active fire was detected inside the boundary of a field, then the whole field was considered to have burned completely. If there were several fire points within one field over a short time period, the fire was regarded as a single burn event and burned area was the same as the area of that single field. If the fire point distance was less than 500 m from a field, it was considered a single burned field. If several fire points were situated in a line along over cropland fields over a short time period, it was not considered as an agricultural burning due to the possibility of this being a sensor anomaly or the spread of a non-cropland fire into cropland areas.

2.4 Emission Calculations

Figure 3 illustrates of the bottom-up approaches to BC emission estimations for the Active Fire based on land cover products (Sect. 2.2) and GIS filed maps (Sect. 2.3) and Agricultural Statistics analyses.

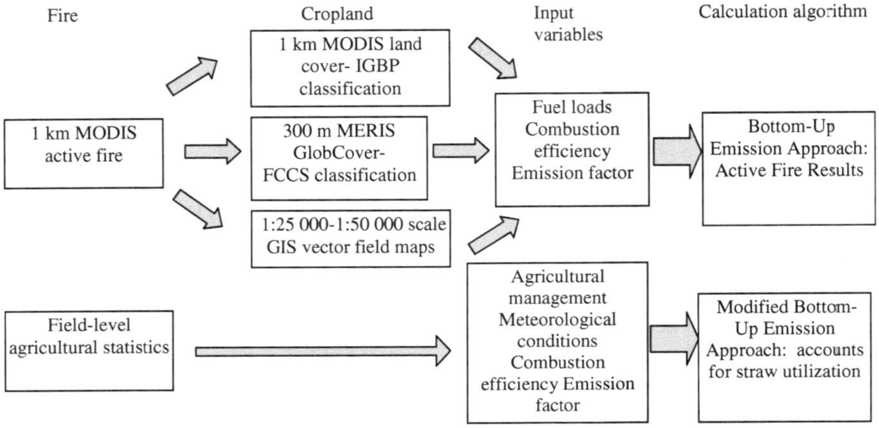

Fig. 3 Basic workflow of BC emission calculations utilized in the analysis

The estimation of possible amounts of straw burning based on the agricultural statistics approach and the active fire methodology is described in McCarty et al. (2012). A set of standard assumptions has been used in the active fire emission estimates, including the assumption of fires in those agricultural fields where cereal crops are cultivated and emission factors BC. Briefly, the equation for the BC emission calculations for the Active Fire Analysis is:

$$\text{Emissions} = A * B * CE * \ Ebc \qquad (1)$$

where A is cropland burned area, B is the fuel load variable (mass of biomass per area), CE is combustion efficiency (fraction of biomass consumed by fire), and Ebc is the BC emission factor (mass of BC per mass of biomass burned). For this analysis, all agricultural burning in Russia was attributed to grain (wheat, barley, rye) production, which accounts for the majority of planted croplands acreages (USDA FAS 2003). Grain yield statistics was used a proxy for fuel load. The Active Fire Analysis utilized the oblast-specific fuel loadings from the official statistics and assumed a yield-to-residue coefficient factor of 1.35 to estimate fuel loads. The Active Fire Analysis used a combustion efficiency value of 0.80.

The equation for the BC emission calculations for the Agricultural Statistics Analysis is

$$\text{Emissions} = C * D * CE * Ebc \qquad (2)$$

where C is annual straw surplus, D is series of correction coefficients used to account for agricultural management and agrometeorological conditions (i.e., temperature and precipitation) that can impact annual crop yield and thus straw burning. C is calculated from grain yield statistics within individual federal administrative units according to a methodology published in Derevyagin (1987).

Throughout this the manuscript, winter is defined as January through March, spring is April through June, summer is July through September, and fall is October through December, respectively.

3 Results

3.1 Emissions Based on the Agricultural Official Statistics Approach

The regions with the highest BC emission estimates can be identified with statistics approach. High values are indicative for the main grain-producing regions with the highest straw surplus. Based on this analysis, straw burning in the above regions could be a source of more than 50 % of total BC emissions.

Direct comparison of the agricultural statistical approach with the remote sensing-based fire radiative energy (FRE) and burned area analyses presented in McCarty et al. (2012) have some restrictions, as part of the fuel load is expected to be burnt next spring and seasonal contribution in straw burning have a considerable variation, both regional and interannual. However, the same trends in BC emissions are visible as in burned area analysis based on definition of croplands from the IGBP (1 km MODIS Land Cover) and LCCS (300 m MERIS GlobCover) land cover classification schemas. Estimations peaked in year 2008 with 36 % higher for statistical approach than the average emission values from 2003 through 2010, mainly as a result of 32 % increase of the total output crop straw.

Uniform high values in 2003–2009 were typical for European Russia regions: Rostovskaya oblast, Krasnodarskiy and Stavropol'skiy Krai (South Federal District), Volgogradskaya and Orenburgskaya oblasts, Republics of Tatarstan and Bashrortostan (Volga Federal District), Voronezhskaya oblast (Central Federal Distict). The highest possible emissions calculated with statistics approach from one region were found to be 6.7–9.9 % of total emissions (Stavropol'skiy Krai). Across the 7 years the highest possible emissions were calculated also for Saratovskaya oblast, Tambovskaya oblast, Samarskaya oblast, Orlovskaya oblast, Lipetskaya oblast (European Russia), Krasnoyarskiy Krai (East Siberia), Omskaya oblast, Altaiskiy Krai, Chelyabinskaya oblast, Kurganskaya Oblast, Novosibirskaya oblast (West Siberia). In East and West Siberia calculated peak of annual BC emissions from crop burning was 7.0 and 4.0 %, respectively.

Straw burning consumption is estimated at 20–27 % of the total output crop straw in 2003–2010. The last range is quite consistent in calculations since 1997. The largest contributions are European Russia oblasts (61–78 % of total burning) with less than 20 % of burning coming from West Siberia and Far East.

3.2 Emissions Based on the Active Fire Analysis

According to the Active Fire Analysis, intra-annual variations for spring, summer and fall burning were 33–70, 9–38, and 11–33 %, respectively. A slightly higher summer cropland burning in comparison with spring was calculated for years 2005 and 2007 within the LCSS cropland definition from the 300 m MERIS GlobCover product while the IGBP cropland definitions in the 1 km MODIS Land Cover product produced an equal contribution of summer and spring burning in total emissions (Table 1). Winter burning contributed for less than 0.5 % of total annual BC emissions.

The highest spring cropland burning from the active fire analysis was calculated for years 2003, 2004, 2006, 2008, 2009 and 2010. The highest annual BC emissions were calculated for years 2008 and 2009 while the lowest BC emissions for 2003 within the 8-year study period. The average BC emissions were 7.7–8.4 Gg for IGBP-Croplands, IGBP-Agriculture, LCCS-Croplands in comparison with the Agricultural Statistics Approach result of 8.5 Gg for years 2003–2010. The same is true for the interannual dynamics, with correlation coefficient of not less than 0.9 (Table 2). More than 60 % of all BC emissions, an average of 64 % for all years, occurred in European Russia. Approximately 50–79 % of all federal subjects with greater than 1 % of total BC emissions were within European Russia. On average, West Siberia was the source of 20 % of total cropland burning BC emissions, Far East Russia accounted for 9 %, and East Siberia accounted for 7 %.

The highest emissions calculated from Active Fire Analysis were found to be 7.1–15.0 % of total emissions (Krasnodarskiy Krai). The other regions which produce the highest values of emissions are Stavropol'skiy Krai as well as Rostovskaya oblast (highest among the European territories in all years except 2003, 2008 and 2009) as well as Omskaya and Novosibirskaya oblasts and Altajskiy Krai in West Siberia—vast territories with quite intensive grain production. These source regions were different than those highlighted by the Agricultural Statistics approach. In East and West Siberia calculated peak of annual BC emissions from crop burning as detected by the Active Fire Analysis was 4.3 and 9.4 %, respectively, while the highest BC emissions outside of European Russia were calculated for West Siberia only according to the Agricultural Statistics approach.

3.3 Comparison of Fire Data with Straw Surplus Statistics

For 13 oblasts of Russia which have the biggest MODIS agricultural fire counts the possible effect of air pollution and BC emission was assessed in 2003–2009. These oblasts are specialized in wheat growing. Straw surplus here was 10–15 % higher than average for all Russian grain production regions. Nevertheless, only for a few oblasts the significant relationship between straw burning estimates based on agricultural statistics and MODIS active fire counts was found. Based on these

Table 1 Average seasonal BC emissions in the Russian Federation from cropland burning as detected from Active Fire Analysis; emissions reported by four definitions of croplands, all emissions reported in Gg

Year	Season	IGBP-Agriculture	IGBP-Croplands	LCCS-Agriculture	LCCS- Croplands
2003	Winter	0.02	0.02	0.01	0.01
	Spring	2.95	2.74	2.08	0.67
	Summer	0.59	0.56	0.57	0.25
	Fall	1.27	1.23	1.31	0.50
2004	Winter	0.01	0.01	0.01	0.00
	Spring	3.34	3.01	3.23	0.72
	Summer	2.43	2.42	2.44	1.23
	Fall	1.06	1.05	0.88	0.40
2005	Winter	0.11	0.11	0.10	0.05
	Spring	2.05	1.95	1.57	0.48
	Summer	3.24	3.23	3.26	1.65
	Fall	2.27	2.23	1.99	0.90
2006	Winter	0.01	0.01	0.01	0.00
	Spring	5.22	4.79	4.91	1.24
	Summer	2.57	2.54	2.65	1.26
	Fall	1.41	1.39	1.28	0.61
2007	Winter	0.08	0.08	0.07	0.03
	Spring	2.44	2.32	1.71	0.61
	Summer	3.09	3.06	3.22	1.53
	Fall	1.22	1.19	1.04	0.46
2008	Winter	0.08	0.08	0.07	0.03
	Spring	5.77	5.36	4.74	1.15
	Summer	5.59	5.57	5.49	2.96
	Fall	2.53	2.51	2.31	1.17
2009	Winter	0.03	0.03	0.02	0.01
	Spring	7.45	6.94	6.71	1.91
	Summer	2.34	2.32	2.34	1.25
	Fall	1.43	1.42	1.34	0.65
2010	Winter	0.02	0.02	0.02	0.01
	Spring	3.18	2.93	2.78	0.67
	Summer	2.00	1.90	2.51	0.83
	Fall	1.01	0.98	0.98	0.38
Average	Winter	0.05	0.04	0.04	0.02
	Spring	4.05	3.75	3.47	0.93
	Summer	2.73	2.70	2.81	1.37
	Fall	1.53	1.50	1.39	0.63

data, in Rostovskaya oblast, Krasnodarskiy Krai and Stavropol'skiy Krai 11–16, 10–14 and 20–26 % of total straw yield can be burned in 2001–2007, respectively, with 0.3–0.7 Gg annual BC emission, $R = 0.74–0.84$ (Fig. 4). With uniform agricultural practice and grain yields within the regions statistical data here can be a reliable source of BC emission estimation. But for the biggest part of Russian regions straw burning does not appear to be the main source of BC emissions, according to this comparison.

Table 2 Annual BC emissions in the Russian Federation from cropland burning as detected from Active Fire Analysis; emissions reported by four definitions of croplands; all emissions reported in Gg

Year	IGBP-Agriculture	IGBP-Croplands	LCCS-Agriculture	LCCS-Croplands	Agricultural statistics
2003	4.82	4.55	3.97	1.43	6.57
2004	6.84	6.49	6.55	2.35	7.27
2005	7.67	7.52	6.92	3.08	7.50
2006	9.22	8.74	8.84	3.11	8.40
2007	6.82	6.65	6.04	2.64	9.08
2008	13.97	13.51	12.61	5.31	12.11
2009	11.25	10.71	10.41	3.81	11.32
2010	6.21	5.83	6.29	1.88	5.97
Total	66.80	63.99	61.64	23.61	68.25
Annual Average	8.35	8.00	7.70	2.95	8.53

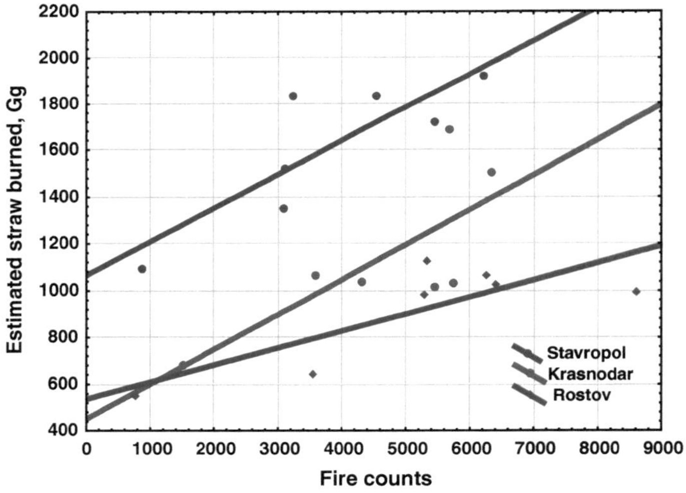

Fig. 4 Comparison of straw burning statistical estimates with MODIS active fire counts for Rostovskaya Oblast, Stavropol'skiy and Krasnodarskiy Krai in 2003–2009

This difference between the regions and their relative role as sources for BC emissions cropland burning was the reason for validation of the cropland mask, which used the cropland and cropland/natural vegetation mosaic classes both in IGBP and LCCS land cover classification schemas.

3.4 Manual Identification of Croplands

Incorrect classification can be connected with remote sensing estimates of fire point to cropland when in fact burned area is not an agricultural one (Figs. 5, 6) or missing the agricultural burned area based on the 1 km MODIS Land Cover dataset. In 2009 incorrect estimates of cropland fires with automatic pixel-based classification of land covers were 0.227, 0.421 and 0.388 for Rostovskaya, Moskovskaya and Kostromskaya oblasts, respectively (Table 3). These estimates include Active Fire points in IGBP-Agriculture land cover classification schema that do not belong to fields in accordance with field mask, Active Fire points in the fields of field mask do not belong to IGBP-Agriculture and the number of duplicate Active Fire points in the fields of field mask, so the total sum can exceed 1.0. Incorrect estimates were higher in 2010, especially for Moskovskaya and Kostromskaya oblasts. This increase in incorrect assignments of land cover classes is connected with increasing total number of fires detected in 2010, but still accuracy is not the same for the 3 regions.

Application of the 300 m MERIS GlobCover dataset did not change the fire classification results significantly. 17.5 % of the area of the region (28.8 % of the arable area) was mistakenly classified as croplands for Rostovskaya Oblast, and 14.3 % of the area of the oblast (23.6 % of the arable area) was mistakenly classified as non-arable lands; the resulting classification error for farmlands identified by MERIS was 52.4 %. 11.5 % of the area of the region (36.0 % of the

Fig. 5 Active fire points estimated as cropland burnings (*red dots*) and natural vegetation burnings (*yellow dots*) in peat bogs based on automatic land cover classification (Moskovskaya oblast)

Fig. 6 The same points from Fig. 5 on a topographic map

arable area) was mistakenly classified as croplands for Moskovskaya Oblast and 12.7 % of the area of the region (39.8 % of the arable area) was mistakenly classified as non-arable lands, the resulting error of the farmlands identified by MERIS was 75.8 %. 0.8 % of the area of the region (6.1 % of the arable area) was mistakenly classified as croplands for Kostromskaya Oblast and 10.8 % of the area of the region (84.0 % of the arable area) was mistakenly classified as non-arable lands, the resulting error of the farmlands identified by MERIS was 90.1 %.

3365 MODIS active fire points were analyzed for the Rostovskaya Oblast in 2010, of which croplands accounted for 76 % of fires. Burned area calculated using the high resolution field mask produced an estimate of 269.9 thousand hectares. For 2009, burned area was calculated using the same method for approximately 440 thousand ha, with croplands contributing as much as 81 % of total active fire detections. For Rostovskaya oblast, an automatic identification of land cover classes from the 300 m MERIS GlobCover product (Fig. 7) exhibits errors of commission, illustrated in Fig. 8.

3804 MODIS active fire points recorded were analyzed for the Moskovskaya oblast for 2010, of which croplands accounted for 23 % of fires. Burned area calculated using the high resolution field mask produced an estimate of 48.2 thousand hectares. In 2009, of the 1839 MODIS satellite detections of actively flaming fires, croplands accounted for 80 % of fires with a total burned area of 95 thousand ha. For Moskovskaya oblast a 2.5 increase of uncertainty in 2010 (Table 3) was connected with incorrect classification of peat and bog lands as well as grasslands as arable fields. This feature of incorrect classification is typical for land cover data created by automatic pixel-based land cover classification

Table 3 Active fire points distribution in the pilot areas

Federal Subject Name	Total amount of fire detections	Active fire points in IGBP-Agriculture, F	Active fire points in IGBP-Agriculture that do not belong to fields in accordance with field mask, A	Active fire points in the fields of field mask	Active fire points in the fields of field mask do not belong to IGBP-Agriculture, B	The number of duplicate active fire points in the fields of field mask, C	Burned fields amount	Burned area, ha	Incorrect estimates of cropland fires, (A+B+C)/F
2009 year									
Kostromskaya oblast	142	114	11	121	18	15	106	8002	0.386
Moskovskaya oblast	1839	1398	48	1707	357	221	1486	95010	0.448
Rostovskaya oblast	4678	4426	150	4445	169	664	3781	440002	0.222
2010 year									
Kostromskaya oblast	249	126	15	162	51	27	135	7477	0.738
Moskovskaya oblast	3804	1111	418	1187	494	307	880	48163	1.097
Rostovskaya oblast	3365	3200	237	3061	98	509	2552	269941	0.264

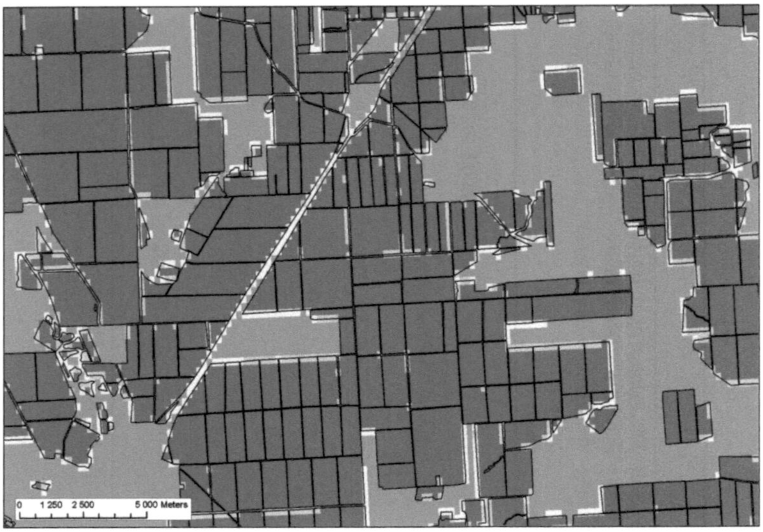

Fig. 7 Field map fragment for Rostovskaya oblast

Fig. 8 MERIS 300 m GlobCover overlaid with the same fragment of field map (Fig. 7)

techniques used for global and regional satellite products produced by USGS, NASA, IIASA, Space Research Institute (IKI), etc. There was a relationship between peak of fire activity on agricultural lands and total fires in April. In year 2010, when total number of fires increased roughly twice with a 50 % decrease of

Table 4 Black carbon emission analysis in the pilot areas

Federal Subject Name	Burned area, ha	Grain yield, dt/ha	Coefficient factor of grain yield in the straw	The BC emission factor g/kg	Combustion efficiency value	BC emission, Gg		
						Calculated from the burned area	Calculated from active fire points	
							IGBP-Agriculture/ IGBP-Croplands	LCCS-Agriculture/ LCCS- Croplands
2009 year								
Kostromskaya oblast	428 611.0	17.0	1.35	0.4	0.8	0.006	0.005/0.006	0.001/0.005
Moskovskaya oblast	93 654.4	23.0	1.35	0.4	0.8	0.093	0.153/0.170	0.048/0.173
Rostovskaya oblast	7 853.9	33.0	1.35	0.4	0.8	0.611	0.471/0.472	0.254/0.482
2010 year								
Kostromskaya oblast	269 941.0	12.1	1.35	0.4	0.80	0.004	0.005/0.007	0.001/0.004
Moskovskaya oblast	48 163.0	21.6	1.35	0.4	0.80	0.045	0.087/0.104	0.017/0.179
Rostovskaya oblast	7 477.0	24.6	1.35	0.4	0.80	0.287	0.339/0.340	0.169/0.343

cropland burned area, the biggest number of incorrectly classified fires was connected with a temporarily dried bog/peatland areas.

The similar situation in 2010 was in Kostromskaya oblast, despite the fact that here boggy areas are mainly situated under natural forest canopies. 249 MODIS Active Fire points were analyzed for Kostromskaya Oblast in 2010, of which croplands accounted for 54 %. Burned area calculated using the high resolution field mask produced an estimate of 7.5 thousand hectares. In 2009, of the 142 MODIS Active Fire points, croplands accounted for 75 % of all fires and with a burned area of 8.0 thousand ha.

Comparative estimation of BC emissions for the 3 pilot areas (Table 4) demonstrate consistent results based on the field masks and IGBP land cover classification schemas in Kostromskaya oblast in 2009. The same calculation in 2010 reveals 20 and 60 % overestimation based on IGBP-Croplands and IGBP-Agriculture land cover classification schemas, respectively. For Rostovskaya oblast, IGBP land cover calculations provide more than 20 % underestimation in 2009 and 18 % overestimation in 2010. The highest errors were found in Moskovskaya oblast. Interannual variability in uncertainty is a subject of special research but it is clear that with the automatic pixel-based classification of land covers provided by the MODIS and MERIS products is not less than 30 % for the three pilot oblasts.

4 Conclusions

The BC emissions results of Active Fire Analysis are fairly consistent with the estimates from the Agricultural Statistics Approach. The highest BC emission values are indicative of the main grain-producing regions. With uniform agricultural practice and grain yields within the region statistical data, provides a reliable source for BC emission estimation.

Straw burning can be a source of at least 1/3 of total BC emissions from agriculture and grassland fires. For the biggest part of Russian regions, straw burning does not appear to be the main source of BC emissions. The peak fire activity after harvesting is indicative for the main grain-producing regions with high straw surplus. For most Russian regions, springtime straw burning shows high variability from year to year. According to Active Fire Analysis intra-annual variation for spring contribution in total annual BC emissions was 33–70 %. Substantial spatial variation of agricultural fires may be a result of different specialization, intensity of technologies, as well as percent of plowing land.

For regions with small number of cropland fires, the accuracy of existing land cover classifications made on the basis of remotely sensed data is insufficient for reliable allocation of agricultural fires. This study shows that it is preferable to make BC calculations from possible burned areas based on field masks and ground validation of fire sources on croplands.

For improving results of remote sensing monitoring of agricultural fires on the territory of the Former Soviet Union the following steps are important:

1. Delineation of uniform territories with similar strategies of using crop residues and groundtruthing of agricultural fire counts in separate fields for checking which crop residues are the main source of agricultural fires, estimate fuel load and combustion efficiency values as well as role of set-aside or fallow agricultural lands as a source of fires in open territories.
2. Electronic map of croplands as well as set-aside or fallow agricultural lands needs to be created. Otherwise incorrect classification of agricultural fires may exceed 25 %, increasing for the northern part of Russia where forest territories are predominant. As the first step this work can be performed for federal subjects in the Russian Federation with greater than 1 % of total annual BC identified by the different approaches highlighted in McCarty et al. (2012). Note that the importance of this work is higher than simply monitoring agricultural fires as it will provide more accurate burned area and subsequent BC emission estimates.
3. High uncertainty is also connected with absence of data about yield variations within the specific regions. It is impossible to make BC calculations based on average yield for the country because data from region to region can differ by more than a factor of 5. To decrease uncertainty in emission estimates, crop yield data and cropping patterns for administrative districts are necessary.

Acknowledgments This work was funded by support from the Clean Air Task Force (http://catf.us/) and the Bellona Foundation of Russia (http://bellona.org/).

References

Cao GL, Zhang XY, Wang YQ, Zheng FC (2008) Estimation of emissions from field burning of crop straw in China. Chin Sci Bull 53:784–790

Derevyagin VA (1987) Practical guidance on straw utilization as fertilizer, all-union research and development, pp. 1–11 (in Russian, translated into English by Bellona Foundation). www.bellona.org. USSR State Agro-Industrial Committee on Design and Technological Institute for Organic Fertilizers and Peat, Moscow, Russia. Available from: http://quickr.mtri.org/croplandburning

Giglio L, van der Werf GR, Randerson JT, Collatz GJ, Kasibhatla P (2006) Global estimation of burned area using MODIS active fire observations. Atmos Chem Phys 6:957–974

Korontzi S, McCarty J, Loboda T, Kumar S, Justice CO (2006) Global distribution of agricultural fires from three years of MODIS data. Global Biogeochem Cycles 20(6): 2021. http://dx.doi.org/10.1029/2005GB002529

Lyuri DI, Goryachkin SV, Karavaeva NA, Denisenko EA, Nefedova TG (2010) Dynamics of agricultural lands of Russia in XX Century and postagrogenic restoration of vegetation and soils. Geos, Mosco, pp. 420 (in Russian)

McCarty JL, Ellicott EA, Romanenkov V, Rukhovitch D, Koroleva P (2012) Multi-year black carbon emissions from cropland burning in the Russian Federation. Atmos Environ 63: 223–238. http://dx.doi.org/10.1016/j.atmosenv.2012.08.053

Rusakova IV (2006) Straw. In: Yes'kov AI (ed) Organic fertilizer resources in Russian agriculture: information analysis reference book, All-Russian Institute of Organic Fertilizers and Peat. Russian Federation, Moscow, pp. 113–119 (in Russian, translated into English by Bellona Foundation). www.bellona.org. Available from:http://quickr.mtri.org/cropland burning

Soja AJ, Cofer WR, Shugart HH, Sukhinin AI, Stackhouse Jr, PW, McRae DJ, Conard SG (2004) Estimating fire emissions and disparities in boreal Siberia (1998-2002). J Geophys Res 109:D14S06. http://dx.doi.org/ 10.1029/2004JD004570

Stohl A, Berg T, Burkhart JF, Fjraa AM, Forster C, Herber A, Hov Ø, Lunder C, McMillan WW, Oltmans S, Shiobara M, Simpson D, Solberg S, Stebel K, Ström J, Tørseth K, Treffeisen R, Virkkunen K, Yttri KE (2007) Arctic smoke - record high air pollution levels in the European Arctic due to agricultural fires in Eastern Europe in spring 2006. Atmos Chem Phys 7:511–534

USDA FAS (2003) Russia: agricultural overview. United States Department of Agriculture Foreign Agricultural Service Production Estimate and Crop Assessment Division. Available from: http://www.fas.usda.gov/pecad2/highlights/2005/03/Russia_Ag/index.htm

Non-Linear Approaches to Assess Water and Soil Quality

Gunnar Lischeid

Abstract Systematic monitoring is indispensable for a thorough water and soil management. However, large data sets with many variables, natural heterogeneities, and a variety of (possible) influencing factors require new approaches for processing and visualization of the data. A variety of advanced techniques has been developed recently in different disciplines. Some of them have been tested for application in water and soil resources management and exhibited very promising results. Two out of these approaches are presented here by application to a data set of shallow groundwater quality that has been complied during a five years period in a small catchment in Northeast Germany. Measured variables of soil or water quality usually reflect effects of various processes. On the other hand, single processes usually affect more than one variable and thus generate a characteristic "fingerprint" that can be used in an inverse approach to identify this process based on observed measured variables. Other processes differ with respect to their "fingerprints" and thus can be differentiated in a large data set. This is the basic idea of applying dimensionality reduction approaches. Every single sample can be ascribed a score of a component that is a quantitative measure for the impact of the respective process on the given sample. Usually, a small number of components (or processes, respectively) accounts for a large fraction of the variance in a data set with many variables. This "dimensionality reduction" helps a lot to gain better understanding of the prevailing processes, of spatial and temporal patterns, and of the reasons for conspicuous data. The larger a given data set, and the larger the number of variables, the more advanced methods of data visualization are required. Modern visualization techniques pave the way for efficient use of the most powerful interface between data stored on a computer and the human brain. A single non-linear projection of high-dimensional data on a two-dimensional graph provides comprehensive information about outliers, clusters, linear and non-linear relationships, spatial patterns, multivariate trends, etc. in the data. This

G. Lischeid (✉)
Leibniz Centre for Agricultural Landscape Research (ZALF) e. V.,
Eberswalder Strasse 84, 15374 Muencheberg, Germany
e-mail: lischeid@zalf.de

L. Mueller et al. (eds.), *Novel Measurement and Assessment Tools for Monitoring and Management of Land and Water Resources in Agricultural Landscapes of Central Asia*, Environmental Science and Engineering, DOI: 10.1007/978-3-319-01017-5_21,

approach could usefully be combined with other dimensionality reduction techniques. This chapter can serve only as an appetizer. A variety of sophisticated new methods exist. These techniques still are not part of textbooks of hydrology or soil science. They require an open mind and some initial training. Then a wealth of powerful tools are at hand as a base for thorough water and soil resources management.

Keywords Water quality · Soil quality · Statistical methods

1 Motivation

Scarce resources need to be managed very carefully. Water quantity, water quality and soil quality certainly are scarce and valuable resources in most parts of the world, and are intimately intertwined. None of them could be replaced by any technical means at affordable prices and at the required scale. Water and soil resources management has to grasp the actual state, current and future threats and options, has to implement and perform measures, and, last but not least, has to evaluate the effects and possible harmful side-effects of the adopted measures as well as the efficiency of the management. Thus a well-designed monitoring scheme is indispensable for providing the required data with the necessary spatial and temporal resolution (Muller 2012).

Natural systems usually exhibit considerable heterogeneity in space and time which any monitoring has to account for. This requires a large number of replicates, yielding large data sets that are not easy to handle. Moreover, usually a multitude of different processes act in parallel and at different scales and have an impact on the observed variables. In return, single measured variables usually are prone to various processes that will be reflected by the measurement data, although to unknown quantities. Last but not least, the monitoring scheme itself needs to be evaluated and to be checked with respect to efficiency as well as to possible errors etc.

2 Example Data Set

Here two new techniques will be presented that have been developed recently and have proved successful in a variety of studies on soil and water quality. Both methods explicitly allow for non-linearities that are to the author's experience more the rule rather than the exception for real world data. Statistical analyses and graphs have been performed using the R software environment (R Development Core Team 2006) which is freely available at http://www.r-project.org.

Table 1 Pearson correlation between the measured variables

	NH_4	NO_3	P	Cl	SO_4	pH	Redox potential	EC
NH_4		−0.2	0.06	0.03	−0.1	−0.08	0.12	−0.01
NO_3	−0.2		−0.14	−0.15	−0.19	−0.08	0.1	−0.1
P	0.06	−0.14		0	−0.08	0.19	−0.12	−0.05
Cl	0.03	−0.15	0		0.31	0.08	0.02	0.62
SO_4	−0.1	−0.19	−0.08	0.31		0.04	−0.07	0.84
pH	−0.08	−0.08	0.19	0.08	0.04		−0.01	0.01
Redox potential	0.12	0.1	−0.12	0.02	−0.07	−0.01		−0.04
EC	−0.01	−0.1	−0.05	0.62	0.84	0.01	−0.04	

Both methods have been applied to a data set that was kindly provided by Dr. Uwe Schindler, Leibniz Centre for Agricultural Landscape Research, Müncheberg, Germany. It comprises data from 718 samples of shallow groundwater, sampled between March 2000 and May 2004 by piezometers at 2–4 m depth at seven different sites in the Quillow catchment, about 100 km north of Berlin, Germany. Land use was arable land mostly, but included grassland and forest sites as well. In total, 41 different pipes have been sampled. Number of temporal replicates per pipe varied between 1 and 38.

Eight variables have been determined: Concentration of NH_4, NO_3, P, Cl, and SO_4, pH value (pH), redox potential, and electric conductivity (EC). The data were tested for plausibility prior to the analysis. Values below the level of quantification were replaced by half the value of quantification level. To compensate for different data ranges for different variables, the data were z-normalized for each variable separately (values divided by the respective mean value, and then divided by the standard deviation).

Most variables are only slightly correlated with each other. Only SO_4 and Cl concentration are clearly correlated with electric conductivity and are closely related to each other (Table 1). This data set is rather small with respect to the number of samples as well as to the number of measured variables compared to some recent applications (Lischeid and Bittersohl 2008; Lischeid 2009; Schilli et al. 2010). Thus it is well suited to serve as an example. A good textbook that provides in-depth information about these methods and many more related approaches is given, e.g., by Lee and Verleysen (2007).

3 Dimensionality Reduction and Process Identification

Measured variables of soil or water quality usually reflect effects of various processes. E.g., nitrate concentration is affected by fertilisation, uptake by plants, nitrogen release by decomposition of organic matter, oxidation of ammonium, leaching, denitrification, etc. On the other hand, single processes usually affect more than one variable. E.g., denitrification requires reduced oxygen availability and low redox potential, which often comes along with enhanced Fe(II) concentration or sulphate concentration due to oxidation of sulphides. This interplay generates a characteristic "fingerprint" that can be used in an inverse approach to identify this process based on observed measured variables. Other processes differ with respect to their "fingerprints" and thus can be differentiated in a large data set. This is in fact the basic idea of applying dimensionality reduction approaches.

The term "dimensionality reduction" reflects another aspect of these approaches: Often only a very small number of processes prevail in a given data set. Thus, a small number of "components" suffices to explain a large fraction of the variance of a data set with many variables, that is, a high-dimensional data set. Thus application of dimensionality reduction techniques facilitates interpretation and understanding the basic features of large, high-dimensional data sets.

A well known approach is the Principal Component Analysis (PCA). In a mathematical sense it performs an Eigen value decomposition of the correlation matrix of the measured variables. It is the most efficient approach as long as the data are Gaussian distributed and only linear relationships have to be accounted for. It has often been applied to large multivariate data sets of water quality (Haag and Westrich 2002; Thyne et al. 2004; Fernandes et al. 2006) as well as in soil science and geochemistry (Gupta et al. 2006; Zhang et al. 2007; Weyer et al. 2008; Langer and Rinklebe 2009).

However, the PCA is neither well suited for non-Gaussian distributed data nor for data sets that exhibit pronounced non-linear correlations. To that end a variety of non-linear methods of dimensionality reduction approaches have been developed (Lee and Verleysen 2007). Unlike for the linear case, there is no clear advice when which approach would be superior to others, and a trial-and-error procedure is suggested (Lee and Verleysen 2007). One of these approaches is presented here. The Isometric Feature Mapping (Isomap) has been introduced by Tenenbaum et al. (2000). In a certain sense it is a non-linear extension of the linear PCA approach (Gámez et al. 2004). The theory of that approach is not presented here. To that purpose, the reader is referred to Tenenbaum et al. (2000), Gámez et al. (2004), Mahecha et al. (2007) or Lee and Verleysen (2007).

Here the potential of that approach for soil and water quality monitoring is illustrated. Compared to the linear PCA, the first two Isomap components explain a substantially larger fraction of variance of the data set (82 % compared to 70 %) (Fig. 1). This provides some evidence that non-linear relationship play a role in the given data set. Including the third component yields about 90 % of the variance both for Isomap as for PCA. Additional components increase that fraction only marginally. It can thus be concluded that only three different processes play an important role for the given data set. In the following only the Isomap results will be considered.

The next step aims at identifying these three processes. The Isomap analysis ascribes a value for each of the eight components to each of the 718 samples of the data set. These component values, called "scores", can be considered a measure for the strength of the effect of the respective (yet unknown) process. The relationship of the measured variables with the Isomap scores reveals the "fingerprint"

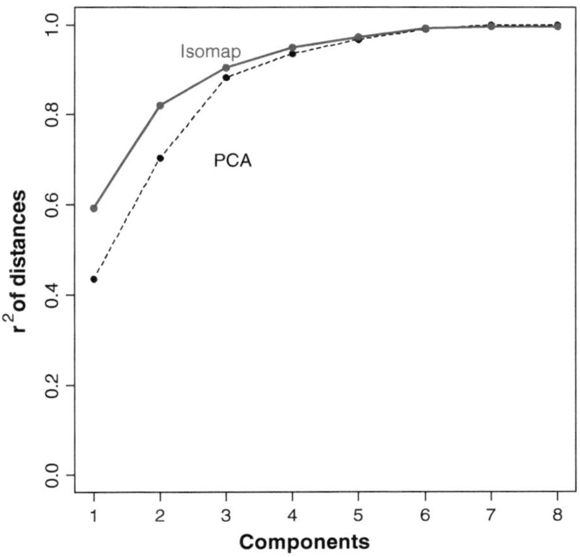

Fig. 1 Cumulative variance explained by eight principal components (*PCA*) and isometric feature mapping (*Isomap*) components

of the respective process. This relationship might be blurred due to the fact that other processes have an impact on the measured variables as well. Those effects can be subtracted from the measured data as follows: A multivariate linear regression of the measured values with all but one Isomap component gives the fraction of the variance of the measured data that is explained by these components. Then the residuals of that regression are compared with the scores of the remaining component.

Figure 2 gives the results for the first component which depicts 59 % of the variance of the data set. It is positively correlated with SO_4 and Cl concentration, pH value and electric conductivity. Negative and partly non-linear correlations are observed for NH_4, NO_3 and P concentration and redox potential. Figure 3 shows that component scores are close to zero at most sites. Only four sites exhibit very large values and substantial temporal variance. These results were unexpected prior to the study. However, both the "fingerprint" as well as the spatial pattern provide strong evidence that upwelling saline groundwater plays a major role at these sites: very high SO_4 and Cl concentration, pH value and electric conductivity are indicative for tertiary deep groundwater that is known to upwell at many sites in East Germany. Low redox potential, and hardly detectable agricultural contaminants like NH_4, NO_3 and P support this hypothesis. Only few piezometers at two different sites were affected. It has been found at other sites that plumes of upwelling saline groundwater can have in fact very small extent.

The second component accounts for 23 % of the data set's variance. The scores of that component increase with NO_3 concentration and redox potential and

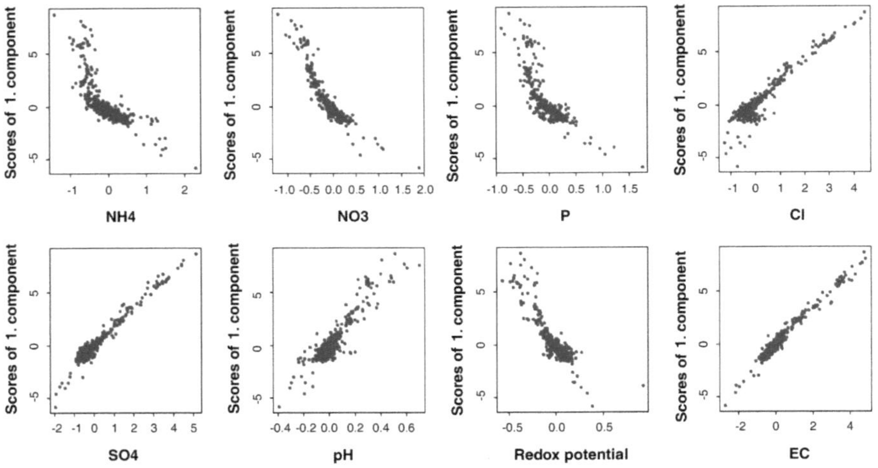

Fig. 2 Scores of the first component versus the residuals of the measured variables

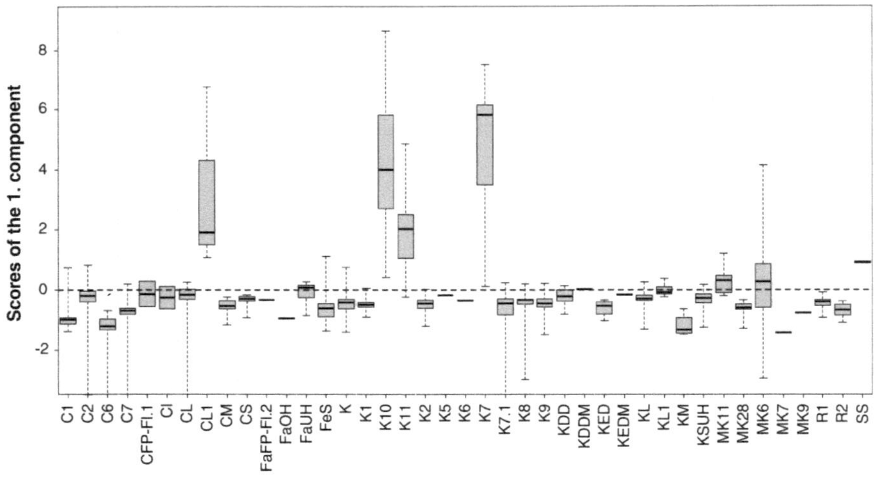

Fig. 3 Scores of the first component for different piezometers. The *boxes* give the 25th, 50th and 75th percentiles, and the *whiskers* the range of the Isomap scores

decrease with NH_4 and P concentration and with pH value. The remaining variables do not show any clear relationship with the second component (Fig. 4). This pattern is consistent with the well known sequence of redox processes. High scores reflect oxic conditions where NO_3 is stable. Being the anion of a strong mineral acid, high NO_3 concentration reduces the pH value. In contrast, low scores of the second component characterise hypoxic conditions where NO_3 is partly denitrified, partly reduced to NH_4, and P is released from sorption to Fe oxides and

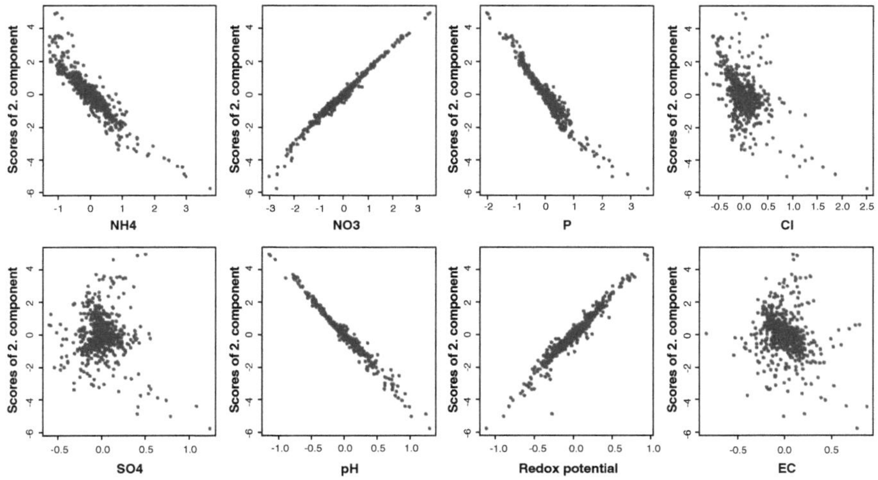

Fig. 4 Scores of the second component versus the residuals of the measured variables

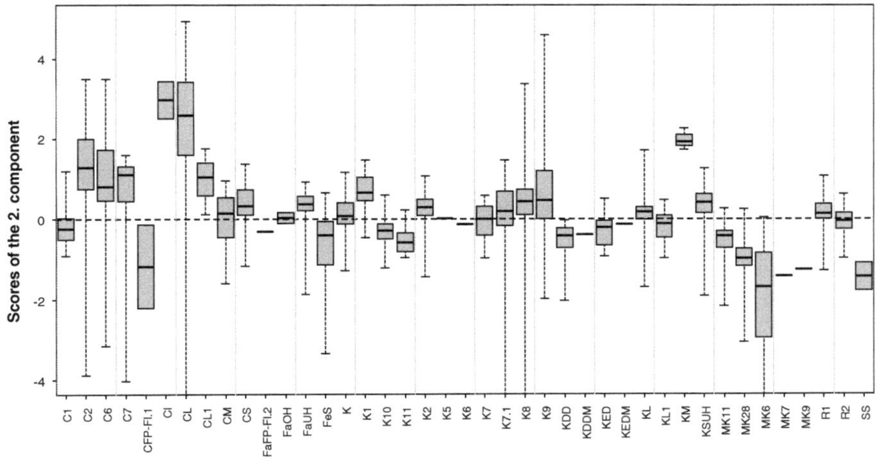

Fig. 5 Scores of the second component for different piezometers. The *boxes* give the 25th, 50th and 75th percentiles, and the *whiskers* the range of the Isomap scores

hydroxides, that become reduced and dissolved under hypoxic conditions. That interpretation is confirmed by the fact that very low scores have been found at the wetland site (MK6, MK28) (Fig. 5).

This component exhibits consistent temporal patterns at many sites. Figure 6 gives time series of the component scores at selected sites. Many sites show very low values during winter 2000/2001, and in the second half of 2002. Precipitation in September 2001 was about four times that of mean monthly precipitation, and

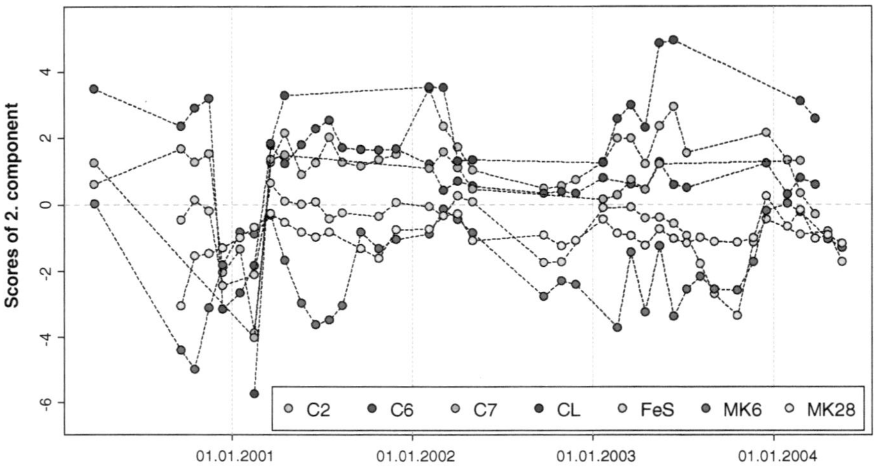

Fig. 6 Time series of scores of the second component for selected sites

continued to exceed the mean until the end of 2002. Thus, soil saturation was higher as usual, which might have promoted oxygen consumption and a decrease of redox potential. It is remarkable that other components exhibited less clear and consistent temporal patterns in this study (not shown).

Another 8.5 % of the variance of the data set is explained by the third component. It is positively correlated with NO_3 and P concentration and pH values, and negatively with NH_4 concentration and redox potential (Fig. 7). Thus this component is determined by the same variables compared to the second

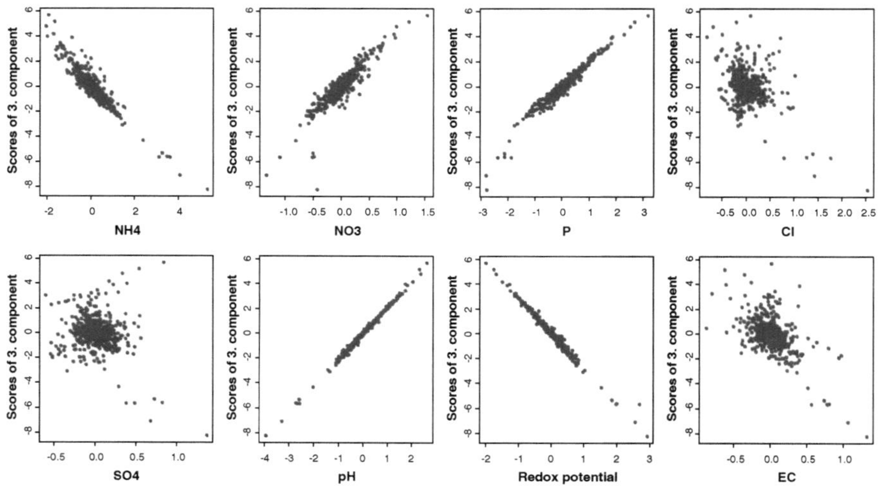

Fig. 7 Scores of the third component versus the residuals of the measured variables

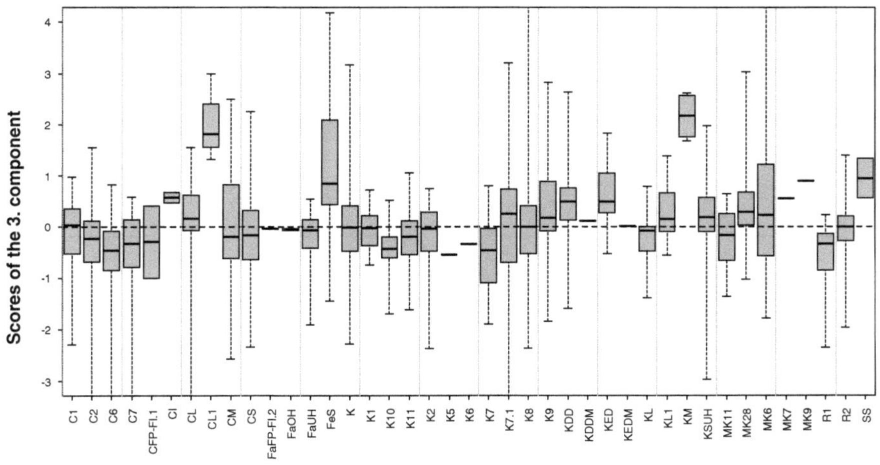

Fig. 8 Scores of the third component for different suction cups. The *boxes* give the 25th, 50th and 75th percentiles, and the *whiskers* the range of the Isomap scores

component, although partly with inverted sign. The third component explains some scatter in the data that has not already depicted by the second component. As will be shown later (Fig. 10), this component mainly differentiates between samples with low scores of the second component, that is, reduced groundwater. Samples with high scores of the third component exhibit higher NO_3 and P concentration and pH value, and lower NH_4 concentration and redox potential than would be expected based on the scores of the second component only. This might be either due to kinetic constraints, or to site-specific peculiarities. In fact, temporarily high scores of the third component have been observed at three sites only (Fig. 8).

Statistics cannot replace the expert's knowledge about biogeochemical processes. The interpretation of the Isomap results might be matter of debate, or might need to be refined and proved by additional measurements. E.g., nitrate isotope data can be used as an independent assessment of denitrification processes. But the Isomap analysis does reveal relationships within the given data set that both challenge and deepen our understanding of the prevailing processes.

4 Visualization

Visualization means more than generating fancy graphs for presentations or papers. Visualization makes use of the most powerful interface between the data stored in a computer and the human brain. The latter is very good in extracting relevant information from large data sets. Within less than a second information from 250 million sensory cells in the man's eyes, that is, a 250 million dimensional

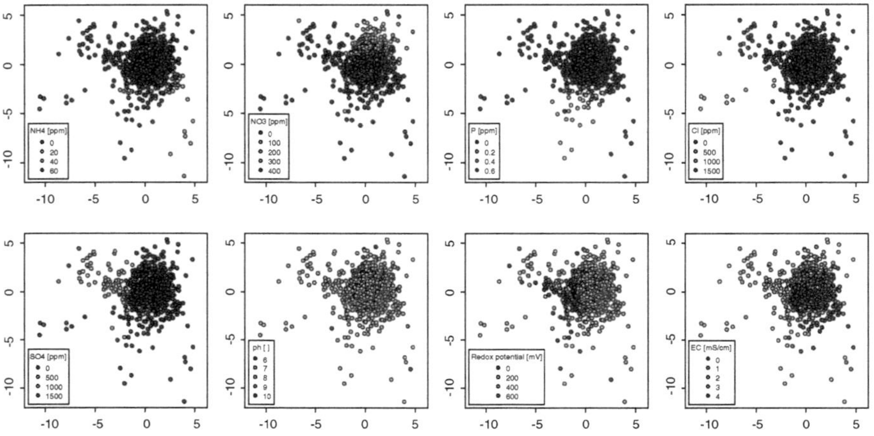

Fig. 9 Measured values of the samples, represented by the SOM-SM

data set, is reduced to a handful of relevant objects. Without this skill man would not have been able to escape predators, enemies, or other road users, and to hunt prey or to find food.

Efficient visualization can help a lot in identifying clusters and outliers, in analysing multivariate trends, or to detect any peculiarities in large multivariate data sets. Here the SOM-SM approach is presented and is applied to the same data set as the Isometric Feature Mapping. It is a combination of Sammon's Mapping (SM; Sammon 1969) and Self-Organizing Maps (SOM), which have been termed Kohonen Feature Mapping as well (Kohonen 2001). Again, the reader is referred to the literature (Sammon 1969; Kohonen 2001; Lischeid 2009) for further details. Here only the application to environmental monitoring data is demonstrated.

The SOM-SM approach yields a two-dimensional graph where every symbol denotes a single sample of the data set. The same graph is used in various ways, where colours indicate different features of the data (Fig. 9). In contrast, the location of the symbols remains the same. The symbols have been allocated in an iterative procedure to ensure that the distance between any two symbols in the graph is proportional to the similarity of the respective two samples with respect to all measured variables. In this study, the two-dimensional graph depicts 89 % of the variance of the data set with eight measured variables. Please note that the scaling at the axes is given for orientation only but has nothing to do with measured data, location of the sampling site in the Quillow catchment, or alike. It is an entirely arbitrary configuration, except for the above stated correlation between distance and similarity.

In a certain sense, any of the graphs given in Fig. 9 can be interpreted analogously to a topographical map where the colour indicates the altitude, with numerous mountain tops and valleys. Similar to a topographic map, a rather smooth "landscape" develops where neighbour points exhibit roughly the same values. However, in these graphs measured values for the samples are given rather than topographic height. Moreover, different "maps" are shown for different measured variables but for the same symbols, that is, the samples of the data set (Fig. 9).

Ammonium concentration is very low and close to the detection level for most samples (Fig. 9). Only samples that plot in the lower right corner exhibit higher NH_4 concentration values. In contrast, NO_3 concentration of these samples is very low, and the highest NO_3 concentration is found in the upper right corner of the graph. Samples with enhanced P concentration plot close to those of high NH_4 concentration, but are not identical. Similarly, Cl and SO_4 concentration and electric conductivity exhibit similar patterns in the SOM-SM with the highest values on the left side (Fig. 9). The pattern is less clear for pH values. There is a slight tendency of increasing pH values from the top to the bottom of the graph. For the redox potential values increase from the left to the right.

These patterns reflect the correlation structure of the data set: Variables with similar patterns, that is, same directions of steepest increase of measured values, are positively correlated, like Cl, SO_4 and EC (cf. Table 1). In contrast, the directions of steepest increase of P and Cl are approximately perpendicular to each other, that is, the correlation is close to zero (Fig. 9, cf. Table 1). The gradients for P and NO_3 are inverse to each other, which is symptomatic for a negative correlation.

Similar to the measured variables the Isomap scores can be represented in the SOM-SM graph (Fig. 10). There is a more or less monotonic increase from the bottom right to the upper left corner for the first component, which corresponds roughly to that of SO_4, Cl and EC which in fact are closely related to this component (Fig. 2). In contrast, the scores of the second component increase from the bottom to the upper right for the second component, which is parallel or inverse to those of the redox-dependent variables (Fig. 4). Please note that the Isomap scores have not been used for generating the SOM-SM. Rather, organisation of the samples within the SOM-SM makes use of the internal relationships of the data set, similar to that of the Isomap analysis.

The SOM-SM graph can be used for analysing spatial and temporal patterns as well. In Fig. 11 samples from single sites are colour-coded. Although there is some

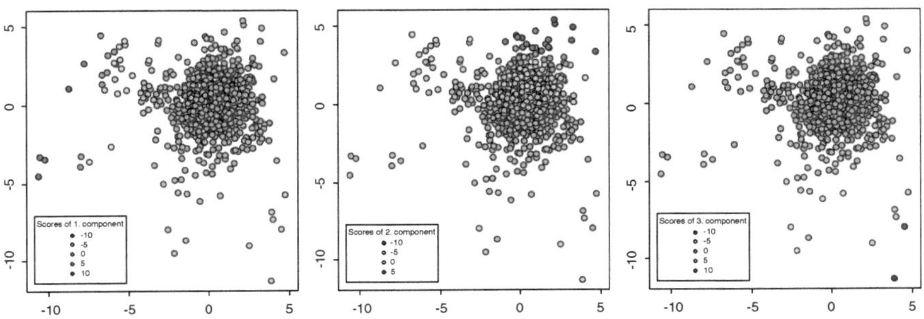

Fig. 10 Isomap scores of the samples, represented by the SOM-SM

overlap, samples of different sites usually cluster in different sub-regions of the graph. E.g., the C6 samples are located in the upper part of the graph, those of site FeS more to the lower right, and those of K10 in the upper left corner. These patterns can be related to those of the measured variables (Fig. 8) or of the Isomap scores (Fig. 9). Moreover, the temporal variance can be investigated for single sites separately. For example, the samples from the MK6 site spread over a larger range compared to those of C6 (Fig. 11). In addition, the MK6 samples scatter more in a direction parallel to the gradient of the second Isomap component rather to that of the first component (cf. Fig. 9). Thus, for this site changes of the redox conditions play a larger role compared to the process which is represented by the first component.

At the K8 and at the C6 site some samples plot far away from the centre of the respective clusters. These "outliers" should be checked for possible erroneous assignment to the sampling site. As an alternative, these deviations from the centre of the respective clusters could be due to extreme conditions during or immediately prior to sampling. This is a common phenomenon especially for stream and river water samples taken during major floods, where deviations from the respective clusters occur roughly in parallel at different sites.

Colour coding in Fig. 11 indicates the year of sampling. This allows investigating trends in the SOM-SM graph, that is, systematic shifts over time. Please note that the SOM-SM represents information about all measured variables. Thus any clear shift in the graph usually is due to a (possibly non-linear) trend of more than one variable. Figure 11 indicates trends only at some of the sites, and none of these is linear. At K8 and K9, the first samples plot close to the right border, those of subsequent years more to the upper left, and those of 2004 more close to the

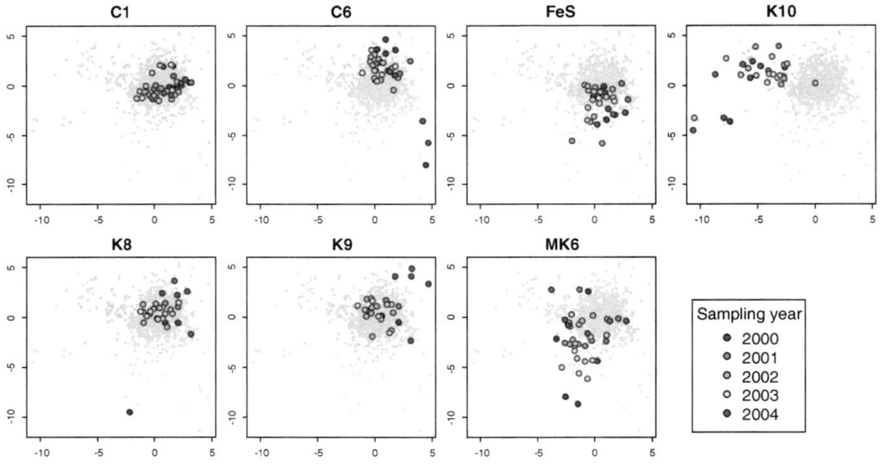

Fig. 11 Time series of soil solution quality at selected sites, represented by the SOM-SM. Samples of the respective site as indicated in the title are *coloured* according to sampling year, all remaining samples are given in *grey*

upper right corner. Thus, this non-linear trend firstly follows an increase of the scores of the first component, and then an increase of those of the second component (cf. Fig. 9), indicating less reduced conditions. In contrast, samples at C6 follow more a spiral from the lower right in anti-clockwise direction and with decreasing radius towards 2004. Correspondingly, the data could be tested for seasonal patterns etc. (not shown).

5 Conclusions

Managing water and soil resources requires sound data that can only be provided by systematic monitoring (Muller 2012). Advanced statistical approaches have been developed to make more efficient use of large data sets in spite of substantial spatial and temporal heterogeneities that are immanent to natural systems. Here only two out of a vast zoo of different approaches have been presented. Natural water resources management as well as science could and should benefit a lot from a variety of approaches that have been developed in other disciplines.

Dimensionality reduction techniques have been applied to time series of groundwater head, lake and river water levels, and stream discharge (Longuevergne et al. 2007; Lewandowski et al. 2009; Lischeid et al. 2010; Lischeid and Kalettka 2012; Thomas et al. 2012). They were successful in differentiating between different effects on the observed dynamics at minor data requirements. The author is firmly convinced that much more can be learned from data by advanced data analysis techniques and will be happy to assist in any related efforts.

References

Fernandes PG, Carreira P, da Silva MO (2006) Identification of anthropogenic features through application of principal component analysis to hydrochemical data from the Sines coastal aquifer, SW Portugal. Math Geol 38:765–780

Gámez AJ, Zhou CS, Timmermann A, Kurths J (2004) Nonlinear dimensionality reduction in climate data. Non-linear Process Geophys 11:393–398

Gupta AK, Sinha S, Basant A, Singh KP (2006) Multivariate analysis of selected metals in agricultural soil receiving UASB treated tannery effluent at Jajmau, Kanpur (India). Bull Environ Contam Toxicol 79:577–582

Haag I, Westrich B (2002) Processes governing river water quality identified by principal component analysis. Hydrol Process 16:3113–3130

Kohonen T (2001) Self-organizing maps. Springer Series in Information Sciences, vol 30, 3rd edn. Springer, Berlin

Langer U, Rinklebe J (2009) Lipid biomarkers for assessment of microbial communities in floodplain soils of the Elbe River (Germany). Wetlands 29:353–362

Lee JA, Verleysen M (2007) Nonlinear dimensionality reduction. Information science and statistics. Springer, Berlin

Lewandowski J, Lischeid G, Nützmann G (2009) Drivers of water level fluctuations and hydrological exchange between groundwater and surface water at the lowland River Spree (Germany): field study and statistical analyses. Hydrol Process 23:2117–2128. doi:10.1002/hyp.7277

Lischeid G (2009) Non-linear visualization and analysis of large water quality data sets: a model-free basis for efficient monitoring and risk assessment. Stoch Env Res Risk Assess 23:977–990. doi:10.1007/s00477-008-0266-y

Lischeid G, Bittersohl J (2008) Tracing biogeochemical processes in stream water and groundwater using nonlinear statistics. J Hydrol 357:11–28. doi:10.1016/j.jhydrol.2008.03.013

Lischeid G, Kalettka T (2012) Grasping the heterogeneity of kettle hole water quality in Northeast Germany. Hydrobiologia 689(1):63–77. doi:10.1007/s10750-011-0764-7

Lischeid G, Natkhin M, Steidl J, Dietrich O, Dannowski R, Merz C (2010) Assessing coupling between lakes and layered aquifers in a complex Pleistocene landscape based on water level dynamics. Adv Water Resour 33:1331–1339. doi:10.1016/j.advwatres.2010.08.002

Longuevergne L, Florsch N, Elsass P (2007) Extracting coherent regional information from local measurements with Karhunen-Loève transform: case study of an alluvial aquifer (Rhine valley, France and Germany). Water Resour Res 43:W04430. doi:10.1029/2006WR005000

Mahecha M, Martínez A, Lischeid G, Beck E (2007) Nonlinear dimensionality reduction as a new ordination approach for extracting and visualizing biodiversity patterns in tropical montane forest vegetation data. Ecol Inf 2:138–149. doi:10.1016/j.ecoinf.2007.05.002

Muller M (2012) From raw data to informed decisions. In: WWAP (World Water Assessment Programme): The United Nations World water development report 4: Managing water under uncertainty and risk. Paris, UNESCO, Chap. 6, pp 157–173

Sammon JW (1969) A nonlinear mapping for data structure analysis. IEEE Trans Comput C-18/5:401–409

Schilli C, Lischeid G, Rinklebe J (2010) What processes prevail? Analyzing long-term soil-solution monitoring data using nonlinear statistics. Geoderma 158:412–420. doi:10.1016/j.geoderma.2010.06.014

R Development Core Team (2006) R: A language and environment for statistical computing. R Foundation for statistical computing. Vienna, Austria. ISBN: 3-900051-07-0, http://www.Rproject.org

Tenenbaum JB, de Silva V, Langford JC (2000) A global geometric framework for non-linear dimensionality reduction. Science 290:2319–2323

Thomas B, Lischeid G, Steidl J, Dannowski R (2012) Regional catchment classification with respect to low flow risk in a Pleistocene landscape. J Hydrol 475:392–402. doi:10.1016/j.jhydrol.2012.10.020

Thyne G, Guler C, Poeter E (2004) Sequential analysis of hydrochemical data for watershed characterization. Ground Water 42:711–723

Weyer C, Lischeid G, Aquilina L, Pierson-Wickmann A-C, Martin C (2008) Mineralogical sources of the buffer capacity in a granite catchment determined by strontium isotopes. Appl Geochem 23:2888–2905

Zhang HB, Luo YM, Wong MH, Zhao QG, Zhang GL (2007) Concentrations and possible sources of polychlorinated biphenyls in the soils of Hong Kong. Geoderma 138:244–251

Using Soil–Water–Plant Models to Improve the Efficiency of Irrigation

Rickmann Michel and Ralf Dannowski

Abstract Crop irrigation is just as essential to agriculture in Central Asia as it is in parts of Eastern and Central Europe. Model-based tools for computerised irrigation scheduling have been under development for several decades in northeast Germany. Having been tested, they are now operational as part of the on-site farm irrigation management process. One of these, the ZEPHYR soil–water–plant model, is presented here and is evaluated against the practical requirements for improving irrigation water use efficiency, meeting crop demand for water and preventing soil salinisation in the context of an irrigated barley crop in southern Kazakhstan.

Keywords Soil · Plant · Irrigation · ZEPHYR model

1 Introduction

Rainfall during the growing season is a crucial parameter for rainfed agriculture worldwide. As a rule of thumb, 20 in. (508 mm) of rain per year is regarded as the minimum requirement for non-irrigated field crops. This threshold differs widely depending on factors such as rainfall patterns, latitude, landscape morphology, geological conditions, soils, farming systems amongst others. The average amount of rainfall over the growing season, however, from the beginning of March

R. Michel (✉)
Ingenieurbüro BODEN und BODENWASSER, Gesundbrunnenstraße 24, 16259 Bad Freienwalde, Germany
e-mail: rjmichel@t-online.de

R. Dannowski
Leibniz Centre for Agricultural Landscape Research (ZALF) e. V., Eberswalder Strasse 84 15374 Muencheberg, Germany
e-mail: danno@zalf.de

L. Mueller et al. (eds.), *Novel Measurement and Assessment Tools for Monitoring and Management of Land and Water Resources in Agricultural Landscapes of Central Asia*, Environmental Science and Engineering, DOI: 10.1007/978-3-319-01017-5_22,

through to the end of September, contrasts quite dramatically across Eurasia, from between less than 100 mm in some regions of southern Kazakhstan to about 350 mm, for example, in the Berlin-Brandenburg region of northeast Germany. A similar range is also observed for the rates of potential evapotranspiration during the growing season (nearly 1,000 and 500 mm, respectively), thus sharpening the contrast mentioned in terms of water availability for agriculture. It follows that the interpretation of irrigation in plant production must also differ greatly between these two parts of Eurasia. Consequently, irrigation can be regarded as being essential to the very survival of agriculture in southern Kazakhstan, whereas, in northeast Germany it can bring about improvements in yield stability and crop productivity.

Furthermore, because of the impact of climate change, agricultural production is likely to be increasingly under the threat of drought over the next decades in both Kazakhstan and Germany. Even lower rainfall levels and even higher evapotranspiration could lead to an additional demand for water. It has been estimated that up to the year 2055, at two investigated potato crop sites in north Germany, for example, this could probably equate to between 24 % (24 mm) and 49 % (53 mm) (Müller 2007). This would increase the demand for potato irrigation water from what is currently about 100–110 mm per year up to 125–160 mm per year in 2055. At the same time, it is anticipated that the amount of water available for irrigation will decrease due to climate change in future.

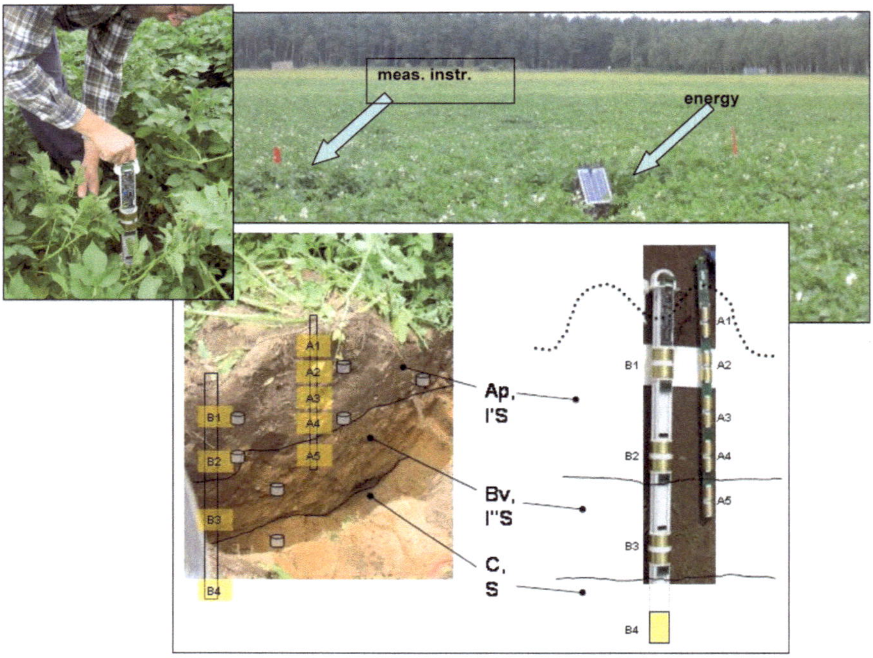

Fig. 1 Measuring the soil water content (*example* sandy soil, potatoes)

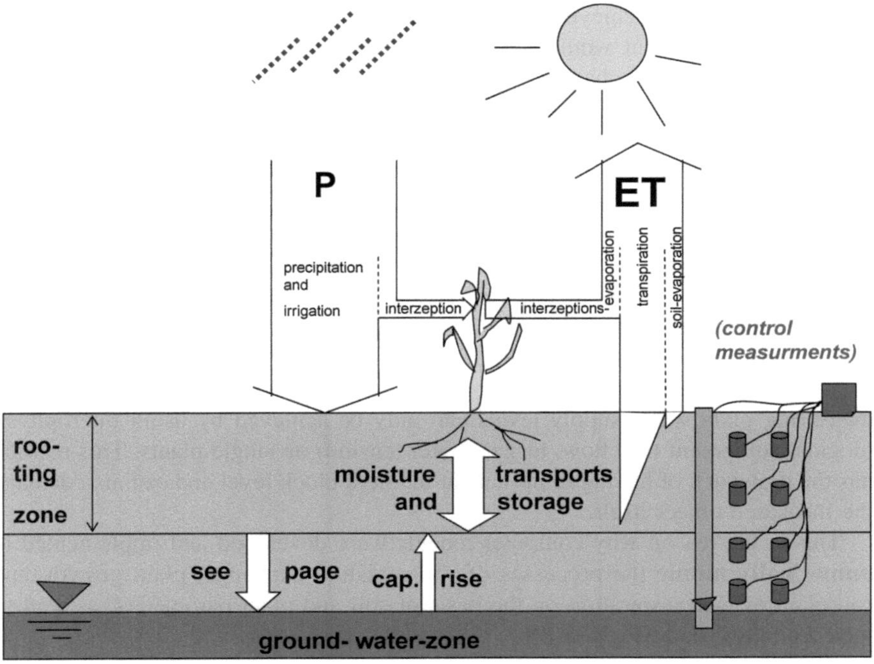

Fig. 2 Principle of modelling the site-specific exchange of water using numerical models for soil–water–plant systems

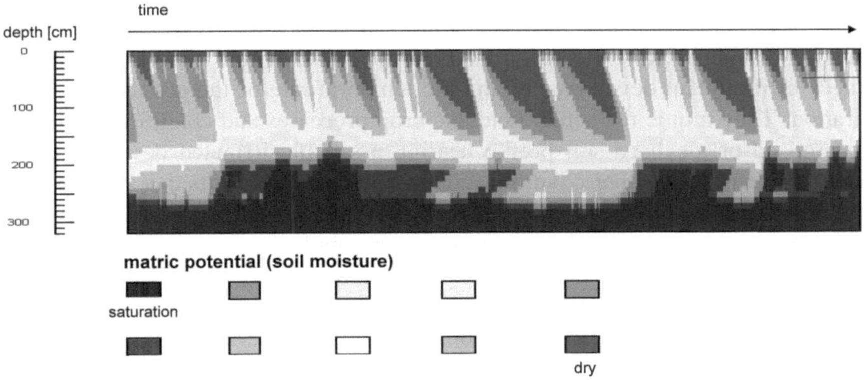

Fig. 3 Propagation of the soil moisture over time, calculated using a numerical model for soil–water–plant systems

Therefore, for both Kazakhstan and Germany, there is an increasing urgency for less water to be used in irrigating field crops by upgrading water use efficiency. As a precondition, the amount of irrigation water must be defined as that which is needed to achieve acceptable yields and that exceeding it would not increase the yields but would merely stress the water resources. This threshold value is dynamic and changes according to the moisture of the soil and plant development over the vegetation period. Thus, the level of demand for water over the growing season needs to be continuously updated. This means that it is always necessary to know the soil moisture status and the level of plant water supply.

In principle, there are two types of methods for monitoring soil moisture levels. Those that **measure** the soil water content or the soil matric potential have the advantage that these parameters can be directly observed in the field (Fig. 1). However, practical experience has revealed that they have some shortcomings such as inaccurate instruments; long-term instabilities; spatial variations in soil conditions and external influences around the probes. However, the alternative of measuring plant water supply levels can only be achieved by using microphysiological equipment (sap flow, turgor, water tension) on single plants. This method has the drawback of having to upscale at the field block level and can also damage the measured object itself.

This is the reason why computer models were developed and implemented to **numerically mimic** the processes of soil moisture exchange, plant growth, and soil/plant evapotranspiration on the basis of soil and plant parameters, as well as meteorological data (Figs. 2, 3).

2 The ZEPHYR Model for Soil–Plant–Water Exchange

Numerical water exchange models for the soil–plant system are capable of calculating or interpolating the spatial and temporal distribution of soil moisture, the stages of root/plant development, and the temporal rate of the actual evapotranspiration. They mimic the actual field and crop processes and conditions through a generalised or idealised soil profile bearing a specific irrigated developing crop under the current weather conditions. Thus site-specific results can be calculated and compared as a proxy for the actual irrigated plot using the criteria of optimum and critical boundaries of the irrigation water supply, depending on the specific soil properties and the plants' actual demand for water. The ZEPHYR model (Michel et al. 2009) presented here was specially developed for on-site farm irrigation scheduling. It takes into account the soil characteristics; the development of the plant cover (from the pool of plant-specific data, interactively adjusted to actual conditions); daily meteorological data; irrigation activities and the depth of the groundwater table.

The basis for modelling soil water exchange, movement and storage is derived from the well-known Richards equation for soil moisture dynamics in the unsaturated zone. The effects of plant roots and transpiration on the vertical moisture distribution in a soil column are calculated using the approach by Koitzsch and Günther (1990). The temporal advance of the transpiration coefficient, the rooting

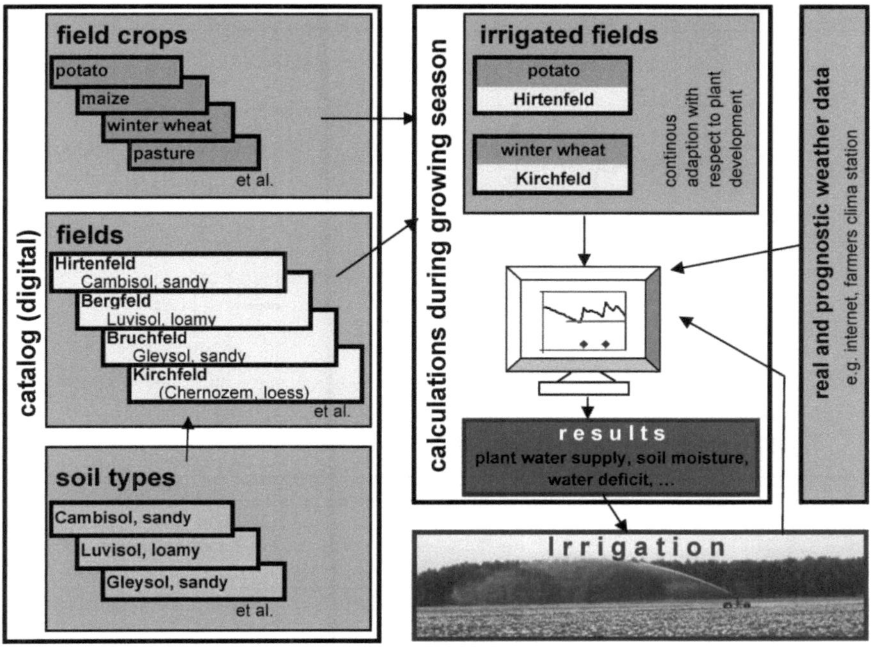

Fig. 4 Principle of calculating the soil moisture dynamics, water deficit and plant water supply in a soil profile with models of soil–plant–water exchange in the context of on-farm irrigation management (ZEPHYR model)

depth and the areal ratio of the plant cover are taken into account. Potential evapotranspiration is calculated after Penman, an equation that has been accepted internationally. It is interesting to note that because of this, Allen et al. (1998) have suggested Penman–Monteith as the FAO-56 PM standard method, as adapted by Wendling for northeast Germany (Wendling et al. 1991).

The ZEPHYR model was thoroughly validated, tested and successfully applied to a variety of cropping systems under sprinkler and droplet irrigation and to different types of farms under German climatic and agricultural conditions. Both farmers and consultants giving advice on farm irrigation management can use it easily. Figure 4 demonstrates the sequence of calculations and the implementation of the model into irrigation scheduling and management.

The main output of the model comprises the following information which is useful for determining the soil water status and irrigation requirements (see also Fig. 5):

- the soil water content of different soil layers and the rooting depth as a whole (current and prognostic);
- the matric potential at different soil layers;
- characteristics of the physiological stress for plants from evapotranspiration;

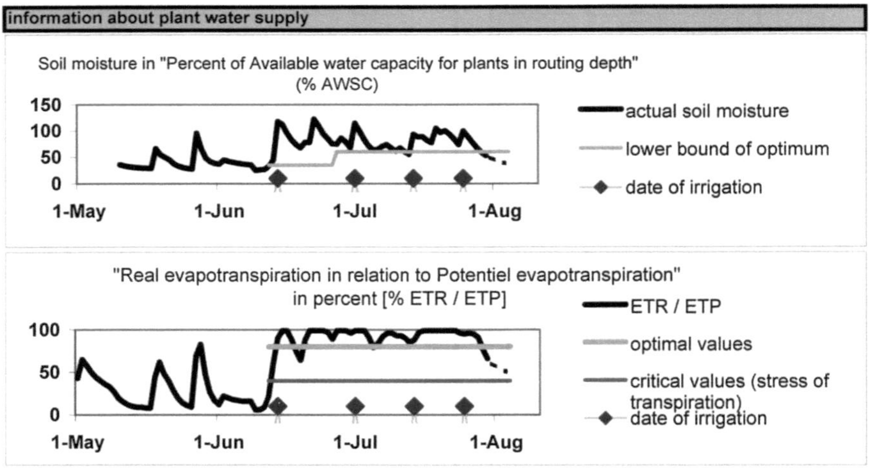

information for planing amount of irrigation water

			amount of irrigation water sufficient to obtain a plant available water capacity in routing depth of ... 80% AWSC	100% AWSC
actual	today	30-Jul	18 mm	31 mm
prognostic	tomorrow	31-Jul	22 mm	35 mm
	day after to.	01-Aug	25 mm	38 mm
	in 3 days	02-Aug	27 mm	40 mm
seepage downward out of routing zone today			0,3 mm/ day	

Fig. 5 Selected results of ZEPHYR calculations for assessing the site-specific soil moisture status and to derive recommendations for farm irrigation scheduling and water quantities

- the total deficit of water in the root zone to recommend the optimum amount of water for irrigation;
- capillary rise from groundwater and the seepage rate (criteria for determining the water use efficiency).

3 First Experience of the ZEPHYR Application in Kazakhstan

The conditions at the site in Kazakhstan are quite particular and are very distinct from those in northeast Germany. This meant that the ZEPHYR model had to be thoroughly tested for this environment prior to its practical application. As mentioned above, in the period between 1 March and 30 September, rainfall in southern Kazakhstan is only about 85 mm; in contrast, the potential evapotranspiration sum rises to around 1,100 mm.

In the first test, the temporal advance of the soil moisture content, actual evapotranspiration and the soil water deficit were calculated for an optimum application of irrigation water for a number of crops, as shown in the example for spring barley in Fig. 6.

Sprinkler irrigation is, however, just one of the standard irrigation techniques used in Kazakhstan. Another method is furrow irrigation, practised by flooding the field in stages. With the aid of a computer model, it is possible to carry out prognoses of water behaviour while the field is flooded. This may help to optimise the flooding regime in order to avoid or reduce water loss both by runoff and leaching (seepage). Figure 7 demonstrates the calculated water balance levels during an irrigation event of 150 mm on the first day when the ground was inundated.

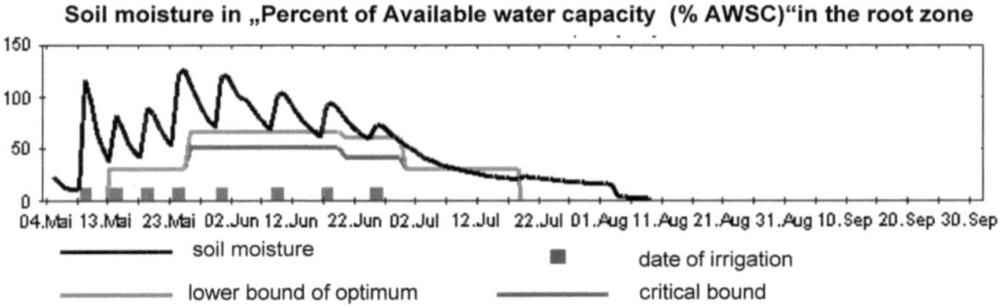

Fig. 6 Advance of the soil moisture content under a crop of spring barley in south Kazakhstan

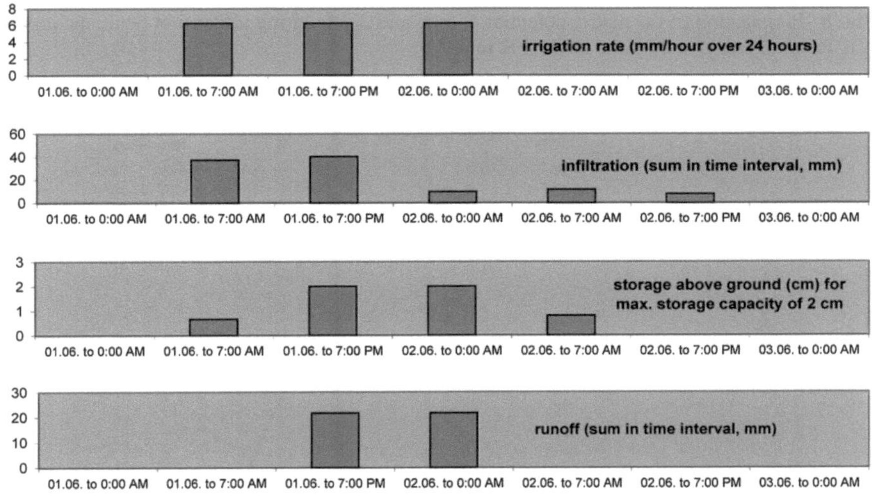

Fig. 7 Water balance levels (infiltration, storage above the ground, runoff) during an irrigation event with an extremely high initial irrigation rate (150 mm/day)

4 Controlling the Tendency of Soil Salinisation

In regions characterised by particularly high evapotranspiration rates, such as south Kazakhstan, irrigation is often connected with the problem of salinisation of soils. This explains why there is a need to control the direction of water movement through the soil. Downward water movement should unconditionally prevail to avoid capillary rise, which otherwise would inevitably be coupled with salt uplift. Figure 8, produced using ZEPHYR again, shows the propagation of the matric potential within the soil during infiltration in a period of flood. The goal of stabilising the predominantly downward water movement during the whole process was achieved.

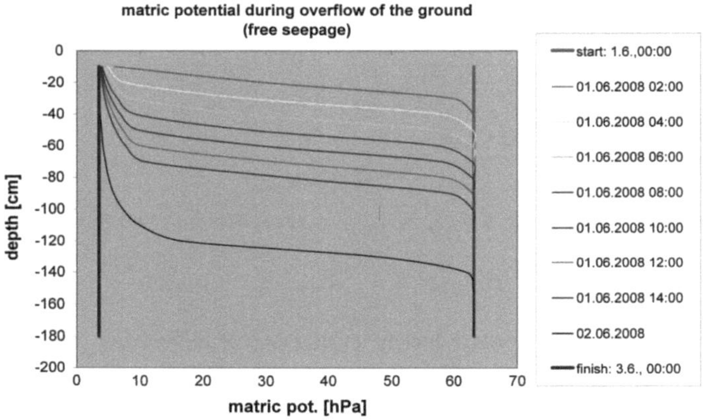

Fig. 8 Propagation of the matric potential in time and depth during infiltration while the ground was flooded, calculated by the ZEPHYR model

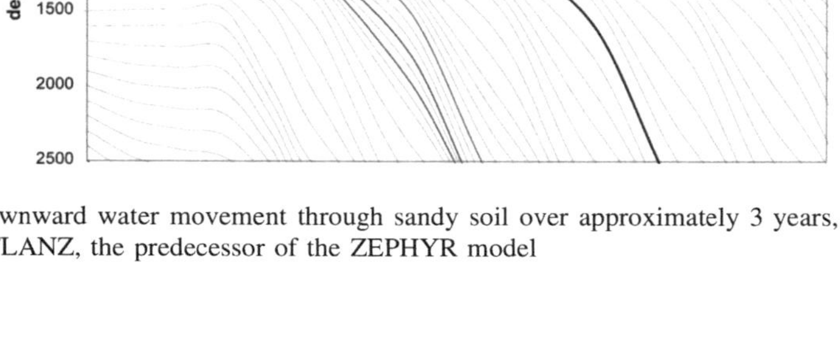

Fig. 9 Downward water movement through sandy soil over approximately 3 years, calculated using SIPFLANZ, the predecessor of the ZEPHYR model

However, more important than observing short-term processes is assessing the long-term behaviour of the soil water in order to identify the salinisation risk and to prevent this phenomenon from occurring. Soil–plant moisture exchange models are capable of calculating the direction of water movement within the soil. Thus, the infiltration fronts presented in Fig. 9, for example, also confirm the prevailing downward water movement through the soil.

5 Conclusions

Soil–water–plant models such as the ZEPHYR model described here are useful tools for improving irrigation water efficiency as they continuously mimic the soil water status under irrigated crops. In this study, comparisons based on regional conditions and field characteristics, have been drawn between sites in south Kazakhstan and northeast Germany and conclusions have been made. In these projects, spring barley was the crop and meteorological data was measured as input data. Recommendations for assessing the required amounts of water; on-site irrigation scheduling and evaluating water use efficiency of furrow irrigation have been made. In addition, as the first results for controlling salinisation were described, discussed and evaluated in anticipation of worsening conditions of severe water scarcity in large parts of Eurasia. The tried and tested technique of computer-aided irrigation scheduling is also regarded as being very promising for the arid southern part of Kazakhstan, provided the required digital information is made available from field surveys, as well as from continuous and reliable meteorological measurements.

References

Allen RG, Pereira LS, Raes D, Smith M (1998) Crop evapotranspiration: guidelines for computing crop water requirements. Irrigation and Drainage Paper No. 56. Rome (Italy): FAO

Koitzsch R, Günther R (1990) Modell zur ganzjährigen Simulation der Verdunstung und der Bodenfeuchte landwirtschaftlicher Nutzflächen mit und ohne Bewuchs. Arch. Acker- u. Pflanzenbau u. Bodenkd., Berlin 34/12, 803.810

Michel R-J, Pfaff M, Müller J, Hoffmann F (2009) Möglichkeiten und Grenzen, sowie Anforderungen bei der Modellierung des Wasserhaushaltes von Pflanzenstandorten. In: Kage H et al. "Modellierung des Systems Nutzpflanze–Boden – Herausforderungen des 21. Jh". Ber. Ges. Pflanzenbauwissenschaften, 4(2009), pp 13–17

Müller J (2007) Development of irrigation demand of potato in Northern Germany due to climatic changes. Rostock University, p 79

Wendling U, Schellin H-G, Thomä M (1991) Bereitstellung von täglichen Informationen zum Wasserhaushalt für die Zwecke der agrarmeteorologischen Beratung. Z Meteorol 41:468–475

MONICA: A Simulation Model for Nitrogen and Carbon Dynamics in Agro-Ecosystems

Claas Nendel

Abstract Process-based simulation models that predict crop growth, evapotranspiration, nitrate leaching or other environmental variables are commonly applied for impact assessment on agricultural crop production or the environment. Algorithms of the dynamic, process-based simulation model MONICA are presented, which was developed for demonstrating the climate and management impact on crop yields and environmental variables on the plot scale and in smaller regions in Central Europe. A legal successor of the HERMES model, it maintains the simple and robust philosophy of its progenitor and adds a full carbon cycle model to it, including the feedback relations between atmospheric CO_2 concentration and other environmental variables on crop growth and water use efficiency. MONICA is the central part of a web-based decision support system that helps farmers and other stakeholders in Germany identifying management options to mitigate the impact of the expected climate change on their business. MONICA has the potential to assess the impacts of climate change and land management on crop yields, carbon balance and nitrogen efficiency in Central Asia.

Keywords Agro-ecosystems · Carbon · Nitrogen · MONICA model

1 Objectives

In many regions of the world, soil and water are scarce resources. The productivity of crops for food or biomass for grazing and the provision of clean water for consumption and irrigation rely on the availability of manageable soils and fresh water input via precipitation or rivers. The protection of these resources is or

C. Nendel (✉)
Leibniz Centre for Agricultural Landscape Research (ZALF) e. V., Eberswalder Strasse 84
15374 Muencheberg, Germany
e-mail: nendel@zalf.de

L. Mueller et al. (eds.), *Novel Measurement and Assessment Tools for Monitoring and Management of Land and Water Resources in Agricultural Landscapes of Central Asia*, Environmental Science and Engineering, DOI: 10.1007/978-3-319-01017-5_23,

should be first priority. Soils are threatened by erosion, compaction, salinisation, contamination and land-use change; factors, which in many cases also affect the availability of water. The projected climate change is expected to have a considerable impact on agricultural crop production (Lobell et al. 2008) and makes future food security for the steadily growing population uncertain. The nature and dimension of the impact, however, are expected to differ greatly in many regions across world (IPCC 2007). Global circulation models are applied to make climate predictions (Lucarini et al. 2007) and down-scaling methods were developed that determine local weather formation (Christensen et al. 2007; Déqué et al. 2007; Enke and Spekat 1997). Finally, statistical or process-based simulation models are used to assess the impacts of climate and management on crop growth, yield formation or other environmental variables, such as nitrate leaching or soil organic matter development (Reidsma et al. 2010). The variables are chosen depending on the size of the area to which they are applied and the issues they are expected to shed light on (Trnka et al. 2004; Wessolek and Asseng 2006). Statistical models that apply regression (Bachinger and Reining 2009; Lobell et al. 2008; Paeth et al. 2008; Potgieter et al. 2005), fuzzy rules (Bardossy et al. 2003; Shrestha et al. 2007) or neural network (Alvarez 2009; Cossani et al. 2007; Wieland and Mirschel 2008) approaches often yield robust and reliable results, capturing most of the local weather—crop relations. The predictions they yield are trustworthy as long as future processes are likely to be governed by the same relations. As soon as feedback processes lead to deviating crop responses to weather, statistical models will fail. Another drawback is that almost all statistical models offer only one output variable (e.g. yield), which then cannot be evaluated in the light of another. Process-based models simultaneously explain the dynamics of a large set of variables and their interactions, and account for feedback relations. Process-based models, on the other hand, are often site-sensitive and thus prone to produce unrealistic results when applied to different locations without prior calibration (Asseng et al. 2012). Many process-based simulation models are available, each of which has its own different strengths and weaknesses (Kersebaum et al. 2007; Rötter et al. 2012).

In this chapter, we present the algorithms of the dynamic, process-based simulation model MONICA (Nendel et al. 2011), developed for demonstrating the climate and management impact on crop yields and environmental variables on the plot scale and in smaller regions in Central Europe; a task recently identified by Reidsma et al. (2010) or Falloon and Betts (2010) as being not yet sufficiently pursued. Since the model is complex, we decided to highlight only the most significant changes to its progenitor model, HERMES (Kersebaum 1989). For all sets of algorithms described in detail elsewhere and borrowed without major alteration from its original source, the reader is referred to the respective publications.

2 Description of the Model

2.1 The Model's Development

The MONICA model (**MO**del for **NI**trogen and **C**arbon dynamics in **A**gro-ecosystems) is the latest generation of HERMES (Kersebaum 1995, 2007) model versions. Extended for the carbon cycle in soil and plant, the recent software marks a principal alteration to the model's code and functionality, hence its new name. Like its progenitor, the model was designed to simulate crop growth, water and nitrogen uptake, and the matter dynamics in the soil for applied purposes. The newly introduced carbon algorithms now allow for simulations of CO_2 impact on crop growth, the effects of climate and management on long-term organic matter dynamics in soil and, in turn, feedback signals to soil physical properties and conditions for crop production. The simple and robust philosophy followed by the HERMES model was maintained, enabling MONICA to operate under restricted data availability. MONICA is a 1D point model which works on a pedon with variable depth (default: 2 m). Results are interpreted for an area of 1 m^2. The temporal resolution for MONICA is 1 day, which is automatically split into smaller time steps when high water and nutrient fluxes occur to ensure stable numerics. The overall module concept is presented in Fig. 1.

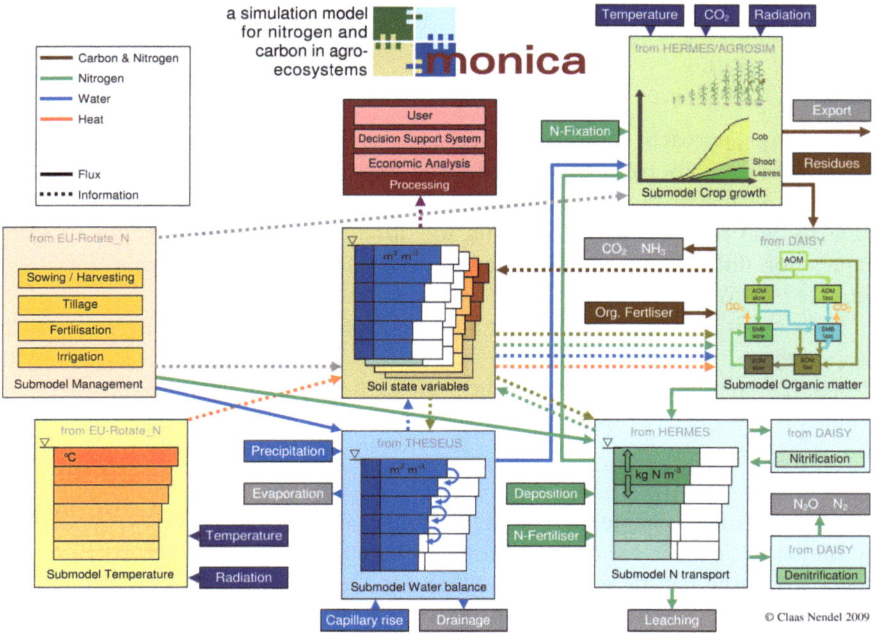

Fig. 1 The MONICA model concept

2.2 Soil Temperature

The soil temperature model was borrowed from Lasch et al. (2002). The approach uses an empirical function for heat conductivity λ_h originally developed by Neusypina (1979).

$$\lambda_h = \frac{3.0 \cdot \rho_B - 1.7}{1.0 + (11.5 - 5.0 \cdot \rho_B) \cdot e^{\left(\frac{-50 \cdot \theta}{\rho_B}\right)^{1.5}}} \cdot 418.4, \tag{1}$$

where the soil bulk density is ρ_B in [1,000 kg m^{-3}] and the constant soil moisture content is θ.

Soil surface temperature T_s is calculated from the minimum (T_{min}) and maximum (T_{max}) air temperature and global radiation (R_g) according to Williams et al. (1984).

$$\begin{aligned} T_s(t) = &(1 - \alpha(t)) \cdot \left(T_{\min}(t) + (T_{\max}(t) - T_{\min}(t)) \cdot \sqrt{0.03 \cdot R_g(t)}\right) \\ &+ (\alpha(t) \cdot T_s(t - \Delta t)), \end{aligned} \tag{2}$$

where the surface albedo is $\alpha(t)$ at time step t and the soil surface temperature at the previous time step is $T_s(t\text{-}\Delta t)$. The impact of R_g is limited to values above 8.33 MJ m^{-2} d^{-1}.

2.3 Soil Water

A capacity approach was used to describe soil water dynamics (Wegehenkel 2000). The capacity parameters are derived from the soil texture, and modified by soil organic matter content and bulk density. Water contents at saturation, field capacity and permanent wilting point for different bulk densities and correction values for different soil organic matter classes (Ad-hoc-AG Boden 2005) are provided by Wessolek et al. (2009) and stored in the database.

If a crop is present, precipitation is partly intercepted and evaporates from the crop surface.

The percolation of water volumes above field capacity is governed by an empirical, texture-dependent rate coefficient (λ).

$$\lambda = 1.15 \cdot f_S^2 + 0.1 \cdot f_C + 0.35 \cdot f_U, \tag{3}$$

where f_S, f_C and f_U denote the sand, clay and silt fractions of the soil texture. If the groundwater level is located within the simulated soil profile, constant groundwater discharge can be adjusted to allow for the rising and falling groundwater level, depending on the soil water balance. The capillary rise from groundwater is considered according to empirical ascent rates (Ad-hoc-AG Boden 2005). The groundwater level oscillates between given maximum and minimum levels with a period of one year.

Reference evapotranspiration ET_0 is calculated using the Penman–Monteith method, according to Allen et al. (1998). This method requires the diurnal minimum and maximum temperature, the water vapour pressure deficit, wind velocity and total global radiation.

$$ET_0 = \frac{0.408 \cdot \Delta \cdot (R_n - G) + \gamma \cdot \frac{900}{T+273} \cdot u_2 \cdot (e_s - e_a)}{\Delta + \gamma \cdot \left(1 + \frac{r_a}{r_s}\right)}, \tag{4}$$

where

$$\gamma = 6.65 \cdot 10^{-4} \cdot P \tag{5}$$

and Δ denotes the slope of the vapour pressure curve, R_n is the net radiation at the crop surface, G is the soil heat flux density, T is the mean daily air temperature at 2 m height, u_2 is the wind speed at 2 m height, e_s is the saturation vapour pressure, e_a is the actual vapour pressure, γ is the psychrometric constant, r_s is the surface resistance, r_a is the atmospheric resistance and P is the atmospheric pressure. The surface resistance for the reference evapotranspiration assumes a 12 cm cut grass crop, and is calculated using

$$r_s = \frac{r_m}{1.44}, \tag{6}$$

where r_m is the stomata resistance (Allen et al. 1998). The latter calculated in the crop growth module (see below). For all other variables in Eq. 4, the reader is referred to Allen et al. (1998).

Crop-specific potential evapotranspiration is calculated using crop-specific factors (K_c) during the growing season and bare soil factors between harvest and crop emergence. The K_c factors are linked to the crop's developmental stages. Partitioning between evaporation and transpiration is a function of the degree of canopy closure β, which in turn is calculated from LAI.

$$\beta = 1 - e^{-0.5 \cdot LAI} \tag{7}$$

The calculation of actual evaporation and transpiration considers the soil water status and vertical root distribution (Pedersen et al. 2010).

2.4 Snow Cover and Frozen Soil Water

Snow covers are simulated following Riley and Bonesmo (2005). Soil freezing algorithms were taken from Olsen and Haugen (1997), assuming uniform thermal properties throughout the profile. Heat conductivity is calculated using a modified version of Eq. 1. Parameter values are taken from the SOIL model (Jansson 1991).

The module requires the input of the surface temperature, as modified by the snow pack. The approach used for thawing is taken from the ECOMAG model (Motovilov et al. 1999). Both freezing and thawing processes have been validated for conditions in Norway and Germany. For snow and frost calculation routines it is assumed that infiltration ceases when the soil freezes. If the soil surface is frozen, it is assumed that precipitation is either stored in the snow pack, if present, or lost to surface runoff. During snowmelt and soil thaw, a quantity of melt-water equal to the difference between field capacity and saturation is stored for later infiltration; the remainder passes to surface runoff. When a complete soil thaw occurs, the stored melt-water passes through the profile.

2.5 Organic Matter Turn-Over

The calculation of organic matter turn-over is based on the routines used in the DAISY model (Hansen et al. 1991). Carbon dynamics in the soil are described by three pairs (slow or rapid decomposition) of conceptual pools (soil organic matter, soil microbial biomass and added organic matter). Decomposition rate coefficients are temperature- and moisture-dependent, and reflect the environmental conditions of the simulated site; decay and respiration rates of soil microbial biomass are further influenced by the soil clay content. Efficiency parameters determine the loss of CO_2 during single turnover processes. N released as NH_4^+ is a consequence of C lost as CO_2 from the system that maintains fixed C-to-N ratios in the different pools. The pool concept is outlined in Fig. 2.

Residues of crops simulated using the crop growth module enter the mineralisation routine with a dynamic C-to-N ratio determined by the residue amount, crop total N content and a factor determining the N content in crop residues relative to the N content in the harvested crop parts, reflecting the growth conditions of the crop during the season with respect to N supply and emanating from the EU-Rotate_N model. A fixed C-to-N ratio is assigned to the slow decomposing part of the material; the C-to-N ratio of the rapidly decomposable part, however, will vary depending on the total N content in the plant material. C-to-N ratios and partitioning coefficients for crop residues are derived from stepwise chemical digestion experiments (Jensen et al. 2005). Parameters for the release of N from manure and slurry were taken from the DAISY model (Abrahamsen and Hansen 2000). Residues with a wide C-to-N ratio cause soil mineral N to be immobilised by soil microbial biomass.

N volatilisation from applied manures and slurries are described using an empirical relation implemented in the ALFAM model (Søgaard et al. 2002). Hydrolysis of, and gaseous N loss from applied urea fertiliser is calculated based on routines of the AMOVOL model (Sadeghi et al. 1988). The processes of nitrification and denitrification are modelled explicitly in the style of DAISY (Abrahamsen and Hansen 2000).

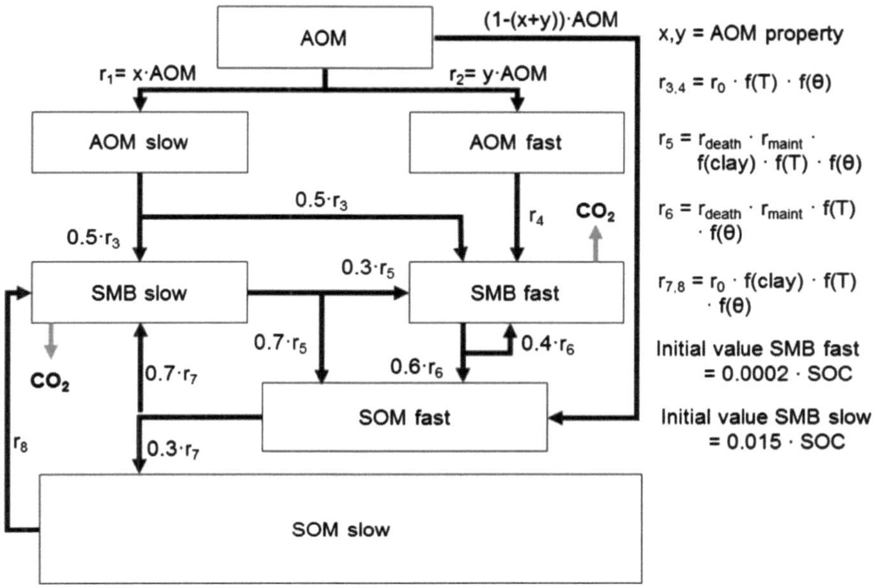

Fig. 2 The pool concept and turn-over rates for organic matter decomposition in MONICA, as adapted from Abrahamsen and Hansen (2000)

2.6 Crop Growth and Development

Crop growth follows a generic approach, based on the SUCROS model (van Keulen et al. 1982). Daily net dry matter production by photosynthesis and respiration is driven by global radiation and temperature. [CO_2] affects the crop's maximum photosynthesis rate and stomatal resistance, which in turn influences transpiration (Nendel et al. 2009). For the former, Mitchell et al. (1995) presented a set of algorithms based on the ideas of Farquhar and von Caemmerer (1982) and Long (1991) for the calculation of the maximum photosynthesis rate

$$A_{\max} = \frac{(C_i - \Gamma *) \cdot V_{c\max}}{C_i + K_c \cdot \left(1 + \frac{O_i}{K_o}\right)}, \tag{8}$$

where C_i is the intercellular concentration of CO_2, $\Gamma*$ is the CO_2 compensation point of photosynthesis, referring to C_i in absence of dark respiration, O_i is the intercellular concentration of O_2, V_{cmax} is the maximum Rubisco saturated rate of carboxylation, and K_c and K_O are the Michaelis–Menten constants for CO_2 and O_2. The temperature dependencies of C_i, O_i, K_c, K_O and V_{cmax} and their parameters were described by Long (1991).

The algorithms for subsaturating light intensities proposed by Mitchell et al. (1995) were not applied. Instead, A_{max} is adjusted for light interception according to Goudriaan and van Laar (1978). For the transition between photosynthetic quantum use efficiency and light saturated photosynthesis (ε_L), Mitchell et al. (1995) suggested using

$$\varepsilon_L = \frac{0.37 \cdot (C_i - \Gamma*)}{4.5 \cdot C_i + 10.5 \cdot \Gamma*}. \tag{9}$$

Stomata resistance r_m was calculated according to the suggestion by Yu et al. (2001) as

$$r_m = \frac{C_s\left(1 + \frac{e_a}{e_s}\right)}{a_S \cdot A_g}, \tag{10}$$

where a_S is a constant, A_g denotes the gross photosynthesis rate, $e_a\ e_s^{-1}$ describes the air water vapour deficit and C_s is the ambient CO_2 concentration at leaf level, which in this case was set equal to C_a. If water is not sufficiently available in soil to satisfy potential transpiration, gross CO_2 assimilation is reduced by a factor calculated from actual transpiration divided by potential transpiration. A crop-specific threshold value defines the crop's sensitivity to drought stress.

Maintenance respiration is calculated separately for day and night periods using AGROSIM algorithms. It is satisfied with priority from assimilates provided by photosynthesis. Thereafter, growth respiration limits the biomass growth increment. Assimilates are assigned to the different organ compartments according to a partitioning matrix, dependent on the crop development stage (Kersebaum 1995). Crop development, in turn, is calculated from a thermal sum (degree-days) and modified, when appropriate, for each stage by day length and vernalisation (Kersebaum 1995).

Root dry matter is distributed over depth according to Pedersen et al. (2010), with the rooting depth increasing linearly with the thermal sum, using crop-specific root growth rates from the EU-Rotate_N model (Rahn et al. 2010). The crop-specific maximum rooting depth is modified by a soil-specific maximum rooting depth, which is calculated from soil sand content and bulk density.

Crop growth is limited by water and N stress. Drought stress is indicated by the ratio of actual-to-potential transpiration. Water and nitrogen uptake is calculated from the potential evaporation and crop N status, depending on the simulated root distribution, the availability of water and N in different soil layers (Kersebaum 1995). Potential N uptake is reduced to actual N uptake if the soil mineral N supply is limited by a maximum uptake rate per root length. Nutrient transport to the roots is simulated using a simplified radial convection and diffusion approach (Kersebaum 1995). The concept of critical N concentration in plants as a function of crop biomass (Greenwood et al. 1990) is applied to assess the impact of N shortage.

$$N_{crit} = a \cdot \left(1.0 + b \cdot e^{-0.26 \cdot W}\right), \tag{11}$$

where W is the aboveground crop dry matter in [t], and a and b are crop-specific parameters (Rahn et al. 2010). The target N concentration is achieved by multiplying the critical N concentration (N_{crit}) by a luxury N factor (k_{lux}), obtained from the EU-Rotate_N model (Rahn et al. 2010). Acceleration of crop ontogenesis by water and nitrogen stress is also considered for specific development stages.

2.7 Automatically Triggered Nitrogen Fertiliser Application and Irrigation

To ensure optimum water and N supply to the crop during long-term scenario calculations, irrigation and N fertilisation can be provided automatically. For N fertilisation, the time slot for reasonable fertiliser application and the type of fertiliser need to be specified. The model determines the first suitable day for fertilisation in the given slot from top soil moisture conditions and calculates the soil mineral N content down to a crop-specific soil depth. Two crop-specific target values for N fertilisation (for the given soil depth and for 30 cm) from a database (Feller et al. 2007) are used to calculate the N deficit. This deficit will be applied by the model (Nendel 2009). This approach mimics the N_{min} method of Wehrmann and Scharpf (1979), which is widely used in Central Europe for N fertiliser recommendations in spring.

A similar method is used for irrigation, where a user-defined soil moisture deficit triggers the application of a user-defined amount of water. Both methods are based on ideas implemented in the EU-Rotate_N model (Rahn et al. 2010). They reflect the farmer's behaviour under changing environmental conditions in a long-term scenario in which he always opts for a sufficient supply of water and N to the crop.

2.8 Data Requirements

MONICA requires a number of different parameters and variables. Parameters describing crop physiology, habitus and phenology, as well as soil processes and more general settings, are stored in a MySQL database, and are accessible only from the science front end. Driving variables describing the site, cropping details and daily weather are listed in Table 1. Their number and format were chosen in accordance with standard data sources available in Europe.

3 Model Testing and Application

The MONICA simulation model continues writing the history of the HERMES model, which originally has been designed for the assessment of N leaching and corresponding groundwater pollution from agricultural production. Equipped with

Table 1 MONICA input variables

Minimum requirement		Optional input	
Variable	Unit	Variable	Unit
Site			
Latitude	°	N deposition	kg N ha^{-1} a^{-1}
Altitude	m	Soil stone content	kg kg^{-1}
Horizon boundaries	m	Soil pH	
Soil sand content	kg kg^{-1}	Max. groundwater table	m
Soil clay content	kg kg^{-1}	Min. groundwater table	m
Soil bulk density	kg m^{-3}	Slope	m m^{-1}
Soil organic matter content	kg kg^{-1}	Texture class[$]	
Crop			
Sowing date	ddmmyyyy	Irrigation dates	ddmmyyyy
Crop type		Irrigation amount	mm d^{-1}
Tillage dates	ddmmyyyy		
Tillage depth	m		
N fertilising dates	ddmmyyyy		
N fertiliser type			
Harvest date	ddmmyyyy		
Fraction residue take-off	%		
Weather			
Precipitation	mm	CO_2 concentration	ppm
Min. air temperature 2 m	°C	Sunshine hours[§]	h
Max. air temperature 2 m	°C		
Relative humidity	Pa Pa^{-1}		
Global radiation[§]	MJ m^{-2}		
Wind speed	m s^{-1}		

[$] The soil texture class following the German classification system (Ad-hoc-AG Boden 2005) may be used if particle size fractions are not available
[§] Either global radiation or sunshine hours is required

functions to describe crop growth response to atmospheric CO_2 concentration and stress induced by extreme temperatures the MONICA model is well suited to assess climate change impact on crop production. It has been calibrated using data from a crop rotation grown under elevated atmospheric CO_2 concentration and different N fertiliser regimes. The rotation included winter wheat, winter barley, sugar beet and maize—typical crops for central Europe (Nendel et al. 2011). MONICA was tested against similar crop rotations from other locations in Germany and evaluated for different variables describing crop growth and development as well as soil water and nitrogen dynamics (Fig. 3). Further calibrating and testing also included data from other crops, such as sorghum, sudan grass, ryegrass, phacelia, oil radish, triticale, oat and field pea. Currently, MONICA is used to simulate yields for maize, soybean, sunflower, cotton and sugar cane in Mato Grosso, Brazil, under different climate and land use scenarios. Its results feed into simulation models for farmer's behaviour and land cover change and back into

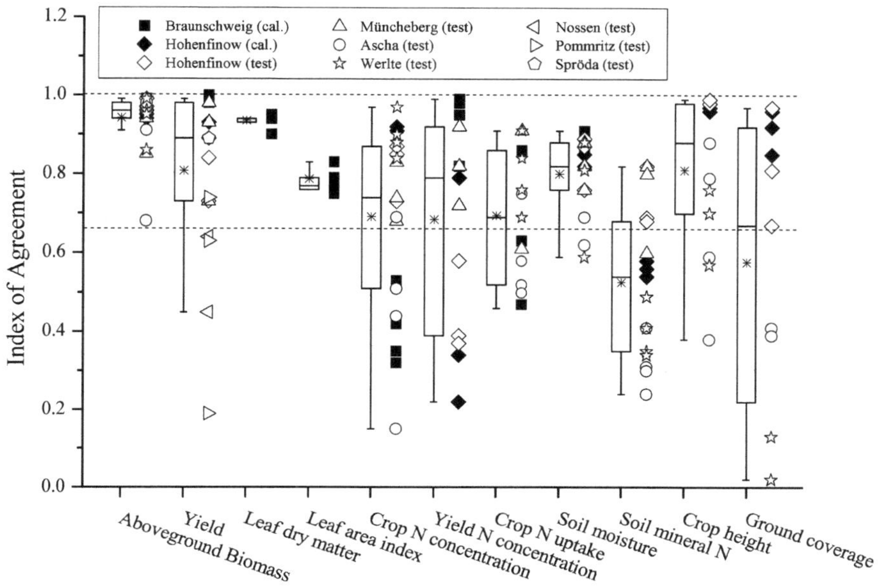

Fig. 3 Overview of results for MONICA simulations of experimental crop rotations at various sites across Germany. Willmott's Index of Agreement (d, Willmott and Wicks 1980) for different soil and crop variables from calibrated (closed symbols) and uncalibrated (open symbols) model runs. The dotted lines contain the acceptable value range for *d* (Nendel et al. 2011)

climate models. Simultaneously, MONICA enables projections on how climate and land use change will impact on greenhouse gas emissions from soil and how soil carbon stocks will develop over time. This feature is used to assess farmer's possibilities to sequester carbon and increase soil quality by adapting alternative management options. In this context, MONICA was also tested against net ecosystem carbon exchange data from a maize experiment on different topography situations (Chatskihk et al. 2012) to verify the effect of erosion or deposit accumulation on the long-term carbon balance in the soil-crop-atmosphere system.

An ensemble of crop models, which included both HERMES and MONICA, was used to simulate spring barley yields under uncalibrated conditions across Europe (Rötter et al. 2012). A similar, albeit larger, ensemble of models was used to simulate the growth of winter wheat at four locations across the globe (Asseng et al. 2013). The range of yields simulated by the models helped to assess the uncertainty that is added by crop models when projecting the impact of a changing climate on food security (Rosenzweig et al. 2012). It also demonstrated that MONICA is capable of simulating agro-ecosystems across a range of different climates and soils. The MONICA ancestor, the HERMES model, was also successfully applied in the cold continental climate of Saskatchewan, Canada (Kersebaum et al. 2008), which underlines the potential of MONICA to address impacts of climate change and land management on crop yields, carbon balance and nitrogen efficiency in Central Asia.

Furthermore, MONICA was used to test the applicability of different spatially resolved sets of weather data for a region in Germany. One of the sets required a high-resolution application of the model, computing 1.6 million data points on a 100 × 100 m grid using a 48 core computer cluster (Fig. 4). This exercise helped to identify the bias that needs to be corrected for when using the point model MONICA for yield simulations in a larger scale context (Nendel et al. 2013).

Within the interactive LandCaRe Decision Support System (Wenkel et al. 2013), MONICA is offered for plot-scale simulations of the user's own scenarios (Fig. 5). Merged with a farm economy coefficient generator (Münch et al. 2013), MONICA outputs can be translated into monetary terms. For some stakeholders costs and returns are better graspable outputs than yields, fertiliser use and irrigation water consumption. Evaluating these outputs, the user can compare MONICA runs with different inputs (weather, crop rotations, soils, management) and improve understanding of the systems behaviour when certain conditions change (Fig. 6). Incidental learning is a secondary benefit of using MONICA and the DSS. However, the provision of answers on what-if questions on possible climate change effects on agricultural production is the primary goal.

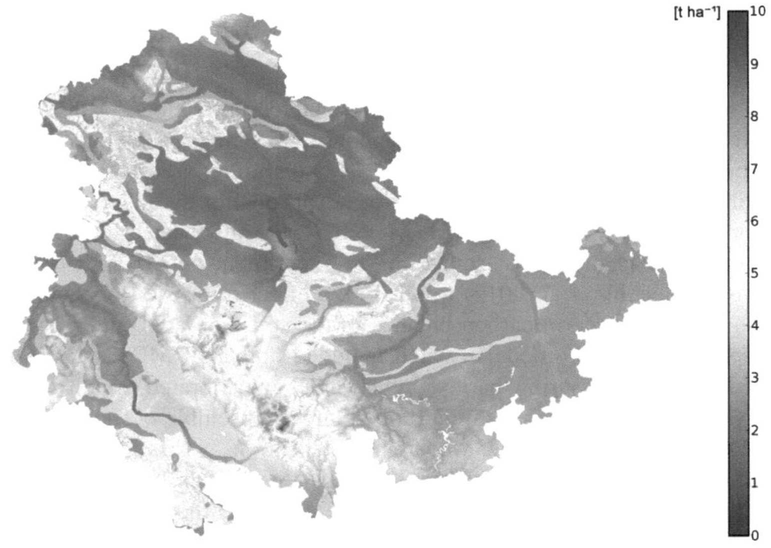

Fig. 4 100 × 100 m MONICA simulations of average winter wheat yields for the years 1992–2010 in Thuringia, Germany (Nendel et al. 2013)

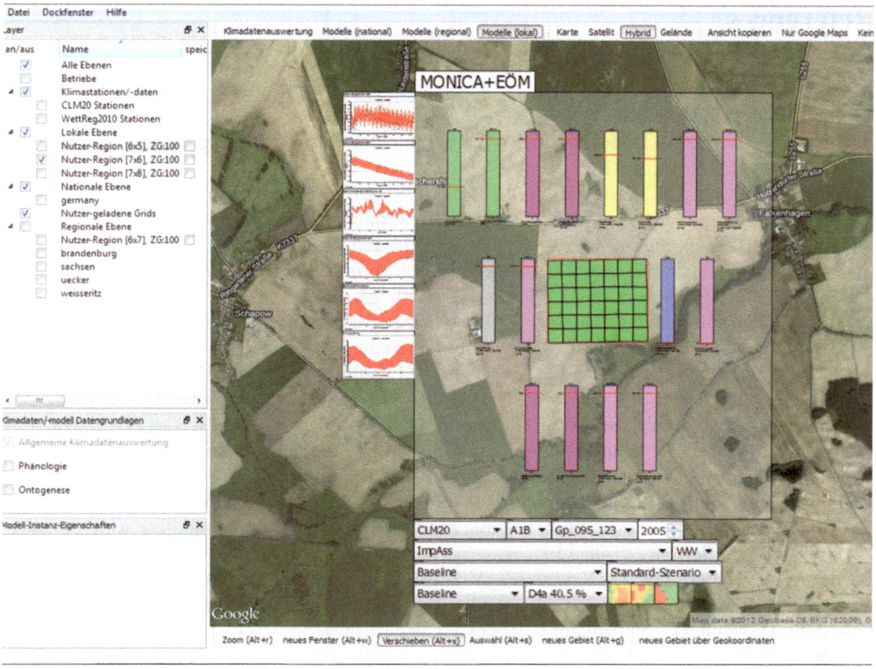

Fig. 5 User interface of the LandCaRe Decision Support System with MONICA plot-scale simulation results displayed in a result panel. Zooming into the panel reveals more details of the simulation results

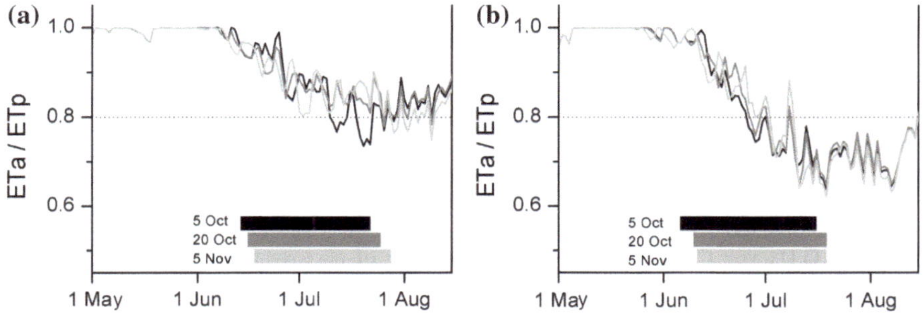

Fig. 6 Effect of winter wheat sowing date on the grain filling phase (bars) at Müncheberg, Germany, in relation to the drought risk indicator ETa/ETp (*lines*). Mean of MONICA simulations for 30 years around 2010 (**a**) and 2070 (**b**). Threshold for drought stress is at ETa/ETp < 0.8 (Nendel et al. 2012)

References

Abrahamsen P, Hansen S (2000) Daisy: an open soil-crop-atmosphere system model. Environ Mod Softw 15:313–330

Ad-hoc-AG B (2005) Bodenkundliche Kartieranleitung. E. Schweizerbartsche Verlagsbuchhandlung, Hannover, p 438

Allen RG, Pereira LS, Raes D, Smith M (1998) Crop evapotranspiration. Guidelines for computing crop water requirements. FAO Irrigation and Drainage Paper 56, Roma

Alvarez R (2009) Predicting average regional yield and production of wheat in the Argentine Pampas by an artificial neural network approach. Eur J Agron 30(2):70–77

Asseng S, Ewert F, Rosenzweig C, Jones JW, Hatfield JL, Ruane AC, Boote KJ, Thorburn P, Rötter RP, Cammarano D, Brisson N, Basso B, Martre P, Aggarwal PK, Angulo C, Bertuzzi P, Biernath C, Challinor A, Doltra J, Gayler S, Goldberg RA, Grant R, Heng LK, Hooker J, Hunt T, Ingwersen J, Izaurralde RC, Kersebaum KC, Müller C, Maresh Kumar S, Nendel C, O'Leary G, Olesen JE, Osborne T, Palosuo T, Priesack E, Ripoche D, Semenov MA, Shcherbak I, Steduto P, Stöckle CO, Stratonovitch P, Streck T, Supit I, Tao FL, Travasso M, Waha K, Wallach D, White JW, Williams JR, Wolf J (2013) Uncertainty in simulating wheat yields under climate change. Nature Climate Change [online]. doi:10.1038/nclimate1916

Bachinger J, Reining E (2009) An empirical statistical model for predicting the yield of herbage from legume-grass swards within organic crop rotations based on cumulative water balances. Grass Forage Sci 64(2):144–159

Bardossy A, Haberlandt U, Krysanova V (2003) Automatic fuzzy-rule assessment and its application to the modelling of nitrogen leaching for large regions. Soft Comput 7(6):370–385

Chatskihk D, Nendel C, Hagemann U, Specka X, Augustin J, Sommer M, van Oost, K (2012) An approach to assess NEE and C costs associated with an energy crop production at different erosion-induced transient states in a typical North-Eastern Germany landscape using process-based agroecosystem modelling. Geophys Res Abstr 14 (EGU2012-13610). http://publ.ext.zalf.de/publications/c3de3ee5-aa90-46cc-aab6-68883c946e44.pdf

Christensen JH, Carter TR, Rummukainen M, Amanatidis G (2007) Evaluating the performance and utility of regional climate models: the PRUDENCE project. Clim Change 81:1–6

Cossani CM, Savin R, Slafer GA (2007) Contrasting performance of barley and wheat in a wide range of conditions in Mediterranean Catalonia (Spain). Ann Apl Biol 151:167–173

Déqué M, Rowell DP, Luthi D, Giorgi F, Christensen JH, Rockel B, Jacob D, Kjellstrom E, de Castro M, van den Hurk B (2007) An intercomparison of regional climate simulations for Europe: assessing uncertainties in model projections. Clim Change 81:53–70

Enke W, Spekat A (1997) Downscaling climate model outputs into local and regional weather elements by classification and regression. Clim Res 8(3):195–207

Falloon P, Betts R (2010) Climate impacts on European agriculture and water management in the context of adaptation and mitigation—The importance of an integrated approach. Sci Total Environ 408:5667–5687

Farquhar, G.D., von Caemmerer, S. (1982) Modelling of photosynthetic response to environmental conditions. In: Lange, O.L., Nobel, P.S., Osmond, C.B., Ziegler, H. (Eds.), Encyclopedia of plant physiology. New series. Volume 12B. Physiological plant ecology. II. Water relations and carbon assimilation. Springer, Berlin, pp. 549–587

Feller C, Fink M, Laber H, Maync A, Paschold PJ, Scharpf HC, Schlaghecken J, Strohmeyer K, Weier U, Ziegler J (2007) Fertilization of field vegetables—a comprehensive database. Institute of Vegetable and Ornamental Crops, Großbeeren, p 266

Goudriaan J, van Laar HH (1978) Relations between leaf resistance, CO_2 concentration and CO_2 assimilation in maize, beans, lalang grass and sunflower. Photosynthetica 12(3):241–249

Greenwood DJ, Stone DA, Draycott A (1990) Weather, nitrogen supply and growth rate of field vegetables. Plant Soil 124(2):297–301

Hansen S, Jensen HE, Nielsen NE, Svendsen H (1991) Simulation of nitrogen dynamics and biomass production in winter-wheat using the Danish simulation-model DAISY. Fert Res 27(2–3):245–259

IPCC (2007) Climate change 2007: the physical science basis. Contribution of working group I to the fourth assessment report of the intergovernmental panel on climate change. Cambridge University Press, Cambridge, United Kingdom and New York, NY, USA

Jansson PE (1991) The SOIL model—Users manual. Swedish University of Agricultural Sciences, Uppsala

Jensen LS, Salo T, Palmason F, Breland TA, Henriksen TM, Stenberg B, Pedersen A, Lundström C, Esala M (2005) Influence of biochemical quality on C and N mineralisation from a broad variety of plant materials in soil. Plant Soil 273:307–326

Kersebaum KC (1989) Die Simulation der Stickstoff-Dynamik von Ackerböden. PhD Thesis, University of Hanover, pp 141

Kersebaum KC (1995) Application of a simple management model to simulate water and nitrogen dynamics. Ecol Mod 85:145–156

Kersebaum KC (2007) Modelling nitrogen dynamics in soil-crop systems with HERMES. Nutr Cycl Agroecosys 77(1):39–52

Kersebaum KC, Hecker J-M, Mirschel W, Wegehenkel M (2007) Modelling water and nutrient dynamics in soil-crop systems: a comparison of simulation models applied on common data sets. In: Kersebaum KC, Hecker J-M, Mirschel W, Wegehenkel M (eds) Modelling water and nutrient dynamics in soil crop systems. Springer, Stuttgart, pp 1–17

Kersebaum KC, Wurbs A, De Jong R, Campbell CA, Yang J, Zentner RP (2008) Long term simulation of soil-crop interactions in semiarid southwestern Saskatchewan, Canada. Eur J Agron 29:1–12

Lasch P, Badeck FW, Lindner M, Suckow F (2002) Sensitivity of simulated forest growth to changes in climate and atmospheric CO_2. Forstwissenschaftliches Centralblatt 121:155–171

Lobell DB, Burke MB, Tebaldi C, Mastrandrea MD, Falcon WP, Naylor RL (2008) Prioritizing climate change adaptation needs for food security in 2030. Science 319:607–610

Long SP (1991) Modification of the response of photosynthetic productivity to rising temperature by atmospheric CO_2 concentrations—Has its importance been underestimated. Plant, Cell Environ 14(8):729–739

Lucarini V, Calmanti S, Dell'Aquila A, Ruti PM, Speranza A (2007) Intercomparison of the northern hemisphere winter mid-latitude atmospheric variability of the IPCC models. Clim Dynam 28(7–8):829–848

Mitchell RAC, Lawlor DW, Mitchell VJ, Gibbard CL, White EM, Porter JR (1995) Effects of elevated CO_2 concentration and increased temperature on winter-wheat—Test of ARC-WHEAT1 simulation model. Plant, Cell Environ 18(7):736–748

Motovilov YG, Gottschalk L, Engeland K, Belokurov A (1999) ECOMAG: regional model of hydrological cycle—Application to the NOPEX region. University of Oslo, Department of Geophysics Report Series 105, Oslo. http://folk.uio.no/kolbjoen/nygen/ECOMAG-REPORT.pdf

Münch T, Berg M, Mirschel W, Wieland R, Nendel C (2013) Considering cost accountancy items in crop production simulations under climate change. Eur J Agron. http://dx.doi.org/10.1016/j.eja.2013.01.005

Nendel C (2009) Evaluation of Best Management Practises for N fertilisation in regional field vegetable production with a small scale simulation model. Eur J Agron 30:110–118

Nendel C, Berg M, Kersebaum KC, Mirschel W, Specka X, Wegehenkel M, Wenkel KO, Wieland R (2011) The MONICA model: testing predictability for crop growth, soil moisture and nitrogen dynamics. Ecol Mod 222:1614–1625

Nendel C, Kersebaum KC, Mirschel W, Manderscheid R, Weigel HJ, Wenkel K-O (2009) Testing different CO_2 response algorithms against a FACE crop rotation experiment. NJAS—Wageningen J Life Sci 57(1):17–25

Nendel C, Kersebaum KC, Mirschel W, Wenkel KO (2012) Testing farm management options as climate change adaptation strategies using the MONICA model. Eur J Agron, http://dx.doi.org/10.1016/j.eja.2012.09.005

Nendel C, Wieland R, Mirschel W, Specka X, Kersebaum KC (2013) Simulating winter wheat yields using input data with different spatial resolutions. Field Crop Res 145:67–77. http://dx.doi.org/10.1016/j.fcr.2013.02.014

Neusypina TA (1979) Rascet teplovo rezima pocvi v modeli formirovanija urozaja., Teoreticeskij osnovy i kolicestvennye metody programmirovanija urozaev. Leningrad, pp 53–62

Olsen PA, Haugen LE (1997) Jordas termiske egenskaper. 8, Ås

Paeth H, Capo-Chichi A, Endlicher W (2008) Climate change and food security in tropical West Africa—A dynamic-statistical modelling approach. Erdkunde 62(2):101–115

Pedersen A, Zhang KF, Thorup-Kristensen K, Jensen LS (2010) Modelling diverse root density dynamics and deep nitrogen uptake—A simple approach. Plant Soil 326(1–2):493–510

Potgieter AB, Hammer GL, Doherty A, de Voil P (2005) A simple regional-scale model for forecasting sorghum yield across North-Eastern Australia. Agric For Meteorol 132(1–2):143–153

Rahn CR, Zang K, Lillywhite RD, Ramos C, de Paz JM, Doltra J, Riley H, Fink M, Nendel C, Thorup-Kristensen K, Pedersen A, Piro F, Venezia A, Firth C, Schmutz U, Rayns F, Strohmeyer K (2010) EU-Rotate_N —a European decision support system—to predict environmental and economic consequences of the management of nitrogen fertiliser in crop rotations. Eur J Hort Sci 75(1):20–32

Reidsma P, Ewert F, Lansink AO, Leemans R (2010) Adaptation to climate change and climate variability in European agriculture: the importance of farm level responses. Eur J Agron 32(1):91–102

Riley H, Bonesmo H (2005) Modelling of snow and freeze-thaw cycles in the EU-Rotate_N decision support system. Grønn Kunnskap 9(112):0

Rosenzweig C, Jones JW, Hatfield JL, Ruane AC, Boote KJ, Thorburn P, Antle JM, Nelson G, Porter CH, Janssen S, Asseng S, Basso B., Ewert F, Wallach D, Baigorria G, Winter JM (2012) The agricultural model intercomparison and improvment project (AgMIP): protocols and pilot studies. Agric For Meteorol, http://dx.doi.org/10.1016/j.agroformet.2012.09.011

Rötter RP, Palosuo T, Kersebaum KC, Angulo C, Bindi M, Ewert F, Ferrise R, Hravlinka P, Moriondo M, Nendel C, Olesen JE, Patil R, Ruget F, Tacáè J, Trnka M (2012) Simulation of spring barley yield in different climatic zones of Northern and Central Europe: a comparison of nine crop models. Field Crop Res 133:23–36

Sadeghi AM, McInnes KJ, Kissel DE, Cabrera ML, Koelliker JK, Kanemasu ET (1988) Mechanistic model for predicting ammonia volatilization from urea. In: Bock BR, Kissel DE (eds) Ammonia volatilization from urea fertilizers. National Fertilizer Development Centre, Tennessee Valley Authority, Muscle Shoals, Alabama, pp 67–92

Shrestha RR, Bardossy A, Rode M (2007) A hybrid deterministic-fuzzy rule based model for catchment scale nitrate dynamics. J Hydrol 342(1–2):143–156

Søgaard HT, Sommer SG, Hutchings NJ, Huijsmans JFM, Bussink DW, Nicholson F (2002) Ammonia volatilization from field-applied animal slurry—the ALFAM model. Atmos Environ 36(20):3309–3319

Trnka M, Dubrovský M, Žalud Z (2004) Climate change impacts and adaptation strategies in spring barley production in the Czech Republic. Clim Change 64(1–2):227–255

van Keulen H, Penning de Vries FWT, Drees EM (1982) A summary model for crop growth. In: Penning de Vries FWT and van Laar HH (eds) Simulation of plant growth and crop production. PUDOC, Wageningen, pp 87–97

Wegehenkel M (2000) Test of a modelling system for simulating water balances and plant growth using various different complex approaches. Ecol Mod 129(1):39–64

Wehrmann J, Scharpf HC (1979) Mineral nitrogen concentration of soil as a gauge of need for nitrogen-fertilizer (N-min method). Plant Soil 52(1):109–126

Wenkel KO, Berg M, Mirschel W, Wieland R, Nendel C, Köstner B (2013) LandCaRe DSS –An interactive decision support system for climate change impact assessment and the analysis of

potential agricultural land use adaptation strategies. J Environ Manage. http://dx.doi.org/10.1016/j.jenvman.2013.02.051
Wessolek G, Asseng S (2006) Trade-off between wheat yield and drainage under current and climate change conditions in northeast Germany. Eur J Agron 24(4):333–342
Wessolek G, Kaupenjohann M, Renger M (2009) Bodenphysikalische Kennwerte und Berechnungsverfahren für die. Praxis 40:1–80
Wieland R, Mirschel W (2008) Adaptive fuzzy modeling versus artificial neural networks. Environ Mod Softw 23:215–224
Williams JR, Jones CA, Dyke PT (1984) A modeling approach to determining the relationship between erosion and soil productivity. Trans ASAE 27(1):129–144
Willmott CJ, Wicks DE (1980) An empirical method for the spatial interpolation of monthly precipitation within California. Phys Geogr 1:59–73
Yu Q, Goudriaan J, Wang TD (2001) Modelling diurnal courses of photosynthesis and transpiration of leaves on the basis of stomatal and non-stomatal responses, including photoinhibition. Photosynthetica 39(1):43–51

Integrated Decision Support for Sustainable and Profitable Land Management in the Lowlands of Central Asia

Nodir Djanibekov and Rolf Sommer

Abstract Land use and crop production in the Khorezm region in western Uzbekistan, exemplarily for the irrigated lowlands of Central Asia, are challenged by the excessive, non-sustainable use of irrigation water and repeated water shortages on one hand, and soil degradation by secondary salinization on the other hand. One of the research objectives of the German-Uzbek Khorezm project, funded by the German Ministry for Education and Research (BMBF) and led by Center of Development Research (ZEF) of Bonn University, was to better understand options for land use and choice of technology at the farm level in order to evaluate and propose technological alternatives and agricultural policy options for sustainable land use. To address these issues, a Farm-Level Economic Ecological Optimization Model (FLEOM) was developed as an integrated decision-support tool capable of optimizing land and resource use at the level of large farms and water user associations, while at the same time assessing the respective economic and environmental impacts at the micro-scale. This chapter provides a brief introduction to characteristics of FLEOM and the necessary steps and studies related to the generation of its database. Based on a comprehensive system-understanding composed by a range of agronomic and socioeconomic studies carried out in the project, the model captures the basic features of the regional agriculture and the interrelations of production activities most prevalent to the local farmers. It comprises regionally specified input–output parameters and conditions of land and water use. The flexibility of the model allows the addition of new crops, the modification of socio-economic parameters in a user-friendly way, such as to simulate changes in socio-economic and production environments,

N. Djanibekov (✉)
Leibniz Institute of Agricultural Development in Central and Eastern Europe (IAMO), Theodor-Lieser-Str. 2 06120 Halle (Saale), Germany
e-mail: djanibekov@iamo.de

R. Sommer
International Center for Tropical Agriculture—CIAT, ICIPE Complex, off Kasarani Road, Nairobi, Kenya
e-mail: r.sommer@cgiar.org

L. Mueller et al. (eds.), *Novel Measurement and Assessment Tools for Monitoring and Management of Land and Water Resources in Agricultural Landscapes of Central Asia*, Environmental Science and Engineering, DOI: 10.1007/978-3-319-01017-5_24,

securing a wide range of its possible uses in the future. FLEOM builds on linear programming optimization routine and a comprehensive agronomic database established with the cropping system simulation model (CropSyst) using field experience and knowledge of a range of agronomic and hydrological studies of the project. A graphical user-interface programmed in Java provides the model's easy usability, by which settings and results are visualized in tables and figures or as maps via a GIS-environment. The model works in three steps: first, the model sets and imports an agronomic database. Next, the agronomic database together with socio-economic information goes through an optimization process in the General Algebraic Modeling Software (GAMS). Finally, the model transfers the results into easily understandable tables, figures and maps.

Keywords Land management · Agronomic data base · Farm level · FLEOM model · Central Asia

1 Introduction

Agriculture in the lowlands of Central Asia has been challenged by a harsh environment with extremely cold winters and hot summers and little to no precipitation. Growing crops like winter-wheat, cotton, maize, rice, sorghum and potatoes is only feasible by irrigation with water withdrawn from the two major rivers, the Syrdarya and Amudarya. Since the introduction of large-scale infrastructure during the Soviet times, the region is suffering from two major problems: (1) the excessive, unsustainable use of irrigation water and at the same time (2) soil degradation by secondary salinization in response to (man-made) shallow groundwater levels and upward movement of water and salts in the soil (Ibrakhimov 2004). An economic-ecological restructuring of land and water use in the region is inevitable, if the mid- to long-term sustainability and profitability of agriculture in this region is supposed to be maintained.

On example of the Khorezm region located in the downstream of the Amudarya Basin of Uzbekistan, this issue was addressed by the Center for Development Research (ZEF) of Bonn University which with local and international partners led the German-Uzbek development research project (ZEF-project; www.khorezm.zef.de) initiated in 2001 with funding from the German Ministry for Education and Research, BMBF (Martius et al. 2012). The restructuring options were investigated through interdisciplinary studies of economic and ecological processes, such as regional or farm economy and regional hydrology or field-level agronomy/crop growth. One of the research objectives of the project was to better understand the options of land use decisions and choice of technology at the farm level in order to evaluate technological alternatives and policy options for sustainable land use in Khorezm. A series of economic factors (e.g., existing policies, and input and output prices), as well as biophysical factors (e.g., environmental conditions, production technologies, land characteristics) of a farming system decide farm management decisions regarding land use. The physical relationships between biophysical attributes of land and farm management practices co-determine agricultural and environmental outcomes. In this respect, an optimization model allows combination of biophysical and socioeconomic information at different scales for farm-level research in developing country agriculture. A proper specification of the modeling framework requires detailed information of realistic technical ratios and biological relationships between inputs and outputs. Cooperation of agricultural economists and agronomists is key for successfully addressing the issue and for the development of models reflecting the decision environment faced by farmers. In this respect, the role of farm-level modeling within a policy formulation is to qualitatively understand the likely responses of producers to various market conditions and policy provisions. Accordingly, such decision-support tools should be flexible to incorporate alternative crops using renewed version of crop-growth component over a multi-period horizon with structure and database transferable between modeled objects and to users for analytical purposes. Additionally, being tested for reliability such model must be able to incorporate various governmental policies and to simulate the cash-flow process in farm based on some type of decision-making routine and farm-level information. Although the environmental and crop process models work at the plot level, the decisions take place at aggregated scales such as a single farm or a group of farms. Hence, biophysical economic (or in short bio-economic) models based on components at micro-level scales can be further aggregated to regional and national levels depending on the problem scale.

To address these issue, the so-called site-specific integrated Farm-Level Economic-Ecological Optimization Model (FLEOM) was developed as a land-use planning support tool for decision-making at the level of farms and Water User Association (WUA), aiming at providing coupled ecological-economic optimization of land allocation. The model was developed such as to optimize farm-level land and resource use while at the same time assessing the respective economic and environmental impacts. The model incorporates the basic features of the regional agriculture and the interrelations of production activities most prevalent to the local farmers and allows for various changes in production factors and input-output relationships that feature economic, resource and policy environment of the modeled system. Finally, the model integrates field-level management decisions on the allocation of inputs and resources taking into account respective ecological and economic consequences in a spatially explicit way.

Potential users of FLEOM are medium-level stakeholders such as WUA representatives and the local water authority. Besides, FLEOM is intended to be a tool for scientists and it may have an important role in university education. Therefore, user-friendliness and easy usability were important aspects when building FLEOM under the following objectives:

- to show the potentials and constraints of alternative land- and water-management options that can increase rural incomes while maintaining agricultural production with at least a medium-term ecological sustainability,
- to design new plans for sustainable development and to check if selected adaptation measures are feasible,
- to develop and suggest to stakeholders optimal land use options under alternative environmental conditions, e.g., water scarcity
- to assess the respective impacts of various external shocks on farm income, crop and animal production, cropping pattern, water and resource use and soil attributes.

To simulate an environmentally friendly development, FLEOM should allow for setting limits to environmental hazards, such as irrational irrigation, pollution of groundwater by excessive nitrate leaching, and deterioration of land by secondary soil salinization.

2 Conceptual Framework

The conceptual layout of FLEOM is the result of a number of interactive (learning) processes. This involved a series of transdisciplinary workshops and numerous meetings, in which the basic systematic concept of FLEOM was summarized in use-case diagrams and a use-case narrative as well as an activity diagram following the notation of the Unified Modeling language (UML). Using UML tools was a worthwhile exercise to streamline and pinpoint the different views and opinions of the project members and agreeing upon a common understanding of the functionalities of FLEOM. However, the established diagrams and narrative were of limited use for the software development process per se. Here, creativity and freedom of choice was in one or the other way hampered by the software tools and modeling frameworks already in use (e.g., GAMS and CropSyst), the in-house software programming expertise, time limitations, as well as anticipated follow-up (software licenses) and maintenance costs ("keep things simple and cheap").

FLEOM represents a stand-alone software that comprises several sub-components (Fig. 1): a graphical user interface (FLEOM-GUI) programmed in Java (NetBeans), a GIS visualization component realized with the Open Systems Mapping Technology (OpenMap, http://openmap.bbn.com/), and an economic-ecological optimization routine written in GAMS. The optimization component that solves using CONOPT solver is connected to the agronomic database stored in a Microsoft Access file. The agronomic database was established with the cropping system simulation model CropSyst (Stöckle et al. 2003).

Arrows in Fig. 1 detail the flow of data or information. In the GUI, a user has access to the database and can read or modify it (e.g., commodity prices), or add new data (e.g., simulation results). The OpenMap GIS interface shows existing farm attributes and provides several options for user-specified visualization or for

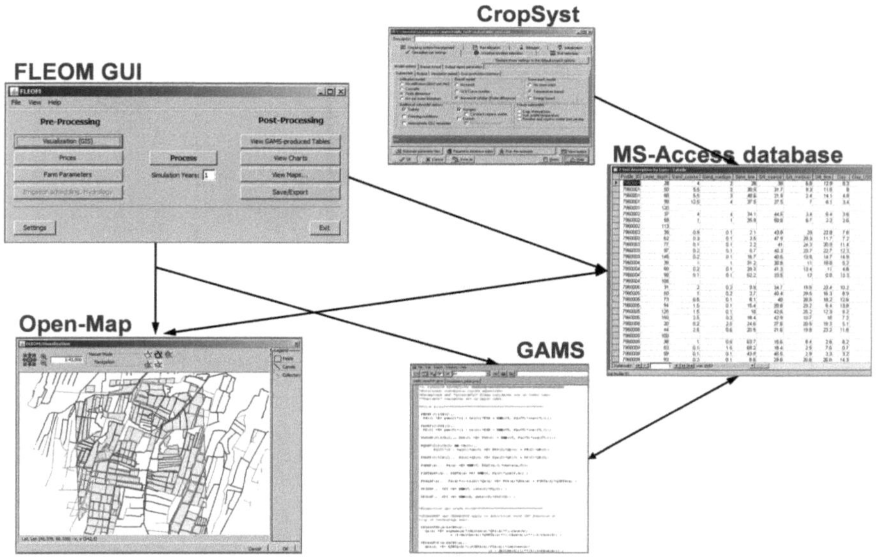

Fig. 1 Components of FLEOM and their links

highlighting/excluding certain fields or attributes. It also allows the display of model results in form of tables and figures as well as their export in a pdf-summary file. This approach was chosen as it is less demanding in software-technical manpower, and at the same time a higher public awareness, availability and professional support (if required) of the used tools outweigh potential disadvantages, such as a missing generic plug-in-pull-out functionality or limitations regarding a potential web-based implementation of such decision-support tool.

Figure 2 illustrates the GAMS-optimization workflow. After setup and start of the optimization analysis, the GUI launches the optimization routine in GAMS which is connected to the agronomic database. This sub-model imports all necessary information from the database. The GAMS routine itself is a slimmed-down version of the optimization routine developed to describe, simulate and optimize the farm production decision. The model results are then exported to the database, from where they used for visualization or export to other formats for further evaluation.

One of the advantages of using OpenMap™ package for the FLEOM GUI is that it allows the visualization of maps based on GIS methodologies. The following assets were complemented with features of OpenMap such as selection, deselection, adding and removal of individual and multiple elements of a layer by click or using a query. In addition, this allows for more flexible layer symbology: the visualization of layer elements was improved by using different symbology to different classes of layer elements. Visualization options provided in ESRI ArcGIS software package were used as a template in this case for category- and quantity-

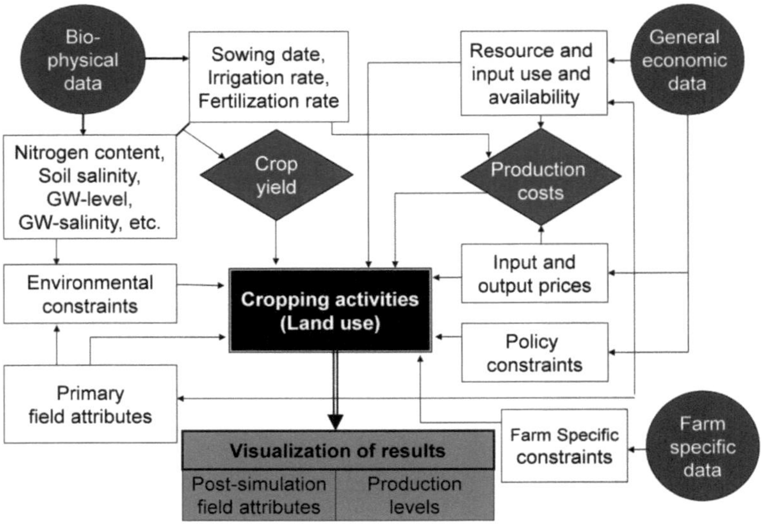

Fig. 2 Workflow of the optimization sub-routine of FLEOM

based classification. Information about the symbology used by each layer of the map appeared in a map legend. This set of GIS functionality responded to the present demands of FLEOM. A full-fledged GIS analysis in FLEOM is not included. However, if desired, FLEOM results can be exported and analyzed in external GIS software.

3 Model Description

The operationally manageable size of a benchmark study area for FLEOM lies in the range of a water user association of a size of up to ~1500 ha. Pakhlavan Makhmud Water User Association (PM-WUA) located in the Khiva district of the Khorezm region, namely its central part (Fig. 3) that covers 822 ha, was chosen as first test site. The modeled area consists of 227 distinct farm fields with largest ones of about 15 ha. Using the biophysical field attributes and further classifications outlined in the next section, these distinct fields can be organized into 45 distinct field types. The fields are grouped into seven cotton-grain growing farms. The sizes of the modeled farms range from 83 to 161 ha. Each farm is distinguished by field-level soil typology and location within the irrigation canal network. Despite being based on the level of one WUA, FLEOM can handle individual fields, group of fields, and individual farms of different sizes thus reflecting actual conditions, in which farms take their decisions somewhat independent of higher-level goals.

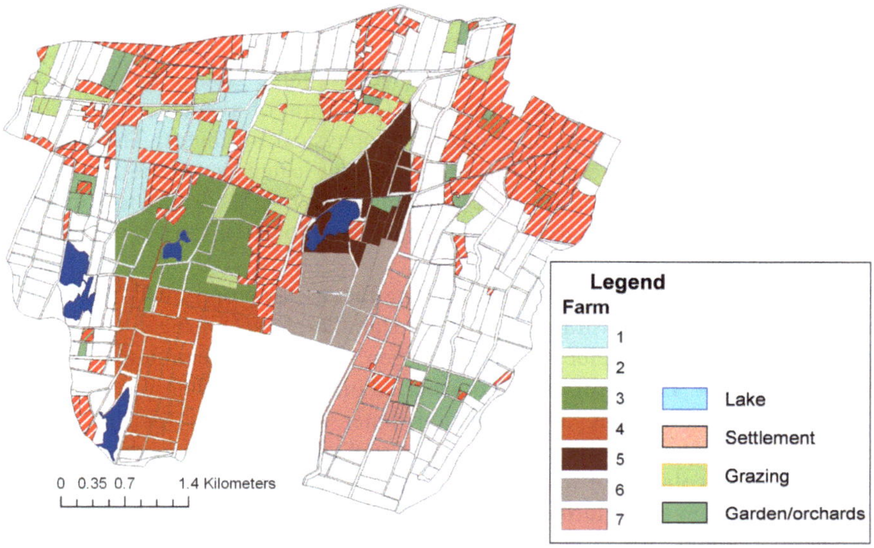

Fig. 3 Boundaries of seven modeled farms, and non-agricultural land use in Pakhlavan Makhmud Water User Association

At the core of FLEOM a linear programming approach was utilized that solves problems defined in terms of an objective function to be maximized/minimized, a set of possible technologies/activities, and a set of constraints of available resources (cf. Hazell and Norton (1986) for presentation of a basic farm-level optimization model). In other words, the model selects the optimal combination of activities that best fulfill assigned objectives and are feasible in terms of physical, human and capital conditions. The optimization model can be used for joint examination of farm activities within the range of resource constraints and costs and returns of modifications in production technologies or any other exogenous change. Moreover, by optimizing a specified objective function, the model can attempt to replicate and show the trends of how a rationally behaving farmer would adjust his production activities to a technological advancements/innovations or any other notable changes in on-farm and external conditions.

The FLEOM model finds an optimal set of production activities and management variants at farm- or WUA-level by maximizing total farm gross margins taking into account product prices, outputs and production costs, soil attributes of each farm field, water and nitrogen fertilizer application techniques presented via Leontief technology coefficient matrix, as well as policy and environmental constraints. The production activities in FLEOM comprise four major crops: cotton (*Gossypium hirsutum* L.), maize (*Zea mais* L.), winter wheat (*Triticum aestivum* L.) and rice (*Oryza sativa* L.) that in total occupied more than 76 % of the sown area and required 82 % of total irrigation water in the Khorezm region in 1998–2006. Particularly, wheat supplies about 60 % of total annual food energy supply in Uzbekistan, while rice generates up to a half of profits of crop-growing farms in the region (Djanibekov 2008).

The constraints in FLEOM can be classified into three groups such as resource availability, state policies and environment. Constraint of available arable land is specified for each field with their distinct biophysical attributes. The fields are further aggregated into seven farms. The cropping calendar covers the regional agriculture characteristic of a double cropping schedule, according to which two crops can be grown and harvested from the same location during the assumed agricultural period.

Since nitrogen is among the main limiting factors for crop growth in the study region, the fertilizer constraint is presented as the annual nitrogen required for cropping activity. The values of each nitrogen application technology are generated in CropSyst. The fertilizers are transferred into elementary nitrogen according to their nitrogen content. The irrigation water requirements for the modeled cropping activities are also generated by CropSyst. The water endowments are fixed at annual volumes of water in the irrigation network. According to this, each farm field is aggregated further into a certain irrigation network area.

Diesel required for crop production indicates fuel costs for machine field operations and water pumping. The diesel requirements for the modeled crops include diesel used for output marketing, as well as for irrigation. Although labor is not reported as a constraint in the agricultural production in Khorezm, the significance of including labor as a constraint in this study is to analyze the potential employment creation by agricultural production constraints in each farm aggregate. The labor requirements for the cropping activities include labor used for all operations in the field and proportional to the amount of water and nitrogen fertilizer application rates and crop harvest.

The cropping activities are also restricted by types of fodder produced by the modeled crops. In this way, fodder production has to match the amount required for feeding the number of livestock in the modeled farms, whereas additional fodder can be purchased. Cash-flow constraint defines that total amount of funds available to each farm, amount of credits and total value of marketed crops is not less than the costs generated during the agricultural activities, amount of purchased inputs, fodder, and hired labor. Policy instrument constraints relate to government policy objectives for regional and national development and are similar to target variables. For instance, the policy instrument constraints of the model can require that production of a certain crop (e.g., cotton) is not less than its assigned amount.

Rotation constraints are related to the maximum number of crops to be cultivated during one season on one field. For agronomic reasons rotation restrictions were set for individual crops as for groups of crops. Two environmental issues are included through constraints for soil parameters, namely (1) the maximum level of nitrogen leached, and (2) the maximum level of soil salinity. The soil fertility of fields depends on cropping activities of a farm. These constraints depict the levels of specific soil parameters of each field modified after certain crop cultivation. The maximum levels can be adjusted by a user depending on his environmental strategy, such as to improve, maintain or ignore soil quality of the fields.

3.1 Database Generation

FLEOM utilizes an extensive agronomic input-output database established with CropSyst using datasets, field experience and knowledge of a range of agronomic and hydrological studies on irrigation and fertilizer responses, planting dates, tillage and residue management carried out in the ZEF-project (Ibrakhimov 2004; Akramkhanov 2005; Sommer et al. 2008; Forkutsa et al. 2009a, b; Djumaniyazova et al. 2010; Kienzler 2010; Kienzler et al. 2011). The following sections will present the details of the agronomic data generation (see Sommer et al. (2010) for detailed description). The database on rice is based on the socio-economic evaluation done by Djanibekov (2008). The socio-economic database comprises output and input prices compiled from survey data of 80 randomly selected farmers conducted in Khorezm in 2007 and 2010. The surveys also provided information on crop labor requirements, diesel use, working hours of combine harvesters, costs at different field operations for four modeled crops and transportation costs. The crop gross margins were calculated as output value per unit of activity less the sum of imputed costs such as seeds, fertilizers, pesticides, labor and machinery costs, land tax, and other fixed costs.

3.1.1 Agronomic Database

In the following details on application of CropSyst to produce the required agronomic database are provided. Since the CropSyst simulations are based on the characteristics of the benchmark study area, we also present its attributes.

Agronomic management of crops comprises a range of issues. Most important parameters for modeling crop growth with CropSyst, or any other biophysical crop model, are the phenological and eco-physiological characteristics of the chosen crop variety, the time of planting, the timing and amount of fertilization and irrigation and the type of tillage operations that are undertaken. This means that for example fertilizer management and crop nutrition beyond nitrogen is rarely simulated, and the same is true for the impact of pesticide and herbicide application, respectively crop pests and diseases. To produce the input-output table for FLEOM with CropSyst, a selection of management practices had to be made allowing on the one hand for a sufficient, representative overview of agronomic management in Khorezm while keeping the total number of simulations at a manageable size, mostly because of limitation given by computing time as well as handling resulting datasets of several gigabytes. The latter database constraint diminishes as computing power increases.

The present version of FLEOM includes biophysically-based input–output tables for cotton, maize and winter-wheat. Only for these crops experiments of sufficient detail were carried out within the project until 2010; data of which could serve for CropSyst calibration. For cotton the variety Khorezm-127 was simulated, as this has been the most commonly used variety in Khorezm in the last years

(50–60 % of the cotton growing area in 2005–2007) and intensively studied by Kienzler (2010). The same was true for the winter-wheat variety Kupava, which was also subject of some intensive agronomic studies (Djumaniyazova et al. 2010). Maize yield response of the variety M-215 (from Moldava) was studied by the project in a crop-rotation fertilizer-response trial carried out in 2005–2006. The dataset for modeling however merely comprised biomass, yield and leaf area index data for maize grown under optimal conditions. In Khorezm, optimal timing of planting is important to achieve best possible yields by avoiding temperature (early and late season) stress or intermediate secondary re-salinization of top soils; to allow for harvest in good time and optimal planting time of the subsequent crop and make best profits. To account for this, CropSyst simulations comprised three different planting dates for cotton.

3.1.2 Agronomic Management Scenarios

Irrigation of annual crops in an arid and salinity-affected region like Khorezm has to consider a range of aspects. These are first of all the actual crop water requirements, which have to be balanced against issues such as soil salinity management and irrigation system specifications. The necessity of pre-season salt-leaching adds to complexity of irrigation management in Khorezm. In view of this, in simulations with CropSyst, irrigation was probably the factor that had to be simplified to the largest extent. Irrigation amount and timing was merely set to comply with crop water requirements. Therefore, the well-established Uzbek Hydro-Mod(ule) scheme (SoyuzNIHI 1992) for irrigation recommendations was used as baseline providing (furrow) irrigation recommendations according to climate, crop, maximum rooting depth, soil texture, groundwater depth and field efficiency.

Four different fertilization scenarios were simulated for each crop, namely standard, medium, low and zero levels of application. Standard fertilization amounts and timing followed Uzbek norms for the crops (see Kienzler (2010) for details). Medium and low application corresponded to 60–20 % of the full amounts. Contrary to irrigation, single fertilizer applications were not pooled. In all simulations, N-fertilizer was broadcasted as ammonium-nitrate.

Only standard tillage practices were simulated for each crop in accordance with common practice: one pass with a moldboard-plow (NRCS implement no. 150) and a spike-tooth-harrow (NRCS implement no. 70) one week before planting. Therewith 99 % of all surface residues of a previous crop were buried to a maximum depth of 20 cm. In this regard, cotton and winter wheat were assumed to be cropped after cotton, and maize after winter wheat. Since Khorezm farmers remove all cotton residues and wheat straw, we set up CropSyst simulations to keep only 5 % of the remainders of the last cotton as surface residues, and 10 % of straw of the last wheat crop as standing stubbles.

3.1.3 Climate, Soils and Calibration Datasets

Given the high contribution of advective energy (dry air blowing in from the surrounding deserts) to evapotranspiration (ET), we used the FAO-56-Penman-Monteith equation (Allen et al. 1998) to simulate actual ET in CropSyst. The required daily time-step weather data precipitation, solar radiation, minimum and maximum air temperature, minimum and maximum relative humidity and wind speed were measured in situ starting from May 2004.

CropSyst, like any other full-fledged crop model, requires three different types of soil information: physical and hydrological description (texture, bulk density, water retention and hydraulic conductivity); initialization of state variables (Soil Organic Matter (SOM), soil mineral N contents (N_{min}), soil salinity, water content); and temporal description of the model system's lower boundary, i.e., in case of the influence of a shallow groundwater table the time and depth of groundwater and salinity. The description and later classification soils is largely based on a comprehensive database of soils in Khorezm from the Uzbekistan Soil Science and Agrochemistry Research Institute and, in the case of SOM, supported by a comprehensive literature study carried out by Kuziev (2006). Soils of PM-WUA were described in detail by Akramkhanov (2005). Groundwater dynamics and salinity in Khorezm were analyzed by Ibrakhimov (2004). Topsoil (0–30 cm) texture was determined by Akramkhanov (2005) for the center region of the PM-WUA. Four soil textures prevailed: sand, loamy sand, sandy loam and loam. Lighter soils were generally found in the south and heavier soils in the north of the PM-WUA (Fig. 4).

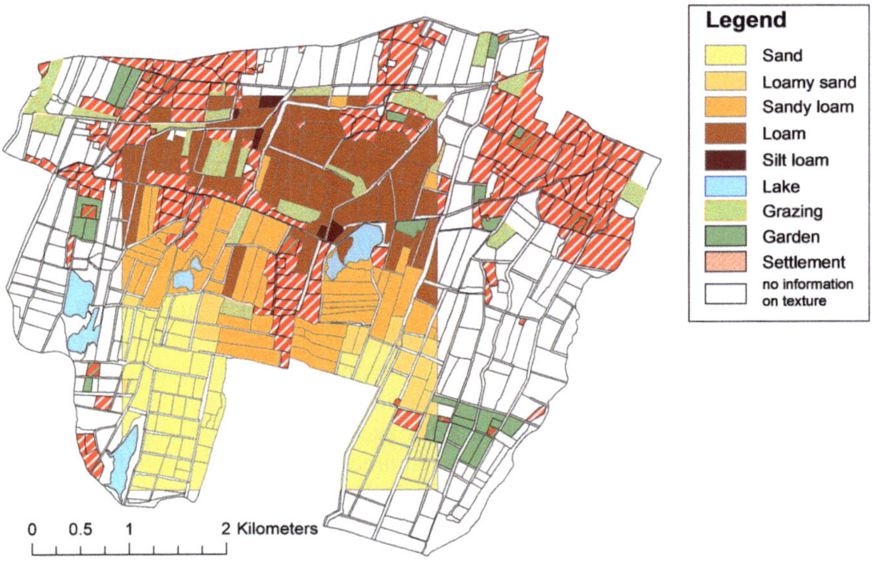

Fig. 4 Soil texture distribution in Pakhlavan Makhmud Water User Association

A detailed description of soil hydraulic properties was not available for a larger dataset; neither for Khorezm as a whole nor for the PM-WUA. In depth studies in this regards in Khorezm were only carried out by the ZEF-project for the first time. Forkutsa et al. (2009a, b) described in detail the soil water retention and hydraulic conductivity of a sand and sandy loam. In 2005, Kienzler (2010) studied crop growth and fertilizer-response of cotton on a loam (gleyic Arenosol) located in the Urgench district of Khorezm (41° 36′ 13″N 60° 30′ 49″E). Winter-wheat followed in 2006 (Djumaniyazova et al. 2010) on a directly neighbored site (same texture). Water dynamics were studied in both years on these fields by simultaneous automated measurements of soil moisture, which served as calibration datasets for the bio-physical simulations of crop-growth and yield, and salt, water and N-dynamics with CropSyst.

3.2 Simulation Scenarios

Based on the different agronomic management scenarios and the potential differences in field/soil attributes, CropSyst simulation scenario files (*.scn) were established, each of which covering one of the multitude of possible combinations. The amount of possible combinations can be calculated by multiplying the number of levels defined for each management and field/soil attribute, as there are for each crop:

Management	Field attributes
3 planting dates for cotton;	4 soil textures
1 planting date for maize and wheat	3 levels of soil organic matter
3 irrigation levels	3 levels of initial soil mineral N
4 fertilization levels	3 levels of soil salinity
	3 groundwater levels
	2 groundwater salinity levels

To model the growth of cotton, a full year (1/Jan/–31/Dec/2005) was simulated. Winter wheat simulations comprised ten months (1/Sep/05–31 Jul/06) and maize simulations six months (1/Jun–1 Dec/06). Altogether, for cotton 16,200 combinations had to be simulated, for winter wheat and maize these amounted to 5,400. For the creation of the 16,200 simulation-scenario files a small program was written to automatically create these scenario files of the feasible combinations, reading-in the necessary data from a MS-Excel spreadsheet. A single simulation run of cotton, winter wheat and maize required around 30, 25 and 15 s of CPU time (Intel Pentium-IV 2.5 GHz, Microsoft XP-Professional OS), respectively. Each of these simulations created about 2.6 MB of results. To compile all necessary results for the completed input-output table 195 h of CPU-time and 69 GB of hard-drive storage capacity were required.

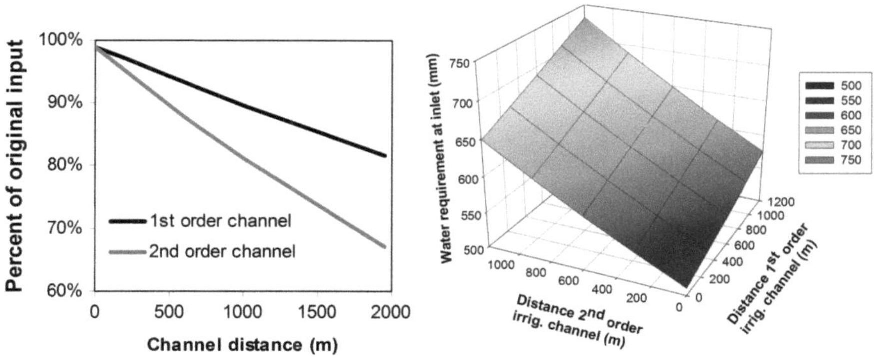

Fig. 5 Loss of irrigation water by seepage as a function of canal length (*left*), and individual percentage losses for first and second order canals (*right*), combined absolute effect assuming exemplarily a field irrigation requirement of 500 mm

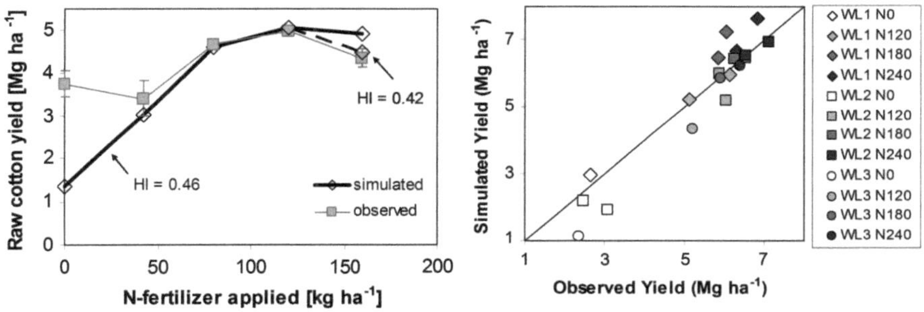

Fig. 6 Observed and simulated raw cotton yield in response to different levels of N-fertilizer applications (*left*), and simulated winter wheat yield plotted against observed yield (*right*); bars denote the standard error of the mean, WL water level, N nitrogen, figures in kg/ha total N application [after Djumaniyazova et al. (2010) and Kienzler (2010)]

In addition to classifying fields in view of their biophysical attributes, each farm field eventually was unique regarding its location within the irrigation canal network that determines amount of seepage loss in the canals (Fig. 5). In this regard, actual water required for field irrigation of each field was calculated based on the field distance to irrigation canals. This procedure accounts a surface water allocation problem by penalizing non-productive use of water by seepage over longer conveyance distances. For instance, if saving irrigation water is a goal, water-demanding crops like rice would better be grouped closer to the main water inlet.

Besides gathering field data for this purpose, CropSyst model calibration was the most labor- and time-intensive part. CropSyst calibration results are exemplarily shown for cotton and wheat, comparing observed and simulated cotton growth response to different levels of N-fertilizer applications and for 160 kg N ha^{-1} with two different harvest indices (HI; Fig. 6) Furthermore, the observed and simulated yields of winter wheat across all irrigation and fertilizer levels are compared.

4 Conclusions

Designed at a WUA level, the model allows for a quantitative analysis of agricultural policies in Uzbekistan. FLEOM captures the basic features of the regional agriculture, as well as the interrelations of production activities most prevalent to the local farmers. It integrates, aggregates and optimizes field-level management decisions on the allocation of water, inputs and labor in a spatially explicit way. It furthermore relates farm-level decisions with constraints or (optimization) goals of networks at the next higher level, such as WUAs or farmer associations. Sommer et al. (2010, 2012) demonstrated scenario simulations to provide first insights into the potentials of FLEOM for assessing production systems of a single WUA in Khorezm.

The developed version of the FLEOM model has a flexible and user-friendly interface and allows for one-click scenario formulation. However, up-to-dateness of integrated simulation models is a constant concern, especially when development of tools like that takes several years as in our case. This is even more so true for the current situation of agriculture and farm enterprises in Uzbekistan that underwent considerable institutional change in the past and most likely will experience further changes in the near future. Of major concern for a tool like FLEOM are the level of ownership and accountability (farm versus WUA), as well as farming obligation such as production targets or maximum allowance of withdrawal of irrigation water. Apart from the total availability of irrigation water, FLEOM also allows setting limits to other environmental hazards, such as the pollution of groundwater by excessive nitrate leaching or the deterioration of land by secondary soil salinization.

Probably one of the biggest advantages of FLEOM is that the GUI is programmed in a way that allows for a multitude of changes for possible scenario analyses including the simulation of some possible changes in the production environment, and thus convenient updating of the agricultural system under study. This can be done by directly adding additional crops and/or modifying socio-economic parameters. Some cases such as an institutional change of the production framework conditions could be coped with by a user via adjusting the model settings through GUI (e.g., by merging small farms to larger enterprises), but other cases (e.g., maximum allowance introduced for water use for a particular crop) might require changes in the GAMS code.

Equally important is FLEOM's generic character that allows simulating any farming environment for which bio-physical data and maps are available. For this, great care was given to the detailed setup and description of the bio-physical settings for the multiple CropSyst simulations that build the basis of the input-output table. Also bio-physical model calibration and validation received special attention. The importance of these underlying datasets cannot be overestimated, as all further integrated simulations do rely thereon. Our experience shows that putting together all necessary data for a single WUA requires considerable time. A current bottleneck still is computing time: processing a single simulation takes between 5 and 10 min depending on CPU power. However, the fast progress in computing power in general will most likely alleviate this bottleneck.

Another aspect is the educational potential of FLEOM, i.e., for gaining new theoretical insights into the interdependencies within the simulated agricultural system, that might be of utmost interest for a user rather than the exact quantities produced for parameters x, y, z under some specific situation.

The complexity of the modeling environment can create completely new challenges. The present experiences with FLEOM development and application showed that simulation results can be difficult to comprehend at first sight necessitating multi-disciplinary team discussions for a complete understanding of causalities.

Acknowledgments This study was funded by the German Ministry for Education and Research (BMBF; project reference number 0339970A). The user interface was programmed by Omonbek Salaev (e-mail: omonbek@gmail.com).

References

Akramkhanov A (2005) The spatial distribution of soil salinity: detection and prediction. ZEF—Ecology and Development Series No. 32, Cuvillier Verlag, Göttingen, p 120

Allen RG, Pereira LS, Raes D, Smith M (1998) Crop evapotranspiration: guidelines for computing crop water requirements. FAO Irrigation and drainage paper 56. FAO, Rome, Italy

Djanibekov N (2008) A micro-economic analysis of farm restructuring in the Khorezm region, Uzbekistan. Dissertation, ZEF, University of Bonn, Germany, p 182

Djumaniyazova Y, Sommer R, Ibragimov N, Ruzimov J, Lamers JPA, Vlek PLG (2010) Simulating water use and N response of winter wheat in the irrigated floodplains of Northwest Uzbekistan. Field Crops Res 116(3):239–251

Forkutsa I, Sommer R, Shirokova Y, Lamers JPA, Kienzler KM, Tischbein B, Martius C, Vlek PLG (2009a) Modeling irrigated cotton with shallow groundwater in the Aral Sea Basin of Uzbekistan: I Water dynamics. Irrig Sci 27(4):331–346

Forkutsa I, Sommer R, Shirokova Y, Lamers JPA, Kienzler K, Tischbein B, Martius C, Vlek PLG (2009b) Modeling irrigated cotton with shallow groundwater in the Aral Sea Basin of Uzbekistan: II Soil salinity dynamics. Irrig Sci 27(4):319–330

Hazell PB, Norton RD (1986) Mathematical programming for economic analysis in agriculture. MacMillan, New York, p 432

Ibrakhimov M (2004) Spatial and temporal dynamics of groundwater table and salinity in Khorezm (Aral Sea Basin), Uzbekistan. ZEF—Ecology and Development Series No. 23, Cuvillier Verlag, Göttingen, p 175

Kienzler K (2010) Improving the nitrogen use efficiency and crop quality in the Khorezm region, Uzbekistan. ZEF—Ecology and Development Series No. 72, Cuvillier Verlag, Göttingen, p 238

Kienzler K, Djanibekov N, Lamers JPA (2011) An agronomic, economic and behavioral analysis of N application to cotton and wheat in post-Soviet Uzbekistan. Agric Syst 104(5):411–418

Kuziev R (2006) Soil evolution in Khorezm region. Unpublished Report, Urgench, p 47

Martius C, Rudenko I, Lamers JPA, Vlek PLG (eds) (2012) Cotton, water, salts and soums—economic and ecological restructuring in Khorezm, Uzbekistan. Springer, New York, p 434

Sommer R, Kienzler K, Conrad C, Ibragimov N, Lamers JPA, Martius C, Vlek PLG (2008) Evaluation of the CropSyst model for simulating the potential yield of cotton. Agron Sustain Dev 28(2):345–354

Sommer R, Djanibekov N, Salaev O (2010) Optimization of land and resource use at farm-aggregated level in the Aral Sea Basin of Uzbekistan with the integrated model FLEOM—model description and first application. ZEF-Discussion Papers on Development Policy No. 139, Center for Development Research (ZEF), Bonn, p 102

Sommer R, Djanibekov N, Müller M, Salaev O (2012) Economic-ecological optimization model of land and resource use at farm-aggregated level. In: Martius C, Rudenko I, Lamers JPA, Vlek PLG (eds) Cotton, water, salts and soums—economic and ecological restructuring in Khorezm, Uzbekistan. Springer, New York, pp 267–283

SoyuzNIHI (1992) Methodology for hydro-module zoning. Tashkent. Fan publisher (in Russian)

Stöckle CO, Donatelli M, Nelson R (2003) CropSyst, a cropping systems simulation model. Eur J Agron 18(3–4):289–307

Efficiency of Duckweed (*Lemnaceae*) for the Desalination and Treatment of Agricultural Drainage Water in Detention Reservoirs

Dagmar Balla, Mohie Omar, Sebastian Maassen, Ahmad Hamidov and Mukhamadkhan Khamidov

Abstract High levels of soil salinity and water mineralization are one of the major problems in arid and semi-arid areas, in particular in Central Asia and North Africa. Besides technological water management solutions, accompanying methods of phytoremediation have been scientifically proven in detention ponds which are integrated into irrigation drainage systems. We report on investigation methods concerning the salt uptake potential of the aquatic macrophytes *Lemnaceae* (duckweed), and regarding technical construction requirements to enhance the purification of drainage water in water reservoirs. The assessment of this method was carried out under laboratory conditions at the Leibniz Centre for Agricultural Landscape Research (ZALF). The results demonstrate that the salt uptake behaviour of duckweed in the form of enclosure in its tissues depends on the degree of salinisation and the initial biomass density. The uptake effect can be

D. Balla (✉) · S. Maassen
Leibniz Centre for Agricultural Landscape Research (ZALF) e. V., Eberswalder Strasse 84
15374 Muencheberg, Germany
e-mail: dballa@zalf.de

S. Maassen
e-mail: Sebastian.Maassen@zalf.de

M. Omar
Nile Research Institute, National Water Research Center, Ministry of Water Resources and Irrigation, El-Quanater El-Khairya, Egypt
e-mail: mohie.omar@hotmail.com

A. Hamidov
Humboldt University of Berlin, Faculty of Agriculture and Horticulture, Invalidenstrasse 42
10117 Berlin, Germany
e-mail: ahmad.hamidov@agrar.hu-berlin.de

M. Khamidov
Bukhara Branch of Tashkent Institute of Irrigation and Melioration, Gazli street 32, Bukhara
200100, Uzbekistan
e-mail: khamidov_m@mail.ru

L. Mueller et al. (eds.), *Novel Measurement and Assessment Tools for Monitoring and Management of Land and Water Resources in Agricultural Landscapes of Central Asia*, Environmental Science and Engineering, DOI: 10.1007/978-3-319-01017-5_25,

described as a first-order decay reaction. The decay/uptake depends on the residence time within the water body. To maximize the detention time in reservoirs the effect of baffles and inlet constructions in the reservoir are investigated using 2D hydraulic modelling with the model package TELEMAC. Therefore, this method is proposed for Central Asian regions as a measure to be taken against high levels of salinity in water bodies.

Keywords Water quality · Phytoremediation · Duckweed · Desalinisation · Model TELEMAC

1 Introduction

In arid and semi-arid areas, salinisation is one of the major problems leading to the deterioration of environmental conditions and the decline of agricultural products. Salinisation affects about 30 million ha of the world's 260 million ha of irrigated land (FAO 2003). Salinisation problems are mainly found in Central Asia and North Africa.

The major rivers of Central Asia (Amudarya, Syrdarya, and Zarafshan) were continuously used for cotton production by the former USSR, which resulted in rising water tables and increasing saline land. Secondary salinisation is rapidly increasing and crop productivity has declined. In the southern part of Uzbekistan, more than 65 % of the area is variously salinised and waterlogged (Toderich et al. 2009). In the arid basin of the Aral Sea, the land has been cultivated for the last 2000 years, which has only been possible through the development of irrigation systems. Irrigated agriculture has spread from river regions (e.g. the Fergana valley and Tashkent region) deep into deserts to form oases (e.g. Khorezm and Bukhara). Nowadays, 10–11 % of the territory of Uzbekistan is irrigated for cotton and wheat production (Gintzburger et al. 2003). For irrigated agriculture an extensive network of irrigation and drainage canals has been developed, which also includes reservoirs and lakes for the storage of residual drainage water. During irrigation the water seeps into the soil and enters the groundwater, where it is salinised. The salinised groundwater is drained by deep drainage canals and reaches rivers, lakes or reservoirs for residual waters. Due to the mixing of salinised return water with better quality river water, irrigation farming can be regarded as the main source of water pollution. In rivers and reservoirs the dissolved salt concentration reaches 0.5–2.5 g/l and in lakes for residual storage it even reaches 4–18 g/l. Fresh water with salt concentrations less than 1 g/l is available only in upstream areas without inflows of return water, such as the upper catchments of rivers and tributary streams or the upper parts of the middle course of major rivers. Consequently, from 1969 to 1995 water salinity in the lower reaches of the Amudarya increased from 0.5 to 1.2 g/l, which exceeds the salinity limit for irrigation water of 0.5 g/l (FAO 2006). Recent measurements in the lower reaches of the Amudarya, in the Bukhara region, showed that on average the salt content in irrigation canals ranged between 1.0 and 1.2 g/l, whereas drainage canals had salinity levels up to 4 g/l (SALUS Workshop 2012).

The seriousness of the environmental situation in Uzbekistan directly affects politics, socio-economics, and human welfare. Lowering the salinity levels of soil and water is absolutely necessary to improve socio-economic and environmental conditions in the region. Desalinating drained groundwater before it enters the river could help delay a further salinisation increase in the main water courses in irrigated regions, such as the Bukhara oasis.

Desalination removes dissolved salts from water to a certain extent depending on the method applied: distillation in evaporators (thermal method), the ion-exchange method, electro-dialysis, the reverse osmosis (membrane) method. Besides these technological, cost-intensive methods for water purification, it may also be posssible to adapt some low-energy measures (Qureshi et al. 2007). Methods of phytoremediation through plants' nutrient uptake are used throughout the world, mainly in natural and constructed treatment wetlands (Andersson et al. 2007; El-Shafai et al. 2007; Kadlec and Wallace 2008; Bal Krishna and Chongrack 2008). Examples of the main groups of plants used for these ecotechnologies are submerged plants such as algae, surface-floating plants such as duckweed (*Lemnaceae*) (Fig. 1) and water hyacinths (*Eichhornia* sp.), emerged plants such as reeds *(Phragmites australis)* and bulrushes (*Typha latifolia)* or, in the sub-tropical and tropical zones, papyrus sedge *(Cyperus papyrus)* (Al Nozaili 2001).

Where sewage treatment has been effective, during the last decades the small aquatic plant duckweed (*Lemnaceae*) has undergone a revival. There is a wealth of literature about its purification behaviour (e.g. nutrients, trace metals, toxic substances), and technical solutions for the adaptation of duckweed for waste water treatment (e.g. Al-Nozaily 2001; Journey et al. 1993). The advantages of duckweed are its fast growth, worldwide occurrence, ease of maintenance and high protein content as fodder.

With respect to combined small-scale purification reservoirs or detention ponds within irrigation systems to treat drainage water under saline conditions, we have proven how duckweed grows in the case of higher salinities, that it can be used for

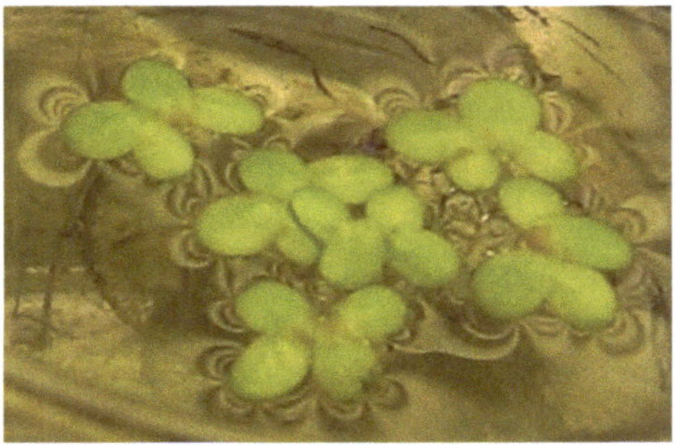

Fig. 1 Six duckweed plants with 4 fronds each (photo: M. Omar)

desalination, and how it responds to technological requirements. The investigations have been performed as part of a doctoral programme in Germany, at the Technical University of Berlin and at ZALF in Müncheberg. Outdoor experiments under mainly humid conditions in Germany are to be transferred and adapted to arid conditions in Uzbekistan (Omar 2011).

2 Materials and Methods

2.1 Lemna *Growth Inhibition Test*

This test is generally designed to assess the toxicity of substances dissolved in water and to quantify how they inhibit the growth of the *Lemna* duckweed species. The test was used in this work based on the existing guidelines provided by the OECD (2002) but many modifications were included to match the objectives of this work. *Lemna* was tested to assess the effect of salinity on the growth rate of *Lemna minor*.

Lemna minor was collected from a pond near ZALF at Müncheberg, where the experimental investigations were performed. Green fronds without visible lesions were washed with deionised water and cultured in a modified Swedish Standard (SIS) medium (OECD 2002). The culture was incubated for one week in a climate chamber with a temperature of 25 °C, light intensity of 8,000 lux and photosynthetically active radiation of 100 μEm2/s. Eleven fronds from the culture were transferred systematically to test vessels with different NaCl concentrations with three replicates each. Two tests were carried out under a large range of NaCl concentrations (0–10,000 mg/l), and two tests under a low range of concentrations (0–500 mg/l). The culture solution was used as a control and to dilute solutions.

The initial dry weight was measured after sub-samples of the fronds were rinsed with deionised water, kept on absorbent paper for 15 min to remove the water, moved to a ceramic dish and dried for 24 h at 60 °C. After the test was finished, the final fronds were counted and the final dry weight was measured.

The response of *Lemna minor* to different NaCl concentrations was assessed using the following parameters obtained according to OECD (2002):

1. Specific growth rate (μ) for each test concentration or control group:

$$\mu = \frac{\ln(N_7) - \ln(N_0)}{T_7 - T_0} \tag{1}$$

where $\ln(N_0)$ is the natural logarithm of the number of fronds in the test or control vessels at the beginning of the test, $\ln(N_7)$ is the natural logarithm of the number of fronds in the test or control vessels at the end of the test, T_0 is the time the test started (d), and T_7 is the time the test ended (d)

2. Inhibition of growth rate in terms of the number of fronds (Ir) for each test concentration:

$$I_r = \frac{(\mu_C - \mu_T)}{\mu_C} \times 100\,(\%) \quad (2)$$

where μ_C is the value for μ in the control group (%), and μ_T the value for μ in the treatment group (%)

3. Inhibition of growth rate of dry biomass (I_b) for each test concentration:

$$I_b = \frac{(b_C - b_T)}{b_C} \times 100\,(\%) \quad (3)$$

where $b_C = \ln(N_7) - \ln(N_0)$ for the control group, and $b_T = \ln(N_7) - \ln(N_0)$ for the treatment group

4. NaCl concentration causing 50 % of growth inhibition (EC_{50}).
5. The lowest observed effect concentration (LOEC): the lowest concentration at which the substance had a statistically significant reductive effect on growth.
6. No observed effect concentration (NOEC): the concentration immediately below the LOEC which, compared with the control, had no statistically significant effect.

3 Investigations Under Naturally Humid Climate Conditions

Six tanks were filled with water to a depth of 10.6 cm resulting in approx. 20.5 l. Two control tanks with an electrical conductivity of EC = 930 μS/cm (salinity S = 0.6 g/l), were not charged with extra salt. The next two tanks were loaded with extra NaCl up to EC = 2,280 μS/cm (S = 1.6 g/l) and the last two tanks up to EC = 3,040 μS/cm (S = 2.1 g/l). For each salinity, one duckweed-covered tank and one open-water tank (without duckweed) were prepared. Two experiments were carried out with initial duckweed biomasses of 50 and 30 g resulting in duckweed densities of 260 and 160 g/m^2, respectively. The EC, water level, and precipitation were measured daily.

The processes affecting the matter concentration are presented in Fig. 2 as follows: dilution and enrichment due to precipitation and evaporation, sedimentation of dead biomass or particles, transformation by biogeochemical processes, and accumulation in duckweed biomass.

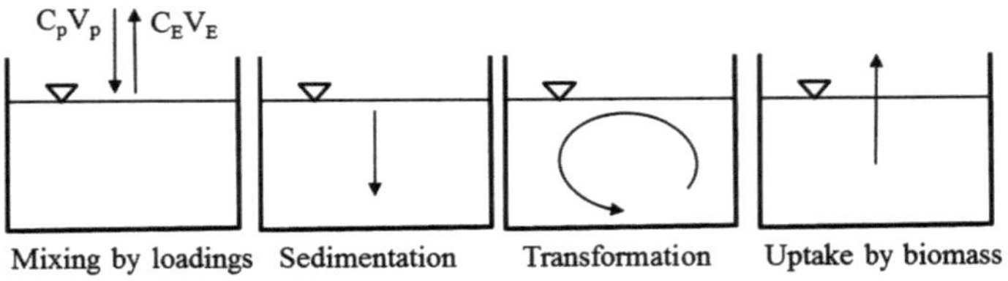

Fig. 2 Main processes of changes in water quality in a tank

The general salt mass balance according to Kadlec and Wallace (2008) is:

$$\frac{d(V \times C)}{dt} = A \times P \times C_P - \Delta V_S C_S - \Delta V_T C_T - A \times J\,(\mathrm{g/d}) \tag{4}$$

where V is the water volume in the container (m^3), C is the salt concentration of water (g/m^3), A is the surface area (m^2), P is the precipitation rate (m/d), Cp is the salt concentration in precipitation (g/m^3), ΔVSCS is the salt mass in sediment (g/d), ΔVTCT is the salt mass transformed by biological and chemical reactions (g/d), and J is the salt removal rate by duckweed (g/m^2*d)

There was assumed to be no salt loss through sedimentation, transformation or evapotranspiration in the tanks. The change in the salt concentration was caused only by changes in water volume due to precipitation and evapotranspiration, and salt accumulation in duckweed tissue. The mass balance can be simplified to:

$$\frac{d(V \times C)}{dt} = A \times P \times C_P - A \times J\,(\mathrm{g/d}) \tag{5}$$

The salt removal for the duckweed-covered tanks has been determined as:

$$(A \times J)_D = A \times P \times C_P - \frac{d(V \times C)}{dt}\,(\mathrm{g/d}) \tag{6}$$

The changes in the concentration due to evapotranspiration have been obtained by comparing this result with the salt removal of open-water tanks:

$$(A \times J) = A \times P \times C_P - \frac{d(V \times C)}{dt}\,(\mathrm{g/d}) \tag{7}$$

The actual salt removal (ΔA × J) was determined as the difference between the salt removal in the duckweed-covered tank in Eq. (6) and that in the open-water tank in Eq. (7)

$$\Delta A \times J = (A \times J)_D - (A \times J)\,(\mathrm{g/d}) \tag{8}$$

The evapotranspiration water loss was determined as the difference between precipitation and measured changes in storage.

$$ET \times A = P \times A - \frac{dV}{dt} \ (\mathrm{m}^3/\mathrm{d}) \tag{9}$$

4 Numerical Simulation of a Pond

An actual pond in East Brandenburg, Germany was investigated using numerical flow and transport simulations. 2D simulations were chosen for simplicity because the water depth in the pond was small (maximum = 0.55 m), the hydrostatic pressure was low and only the horizontal flow velocity was considered. The simulations were carried out considering the following aspects:

- Description and understanding of the hydrodynamic and transport processes in the pond as well as an estimation of the hydraulic residence time.
- Assessment of the influence of length, number and position of baffles on the hydraulic properties and efficiency as well as salt removal efficiency.

The study pond is a detention pond established in North-East Germany (Steidl et al. 2008). The ponds have been created to reduce nutrients from agricultural drainage systems. A drainage system for an agricultural area covering 5.1 ha discharges into the 0.05 ha pond. The pond discharges into the downstream receiving water. The bottom soil of the pond is loam (Fig. 3).

The discharge to the pond flows via a subsurface pipe and is controlled by a V-notch weir. The inflow enters the pond near to the bottom. The height of the inlet weir is 10 cm, resulting in discharges at 1–3 l/s. A similar V-notch weir is installed at the outlet. Surveys have been carried out previously and the resulting bathymetry was used in the pre-processing phase of modelling.

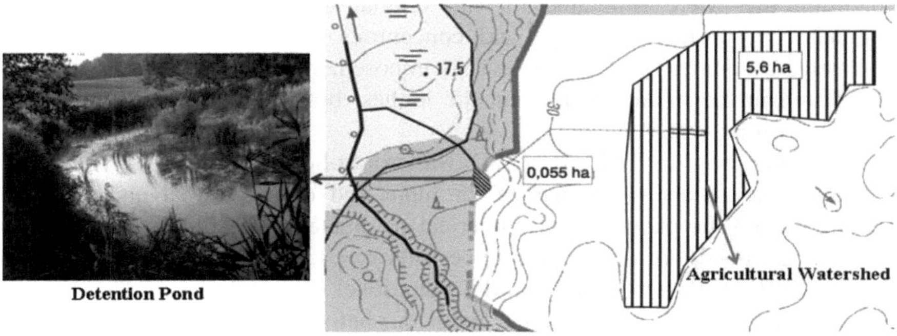

Fig. 3 The detention pond and its drained agricultural watershed (according to Steidl et al. 2008)

The modelling system TELEMAC-2D was used for simulation and to optimise the pond. This program solves the depth-averaged free surface flow and transport equations using the Finite Element Method (FEM) (Hervouet 2007; Hinkelmann 2005). The computational domain in the FEM is subdivided into many small finite elements with nodes. The main results at each node of the mesh are depth of water, depth-averaged velocity components and tracer concentrations.

The equations for the flow and transport processes in surface-water systems are:

1. Flow equations (2D):

$$\frac{\partial h}{\partial t} + \underline{v}\ grad\ h + h\ div\ \underline{v} = q_w\,(\mathrm{m/s}) \tag{10}$$

$$\frac{\partial \underline{v}}{\partial t} + \underline{v}\ grad\ \underline{v} - div(\underline{\underline{\upsilon}}_w\ grad\ \underline{v}) = \frac{1}{\rho_w}(\underline{f} - grad\ p)\,(\mathrm{m/s^2}) \tag{11}$$

where h is the water depth, $\underline{v}$ is the velocity vector of the free-surface flow, q_w is a sink or source term of water, $\underline{\underline{\upsilon}}_w$ is the turbulent viscosity, ρ_w is the density of water, p is the pressure, and f is a momentum source term

2. Transport equation (2D) for a hypothetical tracer S:

$$\frac{\partial S}{\partial t} + \underline{v}\ grad\ S - div(\underline{\underline{\upsilon}}_t\ grad\ S) = q_S\,(\mathrm{mg/l} * \mathrm{s}) \tag{12}$$

where $\underline{\underline{\upsilon}}_t$ is the turbulent diffusivity tensor, and q_S is a tracer sink or source term.

The work was done in three steps: pre-processing, processing and post-processing. The MATISSE module was used for the pre-processing, the TELEMAC-2D module for the processing, and the RUBENS module for the post-processing. In the pre-processing stage, MATISSE was applied to introduce the boundaries and bathymetry of the model. The computational mesh consisted of triangular elements and was refined around the inlet and outlet. The geometry and boundary condition files were generated for the simulation modules. On the upstream boundary, a constant discharge and concentration are prescribed and on the downstream boundary, the water level is imposed. The deepest bottom elevation was found in the middle (0 m) and the highest bottom elevation (0.65 m) was around the inlet (Fig. 4).

The processing in TELEMAC-2D was controlled by a steering file which contained data such as file names for input and output, initial and boundary conditions and the physical and numerical parameters. The time step was 0.3 s, which corresponds to a Courant number of Cr = 1. As an initial condition, the flow velocities were zero and the initial water level was set to 0.9 m. The "general case" of the pond refers to its actual behaviour before its dominant parameters are changed or the baffles configured. For the general case, the inflow discharge was

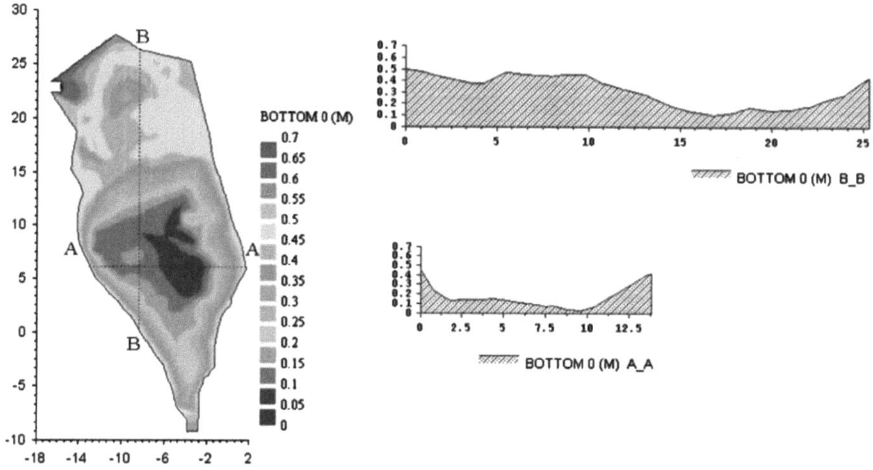

Fig. 4 The topography of the pond

imposed with a constant value of Q = 0.003 m^3/s. Manning's friction coefficient was n = 0.05 s/m1/3. A constant turbulent viscosity was chosen with a value of $\nu = 0.01\ m^2/s$.

Tracer transport simulations were carried out to investigate the spreading of a conservative tracer. Tracer simulations were used in this study to elucidate the degree of apparent mixing. Residence time distributions, RTDs, were generated from tracer simulations to better understand the mixing and the hydrodynamics. The tracer source was chosen as a source term near to the inflow and its flow duration was 10 min. The viscosity and the diffusivity were set as equal ($\nu = \nu_T = 0.01\ m^2/s$) (Fig. 4).

According to Shilton and Harrison (2003) and Werner and Kadlec (1996) the theoretical residence time, tn, can be calculated simply thus:

$$t_n = \frac{V}{Q}\ (\mathrm{d}) \tag{13}$$

where V is the pond volume (m^3), and Q the average flow rate (m^3/d)

The mean residence time (tm) was defined as:

$$t_m = \frac{\sum_{i=1}^{i=n} t_i \times C_i \times \Delta t}{\sum_{i=1}^{i=n} C_i \times \Delta t}\ (\mathrm{s}) \tag{14}$$

where ti is the time at ith time increment (s), Ci is the tracer concentration at ith time increment, n is the number of time steps and Δt is the time increment (here 2000s).

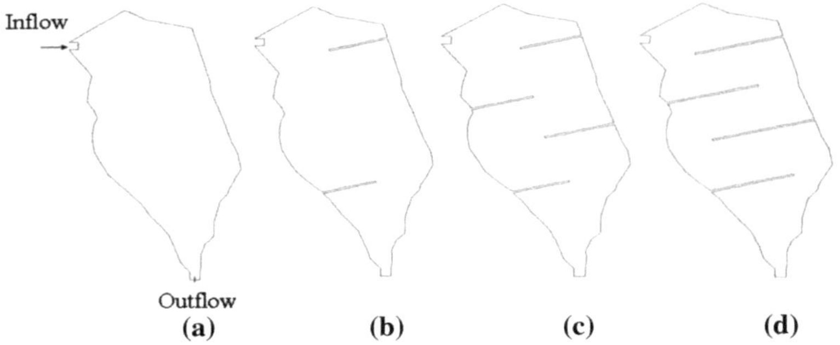

Fig. 5 Unbaffled pond (a), pond with two baffles of 50 % width (b), four baffles of 50 % width (c), and four baffles of 70 % width (d)

In line with Thakston et al. (1987) the hydraulic efficiency (e) was obtained as follows:

$$e = \frac{t_m}{t_n} \tag{15}$$

Finally, the index of short circuiting (I) expressing the amounts of water that flow very quickly to the outlet was defined as:

$$I = \frac{t_o}{t_n} \tag{16}$$

where to is the time at which the tracer first appeared.

The mean residence detention time estimated from Eq. (15) was determined with a first-order kinetic salt removal reaction in duckweed using the following formula:

$$\ln S = \ln S_o - K_r \times t_m \, (\mathrm{r}^2 = 0.95) \tag{17}$$

where S is the resultant outlet salinity (mg/l), So is the initial salinity in the pond = 650 (mg/l), Kr is the first-order salt removal rate of duckweed (1/d), t is the mean residence time (d) and r is the correlation coefficient ($r^2 = 0.95$ indicated that the correlation was very good).

The values resulting from Eq. (17) were only used to compare relative changes in removal efficiency for different pond configurations representing different scenarios of baffles (Fig. 5). However, in practice many factors affect the removal, such as temperature or the existence of other plants that accumulate salts.

5 Statistical Methods

During the *Lemna* test, the influences of NaCl concentrations on both duckweed growth parameters and removal efficiency were statistically assessed (Pearson correlation). During the investigations under natural climate conditions, the salt removal of duckweed was analyzed. The salt removal of duckweed was calculated as the difference between the salt concentration in duckweed-covered and open-water tanks as in Eq. (8). The integral method according to Chapra (1997) was used to determine whether the kinetic salt removal reaction was zero-, first-, or second-order. For a zero-order reaction, the plot should be linear for salt mass (SM) versus time (t), for a first-order reaction it should be linear for lnSM versus t, and for a second-order reaction it should be linear for 1/SM versus t. Each plot includes the best-fit line developed using linear regression.

6 Results

6.1 Lemna *growth inhibition test for salinity tolerance of duckweed*

Tests 1 and 2, with NaCl concentrations from 0 to 10,000 mg/l, obtained the following results (Table 1).

For both concentrations 500 and 1,000 mg/l, the inhibition growth rate had a negative sign compared to a zero concentration (Fig. 6).

When the NaCl concentration increased, the duckweed growth parameters such as dry weight and specific growth rate decreased significantly in Tests 1 and 2 ($p < 0.05$) (not shown). However, an increase in growth parameters was found at low salinities (up to 500 mg/l in Test 1 and up to 1,000 mg/l in Test 2).

Table 1 Growth inhibition in terms of the number of fronds and the quantity of dry biomass for two tests

Pearson correlates with NaCl conc.	Test 1		Test 2	
	Growth inhibition by number of fronds (I_r) $r = 0.85$ ($p < 0.05$)	Growth inhibition by dry biomass (I_b) $r = 0.87$ ($p < 0.05$)	Growth inhibition by number of fronds (I_r) $r = 0.92$ ($p < 0.01$)	Growth inhibition by dry biomass (I_b) $r = 0.99$ ($p < 0.01$)
NOEC	0.5 g/l	0.5 g/l	0.5 g/l	0.5 g/l
LOEC	1.0 g/l	1.0 g/l	1.0 g/l	1.0 g/l
EC50	1.8 g/l	1.6 g/l	1.5 g/l	1.3 g/l

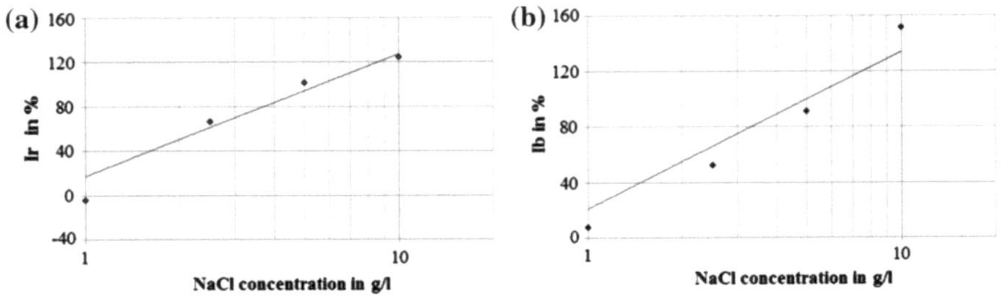

Fig. 6 Influence of NaCl concentrations (0–10,000 mg/l) on growth inhibition of frond count (a) and dry biomass (b) at temperature 25 °C in test 1

Under a low concentration range (0–500 mg/l) in Tests 3 and 4, EC50 lay outside the range. Unlike the previous results, there was no significant correlation ($p > 0.05$) between NaCl concentrations and growth parameters.

7 Investigations Under Humid Climate Conditions

7.1 Salt Uptake

Figure 7 shows plots to evaluate the reaction order for salinities of 0.6 and 1.6 g/l according to Chapra (1997). Both the reactions were first-order since the plots of lnSM versus time (t) most closely approximated a straight line. Table 2 shows the rates of the salt decrease for salinities of 0.6, 1.6 and 2.1 g/l for initial duckweed biomass densities of 260 and 160 g/m^2. A salt decrease rate of 0.122 d$-$1 means that the salinity decreases by 12.2 % per day.

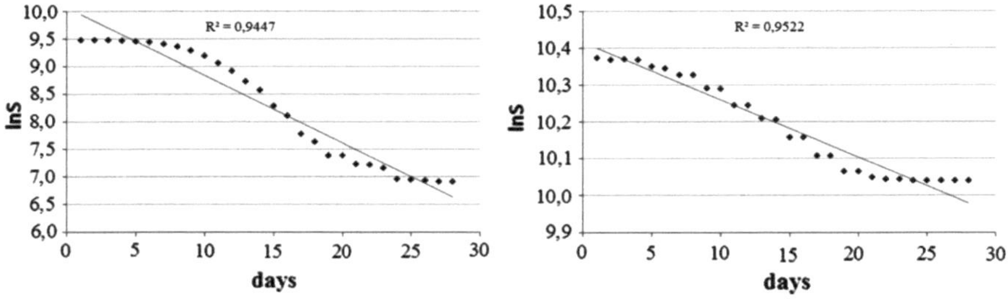

Fig. 7 Plots to evaluate the reaction order for water salinities of 0.6 g/l and 1.6 g/l (duckweed biomass density of 260 g/m^2)

Table 2 Salt decrease rate K_r (d^{-1}) in dependence on salinity and duckweed biomass density

Salt decrease rate K_r (d^{-1})		Initial duckweed biomass density (g/m^2)	
		260 g/m^2	160 g/m^2
Salinity	0.6 g/l	0.122 d^{-1}	0.040 d^{-1}
	1.6 g/l	0.015 d^{-1}	0.005 d^{-1}
	2.1 g/l	0.008 d^{-1}	0.014 d^{-1}

8 Evapotranspiration and Evaporation

With a salinity of 0.6 g/l and an initial duckweed density of 260 g/m^2, the total water loss by evapotranspiration (ET) over the duckweed-covered tank was 10,450 cm^3 and the loss by evaporation (E) over the open-water tank was 12,247 cm^3 (ET = 0.85 E) (Fig. 8). Table 3 shows the results for the different salinities and initial duckweed densities.

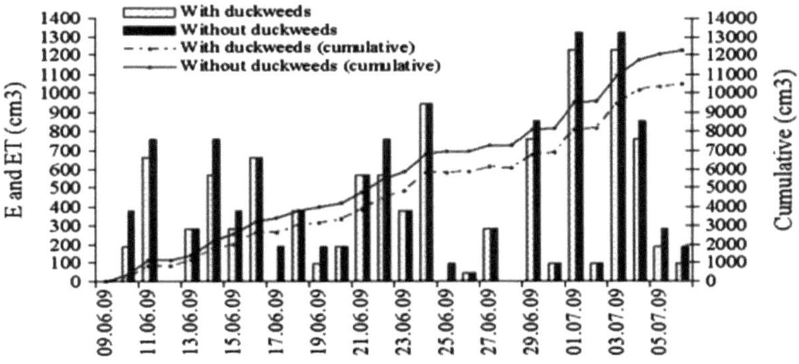

Fig. 8 Daily and cumulative evapotranspiration (ET) and evaporation (E) for a salinity of 0.6 g/l when duckweed intensity = 260 g/m^2

Table 3 Relation between evapotranspiration (ET) and evaporation (E) in dependence on salinity and duckweed biomass density

		Initial duckweed biomass density (g/m^2)	
		260 g/m^2	160 g/m^2
Salinity	0.6 g/l	ET = 0.85 E	ET = 0.94 E
	1.6 g/l	ET = 0.75 E	ET = 0.90 E
	2.1 g/l	ET = 0.75 E	ET = 0.87 E

9 The 2-D Numerical Simulation of the Pond

The hydrodynamic simulation of the unbaffled pond showed a bad velocity distribution with dead zones (Fig. 9). Four baffles of 70 % width showed a superior improvement in velocity distribution that reduced the hydraulic problems. Another hydraulic problem in the unbaffled pond was the swirling around the inlet (Fig. 10). This swirling almost disappeared with four baffles of 70 % width.

Equations (13), (14), (15) and (16) were used to obtain the mean residence time (t_m), theoretical residence time (t_n), hydraulic efficiency (e) and index of short circuiting (I). The hydraulic efficiency in the scenarios with baffles was higher than in the unbaffled pond. Increasing the number of baffles from two to four provided a higher hydraulic efficiency. The maximum hydraulic efficiency was achieved at the pond with four baffles of 70 % width.

The t_m values were integrated with the first-order salt removal rate obtained from the outdoor experiments to calculate the water salinity at the pond outlet. The

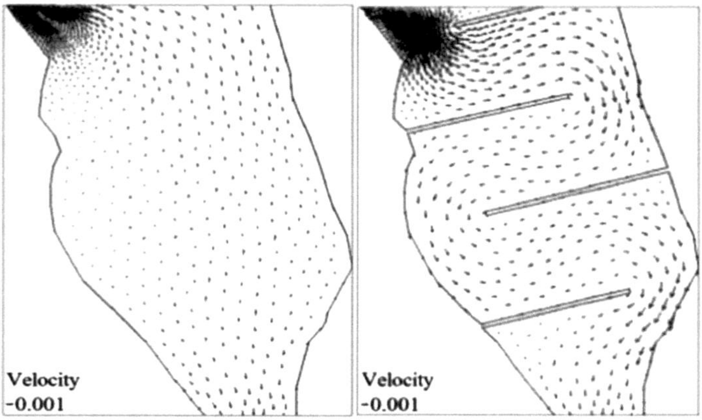

Fig. 9 Velocity vectors (m/s) in the unbaffled pond (left) and in the pond with four baffles of 70 % width (right)

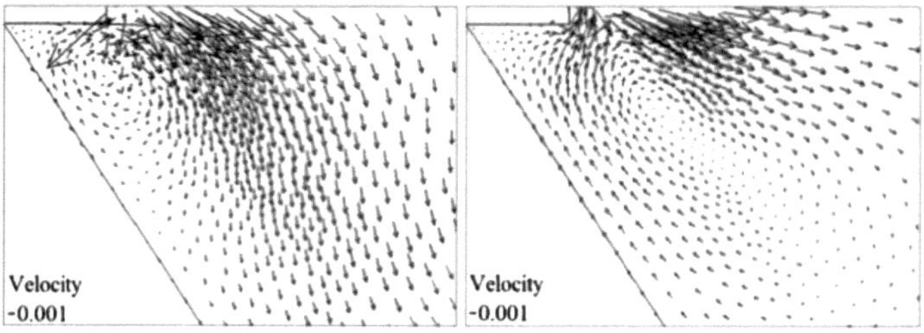

Fig. 10 The swirling around the inlet in the unbaffled pond (left) and its disappearance in the pond with four baffles of 70 % width (right)

value for the first-order salt removal was 12.2 % per day in the case of a water salinity of 0.6 g/l and a duckweed density of 260 g/m^2. The inflow salinity was 650 mg/l and the outflow salinity decreased from 611 mg/l in the unbaffled pond to 600 mg/l in the pond with four baffles of 70 % width.

10 Discussion

Lemna minor growth, under controlled climate conditions, is inhibited when NaCl concentrations increase in the range from 0 to 10,000 mg/l (total salinity from 0.2 to 11.5 g/l). In addition, all duckweed growth parameters such as dry weight, frond count and specific growth rate (μ) decrease significantly ($p < 0.05$). However, the growth does not depend on NaCl concentrations in the range from 0 to 500 mg/l (total salinity from 0.2 to 0.8 g/l). Moreover, low salinity promotes *Lemna minor* growth according to the following observations:

- The observed higher growth parameters at all NaCl concentrations up to 1,000 mg/l compared to those at zero concentration.
- The observed negative values of growth inhibition rates at concentrations up to 1,000 mg/l.

The results of outdoor investigations under naturally humid climate conditions prove that the salt removal of duckweed is first-order reaction kinetics. The first-order salt removal depends mainly on the ratio between duckweed biomass and water salinity. The daily first-order removal in this study is between 0.5 and 12.2 % per day in water at temperatures between 15 and 32 °C. A higher duckweed biomass (duckweed density = 260 g/m^2) achieves higher salt removal. Duckweed accumulates salts in its tissues and can thus desalinate the water.

Duckweed can reduce the evaporation water losses and save up to one quarter of the water volume lost by evaporation. Those results are corroborated in the studies of Ramey and Schardt (2004) and Oron et al. (1984). Ramey and Schardt (2004) reported that ET water loss by duckweed makes up 85–90 % of water loss through evaporation. Oron et al. (1984) reported that a lake that was covered completely by duckweed lost about a third less water through ET compared to an open-water lake.

Baffles reduce the hydraulic problems in the ponds such as dead zones, swirling and short-circuits and subsequently improve the overall hydraulic and removal efficiency. The pond efficiency depends on the number, lengths, and positions of the baffles. The system of four baffles of 70 % width is the most efficient out of the three tested systems with respect to both hydraulic and removal efficiency.

The results concerning baffles are in qualitative agreement with previous studies. Koskiaho (2003) found that the hydraulic efficiency was highly improved by baffles in two constructed wetland ponds in agricultural watersheds in southern Finland. Watters (1973) found that baffles of 70 % width gave superior performance compared to baffles of 50 and 90 % widths. Shilton and Harrison (2003) found that the highest efficiency was achieved when four baffles were applied.

11 Conclusion

The investigations under controlled climate conditions and under naturally humid climate conditions with different water salinities give rise to the following conclusions:

- The optimal growth of duckweed is achieved at salinities from 0 to 1.6 g/l and a temperature of 25 °C. Duckweed can survive up to a salinity of 3.1 g/l.
- Duckweed accumulates salts in its tissue independently of the water salinity and air temperature.
- The salt uptake rate under natural conditions is first-order reaction kinetics and ranges from 0.5 to 12.2 % at water temperatures between 15 and 32 °C.
- Duckweed can save up to 25 % of water loss through evaporation.
- Higher duckweed densities are preferable since a duckweed density of 260 g/m^2 achieves higher salt removal rates and lower evaporation water losses compared to 160 g/m^2.

The numerical simulations of an actual detention pond in the state of Brandenburg, Germany with the modelling system TELEMAC 2D give rise to the following conclusions:

- The design of detention ponds, which is controlled by the geomorphological conditions and composition of the drainage systems, can be modified by baffles that improve both the hydraulic and treatment efficiency.
- The best hydraulic and removal efficiency out of the three tested systems is achieved with a system of four baffles of 70 % width.

In conclusion, the present study proves that duckweed-covered ponds can be a suitable, effective and sustainable measure for bioremediation even in arid and semi-arid areas. Taking into consideration the high levels of salinity in water bodies in Central Asian regions, this current environmentally friendly technique is essential to increase agricultural productivity and improve the ecosystem. In the meantime, the emergence of climate change in the region, associated with an in the of air temperature and a decrease in rainfall, may lead to the problem of water shortage. Thus, the re-use of irrigation water after desalination is vital to improve rural livelihoods and overcome potential malnutrition.

References

Al-Nozaily FA (2001) Performance and process analysis of duckweed-covered sewage lagoons for high strength sewage. The case of Sana'a, Yemen. UNESCO-IHE, Delft

Andersson J, Wedding B, Tonderski K (2007) Estimating wetland non-point source nutrient removal—a comparison of methods and two Swedish regions. In: 2nd international symposium on wetland pollutant dynamics and control (WETPOL 2007), vol 140, pp 35–37

Bal Krishna K, Chongrack P (2008) An integrated kinetic model for organic and nutrient removal by duckweed-based wastewater treatment (DUBWAT) system. J Ecol Eng 34(3):243–250

Chapra S (1997) Surface water quality modelling. McGRAW-HILL series in water resources and environmental engineering, ISBN: 0-07-011364-5

El-Shafai S, El-Gohary F, Nasr F, Van der Steen N, Gijzen H (2007) Nutrient recovery from domestic wastewater using a UASB-duckweed ponds system. J Bio-source Technol 98(4):798–807

Gintzburger G, Toderich KN, Mardonov BK, Mahmudov MM (2003) Rangelands of the arid and semi-arid zones of Uzbekistan. CIRAD-ICARDA

FAO (2006) Report on agricultural drainage water quality. Food and agricultural organization of the united nations. International Commission on Irrigation and Drainage, Rome

Hervouet J (2007) Hydrodynamics of free surface flows: Modelling with the finite element method. Wiley, NY

Hinkelmann R (2005) Efficient numerical methods and information-processing techniques for modelling hydro- and environmental systems. Lecture Notes in Applied and Computational Mechanics, vol 21, Springer, Berlin

Kadlec R, Wallace S (2008) Treatment wetlands. 2nd ed. ISBN 978-1-56670-526-4

Koskiaho J (2003) Flow velocity retardation and sediment retention in two constructed wetland-ponds. J Ecol Eng 19(5):325–337

OECD (2002) OECD Guidelines for the testing of chemicals: *Lemna* sp. growth inhibition test, Guidelines 221, vol 22 Organisation for Economic Corporation and Development, Berlin

Omar M (2011) Improvement of detention ponds with respect to salinity. Book Series of the Department of Civil Engineering, vol 10. Technische Universität Berlin, Aachen

Oron G, Wildschut L, Porath D (1984) Wastewater recycling by duckweed for protein production and effluent renovation. Water Sci Technol 17:803–817

Qureshi AS, Qadir M, Heydari N, Turral H, Javadi A (2007) A review of management strategies for salt-prone land and water resources in Iran. International Water Management Institute, Colombo, Sri Lanka, p 30 (IWMI Working Paper 125)

Ramey V, Schardt J (2004) Plant management in Florida waters. Center for Aquatic and Invasive Plants, University of Florida, IFAS, Gainesville, and Department of Environmental Protection, Bureau of Invasive Plant Management, Tallahassee. http://plants.ifas.ufl.edu/guide

SALUS Workshop (2012) Desalination of irrigation water by usage of water plants for re-using in irrigation agriculture (SALUS) project, funded by the German Federal Ministry of Education and Research (BMBF). Workshop was held 5–6 Nov 2012 in the Bukhara region of Uzbekistan

Shilton A, Harrison J (2003) Guidelines for the hydraulics design of waste stabilization ponds. ISBN 0 - 473 - 08735 - 9

Steidl J, Kalettka T, Ehlert V, Quast J, Augustin J. (2008) Mitigation of pressures on water bodies by nutrient retention from agricultural drainage effluents using purification ponds. In: Proceedings of the 10th international drainage workshop of ICID working group on drainage, Helsinki/Talllin, 6–11 July 2008, pp 187–194

Journey W, Skillicorn P, Spira W (1993) Duckweed aquaculture. A new aquatic farming system for developing countries. The World Bank, Washington D.C

Thakston E, Shields L, Schroeder P (1987) Residence time distribution of Shallow Basins. J Envir Eng 113:1319–1332

Toderich KN, Shuyskaya EV, Ismail S, Gismatullina LG, Radjabov T, Bekchanov BB, Aralova DB (2009) Phytogenic resources of halophytes of Central Asia and their role for rehabilitation of sandy desert degraded rangelands. Land Degrad Devel. doi:10.1002/ldr.936

Watters G, Mangelson K, George R (1973) The Hydraulics of Waste Stabilization Ponds. Research Report; Utah Water Research Laboratory, College of Engineering, Utah State University; Utah, USA

Werner T, Kadlec R (1996) Stochastic simulation of partially-mixed, event-driven treatment wetlands. J Ecol Eng 14:253–267

Conservation Agriculture for Long-Term Soil Productivity

Mekhlis Suleimenov, Zheksenbai Kaskarbayev, Kanat Akshalov and Nikolai Yushchenko

Abstract Objective of this manuscript is to present a methodology of No-Till studies in multifactorial trials. It includes comparison of No-Till and traditional tillage in various crop rotations, under various sowing dates and varieties. This allowed to make comprehensive assessment of No-Till influence on soil moisture and plant nutrients availability as well as on crop yields. On heavy textured black soil No-Till has advantage over traditional tillage on organic matter conservation and moisture conservation in dry weather conditions. Traditional tillage had advantage in infiltration of snowmelt water into the soil in early spring as well as in nitrates availability to plants. No-Till had advantage in grain yields in drier conditions thanks to better moisture conservation. Major advantage of No-Till is conservation of organic matter for long-term soil productivity.

Keywords Conservation agriculture · Soil fertility · Multifactorial trials · Kazakhstan

M. Suleimenov (✉) · Z. Kaskarbayev · K. Akshalov
Scientific Production Center of Grain Farming Named after A.I. Barayev, Shortandy 021601, Kazakhstan
e-mail: mekhlis@yahoo.com

Z. Kaskarbayev
e-mail: tsenter-zerna@mail.ru

K. Akshalov
e-mail: kanatakshalov@mail.ru

N. Yushchenko
Karaganda Research Institute of Crop Science and Plant Breeding, Karaganda, Kazakhstan
e-mail: 12345680@bk.ru

L. Mueller et al. (eds.), *Novel Measurement and Assessment Tools for Monitoring and Management of Land and Water Resources in Agricultural Landscapes of Central Asia*, Environmental Science and Engineering, DOI: 10.1007/978-3-319-01017-5_26,

1 Introduction

In the mid-1950s, after development of vast areas of virgin lands (25 million hectares) in Kazakhstan for grain production, a spring wheat-fallow system was adopted. After several years of field trials at the Scientific Production Center of Grain Farming (SPCGF) it was recommended to fallow 20–25 % of cropland (Barayev 1960). Around five million hectares of cropland was left fallow each year, being a major source of soil erosion both by wind and by water. Later it was suggested that it would be more economical to reduce the fallow share to 15–20 % (Barayev 1984).

As distinct from Canadian practices, Kazakhstan's grain producers never used alternate fallow-wheat system but practiced longer continuous growing of spring wheat. However, summer fallow was recommended as one of the basic elements of dryland agriculture (Barayev 1960). The research results published by Suleimenov (1988) were the first in the Soviet Union published, proving that continuously growing spring wheat may produce more grain from total cropland than recommended rotation of summer fallow with continuous wheat during 3–5 years.

Tillage by moldboard plows was replaced by conservation tillage using blades and sweeps in the late 1960s (Barayev 1984). The conservation tillage system of that time consisted of main tillage done in the fall right after grain harvest of spring wheat, early spring harrowing and seedbed preparation. The purpose of tillage in the fall done in September–October is to cut weeds, and make the soil loose to prevent early spring run-off of snowmelt water. Snow is a very important source of moisture accumulation. Snow ridging is one of traditional methods of snow management to ensure catching snow drifting by wind during snowfalls. The snow ridging allows the accumulation of snow packs as high as 40 cm as compared to 15–20 cm in control. Depth of the fall tillage depends on soil compaction, soil moisture and landscape and varies from 12 to 14 cm to 22–25 cm. The tillage depth increases on compacted, moist soils and on sloping land. In early spring snow melting begins when soil is still frozen to depth of 1–2 m and snowmelt water will run-off on compacted soils especially on sloping land. In such cases loose soil enables better intake of snowmelt water and prevents soil erosion by water.

No-Till studies in Northern Kazakhstan were initiated in 2003 through FAO supported project conducted by CIMMYT and SPCGF on several farms (Karabayev et al. 2009). The results of these on-farm trials demonstrated the feasibility of direct seeding of spring wheat but it wasn't real No-Till research as it was testing of zero till equipment as compared with traditional cultivator-drills. The No-Till studies first were started at a Karaganda Research Institute of Crop Science and Plant Breeding in collaboration with CIMMYT at the Kostanai Research Institute of Agriculture (Dvurechenskiy 2009) and at SPCGF (Suleimenov 2006). Objective of this paper is to present methodology and some results of recent studies of No-Till at Shortandy and Karaganda sites.

2 Materials and Methods

Northern Kazakhstan is a territory between 50° and 54°N latitude and between 60° and 78°E longitude. It covers an area of 57 million hectares and comprises 4 provinces: Akmola, Kostanay, Pavlodar and North-Kazakhstan. Locations referred to in this paper are as follows: Shortandy (51.7°N, 71°E), Kostanai (53.1°N, 63.8°E), and Karaganda (49.8°N, 72.8°E).

At Shortandy conservation agriculture has been studied in a number of multi-factorial trials. The soil is a Southern Calcareous Chernozem on heavy clay loam with organic matter content of about 3.5 %. in the topsoil. The long-term average annual precipitation is 322 mm. The distribution of precipitation is characterized by monthly falling of 20–25 mm throughout autumn, winter and spring and more rainfall in June (40 mm), July (54 mm) and August (35 mm). The snowfall takes one-third of the total annual precipitation and plays an important role in soil moisture accumulation. The annual average air temperature is 2.1 °C with highest temperatures in July (20.1 °C). This type of weather is typical for all northern part of Kazakhstan and adjacent areas of western Siberia.

The data presented in this paper are based on research conducted during 2009–2011. Out of these three years 2010 was one of the driest in history. It was characterized by almost no rainfall during the vegetation period of spring wheat. In 2009 in the period from sowing till wheat jointing stage the weather was dry and cool. The period between jointing and grain filling was favorable as far as rainfall and air temperatures are concerned. This is why grain yields were above average. In 2011 rainfall distribution throughout the vegetation of spring wheat was one of the best in history of the region. In general record grain yield was obtained in northern Kazakhstan. In this paper No-Till refers to continuous zero tillage over several years, whilst minimum tillage refers to occasional direct seeding in a particular year.

In the first trial the factors were cropping technologies and crop rotations (App. 1). The technologies were: (1) Simple, (2) Traditional and (3) No-Till. Under simple technologies intensification inputs (fertilizers and chemicals) were excluded except occasional application of 2.4-D, and tillage includes three tillage operations in the fallow year and sowing with a cultivator-drill on the stubble land. The traditional technology was based on tillage in the fall as deep as 22–25 cm, early spring tillage by needle harrows and sowing with cultivator-drills. The fallow management consisted of four to five tillage operations to control weeds properly. No-Till consisted of controlling weeds by glyphosate herbicides and direct drilling. The traditional and No-Till technologies included an annual application of fertilizers with seeds in all fields except wheat sown after summer fallow at rates of 30–35 kg of N and 15–20 kg of P_2O_5 ha^{-1} and fungicides and insecticides needed. The snow management was done by snow ridging in case of traditional technology and by leaving tall stubble in case of No-Till. All types of cropping technologies have been applied continuously since 2006 (Fig. 1).

The second factor in the first trial was crop rotations: (1) Grains-fallow, (2) Grains, (3–4) Diversified crop rotations, as well as (5–6) Continuous wheat and

Fig. 1 **a** Field days of research center are important to demonstrate and discuss conceptions and results with farmers and other stakeholders. **b** Kanat Akshalov explains the influence of simple, traditional and no-till technologies, fertilization and seeding rate on crop performance in a new experiment on Shortandy site. Photos courtesy of Tobias Meinel

barley. Traditional 5-year rotation of grains and fallow is "fallow-wheat-wheat-barley-wheat". In the rotation of grains summer fallow was replaced by oats, while in the diversified crop rotations summer fallow was replaced by dry pea or rapeseed. The fallow-grains and grains rotations were fully established in 1991, while the diversified crop rotations were established in 2005. Total number of treatments 22, in three replications. Total number of plots 66. Plot size: width—10 m, length—50 m.

In the second trial three factors were tested: (1) Tillage, (2) Sowing time, (3) Wheat varieties (App. 2). Tillage treatments were: Traditional, Minimum and Zero. Sowing dates were: 19 May, 27 May and 2 June. The wheat varieties: 3 varieties of bread wheat—Astana 2, Tselina 50 and Tselinnaya 2007 as well as 2 varieties of durum wheat—Korona and Damsinskaya yantarnaya. The total number of treatments was 45, plots—135. Plot size: width—4 m, length—50 m.

In the third trial three factors were tested: (1) Tillage, (2) Sowing dates and (3) Barley varieties. The tillage treatments: Traditional, Minimum and Zero. The sowing dates: 20 May, 27 May and 3 June. The barley varieties: Astana 2000, Astana 2007 and Pamyati Raissy. Total number of treatments—27, plots—81. Plot size: width—4 m, length—50 m.

Studies of tillage methods in multifactorial trials give opportunity to make comprehensive assessment of No-Till. The factor of crop rotations was included because there is a trend of switching from traditional fallow-grains rotations to the diversified crop rotations where grains are rotated with oilseeds and pulses. The factor of sowing dates was included because sowing at various dates creates a variety of weather conditions for crop development in one year. The factor of varieties was included to get great number of data for evaluation of tillage methods as well as for a better variety assessment under No-Till.

At the Karaganda site No-Till and traditional tillage has been applied in a five year rotation "summer fallow-4 years of grains" since 2001. The observations have been conducted on three landscape parts: Plateau, Northern and Southern slopes.

Soil moisture was determined by drying off soil samples at a temperature of 105 °C.

Assessment of organic matter content in soil was made by method based on dissolving organic matter to carbonic dioxide and water (Tyurin method modified by TSINAO, standard GOST 26213).

Nitrates content in soil samples was determined by standard disulphurphenol method after Grandval Lyazhu.

3 Results

3.1 Soil Moisture

The soil moisture was affected by preceding field, tillage method and snow management (Table 1).

By the time of sowing spring wheat in late May under simple technology with no snow management and no tillage it was accumulated just 71 mm of available moisture in the 1 m soil layer. In case of traditional technology on stubble land including tillage in the fall with sweeps 22–25 cm deep combined with snow ridging soil water storage increased by 47 mm. Under No-Till snow was trapped by tall stubble and snow pack was about the same in both snow management treatments but water infiltration was better in case of tillage in the fall.

In the summer fallow on average additional 15 mm of available moisture was accumulated as compared to stubble land. During summer period of fallow all rainfall water evaporates irrespective of tillage method. From data presented it is obvious that summer fallow has no sense as method of moisture accumulation as only 15 mm accumulates from 320 mm of annual precipitation.

3.2 Soil Nitrates

Availability of nitrates before sowing of spring wheat is most important. It is affected by decomposition of organic matter during summer fallowing and intensive tillage (Table 2).

Table 1 Soil moisture storage in the 0–100 cm layer (mm) prior to sowing of spring wheat on summer fallow and on stubble land (average for 2009–2010)

Preceding field crop	Technology			Average
	Simple	Traditional	NT	
Summer fallow	87	132	122	114
Stubble land	71	118	109	99

The nitrates availability has been affected mostly by fallowing and tillage intensity. This is why traditionally tilled (4–5 operations) summer fallow has notable advantage over No-Till. Under simple technology and chemical fallow an intensity of organic matter decomposition reduced remarkably. On stubble land after grains nitrates availability reduced 3–4 times with no differences between tillage methods. There is a trend observed to increased nitrates availability on plots of wheat sown after dry pea with No-Till reducing this process.

3.3 Crop Yield

The assessment of crop yields as affected by tillage methods in variety of conditions shows that the transfer from traditional tillage system to No-Till not always ensures a crop yield increase. It depends on many factors including soil and weather conditions.

The simple technology shows lowest level of crop yield of spring wheat when no intensification means have been applied. In any year crop yields might be doubled or even tripled (Table 3).

Traditional tillage had an advantage over No-Till in all three years although a significant difference was obtained only in 2011. The traditional tillage ensures more rapid decomposition of organic matter ensuring better availability of nitrates for feeding crop. An advantage of traditional tillage is more remarkable on stubble fields where deficiency of nitrates has been observed. The largest advantage of traditional tillage was noted in 2011 which was characterized by wet weather in June. This moisture was better infiltrated in loose soil. Besides better developed wheat crop demanded more nitrogen lacking on No-Till. This also reduced continuous wheat yield as compared to wheat grown second year after summer fallow.

Analysis of grain yields of other crops shows that they were also associated with tillage effect on moisture and nitrogen availability. Barley grain yields in all three years had trend to increase under traditional tillage as compared with No-Till with more notable advantage in the dry year 2010. Oat grain yield was the best in 2009 in spite of drought in the first part of vegetation. In that year oat gave the record yield and it was obtained on No-Till. In other two years both dry 2010 and favorable 2011 traditional tillage system had an advantage in grain yield over No-Till. On average the two treatments provided the same oat yields.

Table 2 Nitrates availability (mg kg^{-1}) in 0–30 cm soil layer prior to sowing of spring wheat as affected by preceding field and tillage method (average in 2009–2010)

Preceding field crop	Tillage			Average
	Simple	Traditional	NT	
Summer fallow	50	113	84	82
Grains	27	28	27	27
Dry pea	34	33	25	31

Table 3 Spring wheat grain yields (t ha^{-1}) as affected by cropping technology and place in grains-fallow rotation

Technology	2009	2010	2011	Average
Wheat after summer fallow				
Simple	1.39	0.44	2.15	1.32
Traditional	3.13	1.46	4.18	2.92
No-till	3.03	1.34	4.00	2.78
LSD_{05}	0.41	0.30	0.45	
2nd year after summer fallow				
Simple	1.31	0.42	1.83	1.18
Traditional	3.04	1.44	4.09	2.85
No-till	2.92	1.31	3.48	2.43
LSD_{05}	0.32	0.11	0.42	
Continuous wheat since 1979				
Simple	1.48	0.34	1.36	1.06
Traditional	2.93	1.28	3.51	2.57
No-till	2.86	1.23	3.15	2.41
LSD_{05}	0.32	0.18	0.34	

Dry pea grain yields were affected by tillage methods in the same manner as it was shown for oats. On average the traditional tillage had some advantage over No-Till. The advantage of No-Till in 2009 might be explained by the fact that it was applied during four years following previous long-term traditional tillage.

Rapeseed seed yields were affected by tillage methods in different way. In 2009 rapeseed reacted positively on No-Till while in other two years the yields were not affected by tillage methods. This can be explained by better placement of small seeds in case of No-Till as compared with traditional cultivator-drill. It was not enough however, to overcome the other advantages of traditional tillage described before.

As a whole in this particular trial the traditional tillage has an advantage in two factors: better snowmelt water infiltration and in availability of nitrates to crops. The advantage of summer fallow over stubble crop in crop yield was achieved mostly at the expense of better soil nitrogen regime. Nitrogen fertilizers were applied at rather low rates.

In another trial during the same period comparison of two tillage methods have shown advantage of zero tillage in bread wheat and durum wheat grain yields (Table 4).

Grain yield of two wheat types was higher at delayed sowing date of 26 May. This is explained by more rainfall in July which is better utilized by wheat sown later. At all sowing dates zero tillage has an advantage over traditional tillage thanks to better moisture conservation. The yield gains were higher at optimal sowing dates. Both bread wheat and durum wheat produced more nitrogen in grain under traditional tillage method which once more demonstrated advantage of traditional tillage for organic matter decomposition.

Table 4 Grain yield (t ha^{-1}) of bread wheat (average of 3 varieties) and durum wheat (2 varieties) as affected by tillage methods and sowing dates (average for 2009–2011)

Sowing date	Bread wheat		Durum wheat	
	Traditional	No-till	Traditional	No-till
19 May	1.32	1.38	1.20	1.38
26 May	1.39	1.58	1.46	1.73
2 June	1.32	1.42	1.25	1.42

Barley grain yield was affected by tillage methods in different years and sowing dates. Barley gets ripe faster than wheat proving higher grain yields at delayed sowing dates. Under weather conditions of 2009 with delayed rainfall barley used them most productively at late sowing date. Zero tillage had notable advantage at early sowing dates, when barley performance was affected by drought in June. At later sowing date however, when barley development occurred under more favorable moisture conditions traditional tillage had advantage as compared to zero tillage.

Barley grain yield was very low in 2010 when weather conditions were very dry throughout whole vegetation period with remarkable advantage of zero tillage at all sowing dates.

Under favorable weather conditions of 2011 for barley growth at all sowing dates crop yield was very high with notable advantage of traditional tillage. This data shows that zero tillage demonstrated better barley performance in drier conditions while traditional tillage had advantage under more favorable soil moisture conditions.

3.4 Organic Matter

The effect of No-Till on organic matter content was obtained in the longest No-Till study conducted at Karaganda site in the dry steppe zone of Central Kazakhstan. Studies at this site have shown crop yield differences in accordance with landscape: plateau, northern and southern slopes. The grain yield was lowering from plateau to northern and southern slopes because of differences in soil moisture availability which was always higher on plateau. After ten years of continuous No-Till an increase of organic matter has been observed. The organic matter content increased notably in a top 0–10 cm layer from 3.11 % under traditional tillage to 3.26 % under No-Till as a result of continuous no-till application over ten years in a five year rotation "fallow-4 year grains" at the expense of less organic matter losses through nitrification processes in no-tilled soil. In larger farm fields these losses might be increased by soil losses by run-off water especially on sloping lands.

4 Discussion

When conclusions about the possibility of excluding summer fallow from crop rotations were published for the first time (Suleimenov 1988), they were unanimously rejected by all scientists of the Soviet Union working in dryland agriculture research (Kashtanov 1988; Buyankin and Burakhta 1988; Korchagin 1989; Poluektov 1989; Shiyatiy 1989; Dvurechenskiy 2003). Later on more research was conducted comparing traditional rotation of "summer fallow-4 year grains" with the same rotation but replacing summer fallow with oats and dry pea (Suleimenov and Akshalov 2007; Suleimenov et al. 2010). This comparative study was conducted under three levels of crop management: poor, medium and adequate. The results obtained showed that no-fallow cropping produced more grain from total cropland area under all three levels of crop management. The best result from removing summer fallow, however, was noted under best cultural practices.

A number of studies have been supporting the idea about successful replacement of summer fallow by variety of crops in dryland agriculture by many other research done in Canada (Gan et al. 2003; Larney et al. 2004) and in Russia (Demarchuck 2006).

Comparative studies of No-Till and traditional conservation tillage with sweeps and blades in a number of trails at two research fields at Shortandy site with variety of combinations of crops, varieties, snow management and sowing dates have shown that results obtained gave great number of data allowing careful assessment of tillage treatments.

Studies on No-Till and zero tillage have been conducted at a number of stations and research centers in Northern and Central Kazakhstan. The most successful data in favor of No-Till has been obtained at Kostanai site on Chernozem sandy loam soils (Dvurechenskiy 2009). This can be explained by the fact that there is no problem of snowmelt water infiltration in early spring on light textured soils. Also more reliable data in favor of No-Till at this site have been obtained due to more careful snow management practices (Gilevich 2012). This includes leaving tall stubble and producing plant barriers on chemical summer fallow to trap drifting snow during the second year of fallow period. This kind of practice was found useful on light textured soils while on heavier soils it may cause soil erosion by snowmelt water in early spring.

The effect of No-Till on organic matter content was obtained in the longest No-Till study conducted at the Karaganda site in dry steppe zone of Central Kazakhstan. After ten years of continuous No-Till an increase of organic matter has been observed. In larger farm fields these losses might be increased by soil losses during wind and water erosion.

For agricultural producers most important is an economical assessment of cropping technologies. The results of the studies conducted in the region are controversial. Some researchers conclude that No-till is more economical than traditional tillage (Dvurechenskiy 2009; Karabayev et al. 2009) while the others are making conclusions that traditional tillage is more economical (Dubina 2010). It depends not only on differences in crop yield response on transfer from traditional tillage to No-Till. It also depends a great deal on production cost in which cost of chemicals plays most important role. The cost of glyphosate in Kazakhstan is rather high ($US 10/l) which makes No-Till costly. Cost of equipment is also important as modern imported No-Till seeders are much more expensive as compared to traditional soviet cultivator-drills. The other thing is that in plot trials soil erosion is never taken into consideration while in on-farm conditions soil erosion happens frequently on tilled fields especially on tilled summer fallow which is widespread practice. Soil conservation is the major advantage of No-Till for long-term soil productivity.

5 Conclusions

Traditional tillage has advantage over No-Till in soil nitrates availability for performance of variety of crops in various weather conditions. Traditional tillage has also advantage in snowmelt water infiltration on heavy textured soils in some years.

No-Till has more chances for wide adoption on light textured soils while minimum tillage or seem to be more suitable on heavy textured soils.

In general No-Till was more useful in drier soil conditions thanks to better moisture conservation by mulching while traditional tillage system provided better crop performance under favorable rainfall distribution.

For a successful No-Till adoption on Chernozem soils application of adequate rates of nitrogen fertilizers is needed for grain yield increase and higher protein content in wheat grain.

Long-term No-Till ensures better conservation of soil organic matter. This should be considered as major advantage of No-Till in on-farm conditions.

No-till studies should be conducted in long-term trials on various soil types and landscapes. For more adequate assessment of No-Till it is advisable to conduct trials with tillage treatments under different crop rotations, seeding dates and fertilizer applications. In economical assessment of tillage technologies cost of organic matter should be taken into consideration.

App. 1 Experimental design of trial "Comparative study of cropping technologies in various crop rotations"

CW			CB			Fallow			Oat			Pea			Rapeseed			1W after F			1W after O			1W after P			1W after R		
S	T	NT	S	T	NT	S	T	NT	S	T	NT	S	T	NT	S	T	NT	S	T	NT	S	T	NT	S	T	NT	S	T	NT
2W after F			2W after O			2W after P			2W after R			3B after F			3B after O			3B after P			3B after R			4W after F			4W after O		
S	T	NT	S	T	NT	S	T	NT	S	T	NT	S	T	NT	S	T	NT	S	T	NT	S	T	NT	S	T	NT	S	T	NT
4W after P			4W after R			CW			CB			Fallow			Oat			Pea			Rapeseed			1W after F			1W after O		
S	T	NT	S	T	NT	S	T	NT	S	T	NT	S	T	NT	S	T	NT	S	T	NT	S	T	NT	S	T	NT	S	T	NT
1W after P			1W after R			2W after F			2W after O			2W after P			2W after R			3B after F			3B after O			3B after P			3B after R		
S	T	NT	S	T	NT	S	T	NT	S	T	NT	S	T	NT	S	T	NT	S	T	NT	S	T	NT	S	T	NT	S	T	NT
4W after F			4W after O			4W after P			4W after R			CW			CB			Fallow			Oat			Pea			Rapeseed		
S	T	NT	S	T	NT	S	T	NT	S	T	NT	S	T	NT	S	T	NT	S	T	NT	S	T	NT	S	T	NT	S	T	NT
1W after F			1W after O			1W after P			1W after R			2W after F			2W after O			2W after P			2W after R			3B after F			3B after O		
S	T	NT	S	T	NT	S	T	NT	S	T	NT	S	T	NT	S	T	NT	S	T	NT	S	T	NT	S	T	NT	S	T	NT
3B after P			3B after R			4W after F			4W after O			4W after P			4W after R														
S	T	NT	S	T	NT	S	T	NT	S	T	NT	S	T	NT	S	T	NT												

Legend
CW continuous wheat, *CB* continuous barley, *F* summer fallow, *O* oat, *P* pea, R—rapeseed. 1W—1st year wheat, 2W—2nd year wheat, 3B—3rd year barley, 4W—4th year wheat. Cropping technologies: *S* simple, *T* traditional, *NT* No-Till

App. 2 Experimental design of a trial "Comparative study of tillage methods at three sowing dates of two wheat types"

Traditional tillage						Minimum tillage						Zero tillage					
19 May		27 May		2 June		19 May		27 May		2 June		19 May		27 May		2 June	
BW	DW	BW	DW	BW	DW	BW	DW	BW	DW	BW	DW	BW	DW	BW	DW	BW	DW
Minimum tillage						Zero tillage						Traditional tillage					
19 May		27 May		2 June		19 May		27 May		2 June		19 May		27 May		2 June	
BW	DW	BW	DW	BW	DW	BW	DW	BW	DW	BW	DW	BW	DW	BW	DW	BW	DW
Zero tillage						Traditional tillage						Minimum tillage					
19 May		27 May		2 June		19 May		27 May		2 June		19 May		27 May		2 June	
BW	DW	BW	DW	BW	DW	BW	DW	BW	DW	BW	DW	BW	DW	BW	DW	BW	DW

Legend
Tillage methods: traditional, minimum, zero (No-till)
Sowing dates: 19 May, 27 May, 2 June
Wheat types: *BW* bread wheat, *DW* durum wheat

References

Barayev AI (1960) Cropping structure and crop rotations in grain farming of Northern Kazakhstan. J Zemledeliye, Moscow 7:15–21 (in Russia)
Barayev AI (1984) Further development of conservation agriculture: most important challenge. J Vestnik selskokhozyastvennoi nauki, Moscow 12:18–23 (in Russia)
Buyankin VI, Burakhta SN (1988) Ways to improve soil and management systems in Ural Province. J Zemledeliye, Moscow 11:15–20 (in Russia)
Demarchuk G (2006) Sweet clover: green manure and forage crop. In: Ivanov A, Zhdanov S, Pavlov V, Lushnikov A, Podgorbunskaya P (eds) Role of modern technologies in sustainable development of agroindustry complex. Materials of international conference on dedicated to 110 birthday of T.S. Maltsev. Kurgan, Russia, pp 317–321 (in Russia)
Dubina NE (2010) No-till and crop rotations in conservation agriculture of North Kazakhstan province. J Agrarniy Sector, Astana, Kazakhstan 3:45–49 (in Russia)
Dvurechenskiy VI (2003) Water and soil conserving technologies of grain production in northern Kazakhstan. In: Scientific bases of state program of agricultural production of Republic of Kazakhstan for 2003–2005. Ministry of Agriculture, Astana, pp 84–85 (in Russia)
Dvurechenskiy VI (2009) No-tillage technologies in dry steppe of Kazakhstan. In: Suleimenov M, Kaskarbayev Z, Skoblikov V, Dashkevich S (eds) Proceedings of international conference on no-till and crop rotation: base of agricultural policies supporting conservation agriculture for intensification of sustainable production, 8–10 July 2009. Astana-Shortandy, Kazakhstan, pp 91–96 (in Russia)
Gan YT, Miller BG, McConkey BG, Zentner RP, Stevenson FC, McDonald CI (2003) Influence of diverse cropping sequences on durum wheat yield, and protein in semiarid Northern Great Plains. Agron J 95:245–252 (in Russia)
Gilevich S (2012) Role of summer fallow in conservation agriculture. J Agrarniy Sector, Astana, Kazakhstan 1:28–31 (in Russia)
Karabayev M, Wall P, Brown JY, Morgounov A (2009) CIMMYT main activities on conservation agriculture in Kazakhstan. In: Suleimenov M, Kaskarbayev Z, Skoblikov V, Dashkevich S (eds) Proceedings of international conference on no-till and crop rotation: base of agricultural policies supporting conservation agriculture for intensification of sustainable production, 8–10 July 2009. Astana-Shortandy, Kazakhstan, pp 34–39
Kashtanov AN (1988) To follow work began by academician A. Barayev. J Zemledeliye, Moscow 11:12–15 (in Russia)
Korchagin VA (1989) Summer fallow: base of crop rotation. J Zemledeliye, Moscow 3:12–16 (in Russia)
Larney F, Yanzen H, Smith E, Anderson D (2004) Dryland agriculture on the Canadian Prairies: current issues and future challenges. In: Rao S, Rayan J (eds) Challenges and strategies for dryland agriculture. CSSA Special Publication, Madison, Wisconsin, pp 113–139
Poluektov EV (1989) Summer fallow and moisture supply to plants. J Zemledeliye, Moscow 3:16–21 (in Russia)
Shiyatiy EI (1989) Are there bases for categorical conclusions. J Zemledeliye, Moscow 5:20–26 (in Russia)
Suleimenov M (1988) About theory and practice of crop rotations in Northern Kazakhstan. J Zemledeliye, Moscow 9:14–20
Suleimenov M (2006) Dryland agriculture in Northeastern Europe and Northwestern Asia. In: ASA, CSA, SSSA, Peterson G, Unger PW, Payne WA (eds) Dryland agriculture, 2nd edn. Madison, USA, Agronomy monograph no. 23, pp 625–669
Suleimenov M, Akshalov K (2007) Eliminating summer fallow on black soils of Northern Kazakhstan. In: Lal R, Suleimenov M, Stewart B, Hansen D, Doraiswamy P (eds) Climate change and terrestrial carbon sequestration in Central Asia. Taylor and Francis, pp 267–279

Suleimenov M, Kiyas A, Kaskarbayev Z (2010) Replacement of summer fallow with oats and food legumes on black soils of Northern Kazakhstan. Asian Aust J Plant Sci Biotechnol 4(Special issue 1):81–86 @2010 Global Science Books

Modern Technologies for Soil Management and Conservation in Northern Kazakhstan

Tobias Meinel, Lars-Christian Grunwald and Kanat Akshalov

Abstract Crop production in the agricultural steppe regions of southwest Siberia and Northern Kazakhstan is mostly focused on cropping spring cultures. Extreme temperature amplitudes, few, unevenly appearing precipitation events and high annual evapotranspiration are the major natural limiting factors. These regions mainly feature crop rotation systems with a high amount of cereals. The low effort required to apply mineral granular fertilizer and high tilling intensity in the meaning of weed protection rather than soil-protecting chemical treatment, are the main reasons for low average yields and a high risk of soil damage due to nutrient losses and wind erosion. The major goal has to be to reduce process inputs while stabilising and raising average yields. To achieve this, the priority has to be placed on minimising labour costs. Crop rotation systems have to become more diverse. Oscillating market situations might be dealt with more easily. Wider crop rotations would lead to a better soil structure and nutrient supply. Additionally, plant protection can be achieved in an easier, faster and cheaper way. Using no-till systems, yield can be stabilized and soils can be protected from wind erosion in an effective way. Seeding machines with wide row spacings and narrow single depth adjusted hoe opener systems are able to carry out shallow tillage only in the seeding furrow. They are also able to place the seed exactly to the adjusted depth. The old stubble is conserved for reasons of minimizing over-ground wind speed and evaporation. The extensive tillage during seeding with a narrow hoe opener system cleans the furrow from organic material and creates fine textured, loose soil and optimal

T. Meinel (✉)
Amazone company in Kazakhstan, Shortandy 4740704, Kazakhstan
e-mail: Dr.Tobias.Meinel@amazone.de

L.-C. Grunwald
Amazone Company, 49205 Hasbergen, Germany
e-mail: LarsChristian.Grunwald@bbg-leipzig.de

K. Akshalov
Scientific Production Center of Grain Farming named after A.I. Barayev, Shortandy 021601, Kazakhstan
e-mail: kanatakshalov@mail.ru

L. Mueller et al. (eds.), *Novel Measurement and Assessment Tools for Monitoring and Management of Land and Water Resources in Agricultural Landscapes of Central Asia*, Environmental Science and Engineering, DOI: 10.1007/978-3-319-01017-5_27,

contact between the seed and soil. The faster warming of the furrow after seeding accelerates emergence in cool temperate steppe climates. The additional placement of granular mineral fertilizers into the furrow in dry continental regions is almost the only effective means of fertilization and gives the young crops better growing conditions. The no-till seeding machines Citan Z and Condor, produced by Amazone, are developed for and well tested under high continental steppe conditions. In many trials on a farm scale it has been shown that these machines used in no-till cropping systems are able to save more fuel, time and seeding material than conventional seeding machines with chisel-opener systems. Furthermore a stabilization of the yields was observed due to the better water and nutrient supply.

Keywords Climatic extremes • Crop rotation • No-tillage technologies • Effectivity • Wind erosion • Yield

1 Introduction

The Eurasian Steppe in Southern Siberia and Kazakhstan is one of the world's largest contiguous arable farming areas (115 million acres). It extends in a belt about 500 km in width, from the Southern Urals, the Kurgan region, to the Northern Kazakhstan regions, the regions of Omsk and Novosibirsk and the Altai Krai and Kemerovo (Frühauf and Meinel 2007). These regions will be referred to below as the Eastern Ural or Central Asian Steppes (Fig. 1).

Crop farming in the Eastern Ural Steppes is characterised almost exclusively by summer crops and thus fundamentally differs from the cultivation concepts, including the crop rotation sequences, in the European Steppe regions of the Commonwealth of Independent States (CIS) (Meinel 2001; Rostankowski 1979; Smith et al. 2007).

With these climate and soil conditions, as well as the gigantic area structures, these regions show immense potential for sustainable plant production at a high level.

Consequently, in recent years the German company Amazone has developed special machines and machine systems that are optimally adapted to these conditions.

2 Conditions for Plant Production

In conjunction with an optimal area structure, the soils in the Central Asian Steppes, which are often very good, result in enormous potential for sustainable and effective crop farming (Burlakova 1999). The climate conditions are without

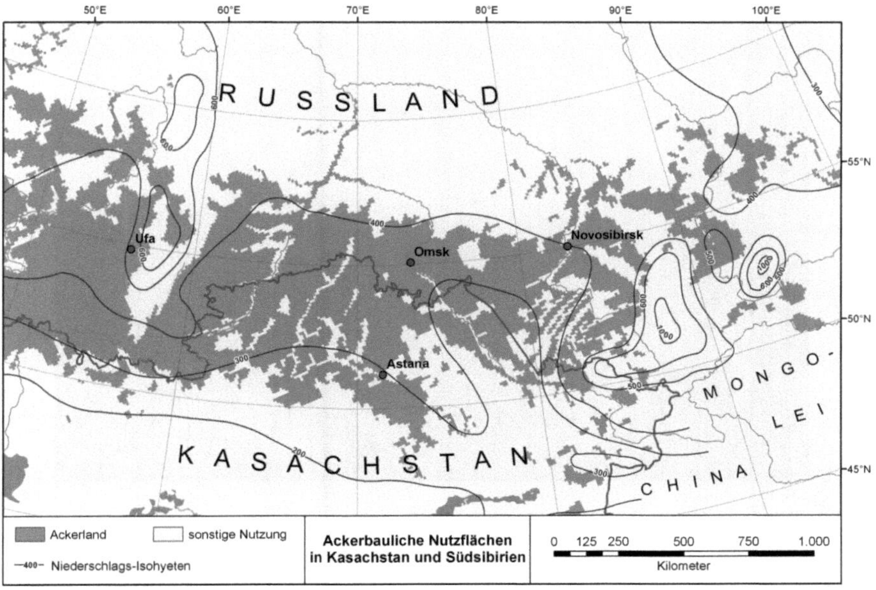

Fig. 1 Arable land in Northern Kazakhstan and South Siberia

doubt the limiting factor. With an annual rainfall from at most 450 mm in the forested steppe areas to a minimum of 250 mm in the dry steppe areas on the border of the rain-fed agriculture areas, the water available for plants is the number 1 minimum factor (Baraev 1976; Meinel 2002; Meye et al. 2008). The temperature regime is an additional difficulty. Summer temperatures of 35 °C are not rare, and high daytime temperatures of 30 °C are reached as early as May (Kharlamova and Revyakin 2006). These conditions cause extremely high evaporation and transpiration of the cultivated plants that far exceed the rainfall; in other words semi-arid conditions predominate (Andreeva and Kust 2003). In winter the low temperatures combined with the low annual snowpack permit almost no cultivation of winter crops (Spaar and Schuhmann 2000). Moreover, the vegetation period from early May to late September, at the latest, is another limiting factor. The extreme exposure of these regions to drought must not be underestimated. Serious droughts occur every two to five years, depending on the location (Bulygina et al. 2007). Another major problem is the danger of wind erosion due to high-intensity soil tilling methods. Various serious wind erosion events have taken place even in recent years (Fig. 2).

On the other hand, in these regions, outdated technology and non-adapted procedures often limit yields and farm results, and above all limit the necessary minimisation of risk in crop farming.

Fig. 2 Wind erosion in the Kostanai oblast, 2005

3 Challenges

Many farmers and farm managers involved with crop farming in the Central Asiatic Steppes are looking for practical reduced procedures. There are many different reasons for this. In these often enormous areas, the priority is to save resources, in order to more efficiently organize cultivation and reduce the risks associated with cultivation. Another priority is to execute work steps more quickly, particularly with regard to tillage, seedbed preparation or sowing, and thus better exploit the optimal time window.

Another current challenge is the extension of the crop rotation sequences. There are many reasons for doing this. On one hand the field tasks can be executed over a significantly longer period of time if different crops are cultivated. On the other hand, there are extremely positive effects in the fight against weeds, diseases and pests (Hudson 1987; McConkey et al. 2003). In addition, the products can be sold with lower risk and higher profits. However, extending the crop spectrum requires high-quality sowing. Particularly in the area of sowing technology, Amazone offers the highest placement quality for sowing grain and small seed crops, and for single-grain sowing.

In addition, for the farmers this means that soil protection concepts are increasing in importance. Only healthy soils have the potential to produce optimal yields, and to compensate for the consequences of negative climatic conditions, such as droughts. And, not least, for procedures that involve reduced tillage, the soil water reserves are used in a better manner and thus higher yields are achieved (Boardman 1990).

To handle these tasks, Amazone offers a broad spectrum of high-capacity, robust machines. In the last decade, machines have been developed and tested on site for these requirements. Amazone not only develops and produces agriculture technology, but also offers the necessary agronomic and technical solutions through extensive testing.

4 Modern Solutions for Sustainable Cropping

Amazone's "Citan Z" and "Condor" drills are designed for specialized, direct seeding on stubble. The openers of these drills are designed to reduce the intensity of their mechanical impact on soil, by wide inter row-spacing, minimal disturbance of stubble and maximum preservation of crop residues on the soil surface. The proposed drills of 12 and 15 m in width can be aggregated with tractors of 220 and 260 horsepower (Figs.3 and 4).

In spring, agricultural producers have to solve difficult issues: the method of planting and getting good seedlings on time. Getting good seedlings depends on the one hand on the moisture content of the upper layer of the soil and on the other hand on the sowing methods and seed placement. Choosing the right seeding rate, combined with maximum germinations, guarantees the most efficient use of soil moisture and precipitation (Fig. 5).

In many cases during spring planting the soil layer quickly dries to 5–7 cm or deeper, and in some arid areas even down to 10 cm. This determines the depth for placing the seed. In this situation, cultivator-type openers can place seeds of crops, particularly spring wheat, at a depth of 7–8 cm, and sometimes deeper. This, of course, affects the duration of the plant-seedling period. Planting to a depth of 7–8 cm and deeper calls for increased fuel consumption and lengthens the sowing

Fig. 3 Seed drill Condor 15001 with K701 (300 hp)

Fig. 4 Condor 15001 seed drill (Photo courtesy of Amazone)

Fig. 5 Differences in germination between old-style air seeder and the single guided tine coulter produced by Amazone Condor

time (Zentner et al. 1998). The main disadvantage of placing crop seeds deep in the soil is that germination is not always uniform and may be delayed, resulting, on one hand, in a shift of the vegetation period and the late maturing of crops and, on the other hand, overgrown weeds in some cases. Many of the cultivator-type openers lead to a high variance in seed placing depth due to higher soil disturbance. This can lead to inhomogeneous emergence and subsequent complications in weed management. In 2009, for example, late-maturing spring wheat led to the deterioration of the quality of grain and in 2011 the harvest was delayed.

One of the requirements for the sowing machines is the placement of crop seeds in moist soil. There is a special requirement for small-seeded crops: they should be placed at a depth of 2–3 cm, and in the wet soil. These requirements are met by the technology of direct seeding. For direct seeding, different companies offer different opener systems, from disks to chisels. However, not every sowing machine opener can place the seeds in moist soil and at a depth of 2–3 or up to 4 cm. In some cases, a disk opener can place the seeds shallowly, in the wet soil, but only on tilled fields. When stubble fields are sown using disc openers problems arise with the depth and placement of seeds no deeper than 2–3 cm for small-seeded crops, and to a depth of 4–5 cm for grain and other crops.

The openers for the "Citan Z" and "Condor" drills for direct seeding are changing the understanding of the optimum depth of sowing seeds for agricultural crops. Seeds are guaranteed to be placed in moist soil and the actual depth of seed placement in the soil is much lower than with cultivator openers. The "ConTec" opener for the "Citan Z" and "Condor" drills is different due to its special ability to penetrate the soil, as well as its accurate seeding depth using a support-press wheel (Fig. 6). This allows good seeds to be given soil contact along with precise incorporation of seed depth.

Getting good and early seedings provides better field germination and plant safety. This, in turn, allows the different seeding rate to get the same crop yield.

Direct seeding drills allow small-seed crops to be sown at a low seed rate: for example the crop of canola can be sown at a rate of up to 360,000 viable seeds per hectare without reducing crop yields due to nearly 100 % germination and preservation (Fig. 7).

Our studies have shown that when sowing with cultivator openers, the average depth of seed placement is about 6.0 cm with a variation from 2.9 to 10.2 cm. When sowing with the "Citan Z" and "Condor" drills, the average depth of seeding is about 3.0 cm with a variation from 2.1 to 4.1 cm. The coefficient of variation of seeding depth by width is less than that of the seeder for sowing seed drills for direct seeding, since each opener can be controlled separately. Spring wheat directly sown by a hoe opener system germinates after 4–5 days, while the

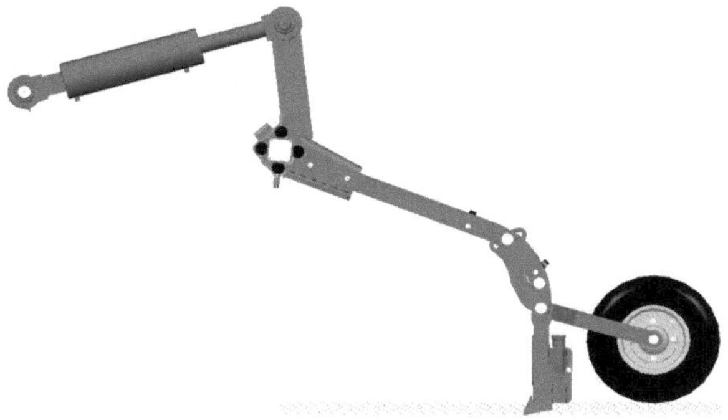

Fig. 6 "ConTec" opener for the "Citan Z" and "Condor" drills (Photo courtesy of Amazone)

Fig. 7 Directly seeded canola with Condor 12001 (Shortandy, June 2011)

full spring wheat seedlings sown by chisel opener systems appear only after 10–12 days.

One of the important agronomic requirements of drills is that they protect the soil from wind erosion. Directly sowing with the "Citan Z" and "Condor" provides straw coverage due to stubble standing in a vertical position. The soil is loosened in the seeding row but nevertheless aggregated. That can reliably protect the soil from wind erosion. The stubble background reduces the wind speed at the surface layer of the soil, reduces evaporation of soil moisture, protects seedlings from dry winds and gives the plant some shade (Fig. 8).

When sowing with the "Citan Z" and "Condor" seed drills with a 25 cm row spacing, the surface of the soil remains untouched between rows, which allows herbicides to be used after sowing. The systematic control of weed infestation and fields reduces the need for herbicides over time.

The more simultaneous the seedings and the drying of the upper soil layer, and the less competition there is for moisture and nutrition, the higher the yield of spring wheat after seeding with the "Citan Z" and "Condor 12001" drills.

Direct seeding has some economic benefits due to good field germination and conservation of plants without declining spring wheat yields at seeding rates down to 100 kg/ha on a calcic Chernozem in southern zones. In the area of chestnut soils the decrease in the seeding rate to 86 kg/ha also does not reduce the yield of spring wheat. Fuel consumption is reduced to 30 % by sowing with no-till drills compared to seeding systems with cultivator openers, with an average flow rate of 3.0–3.5 l on one hectare with no-till drills and up to 5.0 l on one hectare while

Fig. 8 Standing stubble after seeding with the single guide tine coulter of the Condor 12001

sowing with machines with cultivator-type openers. Decreasing the rate of spring wheat seeding from 120 to 100 kg/ha without reducing crop yields can save up to 20 kg/ha of seeds for the "Citan Z" and "Condor" sowing drills, which is equivalent to 4 dollars per 1 hectare or up to 40,000 dollars on 10,000 hectares of sown area. Reducing the consumption of petroleum products by 1.5–2.0 l/ha can save up to 1.0–1.2 dollars per 1 hectare respectively, or 9–12,000 dollars per 10,000 hectares of crops.

One advantage of using drills for direct seeding with the "Citan Z" and "Condor" is that they can be attached to "Kirovets" tractors, which are available to farmers and large agricultural producers.

Tests on farm level have shown that for the "Condor" drill, a tractor with 220 horsepower has enough power. Drilling with "Condor" lowers the fuel consumption to 2.7 l/ha, which is a sensational result. Tests have also shown that a "Condor" drill operating at 8–10 km/h is able to work 120 hectares in one working day (13 h).

In field trials on farm level, yields of spring wheat in direct seeding with the "Condor" no-till drill and an in-row application of 30 kg/ha of granular nitrogen fertilizer levelled out at 2.55 t/ha. Yields of linseed reached 1.75–1.94 t/ha, rapeseed 1.26 t/ha and peas 1.24 t/ha. These drills measured the dose of fertilizers very precisely.

The "Amazone" company's new drills for direct seeding, "Citan Z" and "Condor", give farmers every possibility for the optimal use of the technology of direct seeding.

References

Andreeva IV, Kust GS (2003) Geographical zoning of desertification in the semi-arid and arid zones of Russia. In: Chibilyov AA (ed) Steppes of Northern Eurasia, pp 46–48

Baraev AI (1976) Nautschnoje osnowi poswozaschitnoi sistemu. In: Puti selsko chosjajstwa selinich rajonow. Vaschnil 76, pp 46–63

Boardman JA (1990) Soil erosion and agricultural land. Wiley, NY

Bulygina ON et al (2007) Climate variations and changes in extreme climate events in Russia. Environ Res Lett 2(4):7

Burlakova LM (1999) Potschveno-semelnie resurci. In: Vinokurov UI (ed) Costojanie okruschajuschei sredi v Altaiskom Kraie. OOO NEKSI, Barnaul

Frühauf M, Meinel T (2007) Vom "Neuland unterm Pflug" zum "Dust-Bowl-Syndrom"- der eurasiatische Steppengürtel im Fokus von Global Change – in Planet Erde "Asien"(B 19050-5), Wissenschaftliche Buchgesellschaft, Darmstadt, pp 77–89

Hudson NW (1987) Soil and water conservation in semi-arid areas. FAO Soils Bulletin 57, Rom

Kharlamova NF, Revyakin VS (2006) Regional climate and environmental change in central Asia. In: Environmental security and sustainable land use – with special reference to Central Asia NATO Security through Science Series, pp 19–26

McConkey BG, Liang BC, Campbell CA, Curtin D, Moulin A, Brandt SA, Lafond GP (2003) Crop rotation and tillage impact on carbon sequestration in Canadian prairie soils. Soil Tillage Res 74:81–90

Meinel T (2001) Die Anwendung von Trockenfeldbaumethoden in der Kulundasteppe aus ökologischer und ökonomischer Sicht. Sowremennije problemi geografii i prirodopolsowanija, vol 5–6, Barnaul

Meinel T (2002) Die geoökologischen Folgewirkungen der Steppenumbrüche in den 50er Jahren in Westsibirien. Ein Beitrag für zukünftige Nutzungskonzepte unter besonderer Berücksichtigung der Winderosion. – Diss. Math.-Naturwiss. Fak. Univ. Halle-Wittenberg: p 129, Halle an der Saale

Meyer BC, Schreiner V, Smolentseva EN, Smolentsev BA (2008) Indicators of desertification in the Kulunda Steppe in the south of Western Siberia. Arch Agron Soil Sci 54(6):585–603

Rostankowski P (1979) Agrarraum und Getreideanbau in der Sowjetunion 1948–1985. Duncker and Humblot, Berlin

Smith P et al (2007) Changes in mineral soil organic carbon stocks in the cropland of European Russia and the Ukraine, 1990–2070; comparison of three models and implications for climate mitigations. Reg Environ Change 7(2):106–119

Spaar D, Schuhmann P [Hrsg.] (2000) Natürliche Grundlagen der Pflanzenproduktion in den Ländern der Gemeinschaft Unabhängiger Staaten und des Baltikums. AGRIMEDIA, Bergen an der Dumme

Zentner RP, McConkey BG, Stumborg MA, Campbell CA, Selles F (1998) Energy performance of conservation tillage management for spring wheat production in the Brown soil zone. Can J Plant Sci 78:553–563

Enhancing the Productivity of High-Magnesium Soil and Water Resources in Central Asia

Manzoor Qadir, Frants Vyshpolsky, Khamit Mukhamedjanov, Ussen Bekbaev, Saghit Ibatullin, Tulkun Yuldashev, Andrew D. Noble, Akmal Karimov, Alisher Mirzabaev and Aden Aw-Hassan

Abstract High-magnesium soils and waters are emerging environmental and agricultural productivity constraints. Excess levels of magnesium (Mg^{2+}) in

M. Qadir (✉)
United Nations University—Institute for Water, Environment and Health (UNU-INWEH), 175 Longwood Road South, Suite 204, Hamilton, ON L8P 0A1, Canada
e-mail: Manzoor.Qadir@unu.edu

F. Vyshpolsky · K. Mukhamedjanov · U. Bekbaev
Kazakh Scientific Research Institute of Water Economy, Water Quality Laboratory, 12 Koygeldy Street, Taraz, Kazakhstan
e-mail: mkv_77@mail.ru

U. Bekbaev
e-mail: b_ussen@mail.ru

S. Ibatullin
Executive Committee of the International Fund for Saving the Aral Sea (EC-IFAS), Dostyk Avenue 280, Almaty, Kazakhstan
e-mail: Saghit@inbox.ru

T. Yuldashev · A. Mirzabaev
ICARDA, Central Asia and Caucasus Regional Office, Osiyo 6, Tashkent, Uzbekistan
e-mail: t.yuldashev@cgiar.org

A. Mirzabaev
e-mail: almir@uni-bonn.de

A. D. Noble
International Water Management Institute (IWMI), P.O. Box 2075, Colombo, Sri Lanka
e-mail: A.Noble@cgiar.org

A. Karimov
IWMI—Central Asia, C/O PFU-CGIAR-ICARDA-CAC, Apt 123, Bldg 6, Osiyo str, Tashkent, Uzbekistan100000,
e-mail: A.Karimov@CGIAR.ORG

A. Aw-Hassan
International Center for Agricultural Research in the Dry Areas (ICARDA), P.O. Box 5466, Aleppo, Syria
e-mail: a.aw-hassan@cgiar.org

L. Mueller et al. (eds.), *Novel Measurement and Assessment Tools for Monitoring and Management of Land and Water Resources in Agricultural Landscapes of Central Asia*, Environmental Science and Engineering, DOI: 10.1007/978-3-319-01017-5_28,

irrigation waters and/or in soils negatively affect soil infiltration rate and hydraulic conductivity and ultimately crop growth and yield. Although the levels of Mg^{2+} in irrigation waters and soils are increasing in several irrigation schemes globally, southern Kazakhstan has become a hotspot of such natural resource degradation. The productivity of magnesium-affected soils can be enhanced by increasing the levels of calcium (Ca^{2+}) in the soil to counteract the negative impacts of Mg^{2+}. Studies undertaken on the soil application of phosphogypsum, a major waste product of phosphoric acid factories and a source of Ca^{2+}, have demonstrated beneficial effects of this soil amendment in terms of (1) improved soil quality through a reduction in exchangeable magnesium percentage (EMP) levels; (2) enhanced water movement into and through the soil vis-à-vis increased moisture storage in the soil for use by the plant roots; (3) increased cotton yield and water productivity; and (4) greater financial benefits. In addition to improving crop productivity, these studies demonstrated the beneficial use of an industrial waste material in agriculture. With the aim of addressing the challenge of achieving sustainable agriculture production from magnesium-affected environments, there would be a need for appropriate supportive policies and functional institutions along with capacity building of farmers, researchers, and agricultural extension workers.

Keywords Magnesium to calcium ratio · Exchangeable magnesium percentage · Salt-affected soils · Phosphogypsum · Water quality · Central Asia · Kazakhstan · Cotton

1 Introduction

Magnesium (Mg^{2+}) is necessary for all life. In soils and waters, magnesium is considered to have several common functions with calcium (Ca^{2+}) because of their common divalent cationic forms, i.e. Mg^{2+} and Ca^{2+}, respectively. While this is consistent in several cases, the contrast between these cations stems from the specific effects of Mg^{2+} on soil physical properties. Magnesium may result in deleterious effects on soil structural properties (Oster and Jayawardane 1998; Qadir and Schubert 2002) similar to those caused by sodium ions (Na^+) on soil infiltration rate and hydraulic conductivity (Levy et al. 1988; Keren 1991; Dontsova and Norton 2002). This is particularly important under conditions when water used for irrigation contains higher levels of Mg^{2+} than Ca^{2+}, i.e. Mg^{2+} to Ca^{2+} ratio greater than 1 when ionic concentrations expressed in the same units (Vyshpolsky et al. 2008). Therefore, the ratio between the concentrations of Ca^{2+} and Mg^{2+} is an important aspect for consideration of Mg^{2+} effects on soil structural properties (Qadir et al. 2013).

Based on the typical characteristics and ambient concentrations in soil and/or irrigation water, it has long been anticipated that Mg^{2+} is less efficient than Ca^{2+} in flocculating soil clays (McNeal et al. 1968; Bohn et al. 1985; Rengasamy et al. 1986; Dontsova and Norton 2002). However, it was generally perceived that most agricultural soils and irrigation waters contained much higher levels of Ca^{2+} than Mg^{2+}. Research undertaken in recent decades has provided evidences from several irrigation schemes worldwide (parts of Aral Sea Basin in Central Asia, Cauca River Valley in Colombia, Indus Basin in Pakistan, and Murray Darling Basin in Australia) indicating increasingly high levels of Mg^{2+} in irrigation waters and irrigated soils than those found several decades ago (Garcia-Ocampo 2003; Qadir and De Pauw 2007; Vyshpolsky et al. 2008). The implications are degradation of soil physical properties and marked decrease in crop productivity. This necessitate the synthesis of research-based information on the potential effects of high levels of Mg^{2+} in soils and irrigation waters on soil degradation and crop growth and yield as well as possible interventions that can bring back the magnesium-affected soils to a highly productive state.

With a focus on high-magnesium soil and water resources in Central Asia, this chapter provides an overview of the emerging environmental and agricultural productivity constraints related to these natural resources and their management options through the use of calcium-supplying amendments such as the application of phosphogypsum to soils.

2 High-Magnesium Waters and Soils in Central Asia

Intensive irrigation and excessive leaching of agricultural lands are common practices in many regions of Central Asia. In some regions, such as southern Kazakhstan, irrigation water contains high levels of Mg^{2+} with the Mg^{2+} to Ca^{2+} ratio greater than 1. More than 30 % of the irrigated lands in southern Kazakhstan consist of soils that have exchangeable magnesium percentage (EMP) in the range of 25–45 % (Bekbaev et al. 2005) and in some cases EMP is as high as 60 % (Fig. 1). The corresponding exchangeable sodium percentage (ESP) levels are very low, usually less than 5 %. Such soils are characterized by low infiltration

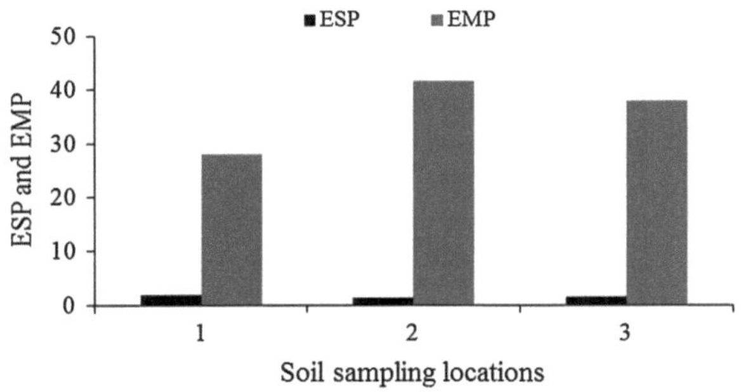

Fig. 1 Exchangeable sodium percentage (*ESP*) and magnesium percentage (*EMP*) levels in soils in the Arys Turkestan area in southern Kazakhstan

Fig. 2 With low infiltration rates and hydraulic conductivities, magnesium-affected soils in southern Kazakhstan form massive clods upon drying after irrigation events, thereby impacting water flow rates along furrows and infiltration. The consequence of using high-magnesium soils and waters in agricultural production systems without suitable management practices has been a gradual decline in crop yield and soil productivity

rates and hydraulic conductivities. The consequence has been a gradual decline in the yield of cotton in a region that relies heavily on this crop for farm and national level revenue (Karimov et al. 2009; Qadir et al. 2013).

An example of soils with high levels of EMP is found in the Arys Turkestan area in southern Kazakhstan where irrigation was initiated in the 1960s (Fig. 2). Since 1965, drainage systems have been constructed to prevent increases in soil salinity. Leaching of soils with excess water application at the rate of 2,500–3,500 $m^3\ ha^{-1}$ had been a common practice for decades, resulting in the removal of excess soluble salts and essential nutrients. The organic matter content has declined by 25 % over the initial levels in the 1960s. The reduction in organic matter content has also affected the cation exchange capacity, in which a 22 % decrease has been determined. Soil bulk densities have increased from 1.33 to 1.45 $Mg\ m^{-3}$ and infiltration rates have decreased by 40–50 %. The levels of gypsum ($CaSO_4\ 2H_2O$), an important source of Ca^{2+} in these soils, have decreased substantially. For example, gypsum content in the top 1 m of the soils in the 1960s was 0.25–0.38 %, but this decreased to 0.03–0.04 % in 2000. This has resulted in a decrease in exchangeable Ca^{2+} levels from 8.8–10 to 4.1–7.5 $cmol_c\ kg^{-1}$ soil, indicating a reduction in Ca^{2+} levels in the range of 25–53 %. The use of high-Mg^{2+} water (Mg^{2+} to Ca^{2+} ratio greater than 1) for irrigation and leaching has increased exchangeable Mg^{2+} levels from 2.0–3.2 to 3.7–6.1 $cmol_c\ kg^{-1}$ soil (85–91 % increase in exchangeable Mg^{2+}) (Karajeh et al. 2004).

The productivity of soils with elevated levels of exchangeable Mg^{2+} can be enhanced by increasing the levels of Ca^{2+} on the cation exchange sites to counteract the impacts of Mg^{2+}. This is accomplished by applying sufficient amounts of

Ca^{2+} to the soil in sufficient amounts (Ghafoor et al. 1992; Karajeh et al. 2004). Phosphogypsum is a source of Ca^{2+} and major waste product of phosphoric acid factories, which use phosphate rock as raw material (Alcordo and Rechcigl 1993). Although phosphogypsum contains variable amounts of ^{226}Ra and other radionuclide concentrations depending on source, studies have shown that its use as an amendment does not produce significant increases in the radionuclide concentrations in soils (Al-Oudat et al. 1998; El-Mrabet et al. 2003). Rather, an application of phosphogypsum could contribute to diluting the radionuclide wastes, with additional value to the farmers (El-Mrabet et al. 2003). Phosphogypsum is available in Kazakhstan and could be used to improve the productivity of high EMP soils. The amendment is also available elsewhere in Central Asia.

3 Productivity Enhancement of High-Magnesium Waters and Soils

The results of a 4-year study conducted in Kazakhstan demonstrated beneficial effects of applying calcium-supplying phosphogypsum to magnesium-affected soils on soil quality and cotton yield (Vyshpolsky et al. 2008). The irrigation water had average values of Mg^{2+} to Ca^{2+} ratio of 1.43 during irrigation period (June, July, and August). In the case of drainage water, the average values of Mg^{2+} to Ca^{2+} ratio were 0.97 during the same period (Table 1). A reduction in the fraction of Mg^{2+} applied in irrigation water points to its adsorption on the cation exchange sites of the irrigated soil with the consequence that less Mg^{2+} appears in the drainage water. Any build up in the fraction of Mg^{2+} on the exchange complex causes the soil to behave like a typical sodic soil, which is characterized by the degradation of soil physical properties. This is a typical case of magnesium-induced soil degradation in many downstream areas in Kazakhstan.

The study site had a smooth landscape with slopes from 0.002 to 0.0035. With high summer temperatures (30–40 °C during June, July and August), annual precipitation ranged from 50 to 250 mm, with 90 % occurring during October through May. The soil was a heavy loam with about 1 % organic matter in the surface layer and a cation exchange capacity (CEC) of 11 ± 1 $cmol_c$ kg^{-1}. Soil bulk density was 1.46 ± 0.01 Mg m^{-3} and pH varied narrowly from 8.1 to 8.2. There were three treatments: control (without phosphogypsum) and two phosphogypsum treatments consisting of soil application of the amendment at the rate of 4.5 and 8.0 t ha^{-1}.

The application of phosphogypsum increased Ca^{2+} concentration in the soil and triggered the replacement of excess Mg^{2+} from the cation exchange sites. After harvesting the first crop, there was an 18 % decrease in EMP of the surface 0.2 m soil over the pre-experiment level in the plots where phosphogypsum was applied at 4.5 t ha^{-1}, and a 25 % decrease in EMP in plots treated with phosphogypsum at

Table 1 Average composition of irrigation water and agricultural drainage water during the cotton growing season in Arys Turkestan zone, southern Kazakhstan (Calculated based on the data from Vyshpolsky et al. 2008)

Parameter	Unit	Canal water[a]	Drainage water[b]
EC	dS m^{-1}	0.73	2.06
pH	–	7.95	8.13
TSS	$mmol_c L^{-1}$	7.30	20.60
Ca^{2+}	$mmol_c L^{-1}$	2.02	6.75
Mg^{2+}	$mmol_c L^{-1}$	2.88	6.59
Na^{+}	$mmol_c L^{-1}$	1.34	5.58
K^{+}	$mmol_c L^{-1}$	0.02	0.03
HCO_3^{-}	$mmol_c L^{-1}$	3.37	7.12
Cl^{-}	$mmol_c L^{-1}$	0.35	2.39
SO_4^{2-}	$mmol_c L^{-1}$	2.95	9.49
$Mg^{2+}:Ca^{2+}$	–	1.43	0.97
SAR	–	0.85	2.15

[a] Average of water samples collected during June, July, and August from Arys Turkestan canal used to irrigate cotton in southern Kazakhstan

[b] Average of drainage water samples collected during June, July, and August in the area irrigated by Arys Turkestan canal

8 t ha^{-1} (Vyshpolsky et al. 2008). The additional beneficial effect of the amendment application resulted in an increase in the soil phosphorus content.

The highest cotton yields were obtained during the first year of the treatment application, which were 2.7 and 3.0 t ha^{-1}, respectively, from the treatments where phosphogypsum was applied at 4.5 and 8.0 t ha^{-1} (Vyshpolsky et al. 2008). Cotton yield in the control plots (1.4 t ha^{-1}) was almost half the yield harvested from phosphogypsum treatments. There was a 93 % increase in cotton yield from the phosphogypsum treatment (4.5 t ha^{-1}) over the control. In the case of the 8.0 t ha^{-1} treatment, cotton yield increased to 114 % over that harvested from control plots. The yield pattern from the control plots for the subsequent years showed a similar response trend (1.3–1.4 t ha^{-1}). In the case of phosphogypsum treatments, cotton yields gradually declined in the subsequent years because of an increase in EMP in the soil as the amendment was applied only at the beginning of the experiment and it was fully utilized in subsequent years. In the phosphogypsum treatment with 4.5 t ha^{-1}, cotton yields decreased from 2.7 t ha^{-1} in first year to 2.2 t ha^{-1} in fourth year. The yield decrease pattern for the 8.0 t ha^{-1} phosphogypsum treatment was from 3.0 to 2.2 t ha^{-1} in four years. The 4-year average cotton yield in the treatments was in the order: soil application of phosphogypsum application at 8.0 t ha^{-1} (2.6 t ha^{-1}) > soil application of phosphogypsum application at 4.5 t ha^{-1} (2.4 t ha^{-1}) > control (1.4 t ha^{-1}) (Fig. 3). Since phosphogypsum was applied once at the beginning of the study, exchangeable Mg^{2+} levels tended to increase after four years of application, particularly in the treatment with 4.5 t ha^{-1} phosphogypsum. This necessitated the need for a booster dose of phosphogypsum to such soils after every 4–5 years to optimize the ionic

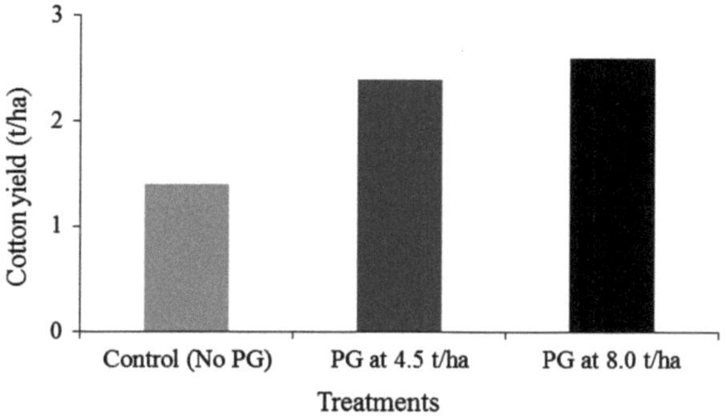

Fig. 3 Four-year average cotton yield as affected by different rates of phosphogypsum application (0, 4.5, and 8.0 t/ha) to a magnesium-affected soil in Kazakhstan (Calculated based on the data from Vyshpolsky et al. 2008)

balance and sustain higher levels of cotton production. The economic benefits from the phosphogypsum treatments were almost twice those from the control.

Another study on a similar type of soil was conducted later to determine the appropriate combinations of the rates and timings of phosphogypsum application to mitigate the effects of high levels of Mg^{2+} in magnesium-affected soils. There were five treatments: (1) control without application of phosphogypsum; (2) soil application of phosphogypsum before snowfall at the rate of 3.3 t ha^{-1}; (3) soil application of phosphogypsum before snowfall at the rate of 8.0 t ha^{-1}; (4) soil application of phosphogypsum after snowfall at the rate of 3.3 t ha^{-1}; and (5) soil application of phosphogypsum after snowfall at the rate of 8.0 t ha^{-1}.

The results of this field study demonstrated the beneficial effects of phosphogypsum application as the phosphogypsum treatments performed better than the typical farming practice in terms of (1) improved soil quality through a reduction in exchangeable magnesium percentage levels; (2) enhanced water movement into and through the soil vis-à-vis increased moisture storage in the soil for use by the plant roots; (3) increased cotton yield and water productivity; and (4) greater financial benefits (Vyshpolsky et al. 2010) (Fig. 4).

In addition to the aforementioned benefits, surface runoff was significantly reduced in the phosphogypsum treatments, which improved the irrigation efficiency and effectively provided more moisture to the root zone through improved infiltration. This in turn increased crop water productivity in the phosphogypsum treatments. Figure 5 presents the water productivity of cotton as affected by different times (before snowfall in January or after snowfall in April) and rates of phosphogypsum application to a magnesium-affected soil. In terms of crop growth and yield, cotton yields in the phosphogypsum treatments were 19–46 % higher than control. Two-year average water productivity in control treatment was 0.32 kg m^{-3} while it ranged from 0.44 to 0.52 kg m^{-3} in the phosphogypsum treatments.

Fig. 4 Patchy germination and growth of cotton on a magnesium-affected soil in Kazakhstan (*above*) and improved germination and growth of cotton on the same soil after the application of phosphogypsum in southern Kazakhstan

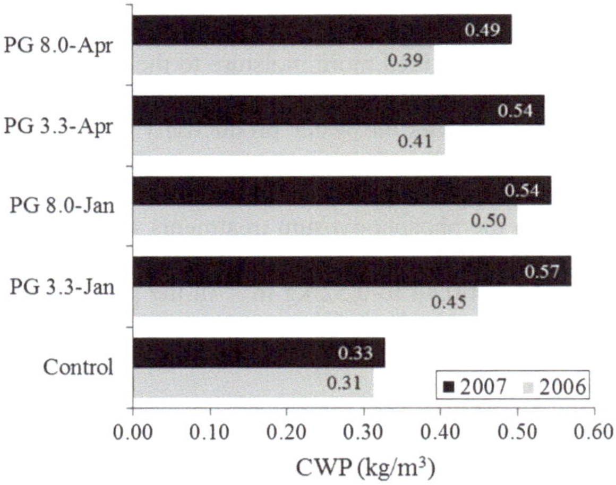

Fig. 5 Effect of different rates and times of phosphogypsum application on crop water productivity (*CWP*) for cotton grown on magnesium-affected soil for the years 2006 and 2007 (Based on the data from Vyshpolsky et al. 2008)

4 Conclusions and Perspectives

Considering declining availability of appropriate quality water and/or highly productive soil resources for agriculture in dry areas, the need to produce more food, forage, and fiber will necessitate the effective utilization of marginal-quality water and soil resources. With the recognition of high-magnesium water and soil resources as environment and agricultural productivity constraints in Central Asia, there is a need to assess the extent of these natural resources along with their management. Recent research and practices have demonstrated that effective utilization of these natural resources in dry areas can improve agricultural productivity per unit area and per unit water applied. The results of these studies demonstrate that application of calcium-supplying phosphogypsum to high-magnesium soils significantly improved agricultural productivity. Furthermore, these interventions were economically viable, revealing that the efficient use of high-magnesium water and soil resources has the potential to improve livelihoods amid growing populations in relevant dry areas whilst reversing the natural resource degradation trend. Restoration of these degraded systems to a level at which they are economically viable is the first step in a process of rehabilitation. Clearly there is a need to rethink the way agriculture is undertaken if we are to manage marginal-quality waters and degraded soils such as high-magnesium soil and water resources.

References

Alcordo IS, Rechcigl JE (1993) Phosphogypsum in agriculture: a review. Adv Agron 49:55–118

Al-Oudat M, Arslan A, Kanakri S (1998) Physical and chemical properties, plant growth, and radionuclide accumulation effects from mixing phosphogypsum with some soils. Comm Soil Sci Plant Anal 29:2515–2528

Bekbaev R, Vyshpolsky F, Ibatullin S, Mukhamedjanov Kh, Bekbaev U (2005) Influence of application of phosphogypsum on infiltration rate of solonetzic soil. Bull Agric Sci Kazakhstan 7:18–20 (in Russian)

Bohn HL, McNeal BL, O'Connor GA (1985) Soil chemistry. Wiley, New York

Dontsova KM, Norton LD (2002) Clay dispersion, infiltration and erosion as influenced by exchangeable Ca and Mg. Soil Sci 167:184–193

El-Mrabet R, Abril JM, Perianez R, Manjon G, Garcia-Tenorio R, Delgado A, Andreu L (2003) Phosphogypsum amendment effect on radionuclide content in drainage water and marsh soils from southwest Spain. J Environ Qual 32:1262–1268

Garcia-Ocampo A (2003) Physical properties of magnesium affected soils in Colombia. Lecture given at the College on Soil Physics, 3–21 March 2003, Trieste

Ghafoor A, Shahid MI, Saghir M, Murtaza G (1992) Use of high-Mg brackish water on phosphogypsum and FYM treated saline-sodic soil. I. Soil improvement. Pak J Agric Sci 29:180–184

Karajeh F, Karimov A, Mukhamedjanov V, Vyshpolsky F, Mukhamedjanov KH, Ikramov R, Palvanov T, Novikova A (2004) Improved on-farm water management strategies in Central Asia. In: Ryan J, Vlek P, Paroda R (eds) Agriculture in Central Asia: research for development. ICARDA, Aleppo, Syria, and Center for Development Research, Bonn, pp 76–89

Karimov A, Qadir M, Noble A, Vyshpolsky F, Anzelm K (2009) The development of magnesium-dominant soils under irrigated agriculture in southern Kazakhstan. Pedosphere 19:331–343

Keren R (1991) Specific effect of magnesium on soil erosion and water infiltration. Soil Sci Soc Am J 55:783–787

Levy GJ, van der Watt HVH, du Plessis HM (1988) Effect of Na/Mg and Na/Ca systems on soil hydraulic conductivity and infiltration rate. Soil Sci 146:303–310

McNeal BL, Layfield DA, Norvell WA, Rhoades JD (1968) Factors influencing hydraulic conductivity of soils in the presence of mixed-salt solutions. Soil Sci Soc Am Proc 32:187–190

Oster JD, Jayawardane NS (1998) Agricultural management of sodic soils. In: Sumner ME, Naidu R (eds) Sodic soil: distribution, management and environmental consequences. Oxford University Press, New York, pp 126–147

Qadir M, De Pauw E (2007) Emerging challenges addressing the characterization and mapping of salt-induced land degradation. Presented at the first expert consultation meeting of the global network on salinization prevention and productive use of salt-affected habitats (SPUSH), 26–29 November 2007, Dubai, United Arab Emirates

Qadir M, Schubert S (2002) Degradation processes and nutrient constraints in sodic soils. Land Degrad Dev 13:275–294

Qadir M, Noble AD, Chartres C (2013) Adapting to climate change through improved water productivity of soils in dry areas. Land Degrad Dev 24:12–21

Rengasamy P, Greene RSB, Ford GW (1986) Influence of magnesium on aggregate stability in sodic Red-Brown Earths. Aust J Soil Res 24:229–237

Vyshpolsky F, Qadir M, Karimov A, Mukhamedjanov K, Bekbaev U, Paroda R, Aw-Hassan A, Karajeh F (2008) Enhancing the productivity of high-magnesium soil and water resources through the application of phosphogypsum in Central Asia. Land Degrad Dev 19:45–56

Vyshpolsky F, Mukhamedjanov K, Bekbaev U, Ibatullin S, Yuldashev T, Noble AD, Mirzabaev A, Aw-Hassan A, Qadir M (2010) Optimizing the rate and timing of phosphogypsum application to magnesium-affected soil for crop yield and water productivity enhancement. Agric Water Manag 97:1277–1286

Advanced Technologies for Irrigated Cropping Systems

Robert G. Evans

Abstract Available supplies of water for irrigation and other uses are becoming more limited around the world, and this trend is accelerating. Emerging computerized precision irrigation technologies will enable growers to apply water and agrochemicals more precisely and site-specifically to match the status and needs of soil and plants as determined by the analysis of data from a variety of sensor networks, wireless communications systems and decision support systems. Speed control and zone control options for site-specific variable rate irritation (SS-VRI) system are currently available, with speed control the most common. Site-specific variable rate sprinkler irrigation systems are wonderful research tools that can provide maximum amounts of information from relatively small areas. A self-propelled SS-VRI sprinkler system has been developed for agricultural research applications and 5 of these systems are now in use. This fully functional research machine has been used from 2005 to 2012 and has been very reliable. These SS-VRI systems offer many benefits for research and they have tremendous potential for a greater use in sprinkler irrigation systems worldwide for both in research and general practice to conserve water, fertilizer and energy.

Keywords Precision agriculture · Decision support · Automation · Irrigation controls · Water conservation

1 Introduction

Production agriculture is facing a series of major challenges in providing sufficient food, fiber, and fuel to support a growing global population while our water

R. G. Evans (✉)
USDA-ARS-NPA, Northern Plains Agricultural Research Laboratory, 1500 North Central Avenue, Sidney, MT 59270, USA
e-mail: bcevans@pocketinet.com

L. Mueller et al. (eds.), *Novel Measurement and Assessment Tools for Monitoring and Management of Land and Water Resources in Agricultural Landscapes of Central Asia*, Environmental Science and Engineering, DOI: 10.1007/978-3-319-01017-5_29,

resources, environmental health, and available arable land decline and climate changes. In agricultural ecosystems, physical and biological processes such as cycling of carbon, water, and nutrients are linked with social and economic processes. To achieve sustainability, it is essential that we understand how these processes interact, and their impacts on the environment through space and time. Innovative approaches will be needed to address these issues and maintain a safe and abundant food supply as the agronomic, economic and social problems facing production agriculture gain in complexity.

Irrigation can be fundamentally characterized as a temporal adaptation to seasonal and annual variation in rainfall (Turral et al. 2010). Irrigation has shaped the economies of many semi-arid and arid areas around the world over many centuries. It has stabilized rural communities, increased income and provided many new opportunities for economic growth. These practices have enabled human habitation, at times quite dense populations, where it otherwise could not exist (Postel 1999). In short, irrigated crop production underpins large sections of current society and lifestyles throughout the world.

Irrigated agricultural activities support diverse components of the world's food chain and provide much of the fruit, vegetables and cereals consumed by humans plus the grain fed to animals that are used eventually as human food. It also provides much of the feed to sustain animals used for work in many parts of the world, and these lands provide considerable sources of food and foraging areas for migratory and local birds as well as other wildlife.

Irrigated land constitutes approximately 18 % of the world's total cultivated farmland, but currently produces more than 40 % of its food and fiber. However, irrigation of lands accounts for over 70 % of fresh water consumed in the world, and is also a major user of energy for farming operations and pumping. Nevertheless, output from irrigated agriculture will have to increase dramatically in the next few decades to help meet the projected world food requirements and to maintain food security and irrigation will necessarily continue to be a major part of world's future agricultural production.

The importance of water and its uneven geographical distribution has made its acquisition and use a matter of great contention between various users and nations. Population growth, rising standards of living and competing uses are forcing major changes in how fresh water is allocated. Agricultural water resources are also coming under increasing pressure from environmental and other social issues, including declining water tables in many areas, industrial and municipal pollution and escalating energy costs. In addition, there is also a persistent loss of arable land to soil salination, soil erosion and urbanization, which collectively emphasize the need to maximize total irrigated food productivity wherever possible. Thus, current water use patterns and practices may not be sustainable in many regions of the world.

Reduced availability of water and energy for irrigated agriculture will require much greater levels of crop and water management than currently in use; but this will be difficult, if not impossible, without advanced irrigation methods. Higher levels of irrigation management and systems that enable greater efficiency of water use can potentially reduce operating costs as well as conserve energy, nutrients and water use. The potential to conserve water depends on the capabilities of the irrigation system and the commitment of the operator to implement timely water-saving practices and technologies (Schütze and Schmitz 2009).

2 Advanced Irrigation Systems

There are basically three types of irrigation systems: surface (which flows by gravity), and pressurized micro and sprinkler systems. Surface (also called gravity) irrigation introduces water at the highest elevation point in a field and the water flows over the soil surface by gravity. Pressurized irrigation systems that include several types of microirrigation and sprinkler systems where water is conveyed by pressurized pipes or tubing to a point in the field where it is applied by various devices. In addition, pressurized systems commonly function as site-specific application methods for fertilizers and other agrochemicals (chemigation), which require appropriate management to avoid environmental contamination. Technology levels in all three types of systems vary widely, and they all have advantages and disadvantages (Hoffman et al. 2007).

The design of a water application system will determine the maximum potential performance level, whereas management dictates the actual benefits realized and the magnitude of any positive or negative ecological impacts. Thus, the amount of water that can be conserved by advanced irrigation systems and practices depends on the ability of a particular type of irrigation system to implement various improved management strategies. Management skills are gained through education and technical assistance, but they are maintained by the irrigator's commitment and economic incentives.

3 Microirrigation Methods

Microirrigation includes drip and microsprinklers. It is a flexible set of technologies that can potentially be used on almost every crop, soil type and climatic zone if justified economically. It is characterized by the applications of water in small amounts using frequent irrigations (i.e. daily). It is also sometimes referred to as localized irrigation and can provide numerous crop production and water conservation benefits that address many of the water quality and supply challenges facing modern irrigated agriculture. Microirrigation acreage is increasing steadily worldwide and is fully expected to continue its rapid growth in the foreseeable future (Lamm et al. 2007). Novel applications for microirrigation systems, such as the reuse of municipal wastewater in turf areas, are providing new opportunities for growth.

These systems can be laid on the soil surface or buried, and can be permanent or temporary installations. They can be used on gardens as well as in fields from 0.5 to more than 50 ha. Because of the relatively high capital cost of these advanced systems and the need for high levels of management, modern microirrigation technologies are typically permanent installations on small blocks of land (e.g. <10 ha) of intensively-managed, high-value specialty crops (e.g. vegetables, grapes, nuts, fruits and berries). Microirrigation could also become more common in perennial forage crops (i.e. alfalfa) and annual row crops such as maize using widely-spaced, buried drip lines in areas with limited water supplies or high water costs.

Greenhouse culture is largely irrigated by microirrigation methods. Year around greenhouse production of fresh vegetables, cut flowers and other high-value crops is rapidly expanding with some of the largest increases occurring around the Mediterranean Basin. All the major factors affecting plant growth can be controlled and managed precisely in these conditions, including the environment, water, nutrients, pests and diseases. Thus, these growing systems provide the ultimate opportunities for the application of many PA concepts.

Microirrigation systems are well suited for automated adaptive control with real-time sensor feedback and decision making capabilities. However, site-specific water management often occurs by default as a function of the scales of underlying environmental variability. Microirrigation systems are generally designed and installed in small blocks to match the needs of particular crop varieties or variations in soil texture and topography across a farm. Thus, additional site-specific irrigation within blocks may not be warranted, although it is technically possible.

A low-cost microirrigation method called pitcher or clay pot irrigation (Siyal et al. 2009) is used on some small fields (e.g. <0.5 ha) in several areas in northern Africa and Middle East. This labor intensive method uses unglazed, porous clay pots which are buried in the ground up to their neck to irrigate individual plants (i.e. fruit and nut trees) that are typically spaced 1 m or more apart. There may be more than one pitcher per plant, which are filled manually with water on at least a daily basis. Water seeps out through the unglazed walls of the pitcher to irrigate the crop.

The above discussion suggests that purposely designed pressurized microirrigation and sprinkler methods are the most suitable for advanced forms of site-specific irrigation. Of these, most of the following discussion will be directed towards self-propelled, continuous move sprinkler irrigation systems, which cover much larger areas than microirrigation methods and their use is rapidly expanding (Sadler et al. 2005; Evans and King 2012).

4 Self-Propelled Sprinkler Irrigation Systems

Self-propelled sprinkler irrigation systems such as centre pivots and linear (also called lateral) move systems have allowed large scale agricultural development of marginal lands that were unsuitable for surface irrigation including areas with light sandy soil and a large variation in topography within the same field. These adaptable systems have experienced tremendous growth around the world over the past 50 years due to their: (1) potential for efficient and uniform water application; (2) high degree of automation requiring less labor than most other irrigation methods; (3) coverage of large areas; and (4) ability to apply water and labeled agrichemicals (chemigation) economically and safely over a wide range of soil, crop and topographic conditions. Similar to microirrigation, these systems apply limited amounts of water during each irrigation event and require frequent applications. A single machine can irrigate fields ranging from about 5 ha to over 200 ha.

These large machines basically consist of a pipeline (lateral) mounted on motorized structures (towers) with large rubber tires. The section between the two towers is called a span, which can vary from about 30 to 70 m in length. Sprinkling devices are mounted on or below the pipe. Maximum application depths of water applied are controlled by varying the speed of machine travel. Field-scale water applications by these systems can be quite uniform.

A center pivot machine basically rotates around a 'pivot' at one end, usually in the centre of the field. These continuously moving systems can irrigate areas ranging from part circle segments to whole circles. Various optional accessories are also available to irrigate portions of field corners. Self-propelled linear move sprinkler systems look and perform almost identically to centre pivot systems except that they move in straight lines to irrigate square- or rectangular-shaped fields. Linear move systems require a global positioning system (GPS) or other methods for physical guidance, and they require considerably greater management and manual oversight than centre pivot systems.

Center pivot and linear move sprinkler systems are designed and generally operated so as to replace the average water used by the crop over the past few days as uniformly as possible across the field. Irrigations are frequent and apply relatively low amounts of water so that soil water is ideally maintained at relatively constant levels. Application rates and base uniformity are primarily established by the sprinkler nozzle package, but the depth of water applied per irrigation with self-propelled center pivots and linear move sprinkler systems is generally controlled by the travel speed of the machine.

Traditionally, irrigation system design and management has strived for maximum uniformity of water applications over the entire irrigated field. However, agricultural fields are never physically or biologically uniform and substantial agronomic and environmental differences can occur. The effects of different spatial and temporal sources of variation on management can be additive, and the stochastic variation of several interrelated factors across a field can affect crop growth, yields and crop quality. Often, the underlying causes of crop performance variation are often not well understood, and they can vary substantially from field to field and year to year. Some sources of variation such as pest problems may have both temporal and spatial considerations.

The high frequency of the irrigations under these machines potentially reduces the magnitude of variability in soil water content in the field. However, stochastic spatial and temporal variability of a number of other interrelated factors (e.g., variations in soil properties, topography, runoff, within field runoff (also called runon), pests, tillage, fertilization, uneven incident precipitation and hail, pesticide carryover effects, and herbicide drift from adjacent fields) across a field can still affect crop growth during the growing season and from one season to the next. These factors can influence management decisions over time, which may also introduce additional in-field variability to crop production. Consequently, the center pivot industry is beginning to market irrigation systems that can adjust for at least some of this spatial and temporal variability, which is typically referred to as site-specific variable rate irrigation (SS-VRI). Manufacturers are just starting to offer site-specific controls for linear move sprinkler systems. Kranz et al. (2012) has summarized characteristics of some of the various currently available commercial site-specific control systems and panels.

SS-VRI can be defined as the ability to vary water application depths spatially across a field to address specific soil, crop and/or other conditions. It is included in the spectrum of precision agriculture technologies because advanced SS-VRI methods can potentially impose treatments in ways that optimize plant responses for each unit of water applied in different areas of the same field. It can also include site-specific applications of water-soluble agrochemicals including fertilizers.

Irrigation management can amplify the negative effects of different sources of variation with a field or it can help to minimize these effects depending on the grower's objectives. Site-specific irrigation management strategies can be justified to account for variation under non-uniform growing conditions. Reducing areas of excess water applications within a field will decrease the potential for runoff, limit the movement of nutrients and agrochemicals below the plant root zone and create conditions for improving crop yield and quality.

Spatial and temporal variation can influence irrigation management decisions, and site-specific managers attempt to characterize the major sources of variation throughout the growing season. However, growers cannot practically manage for all of the many sources of variability. Therefore, they tend to group the most critical properties into relatively homogeneous management zones within a field as much as possible. Thus, different PA technologies may be managed simultaneously at different scales in the same field. For example, the smallest zone that a large scale farmer (e.g. >20 ha fields) prefers to manage for irrigation is probably in the order of about 5 ha even though other PA technologies (e.g. planting and spraying) will often be managed at much smaller scales (e.g. 0.1 ha).

These factors can also accelerate localized leaching of soil nutrients and other agrichemicals past the root zone. Field variability can result in excessive energy use for pumping, affect irrigation system design initially, and, later, management decisions. Furthermore, some management decisions may introduce additional sources of within-field variability.

Self-propelled center pivot and linear move sprinkler irrigation systems are particularly amenable to site-specific approaches because of their current levels of automation and large area coverage with a single lateral pipe. The definition of sprinklers in these applications includes the use of LEPA, bubblers, sprayers, spinners and other related spray techniques to apply water. These devices are usually on drop tubes in or just above the crop canopy. Impact type sprinklers are generally not included because the methods used to vary applied depths of water (e.g., pulse modulation) on commercial systems are not compatible in practice.

The ability to vary water application along the main lateral of a center-pivot system based on position in the field allows the field manager to address specific soil and/or slope conditions and avoid areas of over- or under irrigation, depending on preset management criteria. By aligning irrigation water application with variable water requirements in the field, total water use may be reduced, decreasing deep percolation and surface runoff. Reducing excess water applications will decrease the potential to move nutrients past the plant root zone (Sadler et al. 2000, 2005; King et al. 2009), and fungal disease pressure should also decrease. Site-specific application technologies can be used to treat small areas of a field with simple on/off sprinkler controls in single span-wide treatment areas or to treat the whole field by controlling all spans. Thus, site-specific irrigation can potentially provide water conservation benefits in cases of overirrigation, erroneous irrigation scheduling, in-season precipitation harvesting, or inefficiencies associated with particular crop production practices (e.g., potatoes). Some examples of water conservation strategies under full ET conditions (minimal water stress throughout season) where site-specific applications are potentially able to reduce total water applied include:

- A major cause of excessive water applications can occur when dry areas appear in a center-pivot irrigated field; irrigators tend to run the irrigation systems longer and more often to ensure adequate water across the whole field. These poor management practices result in much of the area being over-irrigated, and large amounts of water are wasted to compensate for suspected drought in relatively small areas. Site-specifically varying water applications can assist in addressing the management decisions that lead to localized, within-field drought and help conserve water, but they do not address the root causes.
- Water can also be conserved by applying less water to hillsides to reduce within-field runoff to low areas, to improve yields by eliminating water ponding in low areas, or to purposefully manage wet areas to reduce the extent of waterlogging problems.
- Water application amounts are generally high for the first span from the pivot point to the first tower even though application rates are low because it moves much more slowly than the outer towers. This area is typically either overwatered by as much as 20 %, because the available nozzle selections do not allow for the correct application of water, or too dry since the nozzles are very small and prone to plugging. The often-wetted foliage in this area also has higher incidences of fungal diseases. One option is to turn off every other sprinkler

along the first span for one or two rotations, and then turn on the sprinklers that are off and turn off the sprinklers that are on for one or two rotations. Another approach to eliminate both the overwatering and plugging problems within the first span area is to use relatively large nozzle sizes and site-specifically pulse the sprinklers on and off, applying the desired amount over time.

- Reduce or eliminate irrigation in wet areas in a field that develop from precipitation runoff, subsurface flow, high water table, springs, or unavoidable irrigation runoff.
- Manage soil water deficits on a spatial basis to maximize the opportunity to capture in-season precipitation in humid climates for subsequent use as crop evapotranspiration.
- The development of control and management technologies that can spatially and temporally direct the amount and frequency of appropriate agrochemical applications by site-specific self-propelled irrigation systems could also be very powerful tools that could increase productivity while reducing total water applications and minimizing adverse water quality impacts.

Evans and King (2012) conducted an analysis of several published simulation studies comparing conventional and SS-VRI management and reported water savings of 0–26 %. Ironically for well-watered crop production, water savings from site-specific irrigation may be greatest in humid climates by spatially optimizing use of non-uniform growing season precipitation. In arid and semi-arid climates, the greatest potential water savings could come from highly managed deficit irrigation strategies (managed drought stress).

5 Site-Specific Sprinkler Irrigation Technologies

Recent innovations in low-voltage sensor and wireless radio frequency (RF) data communications combined with advances in Internet technologies offer tremendous opportunities for development and application of real-time management systems for agriculture. These have enabled implementation of advanced state-of-the-art water conservation measures with self-propelled sprinkler systems such as SS-VRI for economically viable, broad scale crop production with full or limited water supplies.

SS-VRI technologies use many of the same management tools as other precision agriculture technologies, and make it possible to vary water and agrochemical (chemigation) applications to meet the specific needs of a crop in each unique zone within a field. However, site-specific sprinkler irrigation technologies and strategies are not a silver bullet for conserving water and energy; and they must also be integrated with other state-of-the-art farming technologies for maximum potential benefit (Sadler et al. 2005).

Basically, any water application device used on self-propelled sprinkler systems can be utilized for site-specific management of water and agrichemicals applied by the irrigation system. Water application methods commonly used on self-propelled sprinkler irrigation systems are high elevation sprinkler (usually impact style) head applications mounted on the top of the main pipe and medium elevation spray application heads (MESA), low elevation spray application heads (LESA) and low energy precision application (LEPA) methods. MESA is the most common method used on self-propelled irrigation systems in northern Great Plains region. Early work on LEPA was directed towards achieving relatively uniform application depths (Lyle and Bordovsky 1981, 1983, 1995; Bordovsky et al. 2003). Schneider (2000) reported that LEPA could potentially achieve application efficiencies greater than 95 % and that MESA was about 85 % depending on management.

Application depths on self-propelled sprinkler systems are generally controlled by the speed of the machine. However, this is not sufficient under site-specific conditions where variable amounts are needed along the length of the machine. It is possible to control every sprinkler individually, but the management level may increase to the point that the system is not practical because growers probably cannot manage areas less than 0.4–0.5 ha within a field in other cultural aspects of their operation. However, individual sprinkler control would allow more accurate site-specific applications to irregularly shaped areas. Increasing the number of sprinklers per bank would decrease cost, but the control system would lose some ability to adequately match pre-selected treatment areas. Controlling sprinkler heads in groups 10–15 m wide is generally a practical compromise to match operational limits (Evans et al. 2000; Sadler et al. 2000).

Types of SS-VRI Sprinkler Irrigation Systems. In the past few years some commercial companies began marketing center pivot control panels with an option to change center pivot travel speed in increments ranging from 1 to 10° as the machine rotates around the field. This tactic effectively changes application depths in each defined radial sector of the field, and no additional hardware is needed compared to a standard machine (some may need a GPS). This practice is commonly referred to as speed or sector control. It could also be referred to as variable depth irrigation, although some erroneously refer to it as variable rate irrigation. Nevertheless, field variability seldom occurs in long, narrow triangular-shaped parcels and adjusting machine speed may not always be a sufficient level of control because soil and crop conditions often vary substantially in the radial direction.

Consequently, center pivot manufacturers are also offering site-specific variable rate irrigation systems that can differentially apply water site-specifically to irregularly shaped areas or management zones. This is referred to as zone control. Specialized equipment such as control, panels, many valves, supplemental wiring and a GPS are required to control the irrigation in each management zone. Most zone control SS-VRI systems vary water application depths by various forms of pulse modulation (on–off cycling of spray-type sprinkler heads) for a given machine speed. Valves are located on every sprinkler head or groups of heads. Water is then applied to each zone by controlling water output amounts from each group of heads along the length of the machine depending on their location in the field. Zone control has a larger potential for achieving efficient management of water and energy than speed control, and is the general focus of this paper.

The most common site-specific sprinkler irrigation systems in use today are speed control systems, and it is anticipated that much of the short term growth will likely occur with these types of systems. Speed control technologies are probably being used close to their technical capacity to improve water productivity at this time. However, zone control systems can achieve the same effects provided by speed control, but with greater flexibility and provide more management options (Evans et al. 2012, 2013).

Site-specific control systems manage water application amounts by controlling individual or grouped sprinkler nozzle assemblies. These technologies can be used to treat predefined large areas in a whole field or a range of small areas within a field with simple on/off sprinkler controls. Small area (e.g., 0.5–10 ha) site-specific systems can be used to address issues in well defined, localized problem areas where the cost of a full precision irrigation control system may not be justified such as under the first span from the pivot point to avoid overirrigation or minimize foliar disease incidence in that area. These technologies can also be used to treat diverse areas of varying sizes within a large field by variably controlling groups of sprinkler heads across all spans as the machine moves.

SS-VRI Systems for Agricultural Research. There is little doubt that SS-VRI systems are wonderful research tools. They can be used to adjust water applications for artificial variability (e.g., research treatments) on top of natural variability within a field. For example, these systems could serve as an efficient, replicated screening method to identify drought and heat tolerance of various genotypes at a number of different locations to advance crop breeding efforts. SS-VRI systems could also be established to impose a range of abiotic stresses (deficit irrigation) using a line source sprinkler concept to develop water production functions (yields vs. range of available water). Another use would be to evaluate the effects of various degrees of potential climate change in different area on various cropping systems by simulating changing rainfall patterns and other conditions. SS-VRI research systems can be used to develop and test sensor networks that measure crop response to abiotic (e.g., water deficits) and biotic (e.g., insect and disease) stresses for improved management. Data from these types of experiments would also be quite beneficial in developing and testing various models and strategies for automatic control of commercial SS-VRI systems. A case study of a SS-VRI system currently being used for agricultural research is presented below.

6 Case Study of SS-VRI Systems for Agricultural Research

An example of research using SS-VRI system for research is a project conducted on a 4 ha field at the Montana State University (MSU) Eastern Agricultural Research Center) farm (near Sidney, MT (**47.73° N, 104.15° W**) **over five years from 2005 through 2009**. The site-specific irrigation control system was designed to implement research comparing tillage method (strip till vs clean, conventional till) by irrigation method (LEPA vs MESA) in a two year, irrigated crop rotation of sugar beet (*Beta vulgaris* L.) and malting barley (*Hordeum vulgare*). The soil was classified as a relatively heavy Savage clay loam (fine, smectitic, frigid Vertic Argiustolls) with 21 % sand, 46 % silt, and 33 % clay. Average field slope was about 0.5 % to the east.

The research was laid out in fifteen east–west strips parallel to the bi-directional travel of the linear move irrigation system (Fig. 1). Fourteen of these strips were used for research, but all fifteen were capable of site-specific irrigation. Each research strip was divided into four plots with two plots irrigated with MESA and two with LEPA for a total of 56 plots. Each 15 m by 24 m plot including buffers was planted either to sugar beet or malting barley, which alternated from year to year. Half of the plots were irrigated with MESA and the others with LEPA each year. There were 15 m alleys across the middle and ends of the block for turning farm equipment, and 1.2 m wide alleys between the sides of all plots.

All plots were irrigated with a Valley (Valmont Industries, Inc., Valley, NE) 244 m, 5 span, self-propelled linear move sprinkler irrigation system including the cart, which was installed in the spring of 2003. A diesel engine powered an electrical generator (480 V, 3 phase) was located on the cart that provided electricity for the tower motors, cart motors, irrigation water pump, air compressor and control valves. A buried wire alignment system was used with antennas located in the middle of the machine. The linear move machine used a screened floating pump intake in a level ditch as its water supply. Nominal operating pressure was about 250 kPa. Two double direction boom backs were installed at each of the towers (although not at the cart) because the machine irrigated in both directions. Spans were 48.8 m in length except for the center span with the guidance system which was a 47.5 m span. The machine moved at about 2.1 m min^{-1} at the 100 % setting.

A Valley CAMS Pro control panel (Valmont Industries, Valley, NE) was used to turn the machine on or off and control machine ground speed. A separate programmable logic controller (PLC) was designed and fabricated with the purpose of being able to irrigate every plot with either MESA or with LEPA. Individual, pneumatically activated solenoid valves were installed on every sprinkler head and controlled in banks of five MESA or thirteen LEPA heads.

The PLC-based control system activated grouped networks of electric over air-activated control valves. Thirty 15-meter wide banks of sprinklers were controlled with this system (15 MESA banks, 15 LEPA banks.) Both the depth and method of irrigation were varied depending on the location of each plot in the field. When not being used, low-cost pneumatic cylinders lifted the LEPA heads above the MESA heads to avoid spray interference when the MESA is operating over a given plot width and length (Fig. 2).

The amount of water applied was adjusted by pulsing heads on and off (pulse modulation) to achieve a target depth based on a digital map stored in the PLC (or in a remote base computer) of depths for each nozzle location as the machine moved down the field. Water was applied to meet the calculated actual evapotranspiration of each crop using data from a nearby agricultural weather station

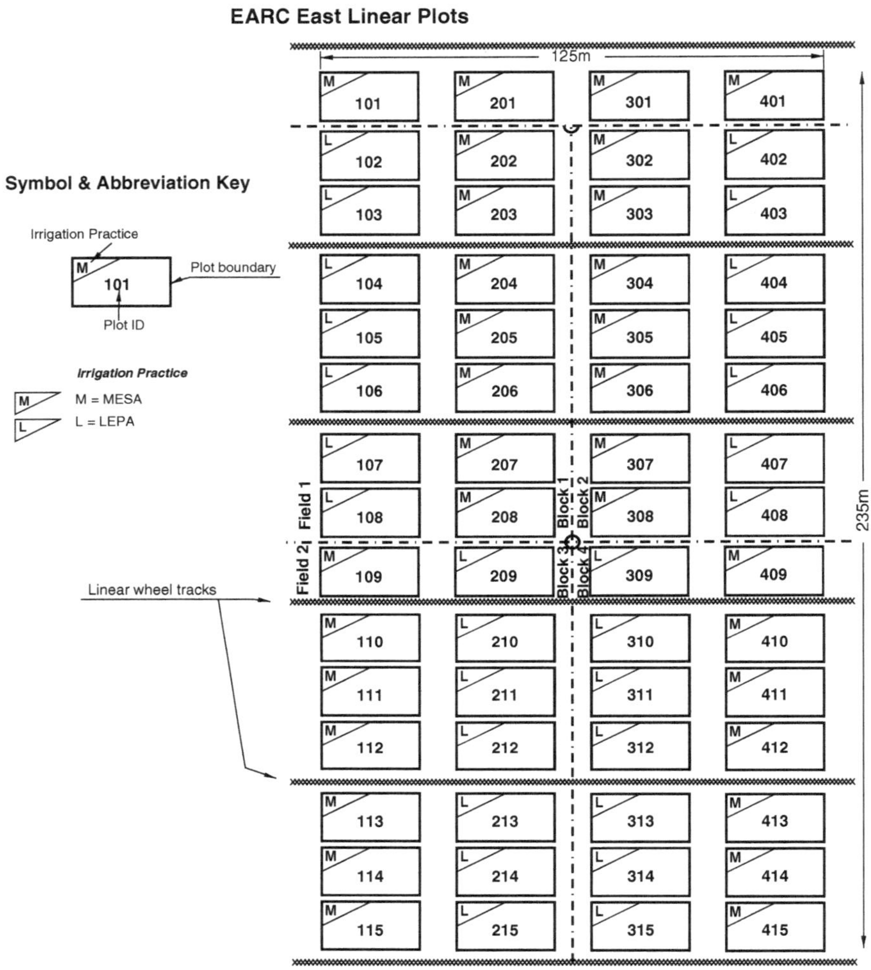

Fig. 1 Plot layout diagram of the field where the site-specific controls were implemented (Evans et al. 2010)

reconciled with weekly neutron probe soil moisture readings. Equivalent depths of water were applied for both irrigation methods for the same crop.

Distributed Sensor Systems and Control. A distributed wireless sensor network (WSN) was integrated into the existing site-specific linear move sprinkler irrigation system described above. Field conditions were monitored by six in-field sensor stations with Campbell CR200 dataloggers (Campbell Scientific, Inc, Logan, UT) distributed across the field based on a soil property map and monitored soil moisture, soil temperature, and air temperature. The CS616 water content reflectometers (Campbell Scientific, Inc, Logan, UT) were calibrated with a

Fig. 2 Photograph of the LEPA heads lifted above the crop canopy and avoiding interference with the MESA spray head water distribution patterns and LEPA heads in operation in the adjacent strip (Evans et al. 2010)

neutron probe and individually identified for their response ranges at each zone. All in-field sensory data were sampled on 10 s intervals. A nearby weather station monitors micrometeorological information on the field, i.e., air temperature, relative humidity, precipitation, wind speed, wind direction, and solar radiation. Communication signals from the sensor network, weather station and PLC irrigation controller to the base station were successfully interfaced using low-cost Bluetooth wireless radio communication.

Results. The overall focus of the project was to assess the environmental impacts of cultural practices and improved management of water, nutrient, and chemical applications as part of a multi-year team project involving several scientists from the location. Practical application of site-specific irrigation technologies with the variability of the research combined with natural variability is certainly more complicated and more challenging than general site-specific field irrigation.

This project illustrates it is possible to effectively install and operate precision site-specific irrigation systems on self-propelled linear move and center pivot systems. The knowledge of soil variability within a field is fundamental to the development of site-specific management areas since different soils have different water holding capacities. The ability to vary water application along the main lateral of the linear move based on position in the field allows researchers as well as producers to address specific soil, crop, and/or special research conditions/ treatments. By aligning irrigation water applications with variable water requirements in the field, total water use may be reduced, decreasing deep percolation and surface run off. Reducing excess water applications will decrease the potential to move nutrients past the plant root zone and fungal disease pressure should also decrease. Cropping systems that more efficiently utilize soil water have been shown to reduce costs and energy use as well as reduce water quality concerns. There is still a need to develop more efficient methods of site-specifically applying crop amendments (e.g., nutrients, pesticides) through self-propelled sprinkler irrigation systems to reduce total amounts applied, improve profit margins and reduce adverse environmental impacts.

References

Bordovsky JP, Lascano RJ (2003) Variable rate irrigation using low energy precision application (LEPA). In: 2003 Proceedings of the Beltwide Cotton Conference, Nashville, Tenn

Evans RG, Buchleiter GW, Sadler EJ, King BA, Harting GB (2000) Controls for precision irrigation with self-propelled systems. In: Evans RG, Benham BL, Trooien TP (eds) Proceedings of 4th decennial national irrigation symposium, ASAE, St. Joseph, MI. pp 322–331

Evans RG, Iversen WM, Stevens WB, Jabro JD (2010) Development of combined site specific MESA and LEPA methods on a linear move sprinkler irrigation system. Applied Eng Agric 26(5):883–895

Evans RG, King BA (2012) Site-specific sprinkler irrigation in a water limited future. Trans ASABE 55(2):493–504

Evans RG, Iversen WM, Kim Y (2012) Integrated decision support, sensor networks and adaptive control for wireless site-specific sprinkler irrigation. Appl Engr Agric 28(3):377–387

Evans RG, LaRue J, Stone KC, King BA (2013) Adoption of site-specific variable rate sprinkler irrigation systems. Irrig Sci. 31(4):871–887. doi:10.1007/s00271-012-0365-x

Hoffman GJ, Evans RG, Jensen ME, Martin DL, Elliott RL (eds) (2007) Design and operation of farm irrigation systems, 2nd edn. ASABE. St. Joseph, MI, p 863

King BA, Wall RW, Karsky TF (2009) Center-pivot irrigation system for independent site-specific management of water and chemical applications. Applied Eng Agric 25(2):187–198

Kranz WL, Evans RG, Lamm FR, O'Shaughnessy SA, Peters RT (2012) A review of mechanical move sprinkler irrigation control and automation technologies. Appl Engr Agric (in press)

Lamm FR, Ayars JE, Nakayama FS (eds) (2007) Microirrigation for crop production. developments in agriculture series, vol 13. Elsevier, Amsterdam, The Netherlands, p 618

Lyle WM, Bordovsky JP (1981) Low energy precision application (LEPA) irrigation system. Trans ASAE 24(5):1241–1245

Lyle WM, Bordovsky JP (1983) LEPA irrigation system evaluation. Trans ASAE 26(3):776–781

Lyle WM, Bordovsky JP (1995) LEPA corn irrigation with limited water supplies. Trans ASAE 38(2):455–462

Postel S (1999) Pillar of sand: can the irrigation miracle last?. Worldwatch Books, W. W. Norton & Co, New York, N.Y, p 325

Sadler EJ, Evans RG, Buchleiter GW, King BA, Camp CR (2000) Site-specific irrigation—management and control. In: Evans RG, Benham BL, Trooien TP (eds) Proceedings of 4th decennial national irrigation symposium, ASAE. St Joseph, MI, pp 304–315

Sadler EJ, Evans RG, Stone KC, Camp CR (2005) Opportunities for conservation with precision irrigation. J Soil Water Conserv 60(6):371–379

Schneider AD (2000) Efficiency and uniformity of LEPA and spray spinner methods: a review. Trans ASAE 43(4):937–944

Schütze N, Schmitz GH (2009) A neuro-dynamic programming approach for optimal scheduling and feedback control of deficit irrigation systems. In: Proceedings of European geoscience union general assembly 2009. European Geosciences Union, Munich, Germany Available at: http://citeseerx.istpsu.edu/viewdoc/download?doi=10.1.1.149.2226&rep=rep1&type=pdf. Accessed 27 June 2011

Siyal AA, van Genuchten MT, Skaggs TH (2009) Performance of a pitcher irrigation system. Soil Sci 174(6):312–320

Turral H, Svendsen M, Faures JM (2010) Investing in irrigation: reviewing the past and looking to the future. Agric Water Manag 97(2):551–560

Multi-Species Grazing on Deer Farms

Axel Behrendt, Andreas Fischer, Thomas Kaiser, Frank Eulenstein, Sylvia Ortmann, Anne Berger and Lothar Mueller

Abstract We developed and tested a specific approach for multi-species grazing on deer farms by combining some principles of contemporary livestock husbandry: the fenced farming of domestic ruminants, deer farming and multi-species grazing. Our approach was designed to optimize the use of extensively managed grasslands. Even in temperate sub-humid climates it is important to prevent grassland degradation due to selective overgrazing and eradication of valuable species. Experimental investigations have focused on determining which processes are self-regulating, and specifically on recognizing which processes are relevant in the interactions between different species of grazing farm animals and wild animals, grassland vegetation and landscapes. The focus of this specific research initiative, run by the Leibniz Centre for Agricultural Landscape Research (ZALF) and the Leibniz Institute for Zoo and Wildlife Research (IZW), is on comparative studies involving fallow deer (*Dama dama*), wild sheep (Mouflon, *Ovis aries*), and sheep, or red deer (*Cervus elaphus*) and Dexter cattle, as well as their specific effects on extensively managed grassland sites. Grazing experiments have shown clear effects and distinct differences in the grazing success between single-species and mixed-group grazing. The initial separation of the individual species in the multi-species groups disappeared over a time period lasting from several days to weeks. The animals grazed peacefully in mixed groups and even rested together in shelters. One particularly positive result was that several species grazed plants that were avoided by other animals. Moreover, weeds and invaders can be controlled

A. Behrendt (✉)
Leibniz Centre for Agricultural Landscape Research (ZALF) e. V., Research Station Paulinenaue, Gutshof 7 14641 Paulinenaue, Germany
e-mail: abehrendt@zalf.de

A. Fischer · T. Kaiser · F. Eulenstein · L. Mueller
Leibniz Centre for Agricultural Landscape Research (ZALF) e. V., Eberswalder Strasse 84 15374 Muencheberg, Germany

S. Ortmann · A. Berger
Leibniz Institute for Zoo and Wildlife Research, Alfred-Kowalke-Straße 17 10315 Berlin, Germany

L. Mueller et al. (eds.), *Novel Measurement and Assessment Tools for Monitoring and Management of Land and Water Resources in Agricultural Landscapes of Central Asia*, Environmental Science and Engineering, DOI: 10.1007/978-3-319-01017-5_30,

by this means. Our experiment was on a wetland site, but the principle of multi-species grazing on deer farms should have potential for other locations as well. We observed clear benefits for the grassland vegetation and ecosystem which are also expected to appear for grasslands in dryland regions. Due to economic constraints (there is a limited market for meat from deer and other animals in this grazing system), the approach remains limited to some innovative farms. It cannot resolve the problems of basic food security and grassland degradation in Central Asia or elsewhere, but a site-specific adaption of this multi-species grazing system could be an interesting option for improving the micro-economy, ecology and cultural lifestyle in some areas.

Keywords Multi-species pasture · Deer farm · Fallow deer · Sheep · Cattle · Red deer · Vegetation

1 Introduction

Global grassland ecosystems are characterized by contrasting systems and tendencies of management which have a critical influence on their function. Breakdowns in pastoral systems and marginalization processes are typical for poor regions and transition phases and countries (Dong et al. 2011). This dominant process is associated with rangeland degradation, a loss of biodiversity and a collapse in ecosystem functions and values (Neely et al. 2009). On the other hand, attempts to achieve viable, sustainable long-term grassland management are typical of many traditional grassland regions, including emerging regions. Systems of "set stocking" or "paddock grazing" are based on fenced grasslands.

Difficulties in achieving sustainable grassland management are typical of public lands (Hardin 1968), particularly in poor regions. Grassland fencing is based on clear property rights and offers a number of innovative options for utilization and management, such as deer farming (Lantz 1908), game farming (Golze 2007) or multi-species grazing (Meyer and Harvey 1985). However, research is required as to whether these systems and their combinations are feasible and environmentally friendly. We combined cattle, sheep, different deer and wild sheep on paddocks and analyzed their behaviour and their impact on vegetation and soil (Figs. 1, 3 and 4).

2 Multi-Species Grazing for Habitat Stability

Herbivores graze using a variety of strategies and are used as self-organized or specifically applied tools in landscape development. They therefore contribute to the diversity of biotope structures and thus to overall biodiversity (Riecken et al. 1998).

Fig. 1 Red deer and Dexter cattle share a pasture, with positive effects on the vegetation and soil

There is increasing demand for the use of ruminants with different dietary requirements in extensively managed grasslands. In the future, those methods and processes, which function in the long term without intensive human input, will become increasingly important both socially and commercially. A case in point is the sustainable use of grassland habits using multi-species approaches.

One fundamental prerequisite for achieving longer-term habitat stability is knowledge about the ecological principles governing the respective habitats. This necessitates an investigation of the interrelationships between the species inhabiting a particular biocoenosis and their relationship to the environment. Knowledge about the functioning of various ecosystems enables appropriate management tools to be developed, for example the selection of grazers to maintain or restore natural habitats.

Many such interrelationships, however, remain unknown. For a process-oriented understanding to be developed, the relevant interactions between the grazer and grazing site need to be identified, analyzed and ultimately generalized. The first step is to determine the specific effect of each ruminant species on the biotope structure. We currently lack a full overview of the effects that arise from the combined use of different herbivore species. One aspect is to quantify and implement stocking densities that are optimally adapted to the seasonally fluctuating yield levels of the respective sites. This also involves incorporating the interspecific relationships between the animal species. This approach can create synergies and help avoid conflict between different species.

One aspect is to recognize the plant sociological, ecological and ethological factors. Another aspect is to make business predictions about the productivity of the experimental variants based on the animals' performance, ongoing costs and work required. The latter aspect has largely been neglected in other studies and is currently an important factor because the economic situation of sheep and deer farming has changed in recent years.

3 Effects on Plant Sociology

The primary expectation from our species combinations in the multi-species groups is that positive synergisms will be achieved for grassland vegetation.

The scientific literature shows that grazing by different animal species creates specific vegetation structures (e.g. spatial arrangement of the vertical and horizontal phytomass; phenological effects) and can change the species compositions. On the one hand, the specific and selective foraging behaviour of grazers promotes the dominance of those individual plant species that are only minimally consumed or generally avoided (e.g. rosette plants, plant species with low-lying vegetation points) or whose substances make them unpalatable or even toxic (Buttenschøn and Buttenschøn 1982). On the other hand, the biomass of strongly preferred food plants is decimated (Noy-Meir et al. 1989; Kratochwil et al. 2004). Some plant species are able to compensate for this and can rebound with considerable phytomass development after being grazed (Belsky et al. 1993; Huhta et al. 2000). The impact of grazers on phytodiversity is interpreted very differently depending, for example, on the grazer, grazing pressure, plant association and climate (Olff and Ritchie 1998). Phytodiversity can increase or decrease after grazing, depending on the productivity of the respective sites (Olff and Ritchie 1998). Moreover, selective grazing alters the competitive situation among the plants themselves (e.g. competition for light), secondarily potentially promoting certain species (Milchunas et al. 1988). In the past, grass-feeding herbivores created open surfaces and therefore played a key role in structuring the landscape (Kaule 1991). Grazing by free-living animals considerably increases the species diversity of an area (Van Wieren 1991).

Our experiments also revealed that, under increasing grazing pressure, the grassland community originally dominated by tufts or top grasses developed into turf-like stands dominated by stolonate bottom grasses and herbage.

The present investigations revealed that fallow deer, mouflon and Skudde (a specific robust sheep breed) tended to select plant species differently at the study site. Methodologically, the comparison was based on a 4-level scoring system developed earlier (Kaiser and Fischer 1997). This involved recording the browsing of major species and problem species in each grazing enclosure on eight 20×20 m test surfaces; unbrowsed plants of the respective species served as a reference. On larger test areas, several browsing levels could occur. Therefore, the percentage surface area of the browsing levels related to the total coverage of a species was estimated for each species. Based on these data, an average browsing level was calculated for each species (Fig. 2). Measurements of plant heights supported the scoring estimates.

In September, fallow deer preferred the high-quality fodder grasses for cattle and only relatively rarely browsed on reed canary grass (*Phalaris arundinacea*) and the hard leaves of tall reed fescue (*Festuca arundinacea*). In contrast, Skudde and mouflon homogeneously browsed the latter two species comparatively close to the ground. The mouflons showed a peculiarity in their browsing behaviour on the creeping thistle (*Cirsium arvense*), a problem species in extensively managed grasslands. As opposed to other ruminants, mouflon browsed not only the apical shoot but also the spiny leaves. By browsing close to the ground, fallow deer were even able to suppress curled yellow dock (*Rumex crispus*), a feared pest on cattle ranges.

Unexpectedly, the wild form of white clover (*Trifolium repens*) was avoided by all three animal species. According to reports in the literature, white clover is

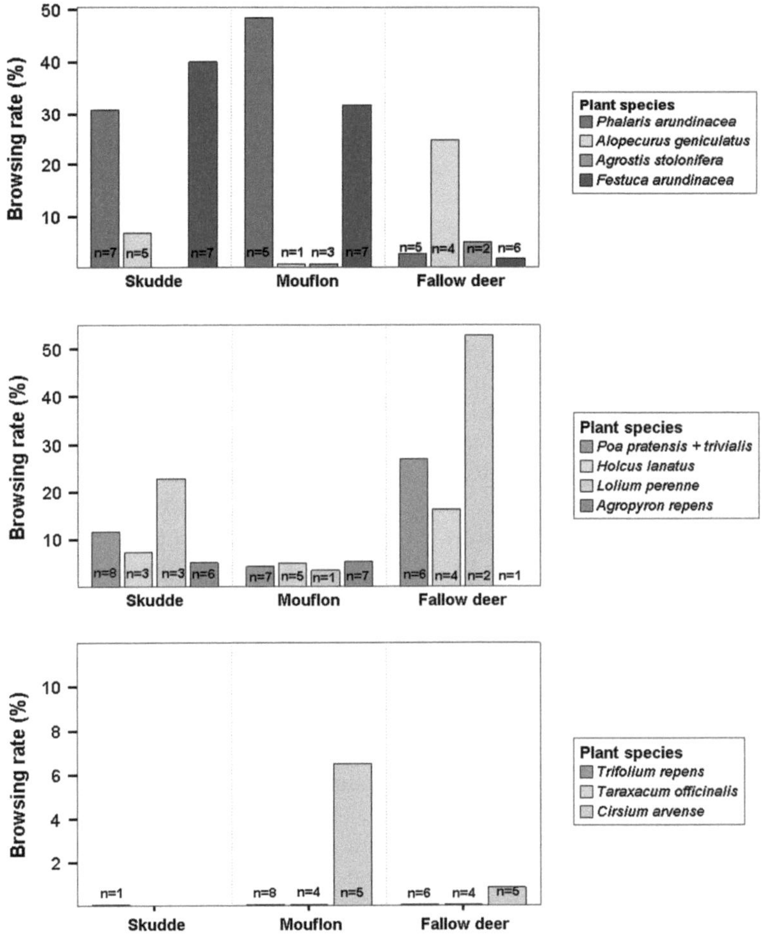

Fig. 2 Browsing ratio of different plant species by the Skudde sheep, mouflon and fallow deer

generally abundantly consumed by cattle, sheep and fallow deer (Rutter 2006). In the case of sheep, however, contradictory selection behaviour has occasionally been reported for white clover (Newman et al. 1992).

In further experiments at the Paulinenaue Research Station we observed that fallow deer preferred to browse on white clover varieties which are low in bitter substances. Wild populations of white clover can contain plants with cyanogenic substances, a feature that has been bred out to varying degrees in today's varieties.

4 Effects on Soil

Positive effects can also be seen on soil protection. Grassland sites, particularly in northeast Germany, are largely lowland moor sites. On these soils, compaction due to anthropogenic activities or trampling by grazers has a protective effect on the moor because it reduces the decomposition of organic material.

Generally, the looser the soil, the more intensive the mineralization of the soil organic matter. This is because aerobic microorganisms play a key role in the transformation processes. Accordingly, mineralization in the organic soil may be reduced by the relatively homogeneous grazing expected when mixed groups are present, as well as by the extensive compaction of the soil by trampling. Moderate soil compaction, as has been clearly demonstrated in experiments by Schalitz and Lehmann (1992), reduces the number of air-filled macropores, which in turn reduces the mineralization of soil organic matter.

Fallow deer, mouflons and Skudde are sure-footed, move relatively evenly across the surface, and have complementary site preferences. Accordingly, moor-protecting effects have been achieved by grazing such mixed groups on boggy sites. The compaction of the topsoil, as measured using cone penetrometers, showed a relatively even surface compaction. The exceptions were the contact points with adjoining enclosures. Here, a greater degree of compaction was visible. This, however, can be explained by the greater use of such areas: the animals seek contact with neighbouring groups, even if the groups are previously unfamiliar with one another.

5 Behaviour of Multi-Species Groups

Herbivores cannot fundamentally hinder natural succession processes, but they can extend early succession stages and influence the species composition of later stages (Davidson 1993). This is because differently specialized animal species—alone and in the complementary interaction—utilize different components of the vegetation. Here, the variously specialized ruminants make up a broad spectrum of types with different ecological orientations. Their feeding strategies are correlated with their spatial distribution, the temporal structure of their behaviour, their flight behaviour and their social structure (Hofmann 1989). Under undisturbed conditions, the different herbivorous species continue to exhibit their unique ecological specialization even today and occupy specific ecological niches (Petrak 1993).

Studies on fallow deer and Skudde under winter grazing conditions have shown that, even when they are held together in the same enclosure, their species-specific feeding behaviours remain intact. Interspecific competition and conflict situations revolving around food sources were recorded in only a few cases. The constellation of fallow deer and sheep represents an example of peaceful coexistence: they were also observed grazing together.

Compared with fallow deer, sheep showed considerably longer feeding activity. The average grazing time of mouflons—5.4 to 5.6 h—was the lowest of the three species. The mouflons generally preferred the mowed half of the pasture for foraging, with the individuals from the separated groups spending more time in the unmowed sector than the animals from the mixed groups. With an average grazing period of 6.6–8.2 h, the sheep showed the peak values among the three species. The mowed half of the enclosure was the preferred feeding site, but the individual sheep groups were also often present in the unmowed sector.

Rainfall and strong wind caused the animals to seek cover under a shelter. When disturbed, the animals reacted with flight or interrupted their respective behaviours. Their susceptibility to disturbance, however, is considered to be moderate. The mouflons behaved indifferently towards the sheep, but negative interactions (e.g. chasing away) were also observed towards the fallow deer (Figs. 3 and 4).

Cattle (Dexter) and red deer also harmonized well (Fig. 5). They grazed and rested together in the shelters. In the first few days of contact, the red deer reacted somewhat sceptically towards the advances of the Dexter, which immediately sought herd contact. After about one week, however, the red deer had already accepted the cattle and the two species complemented one another well in their grazing behaviour.

Fig. 3 Joint grazing of the Skudde sheep breed and fallow deer

Fig. 4 Fallow deer and mouflon in a paddock

Fig. 5 Red deer and Dexter cattle share a shelter

The undemanding Dexter cleanly grazed the fibre-rich reed canary grass (*Phalaris arundinacea*) whilst this plant was typically avoided by red deer.

6 Parasitoses

The performance of grazers can be seriously compromised by endoparasites and hoof diseases. Many studies on various farm animals have shown that parasite infestation negatively affects specific performance features and rearing duration.

This explains the variety of strategic measures used to tackle the problem. Their goal is to comprehensively protect the animals against parasite-related losses with minimal effort and a supra-seasonal effect (Hale 2006; Larsson et al. 2006).

Especially on low-lying grazing sites, where moisture plays a role, the herds must be observed or examined for worm infestation at least in fall and spring. Here, our focus is particularly on calves and lambs. Early action against worm infestation is decisive for rearing success. The most successful strategy is to treat the animals individually. Changing the medications after a certain period has also proven successful in preventing dangerous drug resistances.

7 Economic Considerations

The economy of multi-species grazing as a kind of game farming depends on many factors, such as the macro-economic context and local and national market conditions. It differs even between wealthy countries, such as New Zealand, where deer farming is an export industry, and Germany, where game farming is a niche branch lacking efficiency.

In Germany, the economic situation in sheep-keeping and deer farming is changing rapidly. It is characterized by high revenues from direct payments and by sharply increasing costs, especially for fodder, energy and land leases. Economic calculations by various authors in recent years have questioned the economic viability of game keeping (Fischer 2002; Constantin 2007). This is evident in carcasses with a high fat cover. Our herds do not need the winter fat reserves to the same extent as their wild conspecifics. A reduced supplementary feeding that is adapted to the weather conditions can provide decisive economic advantages. Reduction to the optimal level can therefore increase the effectiveness of this management form. Nonetheless, the profitability of any type of game keeping (and of other sectors) will continue to stand and fall with the business acumen of the producer.

8 Outlook

In many transitional and emerging states of Eurasia, including the states of Central Asia, contrasting processes of both grassland marginalization and intensification will continue. The intensification of grasslands will gain ground. As the economy and the welfare of the population improve this will lead to an increased demand for high quality and special meat of ruminants and deer. This offers options for game farming, multi-species grazing and their possible beneficial effects for the environment as shown above. However, the latter cannot be guaranteed. It seems to be useful to start pilot projects about multi-species grazing in this region. The approach cannot solve problems of basic food security and overall grassland performance. It will remain local, special and in a certain niche. However, it may contribute to improving the micro-economy, ecology and the cultural lifestyle in some areas, setting a new bright spot of achieving sustainable grassland management.

References

Belsky AJ, Carson WP, Jensen CL, Fox GA (1993) Overcompensation by plants—herbivore optimization or red herring. Evol Ecol 7:109–121

Buttenschøn J, Buttenschøn RM (1982) Grazing experiments with cattle and sheep on nutrient poor, acidic grassland and heath. II. Grazing impact. Natura Jutlandica 21, 2, 19–27

Constantin I (2007) Wirtschaftlichkeitsberechnungen zur Landwirtschaftlichen Wildhaltung. Tagungsband der Bundesfachtagung für landwirtschaftliche Wildhaltung 2007 in Kleve, pp 3–12

Davidson DW (1993) The effects of herbivory and granivory on terrestrial plant succession. Oikos 68:23–35

Dong S, Wen L, Liu S, Zhang X, Lassoie JP, Yi S, Li X, Li J, Li Y (2011) Vulnerability of worldwide pastoralism to global changes and interdisciplinary strategies for sustainable pastoralism. Ecol Soc16(2):10. (online) URL: http://www.ecologyandsociety.org/vol16/iss2/art10/, accessed on March 12, 2013

Fischer W (2002) Wirtschaftliche Aspekte der Damwildhaltung, Tagungsband des Oberfränkischen Gehegemanagementseminars für Wildhalter 2002 in Bayreuth, pp 106–122

Golze M (2007) Landwirtschaftliche Wildhaltung. Damwild, Rotwild, Muffelwild, Schwarzwild und andere Wildarten. Ulmer, p 156

Hale M (2006) Managing internal parasites in sheep and goats. ATTRA publication. www.attra.ncat.org/attra-pub/parasitesheep.html, accessed on Jan. 12, 2013

Hardin G (1968) The tragedy of the commons. Science, New Series, 162(3859):1243–1248

Hofmann RR (1989) Evolutionary steps of ecophysiological adaptation and diversification of ruminants: a comparative view of their digestive system. Oecologia 78:443–457

Huhta AP, Hellstrom K, Rautio P, Tuomi J (2000) A test of the compensatory continuum: fertilization increases and below-ground competition decreases the grazing tolerance of tall wormseed mustard (Erysimum strictum). Evol Ecol14(4–6):353–372

Kaiser T, Fischer A (1997) Eignung von Schafen (Skudde und Merinofleischschaf) für die Landschaftspflege auf extensiviertem Niedermoorgrünland. Zeitschrift für Kulturtechnik und Landentwicklung 38:172–176

Kaule G (1991) Arten- und Biotopschutz. 2. Aufl. Verl. Ulmer. Stuttgart, p 519

Kratochwil A, Stroh M, Schwabe A (2004) Restitution alluvialer Weidelandschaften: Binnendünen-Feuchtgebietskomplexe im Emsland (Nordwestdeutschland). Schriften Reihe. Für Landschaftspflege u. Naturschutz 78

Lantz DE (1908) Deer farming in the United States. U. S. department of agriculture. Farmers' Bulletin 330. Issued July 29, 1908. Government printing office, Washington. http://naldc.nal.usda.gov/download/48016/PDF, accessed on March 12, 2013

Larsson A, Dimander SO, Rydzik A, Uggla A, Waller PJ, Höglund J (2006) A 3-year field evaluation of pasture rotation and supplementary feeding to control parasite infection in first-season grazing cattle: dynamics of pasture infectivity. Vet Parasitol 142(3–4)

Meyer HH, Harvey TG (1985) Multispecies livestock systems in New Zealand. In: Proceedings of a conference on multispecies grazing. Winrock International, Morrilton, p 84–92

Milchunas DG, Sala OE, Lauenroth WK (1988) A generalized model of the effects of grazing by large herbivores on grassland community structure. Am Naturalist 132:87–106

Neely C, Bunning S, Wilkes A (2009) Review of evidence on dry lands pastoral systems and climate change implications and opportunities for mitigation and adaptation. FAO Rome, 2009, Online. ftp://ftp.fao.org/docrep/fao/012/i1135e/i1135e00.pdf, accessed on March 12, 2013

Newman JA, Parsons AJ, Harvey A (1992) Not all sheep prefer clover diet selection revisited. J Agr Sci 119:275–283

Noy-Meir I, Gutman M, Kaplan Y (1989) Responses of Mediterranean grassland plants to grazing and protection. J Ecol 77:290–310

Olff H, Ritchie ME (1998) Effects of herbivores on grassland plant diversity. Tree 13(7):261–265

Petrak M (1993) Nischenbreite und Nischenüberlappung bei der Nahrungswahl von Rothirsch (Cervus elaphus Linné, 1758) und Reh (*Capreolus capreolus* Linné, 1758) in der Nordwesteifel. Z Jagdwiss, 39:161–170

Riecken U, Finck P, Klein M, Schröder E (1998) Schutz und Wiedereinführung dynamischer Prozesse als Konzept des Naturschutzes. Schr. f. Landschaftspfl. u. Natursch. H. 56, 7–19, BfN, Bonn-Bad Godesberg

Rutter SM (2006) Diet preference for grass and legumes in free-ranging domestic sheep and cattle: Current theory and future application. Applied Animal Behaviour Science 97:17–35

Schalitz G, Lehmann J (1992) Zum Stellenwert der Bodenverdichtung im Rahmen der ökologiegerechten Nutzung von Niedermoorgrünland. Jahrestagung 1992 der AG für Grünland und Futterbau in der Gesellschaft für Pflanzenbauwissenschaften Stuttgart-Hohenheim, 45–60

Van Wieren SE (1991) The management of populations of large mammals. In: Spellerberg IF, Goldsmith FB, Morris MG (ed) The scientific management of temperate communities for conservation. Blackwell Science Publications, Oxford

Part III
Applications and Case Studies

Assessing the Soil Quality and Crop Yield Potentials of Some Soils of Eurasia

Elena Smolentseva, Boris Smolentsev, Konstantin Pachkin and Lothar Mueller

Abstract The aim of this paper was to quantify soil quality in agricultural landscapes over large regions. For this purpose, representative soil catenas and single soil pits were dug, analysed and classified on test sites in Russia, Kazakhstan and Germany. Soil quality and crop yield potentials were assessed using the Muencheberg Soil Quality Rating (M-SQR) method. This method is based on the rating of indicators relevant to crop yield. The results show that the estimation of all components of the site-specific water balance and drought risk assessment is key for the evaluation of soil functions in agricultural landscapes. We found close correlations between the overall soil quality rating score and grain yields of cereals. The suitability of the M-SQR approach to assess soil quality and crop yield potentials consistently over spatial scales of Eurasia has been confirmed.

Keywords Soil quality · Crop yield potential · Muencheberg soil quality rating · Eurasia

E. Smolentseva (✉) · B. Smolentsev
Russian Academy of Sciences, Siberian Branch, Institute of Soil Science and Agrochemistry (ISSA), Ac. Lavrentieva av., 8, 630090 Novosibirsk, Russian Federation
e-mail: smolentseva@issa.nsc.ru

B. Smolentsev
e-mail: pedolog@ngs.ru

K. Pachkin
Kazakh Research Institute of Soil Science and Agrochemistry named after U. U. Uspanov, Al Faraby Ave 75b 050060 Almaty, Kazakhstan
e-mail: kpachikin@yahoo.com

L. Mueller
Leibniz-Centre for Agricultural Landscape Research (ZALF) e. V., Eberswalder Str., 84 15374 Muencheberg, Germany
e-mail: mueller@zalf.de

L. Mueller et al. (eds.), *Novel Measurement and Assessment Tools for Monitoring and Management of Land and Water Resources in Agricultural Landscapes of Central Asia*, Environmental Science and Engineering, DOI: 10.1007/978-3-319-01017-5_31,

1 Objectives

Many global problems of the 21st century, such as maintaining a supply of energy, water and food, ecological safety, climate change, the protection of biological diversity, or air quality, are closely connected with sustainably functioning soils (Lal 2009). Processes of globalization create needs for international resource quality standards and methods to assess them. The preservation of the soil's productivity is of global significance (Mueller et al. 2010). This key function of soils depends on their properties and also on environmental parameters. Dokuchayev (1951) ascertained that there was a close relation between the properties of soils and the productivity of crops grown on them.

To avoid the degradation of soil resources it is necessary to study their potentials and risks carefully and to develop concepts for sustainable land use (Suleimenov and Oram 2000; Saparov 2013). In most countries in the world, national soil quality assessment systems were created during the twentieth century. However, as their soil and climate conditions are different, and so are the purposes of their soil function assessments, these approaches are different, often not comparable and not transferrable. This dilemma is comparable with that in the field of soil taxonomy.

The basic principles of the old soil bonitation system in the former Soviet Union were published by Gavriljuk (1974). This system is relatively uniform, but differs slightly between individual former Soviet Republics and is not comparable with the soil bonitation systems in other countries of Eurasia. Like old bonitation systems in other countries of Eurasia, it does not meet the criteria of being a contemporary trans-national system of soil quality assessment and evaluation (Mueller et al. 2010).

A standardised methodology for the assessment of soil quality will be demanded by the international community of land users and stakeholders in the near future.

Farmers and peasants manage soils at a local scale. Administrative bodies have to ensure that soil resources are handled sustainably through organizational, financial and legislative decisions at regional and national scales. They all need consistent information on soil quality and crop yield potentials.

At the Leibniz Centre for Agricultural Landscape Research (ZALF), Muencheberg, the Muencheberg Soil Quality Rating (M-SQR) was elaborated (Mueller et al. 2007a). The law of correlations between the properties of soils and the productivity of crop cultivated on them established by V. V. Dokuchayev (1951) served as a theoretical basis for this method. When the properties of soils are estimated this law permits their potential productivity to be assessed at the same time. M-SQR meets most criteria of an overall soil quality rating framework for cropland and grassland, which can operate consistently over spatial scales from field level to a global level (Mueller et al. 2010).

To provide reliable taxonomical and functional diagnostics of the estimated objects and make the results of the assessment comparable in a global context, we used the M-SQR in combination with the international soil classification of the World Reference Base for Soil Resources (2006). It provided comparative results of soil quality assessment in various countries and environments. The purpose of this chapter is to present the results from an assessment of the quality of representative soils which occupy large areas in Eurasia, including the agricultural landscapes of Kazakhstan, the Russian Federation and Germany. Conclusions are to be drawn on the suitability of this method for rating soil quality and productivity potential over large regions.

2 Materials and Methods

2.1 Method of Soil Quality Evaluation

The Muencheberg Soil Quality Rating is based on rating indicators which are relevant to soil productivity and easy to estimate in the field (Fig. 1). Two groups of indicators have been defined. The first group are basic indicators, which characterise the productivity-relevant soil properties of texture, structure and slope at pedon position. Basic soil indicators are scored using scoring tables in a field handbook (Mueller et al. 2007b). Individual scores are on a 5 or 9-point scale ranking from best conditions (2) to worst (0) with possible increments of 0.5 or 0.25. The basic indicators are weighted. Basic soil indicators are estimated entirely in the field. They can be backed by measurements of soil properties. If data regarding texture analysis, organic matter content or dry bulk density are available, they can be used to improve the reliability of the assessment. The final basic soil score ranges from 0 (theoretical minimum, in practice about 15) to 34. It is a measure of soil quality for farming in the field scale. The basic score can be upgraded (Upgraded Basic Score, UBS) to a 100-point scale by multiplying figures by 2.94.

In the second step (Fig. 1, step 2) the rating system also considers "hazard" soil properties and indicators. These properties are so critical for farming that they may limit the total soil quality. Hazard soil properties are the result of extremes of soil forming factors, either in excess or in a minimum. They are often determined by climatic factors. If the latter are present most of them can be identified using field methods or simple field tool kits, by vegetation indicators or the analysis of climate data.

A multiplier is derived from the most critical hazard indicator. The field handbook provides orientation values. Multipliers may rank from 0 to 2.94. Multiplying the basic score with this multiplier yields a final score (SQR score) ranging from 0 to 100. The classes of SQ are <20 = Very poor, 20–40 = Poor, 40–60 = Moderate, 60–80 Good, >80 = Very good.

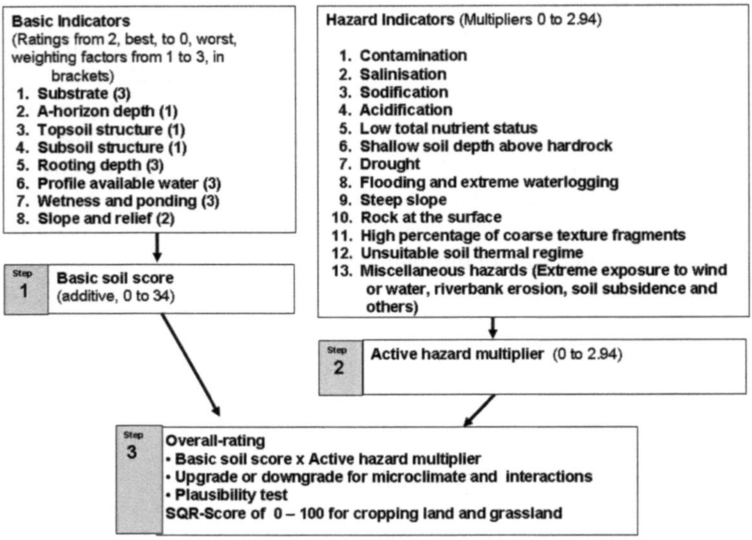

Fig. 1 Indicator scheme of the Muencheberg soil quality rating (M-SQR), (Mueller et al. 2007b)

2.2 Study Sites and Their Basic Data

The method was tested in Eurasia on farmland soils in various climatic zones such as the southern taiga, deciduous zone, northern and southern forest-steppe, steppe and dry steppe. The approbation of the method took place in Russia, in the territory of Western Siberia, in Kazakhstan and in Germany. The characteristics of the study sites and objects of assessment are presented in Table 1.

2.3 Field Procedure

On the sites under study we dug a representative number of soil profiles and performed a common soil survey and classification according to the national soil classification and international guidelines (WRB 2006). Profiles were geo-referenced and documented by photographs. All soil parameters relevant to indicator scorings (Mueller et al. 2007b) were measured and estimated. A monthly climate dataset of temperature, precipitation and potential evapotranspiration was allocated to the soil profile on the basis of a representative climate station. This was important for calculating the climatic water balance, assessing the site-specific drought risk, and estimating the soil thermal regime and the length of the vegetation period. All indicator scores were documented on a form. At the end of our field assessment procedure, each soil was given two names, that of the national soil

Table 1 List of study locations

Point, land, geo-position (Latitude North/ Longitude East)	Zone, subzone	Climate[a]	Soils (WRB)/ Parent material	No. of sampling points
Plotnikovo, RU, (56.886/83.069)	Southern taiga	Humid 482/0.1/ 1.06	Luvisols, Phaeozems/Loess	7
Ust-Kamenka, RU, (55.028/83.888)	Northern forest-steppe	Humid 514/0.1/ 1.11	Phaeozems, Luvisols/Loess	10
Ordinskoje, RU, (54.389/81.954)	Northern forest-steppe	Subhumid 410/0.1/ 0.81	Phaeozems, Chernozems/ Loess	5
Omsk, RU,[b] (55.066/ 73.332)	Southern forest-steppe	Subhumid 390/1.4/ 0.64	Phaeozems/Loess	1
Grushevka, RU, (53.906/77.160)	Typical steppe	Dry subhumid 310/-0.2/0.58	Chernozems, Solonetz/Loess	12
Slavgorod, RU,[c] (53.161/78.722)	Dry steppe	Dry subhumid 313/-0.2/0.59	Kastanozems/Loess	1
Shortandy, KZ,[d] (51.707/71.007)	Dry steppe	Dry subhumid 317/ 2.7/0.59	Chernozems/Loess	1
Besagasch, KZ, (42.801/ 71.406)	Dry steppe	Arid 330/8.9/0.25	Calcisols/ Redeposited loess	4
Dedelow, G, (53.37/ 13.802)	Deciduous forest	Subhumid 498/8.4/ 0.89	Regosols/Moraine	22
Müncheberg, G, (52.517/14.123)	Deciduous forest	Subhumid 540/8.5/ 0.83	Albeluvisols/ Moraine	20
Seelow, G, (52.543/ 14.446)	Deciduous forest	Subhumid 470/8.5/ 0.83	Fluvisols/Alluvium	33
Bad Lauchstädt, G[e] (51.394/11.878)	Steppe	Subhumid 474/8.8/ 0.74	Chernozems/Loess	1

[a] Precipitation in mm/Annual average air temperature in °C/Humidity factor (precipitation-evaporation ratio) according to FAO database New Loc_Clim 1.10 (FAO 2006)
[b] *Estimated from data* Suleimenov et al. (2009)
[c] *Estimated from data* Meinel (2002)
[d] *Estimated from data* Karbozova–Saljnikova et al. (2004)
[e] *Estimated from data* Körschens et al. (2005)

taxonomy and that of the WRB, and soil quality numbers from the M-SQR score. It should be emphasized that the M-SQR rating did not only yield a single bonitation number. It provided information about the reasons behind sub-optimum ratings. This enabled a further evaluation and comparison of the results with those of other soils worldwide. Crop yields of small-grain cereals were taken from research reports of yield plots when the site was a research station. In the case of fields used in practice (Plotnikovo, Ust Kamenka, Ordinskoje, Grushevka) these data were provided by local farmers and managers.

3 Results and Discussion

3.1 Rating Results of All Locations

Table 2 contains the soil quality assessment results in the form of averages of rating scores. Two main scores are presented. The first is the score of the basic rating. It mainly reflects the textural and structural properties of the profile taking into account their importance for the yield formation of small-grain cereals. The second is the overall M-SQR score also considering hazard factors for the yield formation. For a better comparison of the two, we present the upgraded basic score (UBS) in the second column of Table 2. Both are now on a 100-point scale.

The UBS can be evaluated as good or very good on all sites. This is not surprising as all soils are in farmland use or under cropping. Extremely shallow soils, stony soils or steep land were not considered in this study.

Due to the limiting hazard factors such as drought or an unsuitable soil thermal regime, the overall M-SQR score is clearly lower than the UBS. The only sites to have high basic rating scores are several in Siberia and Kazakhstan, and these suffer from limitations caused by hazard factors. Thus, their overall soil quality is poor in most cases. This may change as the climate changes, and the M-SQR methodology could be applicable as a tool for scenarios and monitoring those possible changes.

Table 2 Results of soil assessment, arable land (Smolentseva et al. 2011)

Location	[a]Score of basic rating (UBS)	[b]M-SQR score	Overall soil quality class	[c]Limiting factors
Plotnikovo, RU	87 (Very good)	37 (4)	Poor	TR + D
Ust-Kamenka, RU	88 (Very good)	34 (12)	Poor	TR + D
Ordinskoje, RU	86 (Very good)	29 (10)	Poor	TR + D
Omsk, RU	96 (Very good)	41	Moderate	TR + D
Grushevka, RU	78 (Good)	20 (11)	Poor	D + TR
Grushevka, RU, Solonetz, Solonchak	53 (Moderate)	5 (6)	Very poor	S + TR
Slavgorod, RU	83 (Very good)	28	Poor	D + TR
Shortandy, KZ	89 (Very good)	38	Poor	D + TR
Besagasch 1, KZ, irrigated soils	74 (Good)	57	Moderate	D
Besagasch 2, KZ, dryland	72 (Good)	12	Very poor	D
Dedelow, G	72 (Good)	66 (13)	Good	D
Muencheberg, G	62 (Good)	42 (4)	Moderate	D
Muencheberg, G drained soils	62 (Good)	57 (2)	Moderate	D
Seelow, G	69 (Good)	66 (10)	Good	WL
Bad Lauchstädt, G	93 (Very good)	88	Very good	D

[a] Upgraded Basic score in a 100-point scale (basis rating *2,94) for the comparison with overall soil scores

[b] Overall soil scores average value and standard deviation (in brackets)

[c] Hazard indicators with critically low rating: *TR* thermal regime (cold), *D* drought risk, *S* salinisation, *WL* waterlogging

In the M-SQR score column, the number in brackets indicates the standard deviation at some sites. It differs between landscapes, can be very low in homogeneous loess landscapes under a relatively humid climate (Example: Plotnikovo) and very high under semi-arid conditions (Example: Grushevka). Hummocky landscapes with distinct differences in the soil water regime are a common reason for spatial variability in the soil quality.

The soils in Europe already had lower basic ratings in most cases and larger spatial variability. Soils formed on Pleistocene and Holocene sediments are particularly inhomogeneous. Sandy soils at Muencheberg site are characterised by an unfavourable structure and a low moisture capacity. Low water storage capacity in combination with a clear negative water balance in the vegetation period increases drought at this site. In contrast, Loess soils in Europe, located in a temperate climate (Bad Lauchstädt site), despite having a sub-optimum structure under arable land use, have very high basic and overall ratings and are among the best soils in the world.

3.2 Quality Aspects and Degradation Risks of Soils in Siberia and Central Asia

It was found that all soils which were formed on the loess and loess-like substrata had very high basic rating values (Table 2). These soils are Chernozems, Phaeozems und Kastanozems which are rich in humus and have a mollic horizon. However, these good soils are also prone to degradation. When used as ploughland these soils become substantially degraded because of humus loss and structure change. The crumbly structure of virgin steppe soils (Fig. 2) would be ideal, but this favourable structure cannot be maintained under ploughland. More unfavourable soil structure types, such as coarse granular or even blocky types with lower structure scores (basic indicators 3 and 4), were the result.

Apart from these limitations of structure degradation in the A-horizon, loess soils may provide a deep rooting. Another advantage of loess and loess-like soils consists in their ability to store large quantities of water and nutrients. Under favourable climatic conditions these soils are ideal for agriculture. Their potential water-holding capacity amounts to more than 300 mm in the rooting zone. However, under the current climatic conditions of Western Siberia and Kazakhstan the amount of available moisture at the beginning of the vegetative season is 80–180 mm only, leading to a considerable drought risk.

Loess soils may face further serious degradation risks such as wind, water and tillage erosion. We found clear evidence of wind and water erosion in a number of soil profiles. The M-SQR indicator system reflected this by producing sub-optimum scores.

Fig. 2 Favourable crumbly soil structure of the virgin steppe soils

Water erosion, and snowmelt erosion in particular, led to the local destruction of soil profiles and to a considerable decrease in soil quality. Local variability of basic rating scores of soils at Ust-Kamenka and Ordynskoje is explained by those eroded soil profiles (Fig. 3). Not only soils but also crops in the flow pathways of snowmelt erosion are so heavily damaged in most years that this also needed to be taken into consideration in the hazard rating procedure. Wind erosion led to the

Fig. 3 Shortening of humus horizons in the loess soils under the influence of water erosion. **a** Haplic Luvisol (Siltic), Ust-Kamenka, RU; **b** Haplic Phaeozem (Siltic), Ordinskoje, RU

Fig. 4 Examples of soil degradation due to geochemical and pedogenetical causes: **a** Gleyic Solonchak (Sulphatic, Arenic), **b** Calcic Solonetz (Humic, Siltic)

textural and structural degradation of some steppe soils in Siberia and Kazakhstan. This is a clear result of unsustainable ploughing practices since the second half of the last century. As a result of wind erosion caused by incorrect tillage, soils have gained a more "skeletal" texture and the arable horizon has developed a single-granular-like structure. Consequently, soil quality has become worse and this fact has led to lower ratings. The risk of soil degradation owing to wind erosion in the Kulundinskaya steppe was also recorded by Meinel (2002), Meyer et al. (2008). Loessial soils with insignificant humus content (e.g., Calcisols at Besagash) have slightly lower basic ratings compared with soils having a mollic horizon.

Solonetzic and saline soils at Grushevka are an example of soil degradation with geochemical and pedogenetical causes (Fig. 4). Neighbouring Chernozems under arable land use, located in higher landscape positions along the catena, show clear signs of anthropogenic degradation through wind erosion.

The potential of soil productivity in Siberia is substantially limited by the unfavourable temperature regime and by drought. In the regions of Kazakhstan which are important for cereal cropping, drought is also the most limiting factor.

Assessing all components of the site-specific water balance is of great importance when assessing soil quality and crop yield potentials. On the basis of M-SQR, sustainable management strategies can be established.

Another option for the application of the M-SQR methodology is the creation of soil quality maps of Eurasia. Richter et al. (2009) have already demonstrated this for Germany. In general, all data required are available for Russia and Kazakhstan too (Stolbovoi and Savin 2002; Pachikin et al. 2009). Additional parameters for the assessment of drought and the thermal soil regime can be gained from homogeneous climatic databases, for example from Loc_Clim 1.10 (FAO 2006). The additional use of satellite data will raise the accuracy of soil function maps made on the basis of M-SQR.

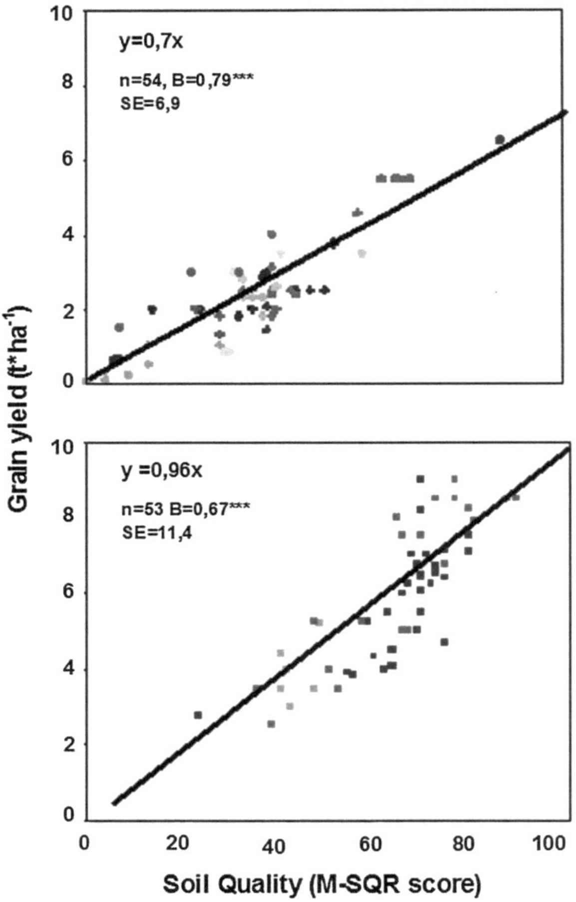

Fig. 5 Correlation between overall soil quality (M-SQR-score) and crop yield (Smolentseva et al. 2011). *Above* Low to moderate input, N-fertiliser <100 kg N ha^{-1}, *below* high input level, N-fertiliser >100 kg N ha^{-1}. Statistical data: *n* sample volume *B* coefficient of determination, *SE* standard error

3.3 Soil Quality and Crop Yields

We found significant correlations between soil productivity potentials in terms of M-SQR scores and yields of small-grain cereals on the basis of our test sites (Fig. 5). As crop yields are also influenced by human activity and the addition of fertilisers in particular, we stratified the dataset. The upper diagram of Fig. 5 shows the regression for locations cereals grown with limited inputs of agrochemicals (N fertilizer <100 kg N ha^{-1}) or in an ecological way. The regression line is linear and has a high coefficient of determination (B) and a relatively low standard error (SE). Using this correlation it is possible to estimate the crop yield on the basis of M-SQR ratings. The lower graph shows the regression at high inputs (N fertiliser >100 kg N ha^{-1}). A linear regression line is also acceptable, but data show there is a tendency towards non-linearity at higher M-SQR scores. This is due to higher inputs on better soils as the crop yield is higher and the risk of crop failure is lower on those soils. Most sites in this diagram are located in Germany.

4 Conclusions

1. The Muencheberg Soil Quality Rating (M-SQR) is a practicable method for detecting and assessing the potential and limitations of soils for their use as cropland or grassland.
2. The final ratings correlate well with grain yields on cropland.
3. This method has potential for the assessment of soils on a global scale.
4. It can be applied as a field method for monitoring agricultural land.
5. It could also be applied for creating uniform soil quality maps of Eurasia.

Acknowledgments Field work in the Novosibirsk region of Russia was greatly supported by Prof. Anatoly Tanasienko (ISSA, Novosibirsk). At the Besagasch experimental station in Kazakhstan, Dr. A. Kalaschnikov, Institute of Water Economy, Taras, guided fieldwork. In Germany, field assessment and scoring was carried out jointly with the managers of the research stations at Dedelow and Muencheberg, with local farmers and with scientists responsible for field experiments. Dr. G. Verch, Dedelow and Dr. D. Barkusky, Muencheberg, provided the most support. Authors gratefully acknowledge the cooperation of all partners and thank them for their collaboration. Work was carried out with partial financial support from the interdisciplinary project of the presidium of the Russian Academy of Sciences (Project No. 14/16) and by the German Federal Agency for Agriculture and Food (BLE, Project No 05/07).

References

Dokuchayev VV (1951) Izbranniye sochineniya, vol 6. Moskau, 515p

FAO (2006) New_LocClim 1.10. Local climate estimator, 2006. Download address ftp://193.43.36.131/SD/Reserved/Agromet/New_LocClim/New_LocClim_V1.10_20060919.zip. Accessed 12 Dec 2012

Gavrilyuk FY (1974) Bonitirovka pochv, Moskva, Vysshaya shkola, 270p

Karbozova–Saljnikova E, Funakawa S, Akhmetov K, Kosaki T (2004) Soil organic matter status of Chernozem soil in North Kazakhstan: effects of summer fallow. Soil Biol Biochem 36:1373–1381

Körschens M, Altermann M, Merbach I, Rinklebe J (2005) Böden als unsere Lebensgrundlage. Schwarzerde ist der Boden des Jahres 2005. Online http://eusoils.jrc.ec.europa.eu/esdb_archive/eusoils_docs/Various/2005_Schwarzerde/A_Texte_Exponate/4_FESTVORTRAG.pdf. Accessed 12 Dec 2012

Lal R (2009) Soils and food sufficiency: a review. Agron Sustain Dev 29(1):113–133

Meinel T (2002) Die geoökologischen Folgewirkungen der Steppenumbrüche in den 50-er Jahren in Westsibirien.- Diss, Halle. Online http://sundoc.bibliothek.uni-halle.de/diss-online/02/03H082/t1.pdf. Accessed on 12 Dec 2012

Meyer BC, Schreiner V, Smolentseva EN, Smolentsev BA (2008) Indicators of desertification in the Kulunda steppe in the south of Western Siberia. Arch Agron Soil Sci 54(6):585–603

Mueller L, Schindler U, Behrendt A, Eulenstein F, Dannowski R (2007a) Das Muencheberger Soil Quality Rating (SQR): ein einfaches Verfahren zur Bewertung der Eignung von Boeden als Farmland—Mitteilungen der Deutschen Bodenkundlichen Gesellschaft 110(2):515–516

Mueller L, Schindler U, Behrendt A, Eulenstein F, Dannowski R, Schlindwein SL Shepherd TG, Smolentseva E, Rogasik J (2007b) The Muencheberg Soil Quality Rating (SQR)—field manual for detecting and assessing properties and limitations of soil for cropping and grazing. Online http://www.zalf.de/home_zalf/institute/lwh/lwh_e/mitarbeiter/mueller_l/pdf/field_mueller.pdf. Accessed 12 Dec 2012

Mueller L, Schindler U, Mirschel W, Shepherd TG, Ball B, Helming K, Rogasik J, Eulenstein F, Wiggering H (2010) Assessing the productivity function of soils: a review. Agron Sustain Dev 30(3):601–614

Pachikin K, Erokhina O, Funakawa S (2009) Properties and distribution pattern of soils in Kazakhstan—Pedologist 53(1):30–37

Richter A, Hennings V, Müller L (2009) Anwendung des Müncheberger Soil Quality Ratings(SQR) auf bodenkundliche Grundlagenkarten—Jahrestagung der DBG: Böden—eine endliche Ressource, Kommission VIII, Bonn: 1–4

Saparov AS (2013) Soil resources of the Republic of Kazakhstan: current status, problems and solutions. In: Novel measurement and assessment tools for monitoring and management of land and water resources in agricultural landscapes of Central Asia (Chapter 2 of this book)

Smolentseva EN, Suleimenov MK, Pachikin KM, Kussainova MD, Bekhbayev UK, Saparov AS, Rukhovich OV, Lukin SM, Schindler U, Mueller L (2011) Assessing agricultural soil quality on a global scale. Soil Sci Agrochemistry (Almaty, Kazakhstan) 4: 81–91 (in Russian)

Stolbovoi V, Savin I (2002) Maps of soil characteristics—CD-ROM *Land Resources of Russia*, Laxenburg. Austria: International Institute for applied systems analysis and the Russian academy of science. CD-ROM. Distributed by the National Snow and Ice Data Center/World Data Center for Glaciology. Boulder

Suleimenov M, Oram P (2000) Trends in feed, livestock production, and rangelands during the transition period in three Central Asian countries. Food Policy 25:681–700

Suleimenov M, Kaskarbayev Z, Chirkov E, Kenzhebekov A, Dvurechenskiy V, Kudaibergenov G, Khramtsov I (2009) Towards minimum and zero tillage in fallow-grains production systems in three Ecoregions of Kasakhstan and Siberia—ISTRO 18th Triennial conference proceedings 1-002, Izmir, Turkey:1–6

WRB 2006 World reference base for soil resources (2006) IUSS working group WRB, 2nd edn. World Soil Resources Reports 103, FAO, Rome, 145

Soils of Kazakhstan, Their Distribution and Mapping

Konstantin Pachikin, Olga Erokhina and Shinya Funakawa

Abstract The territory of Kazakhstan is situated in the centre of the Eurasian continent and extends from 41° to 55° N and from 46° to 87° E. Its landscapes are diverse, with a great variety of climates and forms of relief (from high mountains to desert plains). The soil formation is based on the principles of latitudinal (horizontal) and vertical zonalities. The geographic soil zones in Kazakhstan from north to south are: forest-steppe, steppe, desert-steppe, and desert, where zones are classified according to the types of soil. This article characterises the natural zonality of Kazakhstan's soils and orographic-climatic regions with a spectrum of zones and belts (Altay, North Tien-Shan, West Tien-Shan). For this characteristic of the regions and belts, including the basic climatic indexes, the type of vegetation can be used as a classification criterion. The resulting vertical distribution of soil types and subtypes will be shown. Soil mapping forms a basis for research work in various directions of soil science. Soil maps can be used to ascertain the genesis of soils and create new taxonomic classification approaches. They can also be applied for agro-industrial suitability planning, land reclamation, the estimation of soil degradation etc. Our chapter shows what methods are used to create modern soil maps, based on the application of GIS technologies, field and laboratory analytical work and the utilization of space images for the classification and delineation of soil units. Examples of applied thematic maps are shown, along with the approaches used to create them. Those maps allow terrain and land resources to be estimated and the state of the soil mantle to be monitored as a result of natural and human impacts. They are useful for the quantification of soil functions, such as

K. Pachikin (✉) · O. Erokhina
Kazakh Research Institute of Soil Science and Agrochemistry named after U. U. Uspanov, Al Faraby Ave 75b 050060 Almaty, Kazakhstan
e-mail: kpachikin@yahoo.com

O. Erokhina
e-mail: oerokhina@rambler.ru

S. Funakawa
Graduate School of Agriculture, Kyoto University, Kyoto 606-8502, Japan
e-mail: funakawa@kais.kyoto-u.ac.jp

L. Mueller et al. (eds.), *Novel Measurement and Assessment Tools for Monitoring and Management of Land and Water Resources in Agricultural Landscapes of Central Asia*, Environmental Science and Engineering, DOI: 10.1007/978-3-319-01017-5_32,

productivity potential, or for the allocation of protected areas. All these maps allow measures for the restoration and sustainable development of landscapes to be developed.

Keywords Soil · Mapping · Methods · Soil types · Kazakhstan

1 Objectives

Soils are the natural wealth of every country. They fulfil many functions for society and must be protected and used carefully and sustainably. This requires information on the soil inventory, status and changes over the whole landscape. Different soil maps are important tools for decisions about the handling of soils. The main objective of this chapter is to show some methods for creating basic and thematic maps, and to characterize the distribution of dominant soil types in Kazakhstan.

2 Materials and Methods

The soil mapping methods are based on the principles of soil geography and landscape analyses, and data handling as described in the fundamental works of Isachenko (1980), Korsunov et al. (2002), Sokolov (2004). The soil maps are created using traditional mapping methods (Korsunov et al. 2002), GIS technologies and remote sensing equipment. The mapping process passes through several stages:

1. Work with retrospective data. Previous material from different sources is used. This may include soil maps, thematic (geomorphological, vegetation, soil-meliorative, etc.) and evaluation maps on various scales, material from scientific and planning institutes and other organisations, references, and statistical materials.
2. Structural design and creation of a soil database. The database was created on the basis of concrete soil profiles with geographical coordinates and includes:
    a. Indicators of a soil profile's morphological structure (genetic horizons, colour, moistening, packing, neoformations, inclusions, etc.);
    b. Basic chemical and physical properties (humus, total nitrogen, carbonates, pH, exchanged cations, mechanical structure, compound of salts, etc.);
    c. The characteristic of major factors of soil formation (relief, vegetation, ground water).
3. Work with space images, mainly with satellite data: decoding, separating contours, saturating them whenever possible using data from last year's

research. The basic method for processing space-based information is indirect indicative decoding (Smirnov 2005; Kravtsova 2005) which is based on the determination of soils interconnecting with the components of the landscape, first of all with vegetation and relief. For decoding, mid-scale multispectral Landsat space images were used for a preliminary model of the soil map.

4. On this basis the key or reference sites which characterize the entire variety of soil cover in representative areas are selected for detailed studies of the field conditions.
5. Field research is carried out to update the content of the extracted contours and determine the soil decoding signs. At this stage morphological methods (Rozanov 2004) are applied to ensure that the soil field diagnostics and soil mapping are reliable and valid. Field research is done to obtain samples for laboratory analyses.
6. The detection of the soil taxonomy, compiling of systematic lists and map legends.
7. Extrapolation and final decoding of space images, filling the extracted contours with attributive information, designing the map.

3 Results and Discussion

In Kazakhstan, a variety of climates and relief forms (from high mountains to desert plains) determine the diversity of landscapes and soils whose formation is caused by the laws of latitudinal (horizontal) and vertical zonalities. Soils on the extensive continental plains of Kazakhstan are formed according to a pattern of horizontal zonality. Thus, soils formed on high altitudes, which develop exclusively through atmospheric humidification, are considered zonal soils. The following are the soil geographic zones in Kazakhstan from north to south: forest-steppe, steppe, desert-steppe, and desert, where zones are classified according to the types of soil (Fig. 1). General climatic information on the subzones in continental plains of Kazakhstan is shown in Table 1.

The forest-steppe zone occupies small areas in the northernmost part of the Republic and is indicated by Grey Forest Soils that have developed under grassy birch forests, and sometimes under birch-aspen or pine forests. Grey Forest Soils can be associated with Chernozems, Meadow-Chernozemic Soils, and Solonetzes.

To the south, the forest-steppe zone is replaced by the steppe zone, which is divided into subzones, namely, moderately humidified steppes, moderately droughty steppes, and dry steppes. Ordinary Chernozems are found in a subzone of the moderately humidified steppes; this soil is formed under rich herb-feather grass (Stipa) vegetation and often forms complexes with Solonetz Soils.

Meadow-Chernozemic Soils are formed in relief depressions and lowlands. Level and elevated positions in the droughty steppe subzone are occupied by

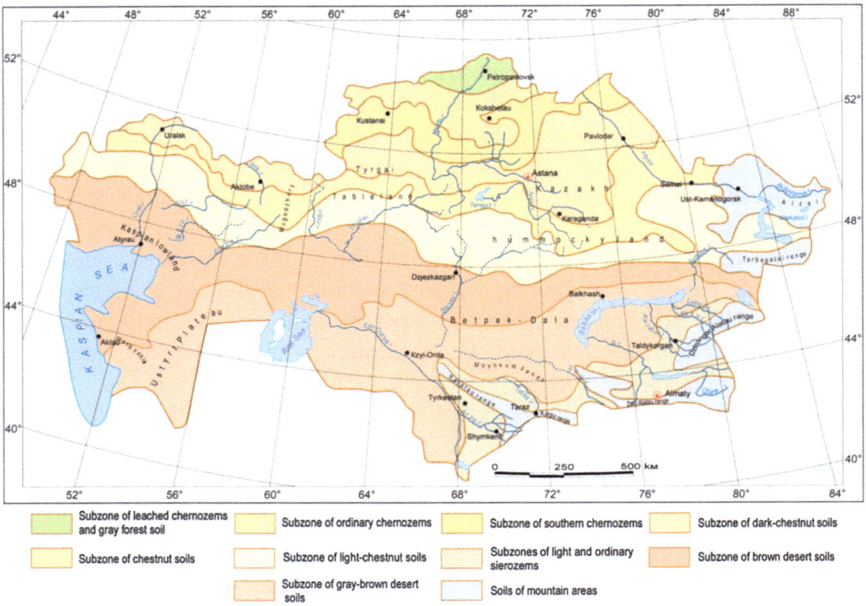

Fig. 1 Scheme of Kazakhstan's soil zones

Southern Chernozems that are also often associated with Solonetz Soils. Dark Chestnut and Chestnut Soils are formed in the moderately dry steppes and dry steppes subzones on relatively elevated positions under feather grass-fescue with poor steppe herb vegetation. The desert-steppe subzone is a transition zone between steppes and deserts. Here, the Light Chestnut Soils are the zonal soil type, and they are formed under sporadic sagebrush-fescue (*Artemisia-Festuca*) vegetation, with the prevalence of Solonetzic Soils in combination with Solonetzes, and in Central Kazakhstan with undeveloped stony soils.

The desert zone is subdivided into the northern and typical (central) desert subzones with Brown and Grey-Brown Desert Soils. Brown and Grey-Brown Soils are characterized by a low organic carbon content, a high quantity of carbonates, and the presence of a superficial porous crust. Often, they form complexes with Solonetz Soils and contain soluble salts in the bottom parts of their profile.

Intrazonal soils are formed in relief depressions under conditions of additional ground or superficial moistening. Semi-hydromorphic soils (Meadow-Chernozemic, Meadow-Chestnut, and Meadow-Brown Soils) are formed in landscape positions that get some additional recharges of surface or subsurface water or even underground water with water table depths of 3–6 m. Various hydromorphic and salinized soils may occur in different zones. They receive more recharge water from shallow (up to 3 m) fresh or salty underground waters and develop in more distinct depressions or lower positions. These soils seldom form homogeneous complexes in combinations with other zonal or intrazonal soils. Intrazonal soil

Table 1 General climatic information on the subzones in the continental plains of Kazakhstan

Subzones	Mean annual temperature (°C)	Cumulative temperature above 10 °C	Thickness of snow cover (cm)	Precipitation (mm)		Hydrothermal coefficient
				Annual	For the period at a temperature above 10 °C	
Ordinary Chernozems in moderately humidified steppes	0.6–0.8	2,070	23–28	320–350	220–230	0.9–1.0
Southern Chernozems in droughty steppes	1–1.5	2,160	21–26	300–330	160–180	0.8–0.9
Dark Chestnut Soils in moderately dry steppes	1.5–2	2,310	20–23	270–300	150–170	0.7–0.8
Chestnut Soils in dry steppes	2–3	2,380	18–23	210–270	130–150	0.6–0.7
Light Chestnut Soils in desert steppes	3–4	2,490	16–21	180–210	90–100	0.4–0.5
Brown Soils in northern deserts	3.5–5	2,800–3,800	10–15	150–180	70–90	0.3–0.4
Grey-Brown Desert Soils in central deserts	5–6.5	4,000–4,100	10–12	90–150	60–70	0.2–0.3

types are represented by Meadow, Alluvial-Meadow, Meadow-Marsh and Marsh Soils, and Solonchaks. These soils can be found in all zones but have some distinguishing features in each zone. Kazakhstan has vast areas of sandy soils, a majority of which are in the desert zone.

Kazakhstan's vertical natural zonality and soil zonality are caused by the height of mountains and their positions in geographically horizontal landscape zones, as well as being influenced by dominating humid (northwest, western) and dry (east, northern) air currents. Simultaneously, this zonality is influenced by the orographic features of the alignment of the main ridges of mountains with the directions of air streams, and seasonal maximum precipitation; an exposition of the slopes of ridges and their dissection. In Kazakhstan, there are three large oroclimatic regions with certain types of vertical zonality, various structures (spectra) of vertical zones and belts: Altay (Altay and Tarbagatay), North Tien-Shan (Zaili Alatau and Dzhungar Alatau ranges), and West Tien-Shan (Table 2, Fig. 1).

In the Altay region, the structure of vertical zonality (from top to bottom) is determined by the following soils: Mountain Tundra, Mountain-Meadow Alpine and Subalpine, Mountain-Taiga acid, Mountain-Forest dark and light Grey Soils (which form combinations with mountain soddy, Mountain Leached and Podzolized Chernozems, Mountain Dark Chestnut, Mountain Light Chestnut, and Foothill Brown Soils in the Zajsan depression).

The North Tien-Shan region is warmer and drier than the Altay and the spectrum of soils in this region is different. From top to bottom, the following types of soils are found: Mountain-Meadow Alpine and Subalpine, Mountain-Forest Dark-coloured, which is formed under mountain *Picea shrenkiana* forest, the Mountain-Forest Dark Grey and Chernozem-like Soils of small-leaved and fruit forests, Mountain Leached Chernozems, Mountain Ordinary and Southern Chernozems, Mountain Dark Chestnut, Mountain and Foothill Light Chestnut, and Northern Ordinary and Light Sierozems, which constitute the first step of vertical zonality.

The West Tien-Shan region is the warmest region with the maximum precipitation during the spring period. It is characterized by a wide distribution of plants that develop according to the spring cycle; ephemers and ephemeroids. Consequently, the soils formed here are completely different from those of the Northern Tien-Shan and Altay regions. A vertical soil zonality appears here with the following types of soils: High-Mountainous Meadow, Meadow-Steppe Alpine and Subalpine, Mountain-Forest Dark Cinnamonic Soils of treelike junipers, Mountain Cinnamonic, Mountain Grey-Cinnamonic, and Foothill Southern Sierozems (Ordinary and Light Sierozems).

Some current problems for Kazakhstan are questions connected with the inventory of lands, an estimation of their possible use, and perspectives of their being used by the agricultural industry for the formation of highly productive adaptive land tenure. Soil mapping is the basis for research in various directions of soil science.

In the 1950s–1970s regional soil maps on a scale of 1:300,000 were created for the whole territory of Kazakhstan. Even now they remain the basis for studying the rules behind the spatial soil distribution pattern in the Republic and provide a great

Table 2 Comparison of soils formed in respective oroclimatic regions with vertical zonality

	Altay		North Tien-Shan		West Tien-Shan	
Alpine and sub-alpine belt	Mountain Tundra Soil		Mountain Meadow Soil		Mountain Meadow-Steppe Soil	
	Elevation (m)	2,000–2,800(N)/ 2,200–3,000(S)	Elevation (m)	2,400–3,400(N)/ 2,500–3,600(S)	Elevation (m)	2,500–3,600(N)/ 2,600–3,800(S)
	MAT(°C)	<0	MAT(°C)	3–1	MAT(°C)	0–5
	AP(mm)	300–600 (summer)	AP(mm)	700–800 (spring–summer)	AP(mm)	600–700
	Vegetation	Lichen-moss-shrub mountain tundra (*Polytrichum, Dicranum, Ctraria, Cladonia, Carex, Betula rotundifolia*)	Vegetation	Coloured grass meadows (*Gentiana, Primula, Saxifraga, Shultsia, Viola*) or short-grass meadows with *Cobresia and Carex*	Vegetation	Coloured grass meadow-steppes (*Festuca, Kobresia, Alopecurus, Poa, Carex, Geranium, Achillea*)
Coniferous belt	Mountain-Forest Acid Podzolized Soil		Mountain-Forest Dark-Coloured Soil		Mountain-Forest Dark Cinnamonic Soil	
	Elevation (m)	1,000–1,600(N)/ 1,400–2,000(S)	Elevation (m)	1,700–2,400(N)/ 2,000–2,500(S)	Elevation (m)	1,500–2,200(N)/ 1,600–2,400(S)
	MAT(°C)	<0	MAT(°C)	3–5	MAT(°C)	6–8
	AP(mm)	600–1,000 (spring–summer)	AP(mm)	800–900 (spring–summer)	AP(mm)	700–800 (winter–spring)
	Vegetation	Fir-Siberian pine-larch herb and green mosses forests	Vegetation	Picea-herb forests	Vegetation	Juniper light forest
Belt of deciduous forests	Mountain-Forest Grey Podzolized Soil		Mountain-Forest Dark Gray Soil			
	Elevation (m)	600–1000(N)/ 1,200–1,400(S)	Elevation (m)	1,500–1,800(N)/ 1,600–2,000(S)		
	MAT(°C)	0–2	MAT(°C)	3–5		
	AP(mm)	600–800 (summer	AP (mm)	700–800 (summer–spring)		
	Vegetation	Aspen, birch and mixed fir	Vegetation	Aspen and birch herb forests		

(continued)

Table 2 (continued)

	Altay		North Tien-Shan		West Tien-Shan	
Mountain forest-meadow-steppe belt (Juniper light forest, shrub high herb semi-savanna in West Tien-Shan)	Mountain Podzolized and Leached Chernozem		Mountain Leached Chernozem		Mountain Leached Cinnamonic Soil	
	Elevation(m)	400–600 (N)/700–1200 (S)	Elevation (m)	1,200–1,500(N)/ 1,300–1,600(S)	Elevation (m)	1,500–2,200(N)/ 1,600–2,400(S)
	MAT(°C)	0–2	MAT(°C)	5–6	MAT(°C)	6–8
	AP(mm)	500–600 (summer)	AP(mm)	700–800 (summer–spring)	AP(mm)	700–800 (winter–spring)
	Vegetation	Forest meadows	Vegetation	Apple tree herb forests, shrub high herb-grass meadows	Vegetation	Shrub high herb semi-savanna with *Crataegus, Elytrigia trichophora, Hordeum bulbosum, Rosa, Spiraea, Ferula, Phlomis*
Feather grass-herb shrub steppes (Shrub high herb semi-savanna in West Tien-Shan)	Mountain ordinary and Southern chernozems		Mountain ordinary and Southern chernozems		Mountain Cinnamonic Soil with carbonates	
	Elevation (m)	300–400(N)/500–700 (S)	Elevation (m)	850–1200 (N)/900–1300 (S)	Elevation (m)	1,000–1,500(N)/ 1,200–1,600(S)
	MAT(°C)	2–3	MAT (°C)	6–7	MAT (°C)	8–10
	AP(mm)	400–450 (summer)	AP(mm)	550–650 (summer–spring)	AP(mm)	600–700 (winter–spring)
	Vegetation	Steppe feather-grass herb (*Stipa, Festuca*) with xerophytic herbs	Vegetation	Steppe feather-grass herb (*Stipa, Festuca*) with xerophytic herbs	Vegetation	High herb semi-savanna (*Elytrigia trichophora, Hordeum bulbosum, Althea, Rosa, Spiraea, Ferula, Phlomis*)
	Altay		North Tien-Shan		West Tien-Shan	

(continued)

Table 2 (continued)

	Altay		North Tien-Shan		West Tien-Shan	
Herb-feather grass-fescue dry steppes	Mountain Dark Chestnut Soil		Mountain Dark Chestnut Soil			
	Elevation (m)	250–300 (N)/300–500 (S)	Elevation (m)	750–850 (N)/800–900 (S)		
	MAT(°C)	3–4	MAT(°C)	7–8		
	AP(mm)	350–400 (summer)	AP(mm)	450–550 (summer–spring)		
	Vegetation	Steppe feather-grass herb (*Stipa, Festuca*) with xerophytic herbs (*Galium, Salvia, Althea*)	Vegetation	Herb-feather grass-fescue (*Festuca, Koeleria, Stipa*)		
Sagebrush-feather grass-fescue desert steppes (High herb semi-savanna in West Tien-Shan)			Light Chestnut Soil		Gray Cinnamonic Soil	
			Elevation (m)	650–750(N)/700–800(S)	Elevation (m)	500–800(N)/600–1,200(S)
			MAT(°C)	8–9	MAT(°C)	10–12
			AP(mm)	300–350 (summer)	AP(mm)	400–600 (winter–spring)
			Vegetation	Sagebrush-feather grass-fescue (*Artemisia, Festuca, Stipa*)	Vegetation	High herb semi-savanna (*Agropyrum trichophorum, Hordeum bulbosum, Ferula, Eremurus, Althea*)
Sagebrush-ephemeroid semi-savanna (Ephemeroid-ephemer low herb semi-savanna in West Tien-Shan)			Northern Sierozem		Southern Sierozem	
			Elevation (m)	450–650(N)/–700(S)	Elevation (m)	350–500(N)/–600(S)
			MAT (°C)	9–10	MAT (°C)	12–14
			AP (mm)	250–350 (summer–spring)	AP (mm)	200–400 (winter–spring)
			Vegetation	Sagebrush-ephemeroid (*Poa bulbosa, Artemisia sublessingiana*)	Vegetation	Ephemeroid-ephemer semi-savanna (*Poa bulbosa, Bromus tectorum, Heteranthelium, Crocus*)

N Northern slopes, *S* Southern slopes
MAT Average long-term air temperature
AP Average long-term precipitation

deal of scientific information. However, most of them have now become considerably outdated. The soil-geography plan studies the territory of Kazakhstan unequally: too little is still known about deserted flat territories and mountain areas. There are demands for more detailed mapping, correcting of zone and subzone boundaries and the revelation of the soil cover structure.

One major trend is creating soil maps using new methods involving remote sensing. This provides more informative soil maps with exact boundaries for soil units and their combinations. Additionally, over recent years we have observed significant changes in the soil cover and the soils, caused by the general aridization of Kazakhstan's territory and negative human impacts.

Since 2000, the Institute of Soil Science and Agrochemistry has begun work on updating existing soil maps. The initial result is that an electronic regional soil map of the southeast of Kazakhstan (scale of 1:500,000) was completed in 2005. Each soil contour on the map contains an attributive data base including three soil components, the structure of the soil cover, a percentage parity of the component's areas within the soil classification units, and also the texture of soils. As a consequence of this work our knowledge about the region's soil cover has been extended. The boundaries of vertical soil zones and belts have been specified, the features of soil cover formation and structure have been defined, and a new systematic list of the

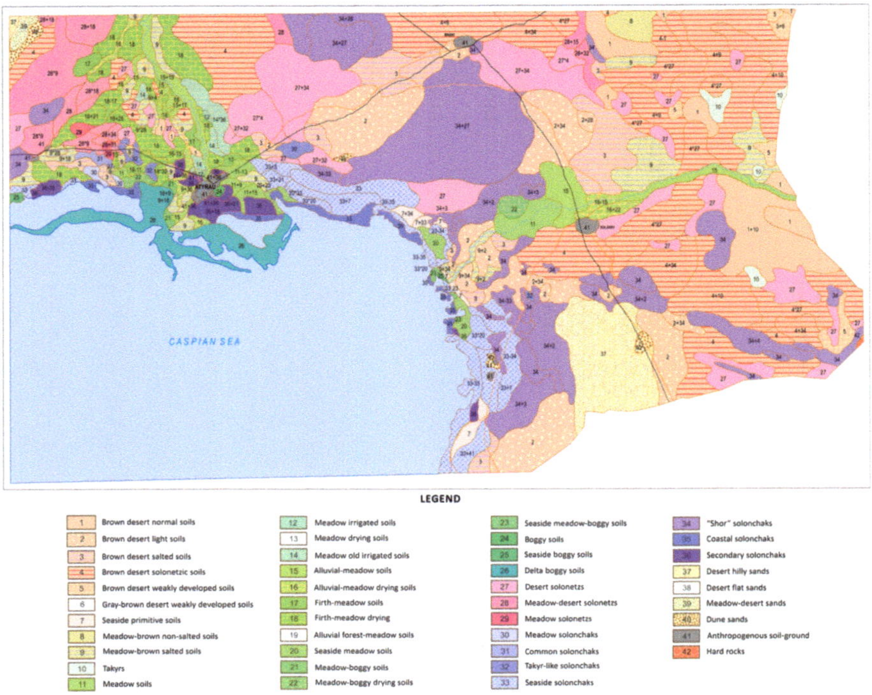

Fig. 2 Soil map of the North-East Caspian Sea region

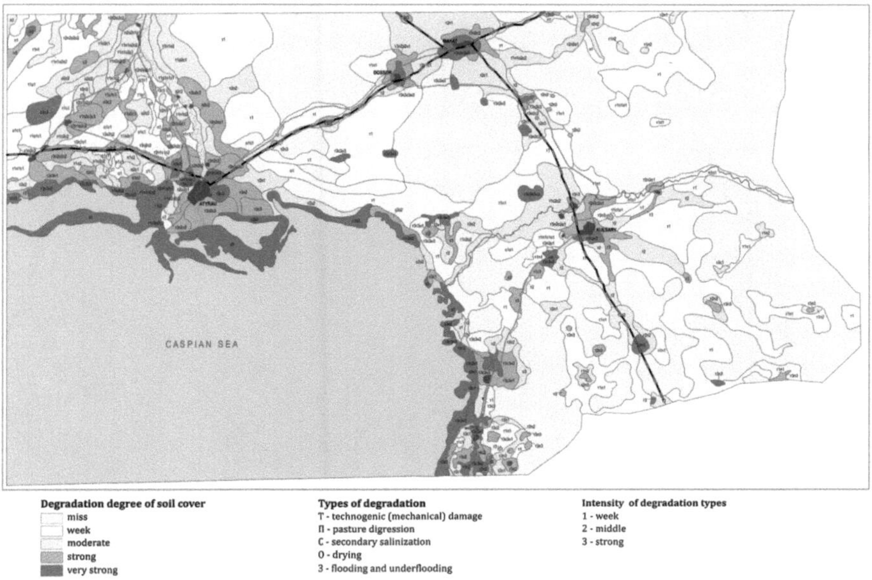

Fig. 3 Soil degradation map of North-East Caspian Sea region

soils of Southeast Kazakhstan with their typical morphogenetic properties has been created. The allocation of some soils which were not marked earlier has been proved. For the first time a soil database including the information on 1896 soil profiles has been created for this region. On the basis of this new soil map, further electronic maps dividing the soil geography into districts, and a thematic map of the soil quality (bonitation) has been created. Similar maps have been made for the territories of the Zhambyl and South Kazakhstan administrative areas.

In recent years great attention has been given to evaluating the potential productivity of soils, to studying the current state of the soil cover, and to working out concepts for the rational use of lands. This work has taken place on test sites in the Karaganda area (the central part of hummocky lands), the South Kazakhstan area (irrigated massifs of old alluvial plains of the Syr-Darya river), territories of rainfed farming in the Western Tien-Shan foothill plains, and also oil-extracting regions of the Caspian Sea region.

As a result of this work it has been possible to study and better understand the rules of soil type distribution over the territory, and of the morphogenetic properties of virgin and anthropogenically destroyed soils. These new soil maps (scale of 1:100,000) were produced using the methods and techniques described above, with use of the modern and retrospective data and GIS technologies.

On the basis of these soil maps various applied thematic maps have been made: maps of soil cover degradation, a map of soil bonitation points and maps of agroindustrial grouping of lands.

Fig. 4 Map of soil bonitation points for the central part of Kazakhstan's hummocky land

Maps of the degradation status of the soil cover bear information about types of anthropogenic degradation such as the overgrazing pattern of pastoral land, the soil fertility depletion of cropping land, technogenic degradation, secondary salinization, petrochemical pollution, waterlogging and bogging, agricultural drought risk, and degradation impacts based on recreational activities. The degree of soil degradation was defined depending on changes in morphological properties (capacity of humus horizon, depth of effervescence, carbonates, water soluble salts) and in chemical and physical–chemical properties (humus content, sums of absorbed cations, absorbed sodium, silt content) on a 5-class scale. Figures 2, 3 show a soil map and map of the soil cover degradation made on this basis for the Northeast Caspian Sea region.

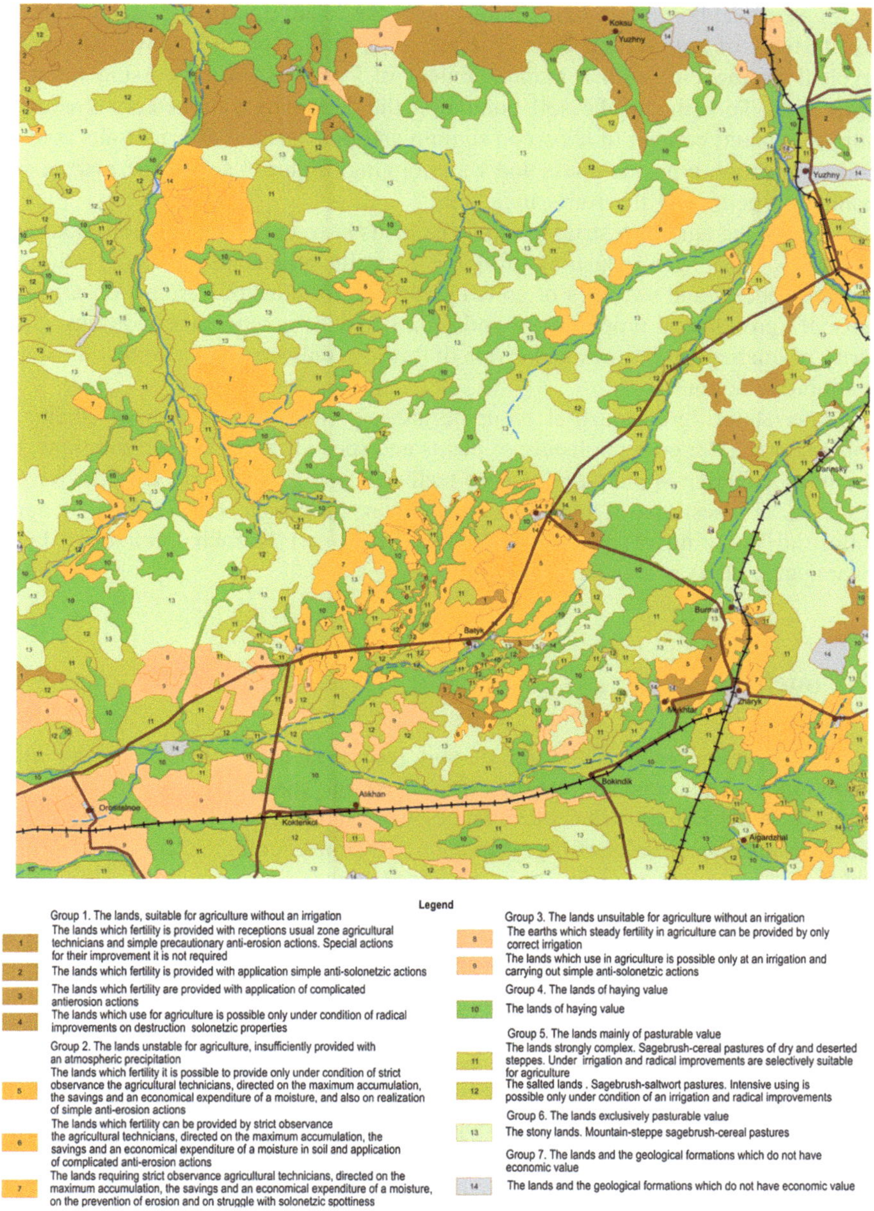

Fig. 5 Maps of agro-industrial grouping of lands

Maps of soil bonitation rating points show the relative quality evaluation of the soil fertility of agricultural land. The reference soil is taken to have 100 points (typical chernozem) with a humus content of 7 % and not possessing negative properties. To estimate the soil bonitation the following parameters were used: humus content; content of absorbed sodium and magnesium; average salts content and salinization type, stoniness and waterlogging. Figure 4 shows the map of soil bonitation points for the Karaganda area.

Maps of the agro-industrial grouping of lands have been created for agricultural territories. The agro-industrial grouping of soils represents the integration of genera, species and varieties of soils in larger agro-industrial groups on a generality of agronomical properties of soils, similarity of ecological conditions, similarity of qualitative features and levels of fertility, uniformity of agrotechnical and meliorative actions.

Materials of agro-industrial grouping are used for the inventory of quality of soil resources, for correct placing of plants and specialization of crop rotations, for the most effective application of agro-technical and ameliorative actions, for the decision of questions of lands transformation. A sample of the map of agro-industrial grouping of the lands of the Central part Kazakh hummocky lands (Karaganda area) is shown in Fig. 5.

4 Conclusions

The preservation and restoration of soil resources is an issue of major state importance. Thus, special urgency is placed on carrying out soil geography research in the Republic to achieve an objective representation of soil resources in Kazakhstan, restore soil fertility, gain control over the ecological conditions of soil cover and forecast tendencies in its transformation.

The introduction of computer technology has essentially expanded the horizons of scientific research in the field of the natural sciences, including soil science. Not only has it affected the development of soil cartography in Kazakhstan from the technical point of view, the methodological plan has also changed both the process of drawing up maps and the ways they can be used. Geographical genetic research on soils in Kazakhstan and on soil mapping are now being successfully developed. Many soil maps, especially for large regions, and also applied maps, have no counterparts. The current primary objectives are further in-depth studies of soil cover of Kazakhstan and its modern state, and the creation of a uniform national soil database.

If these problems are solved, schemes can be put in place for the sustainable development of the region, agriculture management systems and the rational use of the lands for various industrial purposes. Besides this, their solution will form the basis for an estimation of the modern state of soils and their fertility, the major factors behind anthropogenically degraded soils and the degree of their damage that will allow economically effective and ecologically safe technologies to be developed and applied in order to restore such soils.

References

Isachenko AG (1980) Methods of applied landscape researches, 222p. (Методы прикладных ландшафтных исследований. - Л.: Наука, 222 с)

Korsunov VM, Kraseha EN, Raldin BB (2002) Methodology of soil ecological-geographical researches and cartography of soils, 232p (Методология почвенных эколого-географи, ческих исследований и картографии почв. Улан-Удэ: Изд-во БНЦ СО РАН, 232 с.9

Kravtsova VI (2005) Space methods of soils research, 180p (Космические методы исследования почв. М.: Аспект-Пресс, 180 с)

Rozanov BG 2004 Morphology of soils, 432p (Морфология почв. М.: Академический проект, 432 с)

Smirnov LE (2005) Space methods of geographical researches, 348p (Аэрокосмические методы географических исследований. СПб.: Санкт-Петербургский Университет 348 с.)

Sokolov IA (2004) Theoretical problems of genetic soil science, 288p (Теоретические проблемы генетического почвоведения. Новосибирск: Гуманитарные технологии, 288 с.)

Indicators of Land Degradation in Steppe Regions: Soil and Morphodynamics in the Northern Kulunda

Vera Schreiner and Burghard C. Meyer

Abstract As the result of a discussion that started with a debate about the indicators of land degradation and desertification in Central Asia in the context of the United Nations Convention to Combat Desertification (UNCCD), it was realised that the winter cold steppe regions are seldom investigated from a multi-factorial perspective. This article reviews investigations into a framework of indicators developed to assess desertification and land degradation problems as applied to the Kulunda Steppe in the south of West Siberia by focusing on deflation morphodynamics. This article describes preliminary indicators and investigations into problems and indicators of land degradation in the Northern Kulunda at the regional level. An assessment of the deflation problem is illustrated by using the local example of the Ivanovskoe farm in the Bagan district of the Novosibirsk region. Results are presented about (a) the formulation of indicators for land degradation problems; (b) a regional morphodynamics assessment using soil maps and statistics; (c) local soil types and the dominant landscape processes and (d) a deflation assessment example based on the soil degradation indicators. The conclusion presents unresolved research questions about the vegetation-soil–water steppe system in the context of land use and protective measures in a heterogeneous steppe environment.

Keywords Land degradation · Steppe regions · Indicators · Kulunda

V. Schreiner (✉)
Institut für Länderkunde, Schongauerstraße 9, 04329 Leipzig, Germany
e-mail: V_Schreiner@ifl-leipzig.de

B. C. Meyer
Geographisches Institut, Universität Leipzig, Johannisallee 19a, 04103 Leipzig, Germany
e-mail: burghard.meyer@uni-leipzig.de

L. Mueller et al. (eds.), *Novel Measurement and Assessment Tools for Monitoring and Management of Land and Water Resources in Agricultural Landscapes of Central Asia*, Environmental Science and Engineering, DOI: 10.1007/978-3-319-01017-5_33,

1 Introduction

Several steppe regions in Central Asia are characterised by fertile soils which have the economic and ecological potential for dry land agriculture but are prone to land and water degradation. Today's land use is dominated by arable agriculture, and the landscape is structured into large field plots. An enormous number of shallow lakes and the remains of the forest steppe exist. As several of these lakes are affected by the drying out of local rivers, caused primarily by the extensive use of river water by farms, a problematic acceleration of the natural salinisation process is observed. The Kulunda Steppe covers approximately 100,000 km^2. A high seasonal and temporal dynamic and variations in the steppe ecosystem status parameters (temperature, precipitation) are obvious. Morphological activity levels are high and the exposure of the soils to wind erosion is highly problematic in arable agriculture. This was introduced in modern times by the conversion of the grass-steppe-ecosystems into arable farm systems in the Soviet period. Protective measures were applied during the Soviet period in the form of windbreaks and shelterbelts (Meyer et al. 2008).

The United Nations Convention to Combat Desertification (UNCCD) defines the term *desertification* as meaning "land degradation in arid, semi-arid and dry sub-humid areas resulting from various factors, including climatic variations and human activities" (UNCCD 1992). Other authors have researched these major problems of land degradation in Central Asia and partial solutions have been presented. These focus on (a) the applicability of definitions of landscape degradation and desertification especially for the exploration of desertification problems in cold winter regions (Kust 1999) (b) the historic development of land degradation problems on the steppes of Central Asia covering many thousands of years of nomadic land use and the combination of these relatively unknown practices with more recent morphodynamics when the steppes were converted into arable land in the 1950s (Meinel 2002) (c) the long-term changes in land use and land cover in the last millennia including the context of the drying out of lakes as exemplified in the study region by the Lake Chany problem (Bezmaternyh et al. 2007) and (d), the social and economic causes and indicators as major human drivers and the management of protective measures against these processes (e.g. UNCCD 2003).

An explorative research framework was constructed by the authors to differentiate indicators of desertification and land degradation in the Kulunda Steppe in the southern border region of Kazakhstan in Russian Siberia by further developing the research published by Meyer et al. (2008). This article assesses a framework of methods, focusing on the soils and on the impact of morphodynamics on those in the Kulunda. Interpretations and data from the Northern Kulunda are collected by using the example of the Ivanovskoe farm in the Bagan district of the Novosibirsk region as a local case study. The aim is not to provide a quantitative assessment of processes. Instead, it is to supply a general identification of key indicators about land desertification processes in different geographical contexts.

2 Methodology and Data

The article focuses on regional and local investigations by analysing published regional statistics, articles and scientific investigations for Western Siberia and for the Kulunda Steppe. Indicators of different spatial scales are used with the aim of identifying desertification and land degradation. The following approach was taken: (1) selection of general indicators about desertification; (2) development of a regional indicators system and (3) investigation into data availability at different scales.

The local analysis is also based on field survey information and the newly interpreted data by the Siberian Academy of Sciences in Novosibirsk. Applying the UNCCD definition of desertification and land degradation, indicators should include factors influenced by climate and human impacts. It was assumed that, by following the principles of natural processes, human impacts would be relatively rapid and would have a major influence over the actual dynamics started by the conversion of the vegetation cover. Climatic dynamic factors also have an impact, but over a greater period of time. Problems arise when, for example, in the morphodynamics, the quasi-natural processes (e.g. wind or erosive rains) are activated to affect the soils because of the changes in vegetation covering the arable land.

The framework of indicators firstly integrates the following geo-ecological factors: (a) vegetation cover; (b) plant community; (c) plants unusable for pasture farming; (d) biomass; (e) natural production; (f) broadening of C4-plants; (g) biological invasion of species; (h) upper groundwater table; (i) rivers and lakes drying up; (j) soil and water salinisation; (k) deflation; (l) soil erosion; (m) degradation of the upper soil horizon; (n) loss of organic material (humus loss); (o) loss of nutrients and (p) the accumulation of sand and gravel in the topsoil (Table 1). These indicators have been selected to describe different aspects of land degradation. The cited literature should give an introduction to the regional data and to the research methodologies of each of the indicators. It is important to state that overview studies for the Kulunda Steppe are available for each indicator while detailed local investigations in larger areas have not been carried out. The local investigations are based on those performed by Smolentsev et al. (2004) and the expedition report of Koroljuk (2005). Overview studies by other authors that integrate multiple indicators about land degradation have not yet applied to the Kulunda Steppe.

3 Results

3.1 Regional Morphodynamics

Key indications of morphodynamics are the aeolian and fluvial activity of wind and water erosion. Missing or open vegetation cover leads to greater general soil vulnerability. An analysis of diverse public and scientific statistics on the

Table 1 Indicator list, scale of data and availability of studies for the investigation of desertification and land degradation in the south of Western Siberia/Kulunda Steppe and Ivanovskoe farm

Indicator	Available regional data and studies	
	South of Western Siberia/Kulunda Steppe	Local scale (Ivanovskoe)
(a) Vegetation cover	Red List (2008)	Koroljuk (2005)
(b) Plant community	Red List (2008)	Koroljuk (2005)
(c) Plants unusable for pasture farming	Rudskij (1996)	
(d) Biomass	Titljanova (2000)	
(e) Natural production	Titljanova (2000)	
(f) Broadening of C4-plants	Stasova et al. (2007)	
(g) Biological invasion of species	Krasnoborov (2000)	
(h) Upper groundwater table	Osmuschkin (2000); InterNet project (2005)	
(i) Drying up of rivers and lakes		Smolentsev et al. (2004)
(j) Soil and water salinisation	Tanasienko et al. (1999)	Smolentsev et al. (2004)
(k) Deflation	Gos. Doklad (1998); Tanasienko et al. (1999); Vologshina et al. (2003)	Smolentsev et al. (2004)
(l) Soil erosion	Gos. Doklad (1998); Tanasienko et al. (1999)	
(m) Degradation of the upper soil horizon	Rudskij (1996, 2000)	Smolentsev et al. (2004)
(n, o) Loss of organic material (humus loss) and of nutrients	Jushakov (1999); Gos. Doklad (1998); Titljanova (2000)	Smolentsev et al. (2004)
(p) The accumulation of sand and gravel in the topsoil (skeletinisation)		Smolentsev et al. (2004)

vulnerability of the Russian steppe zone reveals a high level of problematic morphodynamic processes. In the Altai region, a tendency to water and wind erosion was assessed by investigating 7,740,200 ha of agricultural land within a total of 10,879,600 ha of land that was officially recorded as being agricultural (Gos. Doklad 1998). In the Altai region, 6,127,500 ha of the arable land (88.5 %) are endangered by deflation or water erosion. Investigations by Jaschutin (1999) showed that in 1998, 16.1 % of the Siberian steppes had been degraded by water erosion and 29.5 % had been caused by deflation. The degradation problem is especially serious in the administrative districts of Bagan, Karasuk, Kupino and Tschistoozérnoe, all of which belong to the Kulunda Steppe. The arable land here is characterised by a degradation of around 30–50 % of the total arable land (Gos. Doklad 1998). Wind erosion in the Northern Kulunda is evident on around 25 % of the total arable land, with a high regional differentiation (of up to 50 % of the arable land; Atlas 2002). Alternative data are summarised by Tanasienko et al. (1999) using soil maps to investigate the regional differentiation in soil erosion and the deflation problems in soils affected by desertification in Western Siberia in Table 2. No original data for these regional assessments are available.

3.2 Local Soils and Dominant Processes

Some background information about regional soil types should be given before describing soil and morphodynamic indicators. The soil cover of the steppe zone in Southern Siberia is heterogeneous. The soil types have developed due to their relief position. Using the IAASA and RAS classification (2002) and by correlating the terms of the Soil Map of Russia with the World Reference Base for Soil Resources (WRB 1998) and the and local Russian nomenclature given by Smolentsev et al. (2004), the zonal soil types are identified as Chernozems and Kastanozems. Both soil types are located in relief sites and are not influenced by ground- or stagnant water and are developed in loess materials. These Chernozems are the typical zonal soil types of the cold winter steppes and are modified in the Kulunda to Southern Chernozems (Haplic Chernozem), which are characterised by a strong humus layer (35–38 cm), an overall low carbon content (up to 2.5 %) and high rates of calcareous in the upper horizons (10–12 %). Kastanozems are only found in the driest parts of the Southern Kulunda Steppe. These soils are characterised by a humus horizon to a depth of 30–35 cm.

As a local example, Ivanovskoe farm possesses a high heterogeneity of soil types which can be differentiated. Dependent on slopes and depressions, several soil types were determined there as well as salt influence and salt dominated soils, e.g. Solonez, Solontshak and Solod soils (Table 3).

Table 2 Regional differentiation in arable soils affected by water erosion and deflation in Western Siberia(after Tanasienko et al. 1999)

Region	Arable land (1,000 ha)	Soils affected by (in 1,000 ha)			Total affected	
		Water erosion	Deflation	Water erosion and deflation	1,000 ha	%
Altai	7,274.7	1,099.9	2,364.5	–	3,464.4	47
Novosibirsk	3,865.4	247.0	251.6	22.0	520.6	13
Omsk	4,453.5	212.9	1,294.6	254.0	1,761.5	40
Altai republic	145.9	86.2	2.4	–	88.6	61
Tomsk	667.0	6.3	–	–	6.3	1
Tumensk	1,740.0	–	–	–	–	–
Kemerovo	1,613.0	144.8	173.6	74.6	393.0	25

Table 3 Local soil types, area, relief position and dominating landscape processes in the area of Ivanovskoe farm in the Northern Kulunda (IAASA and RAS (2002) and local Russian nomenclature in Smolentsev et al. (2004)

Soil subtypes after IAASA and RAS (2002), local Russian nomenclature in Smolentsev et al. (2004)	Area (in ha)	Area (in %)	Relief position	Dominant landscape process
Chernozems ordinary, **Chernozems Chernic (CHch)**	6,659.85	38.8	Upper slopes	Deflation
Meadow-chernozemics, leached; **Phaeozems Haplic (PHha)**	2,503.69	14.6	Medium slopes	Deflation
Meadow-chernozem; **Phaeozems Gleyic (PHgl)**	2,112.52	12.3	Lower slopes	Salinisation
Meadow-boggy; **Gleysols Histic (GLhi)**	2,011.95	11.7	Depressions in hilly reliefs, on lower slopes or around the salt water lakes	Salinisation
Meadows differentiated, (and solodic); **Planosols Luvic (PLlv)**	967.28	5.6		Salinisation
Meadow-boggy solonetzic and solonchakous; **Gleysols Sodic (GLso)**	28.57	0.2		Gleisation
Solonchaks typical; **Solonchaks Haplic (SCha)**	1,028.28	6.0		Salinisation
Solonetzes; **Solonetz Humic (SNhu)**	1,209.34	7.0		Salinisation
Solods; **Planosols Albic (PLab)**	435.90	2.5		Leaching
Alluvials saturated; **Fluvisols Eutric (FLeu)**	217.55	1.3	Along streaming waters	Accumulation
Total	**17,174.93**	**100.0**		

The Chernozems are mainly used for arable agriculture, and the Chernozem meadow soil influenced by water is currently used for pasture; the latter soil type had an intensive deflation and degradation period caused by the steppe clearing and after that was used for arable agriculture for only a short period in the 1950s. Records show that in the year 2000, the Ivanovskoe agricultural business was using the land for arable purposes (44.4 %), meadows (2.8 %) and pasture (27.7 %) and other land uses (25.1 %, including 8.2 % of lakes). The heterogeneity of soils in the water balance and salt problems result in a spatially diverse picture of deflation and wind erosion risk exposure. Little evidence of factors affecting water erosion has been found on the local field investigations at the Ivanovskoe farm.

3.3 Local Deflation Assessment

According to Vologshina et al. (2003), the total area of actual deflated arable soils in 2003 at Ivanovskoe farm was 9,105 ha; of these, 4,064 ha (44.5 %) are classified

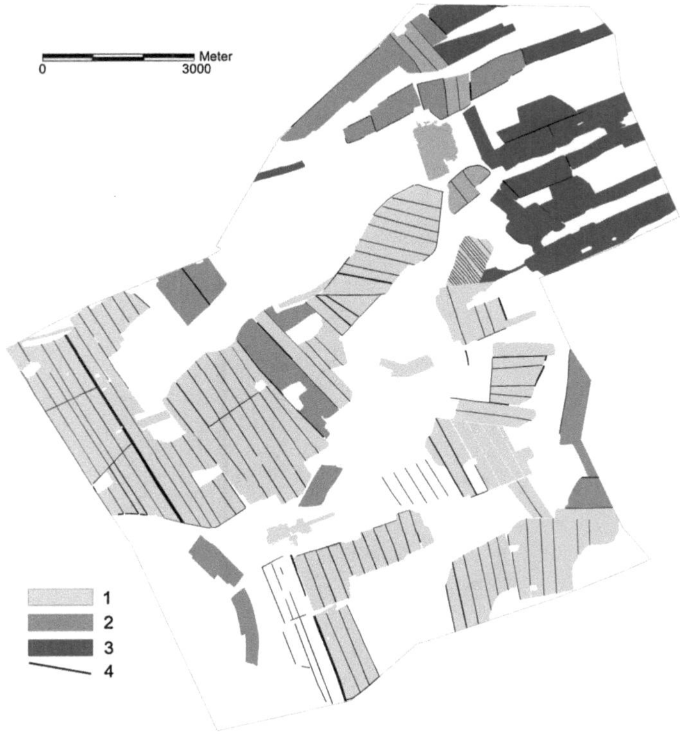

Fig. 1 Actual deflation risk for arable land in the area of Ivanovskoe (Meyer et al. 2008). *1*—Low risk; *2*—medium risk; *3*—high risk; *4*—shelterbelt

as heavily deflated. As protection against deflation, a high number of wind-protection strips (hedgerows and shelterbelts) were established between 1968 and 1970 and beyond. A total of 222 wind-protection strips measured by a GIS-based landscape analysis with a current length of 276.3 km in total are greatly important to soil conservation for the farming area. However, this strategy has not brought the deflation process to a complete halt (Appendix, Photos 1 and 2). The need for protection was assessed using a "soil shelterbelt ratio index method" (Smolentsev et al. 2004). By using information about the soil types' granulometry and the impact of deflation risk and the shelterbelts' "crown base height wind protection capacity," a deflation risk value could be calculated for Ivanovskoe, as mapped in Fig. 1.

3.4 Soil Degradation by Deflation

The actual deflation damage of the arable soils and related soil characteristics were investigated by analysing field data and laboratory results and were used as an

Fig. 2 Land use in the Ivanovskoe area (Meyer et al. 2008); *1*—Arable field; *2*—Pasture and meadows (agricultural grassland); *3*—Lowland peat bog, marshlands; *4*—Natural forest vegetation; *5*—Solonchak; Solonez; Solod; *6*—Lakes; *7*—Shelterbelt and hedges; *8*—River Bagan

indicator assessment in the area of Ivanovskoe. Deflation causes several problems for soil and land usage. Smolentsev et al. (2004, 2006) have characterised the schematic interrelationship between soil deflation and desertification. A key problem caused by deflation is the accumulation of sand on the soil surface (skeletinisation) and the loss of humus und clay fractions in the top soil. Dehumification, soil compaction and the reduction of reserves of productive soil moisture relate to the increase in soil climate aridity and the decrease in productivity of the croplands. The related data used for the assessment of the deflation damage using the parameters humus, clay and sand are summarised in Table 4.

As described in detail in Meyer et al. (2008), the depth of the humus horizon in unploughed Chernozem is up to 38–40 cm. Through active deflation processes, the finer parts of these horizons have been blown away during arable land use measured at a height of 12–22 cm during the degradation process since the 1950s. The humus content for highly deflated soils is significantly degraded by half 1.5–2 % compared with unploughed reference soils. This shows the gradual process of dehumification.

"Accumulation of sand and gravel in the topsoil" is how an indicator describes the recent reduction of clay and silt content in the topsoil and therefore the actual active deflation process. Smolentsev et al. (2004) explain that for the Northern Kulunda "64.6 % of all ploughed soils are loamy soils and 34.2 % are sandy soils (in the horizons up to 1 m depth). Silt content in the latter amounts to less than 6 %, the loam portion to less than 15 %, while the sand portion with 80–85 % dominates".

3.5 Land Use and Protective Measures

As described earlier in this article, a closed vegetation cover is the basis for the protection of land against morphodynamics. In the normal steppe system, only strong and extreme events such as exceptional thunderstorms, heavy rains or very intensive snowmelt processes and long dry periods will have the capability to open the vegetation cover. As analysed in the Kulunda Steppe, it is obvious that the ecosystem compartments are very diverse or heterogeneous. Also, following the conversion of approximately half of the steppe into arable fields, a high proportion of the land is occupied by non-arable land and quasi-natural land use types. Hedges and shelterbelts represent a new land use type introduced after the steppe conversions. The map of land uses in the Ivanovskoe farm area in Fig. 2 shows a diverse mosaic of arable fields, pasture and meadows (agricultural grasslands), lowland peat bogs and marshlands, remains of the natural forest steppe vegetation, solonchaks and salt-dominated lands, lakes, shelterbelts and hedgerows, and also the dried out river bed of the River Bagan (Appendix, Photo 3). A comparison of Figs. 1 and 2 clearly demonstrates the efficiency of the protective shelterbelts and hedgerows in reducing deflation risk. In the eastern part of the maps, the lack of protective measures is obvious and highly degraded soils are found here. A very low rate of productive livestock was observed in the years between 1997 and 2005 of around 0.1 livestock units per ha. However, this could also be explained by higher livestock densities or a local concentration of animals especially in the lake fringes, resulting in the degradation of vegetation, also leading to the activation of deflation processes.

4 Conclusion

The investigation into the application of a framework of indicators to land degradation problems relates to multiple unanswered methodological questions. The overarching research question is how to sustain arable agriculture in the steppe environment without extensive land degradation processes and without losing the soil's capacity for production and to service the landscape. Because of the system's behaviour, the opening of steppe vegetation cover is found to be the key human impact on initiating land degradation processes. In the local test area of

Table 4 Deflation damage of the arable soils and related soil characteristics used for the assessment in the Ivanovskoe area (after Meyer et al. 2008)

	Deflation damage		
	Low	Medium	High
Area (ha) (2000)	3,395	2,390	999
Humus horizon (cm)	28–34	23–27	18–22
Humus proportion (%)	1.7–3.4	1.6–2.6	0.7–1.3
Clay (<0.002 mm) (%)	6.4–14.2	10.7–12.7	6.7–7.6
Sand (0.05–2 mm) (%)	65.2–73.1	64.8–66.9	81.0–82.4

Ivanovskoe, biologically, the opening of vegetation cover has changed and influenced (1) the distribution of plant communities; (2) the distribution of plants unusable for pasture farming; (3) biomass production in general; (4) the broadening of C4-plants and also (5) the biological invasion of species. In terms of water and soil salinisation processes, it has led to (1) changes in the upper groundwater table; (2) the drying up of rivers and lakes; (3) soil and water salinisation. Also, it has induced wind morphodynamics leading to (1) deflation processes followed by (i) the degradation of the upper soil horizon; (ii) the loss of organic material (humus loss) and of nutrients and (iii) the accumulation of sand and gravel in the topsoil and also to (2) soil erosion by water. There was only minor evidence of this latter process in the local study. This local observation confirms the investigations of Meinel (2002) who described only some events of water erosion in the region of Southern Kulunda.

Basic research into the Kulunda Steppe or into comparable fragile and heterogeneous steppe land use systems is needed to extend our knowledge about the behaviour of systems in vegetation-soil–water households in cold winter regions. Little valid data about the process interconnections exists. The short vegetation period, the intensive snowmelt, the high summer evaporation, the significant variation and distribution of annual and inner-annual precipitation and the changing land use system in terms of farm organisation and crop rotation prompt numerous research questions that have yet to be resolved.

In this context, more probing research into land degradation indicators should be linked to climate change impact forecasts and the land use changes observed by the intensification of arable systems and also for the current dry land agriculture systems in the steppes of Central Asia. Further investigation into non-arable land use elements and land uses should be linked to biodiversity. There should be improvements in the development of protective measures based on the natural landscape and ecology of the region. The link between these elements is compatible with predictive modelling techniques that include land use and land cover changes with the aim of reducing wind erosion and water erosion and including the impact of hedgerow systems and other landscape elements. Nevertheless several basic data are missing to apply suitable models.

The salinisation problem seems to be clearly activated by land use changes and the extensive water consumption of agriculture housing, livestock production and by the land use changes from grassland steppe to arable fields when modifying the water household and at total by reducing the run-off of the regional rivers and streaming waters. This aspect should be investigated in the context of the long-term dry-out of the Lake Chany as typical regional problem of Central Asia.

The methodological approach developed to investigate the land degradation using indicators should be further developed for other research areas. The methods are open to be linked to the land use modelling also in other steppe environments.

Acknowledgments The research was funded by the BMBF (WTZ RUS 04/008) in the project "climate change and desertification in model-regions of Russia" in cooperation with the Siberian Academy of Sciences in Novosibirsk. We also like to thank E.N. Smolentseva and B.A. Smolentsev for the intensive and constructive collaboration and all the help in the context of the research and the field trips.

5 Appendix: Photos

Photos 1–3

Photo 1 Multi-row shelterbelt (of ca. 30 years age) on Chernozem in extensively used arable land in the farm area of Ivanovskoe (Photo: B. Meyer)

Photo 2 Layers of accumulated deflation sediments on Chernozem in the same shelterbelt indicating current active deflation events in the area of Ivanovskoe (Photo: B. Meyer)

Photo 3 Dry out of River Bagan in the Northern Kulunda in early summer 2005 near Bagan Town (Photo B. Meyer)

References

Stasova V et al (2007) Anatomitscheskoe stroenie list'ev nekotorich stepnich rastenij juga Sibiri kak pokazatel'tipa metabolizma. In: Opustynivanie i bor'ba s nim. Materialy mechdunarodnoj konferenzii, Abakan, Mai 2006. Rossijskaja akademija sel'skochozjajstvennich nauk, Abakan. 327 S

Atlas (2002) Atlas of the Novosibirsk Region, Novosibirsk

Bezmaternyh D, Dvurechenskaya S, Ermolaeva N, Zarubina E, Kirillov V, Kirillov T, Marusin K, Mitrofanov E, Pestova L, Popov P, Yurlova N, Yanovsky A (2007) Lake Chany. Hydrology, hydrochemy, hydrobiology, ornithology (materials to study). Novosibirsk, IWEP, p 48

Gos Doklad (Gosudarstvennij doklad) (1998) O sostojanii okrushajuschej prirodnoj sredi Rossijskoj Federazii v 1997 godu (About the condition of the environment of the Russian Federation in the year 1997). Moscow, p 606

IAASA and RAS (2002) Land resources in Russia: soil correlation. IAASA and RAS; version 1.1. http://webarchive.iiasa.ac.at/Research/FOR/russia_cd/soil_cor.htm. Accessed 11 Nov 2012

InterNet project (2005) History and culture of the Germans in Siberia. URL: http://museum.omskelecom.ru/deutsche_in_sib/index.htm. Accessed 21 Nov 2012

Jaschutin NV (1999) Potenzial Altana/Altajskij Gosudarstvennij Agrarnij (Agropotential of the Altai). State Agricultural University Altai, Barnaul

Jushakov AI (1999) Consequences of land reclamation of the steppe zone in Siberia. Stepnoj bjulleten 3–4, Novosibirsk

Koroljuk (2005) Field data of the Expedition to Kulunda steppe and the Ivanovskoe farm in the year of 2005 of Botanical Institute of SBRAS (not published)

Krasnoborov I (2000) Po povodu nekotoryx vidov vo flore Altajskogo Kraja. In: Turczaninowia, 3(1), S. 56–57

Kust G (1999) Opustinivanie: prinzipi ekologo-genetitscheskoj ozenki i kartographirovanija. Moskva, p 369

Meinel T (2002) Die geoökologische Folgewirkungen der Steppenumbrüche in den 50er Jahren in Westsibirien. Diss., University of Halle, p 122

Meyer BC, Schreiner V, Smolentseva EN, Smolentsev BA (2008) Indicators of Desertification in the Kulunda steppe in the south of Western Siberia. Arch Agron Soil Sci 54(6):585–603

Osmuschkin V (2000) Engineering-geological evaluation of the Predaltaijsk plane. Reports of the national Technical University of Altai 1, Barnaul

Red List of animals, plants and fungi in the Novosibirsk region (2008) Arta, Novosibirsk. 528 S. http://www.balatsky.ru/NSO/2008_redbook.htm. Accessed 21 Nov 2012

Rudskij VV (1996) Altaj. Ecological-geographical basics of land use. Barnaul, p 238

Rudskij VV (2000) Land use in mountains by the example of Altai and Sajan. Nauka, Novosibirsk, p 207

Smolentsev BA, Smolentseva EN, Fedorov PA (2004) K probleme opustiniwania Severnoi Kulundi. Materiali IV sjezda Dokutschaevskogo obschestva potschwowedow. (The problem of desertification in the northern Kulunda). In: Proceedings of the 4th congress of the Dokuchaev Society of Soil Scientists, vol. 2. Novosibirsk, pp 214–215

Smolentsev BA, Smolentseva EN, Meyer B (2006) Pritschini i indikatori opustinivanija v severnoj Kulunde. Materiali meshdunarodnoj nautschnoj konferenzii po borbe s opustinivaniem. Abakan, pp 73–79

Tanasienko AA, Putilin AF, Artamonova VS (1999) Ekologitscheskie aspekti erozionnich prozessov. Analititscheskij obzor. (Ecological aspects of the erosion processes. Analytic overview). Institute of Soil Science and Agrochemistry. Ekologija 55, Novosibirsk, p 98

Titljanova AA (2000) Land reclamation in the forest steppe zone and steppe zone in Western Siberia increased the carbon emission. Stepnoj bjulleten 8, Novosibirsk

UNCCD (1992) Definition of desertification. http://www.unccd.int/en/about-the-convention/Pages/Text-overview.aspx. Accessed 1 Sept 2012

UNCCD (2003) Subregional action programme for the Central Asian countries on combating desertification within the UNCCD context. Havanna, 21 S. http://www.unccd.int/en/regional-access/Pages/regionaldocuments.aspx?ctx=nap&region=Asia&reportType=subregional. Accessed 11 Nov 2012

Vologshina OW, Smolentsev BA, Ditz LU (2003) Izmenenie strukturi pochvennogo pokrova pod vlijaniem antropogennogo faktora. (Change of the structure of the soil cover by anthropogenic factors). Tomsk, Vestnik TGU 7, pp 57–63

WRB (1998) World reference base for soil resources. World soil resources report 84. FAO, Rome, p 88 (FAO/ISRIC/ISSS)

Erosion Rates Depending on Slope and Exposition of Cropped Chestnut Soils

Dana K. Shokparova, Erkin K. Kakimjanov and Burghard C. Meyer

Abstract The surface runoff erosion of dark and light chestnut soils (Kastanozems) in the Ile Alatau foothills in the Zhambyl/Karasai administrative area and in the western part of Chemolgan near Almaty/Kazakhstan was analysed for use in devising adaptive landscape agricultural systems. 1:25,000 scale maps were modelled by combining geo-information techniques, field work and laboratory analytics. Rain and snowmelt activity depending on exposure on southern and northern slopes of agricultural land was differentiated. Single parameters were measured as snowmelt erosion and water erosion; a runoff coefficient was then calculated to determine the level of soil erosion from the intensity of the erosion processes. The results reveal that dark and light chestnut soils on northern slopes are more resistant to water erosion than those on southern slopes. It was also found that soil erosion processes induced by erosive rain are more intensive than snowmelt erosion. For chestnut soils, water erosion rates ranged from 1.4 to 30.8 t/ha induced by rainfall and from 0.7 to 3.5 t/ha induced by snowmelt, depending on slope inclination and exposure. Greater erosion was detected on southern slopes. No clear differentiation was found when comparing the erosion rates of dark and light chestnut soils.

Keywords Soil erosion · Chestnut soils · Slope · Exposition

D. K. Shokparova (✉) · E. K. Kakimjanov
Al-Farabi Kazakh National University, Al-Farabi Avenue 71 050038 Almaty, Kazakhstan
e-mail: shk.dana@hotmail.com

B. C. Meyer
Geographisches Institut, Universität Leipzig, Johannisallee 19a 04103 Leipzig, Germany
e-mail: burghard.meyer@uni-leipzig.de

L. Mueller et al. (eds.), *Novel Measurement and Assessment Tools for Monitoring and Management of Land and Water Resources in Agricultural Landscapes of Central Asia*, Environmental Science and Engineering, DOI: 10.1007/978-3-319-01017-5_34,

1 Objectives

Only very few practical investigations into the soil erosion exposure of chestnut soils have been undertaken to date. In the foothills to the north of the Ile Alatau Mountains in Kazakhstan, no scientific studies have every been conducted to differentiate erosion rates caused by snowmelt and heavy rain. Little is known about the impact slope exposition has on soil erosion in this part of Central Asia, too. Chestnut soils (Kastanozems) are steppe soils defined by an accumulation of humus in semi-arid and arid environments, normally containing brown coloured upper horizons. These horizons have a high humus content and calcareous or gypseous fractions. The soil is developed mainly in loess materials, as described in the World Reference Base for Soil Resources; IUSS Working Group (2006). The same literature source mentions the key problem of wind and water erosion in Kastanozem soils, especially on fallow land, and the dominant land use of Kastanozem lands by extensive grazing when irrigation and arable land use is infeasible.

According to data obtained from research undertaken at the Institute of Soil Science (Pachikin and Erokhina 2011), approximately 60 % of the soil cover in Kazakhstan is affected by different degrees of soil degradation, differentiated by natural conditions and their soil usage. One of the causes of soil degradation is the extensive development of arable agriculture involving the conversion of steppes into arable land. Further causes are the strong impact of the mining industry and the wide network of former military test sites (from the USSR era). The battle against land degradation is of considerable importance in Kazakhstan. Approximately 43 % of the population (6.47 million) live in rural regions, the majority of whose livelihoods depend directly or indirectly on the agrarian sector and agrarian land use [data provided by Karazhanov et al. (1998)].

The peculiarity of the ecological/geographical situation in Kazakhstan is the generally low resistance of the natural environment to anthropogenic impacts. In data provided by Asanbaev (1998), fragile ecosystem types exist as deserts (45 %) and mountain landscapes (20 %) in the regions where the main pastures are located. Areas suitable for tillage agriculture are located in the chernozemic and dark chestnut soil zone. Chestnut soils are generally located in the dry steppe zone, light Kastanozems in desert steppes; brown Kastanozems in the semi-desert zone and grey-brown Kastanozems in the desert zone.

The research presented here is part of the ALAS (adaptive landscape agricultural systems) approach, based on the agro-ecological land evaluation methodology further developed by Kiryushin (2005). This assessment includes a comprehensive investigation of the territory's geomorphology, lithology, climate and hydrogeological conditions. Following the ideas developed by Kiryushin (2011) about classifications into ALASs and technologies (Editorial Board of Eurasian Soils Science 2011), scientific investigations into problems concerning the practical implementation of these systems have been initiated in several provinces of Russia, including at Kazakhstan Al-Farabi Kazakh National University in Almaty.

Iorgansky (2001) summarises that ALAS increases the production and environmental efficiency of crop production in several regions of Russia. This is achieved primarily by a better differentiation of the agro-ecological site conditions of the land.

In this regard, studies undertaken in Kazakhstan are devoted to the spatial analysis and preparation of landscape classification maps. The work presented here was performed in three stages: (1) territorial analysis of the test area, (2) differentiation and classification of landscapes using morphological characteristics and (3) development of agro-technology measurements for each type of landscape. The bio-physical and geographical analysis presented here was undertaken in the Almaty Oblast administrative region (Zhambyl/Karasai), located on the northern slope of the Ile Alatau, by using a geographical information system (GIS) and combining remote sensing data and field methods.

The aim of this paper is to investigate the impact water erosion processes caused by surface runoff have on dark and light chestnut soils in the foothills of the Ile Alatau Mountains, describing the role played by slope inclination and slope exposure. The analysis described in this paper therefore focuses on the following aspects:

1. Description of the soil characteristics of typical dark chestnut soil profiles,
2. Modelling of agricultural landscapes using GIS in terms of exposure and slope inclination, focusing on northern and southern slopes,
3. Text plot-based measurements for indicators of snowmelt water runoff intensity and their ratios, and indicators of soil erosion intensity depending on steepness and slope exposure for agricultural land,
4. Discussion about the field crop parameters that influence water erosion and water runoff as measures for adaptive landscape agricultural systems.

2 Materials and Methods

The research into dark and light chestnut soils in agricultural areas was undertaken at the foothills of Ile Alatau. The test site for **light chestnut soils** was located on the premises of the K. Mynbaev Kazakh Research Institute of Livestock and Fodder Production (43°13′20″N 76°40′56″E). In this field observatory, soil profiles were analysed on the southern and northern slopes. The sites were investigated using both the genetic-morphological structure as proxies and the field method. The average inclination of the slopes is around 5°. The humus horizon AB is 53 cm at slightly eroded sites; at sites featuring average erosion, this figure is still 38 cm.

The test site for **dark chestnut soil** profiles is located on arable land in “Sholak-Kargaly” village in western Chemolgan (43°22′39″N 76°37′12″E). Here, the humus horizon (AB) is up to 65 cm, and 55 cm with slightly eroded soils

(1–3°); with soils with average erosion (3–5°), the humus horizon is around 45 cm, and up to 35 cm with highly eroded soils (7° inclination and higher).

The erosive activity of surface runoff is measured at selected observation plots (with an area of 75 m^2) for each soil type. The experimental observation sites were established by applying three replications for the measurements of soil erosion intensity (g/l). These observations were made during the spring snowmelt and for heavy precipitation events. The field experiments were conducted in 2010 and 2011.

Table 1 shows the analytics of dark chestnut soils; averages are based on 13–23 replications. The texture is medium loamy and heavy. Due to long-term irrigation, some moving clay fraction is observed from the upper to the lower horizons. Micro-aggregation and water resistance of the macro-structure are relatively high.

The 1:25,000 scale map of the slope inclination was developed using GIS (Arc GIS 9.3 software) and additional functions of 3D visualisation and analysis of building surfaces. The thematic maps were based first on the digitisation of topographic data using analogous maps, which are later used to create a spatial

Table 1 Morphological and analytical characteristics of dark chestnut soils at the Ile Alatau foothills (n—number of soil profiles; max–maximum depth of soil horizon; min–minimum depth of soil horizon; x—average)

Parameters of soil properties	N	Genetic horizons	Parameter		
			Max	Min	X
Thickness of soil horizon, cm	23	A_1	10	6	9.6
		A_2	30	16	25.0
		B_k	60	40	50.3
		C_1	90	70	83.8
CO_2 carbonate, %	13	A_1	4.3	1.3	1.9
		A_2	6.8	1.8	1.9
		B_k	7.1	1.9	4.0
		C_1	8.5	2.6	4.5
Humus, %	17	A_1	4.2	2.0	3.2
		A_2	3.7	1.1	2.3
		B_k	3.6	0.4	1.4
		C_1	1.8	0.3	1.0
Absorbed calcium, mg-eq.	14	A_1	23.9	9.8	14.2
		A_2	16.0	6.2	12.1
		B_k	24.3	4.0	14.2
		C_1	10.7	1.9	5.8
Physical clay (< 0.01 mm), %	23	A_1	52.3	19.0	41.7
		A_2	55.4	23.5	42.7
		B_k	67.6	25.9	45.2
		C_1	62.2	34.9	46.2
Clay (< 0.001 mm), %	23	A_1	18.8	5.6	13.4
		A_2	21.3	3.3	12.1
		B_k	29.6	6.8	16.0
		C_1	29.9	9.8	16.8

model topography (slope maps, slope exposure). The work was performed by constructing a triangulated irregular network (TIN) model and then calculating the slope inclination using the 'Derive Slope' function for the contextual surface mode. Contour lines, roads, land uses, lakes, rivers, elevation points and additional information were added to the digital database for further analysis.

3 Results and Discussion

3.1 Light Chestnut Soils

Figure 1 shows the slope inclination of agricultural landscapes at the **light chestnut soil test site**, comprising 4,776 ha of land. Of this area, 1,893 ha (or 40 % of the territory) have slopes with an inclination of 1–3°; 2,219 ha (or 46 %) have 3–5° slopes; 523 ha (11 %) have 5–7° slopes, and only 141 ha (3 %) have slopes with an even steeper inclination.

The map provides the information required to assess the erosive activity of the landscape. The main factors that could influence the types of water erosion analysed are the physical properties of the soil (soil texture) to analyse the water household (infiltration and runoff). It is important to spatialise the distribution of the soil texture. Other important factors that can be gleaned from the test sites are the surface slope, the snowmelt runoff, the quantity and intensity of rainfall, the depth and extent of soil freezing, the type of land use, and the status of the surface soil, including information about plant residues and the micro-relief. Crops primarily cultivated at both investigation areas are spring barley, winter wheat and alfalfa. Table 2 summarises the observation results of surface runoff and intensity as an example of light chestnut soils.

Snowmelt runoff, storm water runoff and total runoff were categorised into different classes of slope inclination. Thus, the annual runoff of snowmelt and rain water varied depending on the steepness of the slopes with northern exposure from 21 to 67 mm, and erosion from 2 to 24 t/ha/yr. Figures for slopes with southern exposure varied from 24 to 74 mm and 3.9 to 33.9 t/ha, respectively. Higher volumes of runoff and eroded soils are observed on slopes with an inclination of 5–7° and in the class of slopes with an inclination of over 7° with southern exposure. This reflects the greater resistance to erosion of soils located on slopes with northern exposure, which is also confirmed by the lower intensity of erosion varying according to the slopes from 9.5 to 35.8 g/l in the runoff water analysis. The observed soil erosion intensity on the southern slopes is higher, measuring 16.2–45.8 g/l. The coefficient of runoff of snowmelt water was applied using a method developed by Surmach (1969) (Table 3).

Soil erosion at the test site occurs more rapidly after rainfall compared to snowmelt runoff. On the slopes with northern exposure, soil erosion affected by storm water runoff ranged from 1.4 to 20.9 t/ha; snowmelt runoff ranged from

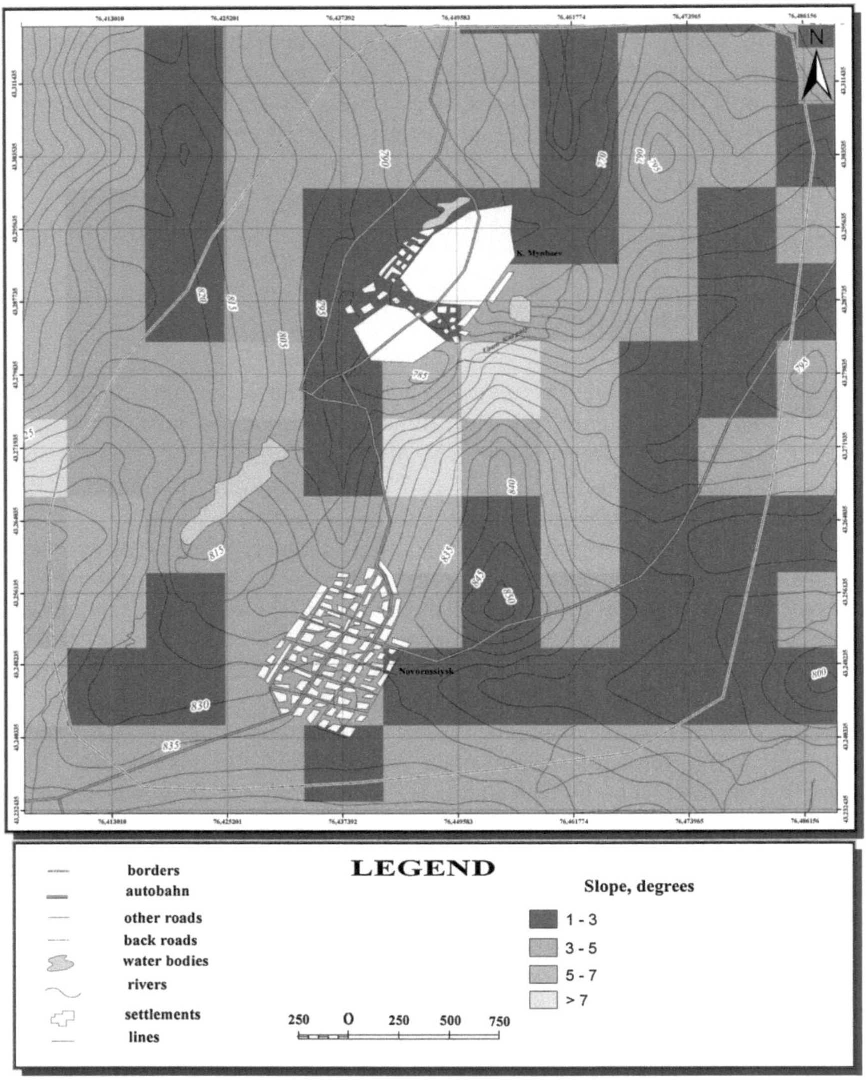

Fig. 1 Map of slope inclination at the light chestnut soil test area, test area K. Mynbaev Kazakh research institute of livestock and fodder production (Karasai/Almaty)

0.7 to 3.1 t/ha. When undertaking the agro-ecological assessment of land, it is also important to consider the soil's resistance to erosion. Resistance can be taken as a measure for assessing economic use, the field's potential crops and the local intensity of rainfall. For a typical rainfall event in May 2010, the results of surface runoff accounting for four elementary runoff test plots showed that the amount of

Table 2 Annual surface runoff and intensity of erosion in light chestnut soils, test area K. Mynbaev Kazakh Research Institute of Livestock and Fodder Production (Karasai/Almaty)

Parameter of erosion	Steepness of slope, degrees			
	1–3	3–5	5–7	>7
Northern exposure				
Snowmelt runoff, mm/year	13	16	20	26
Coefficient of runoff	0.12	0.16	0.20	0.2
Soil erosion, t/ha/year	0.6	0.8	1.9	3.1
Intensity of erosion, g/l	4.6	5.0	8.6	9.6
Storm water runoff, mm/year	8	15	28	41
Soil erosion, t/ha/year	1.4	3.6	11.8	20.9
Intensity of erosion, g/l	17.5	24.0	40.7	52.4
Runoff, mm/year	21	31	48	67
Soil erosion, t/ha/year	2.0	4.4	13.7	24.0
Intensity of erosion, g/l	9.5	14.2	28.5	35.8
Southern exposure				
Snowmelt runoff, mm/year	12	16	21	26
Coefficient of runoff	0.14	0.17	0.26	0.32
Soil erosion, t/ha/year	0.7	0.9	2.2	3.1
Intensity of erosion, g/l	5.8	6.4	10.5	12.9
Storm water runoff, mm/year	12	20	28	48
Soil erosion, t/ha/year	3.2	5.8	13.8	30.8
Intensity of erosion, g/l	26.6	29.0	49	59.2
Runoff, mm/year	24	36	49	74
Soil erosion, t/ha/year	3.9	6.7	16.0	33.9
Intensity of erosion, g/l	16.2	18.6	32.6	45.8

Table 3 Assessment of annual snowmelt runoff intensity

Runoff	Value of runoff, mm	Coefficient of runoff
None	0	0
Very little	Up to 7	Up to 0.05
Measurable	8–20	0.06–0.15
Clearly measurable	21–40	0.16–0.35
High	41–75	0.36–0.65
Very high	76–115	0.66–0.85
Extremely high	>115	>0.85

water runoff and soil erosion on steep slopes (7–8°) varied considerably, depending on which crops grew there (Table 4).

These figures suggest that the anti-erosion stability of light chestnut soils is determined largely by arable crops. It can be significantly regulated by performing crop rotation and planting particular crops. The cultivation of alfalfa and winter wheat in particular enhances anti-erosion stability. The lowest water runoff and soil erosion rates were observed on virgin soil covered by natural vegetation.

Table 4 Surface runoff on light chestnut soils

Date of registration	Steepness of slope degrees	Rainfall in mm	Average intensity of precipitation, mm/min	Indicators of erosion	Land				
					Fallow (uncropped)	Spring barley	Winter wheat	Alfalfa	Natural vegetation on virgin land
2.05.10	7–8	17	0.57	Runoff, m^3/ha %	27 18	23.1	15.1	37.2	10.0
				Soil erosion, t/ha	1.1	0.7	0.4	0.25	0.06
				Erosion, g/l	40.7	30.3	26.4	6.7	6.0

Perennial grasses cultivated on the croplands can also lead to a significant reduction in soil erosion. Storm water runoff measured here was higher than in fields in which winter wheat or spring barley were grown, but there was less soil erosion. The figures obtained are important for designing erosion control measures, for influencing the soil moisture regime and for improving the environmental situation in general.

3.2 Dark Chestnut Soils

Using comparable methods to those described for light chestnut soils, a 1:25,000 scale map was created for the agricultural landscapes of dark chestnut soils (4,361 ha) (Fig. 2). 285 ha (or 6.6 % of the land) belong to the class with a slope inclination of 1–3°; 775 ha (or 18 %) are in the 3–5° slope inclination class; 2,870 ha (66 %) have a slope inclination ranging from 5 to 7°, and 431 ha (9.4 %) of the land feature slopes with an inclination of over 7°.

Table 5 provides information about the surface runoff and intensity of erosion on dark chestnut soils. The quantitative measures of soil erosion with snowmelt and the spring rainfall event show that snowmelt varied significantly with slopes of varying steepness and exposure throughout the investigation, with soil erosion ranging from 3.6 to 31.1 t/ha.

Erosion processes are induced more intensely by early spring rain than by snowmelt; they are also more intense on slopes with southern exposure than on

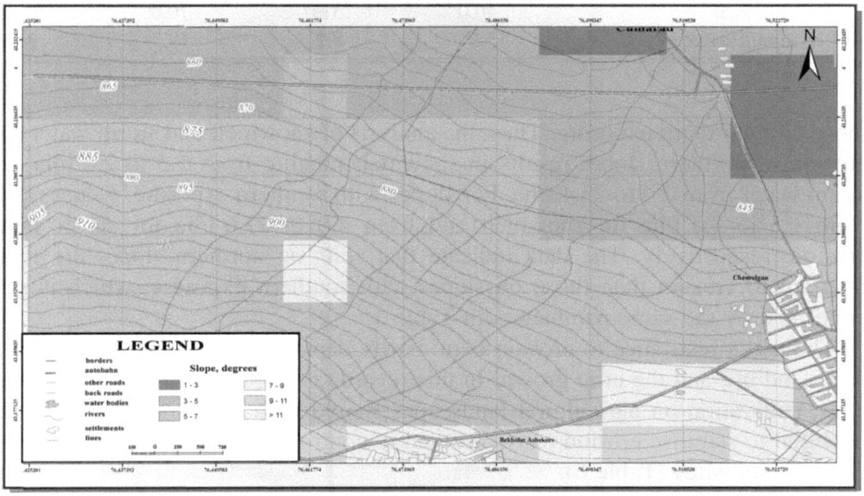

Fig. 2 Map of slope classes in the dark chestnut soil test area, Sholak-Kargaly (Zhambyl/Almaty)

Table 5 Annual surface runoff and intensity of erosion on dark chestnut soils located in Sholak-Kargaly (Zhambyl/Almaty)

Indicators of erosion	Steepness of slope, degrees			
	1–3	3–5	5–7	>7
Northern exposure				
Snow water, mm/yr	99	96	91	86
Snowmelt runoff, mm/year	18	22	26	32
Coefficient of runoff	0.18	0.24	0.29	0.33
Soil erosion, t/ha/year	1.2	1.5	1.9	2.8
Intensity of erosion, g/l	6.5	6.8	7.3	8.7
Rain precipitation, mm/year	128	128	128	128
Storm water runoff, mm/year	14	28	38	45.9
Coefficient of runoff	0.11	0.22	0.33	0.34
Soil erosion, t/ha/year	2.4	4.9	12.3	23.4
Intensity of erosion, g/l	17.1	17.5	32.4	50.9
Runoff, mm	32	50	64	77.9
Soil erosion, t/ha	3.6	6.4	14.2	26.2
Southern exposure				
Snow water, mm/year	80	76	75	72
Snowmelt runoff, mm/year	20	22	24	28
Coefficient of runoff	0.25	0.29	0.32	0.39
Soil erosion, t/ha/year	1.5	1.7	2.4	3.5
Intensity of erosion, g/l	7.5	7.7	10.0	12.5
Rain precipitation, mm/year	128	128	128	128
Storm water runoff, mm/year	20	29	46	50
Coefficient of runoff	0.16	0.23	0.36	0.39
Soil erosion, t/ha/year	4.0	6.5	17.3	27.6
Intensity of erosion, g/l	20.0	22.4	37.6	55.2
Runoff, mm/year	40.0	51.0	70.0	78.0
Soil erosion, t/ha/year	5.5	8.2	19.7	31.1

those with northern exposure. Thus, the analysis of snowmelt and storm water soil erosion of northern slopes results in snowmelt ranging from 1.2 to 2.8 t/ha and storm water from 2.4 to 23.4 t/ha. For slopes with southern exposure, values range from 1.5 to 3.5 t/ha for snowmelt and from 4.0 to 27.6 t/ha for storm water erosion, respectively. The data indicate that soils on northern slopes are more resistant to erosion than those on slopes with southern exposure, which is also confirmed by the same parameters in Table 1.

The soil's resistance to erosion is primarily dependent on the type of agricultural land use and the crop grown. In this regard, as interpreted for light chestnut soils, cultivation on agricultural landscapes should focus on crops that have the potential to stabilise erosion resistance and to control surface water runoff at dark chestnut soil sites (Table 6).

According to Table 6, and taking into account the results of surface runoff at four elementary sites located on slopes with northern exposure and a 7–8° inclination, an average rainfall intensity of 0.57 mm/min was measured. The resulting

Table 6 Surface runoff on dark chestnut soil; Sholak-Kargaly (Zhambyl/Almaty)

Date of registration	The steepness of slope, degrees	Rainfall in mm	Average intensity of precipitation, mm/min	Indicators of erosion	Land		Cropland	
					Spring barley	Winter wheat	Alfalfa	Natural vegetation on virgin land
10.05.11	7–8	17	0.57	Runoff, m^3/ha %	37.5	25.2	48.9	14.7
					17.9	12.0	23.3	7.0
				Soil erosion, t/ha	1.8	0.65	0.34	0.07
				The intensity of erosion, g/l	48.0	25.7	7.0	4.76

flow of rainwater runoff varied strongly depending on the field crops grown. Runoff from spring barley, winter wheat and alfalfa ranged from 25.2 to 48.9 mm. With virgin soil, it was only 14.7 mm. Thanks to ploughing, the cultivation of alfalfa and winter wheat led to more effective erosion control, with soil erosion measuring 0.34 and 0.65 t/ha, respectively. The greatest erosion was detected for spring barley. However, the effectiveness of moisture accumulation on arable land is higher under winter wheat than under alfalfa, and runoff water was eventually almost twice as high. Erosion affects soil most under barley because the late, weak plant development results in the open soil cover of arable land during rainfall.

Compared with arable land, the lowest runoff water and soil erosion was observed on virgin soils covered with natural vegetation. The planning and implementation of measures to combat soil erosion should therefore be developed by farmers in cooperation with nomads and animal husbandry, depending on the crops cultivated on agricultural land. Parakshina et al. (2010) determined that erosion affects major parts of the territory of the Republic of Kazakhstan. Most of the irrigated land and the land used by rain-fed agriculture is located in the piedmont plains, and is endangered by water erosion, as investigations by Dzhanpeisov (1977), for example, clearly show.

4 Conclusion

A 1:25,000 scale map of agricultural landscape slope inclination was developed using GIS technology for dark and light chestnut soil test sites at the foothills of the Ile Alatau, located in the territory of Zhambyl/Karasai administrative area. Arc GIS 9.3 and built-in 3D Analyst were employed to create relief maps of slope inclination and slope exposure. Using experimental plots with different slope situations in the agricultural land, the surface water runoff from storm water and snowmelt and erosion intensity were characterised and typified by indicators for different slope classes (1–3, 3–5; 5–7; > 7). The loss of fertile soil particles by erosion is particularly problematic on the "warm" slopes with southern exposure, where erosion is much higher than that on slopes with northern exposure. This reflects a greater resistance to erosion of soils on slopes with northern exposure, which is also confirmed by the intensity of erosion rates when compared with southern exposed slopes, which are less resistant to erosion. There is the need for further investigations to verify erosion rates, e.g. (1) when applying a larger number of field observations, and (2) applying soil erosion modelling based on GIS systems, including the full range of factors that are known to affect soil erosion due to water. Verification of modelling should be tested on field-scale plots using measurements and modelling for characteristic drainage systems on the landscape scale.

Acknowledgments We thank Mr. A.I. Iorgansky from K. Mynbaev Kazakh Research Institute of Livestock and Fodder Production for providing the field work data used as the basis for calculations in the tables of this article.

References

Asanbaev IK (1998) Anthropogenic changes in soils and their environmental consequences. Ministry of science academy of sciences of the republic of Kazakhstan. Kazakh SRI of soil breeding and agro chemistry named after Uspanov U.U. Almaty, Kylym, 180. ISBN: 5-7667-4645-8

Dzhanpeisov R (1977) Erosion and deflation of Kazakhstan. Nauka, Alma-Ata, p 232

Editorial board of Eurasian soils science (2011) Valerii Ivanovich Kiryushin is 70. Eurasian soil science 44 (8):954–956. doi: 10.1134/S1064229311080072

Iorgansky AI (2001) Actual task of improving soil ecology in agricultural landscapes of Kazakhstan, Problems of agriculture ecology and environment. Proceedings of the 3rd international technical conference (27–29 Sep 2000). Almaty, pp 97–101 (In Russian)

IUSS working group WRB (2006) World reference base for soil resources. World soil resources reports, no 103. FAO, Rome

Karazhanov KD, Akhanov JU, Asanbaev IK, Faizov KSh (1998) Problems of soil ecology in Kazakhstan. In: Akhanov JU et al. (ed) Condition and rational use of soils in Kazakhstan. Institute of soil science, Almaty, Kazakhstan, pp 14–18

Kiryushin VI (2011) The theory of adaptive landscape agriculture and design of agricultural landscapes. M. Izdatelstvo "KolosS", ISBN: 978-5-9532-0779-9 (In Russian)

Kiryushin VI, Ivanov AL (2005) Agro-ecological land evaluation, design adaptive-landscape systems of agriculture and agricultural technology (Methodological Guide in Russian). FSSI "Rosinformagroteh". Moscow, 783

Pachikin KM, Erokhina OG (2011) Geographical and genetic researches for years of independence of Kazakhstan. Soil science and agrochemistry, no 4. Almaty, pp 14–29

Parakshina EM, Saparov AS, Mirzakeev EK (2010) Erosion of Kazakhstan. Almaty, 367. ISBN: 978-601-278-250-9

Surmach GP (1969) Prediction of snowmelt water runoff on chernozem and chestnut soils. Bulletin of agricultural science 12, 53–56. (Прогнозирование стока талых вод с черноземных и каштановых почв//Вестн. с.-х. науки, no 12 С. 53–56)

Methodology of Measuring Processes and Evaluation of Water Resources of the Republic of Kazakhstan

Tursun Ibrayev, Batyrbek Badjanov and Marina Li

Abstract The article highlights the methodology of evaluating the changing processes and water resources of the Republic of Kazakhstan with the help of geo-information technologies, determining their dynamic parameters based on thematic satellite data processing methods and techniques for theoretical and applied purposes. A review of space imaging systems with different technical characteristics was carried out. A technological scheme for making precision mosaics using high-resolution optical electronic space imagery is proposed; this is supplied with mosaics composed of LANDSAT space imagery of the Ertis river basin. A schematic map has been developed showing the water basins of transboundary rivers in Kazakhstan and neighbouring countries with a list of cartographic layers, objects and information, including administrative territorial divisions and terrain (contours, heights, forms of relief), as well as the structure of the GIS database.

Keywords Water resources · Transboundary rivers · Satellite data · Kazakhstan

T. Ibrayev (✉) · B. Badjanov
Kazakh Scientific Research Institute of Water Economy, 12 Koygeldy Street,
Taraz, Kazakhstan
e-mail: tursun_ibraev@mail.ru

B. Badjanov
e-mail: badg_4@mail.ru

M. Li
Kazakh Research Institute of Water Economy, 12 Koygeldy Street, Taraz, Kazakhstan
e-mail: limarina76@gmail.com

L. Mueller et al. (eds.), *Novel Measurement and Assessment Tools for Monitoring and Management of Land and Water Resources in Agricultural Landscapes of Central Asia*, Environmental Science and Engineering, DOI: 10.1007/978-3-319-01017-5_35,

1 Introduction: Managing Water Resources from Transborder Rivers

One of the most important goals of water resource control in Kazakhstan is to develop more precise and detailed methodologies for studying and evaluating water resources, and determining their dynamic parameters based thematic satellite data processing methods and techniques for theoretical and applied purposes.

There are about 260 international river and lake basins in the world, and most of them are located on the territory of several countries. Usually neighbouring countries have problems with protecting their interests regarding the division of water resources from the basin and the deterioration of water quality caused by a neighbouring country. In such cases issues revolving around the rational use of water resources of interstate importance become very pertinent, because there are no real borders for cross-border geosystems and the negative impacts of any party's economic activities have an impact on environmental conditions. This problem is particularly acute for Kazakhstan, because most of the rivers flowing through its territory originate abroad.

To evaluate the water resources and processes occurring in different areas of water collection owned by neighbouring countries, detailed information is needed about the geosystem of the whole territory of the transborder basin. Obtaining such information is possible only if the neighbouring countries have an agreement on regulation of the use and protection of water resources from transborder rivers and lakes.

The issues of using resources from transborder rivers remain unresolved between the countries participating in one of the most influential political organizations in Central Asia: the Shanghai Cooperation Organization (SCO). For many years most of the countries neighbouring Kazakhstan have ignored attempts to make them use transborder water resources in compliance with the principles and norms of international law. At the same time, international law does not foresee any prohibitive or restrictive provisions on the use of water resources from the rivers within a nation's own territory. Water relations are meant to be based solely on the basis of mutual agreements. Currently, Kazakhstan's closest neighbour, China, has not acceded to the two major international agreements: the Convention on the Right of Non-navigational Use of Transborder Water Flows and the Convention on Conservation and Use of Transborder Water Flows and International Lakes, and prefers to solve the issues of regulating transborder waters only through bilateral negotiations, which, in practice, provide only a partial solution to this problem.

The absence of such agreements means an alternative method needs to be found to study the related areas (if possible independently of any political relationships), which will allow a modern information database to be created to monitor and track the changes taking place in the whole water collection area and forecast possible options for their development, depending on the water economy policy of each neighbouring state.

Today, monitoring water resources is the most urgent issue in Kazakhstan. For example, today China has increased its consumption of water at the springhead of the river Ertis. In connection with the increased water consumption, experts in Kazakhstan and Russia forecast that within 15–20 years the Ertis river will become a less navigable river. Urgent issues are also arising on the transborder rivers of Ili and Syrdarya.

2 Key Water Resource Monitoring Issues: Satellite Data and GIS Technologies

Global experience in developed countries (the USA, Canada, the European Union, Russia, Ukraine, etc.) shows that the use of satellite data allows regular evaluations to be conducted of the status of water objects, while the efficiency of water resource management can be significantly improved. Data from the remote sensing of Earth from space (RSE), especially high spatial resolution, is the most valuable material for conducting research in the field of water management.

Kazakhstan has a well-developed scientific and technical potential to solve the tasks of monitoring processes taking place on the surface of river basins under the influence of the weather and human activities. In particular, the development of remote sensing methods to evaluate the variability of different ecosystems in Kazakhstan will allow current issues to be addressed, such as studying the variability of various natural and anthropogenic ecosystems under the influence of weather factors and economic activity or determining long-term changes in ecosystems, including water resources. Based on historical data from ground-space surveys it is possible to evaluate the dynamics of water resource changes in recent years.

In assessing the status of water resources in the RK using GIS technologies in the regions of intensive anthropogenic impact on the environment, the following activities are implemented:

- collecting data on environmental conditions, exhaustion and pollution of surface and groundwater;
- study of the structure of river basins: ranking of water flows and basins; evaluation of the basin structure (areas that are based on water streams of different orders, the ratio of the lengths of water streams of different orders, the ratio of the angles of confluence of water streams and other factors);
- identifying the main factors determining the conditions for the development of negative natural processes in drainage basins of small rivers, as well as indicators of water resource status;
- zoning the territory by making specialized maps of the region's natural and economic conditions (geological and geomorphological, hydrogeological, terrain), as well as the extent of human impacts on water resources (reduction of

river flow, exhaustion of dynamic supplies of groundwater resources, pollution of surface and groundwater);
- formulating mathematical models of the whole set of indicators based on combining specialized maps in a geo-information system;
- outlining main indicators and factors, ranking "input" to the overall degradation of water resources;
- developing mathematical models aimed at outlining river basins with various degrees of disturbance in their functioning;
- designing generalizing maps: zoning based on integral indicators of conditions in which negative natural processes develop; zoning based on integral indicators of water resource deterioration, with the help of GIS technologies;
- analyzing zoning results, developing protection schemes and rational use of water resources.

The information about the structure of river basins and natural and anthropogenic factors is used to make maps, each one describing specific properties of the natural environment (topographic, geological, geomorphological, hydrological, hydrogeological, terrain). The scale of the maps depends on the purposes and objectives of the research and can be used for a general evaluation of large areas; for example, small-scale maps (1:500, 000) are used for administrative regions. In contrast, for small regions where the dynamics of specific processes need to be assessed, it is recommended to use large-scale maps (1:100, 000 and above).

Data on ground-space surveys can also serve as a cartographic base. Space imagery are particularly needed to identify the contours of zones of anthropogenic impacts, and the consequences of human activities.

The geological and geomorphological conditions of the territory are the most important factors forming the structure of river basins. They determine the direction and intensity of natural and anthropogenic processes (nature and intensity of water erosion, transporting capacity of water and transfer of pollutants). The composition of alluvial material and its ability to move depends on the lithologic composition of the relief-forming rocks. One important factor in the functioning of erosion morphosystems is the correlation of water stream slopes of different orders. The study of flow slopes and receiving water flow is necessary to identify zones of accumulation or erosion, which will allow contaminants' migration path to be calculated.

Hydrogeological maps show the general conditions under which the territory is supplied with groundwater and helps find how those waters are linked with surface runoff.

Terrain maps contain information about the structure of the basins as complex systems, the nature and properties of surface water collection, and their resistance to human impacts.

The territory can be zoned on the basis of multi-factor analysis and the classification of natural and economic systems characterized by a large number of factors, using computer programs for data processing and analysis.

Geo-information systems ensure that a comprehensive assessment of the territory can be carried out in terms of the development of negative natural processes in the basins of small rivers; that information is provided for models used for the integrated environmental evaluation of water resources, and that environmental maps can be designed for operative decision making.

Currently the use of space data in the study of various processes on the Earth's surface is a common process all round the world. Historical satellite data have been accumulated for many years and made available since 1972, and they provide both operational checks on current changes in the status of the objects and the possibility of studying ongoing long-term processes, allowing scientists to evaluate the scales and trends in changes to the underlying surface for a period of more than 40 years. In this regard, they are widely used to make key decisions on territory management, the maintenance of the ecological balance, effective use of land and water resources, as well as many other tasks. Studying long-term environmental changes using RSE methods is now the most effective scientific field when it comes to studying processes occurring on the Earth's surface.

At present the developed countries (the USA, Canada, the European Union, Russia, Ukraine, etc.) are focusing on developing methods and technologies for the thematic processing of satellite data for theoretical and applied purposes. Global experience shows that the use of space data will allow regular assessments of rivers, ponds, water systems and facilities to be conducted, and also significantly increase the efficiency of water use for different categories of users. Data from remote sensing of Earth from space (RSE), especially those of high spatial resolution, are the most valuable material for implementing water management activities.

Being a spacefaring power, Kazakhstan has significant scientific and technical potential to monitor the processes taking place in the underlying surface under the influence of weather and human impacts. In particular, the development of remote sensing methods to assess the variability of different ecosystems in Kazakhstan will allow current issues to be addressed such as detecting the variability of natural and anthropogenic ecosystems under the influence of weather factors and economic activity or determining long-term changes in ecosystems, including water resources. In recent years it has become possible to evaluate the dynamics of water resource changes based on historical satellite imagery data.

3 Performance of Remote Sensing Data for Kazakhstan

Satellite images are the most valuable carriers of varied information, which can be analysed to help identify the spatial and temporal interrelations of the system components of water collection, water streams and water reservoirs. Remote sensing data is essential due to the lack of ground-based observations, and often the only available source of information about the status of water objects and their water collection.

Currently, the RSE space systems NOAA, MODIS, IRS, RADARSAT and LANDSAT can be used for monitoring water resources (Table 1).

In addition, the NASA/LANDSAT data archive, which has been maintained since 1972, has been opened up for free access online.

The satellites LANDSAT 4–5, equipped with two types of scanners, Multi-spectral Scanner (MSS) and Thematic Mapper (TM), provide satellite imagery of the Earth's surface with different spatial and spectral resolutions. The MSS data (80 m spatial resolution) are available dating back to 1972, while the TM data (30 m spatial resolution in visible, close and average infrared zones, 120 m in thermal range) goes back to 1982.

The LANDSAT 7 satellite has been in orbit since April 1999. The ETM+ radiometer installed on the satellite is an improved version of the TM scanners. The main significant difference about the device is the presence of a high-resolution panchromatic channel (15 m). The Landsat 7′s range of coverage is 185 km, with a repetition interval of 16 days (233 orbits), 8 channels, and a 30 m spatial resolution (in a thermal range of 60 m).

The imaging device installed on the satellite LANDSAT 7—the Enhanced Thematic Mapper Plus (ETM+) scanning radiometer—provides imaging of the Earth's surface on six channels with a resolution of 30 m and on one IR channel with a resolution of 60 m, along with simultaneous panchromatic imaging with a resolution of 15 m and a width of review for all channels of 185 km. The characteristics of the radiometer, which has 8 spectral channels, are shown in Table 2.

The JSC ("National Center for Space Research and Technology") receives and archives satellite data. Table 3 includes some technical characteristics of the space information received.

Table 4 shows a comparison of the spatial resolutions of the satellite surveys and the scale of the resulting maps.

The basins of the largest trans-boundary rivers such as the Ertis, Balkh-Alakol and Aral-Syrdaria represent the biggest interest in terms of Kazakhstan's water resources (Fig. 1). At present, historical satellite data is being collected and

Table 1 Received RSE data (Report of RCSST 2011)

Spacecraft (country)	Radiometer	Imaging device specifics		
		Spatial resolution (m)	Number of channels (specter, range, mkm)	Range of coverage (km)
NOAA (USA)	AVHRR	1,100	5 (0.58–12.5)	3,000
TERRA, AQUA (USA)	MODIS	250	2 (0.6–0.8)	2,300
		500	5 (0.4–2.0)	
		1,000	29 (0.4–14)	
IRS P6 (India)	LISS-4	5.8	3 (0.52–0.86)	23
	LISS-3	23	4 (0.52–1.7)	140
	AWIFS	56–70	4 (0.52–1.7)	740
RADARSAT-1 (Canada)	SAR	8–100	C-diapason 5.6 cm	50–500

Table 2 Spectral characteristics of ETM radiometer (Report of RCSST 2011)

No. of channel	Spectral diapason (M km)	Spatial resolution (m)
1	0.45–0.515	30
2	0.525–0.605	30
3	0.63–0.690	30
4	0.75–0.90	30
5	1.55–1.75	30
6	10.40–12.5	60
7	2.09–2.35	30
8	0.52–0.90 (panchrome)	15

Table 3 Spatio-temporal characteristics of satellite information (Report RCSST 2011)

Satellite	Spatial resolution (m)	Beginning of the observation period	Receipt frequency
NOAA	1,000	2000	Every day
MODIS	1,000–250	2001	Every day
MOD09 (USA)	250	2000	Every 8 days
IRS LISS	23	2003–2009	Registered
IRS PAN	5.8	2004–2009	Registered
LANDSAT (USA)	30	Beginning of the 1980s	Every 16 days

Table 4 Compliance of spatial resolution of space surveys and scale of resulting maps (Report RCSST 2011)

Range of view (km)	Resolution (pixel size in m)	Map scale
1000–2000	250	1:1000 000
150–200	25–30	1: 100 000
80–120	10–12	1: 50 000
60–70	5	1: 25 000
20–40	2	1: 10 000
10–12	1	1: 5,000
6–8	0.5	1: 2 500

analyzed, techniques are being developed, space monitoring technology for water objects is being developed, and water systems and water use are being analyzed.

Low-resolution satellite data (NOAA, MODIS) can be used to create view maps of the areas studied. Monitoring water objects and medium resolution images (LANDSAT) allows scientists to monitor the area of the water surface of reservoirs, structure and meandering of river beds, and the status of irrigated lands. High-resolution data (less than 6 m) allows them to evaluate indented coastlines, the degree of silting in river beds, etc.

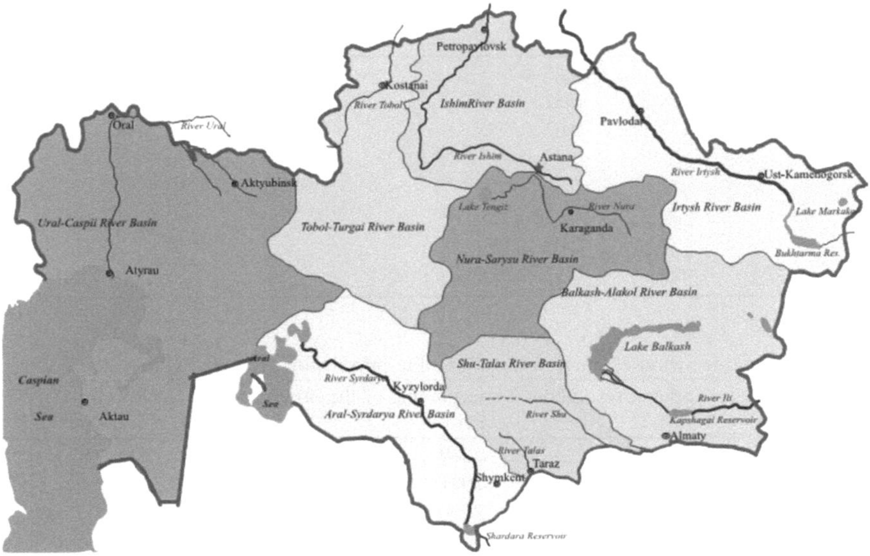

Fig. 1 Map of Kazakhstan river basins

4 Data Processing and Mapping

Improving the image quality by switching to higher-resolution images leads to the need to use a greater number of scenes covering the territory. So, to cover a site 1 using LANDSAT scenes (r. Ertis) 4 scenes are needed: 144/26, 144/27 143/27, 142/27; for site 2 (r. Ili) 6 scenes are needed: 145/30, 146/29, 146/30, 147/29 147/30, 147/31, and for site 3 (r. Syrdaria) 16 scenes: 148/31, 143/31 150/31, 151/32, 151/31, 151/32, 152/31 152/32, 153/31 153/32, 154/31 154/32, 155/31 155/32, 156/31 156/32.

Creating a mosaic of satellite images involves ortho-correcting images, selecting mounted areas and cutting lines, balancing data tone and "stitching" images. In preparing mosaics the satellite imagery archives are used to create integral coverage over large areas. Figure 2 shows a technological scheme for creating high-precision mosaics using optical-electronic high-resolution satellite images.

As shown in the diagram, the first stage is to collect basic data, namely satellite images, totally covering the area required. The next stage is the primary processing of digital information, which includes the binding of a picture according to orbital elements, various kinds of filtration of noise (if any), as well as color levelling and correction of data. The next step in creating a mosaic is to clarify the binding and ortho-transformation of images, which is a mathematically strict transformation of the original image into an orthogonal projection and elimination of distortions caused by terrain, imaging conditions and type of camera.

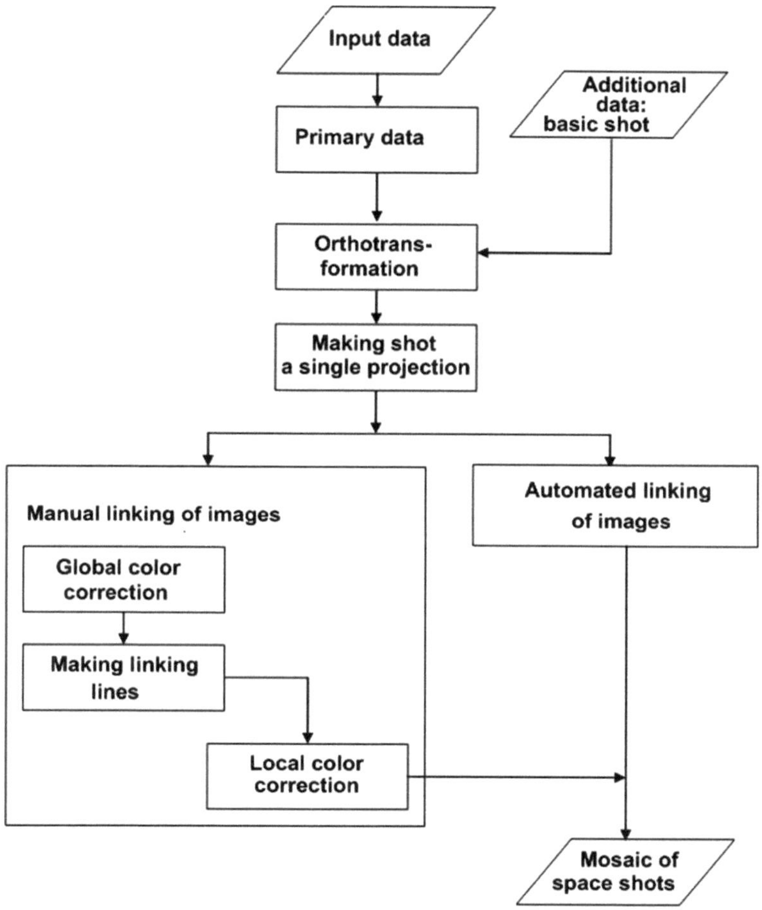

Fig. 2 Technological scheme for making mosaic space images (Based on the report of RCSST 2011)

In ortho-transformation the images with a standard precision of binding, and digital terrain model are used as reference data. At the last stage, the images are linked.

A mosaic for the basins studied is designed for three time periods: the end of the 1980s, the end of the 1990s, and the current status of 2008–2011. Here the basic scale is 1:200,000. Figures 3 and 4 show mosaics made from LANDSAT satellite images of the Ertis river basin. Scenes with a small percentage of cloud cover in the snowless period were selected to make the mosaic.

Figure 5 illustrates how long-term changes in water objects and land use can be evaluated using space images.

Fig. 3 Mosaic of LANDSAT scenes (144/26, 144/27, 143/27, 142/27) of the Ertis river basin on adjoining territory during the period of 1989–1992 (Report RCSST 2011, Courtesy of the Sultangazin—Aerospace Institute Almaty)

Fig. 4 Mosaic of LANDSAT scenes (144/26, 144/27, 143/27, 142/27) of the Ertis river basin on adjoining territory during the period of 1999–2001 (Report RCSST 2011, Courtesy of the Sultangazin—Aerospace Institute Almaty)

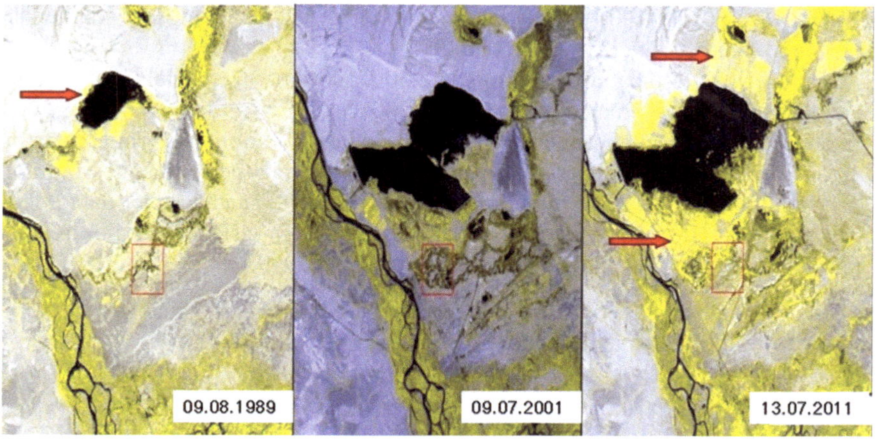

Fig. 5 Changes in land and water use in the Ertis basin (report of RCSST 2011, Courtesy of the Sultangazin—Aerospace Institute Almaty)

Trans-boundary river basins on the territory of neighbouring states were delimited on the basis of available cartographic and satellite data. Eight water-economic basins of trans-boundary rivers were studied: Aral-SyrDaria, Balkhash-Alakol, Ertis, Esil, Zhaiyk-Caspian, Nura-Sarysu, Tobol-Torgay and Shu-Talas.

Map schemes of water basins on the territories of Kazakhstan and neighbouring countries are the basis of the geo-information system "Water resources from trans-boundary rivers of the RK". Currently, a set of map layers include the administrative-territorial division and terrain (contours, eights, forms of terrain). Table 5 shows the structure of the GIS database.

The list of cartographic layers, objects and information content presented in Table 5 is advisory; it does not restrict their extension or additions.

The proposed methodology involves changing the processes and evaluation of water resources in the Republic of Kazakhstan with the help of GIS technologies to determine their dynamic parameters based on the methods and techniques of thematic satellite data processing in theoretical and applied purposes. The method uses the results of historical satellite imagery analysis for the last 30–40 years in the basins of transborder rivers on the territories of the RK and neighbouring states. This satellite imagery from U.S. satellites includes:

- NOAA/AVHRR (1000 m spatial resolution, 5 channels in the visible and infrared bands, range: 2400 km) in 1998;
- TERRA/MODIS (spatial resolution 250, 500, 1000 m, 36 channels in the visible and infrared bands, range: 2400 km) in 2001;
- Landsat 4, 5, 7 (Data Scanner MSS (80 m spatial resolution) are available dating from 1972, scanner TM data (30 m spatial resolution in the visible, near infrared and middle zones, 120 m in thermal range) from 1982, and scanner data ETM+ (spatial resolution 30 m in the thermal range–60 m in visible, near infrared and middle zones) are available dating back to April 1999.

Space imaging materials have been prepared to assess changes in water resources in the 1980s and 1990s, 2000–2005 and 2006–2010.

Table 5 GIS database structure

Water object	List of recommended attributive information	Type of field
River	Name	Symbolic
	Flows into and from what river bank	Symbolic
	Distance from mouth, km	Digital
	Length of water flow, km	Digital
	Water collection, km^2	Digital
	Number of inflows of 10 km long	Digital
	Total length of inflows less than 10 km, km	Digital
Water reservoir	Name	Symbolic
	Watershed or place of origin	Symbolic
	Full capacity on project, mln.m^3	Digital
	Useful capacity on project, mln.m^3	Digital
	NPU water level mark, m	Digital
	UMO water level mark, m	Digital
	NPU mirror area, km^2	Digital
	UMO mirror area, km^2	Digital
Lake	Name	Symbolic
	Total watershed area, km^2	Digital
	Mirror area, km^2	Digital
Irrigation system	Highway, off-farm collector	Symbolic
	Head flow, m^3/day	Digital
	MK length, km	Digital
	MK length in facing, km	Digital
	Crop irrigation method in furrows and bands	Digital
	Crop irrigation method in plots	Digital
	EC system	Digital
Irrigation area	Name	Symbolic
	Irrigation source name	Symbolic
	Irrigation area, thous. ha	Digital

Reference

RCSST Report (2011) (National Center for Space Research and Technology). Development of the complex of space monitoring technology of water objects, water systems and analysis of water use in the basins of transborder rivers. Almaty, p 76

Model-Based Impact Analysis of Climate and Land Use Changes on the Landscape Water Balance

Marco Natkhin, Ralf Dannowski, Ottfried Dietrich, Jörg Steidl and Gunnar Lischeid

Abstract Changes in the water balance in landscapes can be easily observed by measuring water levels, runoff, etc. Determining the causes of changes in the water balance is much more difficult, because of the complex interrelations between the interacting hydrological processes. However, revealing the causes and processes is necessary to understand the source of changes and to evaluate management options. With the help of modelling and scenario analyses, it is possible to differentiate between the effects of, for example, land use/land cover and climate on the water balance, taking the underlying hydrological processes into consideration. Two case studies are presented: the Ngerengere river catchment in Tanzania, Africa and a forested area in North-East Germany. In these areas the observed water balance has changed considerably in the last few decades. The influence of climate conditions and land use change are analysed and determined with the help of the SWAT model, the WaSiM-ETH model and statistical analyses. Both the suitability and the limitations of this methodology of model-based impact analysis are demonstrated.

M. Natkhin (✉)
Johann Heinrich von Thünen-Institut, Institut für Waldökosysteme, 16225 Eberswalde, Germany
e-mail: marco.natkhin@ti.bund.de

R. Dannowski · O. Dietrich · J. Steidl · G. Lischeid
Leibniz Centre for Agricultural Landscape Research (ZALF) e. V., Eberswalder Strasse 84
15374 Muencheberg, Germany
e-mail: danno@zalf.de

O. Dietrich
e-mail: dietrich@zalf.de

J. Steidl
e-mail: jsteidl@zalf.de

G. Lischeid
e-mail: lischeid@zalf.de

L. Mueller et al. (eds.), *Novel Measurement and Assessment Tools for Monitoring and Management of Land and Water Resources in Agricultural Landscapes of Central Asia*, Environmental Science and Engineering, DOI: 10.1007/978-3-319-01017-5_36,

Keywords Landscape water balance · Climate · Land use · SWAT model · WaSiM-ETH model

1 Objectives

Water is a limited resource in Central Asia. Water use for different purposes, above all water abstraction for irrigation, or the artificial drainage of waterlogged and salty soils, can lead to water use conflicts between upstream and downstream users. The energy supply, the protection of ecological functions or flood prevention can evoke a demand for differentiated management. Changes in the water balance are not only caused by direct water use but also by land use and changes in the land cover (Brown et al. 2005). Land cover means the earth's surface including biota, soil, topography, water and anthropogenic structures, and land use is connected with people's intention to change and use the land cover for their purposes. Other main drivers of changes in the water balance are climate conditions and their fluctuations.

Determining the causes of changes in the water balance is important but often not easy, because of the complex interrelations between the interacting hydrological processes. Further, several possible causes simultaneously lead to change, on very different time scales. One method of handling this problem is to use models and scenarios where only one possible cause is changed in one model run, and to analyse the resulting partial effect. Thus, with the help of models and scenario analyses, it is possible to differentiate between the effects of, for example, land use/land cover and climate on the water balance, taking the underlying hydrological processes into consideration. Model-based analyses need data which is often not available as needed for sufficient analysis. Thus, the modelling performance is restricted and the strength of results is limited, but may still be helpful.

This chapter presents two case studies from two completely different geographic and climatic regions and with different models. One study was done for the data-scarce Ngerengere river catchment in Tanzania, Africa. In this catchment an intensification of agriculture has been observed in recent years, with consequences for the flow regime of the rivers. The influence of climate boundaries and land use change in the catchment to the flow regime of the river is analysed with the SWAT model (Arnold and Fohrer 2005). In the second study, changes in groundwater recharge are analysed in a forested area in North-East Germany with recently decreasing water levels. Here the influence of land cover change and changing climate boundaries on groundwater recharge is analysed with the WaSiM-ETH model (Schulla and Jasper 2007). Detailed information about these studies is published in Natkhin et al. (2012 and 2013). This paper shows, based on these two case studies, how hydrological models can help to fill knowledge gaps caused by the complexity of interacting processes. Furthermore, the advantage of scenario analyses is shown, based on process models, for assessing the impact of and evaluating future trends in climate and land use change on the landscape water balance as anticipated for two different regions of the world. Advances in the methodology, and its limitations and concomitants, are discussed against the background of a possible application for Central Asia.

2 Materials and Methods

2.1 Ngerengere Catchment in Tanzania

The Ngerengere River catchment is located in western Tanzania within the longitudes 37°32′ and 38°38′ E and latitudes 6°51′ and 7°09′ S. At the Utari gauge, its size is approximately 2,780 km^2 (Fig. 1). The topography varies from relatively flat land to mountainous areas up to 2,260 m above sea level in the Uluguru Mountains. The geology of the catchment has mainly been created by different plutonic and Neogene events. The groundwater is deep in most parts, but plays a particularly significant role in water storage and as a pathway between the rivers (as water infiltrates) and their floodplains where water is lost by evapotranspiration. The catchment is located in the tropical climate region with two main rainy seasons. The annual rainfall over most parts of the catchment varies between 800 and 1,000 mm/year but increases to over 1,500 mm p.a. in the Uluguru Mountains.

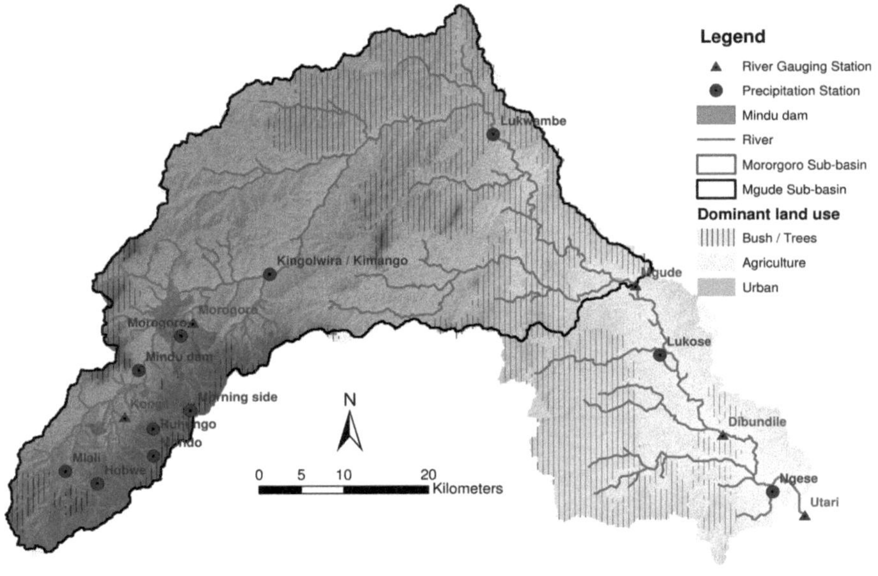

Fig. 1 Study area—Ngerengere river catchment and Mgude and Morogoro sub-catchments with precipitation and runoff gauges

In the hills, especially, the rainfall data indicates high spatial and temporal variability. Normally, potential evapotranspiration is higher than rainfall in the catchment. FAO Penman–Monteith potential evapotranspiration is between 1,500 and 1,700 mm p.a. (FAO 2004). Cultivated areas, most of which are rain-fed, rapidly increased from 16 % in 1995 to 61 % in 2000 (Yanda and Munishi 2007).

2.2 Redernswalde Study Area

The Redernswalde study area is located in North-East Germany, about 50 km northeast of Berlin within the longitudes 13°45′ and 14°00′E and latitudes 53°00′ and 53°05′N. The morphology of this area was formed by the last Pleistocene ice advance. The stratification is marked by up to seven Quaternary aquifers as a consequence of the thickly deposited glacio-fluvial sand, till and clay layers. The water table in the upper, unconfined aquifer varies between the ground surface and more than 40 m below the ground, mostly depending on the surface level. There are a large number of lakes and wetlands intersecting the groundwater surface, but only two natural streams chosen as the boundary of the study area, along with some artificial drainage ditches (Fig. 2). Sandy Cambisols, some clayey, and Cambisols/Podsols are the predominant soils. The forests are dominated by pine monoculture (Pinus sylvestris) as is common in large parts of North-East Germany. This region is located in the transition zone between a maritime and a continental climate, with a long-term (1958–2007) mean precipitation of 529 mm p.a., an air temperature of 8.6 °C and an FAO Penman–Monteith potential evapotranspiration of about 570 mm p.a. In this region a water level decrease has been observed at many lakes and groundwater gauges.

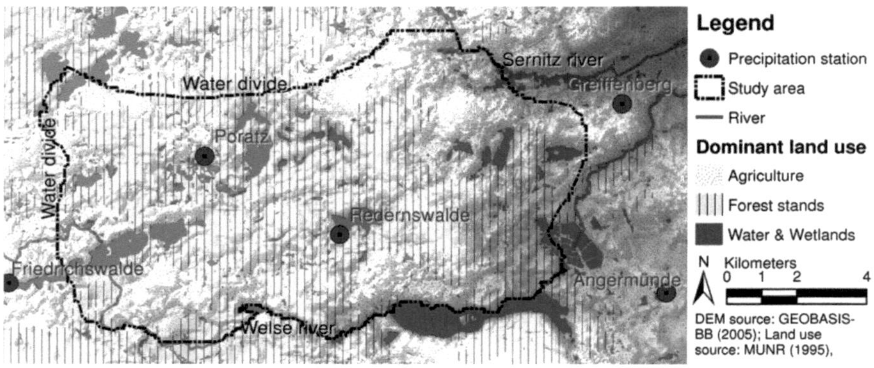

Fig. 2 The Redernswalde study site, with 58 % forested area; and the location of the precipitation gauges used for modelling. Regional groundwater divides and the rivers Sernitz and Welse form the boundaries of the study area

3 Models

One of the benefits of using models is their potential to be linked with scenarios to analyse complex cause-and-effect chains. Different steps in modelling are recommended to obtain trustworthy model results (Fig. 3).

A large number of models have been developed over recent decades in the hydrological community (Gosain et al. 2009). The choice of a model depends mainly on the aim of the modelling, which processes are to be included in the modelling and which data are available. As these studies focus on land use/land cover and climate changes, models should be spatially distributed and rely on a physical base.

For the Ngerengere catchment, where data is scarce, the Soil and Water Assessment Tool (SWAT) is used (Arnold and Fohrer 2005). This model requires the dissection of areas into hydrological response units (HRUs) and is well suited to water catchments. The SWAT model was previously used in Tanzania and East Africa with similar purposes (Mulungu and Munishi 2007; Palamuleni et al. 2011; Ndomba et al. 2005, 2008), but also in arid and semi-arid areas (Wang et al. 2012; Perrin et al. 2012). The Ngerengere catchment is divided for modelling into 124 sub-catchments with 10 land use classes and 6 soil types, resulting in about 280 HRUs. For the calibration and validation process the land use map from 1995 is used. The meteorological time series starts on 9/1/1970. The time until 12/31/1971 is used as a model warm-up. The following three years (1/1/1972–12/31/1974) are used for calibration, and the next 5 years (1/1/1975–12/31/1979) for validation. The model period used for the scenario analysis is from 1/1/1972 to 8/31/2010.

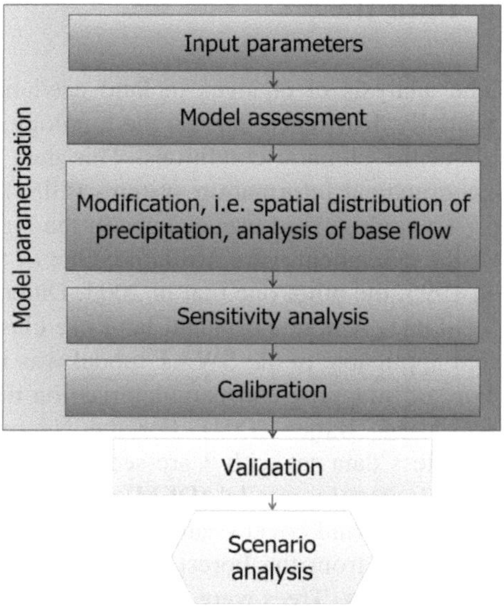

Fig. 3 Flow chart of modelling procedure

The model performance is tested with the coefficient of determination (R^2) and the Percent Bias (PBIAS) between simulated and observed runoff, with treatment of missing values (Yapo et al. 1996). Following Moriasi et al. (2007), the model simulation can be rated as satisfactory as soon as R^2 is greater than 0.5 and the PBIAS is less than 25 %.

The WaSiM-ETH model (Schulla and Jasper 2007) is a widely used water balance model which appeared appropriate to analyse groundwater recharge in the Redernswalde area. This model was applied in different regions with climates ranging from wet alpine to semi-arid (Ahrends et al. 2008). Processes relevant to the water balance are integrated into particular modules. Regarding the available spatial and temporal data and the numerical complexity, the horizontal spatial discretization is carried out using regular grids with a cell extension of 100 by 100 m. The time step length for the result output is 1 day. The model boundaries were chosen along hydrogeologic structures with known flux or known water levels, such as the main groundwater divides and the rivers. The study area is 104 km^2 in size. The model period starts in the year 1951. Until 1957 the calculations are used to stabilize the model run. From 1958 to 2007 groundwater recharge is modelled for analysis. To definitely calculate the rate of groundwater recharge as the amount of water contributing to the groundwater from the unsaturated zone, the groundwater levels must be known, especially for wetland areas with increased evapotranspiration from capillary rise. For this an underlying multilayer 2D groundwater model is included in WaSiM-ETH and was used to design a two-layer model as a simplification of the hydrogeological model. For more parametrisation details see Natkhin et al. (2012, 2013).

4 Data Base

The analysis of changes in land use/land cover at the landscape scale requires spatially distributed data on the topology, geology, soils and land use, and usually about the climatic conditions. This data is supplied and used as digital maps in geographical information systems (GISs).

To analyse land use changes in the Ngerengere Catchment, two land use maps of the catchment were available. One of the maps shows the land use distribution in 1995, the other one that in 2005. Older maps with these spatial details were not available. A new individual land use classification which is also compatible with and applicable to the SWAT model was introduced to homogenise the input data. For soil mapping and parameterisation the SOTER Database for Southern Africa was used (Batjes 2004). For the Ngerengere catchment the SRTM hole-filled seamless data set with 3 arc-second resolution (Jarvis et al. 2006) was used as a digital elevation model (DEM).

For the land cover change of forests in the Redernswalde area the basis was the 2006 data from the Forest Data System of the Federal State Forest Authority of Brandenburg. Trees were assigned here to three age groups (young stands, saplings and mature stands). The vegetation has at most three layers in the model. The first layer is upper-stand trees, the second tree layer is lower understorey stands and the third layer is ground vegetation. For soil mapping and parameterisation for Brandenburg the official generalised 1:300,000 soil map was used supplemented by our own observations and plausibility for land use. Considering the importance of groundwater in the Redernswalde area, the hydrogeological data for the groundwater model is based on hydrogeological maps with parameter ranges of saturated conductivity, layer thickness and effective porosity, as well as Quaternary lithofacies maps, both at a scale of 1:50,000, and some additional drillings of our own. The groundwater and lake gauges used in the study area are described in Lischeid et al. (2010). Elevation data from GEOBASIS-BB (2005) with 25 m resolution was used to create a DEM for the Redernswalde area.

To analyse changes in the climate boundaries and water balance it is necessary to analyse time series. Changes in precipitation are often assumed to be one of the main drivers for changes in the flow regime and water balance. Changes in precipitation in the Ngerengere Catchment are analysed using observed daily data from the precipitation station in Mlali. The analysis of changes in runoff is based on a runoff time series obtained from daily water level measurements taken by the Wami Ruvu Water Basin Office (WRWBO).

Daily values of meteorological data (temperature, air humidity, wind speed and radiation) from the 1970–2010 time series were provided by the Tanzanian Meteorological Agency (TMA) for the Morogoro station. Data gaps were filled by regression and autocorrelation approaches. As precipitation is a main limitation for proper model performance (Mulungu and Munishi 2007 and Ndomba et al. 2008), precipitation was in the focus when the input data were assembled. Additionally to the Mlali Station, 14 TMA and WRBWO precipitation stations were used as a data source, mainly in the upper part of the catchment (Fig. 1), but also outside of the catchment area. Kriging was applied to create daily precipitation maps for the mountains, as precipitation varies widely, especially in mountainous areas. In the lowlands, for which there is mostly no long-term precipitation data, the precipitation is lower than in the mountains. Using average precipitation maps (FAO 1994), the average precipitation station data and the elevation, 80 % of the precipitation of the Morogoro station was assumed to be representative for times and areas with no data. The detailed monitoring programme for the Ngerengere catchment with additional data for the model construction is described in Gomani et al. (2010).

For the Redernswalde area, a full set of daily meteorological data from the Angermünde site was provided by the German Meteorological Service. In addition, precipitation data from four neighbouring sites were used (Natkhin et al. 2012). The daily precipitation was corrected for the local conditions following Richter (1995). The observed discharge data from two gauges in the Sernitz and Welse rivers were used to validate the modelled runoff of the study area. A comparison of long-term water balance averages (1976–2005) following the Glugla-Bagrov approach with other model data is possible using LUA (2009).

5 Analysis of Change

Statistical analysis was used to find significant changes in the time series. The Mann-Kendall test was used to check the significance ($p \leq 0.05$) of a trend. The slope of the trend was calculated applying the Theil-Sen estimator (Sen 1968). The trend of autocorrelated time series was estimated following Yue et al. (2002). Besides a trend analysis the time series were divided into halves to find changes by time. The significance of differences between two time periods with no normal distribution was tested using the Wilcoxon rank sum test ($p \leq 0.05$). This provides the chance to compare long time series and to detect changes even if there is no estimated trend. These statistical analyses were performed using the R software package (R Development Core Team 2006) and with the additional use of the Kendall, nortest and zyp packages.

To find changes in the water balance of the Ngerengere Catchment, indicators are used for precipitation-induced water scarcity, while for floods the 5th percentile of runoff (QP5) is used and for low flow the duration (LFD). The indicators are described in detail in Natkhin et al. (2013).

5.1 Scenario Analysis

The effects of changes in climate and land use/land cover on observed changes in water balance are not easy to determine and to quantify, as more than one possible cause may change at a time and their interactions are complex. One way of handling this problem is to use model scenario analysis, where one boundary condition is changed and the remaining ones are left constant. Comparing different scenarios allows researchers to educe the impact of changes.

For the Ngerengere catchment, the impact of land use change on the flow regime is analysed with the land use maps of 1995 and 2005 under the same climatic boundaries. This approach was used successfully in Githui (2008) and Palamuleni et al. (2011). In the same manner, the impact of climate boundaries on the flow regime is analysed, while land use is not changed. The modelling period is divided into two 19-year periods (1972–1990 and 1991–2009).

In the Redernswalde area, three different scenarios were used for reproducing the development in forests. The first scenario (AGESTAT) assumes that the forest stand structure for the reference year 2006 remains unchanged. Thus, changes in groundwater recharge over time will be induced only by climate fluctuation. The second scenario (AGEDYN) considers changes in the age distribution of trees. Based on the conditions in the reference year 2006 the tree ages are reconstructed in 10-year intervals back to 1956. As a consequence, the percentage of mature forest in the first tree layer is seen to have increased continuously from 1956 to 2006. In the 1980s the fraction of young trees was at its highest. The percentage of tree understorey was smaller in the past. The third scenario (GVDYN) considers changes in ground vegetation in addition to changes in age distribution. To implement a less water-consumptive ground vegetation in the model, it is assumed that none of the pine forests in the study area had an explicit ground vegetation layer before 1981. Thus this scenario is a rough assumption and to be seen as an extremely simplified scenario with its rapid change in ground vegetation simulated in one simultaneous time step for all the pine stands due to a lack of further information.

6 Results and Discussion

Some changes in the Ngerengere catchment can be observed even without modelling. There is intensification in land use change (Fig. 4). Trend analyses of precipitation time series showed a significant extension of drought time (weeks with less than 15 mm precipitation) with an increase of 0.9 weeks per 10 years for the precipitation station in Mlali. To analyse observed discharge changes over time, the two 19-year periods were compared. As an indicator for high-flow events, the 5th percentile value QP5 increases in Morogoro from 2.7 to 3.4 m^3/s (not significant), but decreases significantly in Mgude from 44 to 15 m^3/s. LFD increases at both gauges, but the increase is only significant in Mgude.

The model calibration and validation showed moderate results. The mass balance during calibration and validation fitted very well for Morogoro and Mgude except for a four-month period in the autumn of 1977 with a PBIAS of 0.4 and −1 % for calibration and 1.1 and 9.6 % for validation. The fit of the runoff dynamics, which is represented by R^2, was much worse (0.57–0.31), especially with respect to the amplitude of discharge peaks. This seems to point to inadequate precipitation data quality. The observed and modelled water storage from the Mindu dam, a reservoir, was used to validate the upper catchment runoff, yielding

Fig. 4 Changing land use in the Ngerengere catchment between 1995 and 2005; *source* Sokoine University of Agriculture in Morogoro

Table 1 Modelled effects of changing land use (LU impact; land use scenarios LU1995 and LU2005) and different climate conditions (CC impact; periods of 1972–1990 and 1991–2009) on the 5th percentile of runoff representing floods (QP5) and the low flow duration (LFD) at the gauges in Morogoro and Mgude (*significance)

		Morogoro LU1995	LU2005	LU impact	Mgude LU1995	LU2005	LU impact
QP5	1972–1990	2.94	3.24	10 %*	21.39	24.75	16 %*
	1991–2009	2.7	2.95	9 %*	16.79	19.69	17 %*
	CC impact	−8 %	−9 %		−22 %	−20 %	
		LU1995	LU2005	LU impact	LU1995	LU2005	LU impact
LFD	1972–1990	0.58	0.54	−6 %*	0.43	0.33	−24 %*
	1991–2009	0.64	0.61	−5 %*	0.57	0.48	−16 %*
	CC impact	12 %	13 %		34 %	48 %	

an R^2 of 0.74. To summarize, the modelled values were plausible and the model was considered to be suitable for this study.

Statistically analysed modelling results of the runoff dynamics are presented in Table 1. There is a decrease in overall QP5 over time at the climate boundaries but an increase at Morogoro caused by changing land use. At the Mgude gauge, the same can be observed. Changing land use leads to an increase in QP5. The changes in Mgude are larger than in Morogoro. Changing land use and climate conditions have opposite effects on low-flow days. There is an increase in LFD over time in Morogoro and Mgude at climate boundaries, but the land use change reduces the LFD. As with QP5, changes in Mgude are larger than in Morogoro.

The main model output in the Redernswalde area study is the groundwater recharge. Its measurement on the landscape scale is not possible in a direct way. Thus, the model calibration and validation was carried out in steps (Natkhin et al. 2012). The measured interception loss of trees showed good agreement with the model results. The performance of soil water modelling was tested on a beech stand near the Redernswalde area. Deep seepage of tree stands was compared with literature values of observed stands with comparable boundary conditions. The modelled differences between land use classes are plausible and applicable for scenarios, and they are largely consistent with observed and other modelled values of regional long-term runoff. For more details of calibration and validation see Natkhin et al. (2012).

Changes in the AGESTAT scenario in the modelled groundwater recharge in the Redernswalde area only result from changes in the climate. The annual precipitation figures show an insignificant trend of −1.2 mm p.a.[2] Although there is no significance, the precipitation is 22 mm p.a. lower in the second half of the model period compared to the first half. Evapotranspiration in forest stands increases in the second half of the model period without a significant trend. The air temperature significantly increases in the last two decades from 8.1 (average for 1958–1987) to 9.2 °C (average for 1988–2007). The annual groundwater recharge of the forest stands has a trend with a slope of −1.2 mm p.a.[2] For the entire model period of 50 years this trend results in the groundwater recharge decreasing by

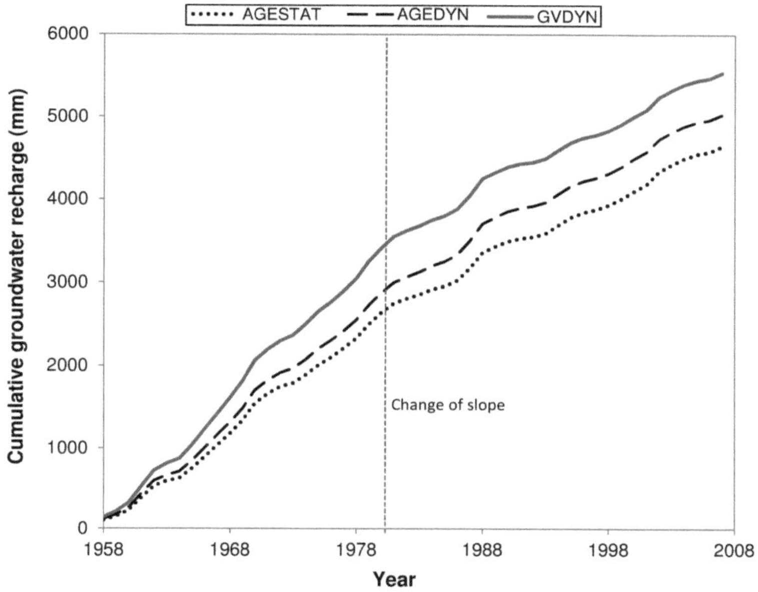

Fig. 5 Cumulative groundwater recharge of forest area in the Redernswalde area modelled for all three scenarios

60 mm p.a. Comparing the second and first halves, there is a decrease of 38 mm p.a. The cumulative groundwater recharge in Fig. 5 shows a change in its slope at the beginning of the 1980s.

The difference in the cumulative groundwater recharge between the AGEDYN and the AGESTAT scenarios for the entire model period is 387 mm (Fig. 5). The biggest differences in tree age distribution and thus in groundwater recharge appear in the first half of the model period. Later on, groundwater recharge values in the two scenarios draw closer, and finally in the last decade they are identical. The annual average groundwater recharge in the first model period is 10 mm p.a. higher compared to the AGESTAT scenario groundwater recharge in the same period (significant). The annual groundwater recharge of the forest stands shows a significant trend, with a slope of -1.6 mm p.a.^{2} For the entire model period of 50 years this trend results in an 80 mm decrease in the annual groundwater recharge. The third scenario, taking into account changes in ground vegetation, leads to higher groundwater recharges until 1982 (Fig. 5). The cumulative difference compared with the AGEDYN scenario in 1982 is 560 mm. Comparing the first half of the model period in this scenario (GVDYN) reveals a higher average groundwater recharge under the forest in the study area, at 22 mm p.a. The annual groundwater recharge of the forest stands shows a significant trend, with a slope of -2.3 mm p.a.^{2} For the entire model period of 50 years this trend results in a decrease of 113 mm p.a. in the annual groundwater recharge.

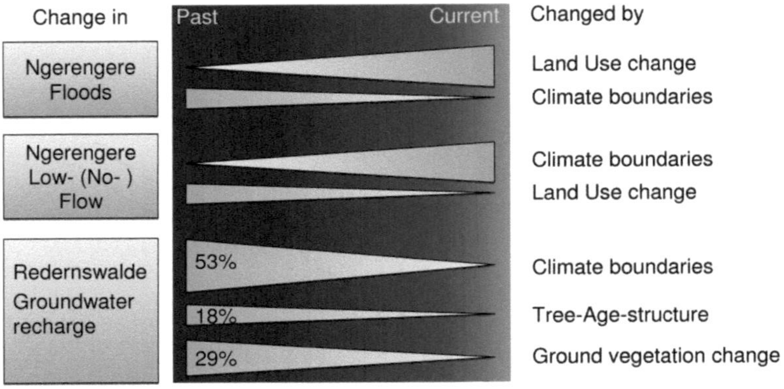

Fig. 6 Subsumed impact of analysed changes on the water balance in the Ngerengere catchment and the Redernswalde area

The effects of climate and land use change for the Redernswalde area can be weighted as it is by Githui (2008) (Fig. 6). The Ngerengere catchment is a data-scarce area. The use of available data results in a discrepancy between the model time and the way that changing land use was monitored. Therefore it is not appropriate to weight the influences here.

7 Conclusions

The use of models permits the analysis of scenarios, which is important in order to examine complex cause-and-effect chains. A number of successfully tested and validated hydrological models are internationally available. However, one basic prerequisite of successful modelling is a sufficient data basis. In data-scarce regions such as the Ngerengere catchment, the model performance is necessarily limited. This means that the modelling results are of limited validity, and in the Ngerengere study we were actually not able to differentiate between the effects of climate and land use change for this catchment. Nevertheless, the model scenarios were useful to analyse changes in the flow regime of the Ngerengere catchment between 1950 and 2010. The use of the model is appropriate to bring together changes observed in the flow regime, climate boundary conditions and land use. Land use change was definitely found to be the main cause of changes in high-flow events in Morogoro. These changes are related to fast flow processes linked with surface runoff. The impact of changing climate conditions dominates the low-flow duration and lowers the frequency of high-flood events in Mgude. The underlying processes are subject to evapotranspiration. Mgude is generally more strongly affected by changes than Morogoro. Land use change and changing climate conditions "work against each other". Changes in the flow regime can also have indifferent characteristics in the catchment, as correspondingly observed in other regions. The main limitations in terms of the model performance are produced by the scarce precipitation input data and discrepancies in the observed data as a whole. In the Redernswalde study the impacts on changing groundwater recharge were successfully analysed, separated and quantified: contributions of changes in climate, forest growth, forest management dynamics, and changes in ground vegetation. To this end a model-based approach was used comparing the groundwater recharge in different scenarios, along with a trend analysis. The WaSiM-ETH model was proved to be an adequate tool to manage this work.

When the processes and interactions have been clarified, as demonstrated in these two studies, it will be possible to use scenario runs to take a look at thinkable futures, which can be important for decision makers, and to find out ways to adapt to probable climate developments. The presented case studies are also believed to be useful in Central Asia as an example of model-based analyses on the hydrological effects of changes in land use and climate.

References

Ahrends H, Mast M, Rodgers C, Kunstmann H (2008) Coupled hydrological-economic modelling for optimised irrigated cultivation in a semi-arid catchment of West Africa. Environ Model Softw 23(4):385–395

Arnold JG, Fohrer N (2005) SWAT2000: current capabilities and research opportunities in applied watershed modelling. Hydrol Process 19(3):563–572

Batjes NH (2004) SOTER-based soil parameter estimates for Southern Africa. ISRIC—World Soil Information. Wageningen

Brown AE, Zhang L, McMahon TA, Western AW, Vertessy RA (2005) A review of paired catchment studies for determining changes in water yield resulting from alterations in vegetation. J Hydrol 310(1–4):28–61

FAO (1994) Map of annual rainfall. FAO

FAO, 2004. Map of Reference Evapotranspiration. FAO

GEOBASIS-BB (2005) Digital elevation model 1: 25 000 (DGM25). Landesvermessung und Geobasisinformation Brandenburg, Potsdam

Githui FW (2008) Assessing the impacts of environmental change on the hydrology of the Nzoia catchment, in the Lake Victoria Basin Department of Hydrology and Hydraulic Engineering Faculty of Engineering, Vrije Universiteit Brussel, Brussels

Gomani MC, Dietrich O, Lischeid G, Mahoo H, Mahay F, Mbilinyi B, Sarmett J (2010) Establishment of a hydrological monitoring network in a tropical African catchment: an integrated participatory approach. Phys Chem Earth 35(13–14):648–656

Gosain A, Mani A, Dwivedi C (2009) Hydrological modelling—Literature review, Climawater, Report no.1

Jarvis A, Reuter HI, Nelson A, Guevara E (2006) Hole-filled seamless SRTM data V3, International Centre for Tropical Agriculture (CIAT). Available at http://srtm.csi.cgiar.org

Lischeid G, Natkhin M, Steidl J, DietrichO, Dannowski R, Merz C (2010) Assessing coupling between lakes and layered aquifers in a complex Pleistocene landscape based on water level dynamics. Adv Water Res 33(SI):1331–1339

Moriasi DN, Arnold JG, Van Liew MW, Bingner RL, Harmel RD, Veith TL (2007) Model evaluation guidelines for systematic quantification of accuracy in watershed simulations. Trans ASABE 50(3):885–900

Mulungu DMM, Munishi SE (2007) Simiyu River catchment parameterization using SWAT model. Phys Chem Earth 32(15–18):1032–1039

Natkhin M, Steidl J, Dietrich O, Dannowski R, Lischeid G (2012) Differentiating between climate effects and forest growth dynamics effects on decreasing groundwater recharge in a lowland region in Northeast Germany. J Hydrol 448:245–254

Natkhin M, Dietrich O, Schäfer MP, Lischeid G (2013) Quantifying the effect of climate and changing land use on the discharge regime of a small catchment in Tanzania. Regional Environmental Change. (online first) doi: 10.1007/s10113-013-0462-2

Ndomba P, Mtalo F, Killingtveit A (2005) The suitability of SWAT model in sediment yield modeling for ungauged catchments: a case of Simiyu river subcatchment, Tanzania. In: 3rd International SWAT Conference, Zürich

Ndomba P, Mtalo F, Killingtveit A (2008) SWAT model application in a data scarce tropical complex catchment in Tanzania. Phys Chem Earth 33(8–13):626–632

Palamuleni GL, Ndomba PM, Annegarn HJ (2011) Evaluating land cover change and its impact on hydrological regime in Upper Shire river catchment, Malawi. Regional Environmental Change

Perrin J, Ferrant S, Massuel S, Dewandel B, Marechal JC, Aulong S, Ahmed S (2012) Assessing water availability in a semi-arid watershed of southern India using a semi-distributed model. J Hydrol 460:143–155

R Development Core Team (2006) R: A language and environment for statistical computing. R Foundation for Statistical Computing. Vienna, Austria

Richter D (1995) Ergebnisse methodischer Untersuchungen zur Korrektur des systematischen Meßfehlers des Hellmann-Niederschlagsmessers. Deutsche Wetterdienst. Berichte des Deutschen Wetterdienstes 194. Offenbach am Main

Schulla J, Jasper K (2007) Model Description WaSiM-ETH. Zürich. Available at http://www.wasim.ch/products/wasim_description.htm

Sen PK (1968) Estimates of the regression coefficient based on Kendall's tau. J Am Stat Assoc 63:1379–1389

Wang ZG, Ficklin DL, Zhang DL, Zhang MH (2012) Impact of climate change on streamflow in the arid Shiyang River Basin of northwest China. Hydrol Process 26(18):2733–2744

Yanda PZ, Munishi PKT (2007) Hydrologic and land use/cover change analysis for the Ruvu river (Uluguru) and Sigi river (East Usambara) watersheds. Dar es Salaam

Yapo PO, Gupta HV, Sorooshian S (1996) Automatic calibration of conceptual rainfall-runoff models: sensitivity to calibration data. J Hydrol 181(1–4):23–48

Yue S, Pilon P, Phinney B, Cavadias G (2002) The influence of autocorrelation on the ability to detect trend in hydrological series. Hydrol Process 16 (9):1807–1829

Biotechnological Restoration Methods of Technogenically Disturbed Soils in Kazakhstan

Farida E. Kozybayeva, Abdulla Saparov, Hasi Dzhamantikov, Gulzhan B. Beyseyeva and Valeria N. Permitina

Abstract This article includes materials on the environmental status of soils as a result of disturbances to the soil surface by resource-extracting industries, and the degradation and desertification of the soil surface in the Aral Sea area. It highlights the methods of field study used in technologically disturbed lands and biotechnological methods of remediating soils which are disturbed and transformed by oil production and aridity. Comprehensive biogeocenologic studies of the initial processes of soil formation in technogenically disturbed ecosystems have been conducted. Integrated approaches have made it possible to develop biotechnological remediation methods. The main aim was to initiate humus formation and soil development by introducing site-adapted vegetation. This was supported by specific biological soil ameliorants and occasionally also by chemical soil ameliorants. *Haloxylon aphyllum* proved a most promising drought-resistant plant for

F. E. Kozybayeva (✉) · G. B. Beyseyeva
Joint Stock Company "KazAgroInnovation", Ministry of Agriculture RK, Kazakh Research Institute of Soil Science and Agrochemistry named after U.U. Uspanov, Al Faraby Ave 75b, 050060 Almaty, Kazakhstan
e-mail: farida_kozybaeva@mail.ru

G. B. Beyseyeva
e-mail: beiseeva2009@mail.ru

A. Saparov
Kazakh Research Institute of Soil Science and Agrochemistry named after U.U. Uspanov, Al Faraby Ave 75b, 050060 Almaty, Kazakhstan
e-mail: ab.saparov@yahoo.com; ab.saparov@mail.ru

H. Dzhamantikov
Joint Stock company "KazAgroInnovation", Ministry of Agriculture RK, Research Institute of Rice Production, Abai Ave, 25b, 120008 Kyzilorda, Kazakhstan
e-mail: pniiaesx@rambler.ru

V. N. Permitina
Institute of Botany and Phytointroduction, Ministry of Education and Science Republic Kazakhstan, Timiryzeva Ave 36d, 050040 Almaty, Kazakhstan
e-mail: v.permitina@mail.ru

L. Mueller et al. (eds.), *Novel Measurement and Assessment Tools for Monitoring and Management of Land and Water Resources in Agricultural Landscapes of Central Asia*, Environmental Science and Engineering, DOI: 10.1007/978-3-319-01017-5_37,

land restoration on many sites. Fencing was very important for the natural restoration of the soil and vegetation.

Keywords Soil · Degradation · Mining industry · Restoration · Biotechnological methods · Kazakhstan

1 Introduction

All over the world we face serious environmental problems relating to land, soils and environmental conservation (Lal 2006). Many scientists have dealt with soil and environmental conservation studies and scenarios for ensuring food security for a growing population. The First Preliminary Conference for the 15th OSCE Economic Forum was devoted to the "Degradation of lands and soil contamination" (OSCE 2006). It was held in Bishkek, Kyrgyzstan. The work of the 5th International Congress of the European Society for Soil Conservation (ESSC), held in Palermo, Italy, and the ISCO's 15th International Congress (2008) in Budapest, Hungary, focused on soil conservation (ESSC 2007; ISCO 2008). The "1st Annual World Congress of Environmental Biotechnology", conducted in 2011, was devoted to biotechnological methods of environmental protection (BIT 2011). All the above-mentioned forums and congresses addressed the protection of soils and the environment. Issues of soil surface ecology and environmental protection are related not only to Kazakhstan, but to the entire planet Earth.

The total area of the Republic of Kazakhstan is 272.5 million (M) ha. 222.5 M ha are in agricultural use, of which 33.7 M ha are arable land, 187.0 M ha are pastures and hayfields, and 1.8 M ha are perennial plants and badlands. About 60 % of the soil surface of the Republic has been degraded to various degrees depending on the characteristics of the natural conditions and economic use.

Degradation in all regions of Kazakhstan is caused by three main factors:

1. Extensive development of agricultural production;
2. Intensive development of resource-extracting industries;
3. The presence of a vast network of former military test grounds (in the period of the USSR).

In the northern regions of Kazakhstan the soils are exhausted, and during almost half a century of virgin land development, 1.4 billion tonnes of humus were destroyed, which is 1/3 of the original stock. Humus reproduction with annual cereal crops is very low, resulting in an imbalance of humus in soils. At the same time the arable land has lost 57 % of humus due to erosion. The area of eroded soils of northern Kazakhstan is 19.1 M ha.

On the territory of the Republic there are soils used in agricultural production which are subject to water and wind erosion. As for the qualitative characteristics of lands, eroded soils in the country occupy an area of 5.0 M ha, of which 1.0 M ha are arable lands. Soils prone to wind erosion occupy 25.5 M ha of which 594.6 thousand ha are arable lands. According to a decree of the Government of Kazakhstan, about 70 % of the territory of Kazakhstan is subject to desertification and degradation of various degrees and needs action.

The Republic of Kazakhstan has the 5th highest amount of pasture land in the world. Rangelands occupy 67 % of the territory. At the same time 125 M ha of pastures are situated in deserts and semi deserts. Historically, the pastures served as the driving force of the country's economy and source of livelihood for the nomads, giving them food, fuel, fodder, medicinal plants, etc. In the mid-1980s, degraded pastures occupied 47 M ha; if we take into account degradation under the influence of technogenic factors, the total area of desertification was 63 M ha of pastures (Arinushkina 1970). From a report of Bekturov et al. (2006), total economic losses in Kazakhstan resulting from the direct and indirect effects of land degradation are estimated at 93 billion Tenge or 6.2 billion USD per year (OSCE 2006).

The condition and use of 59.6 M ha of land in areas suffering from the ecological disaster in the Aral Sea area are a special cause for worry. According to space images, salt and dust streams originating in this region extend in a radius of 150–300 km, and at most 500 km. The area of dust distribution and deposition is about 25 M ha. As a result of the drying up of the Aral Sea, major changes have occurred at the current delta of the Syr Darya and on the dried-up bottom of the Aral Sea.

The Uspanov Institute of Soil Science and Agrochemistry has conducted a number of monitoring studies. The drying up and desertification of hydromorphic soils at the current delta of the Syr Darya river are accompanied by increased salinisation processes, a sharp drop in non-saline soils, an increase in the area of soils whose salinity has changed, and the formation of saline soils. Studies of the soil surface in the Aral Sea area, conducted in 2001, showed that of 1,533,000 ha of soils studied, 50 % are saline (Karajanov and Khaibullin 2005).

In Kazakhstan the soil vegetation surface is heavily disturbed due to the extraction of natural resources. Thus, large areas of disturbed lands are concentrated in the East Kazakhstan region, where ore deposits are situated, while in the west of the country the main disturbances and contamination are caused by the oil and gas industries.

There are various ways to resolve soil ecology problems in Kazakhstan. A special scientifically justified approach needs to be used in each specific region, taking into account the climatic conditions and knowledge of nature regulations.

The aim of this work is to develop methods for restoring technologically disturbed, transformed, degraded and desertified lands using carbon sequestration in order to re-use them in an economical and environmentally friendly manner.

The main objectives are to:

- carry out a reconnaissance study of the disturbed, transformed, degraded and desertified lands;
- define the area, time, extent and forms of negative impacts on soil conditions, defining technogenic terrain forms;

- explore areas with natural vegetation and determine the degree of plant growth under these conditions;
- explore the initial soil-forming processes in natural vegetation and carry out the biotechnological remediation of disturbed and transformed soils;
- compare the recovering landscapes with surrounding undisturbed landscapes (terrain, vegetation and soil surface).

2 Materials and Methods

Studies were conducted on technogenically disturbed lands with polymetal ore deposits in eastern Kazakhstan. The disturbances take the form of industrial waste dumps, open pits and tailings of different ages, forms of terrain and mining methods. At sites of natural vegetation, the degree of the vegetation, the floristic composition of the plants, and their biological productivity were studied. Sites with initial soil-forming processes were outlined depending on the degree of vegetation. In soils at waste rock dumps the microbial fauna were studied which were involved in the primary processes of soil formation in technogenically disturbed ecosystems. Soil-forming processes were studied in terms of the biotechnological restoration of ore deposit rock dumps. Locally adapted grasses, trees and shrubs were planted on various soils.

Due to the major influence of technogenic factors on the oil and gas field, methods for the phytoremediation of technogenically disturbed soils were studied on the Uzen oil and gas field in the Mangistau region. The deposit is an arid-denudation table plateau, characterized by a sharply continental climate with an average annual rainfall of 120 mm. The ground water is 10–15 m deep in the northern part and 30–40 m deep in the southern part. Gray-brown soils of various degrees of salinity and alkalinity, and complexes of gray-brown soils with salt marshes dominate. The soil salinity is 0.5–1.5 and locally, 3.5 g/l or higher. Soils formed from deposits and their vegetation have experienced a technological transformation. Changes in the soil surface occur due to mechanical damage to the upper horizons, penetration of mineralized sewage waters and reservoir fluids (crude oil, oil emulsion).

In order to protect the environment and to create pastures on transformed and saline soils in the oil and gas field, biological remediation was carried out using agrotechnology and phytoremediation.

On lands of the Aral Sea area which are degraded and desertified due to the drying up of the Aral Sea and total aridity, long-term monitoring studies were conducted on soil transformation. On the transformed soils the field biotechnological experiments were carried out to restore their productivity and reduce salt and dust transfer processes caused by wind erosion.

Field research methods on technologically disturbed lands are based on an integrated approach. When ore is mined using open mining methods, the soil plant cover above the ore is disturbed or completely destroyed. After the end of mining, plants from the undisturbed ecosystems begin to grow on disturbed lands. The rate of natural growth of plants on disturbed lands depends on natural climatic conditions, the productivity of the surrounding undisturbed landscapes, and the texture and chemical composition of the substratum. The integrated approach includes methods from soil science, agricultural chemistry, botany, geobotany, soil microbiology and zoology.

In the study of initial soil-forming processes in natural vegetation and the biological remediation of industrial mining dumps, oil and gas fields and soils transformed by aridisation, the method used was that of dividing them into sections and describing the location, terrain, vegetation and morphogenetic characteristics of "young soils" of technogenical terrains and transformed soils by means of sampling.

Before the experiment was planned, a reconnaissance tour of the area was done, and the disturbance of soil and vegetation was mapped, identifying the total areas of disturbed lands, the types and forms of the disturbance, and the technology used for dumping and storing potentially fertile soil layers and rocks. The texture and chemical composition of the substratum were analyzed.

For comparative analysis, technologically disturbed lands were studied by means of the parallel sampling of surrounding undisturbed landscapes. The terrain and vegetation cover were identified, and plants were collected to determine the floristic composition of dominant species. The productivity of the surface (including litter) and underground plant biomass was also determined.

Depending on when the overlay excavation material was dumped (tailings, formed during the technological processing of raw materials and their natural vegetation with plants), the primary processes of soil formation and young soil formation were identified. Plant residues were studied based on their decomposition rate and soil samples were taken to find microbial zoocenoses. The biological productivity of surface and underground plant biomass was determined.

Experimental field methods: The soil amendments studied were biofertilisers and sodium humate. Two types of biofertilisers were applied:

(a) Zher Nury—preparation based on soil nitrogen-fixing bacteria;
(b) MERS from the plant sap on the basis of nitrogen-fixing bacteria and micronutrients.

Sodium humate based on processing balance coal, consisting of a sodium salt of humic acid, which increases organic matter in Takyr soils.

Experiment 1 was on heavy soils. Planting was done in furrows 900 m in length. The distance between the furrows was 5 m and that between plants 1.5 m. The roots of *Haloxylon* were treated using Sh (sodium humate, 2.5 %), Z (Zher Nury, 10 %) and M (MERS, 0.01 %), with these concentrations of the drugs being added to the mix (water + soil) and applied to the roots of the seedlings. Hydrogel (H) was also added to the mix.

Experiment 2 was conducted on the southern side of the Kozzhetpes sand dune on *Haloxylon aphyllum* in two furrows. We used 2–3-year-old naturally growing seedlings.

Hydrogels were applied. They are phosphorus and nitrogen containing polyelectrolytes. The polyelectrolytes which we used increase in size 1,000 times in the presence of moisture, and have the ability to retain moisture and regulate plant water supply.

Laboratory methods: were applied to determine the physical, physical–chemical, chemical and biological soil properties.

Physical–chemical methods: the properties of the soils in the selected samples were determined using methods included in the "Guidelines for soil chemical analysis" (Arinushkina 1970).

- Total humus was detected using Tyurin's method;
- Total nitrogen using Kjeldahl's micromethod;
- Gross phosphorus using that of Ginsburg and Shcheglova;
- Gross potassium using Smith's method;
- Hydrolysable nitrogen using that of Tyurin and Kononova;
- Mobile phosphorus using Machigin's method (calcareous soils);
- Mobile potassium using Machigin's method, as modified by Grabarov;
- Soil pH using the potentiometric method;
- Carbonates using the gasometric method, and the;
- Absorption capacity using Bobko and Askinazi's method as modified by Grabarov and Uvarova.

The botanical and zoological methods used were:

- Methodology for studying the morphology and ecology of the underground part of specific plants and plant communities (Shalyt 1960);
- Methodology for studying underground organs, trees, shrubs, and forest communities in the field of geobotanical studies (Krasilnikov 1960);
- Counting large soil invertebrates (mesofauna) manually on a site of 0.25 m^2, 50 × 50 (Gilyarov 1977);
- Counting with a microfauna elector under a binocular microscope on systematic groups (Gilyarov 1975);
- The study of the soil microflora was carried out using the method of simplified pedoscopes (Myatlikov 1976).

3 Experiments and Results

3.1 Biotechnological Methods for Remediating Disturbed Soils in Ore Deposits

Eastern Kazakhstan is also known as the Ore-Bearing Altai Mountains. This is where the richest ore deposits are concentrated, which were extracted by mining until the end of the 1950s. In the 1960s the ore was extracted in open pits. Large rocks above the ore were taken out of the pit and stockpiled in dumps. The dumps looked like chaotic mounds 50–100 m in height. They consisted of coarsely fragmented, sometimes stony rocks mixed with loam and tertiary red-coloured clays. In addition, during mining the open pits were formed, which were 250–300 m deep.

An industrial town is nearby. Ore is processed here in the processing factories and plants, which give off emissions with a negative impact on the environment. The vegetation around this site has been destroyed and the soil is exposed to erosion.

Biological remediation was carried out at the industrial dumps to protect the environment and improve the sanitary conditions of the industrial town. Mining and biotechnological methods for remediating disturbed landscapes have been developed. Zonal soils, loamy and clay soils were bulked on industrial dumps. These bulk soils had different capacities. Screening layers of gravel, pebbles and technogenic sand (the latter being industrial waste) were used between soils and mouldboard rock to stop the capillary rise of heavy metals from the waste dumps and reduce toxicity at high pH levels. For phytoremediation, zonal trees and shrubs with many seeds were planted along with drought-resistant grasses. The most promising plans are: *Caragana arborescens*, *Betula pendula* and *Populus laurifolia*. From the herbal plants *Psathyrostachys juncea*, *Agropyron cristatum*, and *Onobrychis arenaria* are most promising.

We will present two experimental options for the biotechnological remediation of industrial dumps.

1. Option 1: loamy rock + sand + technogenic sand + dump with phytomeliorants: *Caragana arborescens* and *Populus laurifolia* (Figs. 1, 2).
2. Option 2: a layer of black soil + land-improving plants: *Caragana arborescens*, *Pinus silvestris*, *Medicago sativa*, *Dactylis glomerata* (Figs. 3, 4). This is a bulk layer of black soil, but not a material or humus horizon.

Experiments have shown that *Caragana arborescens*, *Populus laurifolia*, *Betula pendula*, *Pinus silvestris* and *Hippophae rhamnoides* grow better on dumps than other plants; they have good seed resources, reproducing from seeds and root offshoots. *Acer negundo* and other trees and shrubs and herbaceous crops did not grow widely on technogenic dumps. These dumps were remediated 30 years ago.

Fig. 1 Pit. Artificial profile made of Loam + technical sand + dump—site 8

Fig. 2 Phytoameliorants, *Caragana arborescens* and *Populus laurifolia*

Fig. 3 Pit. Bulk layer of black soil chernozem + dump

Fig. 4 Phytoameliorants *Caragana arborescens*, *Pinus silvestris*, *Medicago sativa*, *Dactylis glomerata*

Initial data on the bulk soils show that the humus content is 7 % in black soil, 0.8–1 % in loamy rock and 0.037 % in the rock dump (fine earth). Gross nitrogen makes up 0.38 %, phosphorus 0.34 % and potassium 2.58 %. In the loamy rock gross nitrogen makes up 0.044 %, P-0 12 %, and potassium 2.22 %.

On the remediated dumps where the initial process of soil formation is occurring, the soil profile of "young soils" is formed with an outlined upper humus layer with a capacity of 2 cm. Nutrient accumulation processes were observed (Table 1).

Table 1 Concentrations of nutrient elements in soils of remediated sites

Number of site	Depth, cm	Humus %	Total N %	Gross, %		Mobile, mg/kg		
				P	K	N	P	K
No. 13 bulk layer of black soil + dump	0–6	4.8	0.252	0.15	2.5	98	14	500
	6–15	4.2	0.21	0.13	2.5	84	12	320
	15–33	2.0	0.098	0.05	1.56	Nm	7	140
No. 8 bulk loam layer + technogenic sand + dump	0–2	Nm	0.098	0.05	1.12	72.8	2	80
	2–16	Nm	0.056	0.1	2.25	64.4	13	350
	16–24	Nm	0.098	0.05	1.5	72.8	6	250
	24–34	Nm	Nm	0.05	1.56	53.2	4	220
No. 12 bulk loam layer + technogenic sand + dump	0–2	2.8	0.154	0.09	2.25	58.1	11	430
	2–6	1.8	0.098	0.1	2.19	50.4	6	280
	6–15	1.6	0.098	0.09	2.25	53.2	3	260
	15–25	1.5	0.07	0.08	2.19	39.2	1	190
	25–40	1.3	0.028	0.05	2.06	25.2	1	40

Nm not measured, *N* Nitrogen, *P* Phosphorus, *K* Potassium

The distribution of humus concentration in the profile indicates the differentiation of horizons and transformation of humic substances in the humus bulk layer of soil and loam rock. Thus, in the bulk loam rock the accumulation of organic matter occurs, which does not tally with the original data. In the black soil bulk, the degradation of humic substances takes place. Humus is observed to decrease by 68 % from the original. This decrease in the humus and its distribution across the profile provide a ground for saying that at the initial stages of soil formation the basic soil characteristics are already displayed.

Data on the humus status show the development of the humus formation process and humus accumulation in the upper horizon (0–5 cm). In the primary soil-forming processes a huge role is played by soil biota. As plant food appeared, soil microbial fauna was activated.

On the Ziryan ore deposit on the remediated dumping site No. 13 (dump + – black soil bulk, depth of 0–6 cm), microscopic fungi were found under the microscope which were in the process of reproduction: Cocci (4–6 cells per field of view), single (30–92 cells per field of view), and coupled (2–3 bacillus) dividing cells (on average 3–8 cells). Presumably, the bacillus-type bacteria belong to the genera *Pseudomonas* and *Bacillus*, and the cocci to the genera *Micrococcus*. The bacterial cells shaped like a comma presumably belong to the genera *Clostridium*, and the very big bacillus which are "stuck" to the fungi mycelium are possibly cellulolytic bacteria and bacteria.

At a depth of 5–6 cm microscopic fungi, cocci (3–4 cells per field of view) short bacillus and long single bacillus (1–3 cells per field of view) were found. Presumably, the stick-like bacteria belong to the genera *Pseudomonas* and *Bacillus*, and the cocci to the genera *Micrococcus* (Figs. 5, 6).

On the remediated dump sites the microfauna involved in the initial soil formation processes is represented by microarthropods from different groups of mites (oribatid/Oribatei, trombidiform/Trombidiformes), as well as collembola/

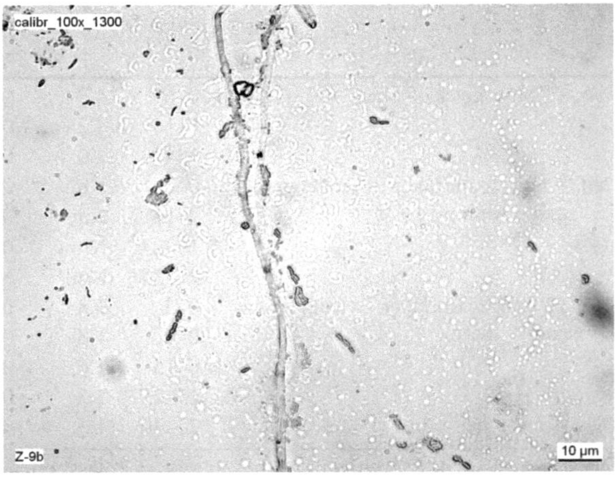

Fig. 5 Site 13. Depth of 0–6 cm *Pseudomonas* and *Bacillus*

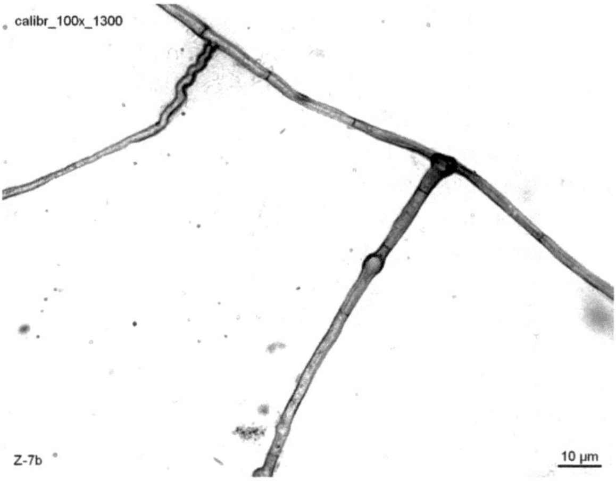

Fig. 6 Depth of 6–10 cm. *Micrococcus, Clostridium*

Collembola. In the case of the oribatid mites representatives of 12 genera are found which belong to 9 families, of which only one genera *Acarus* is a trombidiform from the *Acaridae* family. Collembola are represented by 10 genera belonging to 7 families.

The representatives of mesofauna found included earthworms (*Oligochaeta*), centipedes (*Myriopoda*), adult and larval insects (*Insects*) and spiders (*Arachnids*).

In terms of the composition of components and the quantity of microarthropods among them, the oribatid mites, *Oribatei* (Fig. 7) dominate. They normally comprise 66.8 % (131,600 specimens (sp)/m^2) of the total quantity of microarthropods, trombidiform mites 5.48 % (10,800 sp/m^2) and Collembola 27.6 % (54,400 sp/m^2).

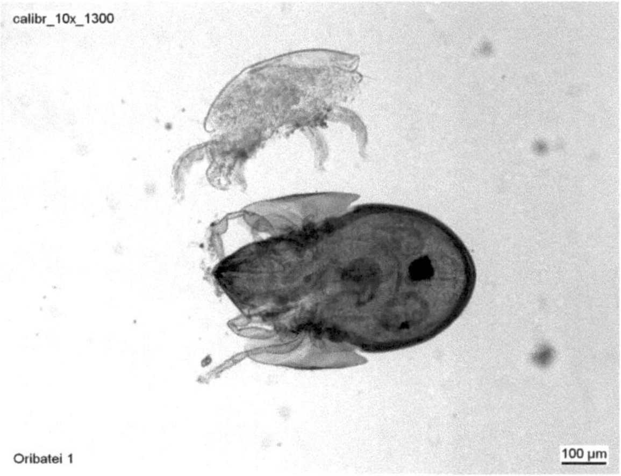

Fig. 7 Site No. 8 (0–5 cm) *Oribatei*

The dominant groups of oribatid mites are representatives of the genera *Nothrus* (18.5 %, 2,440,010 sp/m^2), *Suctobelba* (17.3 %, 22,800 sp/m^2), *Yalumna* (12.7 %, 16,800 sp/m^2) and *Scheloribates* (10.9 %, 14,400 sp/m^2), and the larvae of mites (15.8 %, 20,800 sp/m^2).

There are very few representatives of the genera *Punctoribates, Zygoribatula, Subbelba,* or *Hermanniella.*

Representatives of the genera *Metabelba* are rarely found.

A comparison of the quantity of mites by site and layer showed that the most dense sites were "zonal soil—black soil on undisturbed terrains" (19.4 %, 27,600 sp/m^2), Site 8 (12.6 %, 18,000 sp/m^2) and Site 13 (10.7 %, 15,200 sp/m^2).

Collembola are represented by 10 genera. Springtails or Collembola make up 27.6 % (54,400 sp/m^2) of the total quantity of the microarthropods studied. The predominant components of collembola were representatives of the following genera: *Entomobrya* (27.9 % of the total number of Collembola, 15,200 sp/m^2); *Onychiurus* (19.8 %, 10,800 sp/m^2), *Hypogastrura* (12.5 %, 6,800 sp/m^2) and *Anurida* (11.02 %, 6,000 sp/m^2). There were few representatives of the following genera: *Willemia, Mesaphorura, Folsomia, Jsotoma, Sminturus*. Representatives of the genera *Podura* (Fig. 8) were found very rarely and only on the "dump" site.

During the remediation period the bulk loam and clay underwent a transformation process. During the first 10 years of remediation the bulk layer of black soil degraded, and then its initial restoration process was observed. All the above processes show signs of the initial restoration of the soil's environmental function in the disturbed ecosystems.

Fig. 8 Site No. 13 (0–6 cm) *Collembola* genera *Podura*

3.2 Biotechnological Method of Remediating Soils in the Aral Area which were Transformed by the Aridisation Process

In the areas which were exposed after the drying up of the Aral Sea more than 10 years ago, desert terrains of soil complexes with Takyr soils and salt marshes have formed. A strong Takyr type soil crust has formed on the surface and is the main factor preventing salt and dust from being removed by the wind. *Haloxylon aphyllum* has been planted and there are natural growths of *Tamarix laxa, strobilaceum*, and perennial thistle species (Fig. 9).

To restore the soils of the Aral Sea region which have been transformed by arid conditions, agrotechnical practices were carried out using adaptogens, which stimulate the energy potential of plant, producing increased root formation in adverse environmental conditions. These preparations increase the bio-energy of plant seedlings and strengthen their environmental sustainability and productivity on transformed soils.

Long-term monitoring studies have shown that the Takyr type soils of the Kozjetpes valley are the most suitable for remediation activities on the transformed soils. Kozjetpes Valley is a vast area in the south-eastern part of the Aral Sea region which dried up in the 1990s, with sparse vegetation, sparse shrubs and scattered *Haloxylon* trees, making it suitable for development. Experiments were carried out on Takyr type soils of heavy and light texture.

The experiments revealed various degrees of plant growth on sites with "clay" and "sand" textures. When hydrogel was used on the sandy soil, good results were obtained, but on clay soil the result was only satisfactory.

Fig. 9 Remediated site on transformed Takyr type soil, plantings of *Haloxylon aphyllum*

3.3 Biological Reclamation of Disturbed Areas of Uzen Oil and Gas Deposits

The main types of anthropogenic disturbance to the soil and vegetation on the territory of the deposit are technogenic effects (roads, electric power lines, pipelines, underground communication means, drills, wells, etc.), and petrochemical pollution (accidental spills of crude oil), which lead to the transformation or destruction of vegetation. Limited by drilling and compressor equipment, existing wells, and a branched road network which are constantly subject to technogenic impacts, the restoration of natural vegetation is not observed.

The vegetation cover of the disturbed and contaminated areas is characterized by a change of projective cover, which varies from 10–15 to 40 %, and a low number of species, and depends on the degree of exposure (Dimeyeva and Permitina 2002, 2003). After repeated technogenic impact, the vegetation type which occurs is that of primary succession, with a long-term phase of annual thistle. The floristic composition does not exceed 5–6 species. The introduction of perennial species is slow. Where there is a weak and average technogenic disturbance of the soil surface, the vegetation is made up of indigenous plant species groups: white ground Artemisia (*Artemisia terrae-albae*). Where there is a severe or very severe degree of technogenic disturbance, the first species to appear are weeds: Tatar Atriplex (*Atriplex tatarica*) or Salsola (*Salsola nitraria*).

At the sites with spills of crude oil, the method used is to remove the soil surface. The undisturbed areas remain fragmentary. The plants on undisturbed areas with gray-brown desert saline soils are typically wormwood groups (*Artemisia terrae-albae*) and ephemera (*Lepidium perfoliatum*, *Convolvulus fruticosus*, *Strigosella africana*, *Eremopyrum orientale*), with annual halophytes (*Climacoptera brachiata*, *Salsola nitraria*). The projective cover is 10–15 %. Slopes with removed oil spills, pits and sites with removed soil surface are characterized by sparse groupings of annuals (*Salsola nitraria*, *Senecio noeanus*, *Atriplex tatarica*, *Kochia scoparia*), with southern cane (*Phragmites australis*) and *Tamarix laxa*. Plants are strongly suppressed. The projective cover does not exceed 3–5 %.

On the Uzenmunai gas field in the period between 2000 and 2002, we carried out experiments on remediating vegetation using biological remediation techniques, creating a remediation layer (an artificial bulk layer of a mixture of rocks of light texture with organic fertilizers). Salt-tolerant and drought-resistant species of native flora were used as phytoreclamation plants (Dimeyeva and Permitina 2001, 2006; Dimeyeva et al. 2003, 2007; Permitina 2004, 2010).

The aim of improving the conditions of phytoreclamation plants growth was to change the technogenically conditioned salinisation, alkalinisation and lightening of the texture.

One of the ways of improving living conditions for phytoreclamation plants in terms of nutrition and water–air regime is to create an artificial, potentially fertile layer. In the experiments, furrows were formed with a depth of 25–27 cm using loose rock (quarry near the town of Zhana Ozen) of sandy and loamy texture which has insufficient nutrition elements and has a sulphate salinity in the amount of salts (1 %), but with a low composition of exchangeable sodium in the soil absorbing complex. Organic fertilizers were used to improve the nutrition regime. The plants selected had a wide range of correlation with the degree of salinity.

In preparing sites for planting and sowing, we used different soil treatment methods, aimed at creating a bulk layer with more favourable conditions for the survival of seeds and planting material. The treatment methods were selected based on the type of soil, its texture, degree of alkalinity, salinity and disturbance. For each site an experiment plan was developed, which included the rotation of planting and sowing strips. A technique used on all the sites was cutting furrows to a depth of 20–25 cm using a DT-75 M tractor with a single-hull mounted plough PLN-4-35. On one of the selected sites, Plantation plowing to 25–30 cm in two tracks was carried out using a K-700 tractor with an 8-case plough, followed by harrowing in combination with cutting furrows. The width of the till strip was 10 m. The activities developed for the biological remediation of disturbed lands in the Uzen field included the following:

(a) cutting furrows perpendicular to the prevailing wind direction using a bulk layer of a mixture of light rock texture and organic fertilizer (rotted horse manure). This was used on sites of weak and average technogenical disturbance to the soil surface with partial preservation of vegetation. Soils are gray-brown technogenically saline-alkaline and saline-alkaline;
(b) plantation tilling with turning of the plough layer and simultaneous ploughing of a mixture of sand and organic fertilizers. Before tilling on the site (90 × 10 m) the sand and horse manure were put down in a layer of 10–15 cm, and then it was levelled by bulldozers. After sowing, repeated harrowing and packing was carried out for seeding. This was used on sites with a high, extreme and very extreme degree of technogenical disturbance to the soil surface, with the vegetation completely destroyed. Soils are gray-brown, technogenically disturbed, saline-alkaline;
(c) fencing of disturbed areas was used as a natural method of remediating the vegetation and soil surface.

To carry out the experimental work, the initial status of the sites with various degrees of technogenic soil disturbance was determined (depth of destruction of morphological profile, compaction or spraying the soil surface, decrease of humus content, increase of salinity or alkalinity (Dimeyeva et al. 2007).

Based on the obtained data, a plan was developed of steps to take to improve the physical and chemical properties of soils used for phytomelioration. In the sites with technogenically disturbed soils and polluted with mineralized wastewater, the following agrotechnical techniques were used: cutting furrows to a depth of 25–27 cm, followed by the application of a mixture of loose rocks of light texture (25–30 t/ha) and decomposed manure (10 t/ha). On one of the sites with severe technogenic disturbance plantation tilling was carried out along with the simultaneous application of a mixture of rocks and manure, followed by harrowing.

Fig. 10 Experimental site 1. **a** Initial remediation stage in the year 2000. **b** A monitoring phase after additional material planting in 2002

Experimental plot 1 is located in the western part of Block 3A (Fig. 10). The plot size is 0.62 ha. The terrain is a slightly wavy plain. Soils are gray-brown desert technogenic saline-alkaline. The degree of technogenic disturbance of soil surface is extreme, with the soil profile having been destroyed to a depth of more than 15 cm. The road network covers more than 50 % of the site area. There is a weak level of petrochemical pollution. Oil slicks cover 5 % of the site area. The depth of oil penetration is 3–5 cm (Fig. 10).

As a result of making bulk layers on the furrows, the properties of the substrate changed in terms of a number of indicators. A comparative analysis of the properties of technogenically modified soils on the experimental plots and artificial remediated bulk layers showed positive changes in the conditions for the restoration of vegetation. The sandy loam texture of the substrate helped improve the aeration, the absorption and the filtration capacity of the root zone, as well as the amount of water available for plants. In addition, the type of salinity changed and the degree of technogenic salinity and alkalinity decreased (Table 2).

Table 2 Comparative characteristics of physical–chemical properties of gray-brown technogenic soils and bulk layer. Site 1

Sample depth, cm	Humus, %	Carbonate content, %	pH	Na Exchangeable %	Total salts %	Cl^-/SO_4^{-2}	$\sum$ particles <0.01 mm %
Grey-brown technogenic saline-alkaline soils, 2000							
0–2	0.78	13.3	7.40	10.0	0.25	0.6	33.8
2–10	0.77	11.9	7.15	32.3	0.80	1.3	36.2
10–25	0.38	12.3	7.30	36.1	0.59	1.2	38.7
Remediated bulk layer (furrow), 2001							
0–2	0.5	17.7	8.1	0.7	0.32	0.2	10.4
2–10	0.3	17.6	8.0	1.4	0.49	0.2	10.2
10–25	0.2	17.2	8.0	1.6	0.79	0.2	10.9

Vegetation is represented by thin groups of *Artemisia terrae-albae, Climacoptera crassa, C. brachiata, Atriplex tatarica*, annual salty plants *Salsola nitraria, S. foliosa, S. orientalis, Anabasis eriopoda, Halimocnemis karelinii, Ceratocarpus utriculosus, Lepidium perfoliatum.* The projective cover is 3–5 %.

4 Conclusions

1. Major disturbances to the soil–plant cover and acute ecological conditions in Kazakhstan are caused by the activities of the resource-extraction industries.
2. Soil degradation and the general aridisation of the Aral Sea area has resulted in soil transformation processes and the desertification and increasing of the area of saline soils.
3. An integrated approach to the study of technogenically disturbed lands has made it possible to develop biotechnological methods for remediating industrial waste dumps of ore deposits in eastern Kazakhstan.
4. A technological method of putting soils on dumps and growing locally occurring phytomelioration plants for the biological remediation phase produced positive results. Thus, plants grow on industrial waste dumps through vegetative reproduction and through seeds brought by the wind from test sites.
5. Initial soil formation processes occur at remediated industrial dumps. Soil microbes and plant zoocenoses are actively involved in soil formation processes. The soil microflora is mainly inhabited by bacteria, cocci and microscopic fungi. Microfauna in the "young soils" is represented by hard ticks: *Oribatei* and *Collembola*. Soils deposited on the dumps in layers (black soil, loam rock and technogenic sand) are transformed. A small-scale differentiated profile of "young soil" is formed with a clearly visible upper humus layer of 0–2 cm.
6. Monitoring soils in the Aral Sea area allowed scientists to determine the suitability of Takyr type soils for biotechnological remediation. In the process of these experiments it was determined that the use of remediation (hydrogels, fertilizers and sodium humate) on light soils gave good results, but on heavy soils, the results were only satisfactory.
7. The main agrotechnical practices (cutting furrows, plantation tillage involving turning the layer and simultaneously tilling a mixture of sand and soil, the application of organic fertilizers such as manure) gave positive results when biotechnological remediation was conducted on the Uzen oil and gas field.
8. As a result of experimental studies, the changes in the properties of remediated layer in terms of a number of indicators have been revealed. Thus, the content of exchangeable sodium is seen to have decreased, and the type of salinity to have become mainly sulphate or chloride-sulphate. The texture has lightened to loamy, resulting in the improvement of the hydrophysical properties of the substrate (absorption and filtration). Moisture for the root zone increased and the soil aeration improved.

9. *Haloxylon aphyllum* is the most promising drought-resistant plant. In the drought year, the survival rate of seedlings from the autumn planting was 14 %. In the wet year the survival rate ranged from 12 to 33 %. Fencing disturbed areas stimulated the growth of natural vegetation on the site and promoted the natural restoration of the soil and vegetation.

References

Arinushkina EV (1970) Guidelines for chemical analysis of soils. Moscow State University Press, p 487

Bekturov GB, Levin VG, Alimayev II (2006) Report on "sustainable use of pastures- actual issue for Kazakhstan, In: The first preliminary conference for the 15th economical forum of OSCE "degradation of lands and soil contamination", Bishkek, Kyrgyzstan

BIT (2011) The 1st annual world congress of environmental biotechnology, Dalian, China. Decree of the Government of Kazakhstan, (2005): The program on rational use of agricultural lands for 2005–2007 of the Republic of Kazakhstan, January 5, 2005, No. 3

Dimeyeva LA, Permitina VN (2001) Revegetation techniques at the oil fields of the Eastern Caspian region. In: Conference on extreme conditions and vegetation: climatic change and desertification XVIII Jordanas de Fitocenologia. Leon, Spain, p 67

Dimeyeva LA, Permitina VN (2002) Natural restoration dynamics of vegetation and soil surface on oil spills. In: Results and prospects of development of botanical science in Kazakhstan: materials of international scientific conference devoted to 70th anniversary of the Institute of Botany and Phytointroduction. Almaty, pp 231–233

Dimeyeva LA, Permitina VN (2003) Comparative characteristics of vegetation cover and soils of technogenic ecosystems of Mangyshlak. In: Proceedings of the XI congress of RBO. Barnaul, pp 343–344

Dimeyeva LA, Permitina VN (2006) Ecological restoration of disturbed lands of the Eastern Caspian oil fields. In: Proceedings of the XII congress of Ukrainian botanical society. Odessa, pp 56–57

Dimeyeva LA, Permitina VN, Larikov SN (2003) Features of environmental remediation of oil and gas deposits of Mangyshlak. In: Botanical science for sustainable development in Central Asia: proceedings of international scientific conference. Almaty, pp 124–126

Dimeyeva LA, Permitina VN, Merkuriev VB (2006) Ecological restoration of degraded lands at the oil fields in the Eastern Caspian region. In: Abstracts of international conference on ecological restoration in East Asia. Osaka, p 171

Dimeyeva LA, Permitina VN, Merkuriev VB (2007) Planning stages of environmental remediation in the desert regions of Kazakhstan. In: Biological remediation and monitoring of disturbed lands: materials of the international meeting. Ekaterinburg, pp 202–212

ESSC (2007) The 5th international congress of the European society for soil conservation. Palermo, Italy, 25–30 June 2007

Gilyarov MS (1975) Account of big invertebrates (mesofauna). In: Methods of soil-zoological studies. Moscow, "Nauka", pp 12–29

Gilyarov MS (1977) The study of soil animals. Methods of station studying of soils. M. "Science", p 227–240

ISCO (2008) The international congress of ISCO. Hungary was issued in the book of abstracts

Karajanov KD, Khaibullin AS (2005) Anthropogenic degradation of the aral sea area and scientific bases for their optimal use. In: Proceedings of the Scientific-Technical Society "KAHAK" Almaty, no 2(13) pp 57–62

Krasilnikov PK (1960) Method for studying groundwater organs, trees, shrubs and forest communities in field geobotanical studies. In: Field geobotany. Moscow, Leningrad, issue 2, pp 448–489

Lal R (2006) Soil science in the era of hydrogen economy and billion people. The future of soil science. International Union of Soil Sciences pp 76–79

Myatlikov EA (1976) In: Anikeeva VV, Lukomskaya KA (1983) Guide to practical exercises in microbiology. Moscow, "Education", p 75

OSCE (2006) The first preliminary conference for the 15th economical forum of OSCE "degradation of lands and soil contamination". Bishkek, Kyrgyzstan, pp 16–17

Permitina VN (2004) Evaluation of changes in quality of soils during biological remediation of disturbed lands in Mangyshlak plain. In: Theoretical and applied issues of geography at the turn of the centuries: materials of international scientific-practical conference. Almaty, pp 181–184

Permitina VN (2010) Characteristics of soil and drainage conditions of disturbed areas for bioremediation. In: Proceedings of the National Academy of Sciences of RK. Series biological and medical. # 1. pp 63–66

Shalyt MS (1960) Method of studying morphology and ecology of underground part of specific plants and plant communities. In: Field geobotany. Moscow, Leningrad, issue 2, pp 379–396

Strategy of Sustainable Soil and Plant Resource Management in the Republic of Kazakhstan

Abdulla Saparov

Abstract This chapter presents the research material gained during implementation of the international project led by the International Center for Agricultural Research in Arid Areas (ICARDA) entitled "Sustainable management of land resources in arid regions of Central Asia and Caucasus". This project focused primarily on irrigation and degraded pastures in south and southeast Kazakhstan. Soils and forage crops were studied using conventional, proven methodological approaches and agro-chemical methods. The subjects of the study were: terrains, soil, water, cultivated and pasture plants (rice, triticale and winter wheat, wormwood-ephemeral, grass-shrub and cereal-grass communities). The following aspects were studied and evaluated: soil reclamation and the environmental condition of the soil surface of Shiely irrigation area, fertility levels, soil salinity chemistry and soil degradation under irrigation in the Kyzylorda region, as well as the degradation of pastures in the Sarysu district of the Zhambyl region. The current status of the soil surface of each study object was described. In addition, methods were devised to improve soil reclamation and environmental conditions and conservation, and to increase soil fertility and crop productivity in irrigated saline soils and pastures of the arid zone.

Keywords Soil degradation · Soil reclamation · Kazakhstan

1 Introduction

The Republic of Kazakhstan's soil surface is 272.5 million ha, making it the world's ninth largest soil surface and, per capita, one of the top scorers. Kazakhstan's soils, which develop in arid, extreme conditions, differ to those in

A. Saparov (✉)
Kazakh Research Institute of Soil Science and Agrochemistry named after U. U. Uspanov,
Al Faraby Ave 75b 050060 Almaty, Kazakhstan
e-mail: ab.saparov@yahoo.com; ab.saparov@mail.ru

L. Mueller et al. (eds.), *Novel Measurement and Assessment Tools for Monitoring and Management of Land and Water Resources in Agricultural Landscapes of Central Asia*, Environmental Science and Engineering, DOI: 10.1007/978-3-319-01017-5_38,

other countries (CIS and foreign countries). They are highly vulnerable, have a low resistance to anthropogenic stress, and are subject to degradation and desertification processes.

Anthropogenic impacts, the intensive advancement of degradation and desertification processes, extensive and inefficient land use, violations of the law of return of alienated nutrients and a failure to observe science-based farming systems and recommendations have led to a deterioration of the soil and its ecological status, and a decline in soil fertility. The land is increasingly becoming degraded, saline, secondary saline, slight saline and polluted by oil, oil products, and chemical and radioactive substances.

The current status and outlook of developing agriculture in Kazakhstan are closely linked to regulating the level of soil fertility and managing natural resources. The main strategy of managing natural resource needs to address the priority tasks of using agricultural land rationally and efficiently, and conserving the environment to the greatest extent possible.

In order to make current land cultivation and management sustainable, existing land resource management methods and technologies for the rational use of agricultural land need to be improved, and new methods devised. To achieve this, it is crucial to regulate soil fertility and vegetation.

In Kazakhstan, the fertility of agricultural land has declined by one-third, and by up to 60 % when under irrigation since the past 20 years. One of the main factors for conserving and enhancing the soil fertility of arable land is the use of fertilisers. In Kazakhstan, the volume of mineral fertilisers per hectare of arable land has decreased by a factor of 15 compared with when fertilisers were used intensively (1986), and by a factor of 25 in the case of organically farmed soil, leading to the destabilisation of soil fertility.

According to data provided by the Republic's agrochemical service, most of the area surveyed now has a low humus content, and it has declined by one third. There is therefore an urgent need to take action to regulate and enhance soil fertility by applying organic and mineral fertilisers, introducing scientifically substantiated crop rotations and introducing advanced crop cultivation and diversification technologies, for example.

Kazakhstan's pastures represent the country's major vegetation asset. They cover 67 % of the country's territory, or 187 million ha, about 124 million ha of which are located in the arid zone.

The use of pastures is particularly important in arid conditions characterised by extreme and unstable environmental factors, resulting in the low productivity of pastures and sharp seasonal and annual fluctuations. In addition, the soil surface is characterised by a poor drainage ability and low natural fertility, which fails to act as a buffer against anthropogenic and technogenic loads.

The economic value of pastures depends largely on the species diversity of plant communities, because they should ensure the sustained provision of full-value feed for animals. In addition, plant communities act as an important, permanent soil forming factor and an irreplaceable biological factor preventing the degradation and desertification of pastures. Environmental issues such as conserving plant diversity and preventing soil and plant degradation in pastures are currently the main objectives in developing livestock production in desert areas, as well as developing other branches of the agricultural sector. Economic activities related to land resource management in pastures can only be effective if highly dynamic processes of yield formation in terms of very uneven annual agro-climatic resources in the desert territory are evaluated correctly. It is particularly important to identify the most productive and drought-resistant combinations of forage plants, types of sown pasture, as well as alternative and promising opportunities for flora.

To this end, the international project ICARDA involved conducting research into the sustainable management of land resources in irrigation and pastures in Kazakhstan.

2 Subjects and Methods of Research

Studies were conducted in areas of Kazakhstan focusing on the Shiely irrigation district. The objects of study were terrains, soils, water and plants. New varieties of plants, crop rotations and associations, and reinvented site-adapted natural species were tested. The plants involved were rice, triticale, winter wheat, *Artemísia*-ephemeral, grass-shrub and cereal-grass associations.

In order to achieve our objectives, we used very common, well-proven methodological approaches and methods to comprehensively study field soils: the comparative geographic method and the method of soil keys in studying the specifics of soil formation during the periodic flooding of soils, and morphological profiling. Saline soils were evaluated based on three main criteria: the chemistry (type) of salinity, the salinity rate, and the depth of the saline horizon. The chemistry of saline soils was determined by composing anions and cations. First of all, attention was paid to anions and the extent to which they correlated to water extracts from soils (Bazilevich and Pankov 1968; Meirmanov et al. 1997; Zubairov 2002; Otarov et al. 2007). The agrochemical soil survey was conducted by conventional research using the following methods: humus using the Tyurin method, GOST 26213-91, easy hydrolysable nitrogen using the Cornfield method 1975 (Agrochemical methods 1975), mobile phosphorus compounds and exchangeable potassium using the method of Machigin, GOST 26205-91, and pH-water extraction using GOST 17.5.4.01-84.

The seasonal dynamics of salts was determined by conducting large-scale saline surveys (1:2,000 scale). pH, CO_3, and HCO_3 were analysed by potentiometric, Cl and SO_4 by titration, Ca and Mg by atomic-absorption spectrometry, and K and Na using a flame photometer. Phenological observation of plant communities in rangelands was conducted according to Beideman's methodology (Beideman 1974). All other field and agrochemical analyses were performed using standard methods (Dospechov 1973; Agrochemical methods 1983).

3 Results and Discussion

3.1 Status and Improvement of Irrigated Land

The evaluation of the current soil reclamation status of the Shiely area revealed that all soils have a certain degree of salinity. At the same time, 998 ha (54 %) of the studied test site contain slightly saline rice-swamp soils and moderately and strongly saline soils. Saline soils and deep saline soils make up an insignificant area. Soils are mainly saline on the surface (88 %), the result of the irreversible secondary salinisation of soils, which is currently occurring. Concerning the extent and chemistry of soil salinity, it should be noted that the technical condition of the collector fails to meet design standards, resulting in an increase in groundwater level and soil salinity (Fig. 1).

The results of the salinity survey and comparative analysis revealed that some freshening (flushing) or desalinisation occurs during the growing season. However,

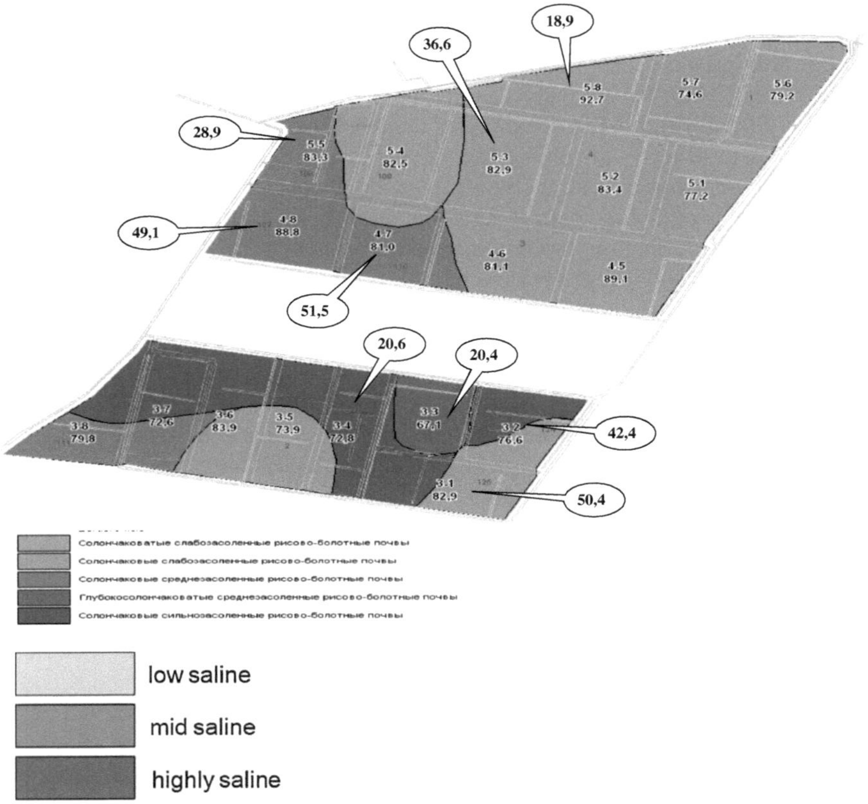

Fig. 1 Map of soil salinity and rice yield capacity

the intensity of the removal of salts from the root zone is insufficient. The yield of rice grown on weakly and moderately saline soils (in rotation with 3-year alfalfa) was 4.2–5.2 t/ha, and only 2.1 t/ha on strongly saline soils. The irrigation norm in a shorter irrigation regime was 24,200, and 25,480 m^3/ha in a constant flooding regime. Since the productivity of crops grown on saline soils depends largely on soil fertility, the presence of organic matter in soils is one of the main aspects affecting the soil fertility of rice fields.

The results of the variation-statistical analysis of the content of humus forms and nutrients in the soil revealed that soils from the test site in Shiely irrigation area have a very low content of common humus (1.1 ± 0.18 %). In this case, the mobile water-soluble form of humus averaged 0.004 ± 0.0005 %. The loss of humus in periodically flooded paddy soils is mainly due to mobile water-soluble forms.

An agrochemical survey was conducted to determine the content of total humus in soils at the experimental site and the extent to which they are supplied with the main nutrient elements at the test site (712 ha). Based on the data obtained, maps were created and the soils were classified into groups of humus content and nutrients, calculating the area of the respective groups. The studies revealed that the entire area under investigation contains groups of soils with a "very low" (95.8 %) and "low" (4.2 %) humus content, i.e., soils at the test site are degraded and subject to the process of dehumification (Fig. 2).

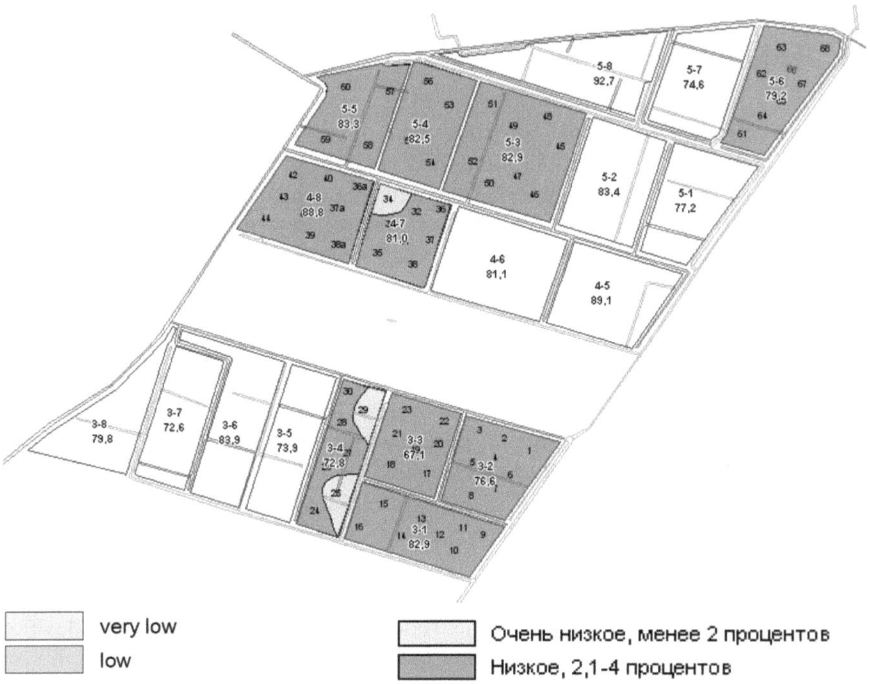

Fig. 2 Map of humus content in soil

Due to the deterioration of soil reclamation status and the intensive development of soil salinity, humus content in soils has decreased; the loss of the mobile water-soluble form of humus in one season was 12–36 % (Otarov et al. 2007). Soils in the surveyed territory are very diverse concerning the availability of mobile phosphorus, namely: very low, low, medium, increased, high and very high. In order to achieve a high yield, 45.2 % of the area requires phosphorus fertilisers. The content of exchangeable potassium was medium and high for over 90 % of the surveyed soils.

Studies on optimising the technology of rice cultivation based on raised bed and minimal-zero till technologies in combination with short or periodical irrigation show that the expenditure of irrigation water can be reduced by 30–40 %, leading to an improvement of the soil's reclamation status and a reduction in the amount of seeds, fuel and lubricants required. With raised bed seed sowing using specialised drills made in India and the equivalent rice yield capacity, the quantity of seeds saved ranged from 100 to 160 kg/ha.

In the Shiely area of irrigation, it was determined that it is possible to introduce a minimal-zero till technique for growing rice that produces the same yield but saves 100 kg/ha seeds compared with conventional rice cultivation techniques.

Based on the experimental studies, we designed soil and terrain maps of the study objects.

3.2 Pasture and Dryland Improvement

The productivity of pastures has decreased sharply in line with the intensive degradation of land. With this in mind, it is very important to develop activities to promote the rational use of land and the enhancement of pastures.

The results of studies on the development of systems to manage pastures, carried out on different soil types and climatic conditions of the Sarysu district of the Zhambyl region, highlighted the need to create sown pastures using plants of the arid zone and to cultivate alternative crops (Figs. 3 and 4).

On light grey soils, where the pasture vegetation cover is 25–40 % and yields are 0.08–0.13 t/ha, the main species were *Calligonum* and *Haloxylon*, *Krascheninnikovia ceratoides* and *Halothamnus*. The plant height of *Kochia prostrata* was 23.6 and 19.5 cm in pure and mixed plantations, respectively, *Krascheninnikovia ceratoides* was 23.1 and 22.8 cm, *Haloxylon* 27 and 25.3 cm, *Calligonum* 43 and 39.2 cm, and *Haloxylon* 23.4 and 19.7 cm. In this case, all of the species featured well-developed lateral shoots and formed symmetrical crowns.

Drought periods in the spring in semi-bound sands with grass-shrub vegetation, with *Artemisia*-ephemeral groups, was reflected to some extent by the development of xerophytes. The height of young plants of *Kochia prostrata, Atriplex* and others was 7–11 cm. Relatively better growth was observed in samples of *Calligonum*, measuring 11–25 cm. In general, the most sustainable plants

Fig. 3 Degraded pastures of Moinkum sands

Fig. 4 Preliminary examination of the "Abylai" farm territory in the Sarysu district (Zhambyl region)

Fig. 5 *Kochia* crops

during the studies were *Sameriaria boissieriana, Kochia prostrata* (Fig. 5), *Krascheninnikovia* and *Halothamnus*, and on sands *Sameriaria boissieriana, Agropyron, Krascheninnikovia, Halothamnus, Calligonum* and *Haloxylon*.

Table 1 Characteristics of wheat plants of the variety vitreous 24 and triticale of the horde variety

Varieties	Development phases							
	Germinations		Waxy ripeness					Full ripeness
	Number of rootlets, pieces	Height, cm	Plant height, cm	Number of shoots, pieces	Number of nods at main stem	Ear length, cm	Number of grains in ear, pieces	Yield capacity, t/ha
Vitreous	3.6 ± 0.4	8.5 ± 0.7	62.6 ± 1.8	1.5 ± 0.2	5.0 ± 0.1	8.0 ± 0.4	36.9 ± 2.3	4.83
Horde	5.6 ± 0.6	11.7 ± 0.6	79.6 ± 0.9	2.1 ± 0.3	5.5 ± 0.1	9.5 ± 0.3	51.0 ± 3.2	5.60

Fig. 6 Comparative indicators of tillering: **a** Winter wheat (variety vitreous 24). **b** Triticale (variety horde)

The cultivation of cereal crops in rain-fed conditions, in particular triticale (variety Horde) and winter wheat (variety Vitreous 24), on ordinary grey soils revealed an advantage of triticale. Plant heights of 79.6 ± 0.9 and 62.6 ± 0.7 cm, respectively, and better tillering were observed. In this case, the biological productivity of triticale was 5.6 t/ha, and the yield of wheat 4.8 t/ha (Table 1 and Fig. 6).

4 Conclusions

- Salinity is a major factor leading to a reduction in soil fertility and crop yield under irrigation conditions.
- The extent and chemistry of soil salinity are important. The salt regime of periodically flooded soils must be controlled by summer washing (when fields are flooded) and, in autumn, after the irrigation period.

- The reclamation and enhancement of the fertility of saline soils is essential for increasing the productivity of crops and land resource management in irrigated agriculture.
- The soil reclamation status should be assessed and soil fertility monitored periodically.
- Based on these survey maps and technology, action must be taken to enhance the fertility of saline soils.
- In order to achieve sustainable pasture management and to ensure fodder production in the arid zone, cultivated pastures must be created by planting natural forage plants and growing alternative crops, particularly triticale, in the ley.

References

Agrochemical methods (1975) Methodological guidance on conducting the agrochemical survey of soils of agricultural land. (Агрохимические методы исследования почв). "Nauka", p 23

Agrochemical methods (1983) Agrochemical methods of soil studies. (Методические указания по проведению полевых опытов с кормовыми культурами). Moscow, "Nauka", p 656

Bazilevich NI, Pankov E (1968) Experience in classifying salinity//Soil Science. (Опыт классификации по засолению//Почвоведение). 11:76

Beideman IN (1974) Methodology for studying plant phenology and plant communities. (Методика изучения фенологии растений и растительных сообществ). Novosibirsk, p 154

Dospechov BA (1973) The field experiment method. (Методика полевого опыта). Москва, p 336

Meirmanov GT, Dzhamantikov H, Dzhamantikova TO, Bekbolat SJ (1997) Environmental impacts of anthropogenic changes in irrigated lands in the aral sea area. (Экологические последствия антропогенных изменений орошаемых почв приаралья//Проблемы экологии АПК и охраны окружающей среды), Issues of ecology and agricultural environment. Almaty, pp 110–111

Otarov A, Ibraeva MA, Saparov AS (2007) Degradation processes and current soil-ecological condition in rice areas of the Republic/environmental bases of formation of soil surface in Kazakhstan in anthropogenesis and development of theoretical bases of reproduction of fertility. (Деградационные процессы и современное почвенно-экологическое состояние рисовых массивов республики/Экологические основы формирования почвенного покрова Казахстана в условиях антропогенеза и разработка теоретических основ воспроизводства плодородия). Almaty, pp 73–105

Zubairov OZ (2002) Reclamation status of irrigated land in the Kyzylorda region/system of agricultural production in the Kyzylorda region. (Мелиоративное состояние орошаемых земель Кызылординской области/Система сельскохозяйственного производства Кызылординской области) Almaty. "Bastau" publishing house, pp 385–387

The Effect of Applying the Microbiofertiliser "MERS" on the Soil Microbial Community and the Productivity of Winter Wheat Under the Conditions of Southeast Kazakhstan

Maira Kussainova, Marion Tauschke and Abdulla Saparov

Abstract Applying microbiofertilisers can be a sustainable alternative to the wide use of chemical fertilisers. They have the potential to reduce the amount of chemical fertilisers applied and thus to minimise environmental pollution, such as nitrogen leaching and gaseous emissions. The aim of this study was to determine the effect of applying the microbiofertiliser "MERS" to the soil microbial community, the yield and quality of winter wheat and soil chemical properties. Over a three-year period (from 2006 to 2009), experiments were conducted at the experimental station of Kazakh Research Institute of Water Resources in the Taraz Zhambyl region. Applying the microbiofertiliser "MERS" to meadow grey soils had an impact on soil properties in all three experimental years. The humus content and content of plant-available N, P and Na were higher in the treated plots than in the control variant. The abundance of the microbial community, in particular heterotrophic bacteria, actinomycetes, yeasts and microscopic fungi, increased for all application rates. The highest increase was found with an application rate of 500 ml/ha. The same application rate had the greatest impact on the yield of the winter wheat cultivar "Almaly" (5.27 t/ha compared to 4.27 t/ha for the control variant).

Keywords Microbiofertiliser · Wheat · Microbial community · Soil fertility

---

M. Kussainova (✉) · A. Saparov
Kazakh Research Institute of Soil Science and Agrochemistry named after U. U. Uspanov, Al Faraby Ave 75b 050060 Almaty, Kazakhstan
e-mail: madgu@inbox.ru

A. Saparov
e-mail: ab.saparov@yahoo.com; ab.saparov@mail.ru

M. Tauschke
Leibniz Centre for Agricultural Landscape Research (ZALF), e. V. Eberswalder Strasse 84 15374 Muencheberg, Germany
e-mail: mtauschke@zalf.de

L. Mueller et al. (eds.), *Novel Measurement and Assessment Tools for Monitoring and Management of Land and Water Resources in Agricultural Landscapes of Central Asia*, Environmental Science and Engineering, DOI: 10.1007/978-3-319-01017-5_39,

1 Introduction

To ensure the stable production of food and feed, Kazakhstan's need for fertilisers in agricultural production grows each year. However, it has to be borne in mind that the global intense use of conventional chemical fertilisers has a major impact on humans, our environment, soil and food quality (Jog et al. 2012). The quality and health of soil, an integral component of the human environment, have declined due to intensive management and the high input of inorganic and organic chemicals, causing changes in physical, chemical and biological properties (Arshad et al. 2002; Doran et al. 2000). The natural resource soil must be used wisely to ensure sustainable development and to meet the increasing demand for healthy, sustainably produced crops and food. It is becoming increasingly important to investigate and develop alternative management and fertilisation strategies (Sadeghi et al. 2012; Abdel-Razzak et al. 2013).

Microbiofertilisers containing living microorganisms or biological substances that promote naturally abundant plant growth-promoting microorganisms can play an important role in maintaining and improving soil fertility in agriculturally used soils (Piotrowska et al. 2012) and in the sustainable production of high yields and high-quality crops (Parr et al. 2002).

Often, the composition of microbiofertilisers is not fully specified, making it difficult for users to evaluate the product and to analyse its effects on the endophytic microbial community and plant growth (Schenck zu Schweinsberg-Mickan et al. 2009).

A wide range of microorganisms are beneficial to plants and are involved in important soil functions. The diversity of the soil microbial community is regarded as critical for maintaining soil health and quality and for ensuring the growth and ecological fitness of their host (Pięta 1999; Raaijmakers et al. 2009; Nihorimbere et al. 2011).

Various interactions take place between different groups of microorganisms and between microorganisms and plants, especially in rhizosphere soil. The composition of microorganisms in the root zone of plants and in bulk soil changes constantly under the influence of different biotic (root exudates) and abiotic (soil, climate) factors (Pięta and Patkowska 2003; Parke et al. 1990; Schoruvitz et al. 1989; Odham et al. 1986).

Regulation of the microbial community, and their life processes and function using microbiofertilisers could play a major role in agrobiotechnologies.

MERS B is a domestic, commercial (Company Ltd. NGO "Ana Zher") product on a natural synthetic basis consisting of chlorophyll-peptide protein compounds combined with 19 macro and micro elements. According to data provided by Usmanov (2008), applying the domestic microbiofertiliser MERS can cause a manifold increase in soil microorganisms, resulting in an increase in humus and nitrate nitrogen content of 1.5–2.0 t/ha and 15.0–30.0 kg soil/ha per year, respectively. Applying the microbiofertiliser means that the amount of fertiliser applied can be reduced by 1.5–3.0 times.

The impact of the microbiofertiliser “MERS” on soil chemical properties, the abundance of the microbial community and the yield and quality of the winter wheat variety “Almaly” were studied in a three-year field experiment. One of the main objectives of the study was to determine the optimal application rate of MERS in winter wheat.

2 Methods

2.1 Study Area

The field experiments were carried out at the research station of Kazakh Research Institute of Water Resources in the Taraz Zhambyl region over a period of 3 years (from 2006 to 2009) (Fig. 1).

The Besagash experimental site is located approximately 30 km south of Taras, at the foot of the Tien Shan Mountains. The following data were taken from a report by Noble (2005). The climate is continentally dry, with hot summers and cold winters. Average air temperature in summer varies from 16 to 19 °C, with an absolute maximum of 43 °C during the summer season. The first frosts usually occur at the end of October. Mean annual precipitation varies between 210 and 380 mm, averaging at 362 mm.

Soil at the Besagash site is a meadow light Sierozem (Calcisol) of silt loam texture. The infiltration rate is relatively low (18.40 mm h^{-1}) due to the high levels of $Mg2^{+}$ on the cation exchange complex. These sodic soils are easily dispersed when wet. The soil bulk density increases with depth, ranging from 1.52 to 1.67 g/cm^3. The depth of the groundwater table varies from 1.5 to 2.8 m, with total dissolved salts ranging from 3,000 to 7,190 mg/l. The soil has a slightly alkaline reaction, with soil pH ranging from 7.60 to 8.4 (Noble 2005).

The potential evapotranspiration at the Besagash site is approximately 1,400 mm/yr. Factors that limit crop yield are drought, alkalinity and sodicity of

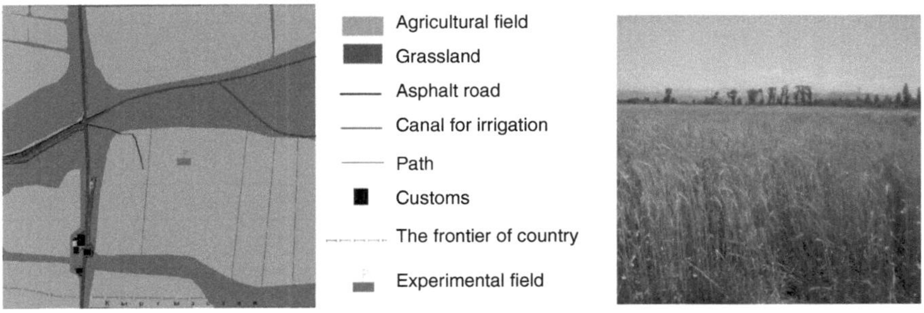

Fig. 1 Experimental field at the research station of Kazakh research institute of water resources near the village of Besagach, Zhambyl

soils. The overall soil quality is very poor without irrigation, and medium to good under irrigation conditions (Smolentseva et al. 2011).

3 Preparation and Application of "MERS"

Winter wheat of the "Almaly" variety was treated with four different concentrations of the microbiofertiliser "MERS" on 20 m^2 plots with three replications and then compared to an untreated control. The MERS solution was prepared in a fertiliser slurry with 30 kg N and P (Table 1).

In order to prepare a "MERS" ready-to-use solution, 30 kg ammonium phosphate was added to 17–25 l water and stirred for 4 h to obtain a homogeneous slurry. The slurry was mixed with 3 kg adhesive protectant ("Dextrin" or any other adhesive, commercially available product) and 100 ml "MERS". Water was added to the working volume of 25–30 l. Before sowing the wheat, seeds were treated with the "MERS" solution.

Wheat plants were additionally sprayed with the appropriate MERS solution at the growth stages of tillering, booting, heading, milk ripe and full ripeness.

4 Soil Sampling and Soil Chemical Analyses

Soil samples for analysing soil parameters were taken at all five growth stages of winter wheat and after fertilisation from the ploughed layer 0 to 20 cm, from 20 to 40 cm and from 40 to 60 cm (Fig. 2).

Soil analyses were carried out using standard methods according to Tyurin (humus), Gedroits (aqueous extract and gypsum), Arinushkina (1970) (total nitrogen: Nt; Kjeldahl, absorbed Ca and Mg and aggregate-size distribution), and Sokolov (1975) (pH - potentiometrically in aqueous suspension). The weight of the solid phase of the soil was determined using the picnometer method; soil moisture was measured by thermal drying to a depth of 100 cm in a 4-fold repetition; wilting point was determined according to Nikolaev (Arinushkina 1970).

Table 1 Scheme of "MERS" application

Variants	Ammonium phosphate (kg)	MERS (ml/ha)
Control	0	0
1	30	50
2	30	100
3	30	200
4	30	500

Fig. 2 Soil sampling with a soil corer

5 Analysis of the Microbial Community

To perform the quantitative analysis of the soil microbial community, soil samples were taken in July each year, before the winter wheat was harvested. A 1 g soil sample mixed from three replicates of the different horizons was used. The soil aggregates were dispersed as follows: the soil was watered and grinded with a mortar for 5 min to obtain a homogeneous soil solution. Soil microorganisms were desorbed from this solution by sequential dilution. The total number of heterotrophic bacteria, actinomycetes, yeasts and microscopic fungi were determined on agar plates using the serial dilution plate method.

The plates were incubated for 5–7 days. The abundance of microorganisms in the different soil layers was calculated according to the following equation:

$$M = M_a + M_b + M_c$$

where M is the number of microorganisms in the appropriate horizon, calculated for 1 cm^2 surface area; M_a is the number of microorganisms in 1 g soil; M_b is the capacity of the soil horizon in cm and; M_c is the relative density of the undisturbed soil.

To obtain an overall number of heterotrophic bacteria, actinomycetes, yeasts and microscopic fungi in the different treatments, the abundance of the respective microorganisms in the different horizons was accumulated.

6 Crop Yield and Grain Quality

For the quality parameters, grain sheafs were manually collected from five different points of each plot before the wheat was harvested. Grain moisture, impurity, full scale mass, vitrescence and gluten content were determined according to GOST 10842-89. The gluten extensibility (quality) was measured using an IDK gluten extensometer.

The wheat was harvested from each plot separately to determine crop yield.

7 Results

In plots treated with the microbiofertiliser "MERS", the humus content and plant-available N, P and Na content were higher than in the non-treated control variant. With the treated variants, the humus content varied between 1.51 % (50 ml/ha MERS) and 1.8 % (500 ml/ha MERS). A humus content of 1.26 % was determined in the control variant. Averaged over the three treatment years, the content of humus in the variant treated with 500 ml MERS/ha was 42 % higher than in the control variant (Table 2).

In addition, plant-available nitrogen was 3.1–4.4 times higher in the MERS treated variants than in the control variant; phosphorus was 3.1–6.4 and potassium approximately 1.1–1.5 times higher (Table 2). The greatest increase amongst all determined soil parameters was in the field plots treated with 500 ml/ha of the microbiofertiliser "MERS".

The abundance of all analysed microbial communities also increased by applying the microbiofertiliser "MERS". However, differences were yielded in the response of the various microbial groups to treatment with "MERS". With the exception of yeasts, the highest increase in microorganisms (heterothrophic bacteria, actinomycetes, microscopic fungi) was detected in the 500 ml/ha MERS application (Table 3).

Table 2 Change of soil properties (average from 2006 to 2009) depending on application of the microbiofertiliser "MERS"

Variants	Ammonium phosphate (kg)	MERS (ml/ha)	Humus content (%)	Plant-available N ($mg \cdot kg^{-1}$)	Plant-available P ($mg \cdot kg^{-1}$)	Exchangeable Na ($mg \cdot kg^{-1}$)
Control	0	0	1.26	22.3	11	225.5
1	30	50	1.51	69.1	31.6	252.2
2	30	100	1.64	84.9	36.0	271.1
3	30	200	1.73	88.7	57.0	282.7
4	30	500	1.80	97.1	61.2	296.8

Table 3 Influence of the microbiofertiliser "MERS" on the abundance of various groups of microflora in meadow-grey soil

	Depth of soil samples (cm)	Heterothrophic bacteria (mln·g^{-1}soil)	Actinomycetes (thous·g^{-1} soil)	Yeast (fungi) (thous·g^{-1} soil)	Microscopic fungi (thous·g^{-1} soil)
Control	0–20	$1.5 \pm 0.1 \times 10^6$	$1.2 \pm 0.3 \times 10^4$	$0.8 \pm 0.3 \times 10^4$	$3.0 \pm 0.4 \times 10^4$
$N_{30}P_{30}$+MERS 50 ml/ha	0–20	$3.3 \pm 0.3 \times 10^6$	$2.2 \pm 0.4 \times 10^4$	$0.8 \pm 0.2 \times 10^4$	$4.7 \pm 0.4 \times 10^4$
$N_{30}P_{30}$+MERS 100 ml/ha	0–20	$3.4 \pm 0.3 \times 10^6$	$2.9 \pm 0.3 \times 10^4$	$1.2 \pm 0.4 \times 10^4$	$4.9 \pm 0.5 \times 10^4$
$N_{30}P_{30}$+ MERS 200 ml/ha	0–20	$3.8 \pm 0.3 \times 10^6$	$3.1 \pm 0.2 \times 10^4$	$1.5 \pm 0.3 \times 10^4$	$5.5 \pm 0.8 \times 10^4$
$N_{30}P_{30}$+ MERS 500 ml/ha	0–20	$3.9 \pm 0.4 \times 10^6$	$4.1 \pm 0.5 \times 10^4$	$1.5 \pm 0.4 \times 10^4$	$5.6 \pm 0.3 \times 10^4$

Compared to the control variant (100 %), the abundance of heterotrophic bacteria in the treatment with 500 ml MERS was 260 %, the abundance of actinomycetes 341 %, the occurrence of yeasts 187 % and that of fungi 186 %. The differences between the various treatments were the most pronounced for the group of actinomycetes. While the number of actinomycetes in 1 g soil treated with 50 ml/ha MERS was $1.2 \pm 0.3 \times 10^4$, nearly twice as many actinomycetes ($4.1 \pm 0.5 \times 10^4$) were isolated in the 500 ml treatment.

In all analysed variants, the abundance of heterotrophic bacteria was highest compared to the other groups of microorganisms (Table 3).

Treatment with the microbiofertiliser "MERS" not only affected the parameters of soil fertility, in particular microbial community abundance, but also increased the productivity and quality of "Almaly" winter wheat in meadow light sierozem soils.

Regardless of the variants, the grain yield of "Almaly" winter wheat varied over the three experimental years (Table 4). Whereas the grain yield in 2007 and 2009 ranged from 3.9 to 5.0 t ha, it exceeded 4.6 t ha^{-1} in all variants in 2008, with a maximum of 5.76 t ha^{-1}. In all 3 years, the highest grain yield was measured in the 500 ml/ha MERS treatment.

Table 4 Influence of the microbiofertiliser "MERS" on the yield capacity of winter wheat of the "Almaly" variety

	Grain yield (t·ha^{-1})			Average yield (t·ha^{-1})	Additional yield (t·ha^{-1})
	2007	2008	2009		
Control	33.9	44.69	33.91	4.17	–
$N_{30}P_{30}$+MERS 50 ml/ha	33.94	55.24	44.39	4.52	0.35
$N_{30}P_{30}$+MERS 100 ml/ha	44.3	55.35	44.51	4.72	0.55
$N_{30}P_{30}$+MERS 200 ml/ha	44.8	55.73	44.91	5.15	0.98
$N_{30}P_{30}$+MERS 500 ml/ha	55.0	55.79	55.03	5.27	1.1

Table 5 Impact of applying the microbiofertiliser "MERS" on quality parameters of winter wheat "Almaly"

	Grain moisture (%)	Impurity (%)	Full-scale mass ($g \cdot l^{-1}$)	Vitrescene (%)	Gluten (%)	Gluten quality
Control	10.3	0.1	780	54	10.4	95
$N_{30}P_{30}$+MERS 50 ml/ha	12.5	0.3	790	60	22.5	96
$N_{30}P_{30}$+MERS 100 ml/ha	13.4	0.3	795	65	24	97
$N_{30}P_{30}$+MERS 200 ml/ha	13.9	0.3	810	75	26.5	98
$N_{30}P_{30}$+MERS 500 ml/ha	14.0	0.3	805	78	27	98

Average yield (from 2007 to 2009) of winter wheat in the control variant was 4.17 t ha^{-1}. Due to treatment with different concentrations of MERS, the average yield increased up to 5.27 t ha^{-1} in the 500 ml/ha treatment. A grain yield surplus was determined in all MERS treatments. The highest surplus (0.98 and 1.1 t ha^{-1}) was observed using 200 and 500 ml/ha of MERS, respectively. Similar positive effects were found for the quality indicators of winter wheat grains.

Gluten content in the grains varied between 22.5 and 27 %; in the untreated control variant, gluten content in the grains was only 10.4 %. In the treated variants, the percentage gluten in grain is at least twice as high as in the control variant. However, no differences in the quality of the gluten were detected. The gluten elasticity was satisfactorily weak in all variants, ranging from 95 to 98 (Table 5).

With the exception of gluten quality and impurity, the other analysed quality parameters (full-scale mass, vitreousity and grain moisture, were influenced positively by treatment with the microbiofertiliser. For all these quality parameters, the highest values were also observed at the highest application rates (200 and 500 ml/ha).

Based on the enhancement in yield and grain quality and depending on the cost of the different MERS application rates, the net income per unit growing area of winter wheat "Almaly" also rose. The highest net income following application of the microbiofertiliser "MERS" ranged from 13,300 to 13,800 tenge/ha or 90.8 to 94.2 USD/ha in the variant with 500 ml/ha MERS.

8 Conclusion

The results of the three-year field study suggest that application of different concentrations of the microbiofertiliser "MERS" can influence soil chemical parameters, the abundance of the soil microbial community, and the yield and quantity of winter wheat of the "Almaly" variety. All analysed parameters increased in the variants treated with the microbiofertiliser. The highest impact was determined for all measured parameters in the variant treated with 500 ml/ha of the microbiofertiliser "MERS". Bearing in mind the cost of applying MERS, the surplus in yield and quality of the winter wheat "Almaly", and the impact on the soil microbial community and soil quality, the optimal microbiofertiliser concentration in this experiment was 500 ml/ha, resulting in a net income of 13,800 tenge/ha. For the widely accepted and effective use of the microbiofertiliser "MERS", further studies are required to evaluate its effect on soil microbial functions such as enzyme activities and its effectiveness on different plant species, soil types and management systems.

References

Abdel-Razzak HS, El-Sharkawy GA (2013) Effect of biofertilizer and humic acid applications on growth, yield, quality and storeability or two garlic (Allium sativum L.) cultivars. Asian J Crop Sci 581:48–64

Arshad MA, Martin S (2002) Identifying critical limits for soil quality indicators in agro-ecosystems. Agric Ecosyst Environ 88:153–160

Doran JW, Zeiss MR (2000) Soil health and sustainability: managing the biotic component of soil quality. Appl Soil Ecol 15:3–11

Jog R, Nareshkumar G, Rajkumar S (2012) Plant growth promoting potential and soil enzyme production of the most abundant Streptomyces spp. from wheat rhizosphere. J Appl Microbiol 113:1154–1164

Nihorimbere V, Ongena M, Smargiassi M, Thonart P (2011) Beneficial effect of the rhizosphere microbial community for plant growth and health. Biotechnol Agron Soc Environ 15(2):327–337

Noble A et al (2005) Enabling communities in the aral sea basin to combat land and water resource degradation through the creation of bright spots. ADB supported project: RETA 6208 (2005–2008). Final report (IWMI, ICARDA, ICBA) http://brightspots.centralasia.iwmi.org/2008.pdf. Accessed on 17 February 2013

Odham G, Tunlid A, Valeur A, Sundin P, White DC (1986) Model system for studies of microbial dynamics at exuding surfaces such as the rhizosphere. Appl Environ Microbiol 52:191–196

Parke JL (1990) Root colonization by indigenous and introduced microorganisms. In: Keister DL, Gregan PB (eds) The rhizosphere and plant growth. Kluwer Academic Publishers, Durdrecht, pp 33–42

Parr JF, Papendick RI, Hornick SB, Meyer RE (2002) Soil quality: attributes and relationship to alternative and sustainable agriculture. Am J Altern Agric 7:5

Pięta D (1999) Initial studies of populations of fungi and bacteria in the soil under influence of the cultivation of spring wheat and winter wheat in a growth chamber. Acta Agrobot 52(1–2):161–166

Pięta D, Patkowska E (2003) The role of antagonistic fungi and bacteria limiting the occurrence of some phytopathogens inhabiting the soybean soil environment. Electron J Pol Agric Univ Horticulture 6:2

Piotrowska A, Dlugosz J, Zamorski R, Bogdanowicz P (2012) Changes in some biological and chemical properties of an arable soil treated with the microbial fertilizer UGmax. Pol J Environm Stud 21(2):455–463

Raaijmakers JM, Paulitz TC, Steinberg C, Alabouvette C, Moënne-Loccoz Y (2009) The rhizosphere: a playground and battlefield for soilborne pathogens and beneficial microorganisms. Plant Soil 321:341–361

Sadeghi A, Karimi E, Dahaji PA, Javid MG, Dalvand Y, Askari H (2012) Plant growth promoting activity of an auxin and siderophore producing isolate of Streptomyces under saline soil conditions. World J Microbial Biotechnol 28:1503–1509

Schenck zu Schweinsberg-Mickan M, Müller T (2009) Impact of effective microorganisms and other biofertilizers on soil microbial charcteristics, organic matter decomposition, and plant growth. J Plant Nutr Soil Sci 172:704–712

Schoruvitz R, Zeigler H (1989) Interaction of maize roots and rhizosphere microorganisms. Z Pflan-zenkrankh Bodenh 152:217–222

Smolentseva E, Sulejmenov MK, Saparov AS, Pachikin KM, Balgabaev NN, Kusainova MD, Bekbaev UK, Rukhovich OV, Lukin SM, Schindler U, Müller L (2011) Ocenka kachestva i potencial'noj urozhajnosti pochv v global'nom masshtabe. Pochvovedenie i Agrokhimija (4):81–91

Sokolov AV (1975) Agrochemical research methods of soil. Publisher's Science (in Russian), 78p

Usmanov S (2008) Agrochemical, biological, ecological and economic efficiency of microfertilizers of series MERS // recommendations on the use of microfertilizers of series MERS in crop cultivation. Almaty Publ House, Olzha, p 36

Water Treatment Systems for Agricultural Water Supply

Valeriy A. Tumlert

Abstract Review of the main water treatment methods recommended for application in water supply processes at rural settlements with a daily water consumption of less than 1,000 m^3, and types of plants where these methods are in current use. One of the factors influencing the productive activities and social development of rural settlements is the water supply. A lot of attention is paid to problems related to the drinking water supply. In Kazakhstan, most water pipes and water treatment plants began operation or were overhauled more than 30 years ago. Outdated water treatment systems do not provide a high-quality water supply. More than 25 % of operational water pipelines do not meet the sanitary requirements. More than 90 % of rural settlements in Kazakhstan are supplied from underground sources (wells, percolation wells, tapping of springs); only the remaining 10 % are from surface sources. The conditions of rural water use (less than 1,000 m^3 a day), the lack of skilled repair personnel and the intermittent electricity supply necessitate the use of compact, energy-saving and nonchemical means of water treatment. Looking through patents and the literature for recent years shows that the basic methods of domestic water improvement are still clearing, decolouration, softening, fluoridation, defluoridation, desalination and disinfection. Thus, taking into account the above review of the main water treatment methods and the types of plants at which reverse osmosis treatment can be applied the following methods are recommended in the water supply systems of rural settlements with a daily water consumption of less than 1,000 m^3. For water clearing and decolouration, low-rate filters should be applied. Local close-burning shungites and zeolite ("Koksu" deposit) can be used as a percolation bed. For hard water with a high fluoride and iron content use the "Struya" water treatment unit. For softening, apply liming, soda lime and the sodium cycle (equipment produced by LLC "Membrane technologies C.A."); for deferrisation and permanganating, filter with preliminary aeration using equipment produced by "Aikos"; for

V. A. Tumlert (✉)
Kazakh Scientific Research Institute of Water Economy, 12 Koygeldy Street, Taraz, Kazakhstan
e-mail: tumlert_e@mail.ru

L. Mueller et al. (eds.), *Novel Measurement and Assessment Tools for Monitoring and Management of Land and Water Resources in Agricultural Landscapes of Central Asia*, Environmental Science and Engineering, DOI: 10.1007/978-3-319-01017-5_40,

defluoridation, apply an ion-exchange pressure filter with an alumina filter bed (Al_2O_3); for disinfection, apply a UV sterilizer (LLC, "Membrane technologies C.A."); for desalination/deionization, apply reverse osmosis modules arranged in the internal chamber of the well casing (LLC "Kazakh Water Economy Research Institute").

Keywords Agricultural water supply · Water treatment · Quality of water

1 Introduction

One of the factors influencing the productive activities and social development of rural settlements is the water supply. A lot of attention is paid to problems related to the drinking water supply. In Kazakhstan, most water pipes and water treatment plants began operation or were overhauled more than 30 years ago. Outdated water treatment systems do not provide a high-quality water supply. More than 25 % of operational water pipelines do not meet the sanitary requirements.

The quality of surface water almost throughout all waterways does not meet water standards. The main waterways of Kazakhstan are polluted by transboundary inflows. There are more than 700 potential pollution sources for subsurface waters within the borders of the Republic. 241 of them have an immediate influence on the hydro-geochemical conditions of subsurface waters.

Thus, the poor condition of water supply plants and systems significantly degrades the water quality. Foodstuffs are exposed to dangerous micro-elements and chemical compounds containing organochlorides, malignant bacteria, toxins and heavy metals.

To solve this problem, events need to take place to promote the building of local water treatment systems with up-to-date equipment. If existing municipal water supply and collecting systems are renovated and new ones constructed, this will cause positive changes to the social, economic, and ecological environment and to public health.

2 Water Treatment Systems for Agricultural Water Supply

More than 90 % of rural settlements in Kazakhstan are supplied from underground sources (wells, percolation wells, tapping of springs); only the remaining 10 % are from surface sources.

The conditions of rural water use (less than 1,000 m^3 a day), lack of skilled repair personnel and the intermittent electricity supply necessitate the use of compact, energy-saving and nonchemical means of water treatment.

One of the most important spheres of public welfare is providing a high-quality water supply. The quality of water supplied to consumers depends on many factors: the operational stability of the water intake facilities, the technical condition of the waterworks and water-distributing systems, the appropriate operation of the water treatment systems, the observance of the water treatment works' running regime and other components of the water supply systems.

Looking through patents and the literature for recent years shows that the basic methods of domestic water improvement remain clearing, decolouration, softening, fluoridation, defluoridation, desalination and disinfection. Table 1 shows in brief the characteristics of the basic water improvement methods recommended for use in water supply schemes at rural settlements with a water consumption of 1,000 m^3 a day, and the kinds of structures on which these pretreatment methods can be used.

Clearing and decolouration. Clearing and decolouration processes are mainly applied for surface waters and involve dredging and removing colloidal ions from influent water. Depending on the concentration of suspended matters and colloidal

Table 1 Basic methods of water pretreatment

Qualitative characteristics	Physical–chemical methods	Kinds of structures and equipment
Turbidity	Curdling, flocculating agents treatment, distilling, filtering	Low-rate filters, clearing reagents, sedimentation tank
Taste and smell	Carbon filtering, preliminary chlorination with pre-ammonation, potassium permanganate treatment, ozone treatment	Contact tank, high-rate filter, wastewater clarifying tank
Bacterial pollution	Chlorination, ozone treatment, UV treatment	Directly in fresh water tank to network through UV treatment plants
Fluorine deficiency (less than 0.5 mg/l)	Fluoridation	Fluoridation plant located forward of the fresh water tank
Fluorine excess (more than 1.5 mg/l)	Defluoridation	Defluoridation plant located forward of the fresh water tank
Excess iron	Aeration, chlorination, recausticising, curdling, potassium permanganate treatment, cation polishing	Air or coagulant supply into filters
Excess hard salts	Decarbonisation, soda lime softening, ion exchange	Batchers, H–Na cation exchange demineralization plants
General salt content above the norm	Ion exchange, electrodialysis, distilling, filtering etc.	Batchers, electrodialysis plants

particles, the water may be treated either in two steps (desilting and filtering) or in one step (filtering). In some cases it is possible to use coagulants (aluminium sulphate, iron chloride) to intensify a process. Flocculation results from coagulation, i.e., fine particles floating in water combine to form larger particles which are able to precipitate. Low-rate filters which meet the demands listed on flow charts for this category of users can be used to clear and decolour water when only marginal volumes are used. As a filter bed, it is possible to use local filtering materials (carbonizing shungite, zeolite etc.) mined in Kazakhstan. Filtering materials are produced by LLC “Shungite” from the Koksu coal field (Work report 2004).

Issues related to drinking water improvement are paid great attention abroad and in CIS countries. Particularly, extensive studies are being carried out by scientists at the “VODGEO” research institute on the effective use of new adsorbents of types AG-3 and AG-5. Advanced specific materials produced under the brands Norit, F and SKKD are also being approved (Zhurba 2004). The technology of bio-sorption after the purification of surface water is a promising new approach.

The “Struya” water treatment unit, meant for surface water, is successfully being applied in the treating systems of rural settlements with low and medium water use. The flow chart involves coagulants, clearing, filtering and disinfection of water. The units are produced with different capacities: 100, 200, 400, 800 m^3 a day. Requirements are influent water: suspended substances between 1,000 and 2,500 mg/l; no limits on chromaticity; less than 5 mg/l of fluorine; less than 50 mg/l of iron; a hardness of less than 15–20 mg-eq./l.

For deep treatment purposes, including organic treatments, it is better to use flotation with polyaluminium chloride as a coagulant. It leads to an increase in plant capacity by up to almost 30 % and a reduction in coagulant by 2–3 times in comparison with aluminium sulphate. The qualitative characteristics of the treated water improve by about 40 % as against desilting (Myasnikov et al. 2004).

Recently, work on the joint application VPK-402 and on coagulants containing mineral iron and aluminium, which was being performed by the Pamphilov Research Institute in Rostov (Russia), has finished (Mikheyeva 2004). The municipal utility “Azovvodokanal” has started for the first time to use the coagulant aluminium oxychloride (produced in Azov) for water treatment. This allows the quality of the supplied water to be improved, significantly reducing the operational work, chemical coagulant feeding and loading. The polyaluminium chloride “Aqua-Aurat™30” showed high clotting properties during pilot research at water treatment plants at an industrial complex in Atchinsk; now it is successfully being applied in other cities such as Kamensk, Shakhty and Novocherkassk (Myasnikova 2004). Besides this, the coagulants CA, KF, AVR, PAX ALG and the flocculating agents Praestol (“Schockghazuen”, “Saitech”) are being tested in a process of water treatment intensification.

Softening. Influent water hardness is caused by salts of calcium and magnesium. The general hardness of water used for domestic needs must be no more than

7.0 mg-eq./l. There are many ways of reducing hardness, but in our conditions the following are more preferable:

- with a view to removing carbonate hardness: limed decarbonisation;
- with a view to removing carbonate and non-carbonate hardness: softening, sodium cycle softening.

The range of limed softening is determined by the degree of water solubility of calcium carbonate and magnesium hydrate. Practically, water softened using the soda lime method has a residual hardness no more than 0.5–1.0 mg-eq./l. The salt contained in influent water is reduced either by lime or soda lime treatment.

The uneven distribution of interflow and sweet underground water in Kazakhstan suppresses economic progress in some provinces. About 3,000 rural settlements in the country use mineralized underground water for their water supply.

Figure 1 shows the distribution of water sources by degree of their mineralization per total amount in percent

Deferrisation and permanganate filtering. Water deferrisation is necessary when water contains more than 0.3 g/l of iron. The deferrisation of underground water is carried out by filtering in combination with a pretreatment method: simplified aeration with use of special means and oxidizing agents. Removing manganese involves transformation into not readily soluble compounds, desilting and filtering through a sand filter bed. In the latter case the sand's thickness must be no less than 1,500 mm.

At present, deferrisation and the permanganate filtering of underground water using filtering methods with preliminary aeration are under further development. In some cases the dual-action technology developed by LLC "Aickos" allows both pollutants to be removed in one technological process.

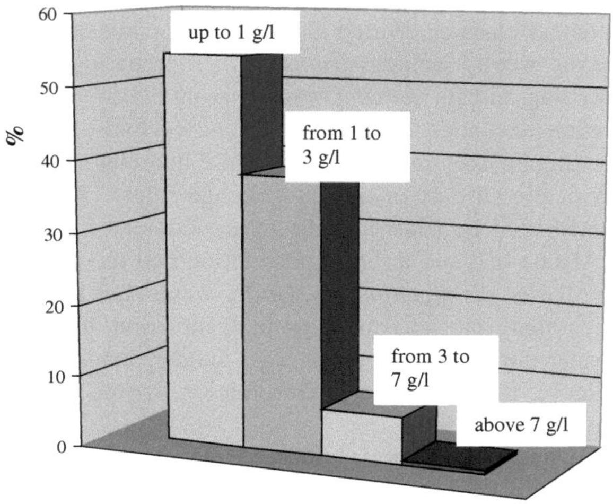

Fig. 1 Distribution of water sources by the degree of their mineralization

Among the methods currently under research for treating underground water containing hard-oxidisable iron and anthropogenic admixtures, some are based on multiple use of the oxidizers H_2O_2, $KMnO_4$, and O_3 in water-bearing beds, or biological methods using biocoenosis and newly designed fibrous-granular bioreactor-filter beds. Particular attention is being paid to water treatment using bioactive substances such as boron, bromine, iodine or silicon. However, because of their cost, labour-intensiveness and operating complexity, they might find little use in rural settlements with a low water consumption and insufficient financial resources; their introduction is rather problematic. Recently, complex water treatment technology based on the primary bioreactor and hydroautomatic inhomogeneous bed filters AFPZ-4 (AFPZ-1), sequential filtering through the boron-selective resins S-108 and AB-17-B and sorption material of activated carbon AG-3 has been developed by specialists at the "VodGeo" Research Institute. Water is cleared of iron, manganese, boron, bromine and nitrates simultaneously. Integration tests on iodine and bromine release methods have been realized in recent years (Fesenko 2004). Thus, a series of modified sorbents are proposed such as KSM silica gel, alumina, Teflon, the anionites AV-17 and AMP and PAN-AV-17 fibrous sorbents. However, while these methods are at the trial stage, they cannot be applied in water treatment practice. Among the metal hydroxide sorbents, one which stands out is aluminium hydroxide, generated from the hydrolysis of the widely used coagulant aluminium sulphate. The oxidation-sorption method, where the oxidizing agent is electrolytic sodium hypochlorite and the sorbent is aluminium hydroxide, allows iodine to be removed completely using a single-step water treatment. The bromine concentration decreases up to 93–98 % during three-step water treatment. During the water treatment process, a sediment forms containing extracted bromine and iodine compound. It can be utilized in iodine-bromine chemical plants.

Fluoridation and defluoridation. Domestic and drinking water fluoridation is necessary when the fluorine content in the water supply source is less than 0.5 mg/l. Fluorine-containing reagents must be introduced either before the filters and contact clarifier or after the treatment plant before disinfection. The fluoridation plant usually includes equipment for preparing the reagent solution, for chemical feeding and for influent water mixing. There are several defluoridation methods: anion exchange (activated alumina or hydroxyapatite), sorption of the emitted sediment (aluminium or manganese hydrates). Water defluoridation is carried out with pressure or open ion-exchange filters. The defluoridation methods may be intensified by trials and the introduction of filtering through an alumina granule (Al_2O_3) bed and an inert percolation bed modified with a coagulant solution.

Water disinfection. As a rule, water that has already been through the other treatment stages has undergone disinfection. In some cases disinfection is the only water purification measure (e.g., underground water). Disinfection may be carried out by means of water chlorination, ozone treatment or bactericidal radiation treatment.

Chloride of lime is used as a disinfectant in processes where the water consumption is low. The world practice of neutralizing water is evidence of the negative influence of active chlorine on a reservoir's biocoenosis in terms of the surface water supply and formation of toxic organochloric compounds and chloramines. In a number of cases chlorine hyperconcentrations have to be kept at a rate of 2–3 mg/l, and in exceptional cases 7 mg/l are necessary to accomplish normative micro-biological characteristics.

The real alternative to chlorination is UV disinfection. The UV method guarantees the effective neutralization of water including chlorine-resistant microorganisms, and the detrimental side products meet all contemporary demands.

Bactericidal radiation treatment is applied to underground water on condition that the physicochemical characteristics are kept permanently at the drinking water standard. At the same time the coli index of treated water must be more than 1,000 un./l, and the iron content less than 0.3 mg/l.

UV disinfection (flow type) and electrodialysis (portion type) "UUFOV" units produced by "Aickos" and "Membrane technologies" (Almaty) have been successfully applied in practice in Kazakhstan.

Ozone treatment is used with a special basis and when the water quality is not at risk of deterioration. The essential dose of ozone for underground water is 0.75–1 mg/l, and for filtered water it is 1–3 mg/l. The ozonators produced by LLC "Membrane technologies" do not require expensive air treatment systems and differ from their foreign equivalents in that prices are lower, maintenance simpler and power inputs lower: the energy intensity for the ozone treatment of water is 0.010–0.020 kWh/ m^3.

Desalination. In a process of desalination all soluble salts are removed from water to a certain extent depending on consumer demands. Desalination is carried out using several methods: distillation in evaporators (thermal method, ion exchange method, electrodialysis, reverse osmosis) and the membrane method.

Natural water desalination using electrodialysis apparatus produced in the former USSR and the mobile water demineraliser "PEDU-100" equipped with the electrodialysis apparatus Э0.400.01 has been investigated for several years at the Water Economy Research Institute (Kazakhstan). The electrodialysis apparatus Э0.400.01 was equipped with membranes modified in various ways and was tested on salted influent water in some regions of Kazakhstan.

Today the domestic companies "Aickos", LLC "Membrane technologies" (Almaty) have succeeded in solving the problems of desalination, softening and disinfection in Kazakhstan.

LLC "Membrane technologies" produces a great range of electrodialysis and reverse osmosis equipment: from domestic apparatus to water demineralization units with high capacity and industrial plants for nonchemical processing of deep desalted water. The company also puts on the market desalting plants with a capacity from 10 to 1,200 m^3 a day installed in heated 20- and 40-feet containers (Fig. 2).

At present the Kazakh Water Economy Research Institute of Taraz city, Kazakhstan is working on technology for an underground water demineralization plant using reverse osmosis modules build into the internal chamber of a well's casing pipe (Patent RK 2010).

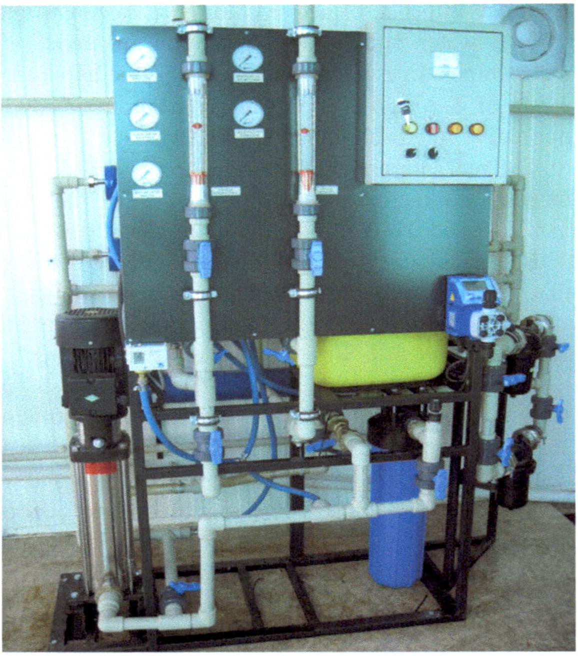

Fig. 2 Reverse osmosis water treatment produced by LLC "Membrane technologies"

The process used in this water demineralization technology is shown in Fig. 3. A high-pressure submersible pump is attached by an outlet pipe to a pumping device divided into suction and feeding zones and the pump is placed in the casing (Tumlert 2009). Connected reverse osmosis membranes of type RE 4040-BLN are attached to the sealed head wall on the well head. The number of membranes depends on the required capacity. Feed water is pumped using the submersible pump from the suction zone to the feeding zone then passed through the reverse osmosis membranes under the above reverse osmosis pressure. The treated water (permeate) is directed to the consumer and brine (concentrate) is utilized.

Advantages of this technology:

- No need for water-lifting pipes as mineralized underground water is pumped along casing;
- No need for high-pressure purge pump for flushing mineralized water through membrane elements; submersible pump is used for this purpose and for water discharge;
- No need for construction and use of water desalination plant in view of reverse osmosis membrane allocation in internal chamber of casing, where ambient temperature is always above +5 degrees centigrade.

The advantages of the reverse osmosis desalination in a well using underground water technology will allow potable water to be produced with minimal inputs. It is of great significance, especially in agriculture where the uneven population

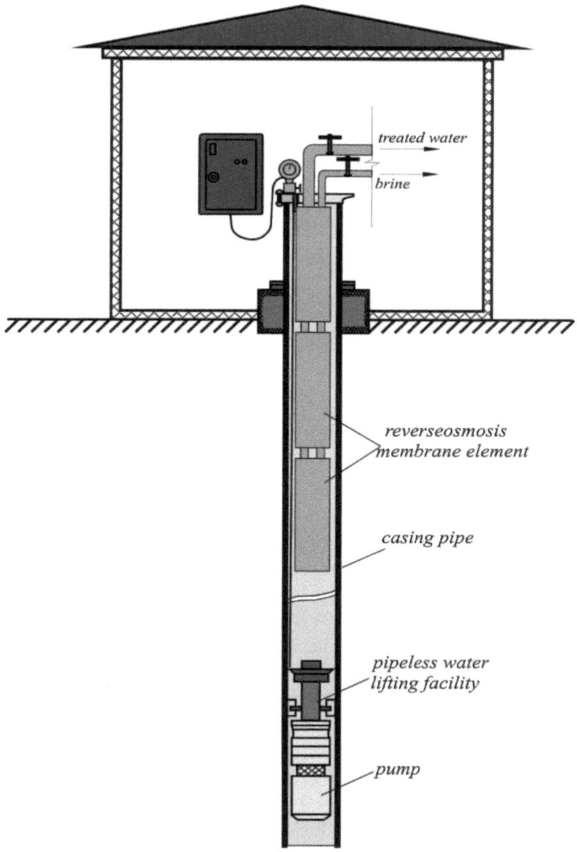

Fig. 3 Method used in underground water demineralization technology (LLC "Kazakh Water Economy Research Institute")

density, limited transmission facilities and lack of experienced technicians make the application of desalination plants difficult.

3 Conclusions

Thus, taking into account the above review of the main water treatment methods and types of plant in which reverse osmosis treatment can be applied, the following methods are recommended for use in water supply processes at rural settlements with a daily water consumption of less than 1,000 m^3:

- for water clearing and decolouration, apply low-rate filters. Local close-burning shungites and zeolite ("Koksu" deposit) can be used as a percolation bed. For hard water with a high content of fluoride and iron, use the "Struya" water treatment unit;

- for softening, apply liming, soda lime and sodium cycle (equipment produced by LLC "Membrane technologies C.A.");
- for deferrisation and de-permanganating, filter with preliminary aeration using equipment produced by "Aikos";
- for defluoridation, apply an ion-exchange pressure filter with an alumina filter bed (Al_2O_3);
- for disinfection, apply a UV sterilizer (LLC "Membrane technologies C.A.");
- for desalination/deionization, apply reverse osmosis modules arranged in the internal chamber of the well casing (LLC "Kazakh Water Economy Research Institute").

References

Fesenko LN (2004) Water treatment by iodine and bromine oxidized –sorption method. Water Supply Sanit Eng 9:9–14

Mikheyeva IV (2004) Conditions of water supply and potable water quality in Rostov province. Water Supply Sanit Eng 1:6–8

Myasnikov IN, Potanina VA, Gorshenin AN, Netrebba MP, Voropayev VA, Fetissov AG (2004) Modernization of potable water treatment by pressure flotation. Water Supply Sanitary Eng 9:7–8

Myasnikova GB (2004) Evaluation of reagent clotting activity in JSC "Achinskiy aluminous industrial complex". Water Supply Sanit Eng 1:14–16

Patent RK (2010) Patent of the Republic of Kazakhstan #23118 B of November 15, 2010, Reverse osmosis in-well underground water desalination method. Inventors: Tumlert VA, Rybintzev, Yu P, Grankin Yu, Ya, Tumlert EV

Tumlert VA (2009) Technology of water –lifting along casing by submersible pump and its effectiveness. Water Resour Water Usage 2(61):25–28

Work report (2004) Report of the research project on "Working out scientific and technical basis of monitoring and arid pastures water supply, rational circuits and water supply technologies", state registration number 0101 RK 00412

Zhurba MG (2004) Water treatment for domestic and potable water consumption. Water Supply Sanitary Eng 2:10–13

Concentration of Heavy Metals in Irrigated Soils in Southern Kazakhstan

Azimbay Otarov

Abstract Currently, one major negative factor in the reduction of soil fertility is contamination with heavy metals, pesticides and other pollutants. In this regard, the main purpose of this work is to study and evaluate soil pollution with heavy metals in irrigated areas of Southern Kazakhstan and to determine the key geochemical parameters of heavy metals. One research object was the soils of large rice farms in the Shiely area of irrigation. The study was conducted using the method of large-scale soil analysis and mapping. Among the environmentally hazardous mobile forms of metals in the soils of the Shiely irrigation area, Pb has the largest share (54 %), then Ni (22 %). Zn and Cu have 12 and 10 % respectively. The share of Cd is only 2 %. Based on the analysis of cartographic material, it has been determined that on the territory investigated, Cu is found in 5 groups of soil which contain more than 3.15 mg Cu per 1 kg of soil. A high concentration of Pb is found in the group of soils, at 9.0–15 mg per 1 kg of soil. Ni is found in high concentration in soils: 4.6–6.0 mg/kg. Thus, based on analysis of the data obtained, we can conclude that for all the mobile forms of the elements studied, one major aspect is the predominance of the accumulative character of distribution in the soil profile is observed, and gross forms are characterized by eluvial-illuvial processes. Environmentally hazardous mobile metal forms available for the plants are subjected to the influence of irrigation to a greater degree than their gross forms.

Keywords Soil · Irrigation · Rice · Heavy metals · Kazakhstan

A. Otarov (✉)
Kazakh Research Institute of Soil Science and Agrochemistry (U. U. Uspanov), Al Faraby Ave 75b 050060 Almaty, Kazakhstan
e-mail: azimbay@bk.ru

L. Mueller et al. (eds.), *Novel Measurement and Assessment Tools for Monitoring and Management of Land and Water Resources in Agricultural Landscapes of Central Asia*, Environmental Science and Engineering, DOI: 10.1007/978-3-319-01017-5_41,

1 Introduction

The key areas with irrigated lands are mainly spread over four regions in the south and south-east of the country. The productivity of irrigated lands in Kazakhstan is 4–6 times higher than that of non-irrigated lands. Irrigated land makes up about 6 % of arable lands and provides more than 30 % of total agricultural production in the country. However, soils in the valleys of major rivers were extensively irrigated without sufficient scientific justification in the Soviet period, and this has resulted in the inefficient use of water resources, their almost complete exhaustion, soil surface degradation (particularly salinisation, waterlogging and simultaneous desertification), and a reduction in the profitability of agricultural production.

Currently, soil contamination with heavy metals, pesticides and other pollutants is one of the main negative factors resulting in the reduction of soil fertility. Usually the irrigated area is located in hydromorphic landscapes which are geochemically subordinate, and therefore, they are inclined to contamination. The results of our research showed that the irrigated soils were contaminated with heavy metals, particularly Pb, Ni and Cu (Otarov et al. 2007). Moreover, the deterioration of soils' environment and drainage conditions has also led to a decrease in their protective capabilities with respect to Pb and Ni up to 3.3 and 4.1, respectively (Otarov 2005).

Water in rivers used as sources of irrigation is becoming more and more polluted, due to the increased anthropogenic pressure on the environment. According to our data, the ecological condition of water in the Kyzylorda region is of particular concern because of the increased content of Pb, and Ni which migrate from irrigation water to the ground water (Otarov et al. 2007). All these problems, as well as a decrease in the level of technological practices, general land cultivation and the level of soil tillage, have finally resulted in the contamination of the raw rice. Rice products contain high Pb and Ni contents (Otarov et al. 2006).

In this context, the main research aim is to study and evaluate soil pollution with heavy metals in irrigated areas of Southern Kazakhstan and identify the main geochemical parameters of heavy metals. In addition, despite the great diversity of soil surfaces in irrigated areas in the Republic, the problem has not been solved of how to measure regional background levels of heavy metals in irrigated soils. With this in mind, the aim is to study and systemize the data on the background content of heavy metals in irrigated soils on a regional level. This is also important and necessary for evaluating the sustainability and stability of irrigated ecosystems in the face of global and regional anthropogenic influences.

The relevance of this work is particularly great in connection with the forthcoming accession of Kazakhstan to the World Trade Organization, which observes the "Agreement on Application of Sanitary and Phytosanitary (SPS) Measures", dated 1994, which regulates the food and agricultural products and is aimed at ensuring food safety. Accession to the WTO assumes the use of internationally recognized and truly objective standards. Such international standards, which are based on the results of scientific analysis of risks to human health, have been adopted by the following organizations dealing with international standards: the "Codex Alimentarius" Commission, implementing the joint FAO and WHO Programme on food safety standards, the International Epizootic Office (IEO), and the World Plant Protection Convention (WPPC).

2 Research Objects and Methods

The research object includes the soils of the big rice-growing farms in the Shiely irrigation area: "Shiely-Avangard" LLP, "Yntymak-Gigant" LLP, "Kaptagay and K" LLP, "Zhana-Bestam", the "Ybyray Zhakay" and "Zhana Akzhol" farms, and small farms (Fig. 1). The coordinates of the centre of Fig. 1 are given in decimal degrees: the longitude is 66.477 E and the latitude is 44.366 N.

In the past, rice was grown on alluvial-meadow, meadow-marsh, marsh and takyr soils in the Shiely area. As a result of the long-term use of these soils for rice

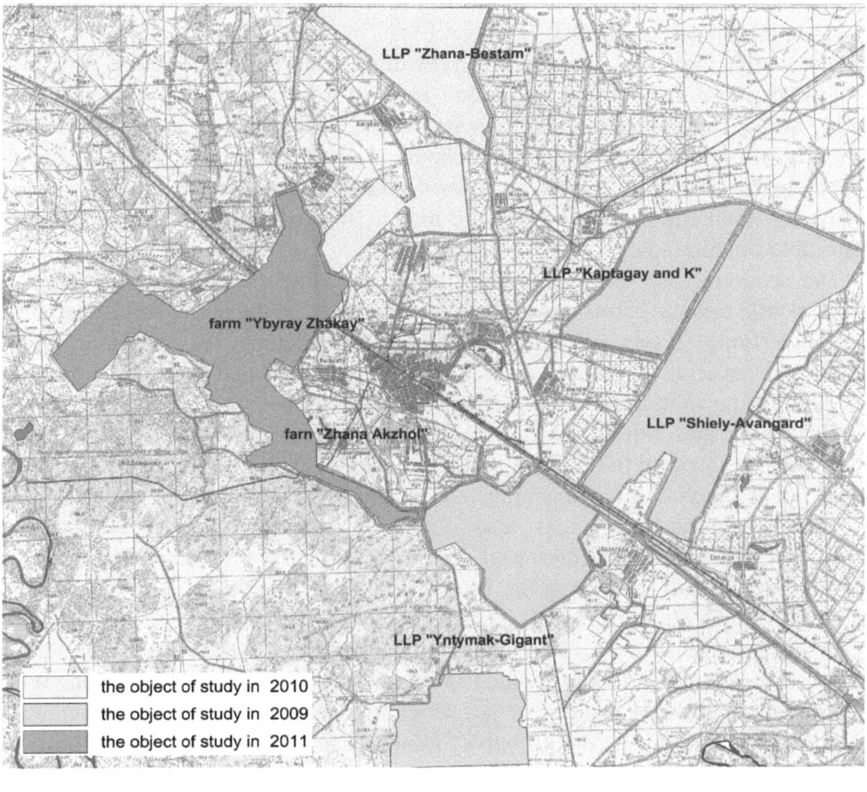

Fig. 1 Research objects

growing, according to the soil classification used by Kazakhstan's soil scientists (Borovsky et al. 1959; Karajanov et al. 1973; Volkov 1983), these soils have been transformed into irrigated (rice) marsh soils. Changes in soils under this crop are associated with specific conditions of its cultivation (permanent flooding).

Rice is a crop which requires water in the plots permanently, throughout the growing season. The duration of continuous flooding of rice is 90–110 days depending on the rice variety. During the rice growing season, large amounts of water—up to 14,000 m^3/ha (Borovsky and Pogrebinsky 1958) are filtered by the soil. This causes fluctuations in the groundwater level from 2.5 m at the beginning of irrigation to joining with irrigation water at normal irrigation, and in the autumn the level falls to its previous depth. A seasonal cycle is observed.

Under periodic flooding, the specific condition is observed which is determined by the characteristics of the soil-forming process, taking place in the flooding and subsequent drying of soils. The constant switching between cycles of flooding and drying, developing contrasting modes, inevitably causes the weakening of specific systems which have formed in the soil. Particularly dramatic changes occur in the oxidation-restoration soil regime, which largely determines the nature of the soil-forming process, in particular the migration of elements in the soil profile, the processes of humus formation, the nutrient regime, etc. (Sharapov 1977; Neunilov 1961; Sheudshen et al. 1998).

Thus, after flooding, as a result of prevailing restoration conditions, two main groups of processes occur which result in the formation of a specific profile in paddy soils. The first is mobilization processes, which lead to an increase in the mobility of organic matter and some chemical elements, including heavy metals, and their transition into low-valence forms. These processes are accompanied by relatively rapid destruction of organic matter, loss of nitrogen and other easily restorable compounds.

The second group is the migration processes which result in the removal of these mobile organic forms and chemical elements from the upper soil horizon into the underlying layers or inter-horizon redistribution. In particular, many researchers, such as Yakupova et al. (1973), have noted that the humus profile is extended somewhat. Significant amounts of humus (about 1 %) can be found at a depth of 1 m and below, which is not typical for other types of delta soil. In addition, fine disperse particles of soil (Borovsky et al. 1958) are also subjected to the process of eluviation (leaching). As a result of mobilization and migration processes, significant changes also occur in the mineralogical composition, properties and profile distribution of the fine disperse fraction, especially in those regions where soil waterlogging is not a natural condition (Kiriyenko 1984; Chizhikova and Zhuravleva 1997; Chizhikova et al. 1997).

In general the process of formation and existence of the rice-marsh soils is characterized by a periodicity which is related to the rice-alfalfa rotation. When rice is in the fields, the process of waterlogging occurs due to the excessive water in these soils, and during the growth of alfalfa or arable crops they transform into meadow soils because of the lack of excessive water on the surface. As we can see, soil-forming processes are very intense in rice-marsh soils, as these soils have a relatively high rate of mobilization and migration processes. Therefore, the level of fertility of paddy soils should be monitored regularly and a wider range of soil properties should be studied. The survey involved widely used and well-proven methodological approaches, and complex methods were used to study the soil. In particular, morphological and profile methods were used, which are the basic methods used in field soil studies. When conducting the soil and environmental surveys, we used "Guidelines for conducting soil and environmental surveys (1979)". The ecological status of soils was determined in accordance with the requirements of soil sampling in general and local pollutions (GOST 17.4.3.03-85 1985; GOST 17.4.3.01-83 1983; Guidelines for conducting field and laboratory studies of soils and plants under the control of environmental pollution by metals 1981; Guidelines for the determination of heavy metals in soils and agricultural plots and crop production 1989). To obtain the gross form of heavy metals we used aqua regia, a mixture of nitric and hydrochloric acid (HCl 3 parts and HNO_3 1 part), and for mobile forms an ammonium acetate buffer (pH 4.8) solution. Heavy metals were determined using the atomic absorption method with the AA - 1200 spectrometer produced by Shimadzu (Japan).

Maps of heavy metal concentration in soils were made using the computer program MapInfo Professional in a GIS environment.

Statistical processing of the obtained data was performed using standard methods of achieving mathematical statistics as described by Dmitriev (1995) and Savic (1972) with the use of the package analysis programs "Excel-97" and "Atte Stat".

3 Research Results and Discussion

In the study of geochemical features of heavy metals in the soils of irrigated areas, the study of their migration and types of distribution in the soil profile are of particular interest. Below is a table showing the distribution of the studied metals in the soil profile in typical sections and some of their geochemical characteristics.

These analytical data show that the mobile forms of all the elements studied are characterized by the prevailing way in which they are accumulated in the soil profile, i.e., all of them are accumulated to various degrees in the plough horizon (Table 1). The accumulative type of distribution in the soil profile is especially seen in copper, lead and nickel.

The process of zinc and nickel being removed from the plough layer and their secondary storage at a depth of 20–50 cm is typical, i.e., an eluvial-illuvial type of distribution is typical for them. The formation of a secondary horizon at the groundwater level is typical for lead and cadmium, i.e., a hydrogenic-accumulation type of distribution is typical for them. The strict uniform-accumulative type of distribution in the profile of the studied soils is typical for copper.

The gross forms of heavy metals have a different type of distribution in the profile of rice-marsh soils than their mobile forms. For all elements investigated the eluvial-illuvial type of distribution is typical, i.e., these soils are seen to be influenced by

Table 1 Distribution of total and mobile forms of heavy metals in the soil profile

No of the section	Depth (cm)	Mobile forms					Gross forms				
		Zn	Cu	Pb	Cd	Ni	Zn	Cu	Pb	Cd	Ni
695	0–2	3.9	1.2	21.1	0.3	3.9	56.8	16	59.2	2	26
	2–22	2.4	1.9	20.3	0.2	9.8	57.6	16.4	54.8	3.2	34.8
	22–45	2.9	2	19.9	0.2	12.9	45.6	10.8	54.8	2.4	36
	45–70	2.9	1.6	21.1	0.4	10	47.6	10.4	56	2.8	32.8
	70–96	2.2	1.4	20.2	0.4	10.2	47.2	11.6	55.6	2.8	38.4

means of irrigation in the form of permanent flooding, and this promotes the removal of heavy metals from the plough horizon to the lower horizons.

Some accumulation in the plough horizon is also typical for copper and cadmium, but this process is weaker than the eluvial-illuvial process. Zinc, lead and nickel form the secondary accumulation horizon.

Based on the available data, it can be concluded that the irrigation of rice-marsh soils by means of permanent flooding has a significant influence on the distribution of mobile and gross forms of the heavy metals studied. Moreover, the environmentally hazardous mobile forms of metals available to the plants are subjected to the influence of irrigation to a greater degree than their gross forms. Usually, in a practical evaluation of the risk of soil contamination by chemicals, their actual concentration is compared with limited permissible concentrations (LPCs), and in their absence, with their regional background content and average content in the soils of the world, as well as their Clarke value. Other aspects taken into account are whether the metal refers to the class of risk, the extent of the soil's buffer capacity with regard to heavy metals (which depends on their basic chemical and physical–chemical properties) and other soil properties. Based on this, the following are presented: the Clarke concentration (CC), the accumulation coefficient (AC) and the rate of migration intensity (Px) of the metals studied in case of rice agrocenosis.

The data obtained show that in the soils studied, the Clarke value of the land crust near the heavy metals under study is excessive. The Clarke value is observed to be especially high in the case of cadmium, the CC of which is above that of the land crust at 15.6 ± 0.82, and the concentration of lead also exceeds that in the land crust, with a CC of 1.9 ± 0.07 (Table 2). It is also necessary to consider that both elements belong to the class of highly toxic elements. Zinc and copper have values which are close to the Clarke value for the land crust, and their CCs are 0.7 ± 0.02 and 0.7 ± 0.03, respectively.

In terms of the Clarke value of the land crust, the metals studied can be arranged in the following decreasing line: Cd > Pb > Zn = Cu > Ni.

All heavy metals studied in the A_{plow} horizon have accumulation coefficients which are close to or higher than one, i.e., all of them are subject to accumulation in the plough horizon (Table 2). Among the metals studied, that with the greatest accumulation in the plough horizon is copper, with an AC of 1.3 ± 0.29, followed by lead and nickel, with Acs of 1.1 ± 0.12 and 1.1 ± 0.19 respectively. Cadmium and zinc are accumulated to a lower extent.

Table 2 Clarke concentration, accumulation coefficients and intensity of heavy metal migration in the horizon of A_{plow} soils in the Shiely irrigation area

Indices	Metals				
	Zn	Cu	Pb	Cd	Ni
CC	0.7 ± 0.02	0.7 ± 0.03	1.9 ± 0.07	15.6 ± 0.82	0.5 ± 0.03
AC	0.8 ± 0.08	1.3 ± 0.29	1.1 ± 0.12	0.9 ± 0.05	1.1 ± 0.19
Px, %	4.9 ± 0.43	9.7 ± 0.83	46.8 ± 2.85	46.9 ± 1.97	25.0 ± 1.94

CC Clarke concentration is the ratio of heavy metal concentration in the soil to the soil Clarke value according to Vinogradov (1957); *AC* Accumulation coefficient is the ratio of heavy metals in soils of a particular genetic horizon, for example in horizon A, to the content of heavy metals in soil-forming rock in horizon C.; *Px* Intensity of migration of heavy metals in soils is the ratio of the mobile form to the gross form, expressed as a percentage, i.e., the share made up by the mobile form

In terms of the rate of migration intensity, cadmium and lead have the highest Px values of 46.9 ± 1.97 and 46.8 ± 2.85 % respectively, while the most passive migrant in periodically flooded soils is zinc. In terms of the intensity of migration in the arable soil horizon in the Shiely area, the elements studied can be arranged in the line: Cd > Pb > Ni > Cu > Zn.

Despite the great variety of soil surfaces in the irrigated areas of the Republic, the problem of determining the regional background levels of heavy metals in irrigated soils still has not been solved. Due to the ever-increasing anthropogenic pressure on the environment, including the soil surface, research into and the systematization of data on the background content of heavy metals in irrigated soils on a regional level is very relevant and important for evaluating the sustainability and stability of irrigated ecosystems in the face of global and regional anthropogenic influences.

Analysing the content and distribution of heavy metals at reference sites helps reveal the features of formation of the regional background level of heavy metals and helps classify the range of concentrations of the metals studied.

In this context, determining the actual levels of heavy metals in soils of reference regions is one of the most relevant tasks of scientific and practical importance.

Statistical data analysis is also very important when studying the characteristics of the content and spatial variation in soil properties or the average content of specific elements in specific soil types or in combination, aimed at increasing the reliability of the data and conclusions. Moreover, the use of statistical analysis also increases the data interpretation possibilities. It should be noted that all conclusions on absolute values of soil properties made by interpreting one or more typical sections without statistical analysis can often be inaccurate and may result in misinterpretation of the data (Sheudshen et al. 1998).

In statistical science, the random variable studied is mainly characterized by two groups of constants. The constants in the first group characterize the average level of the studied value, and those in the second group the degree of variability and changeability. Both groups show the law of distribution of the random variable.

Among the constants characterizing the distribution of various elements in the soils, the main aim of most works is to find the average arithmetic value which characterizes the average level of their concentration and to study their characteristics and determine the true value of the average value, i.e., the true background concentration.

The analytical data obtained on the content of mobile forms of heavy metals in soils in the Shiely irrigation area has been processed in terms of statistical variation and the results listed below. The calculated values of Student's t-criteria show that for all investigated soils the value t_{act} is much higher than t_{tab}. The limits of the confidence interval in almost all heavy metals are relatively narrow, i.e., the obtained average contents of heavy metals in the soils of Shiely area are statistically accurate (Table 3).

With some exceptions the values of their variation coefficients do not exceed the average value. A high degree of variation (>56.1 %) was found only in Cd. Heavy metals decrease in terms of the variation of the coefficient value in the soils investigated as follows: Pb < Cu = Ni < Zn < Cd.

The obtained data was also used to estimate the indices of the concentration of gross forms of the studied heavy metals in soils of irrigation area in Shiely in terms of statistical variation (Table 4).

It also shows that the estimated actual critical values of the Student's criterion are significantly higher than their table values. The limits of fluctuation in the gross forms of heavy metals and the reliable interval of their average value also have significantly narrow intervals.

Table 3 Indices of the concentration of mobile forms of heavy metals in the soil horizon A_{plow} in the Shiely irrigation area in terms of statistical variation

Metals	n	M ± m	Fluctuations	t-criteria		$\pm t_{0.05} * m$	V, %
				t_{act}	$t_{0.05}$		
Zn	162	2.8 ± 0.10	0.3 ÷ 9.2	28.9	2	0.2	44.1
Cu	162	2.5 ± 0.08	0.3 ÷ 5.7	32.4	2	0.2	39.3
Pb	162	13.1 ± 0.25	5.5 ÷ 19.7	52.3	2	0.5	24.3
Cd	162	0.5 ± 0.02	0.1 ÷ 1.2	22.7	2	0.1	56.1
Ni	162	5.2 ± 0.16	0.9 ÷ 9.8	32.4	2	0.3	39.3

n Number of samples. In this case the number of samples analyzed. *M* Arithmetic mean, *m* Error of the mean, *Fluctuations* Limits of fluctuation means the min. and max. concentration of metals in soils within the 162 soil samples, i.e., within n; *t-criteria* The Student's criteria with the help of which the authenticity of the average concentration of metals (M) in soils is estimated. If the calculated or actual value (t_{act}) is greater than the table value ($t_{0.05}$), then the obtained average concentration of heavy metals in soils at a probability level of 95 %, is considered reliable. $\pm t_{0.05} * m$ the limit of the reliable interval. Reliable interval (confidence interval)—the permissible deviation of the observed values from the true ones. The limit of the reliable interval shows the range of the results of sample observations. For example, in our case, the limit of the reliable interval Zn is 0.2–2.8 + 0.2 (mg/kg). The narrower the limit of the reliability interval, the more reliable the average concentration (M) of metal in the soil. *V, %* Variation coefficient. This value shows the variability of the concentration of metals in the variation row or in the study area

Table 4 Indices of the content of the gross forms of heavy metals in the soil horizon A_{plow} in the Shiely irrigation area in terms of statistical variation

Metals	n	M ± m	Fluctuation rates	t-criteria		$\pm t_{0.05}$ * m	V, %
				t_{act}	$t_{0.05}$		
Zn	98	71.9 ± 1.03	48.4 ÷ 91.2	70.1	2	2	14.1
Cu	98	28.1 ± 0.65	16.4 ÷ 46.6	43.5	2	1.3	22.8
Pb	98	26.3 ± 0.76	13.2 ÷ 46.0	34.7	2	1.5	28.5
Cd	98	1.1 ± 0.07	0.4 ÷ 3.2	17.4	2	0.1	56.9
Ni	98	23.2 ± 0.92	8.0 ÷ 46.4	25.2	2	1.8	39.3

The values of their variation coefficients, except for Cd, do not exceed the average value. Cadmium and its mobile forms have a high variation coefficient: 56.9 %. Considering the fact that according to their concentration in terms of weight, heavy metals are microelements, then the relatively high variation coefficients are quite natural.

When the heavy metals studied are placed in decreasing order for their gross forms, the variation coefficient values are slightly different than for those of their mobile forms, and it is as follows: Zn < Cu < Pb < Ni < Cd. Using the analytical data obtained, we calculated the contribution of each metal in the general "metallic" background of the soils in the Shiely area. It was found that the compositions of the mobile and total forms of heavy metals in the soils of the area are substantially different.

Among the mobile forms of the metals studied in the soils of the Shiely irrigation area, lead has the biggest share (54 %), then nickel (22 %). Zinc and copper have 12 and 10 %, respectively. The share of cadmium is only 2 % (Fig. 2).

Among the gross forms of the metals studied, zinc has the largest share (48.0 %), followed by copper (22 %). Lead and nickel take 17 and 15 % respectively. The share of cadmium is only 1 %.

Further, using the analytical data obtained and the computer program MapInfo professional in GIS, maps have been made of the concentration of environmentally significant mobile forms of heavy metals in soils typical in the Shiely irrigation

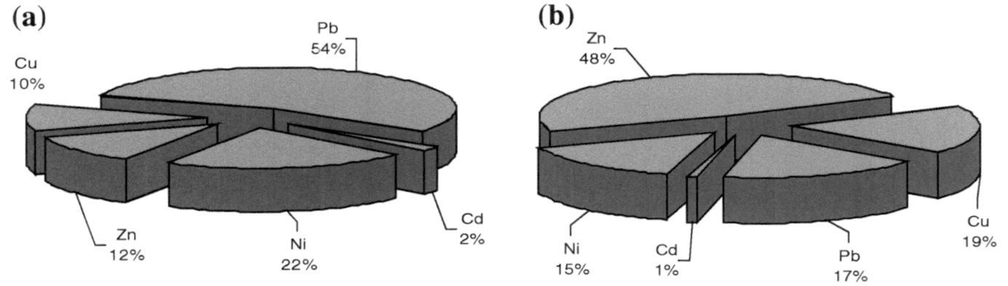

Fig. 2 Share of heavy metals in "metallic" background of soils in Shiely irrigation area. **a** Mobile. **b** Gross forms

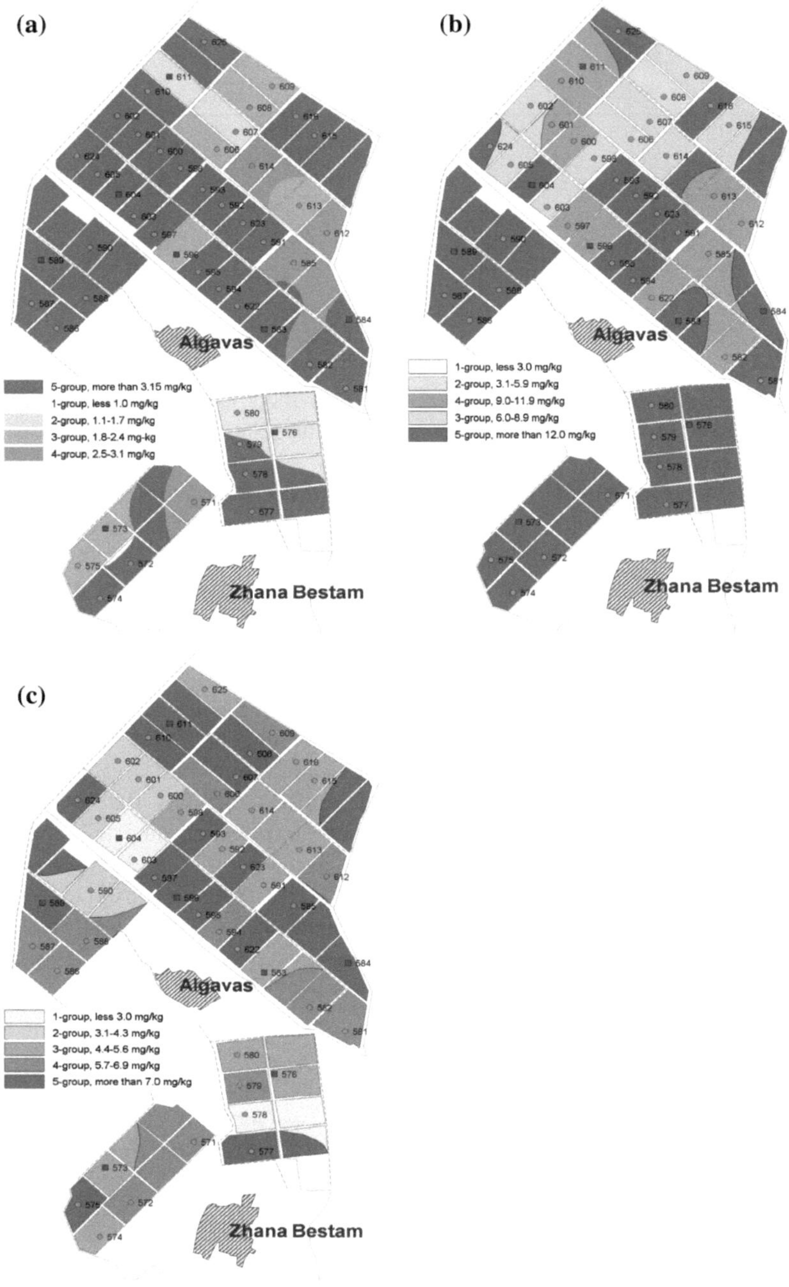

Fig. 3 Map of concentration of heavy metals in soils at LLP "Zhana-Bestam". **a** Copper. **b** Lead. **c** Nickel

area on large rice farms. One example is the map in Fig. 3 of the content of mobile forms of heavy metals in soils at a large farm (LP "Zhana-Bestam") and at smaller farms located on its territory.

These schematic maps containing analytical information allow a number of theoretical and applied tasks to be solved: evaluating the ecological condition of the soils in the irrigated area and categorising them by the level of the content of heavy metals; observing the geochemical migration flows of heavy metals and identifying the areas where they accumulate; justifying measures preventing pollution; using them as a basis for developing recommendations on organizing biogeochemical and ecological-analytical monitoring in the irrigation area.

Based on an analysis of the cartographic materials, we can say that in the case of copper, on the territory of the studied farms, there are 5 groups containing more than 3.15 mg of copper per 1 kg of soil. For lead, the group of soils with increased lead content from 9.0 to 15 mg per 1 kg of soil is observed. As for nickel there is a group of soils with increased content which is 4.6–6.0 mg/kg.

4 Conclusions

Thus, based on the analysis of the obtained data, we can conclude that for all mobile forms of the elements studied, first of all, their distribution in the soil profile is seen to have a predominantly accumulative character, and gross forms are characterized by an eluvial-illuvial process. Environmentally hazardous mobile metal forms available for the plants are subjected to the influence of irrigation to a greater extent than their gross forms.

It is identified that the concentration of cadmium and lead in the soils exceeds the Clarke concentration of the land crust. The CC of cadmium is 15.6 ± 0.82, and that of lead 1.9 ± 0.07. Also it should be noted that both of these elements belong to a class of highly toxic elements. Among the metals studied, that which accumulates the most in the plough horizon is copper, with an AC of 1.3 ± 0.29, followed by lead and nickel, with ACs of 1.1 ± 0.12 and 1.1 ± 0.19 respectively. Cadmium and zinc are accumulated to a lower extent. According to the rate of migration intensity, cadmium and lead have highest mobility; their Px is 46.9 ± 1.97 % and 46.8 ± 2.85 %, while the most passive migrant in periodically flooded soils is zinc.

Among the environmentally hazardous mobile forms of investigated metals in the soils of the Shiely irrigation area, lead has the largest share (54 %), then nickel (22 %). Zinc and copper make up 12 and 10 % respectively. The share of cadmium is only 2 %.

Based on the analysis of cartographic material, we can say that in the studied farms, for copper there are 5 five groups of soils which contain more than 3.15 mg of copper per 1 kg of soil. For lead, there are groups of soils with a high concentration of lead: 9.0–15 mg per 1 kg of soil. Nickel is found in the group of soils which have an increased concentration of 4.6–6.0 mg/kg.

References

Borovsky VM, Pogrebinsky MA (1958) Ancient delta of the Syr-Darya river and northern Kyzyl Kum, vol 1. Publishing House of AS of KazSSR, Alma-Ata, p 512

Borovsky VM, Ablakov EB, Kozhevnikov KY et al (1959) Ancient delta of the Syr-Darya river and northern Kyzyl Kum, vol 2. Publishing House of AS of KazSSR, Alma-Ata, p 418

Chizhikova NP, Zhuravleva MA (1997) Effect of rice cultivation on the composition and properties of fine disperse fractions of some soils of the delta of the Kuban river. In: Proceedings of international conference on "Issues of anthropological soil formation". Moscow, vol 3, pp 55–58

Chizhikova NP, Lebedeva MP, Zhuravleva MA (1997) Mineralogy and micromorphology of soils in rice fields in Kuban. Vestn Mosk Univ. Ser.17. # 4, pp 3–8

Dmitriev EA (1995) Mathematical statistics in soil science. Moscow State University, Moscow, p 320

GOST 17.4.3.01-83 (1983) Nature Conservation. Soils, General requirements for sampling

GOST 17.4.3.03-85 (1985) Nature Conservation. Soils, General requirements for methods of determination of pollutants

Guidelines for conducting field and laboratory studies of soils and plants under the control of environmental pollution by metals (1981) Publishing House "Gidrometeoizdat", Moscow, p 107

Guidelines for the determination of heavy metals in soils and agricultural plots and crop production (1989) USSR Gosagroprom, Moscow, p 62

Guidelines on conducting large-scale soil survey (1979) Alma-Ata, Kazakh SSR, p 137

Karajanov KD et al (1973) Soils of Kazaly area and the prospects for their use. Nauka Kaz SSR, Alma-Ata, p 171

Kiriyenko TN (1984) General regularities of soil evolution in the south of Ukraine under the influence of flooded rice: Abstract. Doctorial Thesis, Kharkov, p 41

Neunilov BA (1961) Increase of soil fertility of rice fields in the Far East. Primorsk Publishing House, Vladivostok, p 59

Otarov A (2005) Protective possibilities of the periodically flooded paddy soil in terms of heavy metals. Status and prospects of development of soil science. In: Proceedings of the international scientific conference devoted to 60th anniversary of the Institute of Soil Science named after U.U. Uspanov. Almaty, "Tethys", pp 131–132

Otarov A, Ibraeva MA, Saparov AS (2006) The influence of ecological condition of rice fields in Kazakhstan on the quality of raw rice and processed products. Grain Processed Prod 3:14–18

Otarov A, Ibraeva MA, Saparov AS (2007) Degradation processes and current soil-ecological condition of rice areas in the Republic: environmental bases for the formation of the soil surface in Kazakhstan in conditions of anthropogenesis and development of theoretical bases of reproduction of fertility. Almaty, pp 73–105

Savich VI (1972) Application of variation statistics in soil science: training manual. Publishing House of the TAA, Moscow, p 103

Sharapov ID (1977) Restoration processes in the rhizosphere of rice and their influence on soil fertility: increase of soil fertility in rice fields. Publishing House Nauka, Moscow, pp 49–70

Sheudshen AH, Bochko TV, Trachov EM (1998) Humus as an integrated indicator of soil fertility and its changes in rice growing: bulletin of Krasnodar research center of Adyge (Cherkess). International Academy of Sciences, Krasnodar, pp 28–32

Vinogradov AP (1957) Geochemistry of rare and trace elements in soils, 2nd ed. Moscow, Academy of Sciences of the USSSR 1957, p 20

Volkov AI (1983) Systematic description of the soils: the soils of the Kazakh SSR, vol 14. Publishing House of Nauka of the AS of Kazakh SSR, Alma-Ata, pp 46–50 Kyzylorda region

Yakupova NY, Karajanov KD, Kievskaya RH (1973) Soils of Kazaly area and prospects of their use. Publishing House of Nauka of Kazakh SSR, Alma-Ata, p 171

Concept and Results of Soil Monitoring in North Kazakhstan

Temirbolat D. Dzhalankuzov

Abstract This concept defines the goals, objectives and areas of work aimed at implementing the public monitoring of agricultural lands and lands used or rented for agricultural purposes within lands of other categories, the data on which are the basis for establishing public information resources on the status and use of these lands. This paper includes the methods for conducting studies monitoring the soils of Northern Kazakhstan, their current status and ways of improving soil fertility in terms of the ordinary and southern black soils of Northern Kazakhstan. It shows the climatic conditions of subzones of ordinary and southern black soils where the studies were conducted, their morphological indices, physical–chemical properties and significant changes in the soil properties due to long-term agricultural use. Use of the results from monitoring studies will allow scientists to forecast the development of negative soil processes, prevent the shrinking of agricultural land, and justify the need for programmes of soil fertility conservation and restoration.

Keywords Soil fertility · Chernozems · Monitoring · Kazakhstan

1 Goals and Objectives of Public Monitoring of Agricultural Land

Public monitoring of agricultural lands is carried out in order to prevent the shrinking of agricultural lands, to encourage their conservation and use in agricultural production, to develop programmes for the conservation and restoration of soil fertility, and to supply public bodies, including executive bodies responsible

T. D. Dzhalankuzov (✉)
Kazakh Research Institute of Soil Science and Agrochemistry (U. U. Uspanov),
Al Faraby Ave 75b, 050060 Almaty, Kazakhstan
e-mail: d.temirbolat@mail.ru

L. Mueller et al. (eds.), *Novel Measurement and Assessment Tools for Monitoring and Management of Land and Water Resources in Agricultural Landscapes of Central Asia*, Environmental Science and Engineering, DOI: 10.1007/978-3-319-01017-5_42,

for the state land control, businesses and individuals, and agricultural producers of all forms of property with reliable information on the status and fertility of agricultural lands and their actual use.

Public monitoring of agricultural lands includes systematic observations of the status and use of field crop rotation, agricultural sites and contours, as well as the parameters of soil fertility and development of their degradation processes (changes in the reaction of the soil environment, the content of organic matter and nutrients, the destruction of soil structure, salinity, alkalinity, waterlogging, excessive watering, the flooding of lands, the development of water and wind erosion, soil contamination with pesticides, heavy metals, radionuclides, industrial, domestic and other waste, and changes in other soil properties). It monitors changes in vegetation cover on arable lands, fallow lands, grasslands and rangelands (change in species composition, yield structure, types and quality of vegetation, degree of resistance to anthropogenic stress).

Public monitoring of agricultural lands will allow the following tasks to be carried out:

- Timely detection of changes in the status of agricultural lands and assessment of these changes;
- Forecasting and making recommendations to improve soil fertility;
- Preventing and eliminating the consequences of negative processes;
- Obtaining data based on systematic surveys of soil fertility;
- Observations of the quality and effective use of agricultural lands as a basic resource of agricultural activity;
- Monitoring of vegetation on agricultural lands;
- Maintaining a soil fertility register for agricultural lands and records of their condition;
- Formation of public information resources on agricultural lands for analysis, forecasting and developing public policy in the sphere of land issues;
- Giving businesses and individuals access to information about the status of agricultural lands;
- Participating in international programmes (ensuring compliance with international obligations).

2 Current Status of Public Monitoring of Agricultural Lands

It is difficult to monitor large areas occupied by agricultural lands, because of the lack of digital maps of agricultural lands indicating the boundaries of crop rotation fields, agricultural sites and contours, because of the undeveloped network of operative monitoring points, surface stations, including meteorological, and because of the lack of air support caused by the high cost of maintenance. In these lands, due to various natural processes and economic activities, the boundaries of areas under crops are constantly changing, as are the vegetation conditions of crops, soil fertility properties, and the development of negative processes.

One effective tool for solving some of the set objectives is the system for the remote monitoring of agricultural lands, combined with ground-based surveys of agricultural lands, which are part of the public information system in agriculture.

To ensure that monitoring works, the following are introduced: new tools and technologies, observation systems, data collection and processing, including those based on remote sensing of land (seen as the most objective and operational in use), which allow scientists to simultaneously monitor the use of land and forecast the development of agricultural crops and potential yields.

Currently, remote monitoring (primarily via satellite) provides objective information on the entire territory occupied by agricultural lands.

The dynamics of land use and status are evaluated based on comparative analyses of multi-temporal cartographic materials. Data from remote sensing of land and ground surveys which aim to identify the processes development scenarios and forecast the situation is based on the use of up-to-date GIS technologies. In addition to monitoring land using remote sensing methods, the Ministry of Agriculture of RK uses centres, agrochemical service stations and centres for chemical treatment and agricultural radiology to monitor the status of soil fertility by performing annual ground surveys of agricultural lands. However, activities carried out for the public monitoring of agricultural lands are mostly fragmentary and done at departmental level. There is no inter-agency coordination and work is not properly organised.

Currently data from the remote sensing of land cannot provide full information without the presence of ground-based measurements (in particular, the characteristics of the species composition and plant biomass), which needs to be used to calibrate satellite data processing algorithms. All these factors lead to a need to conduct ground-based observations, link the obtained data to geographic identifiers and establish public information resources using ground-based observations.

3 The Development of the Public Monitoring of Agricultural Lands

The development of the public monitoring of agricultural lands will require coordinated action by all relevant executive bodies in the following areas:

- Developing and improving the regulatory legal framework, including the procedure of public monitoring of agricultural land and the procedure of public registration of the indicators of soil fertility for these lands, as well as methodological materials on measurement and processing monitoring data;

- Establishing and using a cartographic basis of various scales on the whole territory of Kazakhstan; developing and extensively using remote land sensing methods with different spatial resolutions to monitor the status of agricultural lands;
- Developing a system of indicators for the public monitoring of agricultural lands based on ground surveys and observations, as well as using remote sensing technologies in the case of local governments for different levels;
- Developing ground survey system and observing the status of crops and the condition and use of agricultural lands, the soil fertility of these lands, and the development of negative soil processes;
- Developing a data processing system and analysing soil fertility data, the status of growing plants and the use of agricultural lands;
- Developing and introducing modern technologies, including remote land sensing systems to provide public monitoring of agricultural lands;
- Improving information technologies and information systems providing data for the public monitoring of agricultural lands, and using those data to take managerial decisions at various levels;
- Developing training programmes and professionally training specialists in the field of the public monitoring of agricultural lands.

4 Development and Use of Public Information Resources on Agricultural Lands

Public information resources on agricultural lands, formed by using the data available in the executive bodies, are the centralized resources formed in order to analyze, forecast and formulate public policy in the sphere of land relations and the use of these lands.

The main types of information generated on the basis of public information resources on agricultural lands, using up-to-date information technologies including GIS technology, should be as follows:

- Information about the boundaries of agricultural lands (plots, agricultural sites, contours), their size, shape, form of permitted and economic use, potential productivity;
- Information on agricultural land withdrawn from agricultural use, including the borders, areas, status, type of economic use, year of last use;
- Information on agricultural lands introduced into circulation in the current year and for a given period of observation, including borders, status, type of economic use, potential productivity, length of time that agricultural land has lain fallow during recent years;

- Information about the status of soil fertility, including indicators of morphogenetic properties of soils, their granulometric content, acidity, humus content, macro- and micronutrients, heavy metals and radionuclides, degree of erosion (deflation), waterlogging, salinisation, desertification, stoniness, and characteristics of growing plants by geo-botanical composition and yield capacity, as established during ground surveys.

5 The Importance of Maintaining Soil Fertility

For many years agricultural scientists have studied and explained all aspects of soil fertility. They are primarily related to a soil's ability to provide all factors in the growth and development of green plants through nutrients, sufficient moisture, etc. The plants grow better and faster on fertile soils that provide the basic process of the transformation of solar energy into organic matter. The synthesis of molecules of carbon, oxygen, nitrogen, hydrogen and other elements in the vital amino acids and compounds inherent to living organisms (proteins, carbohydrates, fats, etc.) begins with photosynthesis, which is inherent to green plants.

As civilization develops and the population grows around the globe, two opposite tendencies are accelerating: on the one hand, mankind has come to realize that the soil is an almost non-renewable type of natural resource (like oil, gas, coal, etc.) which, even if the most optimistic forecasts regarding scientific development in the coming century come true, will provide 97 % of the world's population needs for food. On the other hand, the intensified use of soils (uncontrolled irrigation, damage to cropping areas which affects the development of erosion processes and accelerates the mineralization of organic matter, intensive cultivation, the wide use of chemicals) determines the depletion of soil, fertility decline and, eventually, leads to desertification. The introduction of black soils into agricultural use has caused dramatic changes. There is a correlation between almost all soil formation processes and the deterioration of their properties: of the addition of organic matter to the soil has decreased and mineralization of humus has accelerated, the structure has deteriorated, there have been changes in the water regime, etc.

In extensive production systems at the first stage, the black soil evolution process was characterized by gradual degradation (reduction of humus and nutrients, deterioration of water and physical properties). The condition of black soils at the end of the nineteenth century led Dokuchaev (1951) to compare them with a purebred Arabic horse which was exhausted and needed rest and help building up new energy.

The negative trends in decreasing soil fertility are associated with unscientific approaches to the provision of crop rotations with cultivated crops such as sunflower and corn, which activate the mineralization of organic matter.

Unfortunately, the intensification of agricultural production aims to maximize profits and avoid scientifically justified crop rotations as much as possible. Preference is given to the most economically profitable crops, which are mainly sunflower and rapeseed. In our opinion, the government should not allow farming systems of this type, which in practice, leads to the destruction of the best soils in the world. Of the degradation factors, the most painful are the loss of humus and water and wind erosion. Humus is a key indicator of fertility, which determines the productivity of agricultural lands. Its content is interrelated with such properties as structure, water and physical parameters. Humus is an important source of nutrients: in black soils it contains 98 % of the total nitrogen and more than 50 % of the phosphorus. It determines the amount of enzyme activity and is the largest source of accumulation of solar energy, fulfilling one of the most important functions of the biosphere. Although ploughing the straw provides significant opportunities to improve the humus balance in the soil, straw is burnt throughout this area. Besides its ecologically damaging effect (promoting fires, burning forest belts, etc.), burning stubble and straw residues on the fields causes great damage to soils. It has been determined that straw is burned for 30–40 s per square metre, while the surface temperature reaches 360 °C, at a depth of 5 cm: about 50°. The burning of humus was noted in a layer from 0 to 10 cm. At the same time, the hydrophysical properties of soil deteriorate and its biological activity decreases.

During the burning of 4–5 tonnes of straw and stubble, 20–25 kg of nitrogen and 1.5–1.7 tonnes of carbon is lost per hectare.

The situation that occurs when straw is burned can not only lead to a decrease in production volumes but may also cause environmental problems and worsening on a local, regional and even global scale (increased carbon dioxide emissions).

6 Potential for Improvement to the Humus Balance

Since livestock production is a very weak sector, almost non-existing, one of the best sources for replenishing organic matter in the soil is perennial grasses, which enrich the soil through crop residues and roots to 2 tonnes per hectare.

Improvements to the balance of humus in the soil can also be achieved by ploughing all crop residues remaining after the harvest into the soil. When crops are rotated with perennial grasses and the entire mass of by-products is used, humus is regenerated to a greater extent even if organic fertilizers are not used.

The use of soil resources in Kazakhstan should be based on the principles of regenerating soil fertility to a greater extent. This approach will ensure the welfare of the population, and create opportunities to transfer the most fertile black soils to future generations, but not the desert.

The soil surface of Kazakhstan during critical conditions of agricultural production is worsening significantly in terms of physical, physical–chemical, agrochemical and other properties, and the effective fertility range is very wide.

All this gives rise to a need to conduct repeated soil survey mapping based on new methodological principles using remote sensing and geographic information system technologies. This kind of objective information on soil resources can serve as a basis for a qualitative and capital evaluation of land. A lack of this information can cause great damage to the state, because in the absence of proper assessment, the world's best black soils can be sold for very low prices. It will be possible to carry out such activities if a common public service of soil fertility conservation is established in Kazakhstan, which will ensure continuous monitoring of their quality, rational use and cultivation, and of the regeneration of soil fertility.

7 Sites Studied

Monitoring studies on black soils in Northern Kazakhstan started in the late 1960s and early 1970s, when the Institute of Soil Science began conducting station studies of morphogenetic properties and regimes and their changes during long-term agricultural use. To study southern black soils, experimental plots were selected which were typical for the whole sub-zone of southern black soils. The site selected was at the Kostanai regional agricultural research station, 22 km south of the city of Kostanai, south of the village Talapker.

Plots of virgin land were located at: Altitude 187 m. North 53°01′39.7″. East 63°44′12.4″.

Plots of arable land were located at: Altitude 190 m. North 53°01′39.3″. East 63°44′09.2″.

Studies on ordinary black soils were conducted on the territory of the Karabalyk experimental station, located in the area between Uy-Toguzak, in the south-west area of the West Siberian lowland in the subzone of ordinary black soils of moderately dry steppes. The study object is located 150 km to the north-west of Kostanai and 15 km north of the district centre, Karabalyk.

Plots of virgin land were located at: Altitude 201 m. North 53°51′15.6″. East 62°03′08.2″.

Plots of arable land were located at: Altitude 209 m. North 53°51′23.7″. East 62°03′09.9″.

From 1967 to the present time, regular monitoring has been conducted of the morphogenetic, physical–chemical and biological properties of black soils of Northern Kazakhstan which are in long-term agricultural use. Studies have been conducted in the following variants of the experiment: virgin analogue of black soils and arable land, used for growing crops for a long-term period.

8 Changes in Soil Properties

As a result of long-term monitoring, materials have been accumulated on soils' morphogenetic, chemical and physical properties. Their analysis shows that changes in soils' characteristics and properties and the natural process of soil formation are caused by systematic tillage, changes in its natural turf/cereal vegetation caused by annual cultivated plants, and the use of fertilizers.

The climate in sub-zones of ordinary and southern black soils is sharply continental, expressed in extremely cold winters, hot summers and a small amount of precipitation. The discrepancy between the amount of warmth and moisture during the growing season and the excess evaporation in the summer (many times over the amount of rainfall) causes the aridity of the climate.

The climatic conditions of sub-zones of ordinary and southern black soils are as follows (Table 1).

Since the studies were conducted in the sub-zones of ordinary and southern black soils, we will look at their description. Typical ordinary black soils in the Kostanai region occupy the watersheds of Uy-Toguzak, Toguzak-Tobool and form a homogeneous and relatively large array. The underlying rocks are yellow clay and yellowish-brown loam. The most common types of ordinary black soils in the study area are of moderate capacity. Varieties of heavy-loam ordinary black soils are predominant. The morphological characteristics of these soils are characterized by the following indicators: the average thickness of the humus horizon is A + B 60–80 cm, boiling in some places reaches 30–40 cm, solid 40–50 cm, carbonates appear at a depth of 70–100 cm and gypsum is found within the second metre. One distinctive feature is the division of the profile into dark layers (tongues), formed as a result of the cracks filling with material which has been taken off the upper humus horizons by atmospheric and melt water.

Ordinary black soils are richly supplied with organic matter, the quantity of which in the upper layer reaches 6.8–7.7 %. Going deeper, the humus content decreases quite sharply. The soil-absorbing complex is saturated with exchangeable calcium and magnesium. The quantity of exchangeable potassium is very insufficient, making up no more than 1.9 % of the humus horizon. In the soil-forming rock, it is seen to increase by 6–7 %, while magnesium increases to as much as 53–57 %. The reaction of the soil solution changes from neutral in the "A" horizon to alkaline in the "B" horizon and strongly alkaline in the parent rock.

The soils at the station are not saline. The total quantity of water-soluble salts at a depth of 2 m does not exceed 0.11 %, and only at a depth of 230–240 cm is the salt content seen to increase to 0.27 %. Of the water-soluble salts, the sulphates are predominant. Similar studies were conducted in the black soil subzone, where the station observations were made on southern normal heavy loam black soils.

These black soils are formed at a relatively high plains under a mixed vegetation of grass, fescue and feather grass. The soil-forming rocks are yellow–brown diluvial, alluvial-diluvial loams and clays, and loess-like loams. The humus

Table 1 Climatic indices of the subzones of moderate-arid and arid steppes in the Kostanai region

Precipitation, mm		Water supply in snow cover (mm)	Duration of non-frost period (days)	Sum of temperatures higher than 10°	Air temperature (in °C)				Moisture coefficient: correlation of precipitation quantity to evaporation (in mm)
Annual	May–July				Average in July	Average in January	Absolute minimum	Absolute maximum	
Moderate-arid steppe in ordinary black soils									
275–350	120–150	70–80	110–125	2,150–2,200	19.5–20	−18 to −19	−49	39–40	0.25–0.30
Arid steppe in southern black soils									
250–280	105–120	65–75	110–125	2,200–2,400	20–20.5	−17.5 to −18	−46 to −48	39–40	0.20–0.25

horizon thickness at horizons A + B is 55–60 cm on average, while horizon "A" is 15–20 cm thick. The colour of the humus horizon is not uniform due to the tongue profile. This feature of black soils was noted by researchers (Gorshenin 1927; Durasov 1958; Fedorin 1958) as typical of this province. The thickness of horizon "B" ranges from 40 to 50 cm, and its colour is heterogeneous, as the tongue feature is observed in this horizon. The profile of southern normal black soils is strongly cracked. Some cracks reach depths of 100–120 cm. Effervescence after the addition of HCl (boiling) is observed in the lower part of humus horizon at a depth of 35–50 cm, sometimes higher. In the 0–40 cm layer a horizon of carbonate formations is outlined. The apparent display of gypsum is found at a depth of 100–140 cm. One of the characteristics of these soils is the presence of rusty ochre spots at a depth of 2.5–3 m, indicating the passage of the meadow stage.

The soils studied are characterized by their relatively high organic matter content, the amount of which in the upper horizon is 5.8 %, with a sharp decrease down the profile. The amount of bases absorbed is quite high: 35.9 mg/eq per 100 g of soil, and exchangeable calcium (80–90 % of the total amount) prevails throughout the profile. There is less magnesium (up to 27 % of the total amount) and very little sodium (0.4–2.2 % of the total amount). These soils are not saline within one metre. Within the second metre the salt content increases from 0.3 to 0.5 %, in which sulphates prevail.

The results of long-term monitoring observations have shown that long-term treatment leads to significant changes in both the morphological and physical–chemical properties of tilled soils. It should be noted that morphological changes take place immediately after the first tillage (Table 1).

During the process of tilling, the upper layer of virgin soils (Horizon "A") and some of horizon B_1, are destroyed, and thus the arable horizon is formed, which is 30 cm thick with a loose structure combined with fine elements. As a result, its density decreases and total porosity increases, which improves permeability and generally leads to an improvement in the soil water regime. Tillage significantly improves its air regime. Improving the water and air regimes in the arable horizon creates favourable conditions for the development of soil microorganisms and increases their activity. As a result of cultivation, the capacity of the "BC" horizon increases and the lower boundaries of the B_1 and B_2 subhorizons shift down, and in general the thickness of the humus horizon grows. An analysis of Table 1 shows that long-term development (30 years) leads to the restoration of the capacity of sub-horizon B_1. Physical and chemical characteristics are also subject to significant changes, mainly affecting the humus content of the newly ploughed black soils. During the last period (from 1967 to 2011) of cereal cultivation, black soils lost an average of 23–27 % of their natural humus content, the most important indicator of the level and dynamics of soil fertility.

The study results showed that the long-term tillage of black soils leads to significant changes in both morphological and physical–chemical parameters. During the long-term tilling of cultivated soils, an arable horizon of 30 cm of loose fine-grained structure is formed, while there is a decrease in the bulk density and an increase in total porosity. By the 30th year of soil cultivation, the capacity of sub-horizon B1 has been almost completely restored. The final restoration of the capacity of sub-horizon B1 is noted by the 55th year of cultivating black soils. By this time, a sustainable morphological profile has been created of the black soils under long-term tilling (Table 2).

Significant changes also take place in soils' humus content. Over 30 years of cultivation of soils there is a constant decrease in the humus content in the upper plough horizon and only by the 50–55th year of cultivation does a period of relative stabilization occur in the humus content in the plough horizon. A similar pattern is observed in the total nitrogen content (Table 3).

The decrease in the humus content in soils of agricenoses takes place to some extent as the dynamic equilibrium correlation is established between the processes of accumulation of organic matter and new formation of humus on the one hand, and the mineralization of humus and loss of nutrients with the crops on the other.

Table 2 Change in the main morphological parameters depending on the duration of cultivation

Duration of tillage	Genetic horizons					Depth of carbonates (Mycelium spots), cm	Boiling[e], cm
	A[a] (A_{arable})	B_1^b	B_2^b	BC[c]	A + B[d]		
Virgin land	16	21	23	11	60	51	41–49
1 year of cultivation	25	13	22	12	60	55	38–45
25 years of cultivation	28	17	22	18	67	44–46	40–46
30 years of cultivation	28	19	22	20	69	45–63	40–49
55 years of cultivation (old till)	30	21	23	21	75	49–67	40–51

[a] Humus accumulative horizon is indicated by the letter A; [b] illuvial horizon indicated by the letter B
[c] Parent rock by letter C; [d] the presence of a significantly expressed eluvial horizon in the soil is indicated by the letter A_2, and the humus layer is indicated by the letter A_1
[e] Effervescence by application of HCl, indicates carbonates

Table 3 Change in humus content and total nitrogen, depending on the duration of cultivation

Plot	Humus, %				Gross nitrogen, %			
	A (A_{arable})	B_1	B_2	BC	A (A_{arable})	B_1	B_2	BC
Virgin land	7.3	3.9	2.6	1.0	0.39	0.29	0.17	0.06
1 year of cultivation	6.4	3.7	2.6	1.0	0.25	0.20	0.14	0.05
10 years of cultivation	6.2	4.0	2.7	1.3	0.24	0.18	0.12	0.05
25 years of cultivation	5.7	3.9	2.8	1.2	0.26	0.15	0.14	0.06
30 years of cultivation	5.9	4.2	3.1	1.3	0.29	0.21	0.12	0.06
50–55 years of cultivation	5.9	4.3	3.0	1.3	0.29	0.20	0.14	0.08

It has been revealed that the highest humus loss rates occur in the first 5–6 years after cultivation, and this process slows down when the dynamic equilibrium is reached, i.e., a cyclic stabilization of soil fertility takes place, but at a lower level.

A study of changes in humus in southern and ordinary black soils (Tables 4 and 5) shows that in agricultural production, the humus content of ordinary black soils and its distribution in the soil profile is very different from the southern black soils: in the upper horizons of ordinary black soils during their cultivation, the humus content in the plough layer is decreased, while in subsoil horizons a slight increase in humus content is observed, and only in the underlying rocks is their content similar to the content of humus in virgin land analogues.

In the southern normal black soils the distribution of humus is characterized by a steady gradual decline in humus in the soil profile. The different distribution of humus in the soil profile of ordinary and southern black soils is due to their different hydrothermal character and the different means of tillage used on these soils.

A study of the hydrophysical properties of southern and ordinary black soils and those in long-term agricultural use, comparing these soils with virgin lands, is still ongoing. Tables 6 and 7 show the dynamics of some parameters of the hydrophysical properties of southern ordinary soils.

Table 4 Humus content in ordinary and southern black soils

Soils	Plot	Genetic horizons			
		A	A_{arable}	B_1	B_2
Black ordinary normal	Virgin land	7.6		4.5	2.6
	Arable land		6.0	3.8	2.6
Black soils southern normal	Virgin land	5.6		3.4	2.2
	Arable land		4.3	2.3	1.9

Table 5 Changes in humus content in southern and ordinary black soils (long-term average)

Depth, cm	Humus in %			
	Southern black soil		Ordinary black soil	
	Virgin land	Arable land	Virgin land	Arable land
0–10	5.35	3.83	7.7	5.23
10–20	4.68	3.53	5.8	5.04
20–30	3.95	2.97	4.3	4.75
30–40	3.48	2.64	3.3	4.56
40–50	3.04	2.46	2.7	4.47
50–60	2.53	2.03	2.0	4.30
60–70	1.73	1.42	1.6	3.40
70–80	1.20	0.89	1.0	3.10
80–90	0.97	0.88	0.8	1.10
90–100	0.79	0.72	0.6	0.6

Table 6 Hydrophysical properties of southern black soils in the Kostanai region (average over 3 years in the 0–50 cm layer)

Virgin land			Arable land		
Average moisture %	Average soil density, g/cm^3	Moisture supplies, mm	Average moisture %	Soil density, g/cm^3	Moisture supplies, mm
2009					
17.6	1.11	94.1	14.2	1.22	84.7
2010					
13.3	1.13	71.5	14.8	1.11	81.5
2011					
16.9	1.06	85.1	15. 6	1.16	89.8

Table 7 Hydrophysical properties of ordinary black soils in the Kostanai region (average over 3 years in the 0–50 cm layer)

Virgin land			Arable land		
Average moisture %	Soil density, g/cm^3	Moisture supplies, mm	Average moisture %	Soil density, g/cm^3	Moisture supplies, mm
2009					
15.2	1.13	84.5	19.0	1.11	105.8
2010					
17.8	1.09	94.8	15.5	1.10	86.2
2011					
14.0	1.06	85.1	15.4	1.16	89.8

Interesting results were obtained when studying the temperature regime of black soils in the Kostanai region.

This study of the temperature regime of black soils included monitoring the temperature on the surface, in the arable layer and in the deep soil layers. The study provided data which allowed scientists to build thermoizopleths of the process of positive and negative temperatures at different depths. These data made it possible to calculate the number of days with positive temperature at a certain depth. In general, ploughing virgin black soils changes the nature of heat transfer in the soil/atmosphere system during the warm season and especially during the cold season.

9 Conclusion

The implementation of this concept will allow reliable objective information to be obtained about soil fertility and the status and use of agricultural land as a natural resource that is used as the main means of production in agriculture, and establish public information resources, combining the information about these lands collected by various executive bodies, coordinate activities undertaken by executive bodies for the implementation of public monitoring of these lands, and ensure the efficient use of budget funds allocated to the executive bodies for these purposes.

The creation of public information resources on agricultural lands will allow us to:

- Analyze the status and use of lands based on the use of modern information technologies, including geographic information technologies;
- Forecast the development of negative soil processes and the impact of negative natural processes on the state of plant cover (drought, frost, insect infestation, etc.);
- Prevent the shrinkage of agricultural lands, preserve and use these lands in agricultural production;
- Justify the need to develop programmes for the conservation and restoration of soil fertility and assess their feasibility;
- Meet the needs of the State, including the executive bodies responsible for the control of public land, and provide businesses and individuals, including those who deal with land control issues, and agricultural producers of all forms of property with reliable information on the status and fertility of agricultural land, as well as their actual use;
- Conduct an effective public policy in the sphere of land relations in respect of agricultural land;
- Create optimal agro-physical parameters providing the best soil regimes and favourable conditions for the plants in the soil root zone during planting of crops.

The main issues around the conservation of soil fertility are the loss of humus and soil erosion.

The humus balance is currently improved by ploughing the straw into the soil, which enriches the soil due to stubble and root residues to 2 tonnes per hectare.

References

Dokuchaev VV (1951) The study of natural zones and soil classification. v.1. VI. M.-L

Durasov AM (1958) The soils of Northern Kazakhstan. Alma-Ata

Fedorin BV (1958) Black soils carbonate moderate humus in northern Kazakhstan. In: Proceedings of the AS of Kazakh SSR. Botanical series. No. 3

Gorshenin KP (1927) The soils of the black soil belt of Western Siberia. West. Sib. Department of Russian Geographical Society, vol 39. Omsk

Diagnosis and Optimization of Phosphorus Nutrition Conditions of Grain Crops in Northern Kazakhstan

Valentina Chernenok and Dietmar Barkusky

Abstract The phosphorus fertilization of crops is a most important issue because of the limited resources of phosphate rock on the one hand and the possible negative impacts on aquatic and terrestrial ecosystems on the other hand. This chapter deals with the results of long-term field experiments to diagnose and optimize the phosphorus nutrition of cereal crops in Northern Kazakhstan. The efficiency of the phosphorus fertilization was most dependent on the plant-available P content in the soil as analyzed using Machigin's extraction method. We present an approach to evaluate the phosphate status in the soil and to determine the particular requirements of grain crops. The inputs used to calculate fertilizer demand are the optimum content of plant-available phosphorus, its current concentration in the topsoil and the quantity of phosphorus to be fertilized in order to raise the plant-available P_2O_5 by 1 mg kg^{-1} soil. The approach is based on the Machigin method and is currently not transferrable to other regions, where other analytical methods are common. In order to compare this approach with other methods worldwide, methodical comparisons of different analytical methods for assessing the plant available P content in soils are necessary in future. We also require a better understanding and quantification of phosphorus cycling in ecosystems, based on internationally acknowledged analytical methods, models and long-term experiments.

Keywords Plant-available phosphorus · Machigin method · P fertilization · Dark chestnut soil · Northern Kazakhstan

V. Chernenok (✉)
JSC "Kazakh Agro-Technical University named after S. Seifullin", Zhenis-prospect 010000
Astana, Kazakhstan
e-mail: Chernenok2@mail.ru

D. Barkusky
Leibniz Centre for Agricultural Landscape Research (ZALF) e. V., Eberswalder Strasse 84
15374 Muencheberg, Germany
e-mail: dbarkusky@zalf.de

L. Mueller et al. (eds.), *Novel Measurement and Assessment Tools for Monitoring and Management of Land and Water Resources in Agricultural Landscapes of Central Asia*, Environmental Science and Engineering, DOI: 10.1007/978-3-319-01017-5_43,

1 Introduction

Based on common technology and costs, the natural resources for phosphorus fertilizer production in the world will be almost sufficient for the next 120 years. It is predicted that Russia's phosphorus stock will be exhausted within the next 20 years, and that in the U.S.A. within 40 years (Römer 2009). Therefore research into optimizing phosphorus (P) fertilization is very important to secure food production in the future (Dawson and Hilton 2011). On the other hand, long-term phosphorus fertilization has contributed to the eutrophication, pollution and contamination of aquatic and terrestrial ecosystems around the world (Ongley 1996; Schroetter et al. 2005; Smidt et al. 2011). Because of the serious environmental effects of P overabundance in farmland, a reduction or even just a limitation in P fertilizer is being recommended by an increasing number of authors (Guan et al. 2013; Nagumo et al. 2013).

The soils of Northern Kazakhstan feature good physical and agrochemical properties for crop production. They are characterized by a high content of plant-available, mobile nitrogen and potassium in the topsoil. On the other hand, agricultural production suffers from a low content of mobile phosphorus in the soil. As a result of extensive crop production, the phosphorus balance is negative; the amount of phosphorus removed with harvested products exceeds the amount coming into the system through fertilizers or remaining in the field with plant residues. Plants are able to make inert phosphorus forms in the soil accessible to some degree, and these forms are not included in common agrochemical analyses (Römer 2009). Nevertheless, the problem of phosphorus fertilization in Northern Kazakhstan's agriculture has occupied a special position, among other factors, when it comes to determining the level of soil fertility and the resulting crop yield potential. More than two thirds of the soils in this region are characterized by an amount of mobile phosphorus that is insufficient to meet the need of crops.

Because of the deficit in plant-available phosphorus in the soil, crops are not able to use the mobile nitrogen sufficiently. Therefore, under conditions of extensive land use, a steady decline can be observed in the stock of mobile phosphorus in the soil. Its content can only be restored and renewed by means of fertilization, but how should the phosphorus fertilizer be applied, and in what quantity and quality? This question is important and complex. It requires the development of advanced diagnostic techniques.

2 Objects and Methods

2.1 Site Conditions in Northern Kazakhstan

Basic studies were carried out in multi-variant long-term field experiments with different levels of nitrogen and phosphorus nutrition in grain-fallow crop rotations on dark chestnut soils (Kastanozems) in the dry steppe zone in the Akmola and

Karaganda regions and on ordinary black soils (Chernozems) in the forest steppe zone at the Northern Kazakhstan Agricultural Experimental Station. The soils represent more than 60 % of the total arable land in Kazakhstan. They are characterized by relatively favourable physical and physicochemical properties and a high concentration of mobile potassium, but they considerably differ in their humus content, gross forms of nutrients and moisture conditions. Dark chestnut soils contain about three to five percent humus, whereas ordinary black soils contain more than six percent. In dark chestnut soils, the gross content of nitrogen is about 0.15–0.28 % and in ordinary black soils it is 0.30–0.50 %. The gross content of phosphorus in both soils is about 0.10–0.19 % and that of potassium 2 %. Due to an extensive level of crop production, the content of plant-available phosphorus in the topsoil is usually low.

Twelve physical, physicochemical and agrochemical parameters were analyzed to characterize the soil properties in the plots in the field experiments: humus, water-resistant and agronomically valuable aggregates, physical clay, cellulose-decomposing ability, plant-available mineral nitrogen, phosphorus and potassium, calcium, magnesium, pH value and soil moisture. These were measured by means of common methods used in soil science and agricultural chemistry on neutral and calcareous soils (Agrochemical Research Methods 1975; Chernenok 1993). The main factors determining yield production and its quantitative interaction with crop productivity were determined based on multiple regression and correlation analysis.

Under the arid climate conditions in the steppe zone, the rainfall is 200–330 mm a^{-1}. In the forest-steppe zone it is 350 mm a^{-1} and higher. In the capital Astana, which is situated in the area of investigation, the average daily temperature in the year is about 3 °C (Pogodaiklimat.ru 2013). The coldest month is January, with an average daily temperature of −14.2 °C and the hottest July, at 20.8 °C. The vegetation period is about 130 days. It begins in April/May and ends in September/October. The annual precipitation in Astana is 320 mm. The most rainfall occurs in July, with 50 mm on average.

2.2 Analytics of Plant-Available Phosphorus

Phosphorus and potassium were analyzed according to the Machigin method, which is based on the extraction of phosphorus and potassium using a 1 % solution of $(NH_4)_2CO_3$. It was developed in the 1950s under irrigation conditions in Uzbekistan, where the main crop was exclusively cotton (Machigin 1955). A description of this method modified by CINAO is given in the standard ГОСТ 26205-91 published in 1992(ГОСТ 26205-91 2013). The Machigin method has been recommended, inter alia, for carbonate soils in dry step zones. It is the only diagnostic indicator of plant-available soil phosphate conditions in Kazakhstan. The questions of how this method corresponds to the soil and climatic conditions of Northern Kazakhstan and which results can be obtained with other common

Table 1 Classification of the plant-available P_2O_5 in the topsoil according to Machigin in Kazakhstan and according to the VDLUFA recommendation (Kerschberger et al. 1997) in Germany

Machigin method		VDLUFA recommendation	
$(NH_4)_2CO_3$ extraction		Calcium acetate lactate extraction (CAL)	
Content level	mg P_2O_5 100 g^{-1} topsoil	Content level	mg P_2O_5 100 g^{-1} topsoil
1—very low	<1.0	A—very low	≤5
2—low	1.1–1.5	B—low	6–9
3—middle	1.6–3	C—optimum	10–20
4—increased	3.1–4.5	D—high	21–34
5—high	4.5–6.0	E—very high	≥35
6—very high	>6.0		

methods have not yet been acceptably investigated. Table 1 contrasts this method with the CAL method (Schüller 1969); the German standard method. The classifications offer major differences in the absolute content of that part of P which is solubilised by the specific method and then empirically defined as mobile phosphorus in the soil. The Machigin method is suitable for carbonate soils such as dark chestnut and ordinary black soils in Northern Kazakhstan, just as the CAL method in Germany is recommended for soils in Germany with a carbonate content of less than 15 %. The example demonstrates that nothing is comparable and transferrable, and consecutive calculations of the demand for phosphorus fertilizer must also be specific and are not transferrable to sites in other regions.

3 Results

3.1 Phosphorus Balance

Balance methods are crucial and internationally acknowledged tools for process analyses and the management of resources including phosphorus in agro-ecosystems (Fodor et al. 2013; Nagumo et al. 2013). The future need to recycle phosphorus alone (Dawson and Hilton 2011) underpins the crucial role of detailed phosphorus balancing. Balance methods require a basic understanding of nutrient cycling in ecosystems and reliable data on all balanced elements and compartments of the balance. The latter is not yet available for phosphorus balances in Northern Kazakhstan. Thus, balance methods of calculating the mineral fertilizer demand of crops are too uncertain because of unreliable input parameters. Currently, they are not yet a recommended way to optimize phosphorus fertilization in Kazakhstan (Chernenok 2009). Depending on the growth conditions some results of the nutrient removal of 0.10 t grain of wheat can be 2 or 3 times greater than others, while the efficiency of phosphorus in the soil varies 2.5–3 times and that in

Table 2 Variations in phosphorus fertilizer demand for spring wheat calculated using the balance method on the basis of P_2O_5 removal by yield and the efficiency of P_2O_5 in fertilizers and in the soil(*)

	Experimental data	Reference data
Nutrient removal, kg P_2O_5 per 0.10 t ha^{-1} yield	0.4–1.3	1.1
Efficiency of plant-available P_2O_5 in the topsoil %	10–30	10
Efficiency of phosphorus in fertilizers %	0.4–9.3	15
Calculated fertilizer demand, kg ha^{-1} (P_2O_5)	89–500	106

(*) Long-term field experiments, planned yield 2 t ha^{-1}, plant-available nitrogen 40 mg kg^{-1} soil (depth of 0–40 cm) and P_2O_5 20 mg kg^{-1} soil (depth of 0–20 cm), high potassium rate

fertilizers 2.5–20 times. Accordingly, the calculated amount of phosphorus fertilizer has a large range, too (Table 2). The results of the long-term field experiments show that under low and unstable moisture conditions, it is not acceptable to calculate fertilizer amounts based on the balance method because of the high variation of the inputs. Similar results were obtained in field trials with peas (*Pisum sativum* L.) (Chernenok and Nurmanov 2007; Chernenok and Kudashev 2010).

In general, these data demonstrate that phosphorus fertilizers display low efficiency under the soil and climate conditions in Northern Kazakhstan.

3.2 Fertilizer Application Rates, Plant-Available Phosphorus and Crop Yield

Table 3 gives an example of a phosphorus fertilization experiment for a duration of 17 years. The results show that the crop yield increased significantly with enhanced fertilizer application rates, but the variability of the data is high. Over the years, the maximum yield increase was achieved by different fertilizer application rates. This is caused by the different initial content of plant-available phosphorus and nitrogen and their ratio in the soil under the influence of changing soil moisture conditions (Chernenok 1996). The data make it difficult to derive profound conclusions for optimum fertilizer rates.

Furthermore, the results of field experiments indicate that a concentration of no more than 29–30 mg of soluble P_2O_5 per kg soil is suboptimal to achieve high productivity in spring wheat (Table 4). With the help of phosphate fertilizer the yield could be increased even more, by 14–17 %.

The experiments revealed a close correlation between the concentration of plant-available phosphorus in the topsoil according to the Machigin method and the yield in a cereal-fallow crop rotation (Fig. 1). This figure is an example of a Chernozem soil where the highest yield up to 4 t ha^{-1} grains was formed with a concentration of 36 mg of mobile P_2O_5 per kg of soil in the 0–20 cm soil layer. The shape of the curves was similar in other years, but the top of the curve had a

Table 3 Effect of phosphorus fertilizer on the productivity of spring wheat following fallow, example of a long-term experiment

Treatment kg ha^{-1} P_2O_5	Yield without fertilization and yield increase by P fertilizer in different years, t ha^{-1}											
	1	2	3	6	7	8	9	12	13	14	17	Average	
P0	Absolute yield												
	1.02	2.09	1.13	1.48	0.78	1.48	3.02	0.80	0.81	0.94	3.95	1.59	100 %
	Yield increase												
P30	–	0.14	0.29	–	0.25	0.14	0.25	0.40	0.11	0.41	0.48	0.27	17 %
P60	0.06	0.27	0.33	0.36	0.39	0.18	0.26	0.45	0.19	0.58	0.47	0.32	20 %
P90	0.15	0.62	0.44	0.40	0.53	0.27	0.26	0.53	0.36	0.67	0.62	0.44	28 %
P120	0.26	0.67	0.41	0.28	0.74	0.28	0.24	0.55	0.23	0.66	0.69	0.46	29 %
CD at 5 %	0.12	0.27	0.17	0.17	0.21	0.20	0.23	0.21	0.12	0.14	0.13		
mg $P_2O_5^{(*)}$ per kg soil	23.8	29.0	29.4	30.0	22.8	24,8	–	25.7	23.6	26.8	30.3		

(*) Plant-available phosphorus, Machigin method, 0–20 cm soil layer

Table 4 Efficiency of phosphorus fertilizers in 5-field cereal-fallow crop rotation

Plant-available P_2O_5 mg kg^{-1} soil(*)	Applied fertilizer kg P_2O_5 ha^{-1}	Number of crop rotations	Average increase of yield of a crop rotation	
			t ha^{-1}	%
29.9	30	10	0.60	9.3
28.8	60	12	0.83	14.4
28.8	90	12	1.02	17.4
29.2	120	9	0.93	16.5

(*) Plant-available phosphorus, Machigin method, 0–20 cm soil layer

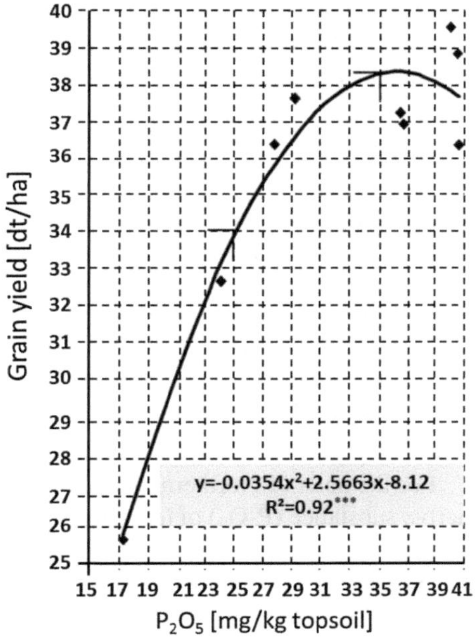

Fig. 1 Correlation between plant-available P and grain yield of spring wheat on a Chernozem

slightly different level. On the average, the highest level of grain yield was achieved at a P_2O_5 concentration of 35 mg kg^{-1} soil at a depth of 0–20 cm. This also holds for the whole crop rotation. The correlation coefficient between the phosphorus content in the soil and the crop yield in any individual year was never lower than 0.70 both in dry and moist years and confirms its high significance. It is important to note, however, that the yield level varied greatly over the years; in wheat, for example, it ranged from 0.8 to 2.6 t ha^{-1}. It is caused by different solubilities and availabilities of water-soluble and citrate-soluble phosphorus salts depending on the moisture conditions in those years and accordingly by varying concentrations of the soil solution.

An increased concentration of plant-available P_2O_5 in the topsoil, rising above 35 mg per kg of soil, had no positive effects or decreased the yield. This indicates that an amount of 35 mg P_2O_5 per kg of soil is the optimum level for spring wheat, which allows its maximum productivity to be achieved with different soil types and under different moisture conditions typical for the region.

3.3 Phosphorus Fertilizer Requirement

Knowing the current and the optimal level of plant-available phosphorus in the soil, as well as the amount of fertilizer needed to increase the value of phosphorus by 1 mg of its plant-available form at a depth of 0–20 cm, it is possible to determine the deficiency of nutrients in the soil. Thus, it is possible to calculate the amount of fertilizer needed in order to optimize phosphorus nutrition, using the following formula:

$$D_f = \left(P_{opt.} - P_{cur.}\right) \cdot Q \left(kg \cdot P_2O_5\, ha^{-1}\right) \quad (1)$$

D_f Calculated amount of fertilizer
P_{opt}. Optimum amount of plant-available P_2O_5 in the 0–20 cm soil layer (Machigin method)
P_{cur}. Current content of plant-available P_2O_5 in the 0–20 cm soil layer (Machigin method)
Q Amount of fertilizer (kg ha^{-1} P_2O_5) to compensate for the deficit of 1 mg mobile P_2O_5 per kg soil

For the soils of Northern Kazakhstan, the parameter Q is calculated at 10 kg of active substance (P_2O_5) of fertilizer. This equation can be seen as versatile for most conditions and crops, for which the optimum amount of mobile P_2O_5 in the soil for the crops and the parameter Q are known. Under soil conditions varying from those in the area of investigation, these two inputs (P_{opt}., Q) may differ.

Studies have shown that plants with different biologies require different levels of plant-available phosphorus saturation in the soil (Chernenok and Nurmanov 2007; Chernenok and Kudashev 2010). For example, the optimal phosphorus content in the soil (depth of 0–20 cm) for pea (*Pisum sativum* L.) ranges from 28 to 30 mg of P_2O_5 per kg of soil. A concentration of less than 28 mg of P_2O_5 per kg of soil is the lowest threshold down to which it is recommended to fertilize with the amount of phosphorus calculated with the help of the equation above.

Knowing the optimal phosphorus concentration in the topsoil for particular plants allows crop rotations to be planned in a more efficient way, by identifying those crops with the greatest need for phosphorus fertilizer, and by placing crops with a lower phosphorus need behind them.

An effective use of phosphate fertilizer just requires it to be applied at a depth of no less than 12 cm to a soil layer with stable moisture.

3.4 After-Effect of Phosphorus Fertilizers, Moisture Conditions and Soil Grading

The main long-term after-effect of phosphorus in crop rotation is its chemical absorption in the soil. However, the best effect of phosphorus fertilization was observed in the first year after fallow due to an accumulation of moisture and mobile nitrogen in soils on fallow land (Table 5). In subsequent years, the increased yield was lower. Nevertheless, the after-effect of applying phosphate is clearly visible from the cumulated yield increase of the first crop rotation. The duration of the after-effect also depends on the level of fertilizing. The input of 30 kg per ha^{-1} of P_2O_5 (P30) ensured an increase in the yield of 0.69 t ha^{-1} in total, over a period of four years up to the end of the first crop rotation. The P60 treatment guaranteed a reliable yield increase cumulated 1.16 t ha^{-1} over five years up to the wheat of the subsequent crop rotation. Treatment with P90 and P120 had an after-effect over 7–8 years with 2.2–2.6 t ha^{-1} total increase in yield.

As can be seen, there is a long after-effect of applied phosphorus, so that a single application of calculated fertilizer amount to the leading crop in the rotation is to be recommended. For the subsequent crop rotation, it is sufficient only to compensate for the total phosphorus removal by plants in the crop rotation, without having to conduct additional surveys. This approach eliminates the present pattern in the application of fertilizers and provides their most efficient use. It enables the optimal phosphorus nutrition of all crops in the rotation with the aim of achieving a maximum yield under unpredictable moisture conditions.

The optimal nutrient regime also provides a relatively efficient water consumption per unit of yield. A quantitative interaction has been identified between the content of plant-available P_2O_5 in the topsoil and the nutrient removal of phosphorus by crops as a result of different moisture conditions (Table 6). Based on this knowledge, it is possible to

- determine the level of mobile P in the topsoil required to achieve the planned yield,

Table 5 The after-effect of P fertilizer applied on fallow land in a five-field crop rotation (fallow, 3 times spring wheat, barley) on dark chestnut soils

Treatment kg P_2O_5 ha^{-1} fertilizer	Spring wheat 2nd field	3rd field	4th field	Barley 5th field	Total
	Absolute yield. t ha^{-1} grain				
PO (control)	0.78	1.87	2.31	1.78	
	Increase in yield, t ha^{-1} grain				
P30	0.25	0.10	0.20	0.14	0.69
P60	0.39	0.13	0.31	0.22	1.05
P90	0.53	0.22	0.34	0.32	1.48
P120	0.74	0.31	0.32	0.33	1.70

Table 6 Interaction of yield with plant-available P_2O_5 in the soil (depth of 0–20 cm)

mg P_2O_5 per kg soil (Machigin method) 0–20 cm soil layer	Spring wheat, t ha^{-1} grain		
	Humidity conditions		
	Very dry	Medium	Moist
10	0.3–0.4	0.7–0.8	1.0–1.2
15	0.5–0.6	1.0–1.2	1.7–2.0
20	0.6–0.7	1.4–1.6	2.2–2.5
25	0.8–0.9	1.7–2.0	2.8–3.0
30	1.0	2.1–2.5	3.3–3.7
35	1.1–1.2	2.5–3.0	3.8–4.0
Nutrient removal, mg P_2O_5 per 0.1 t grain	3.0	1.2–1.4	0.8–1.2

- predict the yield that can be achieved by knowing the content of mobile P in the topsoil,
- calculate the amount of P fertilizer required in order to obtain the planned increase in yield by determining it with the help of the equation or the P_2O_5 removal of yield in years with average moisture.

For example, in order to improve the yield by 0.5 t ha^{-1} it is necessary to increase the quantity of plant-available P_2O_5 by 6 mg per kg of soil by fertilization. The amount of phosphorus to be applied can be calculated in the following way. The planned increase in yield of 0.5 t ha^{-1} multiplied by 1.2 (mg P_2O_5 removal per 0.10 t yields) gives 6 mg of P_2O_5. By multiplying this calculated value with factor "Q" from Eq. 1 ($Q = 10$ for the soils in Northern Kazakhstan), we obtain a quantity of 60 kg per ha^{-1} of P_2O_5 to be applied.

The current restriction to 30 mg of mobile P_2O_5 per kg soil (optimum level according to Machigin grading) does not allow a leading crop such as spring wheat to achieve its potential yield, as can be seen in Table 6. The results allow the zonal grading of plant available phosphorus in the topsoil to be corrected (Table 7).

Table 7 Soil grading for cereals by the content of plant-available P_2O_5 in the soil at a depth of 0–20 cm and nutrient efficiency of phosphorus in fertilizers

Soil grading				Calculated(*) efficiency of P fertilizers
Level		Plant available P_2O_5 mg kg^{-1} soil(**)	Efficiency of P fertilizers	
I	Very low	<15	Very high (30–50 %)	–
II	Low	15–25	High (20–30 %)	24 %
III	Average	25–35	Average (10–20 %)	13 %
IV	Increased	35–45	Low (5–10 %)	6 %
V	High	>45	No	0 %

(*) Based on data from 12 fields in the crop rotation with 60 kg per ha^{-1} of P_2O_5 fertilization
(**) Plant-available phosphorus, Machigin method, 0–20 cm soil layer

4 Discussion

Our long-term results showed a clear empirical relationship between plant-available phosphorus in the soil and crop yields in Northern Kazakhstan. This enabled the development of a simple and practicable method for optimizing the mineral P fertilization in this region. This approach coincides with most common strategies in other countries to use a certain level or range as an orientation value and to apply fertilizer on demand.

On the other hand, this common strategy could not prevent the wasteful handling of P fertilizer and irreversible damage to all ecosystems worldwide. It shows that all orientation values or thresholds for monitoring and controlling ecosystem states, when based on single disciplinary criteria and local empirical methods, face a certain risk. They must be re-evaluated from time to time and must be embedded in more complex decision algorithms.

First of all, the traditional empirical basis of most common methods of analyzing the plant-available phosphorus seems to require international experiments in order to compare and harmonize them.

A number of methods are available for extracting plant-available P (Soinne 2009; Elrashidi 2011; Njukeng et al. 2013).

In Europe, besides mineral P fertilizer, a number of different P-containing fertilizers such as biowaste and sludges are applied, and there are many different and rarely comparable methods for the extraction of soluble P. Janßen (2004) found that chemical bonds are very different in different materials: soil, sludge, biowaste and treated biowaste. The amount of P in the extraction solution of the different extraction methods depends on the extraction power of the methods applied. There is an urgent need for the methods to be harmonised. Janßen (2004) found a preference for the calcium chloride/DTPA method (CAT method, VDLUFA 1997), but discussed possible problems of soils rich in carbonate and having a higher pH. These are precisely the soils we have analyzed in this chapter.

The pH of the extraction solution is a most decisive parameter. It varies widely between methods. For example, the Chirikov method works at pH 2.5, the CAL method at pH 3.8–4.1, the Olsen method at pH 8.5, and the Machigin method at pH 9 (Machigin 1955; Schüller 1969; Janßen 2004; Khristenko and Ivanova 2012).

For soils which are similar to those in Northern Kazakhstan, the Olsen method (Olsen et al. 1954; Sims 2000) is the most recommended method worldwide. It also includes some troubleshooting (Miller and Horneck 2002). There are indications that the Machigin method performs well or even better in those environments (Li et al. 2011). However, this should be tested on a more sophisticated basis.

The empirical formula (1) given in this paper for calculating the fertilizer demand has regional validity for Northern Kazakhstan. Its principle (calculating the amount of fertilizer required to raise the content of plant-available P_2O_5 by 1 mg per 100 g of soil) cannot currently be applied in Germany or elsewhere. It would lead to unrealistically high recommendations for P fertilization in some cases. However, as recommended in this chapter, in Germany it is good agricultural practice to maintain the optimum level of phosphorus concentration in the topsoil, according to VDLUFA grading, and to fertilize with the amount of phosphorus which will be removed by the planned yield.

The formula is very easy in practice but does not consider other P sources, soil conditions and cropping systems. Also, other aspects needing to be taken into consideration are the different types of both mineral and organic P fertilizers that vary in solubility (Chien et al. 2011) and other properties, and finally, reserves of overall P contents in soils and plants.

Sophisticated models and balancing methods, well calibrated and validated to international long-term experiments, will be a more successful way to assess demand for P fertilizers in future.

References

Agrochemical research methods (1975) Moscow, p 655

Chernenok VG (1993) Theoretic basis of optimization and diagnosis of mineral nutrition of cereal crops in dry steppe zone of Northern Kazakhstan. Report of Doctor of Ag.Sc. Omsk, p 54

Chernenok VG (1996) Issue of use of balance methods for determining the doses of fertilizers in low and unstable moisture conditions. Bulletin of Science. Akmola Agricultural Institute. Akmola, pp 58–65

Chernenok VG (2009) Scientific basis and practical methods of management of soil fertility and crop productivity. Astana, p 66

Chernenok VG, Kudashev AB (2010) Coefficients of using nutrients from soil and fertilizers by peas. Bull Sci KazATU S.Seifullin Astana 2(60):28–35

Chernenok VG, Nurmanov ET (2007) Optimization of mineral nutrition of chickpea. Bull Sci KazGATU S.Seifullin Astana 2:69–76

Chien SH, Prochnow LI, Tu S, Snyder CS (2011) Agronomic and environmental aspects of phosphate fertilizers varying in source and solubility: an update review (review). Nutr Cycl Agroecosyst 89(2):229–255

Dawson CJ, Hilton J (2011) Fertiliser availability in a resource-limited world: production and recycling of nitrogen and phosphorus. Food Policy 36, Supplement 1, Jan 2011, pp 14–22

Elrashidi MA (2011) Selection of an appropriate phosphorus test for soils. Soil properties affecting selection of the appropriate phosphorus test and recommended methods. ftp://ftp-fc.sc.egov.usda.gov/NSSC/Analytical_Soils/phosphor.pdf, http://soils.usda.gov/technical/methods/. Accessed on 22 Mar 2013

Fodor N, Csathó P, Árendás T, Radimszky L, Németh T (2013) Crop nutrient status and nitrogen, phosphorus, and potassium balances obtained in field trials evaluating different fertilizer recommendation systems on various soils and crops in Hungary. Commun Soil Sci Plant Anal 44(5):996–1010

ГОСТ 26205-91 (2013) Soils. Determination of mobile compounds of phosphorus and potassium by Machigin method modified by CINAO). http://www.gosthelp.ru/gost/gost10373.html. Accessed on 20 Mar 2013

Guan G, Tu S, Li H, Yang J, Zhang J, Wen S, Yang L (2013) Phosphorus fertilization modes affect crop yield, nutrient uptake, and soil biological properties in the rice-wheat cropping system. Soil Sci Soc Am J 77(1):166–172

Janßen E (2004) Extraction of soluble phosphorus in soil, sludge, biowaste and treated biowaste. http://www.ecn.nl/docs/society/horizontal/hor_desk_25_solubleP_revised.pdf. Accessed on 22 Mar 2013

Kerschberger M, Hege U, Jungk A (1997) VDLUFA-Standpunkt: Phosphordüngung nach Bodenuntersuchung und Pflanzenbedarf. Selbstverlag VDLUFA, Darmstadt, Sept 1997, p 8

Khristenko A, Ivanova S (2012) Improvement of diagnosis accuracy of phosphate status for Ukrainian soils. Better crops, vol 96(2). http://eeca-en.ipni.net/article/EECAEN-2032. Accessed on 22 Mar 2013

Li S-X, Wang Z-H, Stewart BA (2011) Chapter three—differences of some leguminous and nonleguminous crops in utilization of soil phosphorus and responses to phosphate fertilizers. Adv Agron 110:125–249. http://www.sciencedirect.com/science/article/pii/B9780123855312000037. Accessed on 22 Mar 2013

Machigin BP (1955) On effectiveness of not root fertilizing of cotton. Proceedings of the SoyuzNIHI, Tashkent ed. SAGU 45–54

Miller RO, Horneck D (2002) Bicarbonate (Olsen) phosphorus troubleshooting. Online source:http://www.naptprogram.org/files/napt/publications/method-papers/2002-bicarbonate-phosphorus-p-troublehooting.pdf. Accessed on 22 Mar 2013

Nagumo T, Tajima S, Chikushi S, Yamashita A (2013) Phosphorus balance and soil phosphorus status in paddy rice fields with various fertilizer practices. Plant Prod Sci 16(1):69–76

Njukeng NJ, Nkeng EG, Ehabe EE, Schnug E (2013) A comparative study on the use of calcium acetate lactate, calcium chloride and acidic ammonium acetate- ethylene diaminetetra acetic acid (AAAc-EDTA) for the quantification of extractable, P, K and Mg from acidic soils. Int Res J Pure Appl Chem 3(1):22–31

Olsen SR, Cole CV, Watanabe FS, Dean LA (1954) Estimation of available phosphorus in soils by extraction with sodium bicarbonate. USDA Circular 939. 1990. U.S. government printing office, Washington D.C

Ongley ED (1996) Control of water pollution from agriculture—FAO irrigation and drainage paper 55, food and agriculture organization of the United Nations. Rome. http://www.fao.org/docrep/w2598e/w2598e00.htm#Contents. Accessed on 22 Mar 2013

Pogodaiklimat.ru (2013) Климат Астаны. http://www.pogoda.ru.net/climat/35188.htm. Accessed on 27 Mar 2013

Römer W (2009) Ansätze für eine effizientere Nutzung des Phosphors auf der Basis experimenteller Befunde. BMELV Berlin. Berichte über landwirtschaft, band 87, pp 5–30

Schroetter S, Rivas M, Lamas M, Schnug E, Fleckenstein J (2005) Einfluss von Parametern der Bodenfruchtbarkeit und Düngung auf den Uran-Transfer Boden-Pflanze. Landbauforschung Völkenrode, special issue 286, pp 95–105

Schüller H (1969) Die CAL-Methode, eine neue Methode zur Bestimmung des pflanzenverfügbaren Phosphates in Böden. Z. Pflanzenernähr., Bodenkde, vol 123, pp 48–63

Sims P (2000) Soil test phosphorus: Olsen method. In: Pierzynski GM (ed) Methods of phosphorus analysis for soils, sediments, residuals, and waters. Southern cooperative series bulletin, No 396, p 20–21. http://www.soil.ncsu.edu/sera17/publications/sera17-2/pm_cover.htm. Accessed on 22 Mar 2013

Smidt GA, Parks A, Pokharel R, Landes F, Schnug E, Koschinsky A (2011) Indications for phosphorus fertilizer derived uranium mobilization from arable soils to groundwater. Berichte der DBG. http://www.dbges.dehttp://eprints.dbges.de/700/1/Smidt_Berichte_DBG_BERLIN_2011.pdf. Accessed on 22 Mar 2013

Soinne H (2009) Extraction methods in soil phosphorus characterisation—limitations and applications. Pro Terra No. 47, p 49. https://helda.helsinki.fi/bitstream/handle/10138/20847/extracti.pdf?sequence=1. Accessed on 22 Mar 2013

VDLUFA (1997) Schriftenreihe 45, Prüfung und Eignung der Calciumchlorid/DTPA-Methode nach Alt (CAT-Methode) zur Untersuchung von Ackerböden auf pflanzenverfügbare Nährstoffe. VDLUFA Verlag, Bonn

Part IV
Executive Summary

Executive Summary and Conclusions

Lothar Mueller, Abdulla Saparov and Gunnar Lischeid

1 Summary of the Book's Intention and Structure

This book deals with methods which could gain importance for better monitoring and management of water and land resources in Central Asia. Knowledge about the existence, principles and potential benefit of these methods could improve the work of researchers, the authorities, managers, and decision makers. Science teachers at universities have a particular responsibility and interest regarding innovation in the international scientific arena. They have to educate a new generation of academics who will make tomorrow's decisions. This must include profound knowledge about the most crucial resources for the prosperity and welfare of their nations: water and land. Ways of sustainably handling these resources must be initiated very soon.

Our book is intended to support this. It cannot provide full guidance on what steps to take, but can be a source of knowledge and inspiration. Neither the editors nor the authors of this book were seeking all-round approaches or complex solutions. Their intention was to analyse the status of water and land resources, to offer some tools to explain their intended benefits, and in most cases to indicate a framework for their effective functioning.

The book is structured to make it accessible to readers coming from different disciplines, the idea being that readers could pick out their topic of interest first.

L. Mueller (✉) · G. Lischeid
Leibniz-Centre for Agricultural Landscape Research (ZALF) e. V., Eberswalder Str. 84
15374 Muencheberg, Germany
e-mail: mueller@zalf.de

G. Lischeid
e-mail: lischeid@zalf.de

A. Saparov
Kazakh Research Institute of Soil Science and Agrochemistry named after U. U. Uspanov,
Al Faraby Ave 75b 050060 Almaty, Kazakhstan
e-mail: ab.saparov@yahoo.com

L. Mueller et al. (eds.), *Novel Measurement and Assessment Tools for Monitoring and Management of Land and Water Resources in Agricultural Landscapes of Central Asia*, Environmental Science and Engineering, DOI: 10.1007/978-3-319-01017-5_44,

Table 1 Thematic clusters

	Themes	Chapters
1	Status and trend analyses	6
2	Laboratory and field measurement methods	5
3	Methods of resource evaluation, functional mapping and risk assessment	11
4	Remote sensing methods for monitoring and modelling large areas	3
5	Methods of data analysis and ecosystem modelling	5
6	Methods of soil and water bioremediation	5
7	Field soil monitoring	2
8	Methods and technologies for optimising land use systems	6

Thus, the editors have kept the book's structure quite simple and non-hierarchical, with separate chapters by different authors in three sections.

The chapters in Part I, **"Environmental and Societal Framework for the Monitoring and Management of Land and Water Resources"**, analyse the status of land and water resources and the overall agricultural situation in the five Post-Soviet Central Asian States of Kazakhstan, Kyrgyzstan, Tajikistan, Turkmenistan and Uzbekistan. It was the authors' intention to do this from a distant, neutral perspective and to assess data and facts in a dispassionate, critical manner. Part II, **"Novel Methodologies for the Measurement of Processes and Assessment of Resources"**, presented and offered new tools and procedures. In most cases they were developed outside of Central Asia and have not yet been tested or tailored for their application in this region. Part II is the central section of the book. The methods presented in it have a particular degree of novelty as the authors of this section are their inventors, creators or activists. Part III, **"Applications and Case Studies"**, was intended to give examples of the practical application of novel methods presented in Part II. This would have been the ideal case, but it came true in exceptional cases only. Most chapters in Part III support Part I as they provide detailed information about some typical problems of local research. This is important for understanding the relevance and seriousness of the topic and gives inspiration for practical conclusions or applications, or for focusing on deeper research.

The themes of the book fit into 8 thematic clusters (Table 1).

In the following part of this (Part IV) we give a short review of all the individual chapters, analyse their importance for Central Asia in the context of other tools, and characterise the framework for their possible application. This will be done based on the thematic clusters listed in Table 1 above. For didactical reasons, some text is identical with parts of the authors' summaries.

2 Review of Thematic Clusters

Cluster 1, **"Status and trend analyses"**, characterised the current status of water and land resources and the framework conditions for monitoring and management options. This cluster is identical with Part I of the book. It includes the following chapters:

1. Mueller et al.: "Land and Water Resources of Central Asia, their Utilisation and Ecological Status" (Земельные и Водные Ресурсы Центральной Азии, их Использование и Экологическое Состояние"), (Chap. 1)
2. Saparov: "Soil Resources of the Republic of Kazakhstan: Current Status, Problems and Solutions" (Почвенные Ресурсы Республики Казахстан: Современное Состояние, Проблемы и Решения), (Chap. 2)
3. Ibrayev et al.: "Long-Term Monitoring and Water Resource Management in the Republic of Kazakhstan" (Долгосрочный Мониторинг и Управление Водными Ресурсами Республики Казахстан), (Chap. 3)
4. Suleimenov: "Trends in the Agriculture of Central Asia and Implications for Rangelands and Croplands" (Тенденции в Сельском Хозяйстве Центральной Азии и их Последствия для Пастбищ и Пашни), (Chap. 4)
5. Lischeid: "Landscape Hydrology of Rural Areas: Challenges and Tools" (Гидрология Ландшафтов Сельских Территорий: Проблемы и Методы), (Chap. 5)
6. Mueller et al.: "Productivity Potentials of the Global Land Resource for Cropping and Grazing" (Потенциальная Продуктивность Глобальных Земельных Ресурсов для Возделывания Сельскохозяйственных Культур и Выпаса) (Chap. 6).

(1) The analysis by **Mueller et al.** revealed that Central Asia is the global hotspot of a resource nexus. Land, water and food are key issues in this nexus. Food security cannot yet be considered to be ensured in Central Asia. Most problems occur in Tajikistan and Kyrgyzstan, where the overall economy is still poorly developed. Agriculture is crucial for the economies of all Central Asian countries and responsible for about 90 % of their water use. Land and water resources may fulfil their function of regional food supply, but the agricultural productivity of grassland and cropland is low. Agriculture is inefficient and not sustainable. Irrigation causes serious detrimental side effects of soil and water salinisation. Dryland farming, as currently practiced, involves a high risk of wind and water erosion. Rangelands seem to be particularly degraded, but there is a lack of objective criteria. Water bodies, aquatic, agricultural and grassland ecosystems are in a critical state with tendencies towards accelerated degradation and landscape desertification. Past and current industrial activities are a main source of water and land pollution. Competition for water is a permanent issue as all the main rivers in the region are trans-boundary. The downstream riparians Uzbekistan, Turkmenistan and Kazakhstan face increasing risks of water scarcity. Despite all these limitations, agricultural landscapes in Central Asia have great potential for a multi-functional use as a source of income for the rural population, tourism and eco-tourism included (Fig. 1). As a precondition, peaceful development must be provided.

Overcoming the crisis of agri-environmental research and applying advanced methods in science and technology are key issues for further development. Science and technology may provide an overall knowledge shift in assessing processes and initiating sustainable developments.

Fig. 1 Steppe and high mountains with snow and ice. This photo demonstrates the typical situation of the main resources: land and water. Melting water from high mountains is the basis for land use in the lowlands downstream, such as the valleys, steppes and deserts of Central Asia. Photo: Konstantin Pachikin

(2) The results of more detailed studies in Kazakhstan given by **Saparov** confirm the tendency towards intensive land degradation and desertification, salinisation, and soil contamination by oil, petroleum products, chemicals and radioactive substances, humus loss, decreasing soil fertility and the deterioration of the potential productivity and environmental status of soil resources. Sustainably developing agriculture in Kazakhstan and ensuring food and environmental safety are closely related to the rational and efficient use of soils. It is necessary to develop appropriate laws and regulations, programmes and related activities and to take measures to prevent rapid land degradation, desertification and the deterioration of the environmental situation, and to restore soil fertility.

(3) The results of long-term monitoring and water resource management were presented by **Ibraev et al.** They confirmed that the water resource situation in Kazakhstan is critical. Most urgent water management problems result from trans-boundary river constellations. Rivers such as the Irtysh, Ili, Syr Darya and others are required to provide a water supply for the population, agriculture and industry. All water bodies including reservoirs need to be controlled better, water scarcity and catastrophic floods must be avoided. Great progress in monitoring is expected from remote sensing methods and GIS technologies. This may enable water resources to be evaluated more effectively, emergency situations to be identified and action plans to be carried out (Fig. 2).

(4) **Suleimenov** showed that developments in agriculture are largely influenced by decisions of the past. During the first years of independence all Central Asian countries focused their policies on increasing wheat grain production as much as possible. Grasslands including rangelands were disregarded and are in a critical state. Livestock productivity might be improved only if measures to improve rangeland management are undertaken. It is the right time to change policies towards supporting rangeland improvement and integrated crop and livestock production. This has to include long-term soil conservation (Fig. 3).

Fig. 2 Water management and salinity control are essential for the sustainable use of water and land resources

(5) **Lischeid** pointed out the role of water resources for global food security and the development of rural areas. Agriculture inevitably depends on soil quality as well as on water resources. More than 70 % of human water use is for irrigation. In addition, transpiration from rainfed agriculture comprises a substantial part of the earth's water cycle. Thus, land use both highly depends on and affects water availability and is intimately intertwined with water resource management. Even today, groundwater and river water over-exploitation due to increasing irrigation is a matter of concern. In some regions groundwater levels have decreased by 1 m per year during the last few decades. Even large rivers are increasingly falling dry. This has severe implications for water resources further downstream, not to mention biodiversity aspects. Ineffective water management and land degradation are closely

Fig. 3 Flock control by shepherds is a common grazing practice. The regulation of stocking rates must be given top priority to avoid further rangeland degradation and desertification. For the grassland to be maintained and restored, grassland inventories must be carried out and allowable stocking rates must be estimated for each site taking into account productivity and environmental aspects

connected to each other. Climate change is now considered an increasing threat to water resources. Climate change models are fraught with major uncertainties with respect to precipitation. However, experts agree that in general the frequency and intensity of extreme events such as floods and drought are likely to increase, which places agricultural management and soil resources at increasing risk. Besides this, the melting of glaciers in mountainous regions currently increases water availability in the lowlands downstream, whereas the opposite will be true in the long term. Thus there is urgent need for "Integrated Water Resources Management" (IWRM) which also has to include soil quality management. Various ideas, concepts and methods exist. However, local conditions have to be considered. Thus, experience from different parts of the world needs to be shared. This book is intended to contribute to that.

(6) As a precondition for management and action plans, the status of resources must be monitored based on reliable, internationally acknowledged methods. Croplands and rangelands need to be assessed according to objective criteria. **Mueller et al.** conducted a global analysis of soil productivity potentials and methods for recognising them. The land productivity of the Central Asian countries is low compared with other global regions. However, productivity potentials are largely unknown as there are still no objective frameworks or criteria. A comparative analysis of several soil and land evaluation methods revealed the usefulness of indicator-based approaches, reliably but simply and consistently practicable over different scales, from the field level to soil function maps over large regions. Basic soil surveys including visual tactile soil structure assessments were identified as useful diagnostic tools to recognise productivity-limiting soil attributes and estimate indicator values. The Muencheberg Soil Quality Rating (M-SQR) could provide a thorough analysis of the status of land resources for high and sustainable agricultural productivity in Central Asia.

Cluster 2 describes "New Laboratory and Field Measurement Methods". The results of these chapters explain new measurement devices and systems for applications in agri-environmental research and monitoring. This thematic cluster includes the following chapters:

1. Schindler: "A Novel Method for Quantifying Soil Hydraulic Properties" (Новый Метод Количественной Оценки Почвенных Гидрологических Свойств), (Chap. 7)
2. Meissner et al.: "Advanced Technologies in Lysimetry" (Перспективные Технологии в Лизиметрических Наблюдениях), (Chap. 8)
3. Hertel and von Unold: "Third-Generation Lysimeters—Scientific Engineered Monitoring Systems" (Лизиметры Третьего Поколения—Научно Спроектированные Системы Мониторинга), (Chap. 9)
4. Schindler: "A Field Method for Quantifying Deep Seepage and Solute Leaching" (Полевой Метод Количественного Определения Глубокой Инфильтрации и Выщелачивания Растворённого Вещества), (Chap. 10)

5. Mueller et al.: "Simple Field Methods for Measurement and Evaluation of Grassland Quality" (Полевые Экспресс-Методы Оценки Изменения Состояния Травянистых Экосистем), (Chap. 11).

   (1) **Schindler** reported on the development of a novel laboratory method and equipment for measuring a soil's hydraulic properties. Knowledge of a soil's hydraulic properties, such as its water retention curve and unsaturated hydraulic conductivity function, is required for soil water modelling and various soil hydrological studies. The extended evaporation method (EEM, trade name HYPROP) is the most advanced measurement system in the world. It uses new tensiometers which provide delayed boiling and applies the air-entry pressure of the tensiometer's porous ceramic cup as the final tension value, allowing the quantification of both hydraulic functions close to the wilting point. Tailored software provides data recording, calculation, evaluation, fitting and export of the hydrological data. The method is based on soil sampling (Fig. 4).

   (2) The chapter by **Meissner et al.** reviews the application of lysimeters in research and monitoring studies and reports on the author's own new developments. Quantifying soil water flow in situ is a prerequisite for accurately predicting solute transfer within the unsaturated zone. Monitoring these fluxes is challenging because the results have to answer both scientific and practical questions regarding groundwater protection, the sustainable management of agricultural, forestry, mining or set-aside industrial areas, or how to reduce leachate loss from landfills or explain the fate of harmful substances to the environment. In Europe, new lysimeter techniques have been developed. It is now possible to collect large monolithic soil columns and to measure the soil water balance of these monoliths (with a surface area of 0.03–2 m^2 and depths up to 3 m) with high precision (±20 g). Furthermore, progress in lysimetry means that as well as determining precipitation and seepage, experiments can

Fig. 4 Soil sampling for assessing soil quality and measuring soil hydraulic properties in a joint Kazakh-German project

ascertain the mass input of dew and calculate actual evapotranspiration. In addition to gravitation lysimeters, weighable groundwater lysimeters have been developed and successfully tested.

(3) **Hertel and von Unold** reported on special lysimeters for monitoring the hydraulic and thermal conditions of the surrounding field. Different lysimeter layouts were developed for the special requirements and increase the measuring capacity in the respective tasks, e.g. various treatments of fertilisation or irrigation under identical climatic conditions. As an example, the meteorological lysimeter measures precipitation, evapotranspiration and amount of leachate. To achieve this, the lysimeter weighs a certain mass of soil in the gram range with a defined surface over an extended period of time and supplies us with precise data on the input of water such as rain, dew, frost or snow, on water output by evapotranspiration and leachate, and on changes in the soil water content.

(4) **Schindler** developed a cost-effective hydrological field setup which can provide lysimeter-like results for water and solute transport at several sites. The presented method is designed to estimate deep seepage and solute leaching in the field based on soil hydrological measurements below the zero flux plane, based in turn on a calibrated hydraulic conductivity function. The method is simple to apply, flexible and cost saving compared with lysimeters. The required soil hydraulic properties are derived from tension and water content field recordings at the measurement depth. After calibration no further information on soil properties, weather, management and land use or other data are required. Since 1994, the method has been successfully applied at some plots in Northeast Germany. A comparison between lysimeter discharge measurements and discharge calculations confirmed the validity of this method on sandy to loamy soils with a deep water table and a not excessively fluctuating zero flux plane.

(5) **Mueller et al.** reported on simple field analysis methods for measuring and assessing grassland quality. Soils under grassland can be heavily affected by soil and water management. Simple field methods may help to detect properties and processes limiting their use. Methods were presented characterising aspects of the physical, chemical and biological status of grasslands in conjunction with a soil survey. The soil strength, measured with penetrometers, sink cones and shear testers may characterise spatio-temporal changes in soil conditions best. Further important attributes of the soil water status can be measured using TDR probes, field tensiometers and simplified infiltration experiments. A number of expert- based approaches are available to estimate site properties with visual-tactile methods and the carry out the bio-indication of site properties using vegetation analyses. Some of these methods can be applied solely to study particular aspects of the grassland quality such as trampling effects under different animals and grazing systems or the trafficability of meadows, testing the compaction status of the soil, the quality of the soil structure,

vegetation and other factors (Fig. 5). In many cases a set of methods in combination with an overall assessment of the soil quality (Muencheberg Soil Quality Rating) would be useful for a reliable assessment of the status of Central Asian grasslands.

Fig. 5 Grassland monitoring and research for sustainable use as pasture and winter forage (hay, silage). Cattle bred for "Uckermärker" meat are widespread in those parts of Germany having a relatively dry and continental climate. This very robust breed originates from "Fleckvieh X Charolais". Uckermärker are relatively peaceful and often hornless. Bulls weigh up to 1,300 kg and cows about 850 kg. This was one reason to test their grazing pattern and overall action on soils and vegetation at the Paulinenaue Research Station. The suitability of this new breed and other ones should also be tested in Kazakhstan

Cluster 3, "Methods of resource evaluation, functional mapping and risk assessment", includes expert-based assessment methods of resource and process evaluation. It contains the following chapters:

1. Helming: "Impact Assessment for Multifunctional Land Use" (Оценка Воздействия для Обеспечения Многоцелевого Землепользования), (Chap. 12)
2. Mueller et al.: "The Muencheberg Soil Quality Rating for Assessing the Quality of Global Farmland" (Мюнхебергская Система Рейтинга Качества Почв как Метод Оценки Земель Сельскохозяйственных Угодий в Мировом Масштабе), (Chap. 13)
3. Smolentseva et al.: "Assessing the Soil Quality and Crop Yield Potentials of Some Soils of Eurasia" (Оценка Качества Почв и Потенциальной Урожайности Зерновых на Примере Почв Евразии), (Chap. 31)
4. Pachikin et al.: "Soils of Kazakhstan, their Distribution and Mapping" (Почвы Казахстана, их Распространение и Картографирование), (Chap. 32)
5. Hennings: "Use of Pedotransfer Functions for Land Evaluation: Mapping Groundwater Recharge Rates under Semi-Arid Conditions" (Составление Карт Скорости Пополнения Запасов Грунтовых Вод в Полуаридных

Условиях с Использованием Педотрансферных Функций для Оценки Земель), (Chap. 14)

6. Eulenstein et al.: "Nutrient Balances in Agriculture—A Basis for the Efficiency Survey of Agricultural Groundwater Conservation Measures" (Баланс Питательных Элементов в Сельском Хозяйстве как Основа для Оценки Эффективности Мер по Охране Грунтовых Вод), (Chap. 15)
7. Dannowski et al.: "Methods of in situ Groundwater Quality Monitoring—Basis for the Efficiency Survey of Agricultural Groundwater Conservation Measures" (Методы Полевого Мониторинга Качества Грунтовых Вод—Основа Эффективности Сельскохозяйственных Мероприятий по Сохранению Грунтовых Вод), (Chap. 16)
8. Godbersen et al.: "Methods in the Exploratory Risk Assessment of Trace Elements in the Soil-Groundwater Pathway" (Поисковые Методы Оценки Риска в Системе Почва-Грунтовые Воды для Микроэлементов), (Chap. 17)
9. Funk et al.: "Methods for Quantifying Wind Erosion in Steppe Regions" (Методы Количественной Оценки Ветровой Эрозии в Степных Регионах), (Chap. 18)
10. Schreiner and Meyer: "Indicators of Land Degradation in Steppe Regions – Soil and Morphodynamics in the Northern Kulunda" (Индикаторы Деградации Земель Степных Регионов - Почва и Морфо-Динамика в Северной Кулунде), (Chap. 33)
11. Shokparova et al.: "Erosion Rates Depending on Slope and Exposition of Cropped Chestnut Soils" (Скорости Эрозии в Зависимости от Склона и Экспозиции Возделываемых Каштановых Почв), (Chap. 34).

(1) **Helming** explained the need, elements and procedure of Impact Assessment for multifunctional land use. Land is a scarce and limited resource that has to fulfil many functions in addition to the production of food and fibre. It is used for water regulation, and carbon sequestration, and as a biodiversity pool, an area for settlement and infrastructure, human health, recreation, and employment and a base for economic activities. In many regions of the world these functions have to be fulfilled simultaneously in space and time. Ex-ante impact assessment can support decision making and land use management to achieve a balance. To make the concept of multifunctional land use operational for impact assessment and decision making, the Land Use Function (LUF) framework was developed for scientific purposes. This approach was also implemented in a European typology for planning and decision making. The Impact Assessment procedure could be implemented as part of the Central Asian land use strategy.

(2) Sustainable use of soils is a vital issue in the 21st century to meet the global challenges of food security, demands for energy and water, climate change and biodiversity. Tools for reliably, simply and consistently evaluating the status of the soil over a wide range of scales can help assess its suitability for crop growth and yield potentials. **Mueller et al.**

presented a uniform methodology for comprehensive analyses of the status and productivity potentials of land. The Muencheberg Soil Quality Rating (M-SQR) may serve as a framework for evaluating the performance of the land resources of Eurasia for cropping and grazing. Soil properties that limit crop yields and crop productivity potentials can be assessed consistently over large regions. The approach is based on 8 basic indicators and at least 12 hazard indicators. The soil quality is rated during a normal soil survey mainly by applying visual methods of soil evaluation. A field manual provides rating tables based on response curves and thresholds for different hazard indicators (such as risk of drought) (Fig. 6). Finally, overall soil quality rating scores ranging from 0 (worst) to 100 (best) characterise crop yield potentials. The current approach is valid for grassland and cropland. The authors anticipate the creation of comparable soil functional maps of the whole of Eurasia using this method. The method could become a component of the Impact Assessment procedure explained in the chapter by Helming, with a focus on assessing the productivity of land.

(3) **Smolentseva et al.** reported on field tests of the M-SQR method. For this purpose, representative soil catenas and single soil pits were dug, analysed and classified on test sites in Russia, Kazakhstan and Germany. Soil quality and crop yield potentials were assessed. The results showed that estimating all the components of the site-specific water balance and assessing the drought risk is key for the evaluation of soil functions in agricultural landscapes. The authors found close correlations between

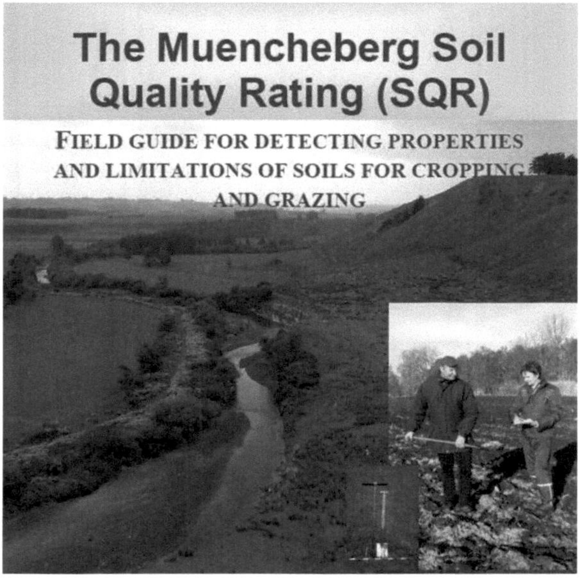

Fig. 6 The field manual of the Muencheberg Soil Quality Rating contains rating tables of soil quality indicators. This new approach enables a rating of soil quality and productivity potentials of cropland and grasslands over Eurasia based on a uniform methodology. The manual can be downloaded from the internet

the overall soil quality rating score and the grain yields of cereals. The suitability of the M-SQR approach for the consistent assessment of soil quality and crop yield potentials on Eurasian spatial scales has been confirmed (Fig. 7).

(4) **Pachikin et** al. reported on their experience with soil and land mapping in Kazakhstan. This task is challenging because of the variety of climates and forms of relief, ranging from high mountains to desert plains. The natural zones and soil zones of Kazakhstan's large orographic-climatic regions with their variety of zones and belts (Altay, North Tien-Shan, West Tien-Shan) need to be considered. The geographic soil zones in Kazakhstan from north to south are: forest-steppe, steppe, desert-steppe, and desert, where the zones are classified according to the types of soil. The regions and belts were characterised according to basic climatic indexes and typical vegetation. Soil mapping provides further research in soil science. The soil map can be used as a basis for issues connected with basic research such as the classification and taxonomy of soils based on the genesis of soils, and with applied research such as their classification regarding agro-industrial suitability, land reclamation, the estimation of soil degradation and more. The creation of modern soil maps is based on field surveys, airborne data and GIS-technology. Some examples of thematic maps are shown. They allow scientists to estimate the

Fig. 7 Extreme soil profiles in steppe regions. **a** Deep (Pachic) chernozem. The soil profile has optimum physical, chemical and biological properties. **b** Solonchak: the extremely high salt content is hostile to vegetation

status of land and soil resources, to analyse the influence of anthropogenic factors, or to decide whether areas should be designated as protected. These kinds of maps will allow measures to be developed for the restoration of physical landscapes and for sustainable land development.

(5) Regional maps of land, water and other resources are important for resource monitoring and land use planning. **Hennings** presented a methodology to derive hydro-pedotransfer functions to estimate annual percolation rates from available information on climate, soil characteristics and land use. The method was developed for the new Hydrological Atlas of Germany and also successfully applied in the Arab region. The approach is based on the FAO56 concept, the CLIMWAT database and the CROPWAT model. The first step was to carry out simulations for eight countries, three kinds of land use and varying soil hydrological properties. The second step, was to use meteorological data from Syria to carry out simulations for six land use scenarios and varying soil hydrological properties. The resulting country-specific regression equations and nomograms are presented as well as the general magnitude of groundwater recharge under typical crops of the Eastern Mediterranean environment. This method has potential for a better estimation of groundwater recharge rates and renewable water resources, especially in Uzbekistan, Kazakhstan and Turkmenistan. Lysimeters as presented by Meissner et al. (Chap. 8) and Hertel and von Unold (Chap. 9), soil hydrological field setups (Chap. 10) and simple tracer applications may provide calibrations and validations of the method in Central Asia.

(6) In the same context of protecting aquifers and open water bodies against pollution by agriculture, **Eulenstein et al.** explained different balancing methods for the identification of nitrate problems and risks to water bodies. This can be carried out by complete operational balance sheets for farm balances or field balances. This evaluation is very important to gain an overview of the nutrient surplus, as the nutrient surplus can contribute to a series of environmental concerns, including water eutrophication, air pollution and the release of greenhouse gases. Other methods which can be applied include applicable measurements and estimation of seepage, groundwater recharge and water quality by deep drills and analyses of pore water in soil samples, the application of tracer methods, or lysimetry. The simulation models presented in Cluster 5 are also recommended. All the methods mentioned and explained in this and in the next chapter are named in the latest methodical guidelines of the German Association for Water, Wastewater and Waste (DWA). These methods must be taken into consideration when controlling agriculture in water protection areas in Germany. They are transferrable to other regions.

(7) **Dannowski et al.** described the technical requirements and methodical guidelines for monitoring the efficiency of agricultural practices for the protection of groundwater resources. The focus is on sampling aquifers.

This holds true especially for areas under predominantly agricultural use, to protect against nitrates from plant production. A catalogue of agricultural groundwater conservation measures is available. It is to be supplemented by a methodology for in situ groundwater quality monitoring as the basis for surveying the efficiency of those measures. General characteristics and the benefits and disadvantages of recent monitoring techniques are presented and summarised under the term "appropriateness for efficiency survey". Against the background of the paramount importance of fresh water resources for human well-being in Europe's industrialised societies, the protection of groundwater bodies is one of the top environmental ambitions of the European Union. The chapter can serve as a basis for assessing whether these methods are also beneficial and feasible for monitoring the groundwater resources in Central Asia.

(8) **Godbersen et al.** analysed soil-groundwater interactions and developed a new risk assessment method for groundwater pollution by trace elements. Trace elements in the geo-hydrosphere are not yet as well understood as nitrate problems. Groundwater protection starts by protecting the soil. When soils are contaminated there is always a risk that the groundwater will be affected by this contamination as well. The goal of exploratory investigations is to investigate the risk for groundwater contamination with the minimum possible effort. Exploratory investigations concerning trace elements can be carried out by analysing soil material, percolation water or groundwater. While percolation water and groundwater analysis can only confirm contamination, information gathered using soil analysis permits the current and future contamination risk for the groundwater to be estimated and additionally allows precautionary measures to be planned to protect the groundwater. The risk of contamination can only be judged when standards are available defining good-quality soil or percolation water. Local or regional background values of trace elements in the soil or percolation water can serve as adequate standards. Analytical methods applied in the exploratory investigations for the risk assessment of the soil-groundwater pathway were compared. Some common methods were introduced to analyse the total content of trace elements in the soil and present conversion functions allowing the results from different digestion agents to be converted. Subsequently an aqueous batch extraction method to routinely estimate the soluble fraction of trace elements in the soil for exploratory purposes is presented and its results are compared with results gained from direct percolation water. The method could be applied for risk assessment in areas of Central Asia that were polluted in the past.

(9) **Funk et al.** presented their method and the latest results of quantifying wind erosion (Fig. 8). Wind erosion has become an important soil degradation process in steppe regions around the world, mainly due to

overgrazing and conversion to arable land. The authors' approach is based on risk assessment and mapping measurements and algorithms. Measurements of horizontal fluxes were used to derive soil losses/dust emissions or the deposition of the transported particles. The thickness and extent of sufficient deposition can be measured to calculate the relocated volume or mass. Losses of fine particles and organic matter can be derived by comparing the grain size distribution and the organic matter content of both the original soil and the deposit. The Fallout Radionuclide Method (FRN, especially ^{137}Cs) is suited to identifying wind erosion and dust deposition patterns at large spatial and temporal scales. Remote sensing and GIS procedures were finally used to present areas for wind erosion and dust deposition for large landscape units. The method had already been tested successfully in Germany, the United States, Argentina and China. It also has potential for application in the deserts and steppe regions of Central Asia. The focus should be on mapping risk areas of deflation and deposition.

(10) **Schreiner and Meyer** defined and characterised indicators of land degradation in steppe regions. Their investigations into the development of a framework of indicators to assess desertification and land degradation problems have been applied in the Kulunda Steppe in southwest Siberia by focussing on deflation morpho-dynamics. The authors presented results about the formulation of indicators for land degradation problems, with a list of indicators that can be used to assess degradation. These are based on vegetation (vegetation cover, plant community, biological invasion of species, biomass), water (upper groundwater table, drying up of rivers and lakes, water salinisation) and soil parameters (deflation, loss of organic material, accumulation of sand and gravel in the topsoil, salinisation). Furthermore, morpho-dynamics were assessed on a regional scale on the basis of soil maps

Fig. 8 Wind erosion in Northern Kazakhstan in September 2011. Ploughing or other tillage that destroys the soil-protecting vegetation and mulch cover is a common reason. Photo: Tobias Meinel

Fig. 9 Soil surface prone to and already damaged by wind erosion. Ploughing steppe soils degrades their fertility and is unsustainable

and statistics in a case study. Soil types and dominant landscape processes were identified on a local scale. The data provided mapping of the actual deflation risk of arable land on a local scale in the Kulunda Steppe. The opening of steppe vegetation cover was found to be the key human impact initiating land degradation processes (Fig. 9).

(11) **Shokparova et al.** measured and modelled water erosion rates on cropland in Kazakhstan. The surface runoff erosion of light and dark chestnut soils was studied and quantified at two sites. Individual parameters such as snowmelt erosion and water erosion were measured, then a runoff coefficient was calculated to determine the level of soil erosion by the intensity of the erosion processes. The results explain that the resistance of the light and dark chestnut soils to water erosion is higher on the northern slopes compared to the southern slopes. It was also found that soil erosion processes induced by erosive rains are more intensive compared to snowmelt erosion. For chestnut soils the water erosion rates range depending on the slope inclination and exposure within a variation of 1.4–30.8 t/ha on different slopes induced by rainfall and 0.7–3.5 t/ha by snowmelt. Maps on a scale of 1:25.000 have been produced by combining geo-information techniques, field work and laboratory analytics.

Cluster 4 deals with "Remote sensing methods for monitoring and modelling large areas". These methods are essential for getting and processing consistent information about land and water resources and their alterations over large regions. The cluster contains three chapters of crucial importance for Central Asia.

1. Klein et al.: "Generation of Up to Date Land Cover Maps for Central Asia" (Создание Современных Карт Типов Наземного Покрова Центральной Азии), (Chap. 19)
2. Ibrayev et al.: "Methodology of Measuring Processes and Evaluation of Water Resources of the Republic of Kazakhstan" (Методология по Измерению Процессов и Оценки Водных Ресурсов Республики Казахстан), (Chap. 35)
3. Romanenkov et al.: "Estimating Black Carbon Emissions from Agricultural Burning" (Оценка Эмиссии Черного Углерода при Сельскохозяйственных Сжиганиях), (Chap. 20)

(1) **Klein et al.** present a new land cover and land use classification approach for Central Asia. A specific classification scheme was developed based on the Land Cover Classification System (LCCS) adopted by the Food and Agriculture Organisation of the United Nations Environment Programme (FAO-UNEP). The classification was performed using a supervised classification method applied to metrics derived from Moderate Resolution Imaging Spectroradiometer (MODIS) data with 250 m spatial resolution. The metrics were derived from an annual time series of red and near-infrared reflectance as well as from the Normalized Difference Vegetation Index (NDVI) and thus reflect the temporal behaviour of different land cover types. The reference data required for a supervised classification approach were collected from several high-resolution satellite images distributed all over the study area. The presented classification approach is a suitable tool for monitoring land cover and land use in Central Asia. This kind of independent information is important for the accurate assessment of water and land resources. The authors documented significant alterations to the landscape, for example the eradication of forest ecosystems in Kazakhstan. This should be an issue of great concern requiring action by the authorities.

(2) The paper by **Ibrayev et al.** highlights the methodology of evaluating the changing processes and water resources of Kazakhstan with the aid of geo-information technologies, determining their dynamic parameters based on thematic satellite data processing methods and techniques for theoretical and applied purposes (Fig. 10). Different techniques for analysing satellite data were tested. The authors propose a technological scheme for making precision mosaics using high-resolution optical electronic space imagery combined with LANDSAT space snapshots of the Irtysh river basin. A schematic map of the water basins of transboundary rivers in Kazakhstan and neighbouring countries has been developed with a list of cartographic layers, objects and information, including administrative-territorial divisions and terrain (contours, heights, forms of relief), as well as the structure of the GIS database.

(3) **Romanenkov et al.** developed a method for estimating losses of black carbon due to agricultural burning (Fig. 11). Black carbon is a threat to the Arctic Region, as it makes ice and snow melt faster. Straw burning and

Fig. 10 Monitoring and managing transboundary rivers based on binding international conventions and treaties is key to the survival of rivers and inland lakes. The photo shows the Charyn in Kazakhstan, just behind the Chinese border. The Charyn is a tributary of the Ili which is the main feeder of lake Balkhash, the largest freshwater lake of Central Asia. The Ili river and Lake Balkhash are candidates for a virtual red list of natural water bodies, e.g. threatened by extinction

possible BC emissions have been estimated at the oblast level in Russia for the years 2003 through 2010 using a comparative study with the 1 km MODIS Active Fire Product and the agricultural statistics approach. Both calculations produced consistent results, including increasing fire activity in the years with additional straw surplus and the highest absolute values for vast territories with relatively intensive grain production. The average BC emissions from cropland burning were 7.7–8.4 Gg for different land cover classifications, compared with 8.5 Gg calculated from statistics. Straw burning may be a source of at least 1/3 of total BC emissions from agriculture and grassland fires. Burning agricultural land is also common in

Fig. 11 Straw burning is still a common practice in some countries but should be banned. It diminishes soil carbon contents and causes greenhouse gas emissions

Northern Kazakhstan. It should be banned as it damages soils and the environment. The method has potential as a permanent system for examining and monitoring agricultural lands and conducting actions against burning.

Cluster 5, "Methods of data analysis and ecosystem modelling", comprises methods of both statistical process analysis and deterministic modelling. The cluster contains the following chapters:

1. Lischeid: "Non-linear Approaches to Assess Water and Soil Quality" (Нелинейные Статистические Методы для Оценки Качества Воды и Почв), (Chap. 21)
2. Natkhin et al.: "Model-based Impact Analysis of Climate and Land Use Changes on the Landscape Water Balance" (Основанный на Моделях Анализ Воздействия Изменения Климата и Землепользования на Водный Баланс Ландшафтов), (Chap. 36)
3. Michel and Dannowski "Improving the Irrigation Efficiency by Use of Soil–Water–Plant Models" (Повышение Эффективности Орошения через Исполь, зование Моделей Системы Почва-Вода-Растения), (Chap. 22)
4. Nendel: "MONICA—a Simulation Model for Nitrogen and Carbon Dynamics in Agro-Ecosystems" (Имитационная Модель Динамики Углерода и Азота в Агроэкосистемах MONICA), (Chap. 23)
5. Djanibekov and Sommer: "Integrated Decision Support for Sustainable and Profitable Land Management in the Lowlands of Central Asia" (Интегрированная Поддержка Принятия Решений для Устойчивого и Рентабельного Землепользования в Условиях Низменностей Центральной Азии), (Chap. 24)

(1) Systematic monitoring is indispensable for thorough water and soil management. However, large data sets with many variables, natural heterogeneities, and a variety of influencing factors require new approaches for processing and visualisation of the data. The method presented by **Lischeid** can detect hidden structures in time series data. Every single sample can be ascribed a component score: a quantitative measure of the impact of a certain process on the given sample. Usually, a small number of components (or processes, respectively) account for a large fraction of the variance in a data set with many variables. This "dimensionality reduction" helps a lot to gain a better understanding of the prevailing processes, of spatial and temporal patterns, and of the reasons for conspicuous data. A single non-linear projection of high-dimensional data onto a two-dimensional graph provides comprehensive information about outliers, clusters, linear and non-linear relationships, spatial patterns, multivariate trends, etc. in the data. This approach could usefully be combined with other dimensionality reduction techniques. These methods are not yet mentioned in textbooks of hydrology or soil science. They require an open mind and some initial training. Then a wealth of powerful

tools are at hand as a basis for thorough water and soil resource management.

(2) The chapter by **Natkhin et al.** combines methods of classical hydrological modelling with the statistical techniques described above. Changes in the water balance in landscapes can be easily observed by measuring water levels, runoff, etc. Determining the causes of changes in water balance is much more difficult, because of the complex interrelations between the interacting hydrological processes. With the help of modelling and scenario analyses, it is possible to differentiate between the effects of land use/land cover and climate on the water balance, taking the underlying hydrological processes into consideration. Two case studies were presented: a river catchment in Tanzania, Africa and a forested area in North-East Germany. In these areas the observed water balance has changed remarkably in the last few decades. The influence of climate conditions and land use change are analysed and determined with the help of the SWAT model, the WaSiM-ETH model and statistical analyses. Both the hydrological models applied and the method of process analysis are practicable under different climate conditions. The re-afforestation of lands, which is an urgent need for some regions of Central Asia, could be better planned and accompanied by those scientific tools.

(3) **Michel and Dannowski** present the ZEPHYR model for irrigation scheduling based on soil water balancing. The model considers all the important elements affecting the soil water balance: soil characteristics, plant development, daily meteorological values, irrigation and the ground water table. Soil water movement and storage in the profile are based on the Richards equation. The effects of plants are calculated using the Koitzsch-Guenther evapotranspiration model. Potential evapotranspiration is computed using the Penman–Monteith equation, modified by Wendling. The ZEPHYR model is practicable for farmers or consultants in irrigation management. It has already proven very efficient for controlling irrigation systems in Germany. Initial calculations for a barley cropped site in Southern Kazakhstan indicate its suitability for other regions. ZEPHYR could also have potential for water-saving irrigation in Central Asia.

(4) **Nendel** reports on the brand-new process-based simulation model MONICA, the legal successor of the HERMES model, which has been successfully applied in research studies worldwide. It maintains the simple and robust philosophy of its progenitor and adds a full carbon cycle model, including the feedback relations between atmospheric CO_2 concentration and other environmental variables on crop growth, nitrate leaching and water use efficiency. MONICA is the central part of a web-based decision-support system that helps farmers and other stakeholders in Germany identify management options to mitigate the impact of the expected climate change on their business. MONICA has the potential to

assess the impacts of climate change and land management on crop yields, carbon balance and nitrogen efficiency in Central Asia.

(5) **Djanibekov and Sommer** present the Farm-Level Economic Ecological Optimization Model (FLEOM), developed as an integrated decision-support tool for optimising land and resource use at the level of large farms or water user associations in the Khorezm province of Uzbekistan. The model builds on the linear-programming optimisation routine and a comprehensive agronomic database established with the cropping system simulation model, CropSyst. The FLEOM works in three steps. First, a user defines farm-specific and economic parameters which are entered into the model's central database along with the agronomic database. Following this, the agronomic database and the socio-economic information are used in an optimisation process in the General Algebraic Modelling Software (GAMS). Finally, the model results are exported into the central database and can be visualised in tables and figures via a graphical user interface or as maps in a GIS environment. The design of the model means one-click scenario formulation can be used to simulate changes in socio-economic and production environments, for instance by modifying socio-economic parameters and adding new crops. The user-friendly interface of the FLEOM and its flexibility secure a wide range of its possible uses in the future here and in other regions of Central Asia.

Cluster 6, "Methods of soil and water bioremediation" presents emerging biotechnologies for the rehabilitation and improvement of soil, plant and water quality (Fig. 12). It includes five chapters:

1. Kozybaeva et al.: "Biotechnological Restoration Methods of Technogenically Disturbed Soils in Kazakhstan" (Биотехнологические Методы Восстановления Техногенно Нарушенных Почв Казахстана), (Chap. 37)

Fig. 12 Biotechnological methods help to create healthy soils, water bodies and plants. The photo shows a new potato variety, bred in the Kazakh Research Institute for Vegetables and Potatoes by means of in vitro cultivation

2. Saparov: "Strategy of Sustainable Soil and Plant Resource Management in the Republic of Kazakhstan" (Стратегия Устойчивого Управления Почвенными Ресурсами и Растительностью Республики Казахстан), (Chap. 38)
3. Kussainova et al.: "The Effect of Applying the Microbiofertiliser "MERS" on the Soil Microbial Community and the Productivity of Winter Wheat under the Conditions of Southeast Kazakhstan" (Эффект Применения Микроудобрения "МЭРС" на Микробное Сообщество в Почве и Производительность Ози, мней Пшеницы при Условиях Юго-Восточного Казахстана), (Chap. 39)
4. Balla et al.: "Efficiency of Duckweed (*Lemnaceae*) for the Desalination and Treatment of Agricultural Drainage Water in Detention Reservoirs" (Еффективность Применения Ряски (*Lemnaceae*) для Отбора Соли и Очистки Сельскохозяйственных Дренажных Вод в Прудах-Очистителях), (Chap. 25)
5. Tumlert: "Water Treatment Systems for Agricultural Water Supply" (Водоподготовка на Системах Сельскохозяйственного Водоснабжения), (Chap. 40)

(1) The chapter by **Kozybaeva et al.** highlights the emerging and most important problems related to soil bioremediation. This includes an analysis of the environmental status of soils as a result of disturbances by resource-extracting industries. The study explains field methods for the biotechnological remediation of soils which have been disturbed and transformed by oil production and aridity. During the remediation of disturbed and transformed soils on ore dumps and oil and gas deposits in the arid zone of the Aral Sea area, biotechnological field experiments were conducted on a number of sites. A number of special plants were successfully tested in these field trials. Biofertilisers, sodium humate, hydrogels and other soil amendments have been successfully tested to stimulate the growth of saksaul (*Haloxylon aphyllum*) and other promising plants for soil remediation. One key aim was to ensure that vegetation grew rapidly and to prevent the transport of hazardous substances to water bodies and other places or into the food chain. Fencing off those contaminated sites and experiments is of high practical importance.

(2) The chapter by **Saparov**, "Strategy for Sustainable Soil and Plant Resource Management in the Republic of Kazakhstan", also includes the results of soil bio-remediation. Research material was presented which had been obtained during the implementation of a cooperation project with the International Center for Agricultural Research in Arid Areas (ICARDA): "Sustainable management of land resources in arid regions of Central Asia and Caucasus", in particular in terms of irrigation and degraded pastures in the south and south-east of Kazakhstan. The objects of the studies were terrain, soil, water, cultivated and pasture plants (rice, triticale and winter wheat, wormwood-ephemeral, grass-shrub and cereal-grass communities). The paper presented the results of soil reclamation and environmental conditions in the Shieli irrigation area. Fertility levels,

soil salinity and soil degradation under irrigation in the Kyzyl Orda region, as well as the degradation of pastures in the Sarysu district of the Zhambyl region were also studied. Methods were developed for improving soil reclamation and environmental conditions, and for conserving and increasing soil fertility and crop productivity in irrigated saline soils and pastures in the arid zone. On silty soils in Southern Kazakhstan the most sustainable plants were *Sameriaria boissieriana*, *Kochia prostrata*, and *Krascheninnikovia* and *Halothamnus*, and on sands *Sameriaria boissieriana, Agropyron, Krascheninnikovia, Halothamnus, Calligonum and Haloxylon* (Fig. 13). For sustainable pasture management and fodder production, it is necessary to create cultivated pastures by planting natural forage plants and growing alternative crops for field feed, in particular triticale.

(3) Micro-biofertilisers could be an option for improving the productivity of agricultural land by means of biological components enhancing the populations and performance of soil organisms that stimulate plant growth using several mechanisms. Just under restricted site conditions, those fertilisers may have some potential. The chapter by **Kussainova et al.** presents research data on the impact of a special micro-biofertiliser (trade name: "MERS") on the fertility of meadow gray soils (Calcisols) and the productivity of winter wheat in south-eastern Kazakhstan. This fertiliser provided slightly higher grain yields of winter wheat ("Almaly" cultivar), at a dose of 500 ml/ha. The yield was 5.3 t/ha, and the control variant without any fertiliser yielded 4.3 t/ha.

(4) **Balla et al.** reported on investigation methods concerning the salt uptake potential of aquatic macrophytes *Lemnaceae* (duckweed). They tested their performance for the purification of drainage water in small water reservoirs. The salinisation of soils and water bodies is one of the major problems in the desert and steppe regions of Central Asia. The results

Fig. 13 Reafforestation of desertified lands with *Haloxylon* or other site-adapted vegetation. Sampling, analysis and mapping of those areas are pre-conditions for success. Soil amendments and bio-fertilisers can support those measures

demonstrate the salt uptake behaviour of duckweed by enclosure in the tissues in dependence on the degree of salinisation and the initial biomass density. The uptake effect can be described as a 1st order decay reaction. The decay/uptake depends on the residence time within the water body. To maximise the detention time in reservoirs the effect of inside baffles and inlet constructions are investigated by 2D hydraulic modelling with the model package TELEMAC. *Lemnaceae* should be tested for improving the quality of salinised open waters in Central Asia.

(5) The chapter by **Tumlert** describes water treatment methods in rural areas of Kazakhstan. Rural populations' supply of clean drinking water is one of the factors influencing livelihoods and social development in rural settlements. Most of the water pipes and water treatment plants are more than 30 years old and do not meet sanitary requirements. More than 90 % of the rural settlements of Kazakhstan are supplied from underground sources. The conditions of rural water use (capacity less than 1000 m^3 a day) are characterised by deficiencies in skilled personnel for maintenance and repair and intermittent faults in the electricity system. Compact and robust solutions are required. The basic methods of water treatment are clearing, decolouration, softening, fluorination, de-fluoridation, de-salination and disinfection. A sufficient number of proper measures are available. For water clearing and de-colouration, low-rate filters are recommended. Local close-burning shungites and zeolites can be used as a percolation bed. Water treatment plants of the "Struya" type improve hard water and produce high contents of fluoride and iron. For softening, the application of lime or lime-soda and the installation of a sodium cycle is recommended. For de-ironing and de-permanganating, filters with preliminary aeration could be installed. De-fluoridation can be provided by applying ion-exchange pressure filters with alumina filter beds (Al_2O_3). For disinfection UV sterilisers are available. For desalination/deionisation reverse osmosis modules have been developed. Technical solutions have been created and tested by Kazakh companies in cooperation with the Kazakh Research Institute of Water Economy.

Cluster 7, "Field monitoring of soils", includes two chapters from Kazakhstan as an example. They are:

1. Otarov: "Concentration of Heavy Metals in Irrigated Soils in Southern Kazakhstan" (Содержание Тяжелых Металлов в Орошаемых Почвах Юга Казахстана), (Chap. 41)
2. Dzhalankuzov: "Concept and Results of Soil Monitoring in North Kazakhstan" (Концепция и Результаты Мониторинга Почв Северного Казахстана), (Chap. 42)

(1) Reductions in soil fertility due to soil contamination with heavy metals, pesticides and other pollutants are of great concern in intensively used and irrigated soils. **Otarov** reported on a study on the measurement and

evaluation of soil pollution with heavy metals in irrigated areas of southern Kazakhstan. The research object included the soils of large rice farms in the Shiely area of irrigation. The study was conducted using the method of large-scale soil mapping. Among the environmentally hazardous mobile forms of metals in the soils, Pb had the largest share (54 %), followed by Ni (22 %). The figures for Zn and Cu were 12 and 10 % respectively. The share of Cd is only 2 %. Cu was found in 5 groups of soil which contained more than 3.15 mg of Cu per 1 kg of soil. High concentrations of Pb were found in the group of soils, at 9.0–15 mg per 1 kg of soil. Ni was also found in high concentration in the soils: 4.6–6.0 mg/kg. Gross forms of elements were characterised by eluvial-illuvial processes. Environmentally hazardous mobile metal forms available for the plants were subjected to the influence of irrigation more than their gross forms. The author concluded that for all mobile forms of heavy metals, first of all, the distribution and accumulation process in the soil profile needs to be ascertained.

(2) **Dzhalankuzov** presented a monitoring concept for soils on the agricultural lands in northern Kazakhstan. He defined the framework in terms of goals, objectives, parameters and application areas. This chapter includes methods for monitoring soils and their current status, and ways of improving the soil fertility. The climatic conditions of the subzones of Ordinary and Southern Chernozems were analysed along with their morphological indices, physical and chemical properties and significant changes in the soil properties due to long-term agricultural use. Using the results of monitoring studies, scientists will be able to forecast the development of negative soil processes, prevent the withdrawal of agricultural land, and justify the need for programmes of soil fertility conservation and restoration.

Cluster 8, "Methods and Technologies for Optimizing Land Use Systems", presents advanced technical solutions which are ready for immediate practical application. The cluster has 6 chapters:

1. Suleimenov et al.: "Conservation Agriculture for Long-Term Soil Productivity" (Ресурсосберегающее Земледелие для Длительного Сохранения Плодородия Почвы), (Chap. 26)
2. Meinel et al.: "Modern Technologies for Soil Management and Conservation in Northern Kazakhstan" (Современные Технологии Рационального Использования и Охраны Почв Северного Казахстана), (Chap. 27)
3. Chernenok and Barkusky: "Diagnosis and Optimization of Phosphorus Nutrition Conditions of Grain Crops in Northern Kazakhstan" (Диагностика и Оптимизация Условий Фосфорного Питания Зерновых Культур в Северном Казахстане), (Chap. 43)
4. Qadir et al.: "Enhancing the Productivity of High-magnesium Soil and Water Resources in Central Asia" (Повышение Урожайности Сельскохозяйственных Культур и Продуктивности Оросительной Воды на Землях с Повышенным Содержанием Магния в Центральной Азии), (Chap. 28)

5. Evans: "Advanced Technologies for Irrigated Cropping Systems" (Новые Технологии Орошения Пахотных Земель), (Chap. 29)
6. Behrendt et al. : "Multi-Species Grazing on Deer Farms" (Многовидовый Выпас Скота в Оленеводческих Хозяйствах), (Chap. 30)

(1) The objective of the chapter by **Suleimenov et al.** was to present the methodology of no-till studies in multifactorial trials. This includes a comparison of no-till methods and traditional tillage in various crop rotations, under various sowing dates and varieties. Crop production in the steppe regions of southwest Siberia and Northern Kazakhstan is mostly focused on cropping spring cultures with a high percentage of cereals in the rotation. Wider crop rotations lead to better soil structure and nutrient supply. Trials provided a comprehensive assessment of the influence of no-till methods on soil moisture and plant nutrient availability, as well as on crop yields. On heavy textured black soil (Chernozems), no-till methods had advantages in terms of organic matter and moisture conservation. Traditional tillage had an advantage in terms of the infiltration of snowmelt water into soil in early spring as well as of nitrogen availability to plants. No-till methods provided higher grain yields in dry years, but their major advantage is the conservation of organic matter for long-term soil productivity.

(2) The chapter by **Meinel et al.** presents the most advanced technologies for productive and sustainable cropping (no till) (Fig. 14). Modern drills such as the "Citan Z" and "Condor 12001" are designed for specialised direct seeding on stubble. The concept of the of the drill openers is designed to reduce the intensity of the mechanical impact on the soil, producing wide inter-row spacing, minimising the disturbance of stubble and maximising the preservation of crop residues on the soil surface. The proposed drills of 12 and 15 m in width can be aggregated with tractors of 220 and 260 horsepower. When sowing with a row spacing of 25 cm, the surface of the soil remains untouched between rows, which allows herbicides to be used

Fig. 14 Experiments at the Shortandy research station revealed the excellent performance of no-till methods under steppe conditions. Photo: Tobias Meinel

after sowing. The systematic control of weed infestation in fields reduces the need for herbicides with time. Direct seeding has economic benefits. In field trials at the farm level, using no-till technology with the "Condor12001" direct seeding drill the yields of spring wheat were at a level of 2.55 t/ha. Yields of linseed were 1.75–1.94 t/ha, that of rapeseed was 1.26 t/ha and that of peas 1.24 t/ha. These drills metered the fertiliser dosage very precisely.

(3) The chapter by **Chernenok and Barkusky** dealt with optimising the phosphorus status of soils on croplands. On soils with a high pH, the phosphorus status and ability for plants can be a limiting factor. On the other hand, phosphorus is a scarce resource, and high doses of fertilisation have led to soil contamination worldwide. The chapter is based on long-term studies of the diagnosis and optimisation of the phosphorus nutrition of cereal crops in northern Kazakhstan and in Siberia. The authors present an empirical formula for the estimation of fertilisation rates. The chapter reveals gaps in our knowledge and understanding of the phosphorus nutrition of plants. It is high time for international comparisons of analytical methods and field trials.

(4) **Qadir et al.** developed a technology for the improvement of soils degraded by high magnesium levels. Excess levels of magnesium (Mg^{2+}) in irrigation waters and/or in soils negatively affect the soil infiltration rate and the hydraulic conductivity and ultimately the crop growth and yield. Southern Kazakhstan has become a hotspot of such resource degradation. These soils require adequate quantities of calcium (Ca^{2+}) to mitigate the effects of excess Mg^{2+}. Phosphogypsum, a by-product of the phosphorus fertiliser industry, is a beneficial ameliorant. Field experiments revealed that the application of phosphogypsum increased Ca^{2+} concentrations in the soil and triggered the replacement of excess Mg^{2+} from the cation exchange complex. Studies have demonstrated the beneficial effects of this soil amendment in terms of improved soil quality, enhanced water movement into and through the soil, increased cotton yield and water productivity, and significant financial benefits. The economic benefits from the phosphogypsum treatments were almost twice those of the control. In addition to improving the crop productivity, these studies demonstrated the beneficial use of industrial waste material in agriculture. With the aim of addressing the challenge of achieving sustainable agriculture production from magnesium-affected environments, there would be a need for appropriate supportive policies and functional institutions along with capacity building for farmers, researchers, and agricultural extension workers.

(5) **Evans** presented a piece of highly advanced irrigation technology. Available supplies of water for irrigation and other uses are becoming more limited around the world, and this trend is accelerating. Increasing the water use efficiency of cropping systems is key to irrigated farming. Emerging computerised precision irrigation technologies will enable growers to apply

water and agrochemicals more precisely and site-specifically to match the status and needs of soil and plants as determined by the analysis of data from a variety of sensor networks, wireless communications systems and decision support systems. Speed control and zone control options for site-specific variable-rate irrigation (SS-VRI) systems are currently available, with speed control the most common. Site-specific variable-rate sprinkler irrigation systems are excellent research tools that can provide maximum amounts of information from relatively small areas. A self-propelled SS-VRI sprinkler system has been developed for agricultural research applications. This fully functional research machine has been used from 2005 through 2012 and has been very reliable. These SS-VRI systems offer many benefits for research and they have tremendous potential for greater use in sprinkler irrigation systems worldwide for both in research and general practice to conserve water, fertiliser and energy. The application of these systems in Eurasia would promise not only resource savings in irrigation, but also benefits for soil and plant quality.

(6) Controlled multi-species grazing, game farming and the combination of both are possible alternatives to existing pastoral systems for a certain proportion of farmers. The chapter by **Behrendt et al.** demonstrates these alternatives for advanced pastoral farms and other landowners to combine entrepreneurial initiatives with significant benefits for grassland quality and the landscape and ecological diversity. A number of interesting options have been tested, such as the combination of cattle and sheep breeds with deer. Considering the rules of good farming practice, animal health, soil and grassland quality can be controlled very well in these fenced systems. The introduction of these systems into Central Asia is not merely an important contribution to food security in the region but can also be seen as improving the lifestyle, stabilising local economies and valorising the landscape and ecosystem quality.

3 Availability of Novel Methods and Conditions for their Application

Table 2 lists and categorises addressees and the availability of novel tools, most of which are presented in Part II of the book. There are various potential addressees for these methods. Most tools could be applied by research and monitoring groups, but landscape planners, consultants, farmers and other people could also benefit from them.

The first group of tools are mainly of a physical nature and are well-engineered commercial products. The measurement devices presented in this group are useful for application in research and monitoring and may provide a knowledge shift in

Table 2 Characteristics of novel methods in Part II, in brief

Tool designation	Authors of chapter	Addressees	Accessibility
Soil hydrological measurement method (EEM)	Schindler	Research laboratories	Commercial product
Weighable lysimeters, different types, high-tech instrumentation	Meissner et al., Hertel and von Unold	Research and monitoring stations	Commercial products
Soil hydrological field-setups	Schindler	Research and monitoring stations	Consists of commercial products
Simple field methods for measuring soil quality of grasslands	Mueller et al.	Consultants, research groups	Consists partly of low-cost commercial products
Method of impact assessment	Helming	Planning bodies, research groups	Method for free, training advisable
Muencheberg Soil Quality Rating, M-SQR	Mueller et al., Smolentseva et al.	Consultants, planning companies and bodies, researchers	Method for free, training advisable
Method of mapping groundwater recharge	Hennings	Monitoring authorities, planning bodies	Method for free, training advisable
Methods of monitoring and assessing groundwater pollution by agriculture	Eulenstein et al. Dannowski et al.	Monitoring authorities, planning bodies	Method for free, training advisable, commercial equipment required
Method of assessing groundwater risk from trace elements	Godbersen et al.	Monitoring authorities, planning bodies	Method for free, training advisable
Method of land use mapping	Klein et al.	Research and monitoring groups, planning bodies	Methods partly for free, but commercial data and equipment required
Method of quantifying wind erosion	Funk et al.	Research and monitoring groups	Method for free, training advisable, commercial measurement devices required
Geostatistical tool	Lischeid	Research groups	Research tool for free, training advisable
MONICA model	Nendel	Research and planning groups, consultants	Method for free, training advisable
ZEPHYR model	Michel and Dannowski	Consultants	Part of commercial irrigation scheduling

(continued)

Table 2 (continued)

Tool designation	Authors of chapter	Addressees	Accessibility
FLEOM model	Djanibekov and Sommer	Consultants, research and planning groups	Method for free, training advisable
Drainage water desalinization by *Lemnaceae*	Balla et al.	Water user associations	Investment in pilot projects
Phosphogypsum application for soil rehabilitation	Qadir et al.	Advanced farms	Technological costs
Technology for conservation tillage	Meinel and Akshalov	Advanced farms	Commercial products
Automatic, intelligent sprinkler irrigation system	Evans	Agricultural research stations, advanced farms	Commercial product
Multi-species grazing systems	Behrendt et al.	Advanced farms, agricultural research stations	Investment in pilot projects (livestock and game species, fencing, ranching)

understanding soil hydrological processes. One typical characteristic of agricultural research institutes in the former USSR was the non-existence of analytics in soil physics and soil hydrology. The application of advanced laboratory methods such as EEM, lysimetry and hydrological field setups could initiate soil hydrological analyses in Research Institutes of Central Asia, dealing with agri-environmental topics.

It should be mentioned that the application of all these instruments requires particular qualification on the part of the people dealing with this. Also, the application of measurement devices and systems needs to be embedded in an advanced environment. For example, lysimeters are excellent research tools, but their potential cannot be used when they are applied as stand-alone devices. Skilled scientific staff and advanced further laboratory equipment will be required for reliable results. Also, the technology for conservation tillage and for automatic water-saving irrigation requires skills, knowledge and the courage to change technologies.

A second important group of methods are not devices, i.e. not of a physical nature. They are virtual tools for the recognition, assessment and evaluation of states and processes in agricultural landscapes. Some addressees of these evaluation methods and models are also researchers, but others are monitoring groups, authorities or advanced farmers. These methods are freely available, but they also require qualifications and a lot of creativity. This is a critical item and sometimes a reason for them being neglected. In many cultures commodities are considered valuable if they are of a physical nature and come at a high price. Virtual goods and goods which are free are underestimated. These attitudes have contributed to the failure or limited success of international cooperation projects in the past. All the tools presented in Table 2, and some others, are worth being examined and tested for their first application in Central Asia or for their broader or general application in this region.

It may be concluded that investment in new devices, instruments, and technological systems can help to overcome the current crisis in agri-environmental research and monitoring. But it is not the key to resolving it or to initiating sustainable resource management. Key to this is mental access to available research tools, including "non-physical" ones. This clearly requires high-level training and qualification in their handling, based on a standard level of scientific education.

4 Proposals for Initiating Sustainable Resource Use

The contributors to this book represent an immense international research network having great potential for supporting activities to improve the situation of agri-environmental research, education and outreach in Central Asia in the coming years. They are willing to take the next step and to train candidates. This requires not only open minds but also thinking about flexible institutional frameworks. Young scientists can already take on internships and train at institutes and companies abroad under current conditions. However, this cannot be the only means of knowledge transfer, and there is a risk that young people do not come back to their home country. Strengthening research and education in Central Asia should be more promising. For example, pilot project groups for field-testing novel methods and technologies could be installed at research institutes, monitoring authorities, water user associations or advanced farms. Their work could be supported by external scientific consultants. In the area of education, "Schools of Environment and Natural Resources" work successfully at some universities in the US and in Europe. They teach students methods of sustainable resource management. Employing external lecturers means that new ideas and methods can have an effect.

The status of the Central Asian grasslands and rangelands seems to be a particularly critical problem, as desertification is probably triggered by grassland degradation. It seems useful to establish an international project group that has to work out comparable methodical guidelines for the evaluation of grassland sites and to prepare a Trans-national Land Inventory of Central Asia (**TraLINCA**).

It is worth thinking about those ways of introducing more durable cooperation between the countries of Eurasia than current short-term, intermittent bilateral research projects provide. In other words, strategies of cooperation in science, technology and education, focused on the performance of water and land resources, must also become more sustainable.

The final motto of this book, "Sharing scientific methodologies is key to sharing prosperity in an intact environment", should be the guiding principle behind a new era of cooperation between research teams in Eurasia.

5 Overall Conclusions

(1) The land and water resources of Central Asia are in a critical state. Some keywords are: grassland degradation, humus loss and wind erosion on cropping land, salinisation of irrigated land, low agricultural productivity and water use efficiency, water scarcity, water pollution.
(2) Objective and novel methods and technologies for monitoring and control of resources, meeting international standards, must be set into operation. The above book chapters present an array of methods which can provide a knowledge shift in agri-environmental research, monitoring and agricultural practice.
(3) Not investment alone, but also educational training about novel methods in the framework of international scientific cooperation and partnership are the key to initiating sustainable developments in resource utilisation.
(4) Pilot projects, permanent think tanks, Scientific-Technical Education Centres, and Schools of Environment and Natural Resources are possible promoters who can carry novel methods into the heads and hearts of people in Eurasia, including Central Asia.

About the Editors

Lothar Mueller (Müller) is a scientist at the Leibniz Centre for Agricultural Landscape Research in Müncheberg, Germany. He studied Agriculture at the Humboldt University in Berlin (Agricultural Engineer, 1970) and graduated from the Academy of Agricultural Science of the GDR (Dr. agr.) in 1974. Since 2013 he is an Honorary Professor of the State Agricultural University of the Kuban in Krasnodar (Russian Federation) and an Honorary Doctor of the Russian Research Institute of Rice in Krasnodar. His main fields of work and expertise are in the areas of soil hydrology, soil quality, agricultural soil and water management, drainage engineering and subsoiling technology.

Lothar Mueller has published more than 350 papers, amongst them more than 100 peer reviewed publications and book chapters. Most of this work was done in cooperation with his colleague Uwe Schindler.

Over the past 15 years Lothar Mueller built up international cooperation with worldwide leading experts in the topic of agricultural soil and water management. Many of them are contributors to this book.

L. Mueller et al. (eds.), *Novel Measurement and Assessment Tools for Monitoring and Management of Land and Water Resources in Agricultural Landscapes of Central Asia*, Environmental Science and Engineering, DOI: 10.1007/978-3-319-01017-5,

Abdulla Saparov has been General Director of the U.U. Uspanov Kazakh Research Institute of Soil Science and Agrochemistry in Almaty, Kazakhstan, since 2003. He graduated from the Labor Red Banner Order Kazakh Agricultural Institute, Faculty of Soil Science and Agrochemistry in 1972, and gained his doctorate in 1997 at the Kazakh National Agrarian University. In 1998 he became a Professor (Agronomy) at the National Academic Center of Agrarian Researches in Kazakhstan. His main specialisation is in Soil Science and Agrochemistry, Agronomy and Ecology.

Abdulla Saparov's current research interests are the efficiency of productivity of crops, their qualities and their safety, and soil fertility depending on the conditions of fertilization.

He has a number of administrative and advisory tasks, is editor of the journal "Soil Science and Agrochemistry", and associate editor of some other scientific agricultural journals in Kazakhstan.

Abdulla Saparov has been the initiator and National Coordinator of many crucial projects about the sustainable management of land resources in Central Asia. He has published more than 280 research papers, including 4 books, and 6 recommendation guidelines. Abdulla Saparov has educated a number of young scientists of Kazakhstan.

Gunnar Lischeid has been the head of the Institute of Landscape Hydrology at the Leibniz Centre for Agricultural Landscape Research in Muencheberg, Germany, and Professor of Landscape Hydrology at Potsdam University since 2008. He studied Agriculture and Geology at the Universities of Bonn and Göttingen. Gunnar Lischeid graduated from the University of Göttingen's Institute for Soil Science and Forest Nutrition in 1995 (Dr. sc. agr.) and from the University of Bayreuth (Dr. habil) in 2004.

His specialist expertise is in the fields of non-linear data analysis, minimizing model structure uncertainty, and groundwater and surface water quality. He has worked on some national and international hydrological projects, and heads a number of current projects. Gunnar Lischeid has published more than 100 scientific papers, the majority of them in peer-reviewed journals. He is active in the international scientific community and a member of numerous scientific organizations.

MIX
Papier aus verantwortungsvollen Quellen
Paper from responsible sources
FSC® C105338

If you have any concerns about our products,
you can contact us on
ProductSafety@springernature.com

In case Publisher is established outside the EU,
the EU authorized representative is:
Springer Nature Customer Service Center GmbH
Europaplatz 3, 69115 Heidelberg, Germany

Printed by Libri Plureos GmbH
in Hamburg, Germany